International Conference on Advances in Earth and Environmental Studies (AEES 2022)

IOP Conference Series: Earth and Environmental Science
Volume 1032

Online
25 – 26 February 2022

ISBN: 978-1-7138-5815-7
ISSN: 1755-1307

Printed from e-media with permission by:

Curran Associates, Inc.
57 Morehouse Lane
Red Hook, NY 12571

proceedings.com

Some format issues inherent in the e-media version may also appear in this print version.

Institute of Physics
Dirac House, Temple Back
Bristol BS1 6BE UK

Phone: 44 1 17 929 7481
Fax: 44 1 17 920 0979

techtracking@iop.org

Additional copies of this publication are available from:

Curran Associates, Inc.
57 Morehouse Lane
Red Hook, NY 12571 USA
Phone: 845-758-0400
Fax: 845-758-2633
Email: curran@proceedings.com
Web: www.proceedings.com

TABLE OF CONTENTS

Preface

Peer Review Statement

Geochemical Evolution of Basaltic Flows from Dongargarh Supergroup, Bastar Craton, Central India ... 1
Tanveer Haidar, Sagar Misra, Neeraj Vishwakarma, K. R. Hari

Utilization of Waste Sole Leather with Fly Ash for Self-Compacting Concrete ... 12
Ankush Thakre, Tarun Kumar Rajak

A Study on the Influence of Vegetation Growth on Slope Stability ... 23
Pooja Naredla, S Sangeetha

Impact of COVID-19 Lockdown on the Risk of Breast Cancer: A Case Study ... 35
Akriti Sharma, Nishtha Hooda, Nidhi Rani Gupta, Renu Sharma

Effectiveness of PNL Technique in Disaster Damage Assessment: Evidence from Selective Case Studies ... 53
P Dave, S Pasari

Life Cycle Cost Analysis of Wastewater Treatment Technologies ... 58
Purva Rathore, D J Killedar, Divyesh Parde, Akansha Sahare

A Concise Air Quality Study on Impact of COVID-19 Lockdown for Two Industrial Cities of Chhattisgarh ... 70
Akansha Sahare, D.J. Killedar, Ajay. V. Ahirwar, Purva Rathore

Machine Learning Algorithms in Geostatistical Data Analysis: Formulation and Observation ... 81
S Pasari, P Asudani, A Mehta

Channel Migration and Consequential Land Use Land Cover Changes of Subansiri River, Assam, North-Eastern India ... 86
Shukla Acharjee, Rajib Tarani Das, Shubhendu Shekhar, Neelratan Singh, Uttam Goswami, M.M.G.S. Dilini, Y.K. Mawale, B.K. Iyengar, R S Negi

A Gis-Based Parametric Analysis of Irrigational Tanks in the Oldest Civilized City in India-Madurai ... 104
J Praveenkumar, Yazhini Malar Pa, K Akash Kannan, B Sathya Bama, S Mohammed Mansoor Roomi

Petrology and Geochemistry of Charnockite Patches and Host Leptynites from Digapahandi Area, EGB: Evidence for Incipient Growth of Charnockites ... 112
Sagar Misra, Varun Kumar Mishra, Rabindra Nath Tiwari, Neeraj Vishwakarma

Water Quality Index Assessment for Drinking and Irrigation Purpose for Wardha District ... 128
Ashay Devidas Shende, M.M. Sahoo, Shrikant Tekade

Footprints of Sedimentation on Loss of Reservoir Life Using Satellite Remote Sensing Technique ... 138
Tulshidas M. Jibhakate, Yashwant B. Katpatal

Estimation of Shape Factor for Irregular Particles Using Three-Axial Measurement Approach ... 145
Mahendra Umare, Swati Ambadkar

Hydro-Geochemical Elucidation and Its Implications in the Wardha Valley Coalfields of Central India 151
P S Ganvir, Rajeeva Guhey

Application of Hybrid Support Vector Machine Model for Streamflow Prediction in Barak Valley, India 163
A Sahoo, D K Ghose

Rice Leaf Disease Classification Using Cnn 173
Pallapothala Tejaswini, Priyanshi Singh, Monica Ramchandani, Yogesh Kumar Rathore, Rekh Ram Janghel

Developmental Strategies for a Water Quality Assessment Model with Limited Datasets – a Case Study from River Bhavani, India 187
A Ravinashree, C Sivapragasam, M Vasudevan

Evaluation of a Preliminary Support Design of Railway Tunnel Adit in Inner Lesser Himalaya India: An Empirical Analysis 198
Md Alquamar Azad, Neelratan Singh, S K Singh, Apurva Alok

A Study of Environmental and Ecological Impacts Due to Construction and Operation of Tehri-Polavaram Dam 209
S K Verma, R T. Sahu, H Singh, A.D. Prasad, M K Verma

Groundwater Quality Assessment and Its Insinuations in Some Parts of Aheri Taluka Dist. - Gadchiroli, Maharashtra 217
S T Peddiwar, D K Ingole, G R Nimbarte

Characterization of Coke Plant Effluent by Lysimeter Test to Evaluate Safe Disposal to Land 227
Nikhil Kumar, H K Naik, Lipsa Mishra, S Jena

Study of Effects of Some Metals and Hydraulic Retention Time on Performance of AMD Treatment by SAPS 235
M. D. Patel

GIS-Based Industrial Land Suitability Analysis for Locating Industrial Parks in Raipur and Nava Raipur 249
Ankur Baghel

InSAR Data Analysis Using Deep Neural Networks 261
Ayush Kumar Agrawal, Himanshu Verma, Sumanta Pasari

Various Types of Constructed Wetland for Wastewater Treatment-A Review 267
Arvind Kumar Swarnakar, Samir Bajpai, Ishtiyaq Ahmad

Kimberlitic Olivine – a Proxy to Kimberlite Petrogenesis and Ascent Process 280
V. K. Mishra, R. N. Tiwari, C. B. Verma, A. Mukherjee, R. K. Dixit, G. Prabhakar

Landuse/Landcover Monitoring and Spatiotemporal Modelling Using Multilayer Perceptron and 'multilayer perceptron'-Markov Chain Ensemble Models: A Case Study of Dausa City, Rajasthan 287
Sangeeta Soni, Harvir Singh, Jameel Qurashi, Mirza Shuja, Manish Pandey, Aman Arora

Challenges and Solutions in COVID-19 Related Pandemic Solid Waste Management (PSWM) - A Detailed Analysis with Special Focus on Plastic Waste 306
Kshitij Upadhyay, Samir Bajpai

Noise Mapping During New Year Eve in Selected Residential Zone of Raipur City 332
Vishal Kumar, A.V. Ahirwar, A.D. Prasad

Estimation of Crop Water Requirements Within Wainganga Sub-Basin for Kharif and Rabi Season Using Spatial Analysis 337
Digambar S Londhe, Mamta D Kapgate, Y B Katpatal, M S Mukesh

Critical Analysis of Hydrological Mass Variations of Northwest India 346
S Mira Shivani, S Srivastava, A Singh

Physically Based Modelling Techniques for Landslide Susceptibility Analysis: A Comparison 353
V Meena, S Kumari, V Shankar

Spatial Variation of Physico-Chemical Parameters and Water Quality Assessment of Urban Ponds at Raipur, Chhattisgarh, India 366
Anjali Tamrakar, Kshitij Upadhyay, Samir Bajpai

Exploring Temporal Variation of $PM_{2.5}$ and PM_{10} and Their Association with Meteorological Data in Raipur, Chhattisgarh 385
P P Khobragade, A V Ahirwar

Kodagu Disaster (Floods - Emphasis on Catchment Fragmentation Index and Unscientific Land Usage) Analysis Using GIS 395
M Kushal, Sangita Deb Barman

Advancement of Environmental Concerns Through Cli-Fi: James Bradley's Clade as a Prototype 410
Anil Manjhi

Coastal Vulnerability Assessment: A Case Study of the Ratnagiri Coast, Maharashtra, India 415
A B Yadav, P C Mohanty, A Singh

3D Monitoring & Its Significance in Tunneling: Special Emphasis on Tunnel T-74R of USBRL Project, India 424
A Ahmed, S Mishra, A K Singh, A Azad

An Experimental Study of Concrete by Partial Replacement of Fine Aggregate with Bakelite Waste 442
R R Bhopi, M Sinha

Retrieval of Aerosols Extinction Coefficient from Calipso Satellite Observations: A Case Study Over Bhubaneswar 452
Sushree Sasmita, Dudam Bharath Kumar

Impact of Directional Effect of Strong Ground Motion on Scenario-Based Earthquake Hazards: Preliminary Results 458
A Kumar, S Pasari, A Mehta, H Verma

Assessment of Active Tectonics in the Siwalik Basin Around the Subansiri River, NE India 463
Archana Singh, Bijoylakshmi Gogoi, Devojit Bezbaruah, Neelratan Singh

Assessment of Air Pollution Before and During COVID-19 Pandemic Lockdown in Chhattisgarh State, India 480
H Rajput, S Barde

Igneous Rock Classification Using Convolutional Neural Networks (CNN) 493
Surendra Patro, D.C. Jhariya, Mridu Sahu, Pankaj Dewangan, P.Y. Dhekne

Classifying Minerals Using Deep Learning Algorithms 506
Tajendra Singh, D.C. Jhariya, Mridu Sahu, Pankaj Dewangan, P.Y. Dhekne

Delineation of Groundwater Potential Zones in Koyna River Watershed, Maharashtra Using Remote Sensing and GIS 521
R C Swain, D C Jhariya, A Swain, Y K Mawale, S Rana

Research on Shear Strength of Geopolymer Concrete by Using Fly Ash 534
Sachin A. Daur, Madhulika Sinha

Development of Different Hydrogeological Influencing Thematic Layers and Groundwater Potential Zones of Deo River Watershed of Mayurbhanj District, Odisha, India 550
N K Mohanta, D C Jhariya, Y K Mawale, M Sahu

Author Index

Preface

International Conference on Advances in Earth and Environmental Studies (AEES 2022)

February 25th - 26th, 2022

National Institute of Technology (NIT) Raipur is delighted to introduce the IOP conference series International Conference on "Advances in Earth and Environmental Studies (AEES 2022)" from 25th to 26th February 2022 in virtual mode. It was jointly organized by the Department of Applied Geology, Department of Information Technology (IT), and Department of Civil Engineering of NIT Raipur.

Earth and Environmental studies are essential as the whole world is facing problems related to different types of pollution. To mitigate the pollution on our Earth, such research will play a vital role consequently, these studies will add a new and better application of tools and techniques in order to mitigate pollution and make our surroundings healthy, clean and green.

The conference aimed to provide a forum for researchers, practitioners, and professionals from the industry and academia to share their research findings and the current trend of research in the field. Topics like Earth Processes, Structure, Tectonics, Crustal evolution, Petrology and Geochemistry, Ore Farming processes, Economic Potential of Rocks, Mineral Exploration, Mineral Processing, Waste Management, Hydrogeology and its Management, Environmental Pollution, Geological processes and Anthropogenic activities affecting Environment, Mining and Geotechnical Engineering, Remote Sensing & GIS Applications, Geoinformatics and Physical Geology, Medical Geology are in the scope of the conference.

This two-day conference held between 25th to 26th February 2022 was highly successful. The program was chaired professionally and efficiently by the session chairmen who had significant experience in the related field. Dr. Sunil Kumar Singh, Director, National Institute of Oceanography, Goa was the Chief Guest in the inaugural ceremony. Valedictory function was presided by Prof Shalivahan, Joint Director at the Indian Institute of Technology (ISM), Dhanbad. Keynote Sessions were organized for both days. The first-day keynote speech was delivered by Dr. Sunil Kumar Singh, Director, National Institute of Oceanography, Goa and Dr. Zoran Stevanovic, Professor, University of Belgrade, Serbia. On the second-day keynote speech was delivered by Dr. Y. S. Katpatal, Professor, VNIT Nagpur and Dr. Anil Kumar, Scientist, IIRS, Dehradun.

There were more than 100 highly gratifying delegates who all participated from different institutes, organizations across the globe. Among the registered participants 50 have shared their research work in online mode in six different tracks, discussing the problems associated in the area of Earth Processes, Waste Management, Environmental Pollution, Mining and Geotechnical Engineering, Remote Sensing & GIS Applications, etc. All the presented papers were evaluated by 18 session chairs 3 in each track from different renowned organizations. This Proceeding includes 49 research papers related to the conference theme, including the paper presented in the conference covering the essential topics.

On behalf of the Conference Organizers, we would like to express our sincere thanks to the IOP for becoming our publication partner. Furthermore, we are grateful to our Institute, "National Institute of Technology Raipur," Chief Patron, Patron, Chairman, Organizing Secretory, Session Chair Persons, Researchers, and last but not least, all the members of our organizing committee who gave their invaluable support and contributions towards the successful completion of this conference.

Editors
Dr. D. C. Jhariya
Dr. Mridu Sahu
Mr. Vishal Kumar

Technical Program Committee

Chief Patron
Dr. A. M. Rawani, Director, NIT Raipur

Patron
Dr. Prabhat Diwan, Dean (R & C), NIT Raipur

Chairman
Dr. Neeraj Vishwakarma, HOD, Applied Geology

Organizing Secretary(s)
Dr. D. C. Jhariya
Dr. Mridu Sahu
Dr. Ajay Vikram Ahirwar

Session Chair/ Reviewer

Dr. Chandan Kumar Singh, NIT Raipur
Dr. Hrushikesh Naik, NIT Rourkela
Dr. Tarun Kumar Rajak, SSIPMT Raipur
Dr. Y. K. Mawale, SGBMA University, Amravati
Dr. Ram Krishna Mishra, BIT Raipur
Dr. Kabita Biswas Sharma, NIT Raipur
Dr. Vineesha Singh, Barkatullah University, Bhopal
Prof. Shailoneil Sahu, NIT Raipur
Dr. Jayant Damodar Supe, RCET Bhilai
Dr. Mridu Sahu, NIT Raipur
Dr. Bhumika Das, MATS University, Raipur
Dr. N. Bodhankar, Pt. RSU, Raipur
Prof. Harel Thomas, Sagar University
Dr. Chandrashekhar Dwivedi, Jharkhand Central University
Prof. Sayon Pramanik, NIT Raipur
Dr. Arvind Chandra Pandey, Jharkhand Central University
Dr. A. V. Ahirwar, NIT Raipur
Dr. Kunamineni Vijay, BEC Bapatla
Dr.Meena Murmu , NIT Raipur
Dr. L. K. Yadu, NIT Raipur
Dr. Sagar Misra, NIT Raipur
Dr. N. Vishwakarma, NIT Raipur

IOP Conf. Series: Earth and Environmental Science 1032 (2022) 011001 doi:10.1088/1755-1315/1032/1/011001

Ms. Manisha Tamboli
Mr.Vishal Kumar
Mr. Kislay Kumar
Mr. Gaurav Mishra
Mr. Rudra Narayan Mishra
Dr. Bhargava Kumar Iyengar
Ms. Pallavi Pradeep Khobragade
Mr. Vinit lambey
Mr. ShashiKant Verma
Ms. Indhulekha K
Dr. D.C. Jhariya

Management Chair

Dr. Sagar Mishra, NIT Raipur
Dr. Bhargava K. Iyengar, NIT Raipur
Ms. Samrudhi Mohdiwale, NIT Raipur
Mr.Vikas Dilliwar, NIT Raipur
Mr. Upendra Chaurasiya, NIT Raipur
Mr. Gaurav Jangde , NIT Raipur
Dr. Sahajpreet Kaur, NIT Raipur
Ms. Indhulekha K. , NIT Raipur
Mr. Gaurav Mishra, NIT Raipur
Ms. Monika, NIT Raipur
Mr. Tanveer Haidar , NIT Raipur
Ms. Priyanka Arya, NIT Raipur
Mr. Aditya Tiwari, NIT Raipur
Ms. Snehal Hishikar, NIT Raipur
Mr. Anshuman Swain, NIT Raipur
Ms. Muskaan Tripathi, NIT Raipur
Ms.Aanchal Sharma, NIT Raipur
Ms. Rajasi Pandey, NIT Raipur
Mr.Vicky Manipuri, NIT Raipur
Mr. Nihar Ranjan Mohanta, NIT Raipur
Ms. Shumela Naz, NIT Raipur
Mr. Neeraj sundrani, NIT Raipur
Ms. Vanshika Gupta, NIT Raipur
Mr. Suresh Kumar Paul, NIT Raipur
Ms. Anushree Shrivastava, NIT Raipur
Mr. Neeraj, NIT Raipur
Mr. Abhijeet Kumar Singh, NIT Raipur
Mr. Vishal Kumar, NIT Raipur

Peer Review Statement

All papers published in this volume have been reviewed through processes administered by the Editors. Reviews were conducted by expert referees to the professional and scientific standards expected of a proceedings journal published by IOP Publishing.

- **Type of peer review:** Single Anonymous
- **Conference submission management system:** Morressier
- **Number of submissions received:** 69
- **Number of submissions sent for review:** 69
- **Number of submissions accepted:** 50
- **Acceptance Rate (Submissions Accepted / Submissions Received × 100):** 72.5
- **Average number of reviews per paper:** 1.98
- **Total number of reviewers involved:** 21
- **Contact person for queries:**
Name: DC Jhariya
Email: dcjhariya.geo@nitrr.ac.in
Affiliation: NIT Raipur - Applied Geology

Geochemical evolution of basaltic flows from Dongargarh Supergroup, Bastar Craton, Central India

Tanveer Haidar[1], Sagar Misra[1], Neeraj Vishwakarma[1*], K. R. Hari[2]
[1] Department of Applied Geology, National Institute of Technology, Raipur 492010, India
[2] School of Studies in Geology and WRM, Pt. Ravishankar Shukla University, Raipur 492010, India

*Corresponding Author

E-mail: nvishwakarma.geo@nitrr.ac.in

Abstract. Composition of basalts in magmatic arcs influenced by the subducting lithosphere, mantle wedge, dehydration of oceanic crust, and/or crustal assimilation beneath the arc. In this paper, we compiled earlier published geochemical data of Dongargarh basalts to decipher the genesis of volcanic rocks. SiO_2 vs (FeO + MgO) plot of basalt suggests the volcanic rocks are tholeiitic in composition. Primitive mantle and REE normalized plots indicate either the source was enriched mantle or a possible interaction of depleted magmatic source with the Paleoarchean continental crust in the Bastar Craton. The primitive mantle normalized diagram shows a negative anomaly of Nb, Ti, and Ta indicates subduction-related magmatism. In addition to the basalt composition, variation diagrams for tectonic settings represent the continental arc-related magmatism. From the available geochemical data of basalts and earlier studies on Dongargarh volcanic, there was an oceanic ridge that was subducted beneath the continental plate. The source of Pitepani basalts was significantly enriched in HFSE and REE as compared to mid-oceanic basalts. Thus the study finds the volcanic rocks are part of enriched mantle source that formed in the subduction-related magmatism.

Keywords: Bastar Craton, Dongargarh Supergroup, Dongargarh volcanics, Basalt, Kotri-Dongargarh Mobile Belt.

1. Introduction

Mafic volcanic are found intercalated with granites in the Bastar Craton. This type of mafic-felsic volcanic intercalations with granitoids was more frequent during the Archean and early Proterozoic periods. The Neoarchean-Paleoproterozoic Kotri-Dongargarh mobile belt (KDMB) in the central part of the Bastar craton is considered to be composed predominantly of bimodal volcanism, volcano-sedimentary units and granitoids [4][23]. Different tectonic models have been proposed by earlier workers and considered these rock suites as a key factor in the construction of geotectonic models of KDMB [2][10][16][19][24][25][34]. An intra-cratonic rift model given for the evolution of Dongargarh Supergroup by Neogi et al. [19] and references therein, a subduction-related arc-magmatism model for KDMB has been proposed by Asthana et al. [3], Khanna et al. [10], and Santosh et al. [25]. The geochemical signature of boninite dykes from the Dongargarh Supergroup also gives the subduction-related characteristics [8]. Large igneous provinces model for the emplacement of Paleoproterozoic lamproites dykes [24] and mafic dyke swarms and mantle-derived mafic magmatism [31] in rift environment which was possibly induced by the plume.

According to proposed tectonic models and earlier published geochemical data, the KDMB may represent a complex tectonic evolution. Geochemical study of the mafic/basaltic flows from the Dongargarh Supergroup may provide crucial evidence in the evolution of KDMB and the Baster Craton. In this study, we assembled earlier published geochemical data of major basaltic flows from the Bastar craton including Pitepani, Sitagota, Mangikhuta and Kotima formation of Dongargarh Supergroup. The study aimed to find a possible magmatic source of basaltic flows in the Dongargarh Supergroup and to propose a possible magmatic source and geochemical evolution of KDMB based on geochemical attributes.

2. Geological setting

Bastar Craton lies in the central part of the Indian shield and the boundaries of the craton are bordered by two mobiles belts and two rifts/grabens. The northwestern and southeastern boundaries of the craton are bounded by Satpura Mobile Belt also known as Central Indian Tectonic Zone (CITZ) and Eastern Ghats Mobile Belt (EGMB). The Mahanadi graben defines the northeastern fringe of the craton and the Godavari graben demarcate the southwestern boundary of the craton. The Bastar Craton is composed mainly of five major tectonic belts: Sausar-Chilpi belt, Bengpal-Sukma belt, Sonakhan belt, Amgaon belt and Kotri-Dongargarh belt flanked the northern, southern, eastern, western and central part of the craton and three supracrustal sequences: Dongargarh suite, Sakoli suite and Sausar suite [24]. The oldest group among five belts is Bengpal and Sukma group [22][26] and the Chilpi group represents the youngest age [17][18]. The gneissic complex is considered as the basement of the craton which essentially comprises tonalite-trondhjemite gneisses (TTG) and granites with paleo-archaean to neo-archaean in age [7][22][27].

The Kotri-Dongargarh belt is the north-south trending tectonic belt located at the central part of Bastar Craton and consisting of various cycles of volcanism and sedimentation [6][28]. The granitic plutons, intercalated volcanic and sedimentary sequences, and gneissic rocks are collectively considered as Dongargarh Supergroup located in the northwestern flank of the Bastar Craton [18]. The rocks of the Dongargarh Supergroup evolved in the Sitagota syncline [27] and the Amgaon group represented as the basement for the volcano-sedimentary sequence [19]. The Dongargarh Supergroup has been sub-divided into two distinct lithostratigraphic successions: Nandgaon Group and Khairagarh Group, and the Dongargarh granites stratigraphically divides the older Nandgaon and younger Khairagarh Groups [3][6][19]. Nandgaon Group is composed of bimodal volcanism: the Bijli rhyolites and Pitepani mafic volcanic. Pitepani volcanic flows are typically composed of basalts, andesite and/or basaltic andesites. Manikyamba et al. [14] proposed that the Bijli rhyolites (ages of 2479 Ma and 2463 Ma), Dongargarh granites (age of 2465 Ma), and Pitepani volcanic (age of 2471 Ma by Asthana et al. [3]) shows a contemporaneous evolution. The Khairagarh group is composed of three volcanic and three sedimentary intercalated sequences. The lowermost sedimentary sequence is the Bortalao formation and is composed of sandstone and/or conglomerates. The volcanic sequence overlying the Bortalao formation is Sitagota basalts. Sitagota basalt is overlain by a sedimentary sequence of quartz arenite named as Karutola formation. Mangikhuta basalt and/or basaltic-andesite formation underlain by Karutola formation. The overlying Ghogra Formation is consisting of sub-arkosic arenite. The uppermost sequence of the Dongargarh Supergroup is the Kotima basalts. The detrital U-Pb zircon maximum depositional ages of sedimentary sequences of the Khairagarh group are reported as 2463 Ma for Bortalao formation, 2453 Ma for Karutola formation, and 2210 Ma for Ghogra formation [10]. The Khairagarh group is unconformably overlain by the Chilpi group of rocks.

3. Result and Discussion

The present study involves the compilation and comparison of previously published geochemical data of basaltic flows from the Northern and Central parts of the Bastar Craton and Central Dongargarh Orogen (CDO). Basaltic flows of Pitepani, Sitagota, Mangnikhutta, and Kotima belong to Central Dongargarh Orogen and the whole rock data are compiled from Khanna et al. [10] and Neogi et al. [19]. Geochemical data of basaltic flows from central parts of Bastar Craton taken from Srivastava et al. [31].

Representative geochemical data of basalts and basaltic-andesites are taken from Srivastava et al. [32] that belongs to the northern and central parts of the Bastar Craton.

Table 1(a). Representative whole-rock analysis of Basalts from Central Dongargarh Orogen, Bastar Craton

	Central Bastar Craton						Northern and Central Bastar Craton		
	RK1	**RK2**	**RK3**	**RK4**	**RK5**	**RK6**	**RS1**	**RS2**	**RS3**
Major oxides									
SiO_2	49.33	49.30	52.02	50.48	48.80	51.32	50.16	48.57	48.84
TiO_2	0.99	0.33	0.58	0.49	0.93	0.57	1.36	1.01	1.15
Al_2O_3	14.99	15.15	13.17	14.35	14.17	11.23	15.46	13.25	13.97
Fe_2O_3	13.48	10.12	11.50	11.25	14.96	13.28	15.28	15.56	15.38
MnO	0.18	0.16	0.17	0.17	0.20	0.18	0.19	0.21	0.21
MgO	6.64	8.45	7.40	8.01	6.64	10.51	3.25	6.70	6.08
CaO	10.62	10.64	9.66	11.40	9.60	8.42	8.92	10.69	10.14
Na_2O	2.15	2.50	2.15	1.72	2.36	1.59	2.81	1.70	2.21
P_2O_5	0.12	0.08	0.08	0.07	0.13	0.12	0.18	0.07	0.12
K_2O	0.44	0.66	0.97	0.44	1.02	1.21	0.46	0.34	0.29
Total	98.94	97.39	97.70	98.38	98.81	98.43	98.07	98.10	98.39
Trace elements									
Cr	200.00	280.00	250.00	130.00	150.00	700.00	30.00	90.00	120.00
Ni	130.00	170.00	130.00	150.00	110.00	200.00	40.00	80.00	90.00
Rb	13.00	27.00	62.00	34.00	75.00	78.00	20.00	21.00	13.00
Ba	91.00	143.00	226.00	108.00	172.00	327.00	160.00	88.00	75.00
Sr	115.00	158.00	131.00	112.00	165.00	187.00	157.00	112.00	119.00
Nb	4.00	3.00	3.00	4.00	4.00	4.00	7.00	2.00	3.00
Ta	0.20	0.10	0.20	0.10	0.20	0.20	0.30	0.10	0.20
Zr	59.00	44.00	55.00	40.00	65.00	56.00	109.00	42.00	73.00
Hf	1.80	1.10	1.60	1.10	2.00	1.60	2.90	1.00	1.60
Y	21.00	10.00	16.00	12.00	23.00	11.00	33.00	15.00	24.00
Ga	14.00	11.00	12.00	11.00	15.00	11.00	21.00	15.00	15.00
Th	0.40	1.60	3.00	1.40	0.80	2.70	1.50	0.50	0.70
U	0.10	0.40	0.80	0.20	0.20	0.70	0.40	0.10	0.20
La	4.60	8.20	9.60	5.90	5.70	9.00	12.40	4.30	5.10
Ce	11.50	16.90	19.30	12.50	13.80	18.50	28.10	9.70	12.00
Pr	1.59	1.85	2.08	1.46	1.89	2.15	3.94	1.36	1.68
Nd	7.60	7.30	8.50	6.10	9.00	8.80	16.50	5.90	8.20
Sm	2.30	1.50	2.00	1.40	2.50	2.00	4.70	1.80	2.60
Eu	0.86	0.47	0.61	0.49	0.89	0.59	1.68	0.70	0.92
Gd	3.10	1.70	2.30	1.70	3.30	2.10	5.90	2.50	3.50
Tb	0.60	0.30	0.40	0.30	0.60	0.40	1.10	0.50	0.60
Dy	3.70	1.70	2.60	2.10	3.90	1.30	6.70	3.00	4.00
Ho	0.80	0.40	0.50	0.40	0.80	0.20	1.40	0.60	0.90
Er	2.20	1.00	1.60	1.30	2.40	1.30	4.00	1.90	2.60
Tm	0.34	0.16	0.25	0.21	0.37	0.20	0.62	0.29	0.41
Yb	2.40	1.10	1.70	1.40	2.50	1.30	3.80	1.90	2.70
Lu	0.39	0.17	0.28	0.22	0.40	0.20	0.62	0.30	0.43

Table 1(b). Representative whole-rock analysis of Basalts from Central Dongargarh Orogen, Bastar Craton

	Northern and Central Bastar Craton			Central Dongargarh Orogen						
	RS4	**RS5**	**RS6**	**KP1**	**KP2**	**KP3**	**KP4**	**KS1**	**KS2**	**KS3**
Major oxides										
SiO_2	49.10	48.83	50.37	51.93	50.72	51.87	48.07	52.20	50.60	50.10
TiO_2	1.25	1.30	1.23	1.15	1.16	1.11	1.27	0.70	0.73	1.07
Al_2O_3	14.02	13.47	13.81	13.41	13.37	11.62	12.36	12.83	13.06	14.07
Fe_2O_3	15.30	15.53	15.10	14.08	15.11	13.22	15.39	12.21	12.74	14.01
MnO	0.21	0.21	0.20	0.18	0.18	0.14	0.17	0.16	0.20	0.16
MgO	5.92	5.11	5.67	8.17	8.03	6.35	7.20	11.97	11.55	9.90
CaO	9.26	8.90	9.60	8.22	8.44	11.89	11.73	7.51	7.63	7.91
Na_2O	2.12	2.33	2.06	1.31	1.53	2.26	1.85	1.63	2.09	1.38
P_2O_5	0.13	0.14	0.15	0.14	0.15	0.14	0.15	0.07	0.08	0.14
K_2O	0.57	0.65	0.54	1.42	1.30	1.40	1.80	0.76	1.28	1.30
Total	97.88	96.47	98.73	100.01	99.99	100.00	99.99	100.04	99.96	100.04
Trace elements										
Cr	120.00	80.00	110.00	62.00	59.00	48.00	58.00	279.00	384.00	210.00
Ni	100.00	60.00	120.00	80.00	82.00	78.00	87.00	115.00	121.00	107.00
Rb	46.00	25.00	41.00	40.00	34.00	43.00	63.00	23.00	38.00	42.00
Ba	140.00	181.00	181.00							
Sr	147.00	154.00	139.00	300.00	346.00	302.00	263.00	126.00	136.00	123.00
Nb	5.00	4.00	4.00	5.40	5.10	6.60	7.00	2.10	1.80	2.90
Ta	0.40	0.20	0.20	0.38	0.36	0.57	0.58	0.18	0.17	0.21
Zr	87.00	90.00	89.00	124.00	119.00	122.00	140.00	49.00	45.00	73.00
Hf	2.30	2.30	2.30	3.10	3.00	3.00	3.50	1.30	1.20	1.90
Y	27.00	25.00	26.00	28.00	27.00	29.00	33.00	15.00	13.00	23.00
Ga	17.00	18.00	15.00	20.00	19.00	20.00	20.00	11.00	10.00	13.00
Th	1.10	1.20	1.30	3.00	2.80	4.20	4.40	1.00	1.00	1.00
U	0.30	0.30	0.40	0.80	0.80	1.20	1.20	0.40	0.40	0.40
La	8.20	9.70	8.50	18.98	18.52	20.25	23.69	5.67	5.48	6.83
Ce	18.50	21.80	19.60	38.20	37.08	40.76	47.58	12.19	12.22	15.61
Pr	2.58	3.04	2.60	4.71	4.56	4.96	5.83	1.58	1.57	2.13
Nd	11.80	12.50	11.40	18.15	17.67	19.11	22.15	6.52	6.17	9.12
Sm	3.40	3.60	3.20	4.02	3.94	4.18	4.94	1.74	1.55	2.37
Eu	1.12	1.23	1.05	1.26	1.22	1.28	1.49	0.57	0.66	0.86
Gd	4.30	4.60	4.00	4.52	4.39	4.49	5.36	2.02	1.87	2.99
Tb	0.80	0.80	0.80	0.73	0.71	0.74	0.87	0.36	0.32	0.53
Dy	4.90	5.20	4.80	4.52	4.37	4.52	5.42	2.35	2.09	3.54
Ho	1.10	1.10	1.00	0.97	0.95	0.97	1.16	0.53	0.47	0.77
Er	3.00	3.20	3.00	2.74	2.65	2.73	3.21	1.40	1.27	2.12
Tm	0.47	0.48	0.46	0.42	0.41	0.42	0.50	0.23	0.20	0.34
Yb	3.10	3.10	3.00	2.82	2.68	2.81	3.25	1.50	1.36	2.36
Lu	0.47	0.49	0.49	0.45	0.42	0.43	0.51	0.24	0.22	0.36

Table 1(c). Representative whole-rock analysis of Basalts from Central Dongargarh Orogen, Bastar Craton

	Central Dongargarh Orogen								
	KS4	**KM1**	**KM2**	**KM3**	**KM4**	**KK1**	**KK2**	**KK3**	**KK4**
Major oxides									
SiO_2	49.20	48.18	49.43	52.65	51.64	52.43	50.97	49.92	52.10
TiO_2	1.19	0.65	0.71	0.78	0.81	0.71	0.65	0.62	0.86
Al_2O_3	12.96	15.69	15.96	16.92	16.27	12.63	12.30	10.78	12.74
Fe_2O_3	15.49	13.78	13.39	10.77	12.04	12.24	13.12	12.42	12.87
MnO	0.20	0.15	0.15	0.13	0.14	0.17	0.19	0.16	0.19
MgO	8.71	11.10	9.74	7.55	7.78	11.99	10.63	9.54	9.27
CaO	9.52	7.61	7.14	7.96	7.89	6.17	8.52	14.63	8.14
Na_2O	2.24	1.33	1.51	1.66	1.68	2.98	2.56	1.21	3.01
P_2O_5	0.12	0.07	0.08	0.10	0.10	0.09	0.07	0.09	0.10
K_2O	0.41	1.44	1.90	1.48	1.57	0.58	0.98	0.64	0.72
Total	100.04	100.00	100.01	100.00	99.92	99.99	99.99	100.01	100.00
Trace elements									
Cr	234.00	946.00	851.00	209.00	196.00	308.00	456.00	777.00	217.00
Ni	179.00	183.00	163.00	74.00	69.00	82.00	130.00	206.00	67.00
Rb	14.00	47.00	57.00	71.00	71.00	67.00	25.00	12.00	13.00
Ba									
Sr	183.00	75.00	110.00	131.00	132.00	173.00	133.00	111.00	195.00
Nb	4.00	5.10	5.80	6.40	5.80	1.70	2.10	2.50	2.70
Ta	0.25	0.35	0.42	0.44	0.42	0.12	0.17	0.17	0.17
Zr	98.00	83.00	92.00	109.00	92.00	39.00	52.00	56.00	65.00
Hf	2.60	2.50	2.70	2.80	2.60	1.20	1.50	1.50	1.60
Y	34.00	16.00	17.00	19.00	17.00	11.00	13.00	12.00	15.00
Ga	18.00	14.00	15.00	15.00	13.00	8.00	12.00	9.00	14.00
Th	2.00	4.00	4.30	4.50	4.90	1.10	1.50	1.00	1.40
U	0.70	1.30	1.40	2.60	2.80	0.30	0.40	0.20	0.40
La	10.91	16.88	17.11	17.90	15.73	7.24	8.08	5.95	8.52
Ce	24.41	33.73	35.51	35.11	32.60	14.61	16.84	12.65	18.43
Pr	3.31	4.06	4.31	4.18	3.95	1.87	2.09	1.65	2.38
Nd	13.96	15.23	16.25	15.00	14.12	7.35	7.88	6.49	9.54
Sm	3.85	3.16	3.41	3.06	2.81	1.60	1.83	1.58	2.21
Eu	1.25	0.74	0.75	0.82	0.76	0.57	0.61	0.52	0.66
Gd	4.65	2.73	2.91	3.34	3.09	1.84	2.17	1.97	2.51
Tb	0.83	0.41	0.44	0.50	0.45	0.29	0.35	0.32	0.40
Dy	5.36	2.42	2.62	3.02	2.82	1.76	2.21	1.98	2.45
Ho	1.23	0.53	0.56	0.65	0.61	0.39	0.49	0.43	0.53
Er	3.41	1.45	1.53	1.72	1.65	1.06	1.32	1.15	1.34
Tm	0.54	0.22	0.23	0.27	0.25	0.16	0.20	0.17	0.21
Yb	3.68	1.55	1.63	1.77	1.68	1.05	1.31	1.15	1.35
Lu	0.58	0.24	0.25	0.27	0.26	0.16	0.22	0.18	0.21

Table 1(d). Representative whole-rock analysis of Basalts from Central Dongargarh Orogen, Bastar Craton

	Central Dongargarh Orogen							
	NP1	**NP2**	**NP3**	**NS1**	**NS2**	**NS3**	**NM1**	**NM2**
Major oxides								
SiO_2	54.87	54.38	51.80	50.77	49.80	50.41	53.09	51.77
TiO_2	0.55	0.55	0.52	0.91	0.87	0.56	0.57	0.58
Al_2O_3	12.69	12.69	12.72	13.97	14.36	14.35	13.22	13.00
Fe_2O_3	8.68	8.70	9.44	11.55	11.60	9.37	9.88	10.65
MnO	0.17	0.16	0.16	0.18	0.18	0.15	0.15	0.17
MgO	7.69	7.95	9.51	6.84	7.05	8.30	8.96	9.67
CaO	10.03	10.06	9.81	9.93	9.52	9.86	6.53	6.62
Na_2O	2.15	1.95	1.73	2.08	2.64	2.47	1.41	1.42
P_2O_5	0.14	0.15	0.06	0.13	0.11	0.06	0.09	0.09
K_2O	1.20	1.16	1.41	0.98	0.34	0.87	1.92	1.28
Total	98.17	97.75	97.16	97.34	96.47	96.40	95.82	95.25
Trace elements								
Cr	474.00	469.00	533.00	240.00	111.00	389.00	512.00	409.00
Ni	101.00	99.00	171.00	120.00	144.00	143.00	184.00	190.00
Rb	44.00	43.00	52.00	25.00	7.00	22.00	54.00	36.00
Ba	318.00	297.00	206.00	176.00	47.00	183.00	410.00	283.00
Sr	296.00	304.00	134.00	132.00	139.00	154.00	109.00	71.00
Nb	6.00	6.00	1.00	4.00	4.00	3.00	7.00	8.00
Zr	92.00	92.00	43.00	73.00	68.00	46.00	86.00	85.00
Hf	2.00	2.00	2.00	2.00	2.00	2.00	2.00	2.00
Y	18.00	16.00	16.00	26.00	25.00	16.00	19.00	20.00
Th	5.20	5.20	1.30	1.30	1.20	1.20	6.20	6.10
U	1.50	1.50	1.50	2.50	0.50	0.50	2.50	2.00
La	21.00	21.00	5.50	8.50	7.50	6.00	18.00	17.00
Ce	37.00	37.00	11.00	17.00	16.00	13.00	32.00	33.00
Nd	13.00	13.00	3.00	9.00	6.00	3.00	11.00	11.00
Sm	3.20	3.20	1.60	2.80	2.40	1.70	2.80	2.80
Eu	0.75	0.80	0.40	0.90	0.90	0.60	0.70	0.60
Tb	0.20	0.20	0.50	0.80	0.40	0.20	0.50	0.80
Yb	1.50	1.70	1.50	2.60	2.40	1.70	1.70	1.80
Lu	0.20	0.20	0.20	0.35	0.35	0.20	0.25	0.25

SiO_2 ranges from 48-54 wt%, Fe_2O_3 from 8-16 wt%, MgO shows a little wide range from 3-15 wt% and Al_2O_3 from 11-15 wt%. Based on the Ti-content of the studied Basalts, these are subdivided into two groups; low-Ti basalts having Ti < 0.9 wt% and relatively high-Ti basalts having Ti > 0.9 wt% [1][10][13]. The plot of compiled data are showing the composition of basalt to basaltic-andesite in composition, lying in the sub-alkaline field in the total alkalis versus silica (TAS) diagram (Figure 1). REE pattern of all samples on chondrite normalized (Figure 3.a) plot shows a significant enrichment of LREE as compared to HREE concentration. A strong negative of Nb, Ta and variable negative P, Ti anomalies are also recognized relative to neighbouring elements on primitive mantle normalized spider plots (Figure 3.b).

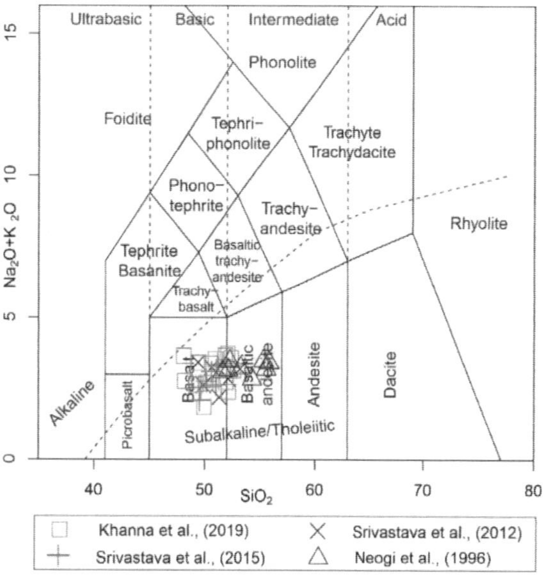

Figure 1. Total alkali versus silica (TAS) diagram [12] showing basalt to the basaltic-andesite composition for the studied basalts.

Basaltic flows in Dongargarh Belt collectively represents tholeiitic composition. Khanna et al. [10] described that Mangikhuta and Kotima basalts are high in Mg#, and low in FeO*, TiO_2, Y and Yb as relative to Pitepani basalts. Asthana et al. [3] categorised the Pitepani suits as high-Mg, low-Ti suits and low-Mg, high-Ti suits. Whereas Sitagota basalts show a wider range in concentration of these elements and can be grouped into high-Ti basalts and low-Ti basalts [1][10][13].

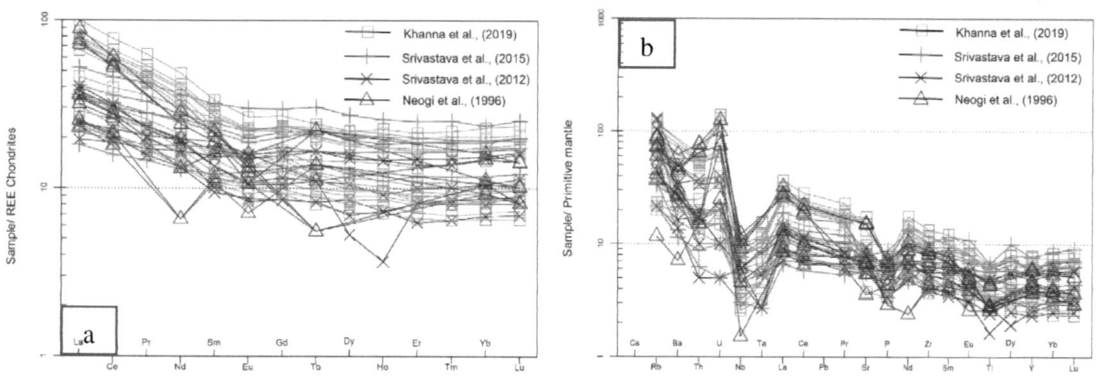

Figure 2. a. Chondrite normalized REE diagram [33], b. Primitive mantle normalized diagram [33] for basaltic flows of Bastar Craton.

Several mafic dykes in the Bastar Craton [30][31] and Dongargarh volcanic magmatism [19] were emplaced in rift setting. Srivastava et al. [31] and Santosh et al. [25] proposed that the emplacement of these mafic dykes in intra-cratonic rift setting were influenced by the plume. Crustal contamination was involved in the generation of some basalts in the Bastar Craton. Initial εNd and εHf values and mantle normalized diagram for basaltic flows of Dongargarh suggest the possible source of basaltic volcanism was generated either through the blending of the depleted mantle and enriched mantle source [10][19] or crustal contamination of melt generated during subduction [10][25]. Low-Mg and high-Ti basalts of Pitepani represents that its magmatic source is crustal contaminated [3].

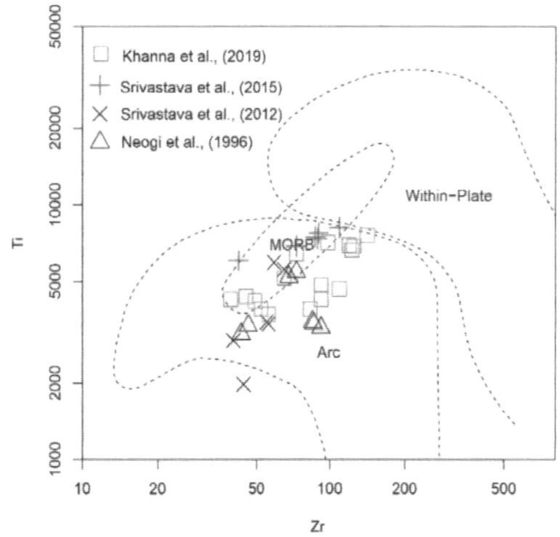

Figure 3. Zr vs. Ti discrimination diagram [20] for compiled basalts collectively representing arc setting.

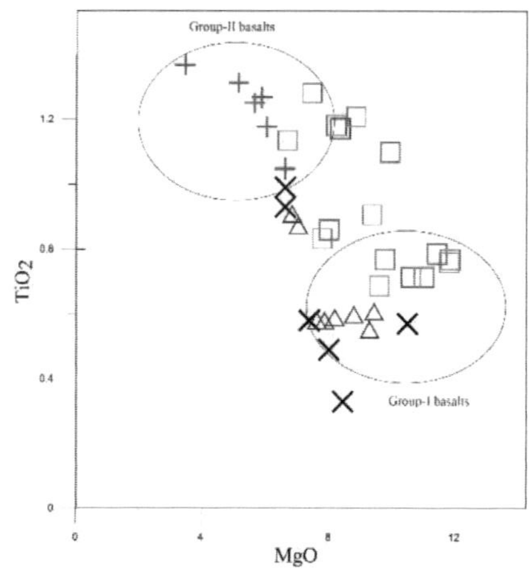

Figure 4. TiO_2 vs. MgO diagram showing high-Ti basalts and low-Ti basalts.

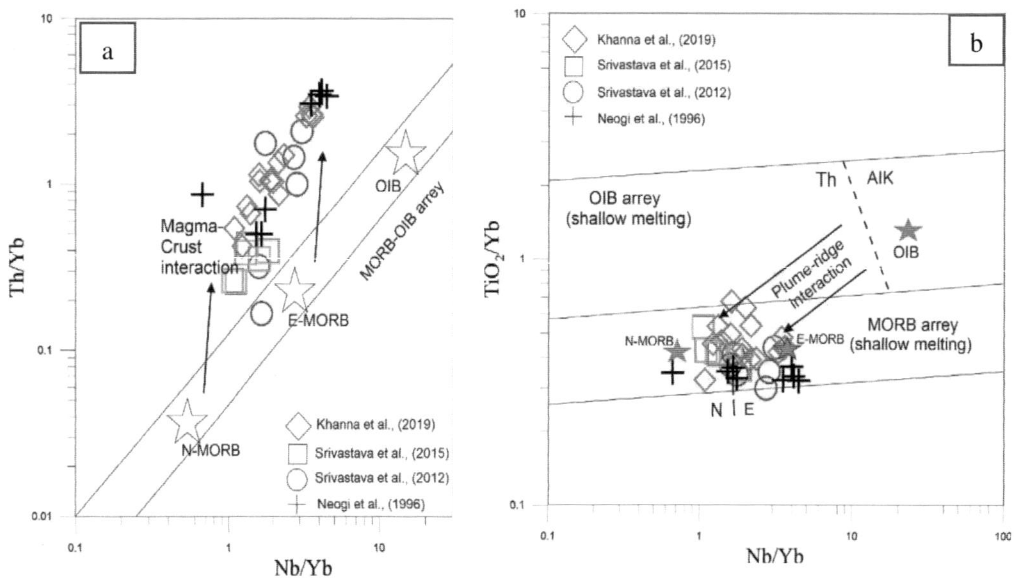

Figure 5. a. Nb/Yb vs. Th/Yb discrimination diagram [21] showing the interaction of magmatic source of studied basalts with crust, b. Nb/Yb vs. TiO_2/Yb discrimination diagram [21] for complied basalts demonstrating a mixed source of N-MORB and E-MORB. Where OIB: Oceanic Island Basalts, N-MORB: Normal Mid Oceanic Ridge Basalt, E-MORB: Enriched Mid Oceanic Ridge Basalt, Th: Tholeiitic, and Alk: Alkaline.

Negative Nb anomaly and low Nb/Th ratio with relatively high LREE in the composition of basaltic rocks despite geological age (e.g., subduction-related basaltic magmatism in Archean time by Smithies et al. [29], Hoffmann et al. [9], Khanna et al. [11], and in Phanerozoic time by Yogodzinski et al. [35] are suggestive of subduction-related arc magmatism. The chondrite normalized and primitive mantle normalized data of mafic and ultramafic volcanic from the Dongargarh belt shows subduction-related magmatism in arc settings [25].

By comparing all the geochemical data of basaltic flows from central Dongargarh Orogen it is observed that strong depletion in Nb, Ta and Ti concentration relative to other trace elements of basaltic samples (K-series, N-series, RK-series and RS-series) on mantle normalized diagram (Figure 3. b) and shows a higher concentration of light-REE over heavy-REE on chondrite normalized diagram (Figure 3. a) marks the subduction-related magmatism for the basaltic flow of Central Dongargarh belt in the Baster Craton. Khanna et al. [10] mentioned that the mantle source of the Dongargarh basalts may modify by subduction derived components. The Nb/Yb and Th/Yb ratios (Figure 5. a) of compiled geochemical data represent that there was possible crustal contamination in the generation of basaltic magmatism of the Dongargarh Belt. Low TiO_2/Yb (< 0.8) and relatively high Nb/Yb (>1.0) values on Nb/Yb vs TiO_2/Yb plot (Figure 5. b) and REE plot represent the Dongargarh basaltic flows were formed from the fertile mantle source. On MgO vs TiO_2 plot (figure 4) all studied basalt samples clustered around two locus. One cluster of basalts showing high-Ti with low-Mg and another cluster of basalts representing low-Ti with high-Mg. Enrichment of Ti in group-II basalts may accompanied by crustal contamination of the basaltic source. As compared to mid-oceanic basalts and island arc basalts the within plate basalts have low concentrations of Mg, Fe, Ti and other compatible elements [5]. The Zr vs Ti plot (Figure 3) for the Dongargarh basalts indicate the samples are fall in the arc and MORB setting but the majority of the samples belong to the arc setting. It suggests that the basaltic magma generated in MOR to arc setting and the basaltic magmatism was not related to intra-plate eruptions.

4. Conclusion

Dongargarh basaltic flows can be categorized into two groups: Group-I basalts having High-Mg and Low-Ti, and Group-II basalts having Low-Mg and High-Ti. REE patterns and primitive mantle normalized plots of complied geochemical data of basaltic flows represents the subduction-related magmatism and enriched mantle source. Zr vs. Ti discrimination diagram and Nb/Yb vs. Th/Yb ratios suggest that the basaltic melt generated in subduction environment and ascending melt may contaminate during magma-crust interaction.

5. References

[1]. Asthana, D., Dash, M.R. Pophare, A.N., S.K., 1996. Interstratified low-Ti and high-Ti volcanics in arc-related Khairagarh Group of Central India. *Curr. Sci.* **71 (4)**, 304-306.

[2]. Asthana, D., Kumar, H., Balakrishnan, S., Xia, Q., Feng, M., 2016. An early Cretaceous analogue of the ~2.5 Ga Malanjkhand porphyry Cu deposit, Central India. *Ore Geol. Rev.* **72**, 1197–1212.

[3]. Asthana, D., Kumar, S., Vind, A.K., Zehra, F., Kumar, H., Pophare, A.M., 2018. Geochemical fingerprinting of ~2.5 Ga forearc-arc-backarc related magmatic suites in the Bastar Craton, Central India. *Journal of Asian Earth Sciences.* **157**, 128-234.

[4]. Asthana, D., Pophare, A.M., Rajalingam, S., Kumar, H., 2014. Neoarchaean Dongargarh Rapakivi A-type Granites and its Relationship to Pitepani Tholeiites. *Gondwana Geol. Mag.* **16**, 25–40.

[5]. Condie, K.C., 1997. Plate tectonics and crustal evolution. *Butterworth-Heinemann. Burlington.* ISBN 0 7506 3386 7.

[6]. Deshpande, G.G., Mohabey, N.K., Deshpande, M.S., 1990. Petrography and tectonic setting of Dongargarh volcanics. *Geol. Surv. India Spec. Publ.* **28**, 260–286.

[7]. Ghosh, J.G., 2004b. Geochronological constraints on the evolution of the Kotri linear belt and its basement. *Geol. Surv. India Rec.* **136 (2)**, 24–26.

[8]. Hazarika, B., Malpe, D.B., Dongre, A., 2019. Petrology and geochemistry of a boninite dyke from the western Bastar craton of central India. *J. Earth Syst. Sci.* **128**, 1–17. https://doi.org/10.1007/s12040-018-1052-y

[9]. Hoffmann, E.J., Munker, C., Polat, A., Konig, S., Mezger, K., Rosing, M.T., 2010. Highly depleted Hadean mantle reservoirs in the sources of early Archean arc-like rocks, Isua supracrustal belt, southern West Greenland. Geochim. *Cosmochim. Acta* **74**, 7236–7260.

[10]. Khanna, T.C., Bizimis, M., Barbeau, D.L., Keshav Krishna, A., Sesha Sai, V. V., 2019. Evolution of ca. 2.5 Ga Dongargarh volcano-sedimentary Supergroup, Bastar craton, Central India: Constraints from zircon U-Pb geochronology, bulk-rock geochemistry and Hf-Nd isotope systematics. *Earth-Science Rev.* **190**, 273–309. https://doi.org/10.1016/j.earscirev.2018.11.014

[11]. Khanna, T.C., Bizimis, M., Yogodzinski, G.M., Mallick, S., 2014. Hafnium–neodymium isotope systematics of the 2.7 Ga Gadwal greenstone terrane, Eastern Dharwar craton, India: implications for the evolution of the Archean depleted mantle. Geochim. Cosmochim. Acta 127, 10–24.

[12]. Le Bas, M.J., Le Maitre, R.W., Streckeisen, A., Zanettin, B., 1986. A chemical classification of volcanic rocks based on the total alkali – silica diagram. *J. Pet. Geol.* **27**, 745–750.

[13]. Longjam, K.C., Ahmad, T., 2012. Geochemical characterization and petrogenesis of Proterozoic Khairagarh volcanics: implication for Precambrian crustal evolution. *Geol. J.* **47**, 130–143.

[14]. Manikyamba, C., Santosh, M., Chandan Kumar, B., Rambabu, S., Li, T., Saha, A., Khelen, A.C., Ganguly, S., Singh, D.Th., Rao, D.V.S., 2016. Zircon U-Pb geochronology, Lu-Hf isotope systematics, and geochemistry of bimodal volcanic rocks and associated granitoids from Kotri Belt, Central India: Implications for Neoarchean–Paleoproterozoic crustal growth. *Gondwana Res.* **38**, 313–333.

[15]. McDonough, W.F., Sun, S. s., 1995. The composition of the Earth. *Chem. Geol.* **120**, 223–253. https://doi.org/10.1016/0009-2541(94)00140-4

[16]. Mishra, D.C., Singh, B.C., Tiwary, V.M., Gupta, S.B., Rao, M.B.S.V., 2000. Two cases of continental collision and related tectonics during the Proterozoic period in Indiain sights from gravity modeling constrained by seismic and magnetotelluric studies. *Precamb. Res.* **99**, 149–169.

[17]. Mohanty, S., 2015a. Palaeoproterozoic supracrustals of the Bastar Craton: Dongargarh Supergroup and Sausar Group. In: Mazumder, R., Eriksson, P.G. (Eds.), Precambrian Basins of India: Stratigraphic and Tectonic Context, 43. *Geological Society of London Memoirs, pp.* 151–164. https://doi.org/10.1144/M43.11.

[18]. Mohanty, S.P., 2021. The Bastar Craton of Central India: Tectonostratigraphic evolution and implications in global correlations. *Earth-Science Rev.* **221**.

[19]. Neogi, S., Miura, H., Hariya, Y., 1996. Geochemistry of the Dongargarh volcanic rocks, Central India: implications for the Precambrian mantle. *Precambrian Res.* **76**, 77–91.

[20]. Pearce, J.A., 1982. Trace element characteristics of lavas from destructive plate boundaries. Thorpe, R.S. (eds): Orogenic Andesites and Related Rocks, *Wiley, London*, 525-48.

[21]. Pearce, J.A., 2008. Geochemical fingerprinting of oceanic basalts with applications to ophiolite classification and the search for Archean oceanic crust. *Lithos* **100**, 14–48.

[22]. Rajesh, H.M., Mukhopadhyay, J., Beukes, N.J., Gutzmer, J., Belyanin, G.A., Armstrong, R.A., 2009. Evidence for an early Archaean granite from Bastar craton, India. *J. Geol. Soc. Lond.* **116**, 193–196.

[23]. Ramakrishnan, M., Vaidyanadhan, R., 2010. Geology of India. *Geol. Soc. India, Bangalore*, **vol. 1**, pp. 1–556. ISBN: 978-81-85867-77-9.

[24]. Santosh, M., Hari, K.R., He, X., Han, Y., Prasanth, M.P.M., 2018. Oldest lamproites from Peninsular India track the onset of Paleoproterozoic plume-induced rifting and the birth of Large Igneous Province. *Gondwana Res.* **55**, 1–20.

[25]. Santosh, M., Tsunogae, T., Yang, C.X., Han, Y.S., Hari, K.R., Prasanth, M.P.M., Uthup, S., 2020. The Bastar craton, central India: A window to Archean – Paleoproterozoic crustal evolution. *Gondwana Res.* **79**, 157–184. https://doi.org/10.1016/j.gr.2019.09.012

[26]. Sarkar, G., Corfu, F., Paul, D.K., McNaughton, N.J., Gupta, S.N., Bishul, P.K., 1993. Early Archean crust in Bastar craton, Central India – a geochemical and isotopic study. *Precambrian Res.* **62**, 127–137.

[27]. Sarkar, S.N., 1957. Stratigraphy and tectonics of the Dongargarh system: a new system in the Precambrians of Bhandara-Drug- Balaghat area, Bombay and M.P. J. *Earth Sci. Eng. Res.* **1**, 237–268.

[28]. Sensarma, S., Palme, H., Mukhopadhyay, D., 2002. Crust–mantle interaction in the genesis of siliceous high magnesian basalts: evidence from the early Proterozoic Dongargarh Supergroup, India. *Chem. Geol.* **187**, 21–37.

[29]. Smithies, R.H., Champion, D.C., Van Kranendonk, M.J., Howard, H.M., Hickman, A.H., 2005. Modern style subduction processes in the Mesoarchean: geochemical evidence from the 3.12 Ga Whundo intra-oceanic arc. *Earth Planet. Sci. Lett.* **231**, 221–237.

[30]. Srivastava, R.K., 2008. Global intracratonic boninite-norite magmatism during the Neoarchean-Paleoproterozoic: evidence from the central Indian Bastar Craton. *Int. Geol. Rev.* **50**, 61-74.

[31]. Srivastava, R.K., Gautam, G.C., 2012. Early Precambrian mafic dyke swarms from the Central Archaean Bastar Craton, India: Geochemistry, petrogenesis and tectonic implications. *Geol. J.* **47**, 144–160. https://doi.org/10.1002/gj.1302

[32]. Srivastava R.K. and Gautam G.C., 2015. Geochemistry and petrogenesis of Paleo-Mesoproterozoic mafic dyke swarms from northern baster Craton, central India: Geodynamic implication in reference to Columbia supercontinent. *Gondwana Res.* **28** 1061–1078.

[33]. Sun S.S. and McDonough W.F., 1995. The composition of the Earth. *Chem. Geol.* **67(5)**, 1050-1056.

[34]. Yedekar, D.B., Jain, S.C., Nair, K.K.K., Dutta, K.K., 1990. Central Indian collision suture. *Geol. Soc. India Spec. Pub.* **28**, 1–43.

[35]. Yogodzinski, G.M., Vervoort, J.D., Brown, S.T., Gerseny, M., 2010. Subduction controls of Hf and Nd isotopes in lavas of the Aleutian island arc. *Earth Planet. Sci. Lett.* **300**, 226–238.

UTILIZATION OF WASTE SOLE LEATHER WITH FLY ASH FOR SELF-COMPACTING CONCRETE

Ankush Thakre[1] & Tarun Kumar Rajak[2]

[1]**PG Student, Department of Civil Engineering, Shri Shankaracharya Institute of Professional Management And Technology, Raipur, CG, India-492001**

[2]**Assistant Professor, Department of Civil Engineering, Shri Shankaracharya Institute of Professional Management And Technology, Raipur, CG, India-492001**

[1]**ankushthakre1996@gmail.com**

[2]**rajak.tarun18@gmail.com**

ABSTRACT

The majority of solid waste from leather industries comes from various processes after manufacturing leather products, and it has been gathered separately from industries. Sole leathers were also emitted in the leather industry. After used, sole leather is thrown at dump yard and the dumping of sole leather on the ground is highly damaging to the environment's health, and the recycling procedure is approximately expensive, therefore this research examines the creation of Self-Compacting Concrete (SCC) mix using Fly Ash and Sole Leathers. This article examines that utilization of waste sole leather including F.A. (Fly Ash) and superplasticizer (S.P.) in self-compacting concrete (SCC) mixtures. fly Ash & sole leather is a novel research concept with high deformability and moderate viscosity, which is required to provide consistent solid particle dispersion during shipping, installation, and until the concrete sets. As per obtained the findings of the current research, the addition with sole leather and fly ash in conjunction with a superplasticizer appears to improve the fresh, hardened characteristics developed strength & proper durability qualities in SCC. The maximum compressive strength was obtained for $SCC_{5\%F.A.+5\%S.L.}$ i.e., 41.10 MPa as compare to other mix proportions i.e. $SCC_{10\%F.A.+7.5\%S.L.}$ (39.95 MPa) & $SCC_{15\%F.A.+10\%S.L.}$ (36.70 MPa).

KEYWORDS: *Self-compacting concrete, Fly Ash, Sole Leather, Superplasticizer*

1. INTRODUCTION

The concrete has a common structural substance in civil engineering projects. Sole leathers can be utilized in concrete, according to preliminary studies; this sort of material has been a prominent study topic in recent years. The lightweight construction materials are seen to be beneficial in encouraging repurposed materials like sole leathers [1]. The majority of sole leather waste is discarded to a landfill after use, and the majority of leathers are not recycled. The findings indicated that, under the tested conditions, sole leathers did not provide complete environmental protection after being dumped into a dump yard to be utilized in concrete mix. SCC (Self-Compacting Concrete) mixtures applied with F.A. (fly ash), Superplasticizer & Sole leathers is frequently used to amend the characteristics strength of fresh and hardened concrete. Many varieties of leather soles are as robust as, if not more so than, metals. Sole leathers are usually corrosion-resistant, letting them to last eternally in SCC mix outside in protected conditions. Sole leathers are adaptable, light, flexible, and long-lasting. It is widely utilized and accounts for the majority of usage worldwide. Leather production factories are quite common these days, and they produce a lot of sole leathers. Sole leather disposal is a significant problem for society and engineers. Due to rising industrialization, population growth, and urbanization, environmental contamination from the sole leather sector is increasing every day. Sole leather manufacturing factories produce the majority of the sole leathers. As a result, sole leathers from these plants are frequently disposed of in huge numbers. The gap in this project, Sole leathers may aid in the reduction of natural resource usage while making self-compacting concrete mixtures. As a result, sole leathers might be used as an alternate material in SCC. A treble leather sole may frequently outlive a rubber sole, thus the more layers of leather included, the exponentially stronger they become. In the present study, the purpose of this study is to obtain the fresh and hardened characteristic strength of various materials & self-compacting concrete characteristics with a mix of SCC-01 ($SCC_{5\%F.A.+5\%S.L.}$), SCC-02 ($SCC_{10\%F.A.+7.5\%S.L.}$) and SCC-03 ($SCC_{15\%F.A.+10\%S.L.}$) where, F.A. represents Fly Ash & S.L. represents as Sole Leather. In this study, self-compacting concrete has made with a mixture of sole leather, fly ash and super plasticizers with a different-different ratio. The sole leather wastes generated as byproduct material during production processes are used in this project. The purpose of this study is to see how shredded 12.5mm sole leather waste affects the qualities of fresh and hardened self-compacting concrete. The percentage of discarded sole leather material in concrete is one of the factors in this study. The waste percentages employed in the concrete mixture were 5%, 7.5%, and 10% by weight of coarse aggregate. Many tests have been undertaken on the workability & characteristic compressive strength of SCC (self-compacting concrete) using with fly ash & sole leather wastes, both fresh and hardened. The characteristics compressive strength of concrete has found decreasing as an amount of fly ash and sole leather waste in the mix increased. The fresh workability of fresh concrete has improved with adding a maximum of 5% sole leather waste. The workability is higher at 5% sole leather, but as the proportion increases to 7.5% and then 10%, the workability decreases. The "Effects of Rubber and Leather Wastes on Concrete Properties" were discussed in his article. He also claimed that the addition of fine and coarse leather particles to fresh concrete improved its workability. The workability of this concrete was better than that of rubber waste concrete. When utilized as coarse & fine aggregates in SCC (self-compacting concrete), leather waste produces a significant loss in compressive strength after 28 days of proper water curing. Fine leather aggregates, on the other hand, have a greater impact on compressive strength than coarse leather aggregates. SCC (Self-compacting concrete) is a non-secede(segregation), extremely permeable (flowable) Every corner of the formwork is filled with concrete consistently & fully with its own weight and encapsulates reinforcement without causing vibration while preserving homogeneity. Compaction only be accomplished by gravity. Such concrete should have a low yield value to assure high flow ability, a moderate viscosity to resist segregation and bleeding, and maintain homogeneity during shipping, placing, and curing to achieve adequate structural performance and long-term durability [2-3]. There is no standard method for SCC mix design, and many academic institutions as well as admixture, ready-mixed, precast and contracting companies have developed their own mix proportioning methods. As per EFNARC recommendation for SCC mix design, the addition of a mineral additive is one of the most significant differences between Self-compacting Concrete (SCC) and traditional concrete [4]. It is expected that the application of sustainable technologies such as supplemental cementitious materials (SCMs) and/or recycled materials will improve the performance of concrete mixtures. SCC is a new concrete substance that has been developed from traditional concrete [5]. In other hand, Waste tyre rubber could also be used to partially substitute finer and gravels in self-compacting concrete, minimizing sand and gravel consumption even while safeguarding natural environment resources [6]. Research of conventional rubberized concrete (CRC) to self-compacting rubberized concrete (SRC) is conducted done (SCRC). In both CRC and SCRC, the substitution of rubber with coarser gravel ranges from 0% to 20%. Rubber bits sized 5mm and 10mm have been used. Rubber shards account for 40% of coarse aggregate replacement in the 5mm size and 60% in the 10mm size [7]. When compared to normal concrete, the rubberized concrete combinations had a reduced unit weight and better workability. When coarse aggregate was replaced instead of fine aggregate, the results of compressive and flexural tests showed that the mechanical properties of rubcrete were reduced further [8]. Tires that have been worn out have accumulated, posing a fire and health risk. An experimental study was done to explore the possibilities of employing tyre chips and crumb rubber as aggregate in Portland cement concrete as a viable a remedy for the issue of dumping of scrap tyres [9]. The paper summarizes the mechanical behavior of specimens composed of concrete that has been filled with tiny crumbled tyre rubber volumes proportions and polypropylene short fibers at 7 and 28 days under static and dynamic loads. The experimental findings are compared to those of concrete specimens with identical characteristics but no fibres or tyres [10]. All the testing of self-compacting concrete done with the help of EFNARC guidelines [11]. Increasing the admixture content beyond a certain level leads to a reduction in strength and increase in absorption [12]

2. MATERIALS

Materials used for the present investigation on self-compacting concrete are:
1) Ordinary Portland Cement (OPC) of Grade-53
2) Fly ash
3) Fine Aggregate
4) Coarse aggregates
5) Super plasticizer
6) Potable Water
7) Sole Leather

In Self-compacting concrete (SCC) mixes with different replacements of Sole leather (S.L.) & Fly Ash (F.A.) were prepared & scrutinize to calibrate the fresh properties of Self-compacting concrete. Beneath tables represents the composition of Self-compacting concrete (SCC) trial mixtures. The replacement was carried out at levels of $SCC_{5\%F.A.+5\%S.L.}$, $SCC_{10\%F.A.+7.5\%S.L.}$ & $SCC_{15\%F.A.+10\%S.L.}$ of fly ash & sole leather content. After iterative trial mixes the water/powder mass ratio (w/p) was selected as 0.40. The final cementitious content was fixed as 340 KG/m^3. The mixes were named using the mix design procedure described in IS 10262-2019. The **M-30** mix proportions were designed.

2.1 Cement

Cement company ACC and grade OPC-53 are using as a cement (binding materials). It confirms the IS 12269:1987 standard. Cement properties are investigated in the laboratory. The outcomes are listed in the Table 1: -

Table 1: Characteristics of Ordinary Portland Cement

Serial No.	Physical requirements	Results	As per IS code
1	Cement Brand Name	ACC Limited	-----
2	Grade of cement	53-OPC	-----
3	IST (Initial setting time) of ordinary cement	138 Minutes	Not less than 30 Minutes
4	FST (Final setting time) of ordinary cement	257 Minutes	Not more than 600 Minutes
5	Comp. strength of cement @ 28 days	58.75 MPa	Not less than 53
6	Specific gravity of cement	3.15	-----
7	Temperature during testing (°C)	28.2 °C	27±2 °C
8	Normal consistency (%)	31 %	-----
9	Fineness of cement (Sieve method)	2.54%	-----

2.2 Fly Ash

The fly ash is employed, according to a requisition from NSPCL Plant, Bhilai, C.G., which confirms to the IS 3812 Part 1(2003) and Part 2(2003) specification. IS 1727-1967 is used to perform and test the physical attributes. It's employed as a fine (as an alternative to cement) in S.C.C. (self-compacting concrete), where it acts as a powder form content. Table 2 & 3 summarizes the physical and chemical characteristics of fly ash respectively.

2.3 Fine Aggregate (Sand)

Fine aggregate also termed as sand is utilized in this construction projects, that meet the IS: 383-1970 specifications. The finer aggregates is collected the location Mahanadi River bed near Dhamtari district, C.G. The physical properties have been performed and tested, and the results are shown in the Table 2.

Table 2: Fine Aggregate (F.A.) Physical Characteristics

Serial No.	Properties	Outcomes
1	Color	Whitish greyish
2	State	Fine Powdered form
3	Specific Gravity	2.01
4	Moisture content	<0.29%
5	Residue on 45 μ sieve (%)	14.9%
6	Fineness (m^2/kg)	359

Table 3: Fly Ash Chemical Composition

Serial No.	Composition of Chemicals	Proportion (%)
1	Silica (SiO_2)	64.9
2	Alumina ($A12O_3$)	27.9
3	Ferric Oxide (Fe_2O_3)	2.9
4	Lime (CaO)	0.9
5	Magnesia (MgO)	0.9
6	Titanium Oxide (TiO_2)	0.6
7	Sulphur Tri-Oxide ($S0_3$)	0.3
8	Loss of ignition	0.3

Table 4: Fine Aggregate Physical Composition

Serial No.	Properties	Outcomes
1	Specific Gravity (S.G.) of F.A.	2.68
2	Water absorption capacity	4.9%
3	Shape of fine aggregate	Circular & angular

2.4 Coarse Aggregates (Gravel)

The size of coarse aggregate has the most impact on the self-compacting concrete's fresh properties. Because the flowability and passing ability of concrete diminishes when the higher aggregates sizes usage of self-compacting concrete (SCC) has increased, segregation increases, Maximum sizes of C.A. (gravel/coarse aggregates) usage in SCC is sets to provide good flowability and passing ability. The Coarse aggregate is collected from the Mandir Hasaud, near district- Raipur, C.G. The tests are performed in the laboratories which is mentioned in the below Table 5.

Table 5: Coarse Aggregate Composition

Serial No.	Properties	Outcomes
1.	Shape of aggregate	Angular
2.	Specific Gravity (S.G.)	2.86
3.	Absorption of Water	0.8%
4.	Maximum Sizes of Aggregates	20 mm & 12.5 mm

2.5 Super Plasticizers

The super plasticizer is used to increase the concrete's workability by lowering the water-to-powder ratio. The compounds **Master Glenium SKY**, which are based on modified **poly carboxylic ethers**, are utilized. The dose is determined by following the recommendations and specifications.

2.6 Water

From IS 456:2000, clean and portable water is used from laboratory.

2.7 Sole Leather

The sole leather wastes generated as byproduct material during production processes are used in this project. The purpose of this study is to see how shredded 12.5mm sole leather waste affects the qualities of fresh and hardened self-compacting concrete. The percentage of discarded sole leather material in concrete is one of the factors in this study. The waste percentages employed in the concrete mixture were 5%, 7.5%, and 10% by weight of coarse aggregate.

3. METHODOLOGY

I.S. & EFNARC guidelines were used to calculate the material mix proportions. It is important to perform a specific test on SCC in order to look into the features of freshness and strength. SCC (Self-compacting concrete's) is to be tested in dual its states of freshness and hardness, with different proportions of fly ash, super plasticizers, and sole leather. Cubes and cylinders are cast using different percentages for different volumes. They are normally cured for up to 28 days before being tested.

To ensure that all of the new property standards were met, the following tests were conducted as per standard testing codes; -

Serial No.	Name of Experiments	Standard Code
1.	Filling ability test that used with T-slump flow test.	EFNARC guidelines [11]
2.	Filling ability test that used with T_{50cm} -Slump flow test.	
3.	Filling ability test that used with V-Funnel test.	
4.	Passing ability test that used with L-Box test.	
5.	The U-Box test for determining passing ability.	
6.	Compression Strength Test	

4. SELF-COMPACTING CONCRETE APPROACH AND WAYS OF APPLYING SOLE LEATHER, FLAY ASH AND SUPERPLASTICIZERS IN CONCRETE

4.1 Approach's for SCC (Self-Compacting Concrete)

A flawless self-compacting concrete mechanism should perceive superior flowability, ability to filling, ability to passing & resistance to segregation, which also can control the additive(admixture) chemical releases. Self-compacting concrete techniques are good approaches for rehabilitation of concrete. The autogenously self-compacting concrete techniques show good results with sole leather, fly ash & superplasticizers.

4.2 Mechanism of Applying Sole Leather, Fly Ash and Superplasticizers in Concrete

As per literature the sole leather, fly ash & superplasticizers is used to prepare the self-compacting concrete. Previous literature revealed that the application of sole leather, fly ash & superplasticizers in self-compacting concrete by directly, it has been revealed that SCC as a good carrier concrete for construction and it has given better results. The application of self-compaction concreting method used for finding out optimum workability for fresh concrete and hardened strength purpose. The sole leather used in this study was obtained from leather industries. The sole leather waste was applied as chipped to replace coarse aggregates in concrete mixture up to a percentage of 10%. The sole leather waste was obtained from shoes industries and used as small as shredded pieces. The workability of fresh concrete & the compression strength properties of hardened concrete cubes(15cm^3) were tested on the self-compacting concrete specimens.

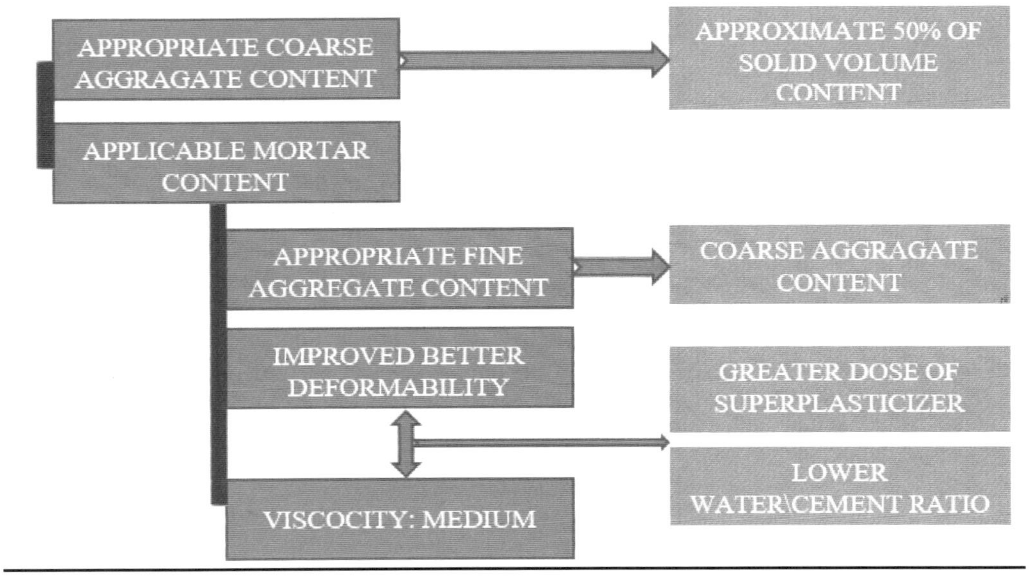

Fig. 1: Methods to achieving self-compaction ability

5. RESULTS AND DISCUSSIONS

5.1 SCC (SELF COMPACTING CONCRETE) FRESH PROPERTIES

5.1.1 TEST FOR T-SLUMP FLOW

SCC's workability criteria are determined using this test. The findings of this test demonstrate that the concrete has high stuffing capacity (filling ability). Values continue to decline as fine aggregate & sole leather material becomes more abundant. Figure 2 depicts overall outcomes of such T-slump flow testing.

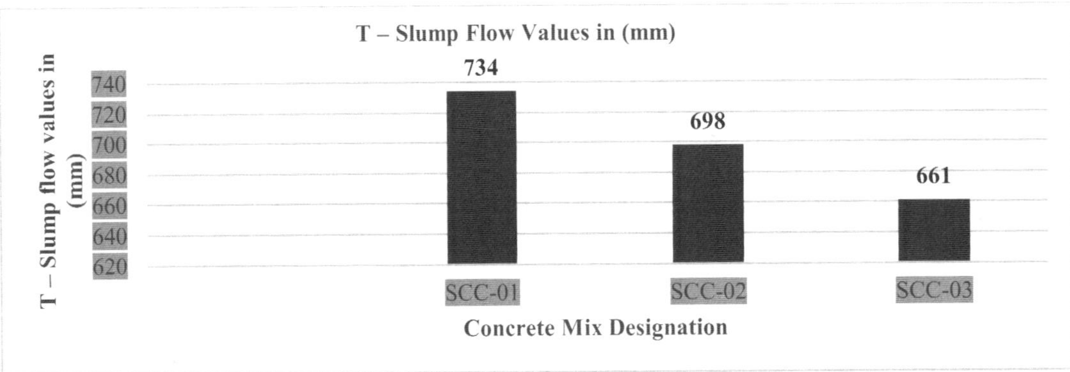

Fig. 2: T-Slump Flow Test Values

This test is intended to determine the self-compacting concrete's filling ability (high stuffing capacity) and workability standards (SCC). The results of this test reveal that the concrete has good filling ability up to $SCC_{15\%F.A.+10\%S.L.}$, beyond which the high stuffing capacity values decrease. It may be due to the coarse material (sole leather & coarse aggregate) contents grow because flowing ability is lesser due to the resistance in materials.

5.1.2 T_{50CM} - SLUMP FLOW TEST

T_{50CM} - slump flow test is performed to ascertain the requirements for workability standards, which has stuffing capacity (filling ability) of SCC (Self-compacting concrete). The obtained outcomes are tallied in-to the beneath. Figure 3 depicts overall outcomes of such T_{50cm} slump flow testing.

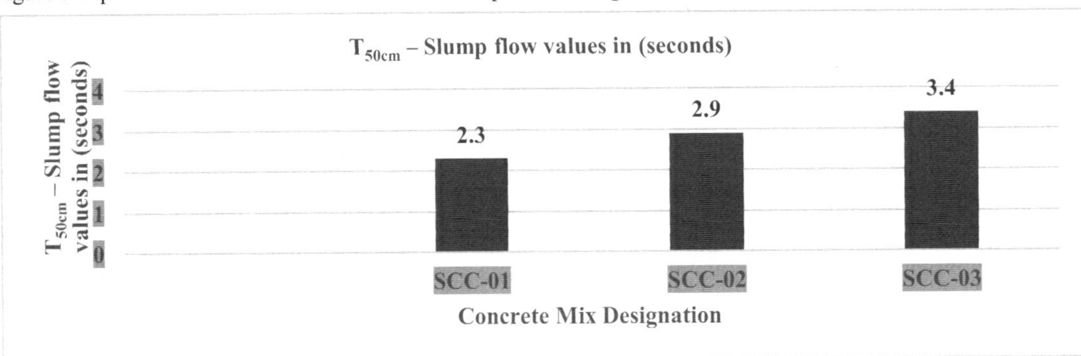

Fig. 3: T_{50cm}– Slump Flows Test Values

In this test, it is used to known the filling ability (high stuffing capacity) and workability criteria of SCC (self-compacting concrete). The result demonstrates that, concrete has good filling ability up to an $SCC_{15\%F.A.+10\%S.L}$ and subsequently the values downs(decreases) as the coarse material (sole leather & coarse aggregate) contents grow because flowing ability is low due to higher resistance.

5.1.3 V-FUNNEL TEST

The workability requirements are determined using this test on SCC. The test results are displayed in Fig 4.

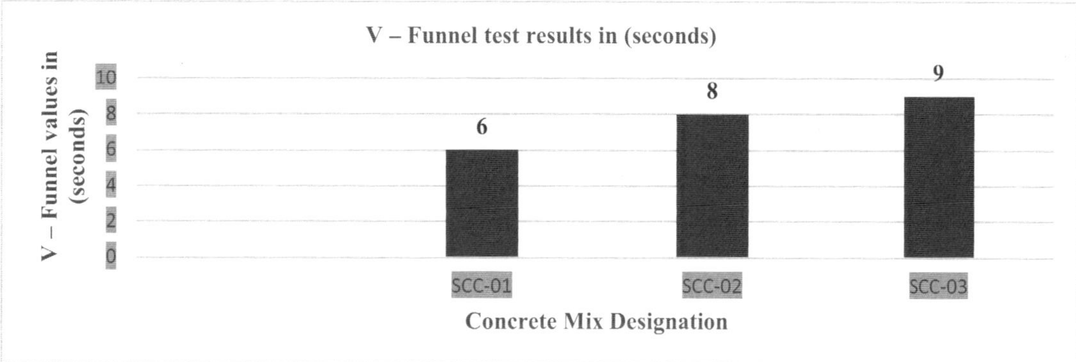

Fig. 4: V – Funnel Test Values

In this test, it is used to known the filling ability (high stuffing capacity) and workability criteria (EFNARC guidelines) of self-compacting concrete (SCC). The outcome of this project indicates that the concrete has a satisfactory filling ability up to a $SCC_{15\%F.A.+10\%S.L.}$ and then the values go on decreasing and segregation & bleeding started with an increase in the fly ash and sole leather contents.

5.1.4 LABORATORY TEST ON L - BOX TEST

L - Box test is intended to appraise both the workability and the ability of pass (passing ability) criteria. Figure 5 depicts overall outcomes of such L-Box testing.

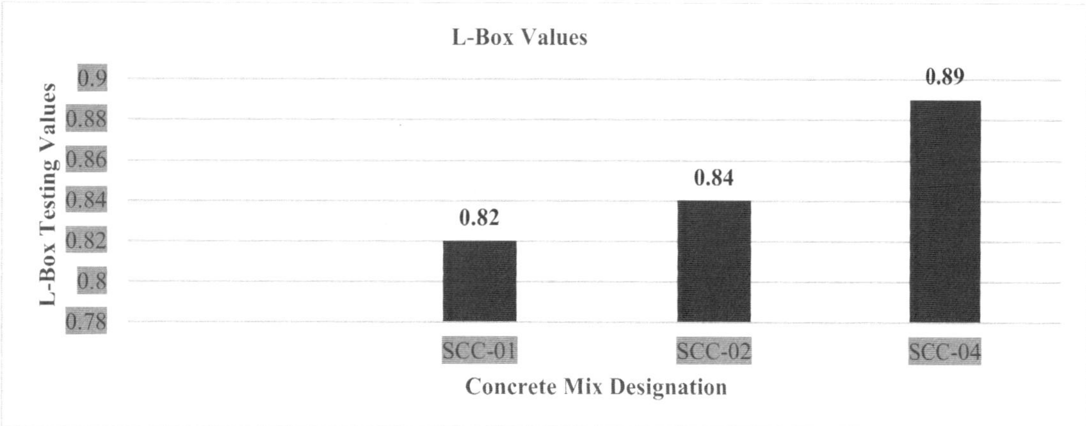

Fig. 5: L – Box Test Values

This test is designed to determine the self-compacting concrete (SCC) passing ability and workability standards. The outcome of this test indicates that the concrete has a strong passing capacity up to a certain point $SCC_{15\%F.A.+10\%S.L.}$ and then the values go on decreasing and segregation & bleeding started with an increase in the fly ash and sole leather contents.

5.1.5 U - BOX TEST

Self-compacting concrete (SCC) uses this test to determine the passing ability & workability requirements, one of which is transitory ability (passing ability). This test results are listed in Fig. 6.

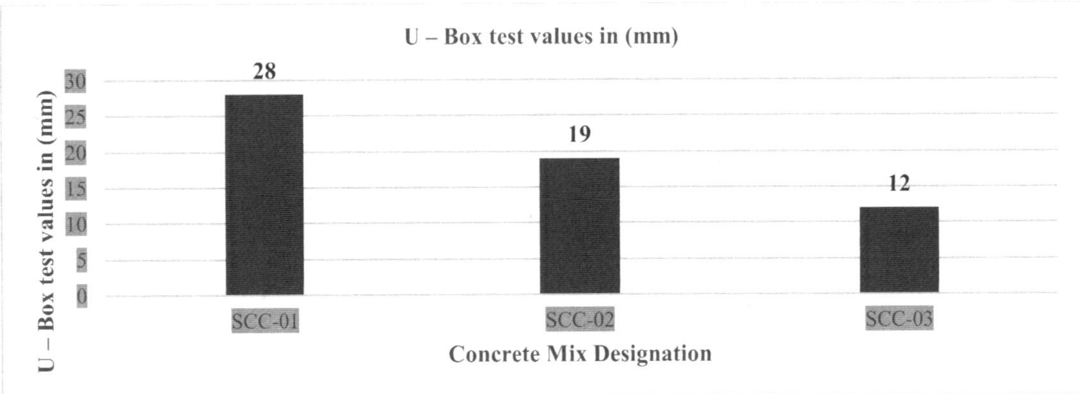

Fig.6: U-Box Test Values

This test is designed to determine the self-compacting concrete (SCC) passing ability and workability standards. The outcome of this test indicates that the concrete has a strong passing capacity up to a certain point $SCC_{15\%F.A.+10\%S.L.}$ and then the values go on decreasing and segregation & bleeding started with an enhance to the fly ash and sole leather contents.

5.2 HARDENED PROPERTIES OF SCC RESULTS

5.2.1 TESTING ON CHARACTERISTICS COMPRESSION STRENGTH

Figure 7 depicts overall outcomes of such Compressive Strength Testing.

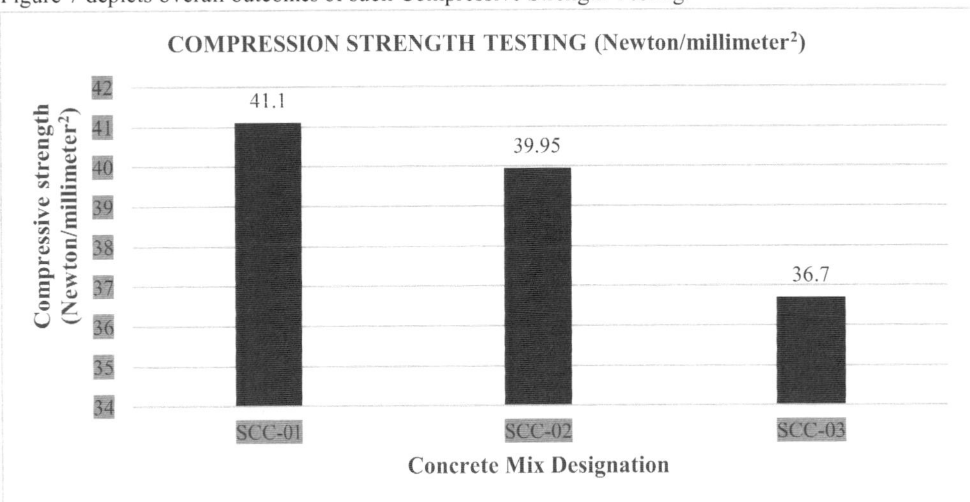

Fig. 7: Compression Strength test @ 28 Days.

The target mean strength (F_m) of 15cm^3 cubes specimens for Self-compacting concrete with $SCC_{5\%F.A.+5\%S.L.}$ replacement is greater i.e. 41.10 MPa, which is compared to other concrete mixes $SCC_{10\%F.A.+7.5\%S.L.}$ & $SCC_{15\%F.A.+10\%S.L.}$ replaced Self-compacting concrete with all the mix proportions. The target mean strength (F_m) values decreases but achieve greater value at $SCC_{10\%F.A.+7.5\%S.L.}$ replacement i.e., 39.95MPa and after that target mean strength (F_m) values get decreased which is $SCC_{15\%F.A.+10\%S.L.}$ replacement i.e., 36.70MPa at 28 days compressive strength.

6. CONCLUSION

Based on the test results, following conclusions are drawn:

a) Based on the T Slump test, the filling ability of SCC-01 mix is higher as compared to other concrete mixes i.e., SCC-02 & SCC-03.

b) Based on the T_{50CM} Slump flow test, the filling ability of SCC-01 mix is higher as compared to other concrete mixes i.e., SCC-02 & SCC-03.

c) Based on the V-Funnel test, the filling ability (high stuffing capacity) & workability of SCC-01 mix is better as compared to other concrete mixes i.e., SCC-02 & SCC-03.

d) Based on the L-Box test, the ability of pass (passing ability) criteria of SCC-01 mix is better as compared to other concrete mixes i.e., SCC-02 & SCC-03.

e) Based on the U-Box test, the passing ability & workability criteria of SCC-01 mix is better as compared to other concrete mixes i.e., SCC-02 & SCC-03.

f) Based on the concrete compression strength test, the target mean strength criteria of SCC-01 mix is much better as compared to other concrete mixes i.e., SCC-02 & SCC-03.

From these experimental studies, SCC (self-compacting concrete) with sole leather resulted in compressive strength close to the values for concrete only, and no significant reduction in concrete compression strength, but the concrete's compression strength was reduced of SCC-02 i.e., 39.95 MPa & SCC-03 i.e., 36.70 MPa in self-compacting concrete mix design. The workability lowered with increasing percentage of fly ash & sole leather. The specific conclusions are drawn from this study for concrete with sole leather waste. The workability was greater than conventional concrete using sole leather waste. Using sole leather waste causes dramatic decrease after 28 days water curing, when used as $SCC_{15\%F.A.+10\%S.L.}$

REFERENCE

[1] Nabil M. A. Al-Joulani, October 9-10, 2013, "Effects of Rubber and Leather Wastes on Concrete Properties", Paper Code. ENV163, (ICEEP III), October 9-10, 2013, Hebron.

[2] Specification and Guidelines for "Self-Compacting Concrete", February 2002 EFNARC, website www.efnarc.org Association House, 99 West Street, Farnham, Surrey GU9 7EN, UK.

[3] S. Mahesh (MTech), "Self-Compacting Concrete and Its Properties", Int. Journal of Engineering Research and Applications ISSN: 2248-9622, Vol. 4, Issue 8(Version 7), August 2014, pp.72-80.

[4] Dhiyaneshwaran, S., Ramanathan, P., Baskar, I. and Venkatasubramani, R., "Study on Durability Characteristics of Self-Compacting Concrete with Fly Ash", Jordan Journal of Civil Engineering, Volume 7, No. 3, 2013.

[5] Snehal Afiniwala Dr. Indrajit Patel, Nirav Patel,"Effect of High-Volume Fly ash on Rheological Properties of Self Compacting Concrete", International Journal of Emerging Technology and Advanced Engineering (ISSN 2250-2459, ISO 9001:2008 Certified Journal, Volume 3, Issue 7, July 2013).

[6] Robert Busic, Ivana Milicevic, Tanja Kalman Sipos, and Kristina Strukar, "Recycled Rubber as an Aggregate Replacement in Self-Compacting Concrete", Materials, 2018, 11, 172, 14 September 2018

[7] M. Mishra and K.C. Panda, "Influence of rubber on mechanical properties of conventional and self-compacting concrete", 14 May 2017

[8] Aiello M, Leuzzi F (2010), "Waste tyre rubberised concrete: properties at fresh and hardened state", Waste Manage 30:1696–1704

[9] Neil N Aldin, Ahmed B Senouci, "Rubber – tyre particles as concrete aggregate" journal of material in civil engineering, Volume 5, Issue 4, November 1993.
[10] Hernandez-Olivares F, Barluenga G, Bollati M, Witoszek B (2002), "Static and dynamic behavior of recycled tyre rubber-filled concrete", Cem Concr Res 32:1587–1596
[11] EFNARC Guidelines, "Specification and guidelines for self-compacting concrete". February 2002
[12] J.M.Khatib, "Performance of self-compacting concrete containing fly ash", Construction and Building Materials Volume 22, Issue 9, September 2008, Pages 1963-1971

A Study on the Influence of Vegetation Growth on Slope Stability

Pooja Naredla[1], S Sangeetha[2]

[1]PG Student, Department of Civil Engineering, VNR VJIET, Hyderabad, India-500090
[2]Assistant Professor, Department of Civil Engineering, VNR VJIET, Hyderabad, India-500090

Email: [1]poojanrdl@gmail.com; [2]mailsangeethasundar@gmail.com

Abstract: Vegetation plays a major role in terms of slope stability. Roots of grass and trees can increase the strength properties of soil which improves the stability of slope. A natural slope consists of grass and infinite number of trees inclined in different directions along with inclined roots. The aim of the paper is to study the effects of vegetation growth on slope stability for different root reinforced soil depth by varying tree inclination (along with roots), root spread and tree spacing for different slope geometry. The study involves determination of safety factor (FOS) of natural slope existing with and without vegetation growth (such as only grass, grass and trees). GeoStudio Slope/w software with limit equilibrium method was used for stability analysis of different slope inclinations. A tree was modelled as a point load and its roots were represented as reinforced soil properties with anchors and only as root reinforced soil. When compared to the bare slope, it was observed that there was an increase in FOS of slope on both the models of vegetation. Variations in FOS by varying root reinforced soil depth, tree inclinations and spacings were reported for different slope geometry and compared. Considering all these parameters, an empirical corelation for computation of vegetation influenced FOS has been developed. This study provides a mechanical accountability of vegetation cover on the stability of slopes.

Keywords: Slope stability, GeoStudio software, Tree spacing and inclinations, Root spread, Slope inclinations, Anchor reinforcement

1. Introduction

Slope stability analysis is an important step in defining slope failure. It is influenced by many factors and vegetation growth is one among them. Vegetation growth affects the stability of the slope hydraulically and mechanically. Roots act as anchor reinforcement and also it provides additional cohesion which enhances the stability of the slope by improving the soil strength.

Natural slopes consist a wide variety of vegetation with different tree inclination, at different spacing, root spread and they will not have a uniform posture of tree and the root. Root reinforces the soil at different depths within the slope and it also plays an important role in strengthening the soil as it has an ability to hold the soil particles tightly. Tree inclination contributes for resisting the forces but it can also affect the slope in a negative way by acting as a driving force.

2. Literature review

The elimination of soil moisture by evapo-transpiration through vegetation implies hydrological effects that can result in an improvement in soil suction or a decrease in pore water pressure, thus an improvement in shear strength. In addition to enhancing the strength of the soil by decreasing its water content, the weight of the soil mass is decreased by plant evaporation. In this study they analysed factor of safety for three tree positions on slope such as at toe, middle and crest of the slope. The presence of tree on slope acts as an additional shear resistance which is greater than shear stress developed in soil that leads to the movement of the slope and when factor of safety was analysed using SLIP4EX for three tree positions on slope such as at toe, middle and crest of the slope, it was observed that stability of slope under the influence of tree increases more when tree is located at the toe of the slope when compared to other locations [1].

Slope stability was found by varying root zone depth 1m, 1.5m, and 2m by considering the vegetation on entire ground, on top of the slope and only on slope and FOS is more in case of vegetation present on entire ground and then followed by vegetation only on slope. It was noted that due to the increase in the penetration of the plant root or increase in the depth of plant root results in the increase of factor of safety. When the trees or forests are cut down then the soil moisture present in that area is more when compared to uncut which results in the failure of slope [2].

Slopes are found to be stable due to the presence of vegetation which increases the shear strength of the soil in terms of cohesion and friction angle [3]. There is more resistance for soils with root compared to soils without roots and soil strength gets increased due to the presence of roots that undergo fewer deformations when subjected to vertical stresses [5]. Analysis for different root architectures like without vegetation, with semi-spherical vegetation, uniform vegetation and cylindrical vegetation was done and it was observed that rooted soil showed much improved values of factor of safety and shear strength than the soil without root [6]. Roots of the vegetation can reach up to the depths of 3m [4, 7, 9]. The zone of Soil with roots is represented in terms of root reinforced soil properties with anchors [8, 10] and only as root reinforced soil [8].

3. Methodology

GeoStudio 2021 slope/w software was used for numerical analysis of slopes with and without vegetation cases by adopting limit equilibrium method. Tree roots were considered as anchors and reinforced soil with additional shear strength properties in terms of cohesion and internal friction angle. The self-weight of the tree was considered as a point load.

For different slope inclinations by varying root reinforced soil depth, tree inclination, tree spacing and root spread angle, slope stability analysis was performed. All these aforementioned parameters were varied and their influence on the stability of the slope was observed. Around 420 slope models were analyzed and an equation was generated for given soil properties. All the parameters considered in the analysis are illustrated in figure 1.

3.1 Input Data

Slopes of different slope inclinations such as 25°, 30°, 35°, 40° were considered. The height of the slope is 10m. Soil properties of two layers such as, the lower layer with an unreinforced soil and the upper layer with root reinforced soil were given in Table 1. The self-weight of the tree and its root (anchor) parameters were given in the Table 2.

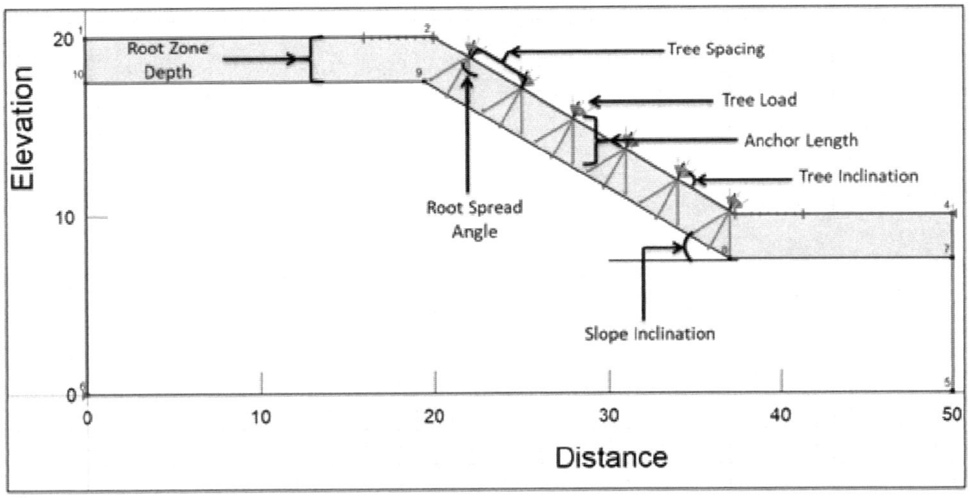

Figure 1. Parameters considered for analysis

Table 1. Properties of soil layer

Properties of soil	Lower layer [11]	Upper layer [8]
Unit weight, γ (kN/m^3)	20	22
Cohesion, C (kN/m^2)	0	25
Angle of internal friction, Ø (°)	25	33.75

Table 2. Tree root reinforcement parameters

Parameter	Values
Tree load	15 kN [8]
Tree inclination	50°, 60°, 70°, 80°, 90°
Tree spacing	2m, 3m, 4m
Root tension	1.5 kN [10]
Root Length	0.5m, 1m, 1.5m, 2m, 2.5, 3m,
Root spread angle	30°, 45°, 60°

4. Analyses

A bare slope was initially modeled as represented in figure 2 for different slope inclinations such as 25^0, 30^0, 35^0, and 40^0 with the properties given in Table 1 and FOS was calculated by limit equilibrium method using GeoStudio slope/w software.

FOS was calculated for all the above slope inclinations with two layers as given in Table 1 by varying the depth of the upper layer (i.e. root reinforced soil depth) such as for the depths 0.5m, 1m, 1.5m, 2m, 2.5m and 3 m as shown in figure 3.

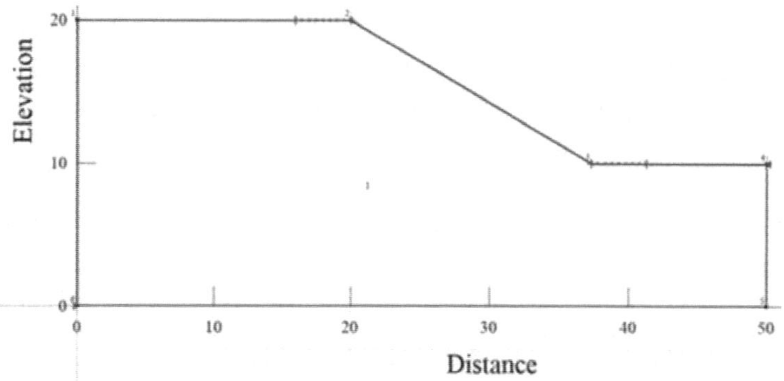

Figure 2. Bare slope of 30^0 inclined slope

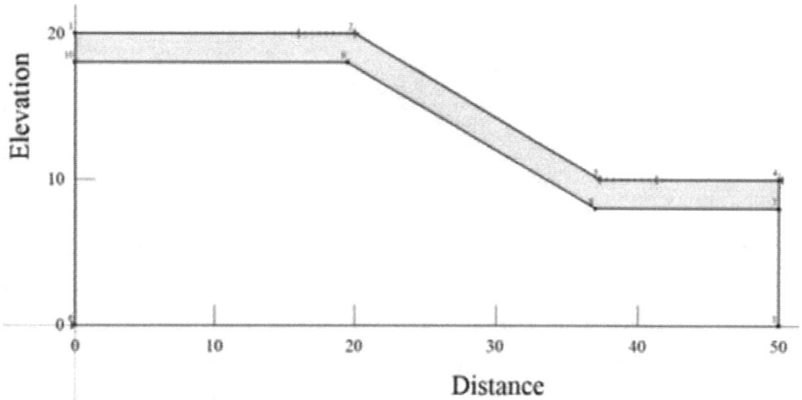

Figure 3. Slope with 2m root zone before analysis of 30^0 inclined slope

In this work, vegetation is represented in two ways, such as in terms of root reinforced soil considering cohesion, angle of internal friction along with anchor and only as root reinforced soil. A two layered slope with trees as point loads inclined with an angle of 50^0, 60^0, 70^0, 80^0, 90^0 in an anti-clockwise direction with respect to the ground, spaced at 2m, 3m, 4m and roots as anchors with different root spread angle of 30^0, 45^0, 60^0 were modeled. FOS was calculated for various combinations of those parameters. FOS was found varying in each case. An example of two layered slope model with tree load, tree spacing, tree inclination, root anchors and root spread angle is shown in figure 4. Around 420 models were analyzed in this study.

Initially a 25^0 slope was modeled and different cases or combinations of every root zone depths, tree inclinations, tree spacing and root spread angle were modelled and then for all those cases FOS was calculated. Similarly for different slope inclinations, FOS was calculated and compared. All the results are graphically represented as shown in figure 10, 11, 12 and 13.

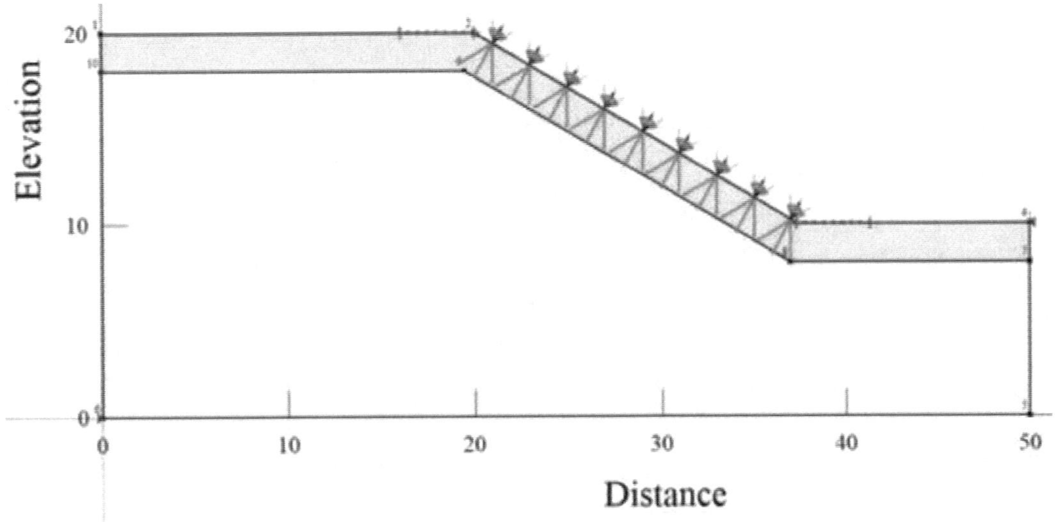

Figure 4. A 30^0 slope with 2m root zone, root anchors, root spread of 60° and 60^0 inclined tree loads spaced at a distance of 2m before analysis

5. Results and Discussion

It was observed that FOS shows much improvement in all cases where the vegetation is present when compared to bare slopes. An example of final output resemblance of FOS after analysis of different cases such as bare slope, slope with only root reinforced soil zone and reinforced soil zone along with root anchors which displays FOS values of 0.809, 1.354 and 1.538 respectively is represented as shown in figure 5, 6 and 7.

When compared between different slope inclinations, 25^0 and 30^0 slopes show higher FOS than 35^0 and 40^0 slope due to its steepness nature as shown in figure 8. It was observed that as the inclination of the tree decreases, FOS increases as it contributes to resisting forces and as the spacing between tree decreases, FOS increases because the interaction between the soil and the roots will be more.

Whereas the change in root spread angle in terms of anchorage has not shown any variation in FOS, hence the obtained curve for all the angles in graph are same as represented in figure 9. But in reality, the extent of root spread creates reinforced soil zone. Here, in this study, it has been considered that reinforced soil is found to exist all along the length of the slope. The root reinforced soil only to different geometry of root zones has confined [6].

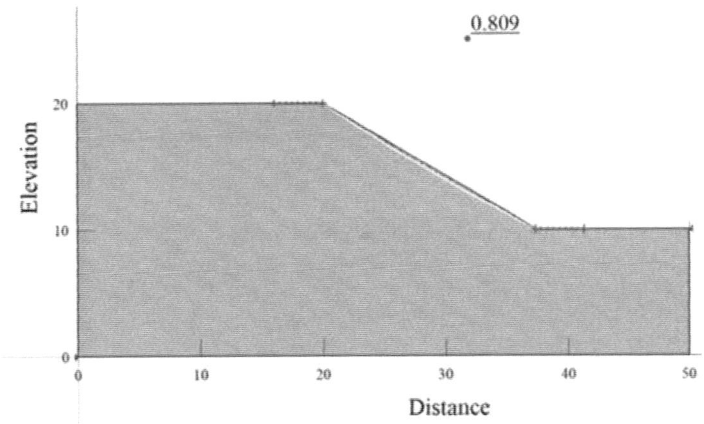

Figure 5. Bare slope of 30^0 inclined slope after analysis

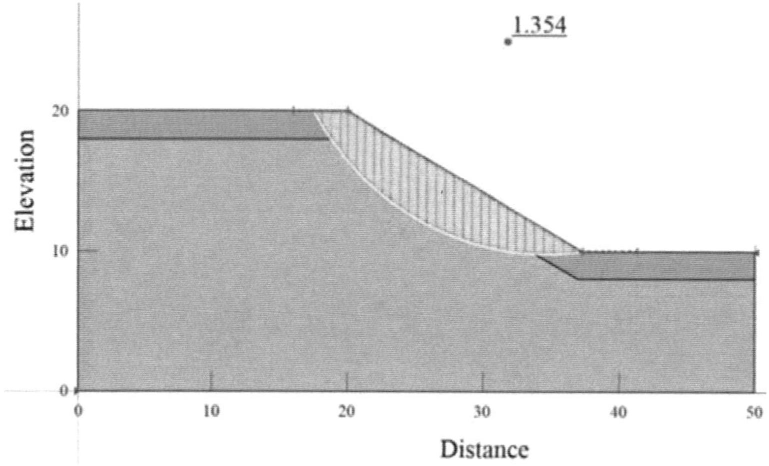

Figure 6. A 30^0 slope with 2m root zone after analysis

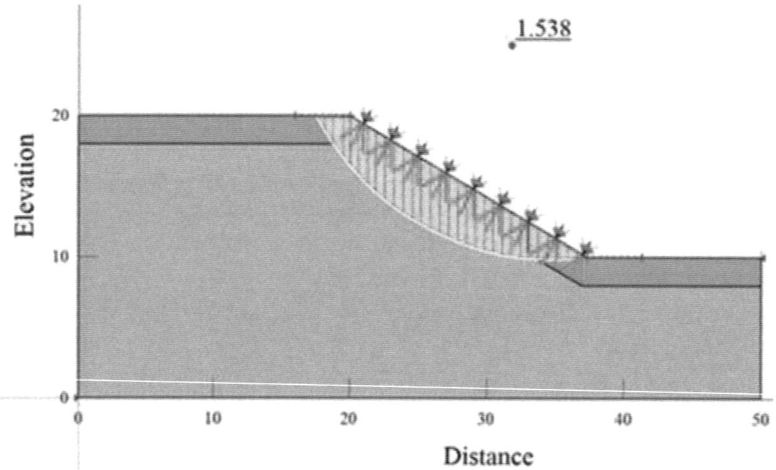

Figure 7. A 30^0 slope with 2m root zone, root anchors, root spread of 60° and 60^0 inclined tree loads spaced at a distance of 2m after analysis

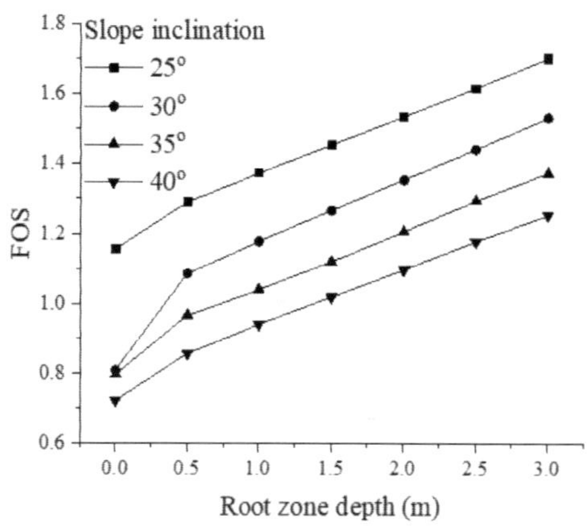

Figure 8. FOS for different slope inclinations at various root zone depths

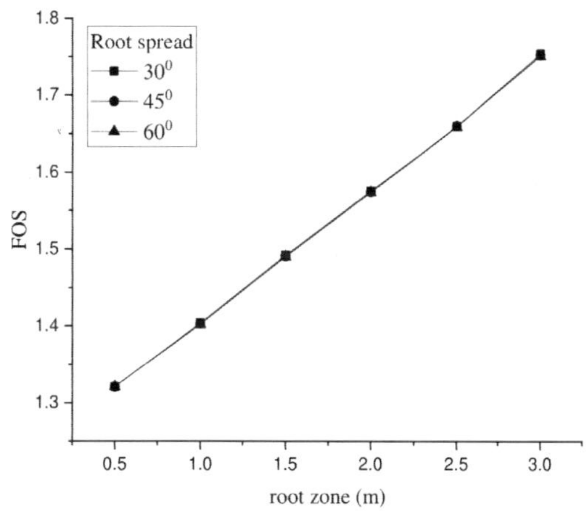

Figure 9. FOS for different root spread angle at varied root zone depths

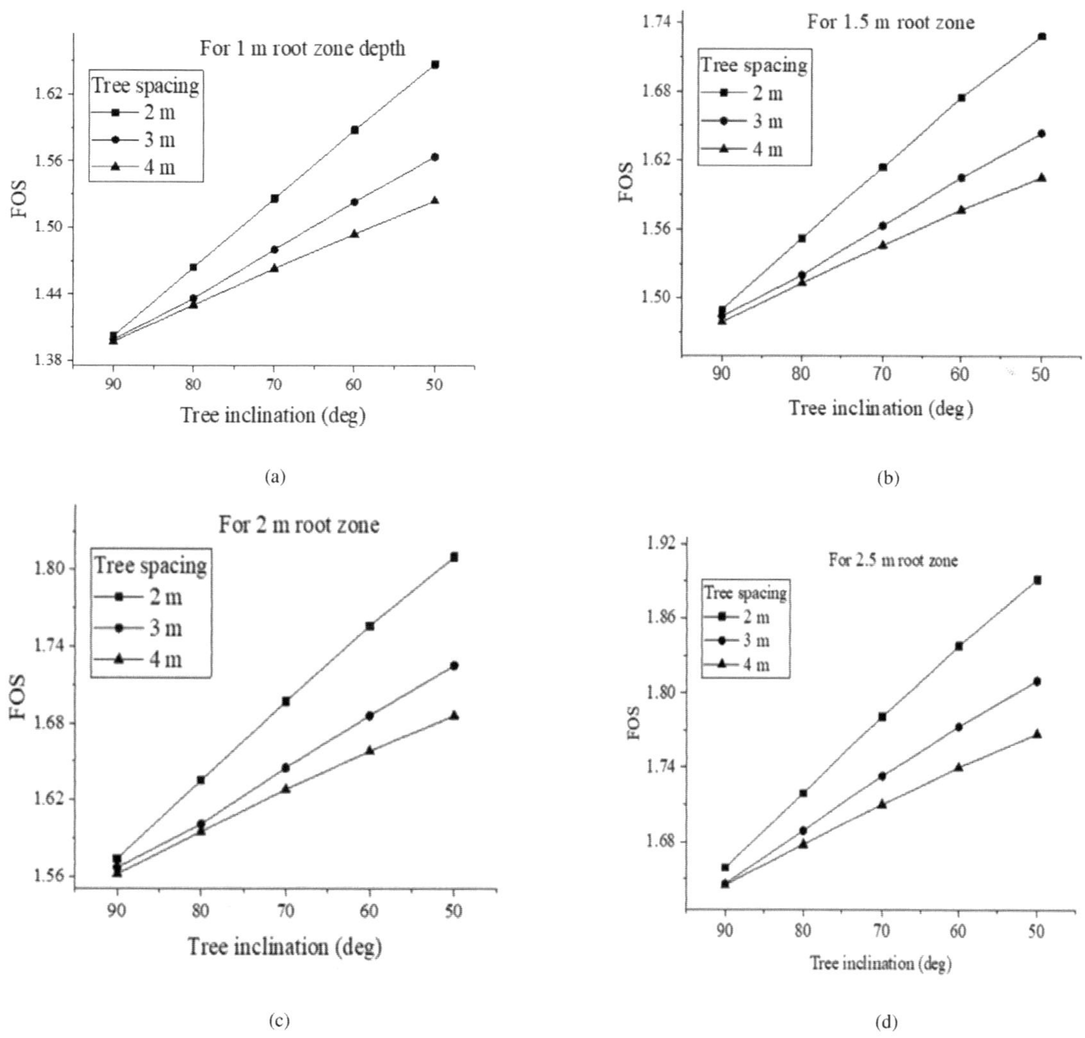

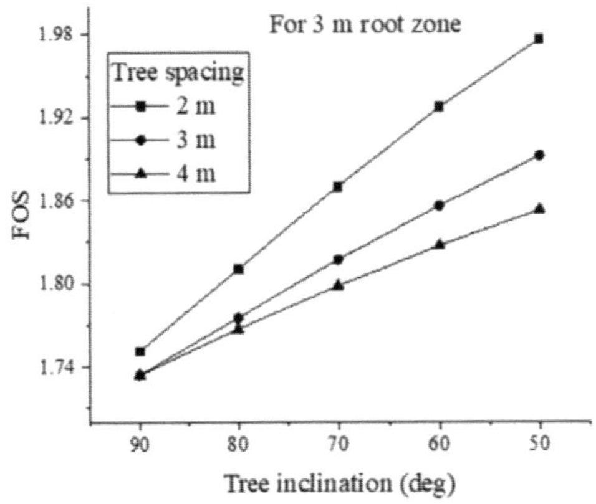

(e)

Figure 10. FOS for different tree inclinations spaced at 2m, 3m, 4m for root zone depths of 1m, 1.5m, 2m, 2.5m, 3m for a 25° slope model

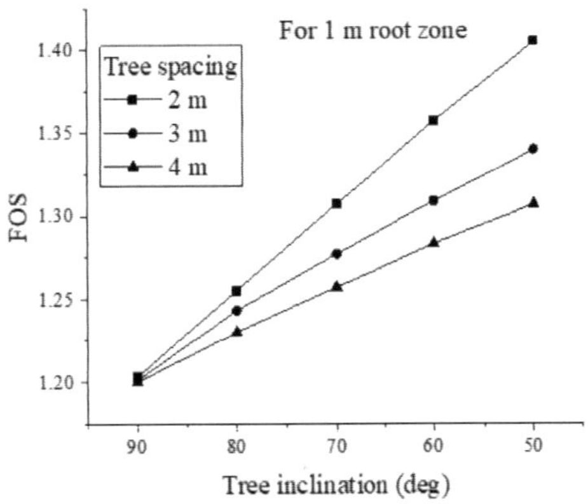

(a)

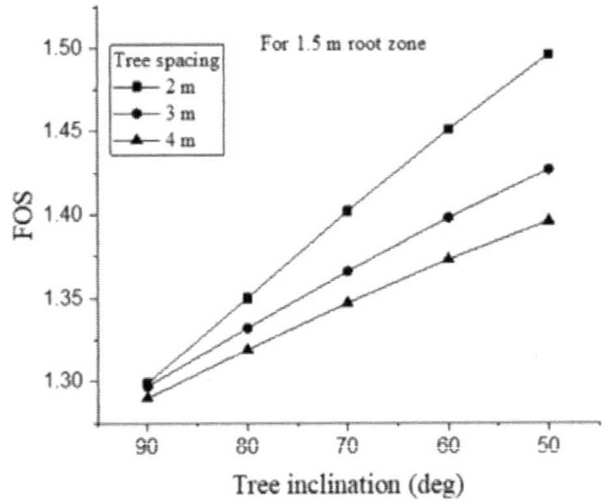

(b)

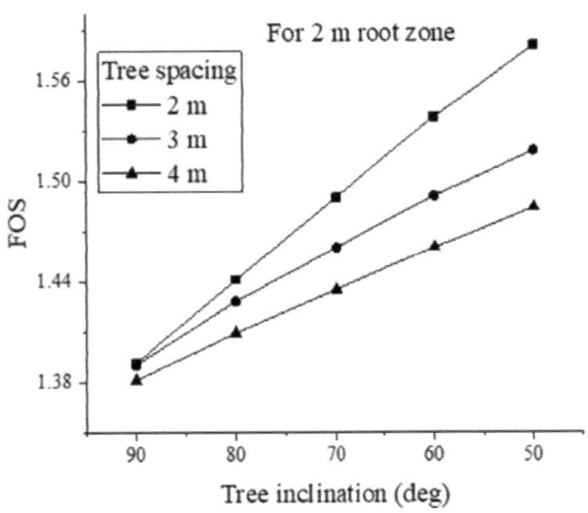

(c)

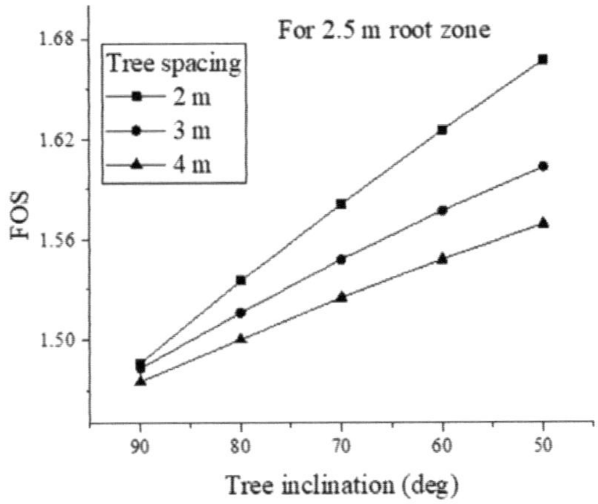

(d)

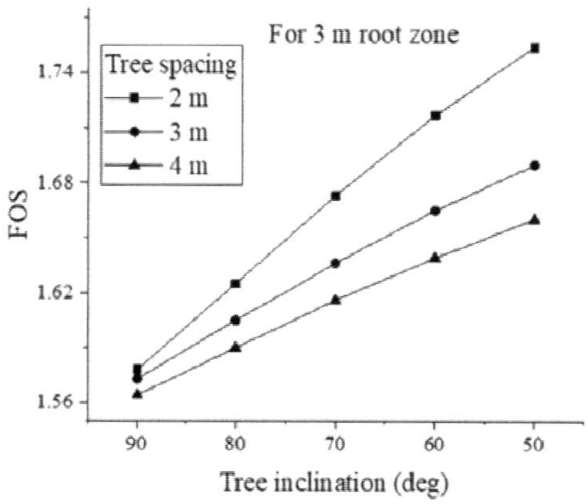

(e)

Figure 11. FOS for different tree inclinations spaced at 2m, 3m, 4m for root zone depths of 1m, 1.5m, 2m, 2.5m, 3m for a 30° slope model

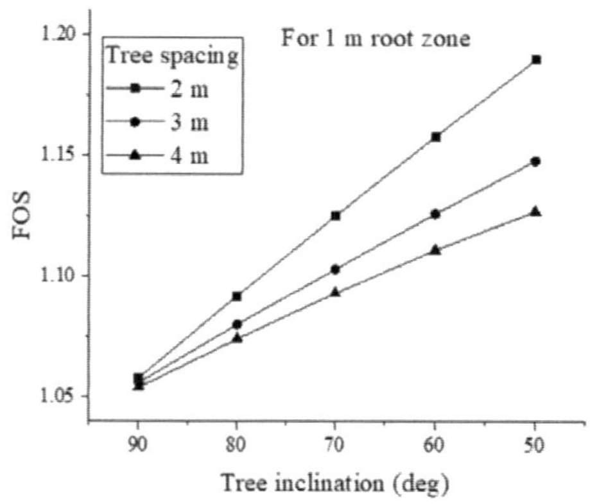

(a)

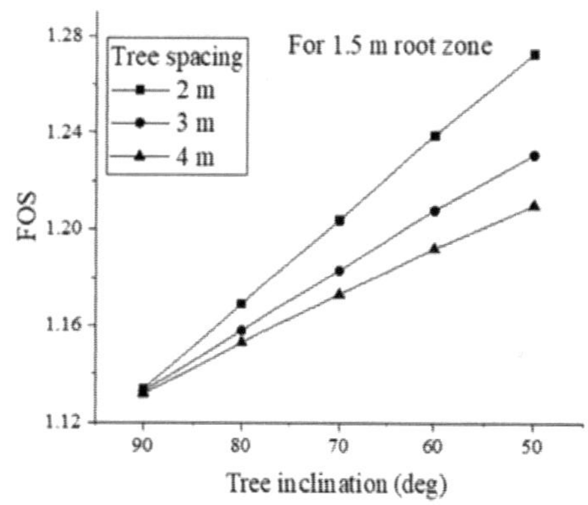

(b)

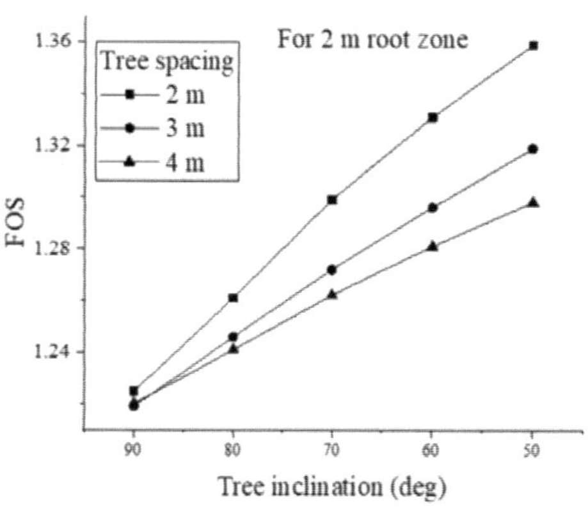

(c)

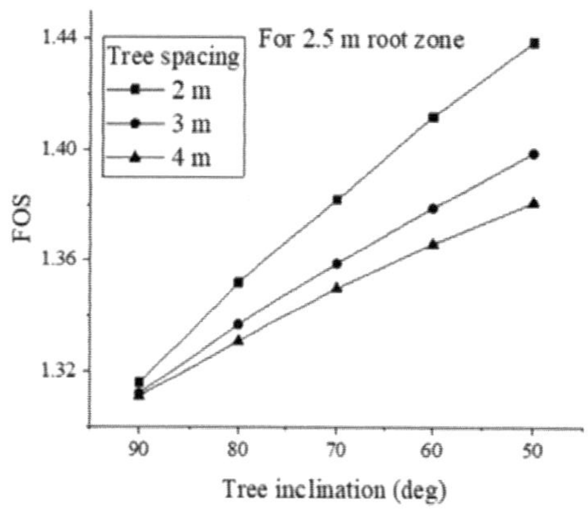

(d)

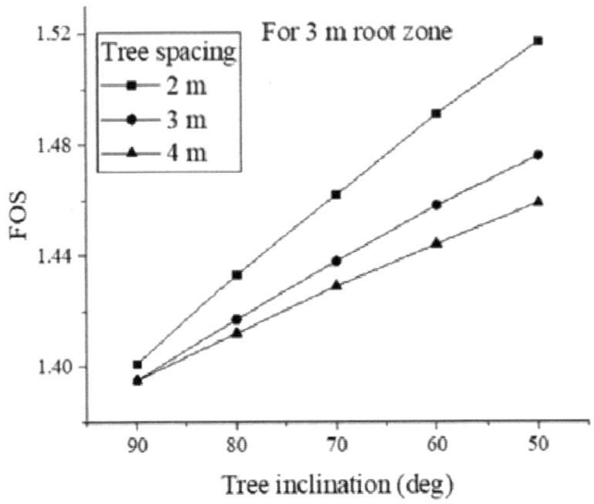

(e)

Figure 12. FOS for different tree inclinations spaced at 2m, 3m, 4m for root zone depths of 1m, 1.5m, 2m, 2.5m, 3m for a 35° slope model

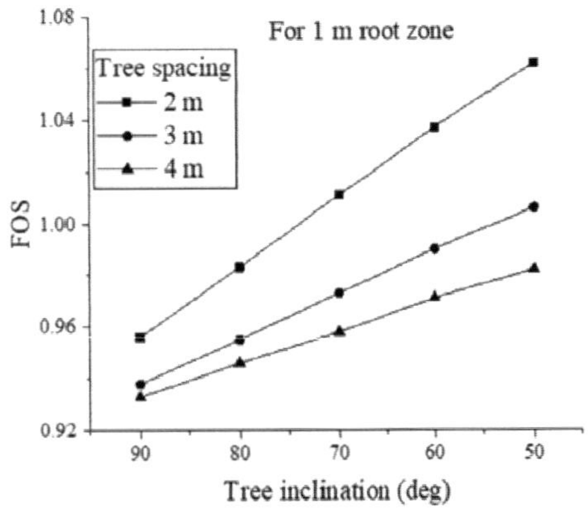

(a)

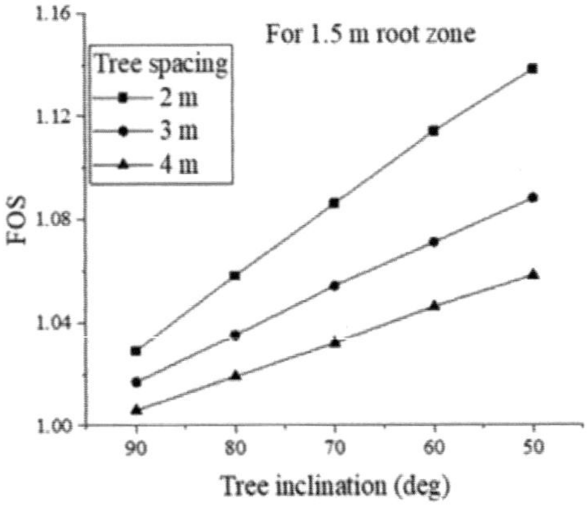

(b)

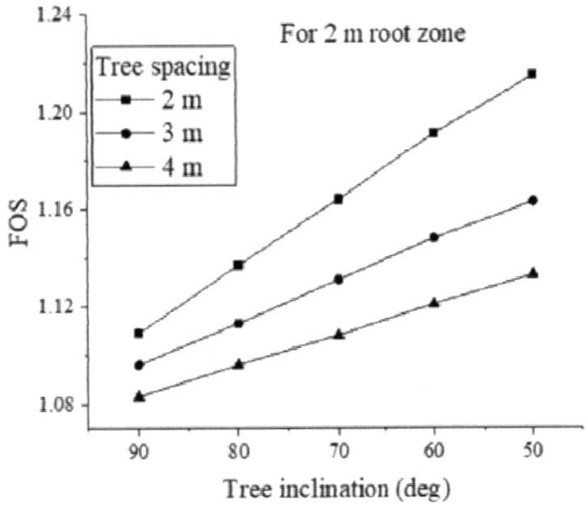

(c)

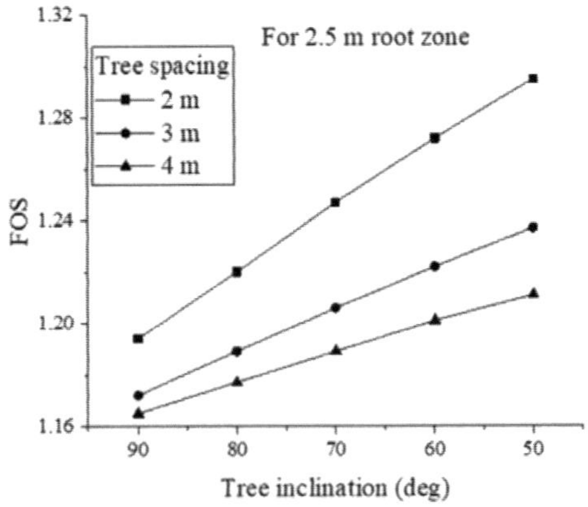

(d)

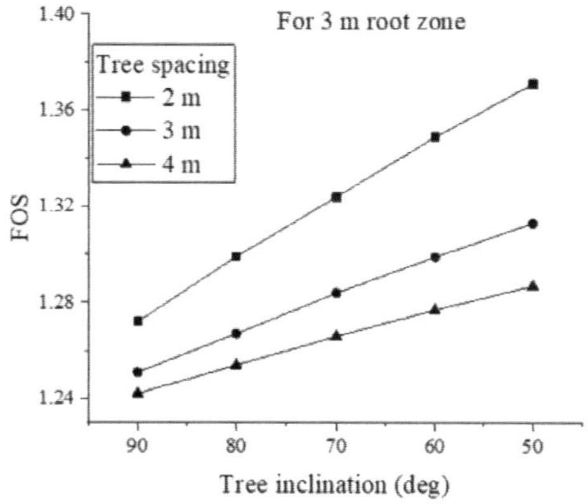

(e)

Figure 13. FOS for different tree inclinations spaced at 2m, 3m, 4m for root zone depths of 1m, 1.5m, 2m, 2.5m, 3m for a 40° slope model

The maximum FOS observed was 1.976 in case of 25^0 slope with 3m root zone depth along with anchors and tree inclination of 50^0 spaced at 2m distance. The minimum FOS observed was 0.857 in case of 40^0 slope with only 0.5m root zone depth.

An equation for FOS was generated using multiple regression analysis for given C, Ø parameters by analyzing all the models with a variation of slope inclination(i), root zone depth(Z_R), Anchor length(A_R), tree load(T_L), tree inclination(T_i), tree spacing(S_T) as represented below.

$$FOS = (-0.0343)\, i + (0.1866)\, Z_R + (-0.0210)\, A_R + (0.0288)\, T_L + (-0.0031)\, T_i + (-0.0279)\, S_T \ldots\ldots (1)$$

$R^2 = 0.981$ (or 98.1%), Standard error = 0.03

R^2 and standard error were obtained through regression analysis by incorporating various aspects of vegetations and slope geometry considered in the 420 models analysed in the present study. R^2 represents the accuracy of FOS obtained from the (1) by comparing the FOS obtained from numerical analysis. And the standard error represents the variation between FOS values obtained through numerical analysis and eqn. (1).

FOS obtained for 420 models with varying vegetation and geometry parameters were tabulated and the multiple linear regression analysis has been performed. The mathematical equation was obtained. The obtained equation may be used to determine FOS, if prevailing vegetation conditions are known.

6. Conclusions

The study has involved in quantifying the effects of vegetation by considering various parameters like slope inclination, root zone depth, root spread angle, Anchor length, tree load, tree inclination and tree spacing. The effects are as follows

- As the slope inclination decreases, FOS increases. This is because as the steepness of the slope decreases, the slope gets more stable.
- Maximum FOS was found at a greater root zone depth and anchor length as it is found that at deeper root zone, higher shear strength to soil is induced.
- As tree inclination decreases towards the slope with respect to the ground it contributes to resisting forces, hence FOS increases

- Slope becomes more stable as the spacing between the trees are less as the number of anchorages along the slope increases.
- Change in Root spread angle has shown no influence on slope stability in all different slope models. But in the actual field condition Root spread angle decides the horizontal spread of reinforcement.
- Even though there is an increment in FOS due to reinforcement, the tension offered by the root anchors has minor effect, whereas the contribution of reinforced soil is more when compared.

Therefore, all the cases with the presence of vegetation showed good increment FOS when compared to slopes without vegetation cases.

7. Future scope

Additional effects of construction of roadway in between the natural slopes, trees uprooting on slope stability can be analyzed. By varying the cohesion and internal friction angle along with other parameters, an equation including the effect of shear parameters can also be obtained.

References

[1] Ali N, Farshchi I, Mu'azu M, and Rees S W 2012 Soil-root interaction and effects on slope stability analysis *EJGE* **17** 319-328.
[2] Noroozi A G, Hajiannia A 2015 The Effect of Vegetation on Slope Instability as Predicted by the Finite Element Method *EJGE* **20(28)**
[3] Badhon F F, Islam M S and Islam M A 2021 Contribution of Vetiver Root on the Improvement of Slope Stability *Indian Geotechnical J* **51(4)** 829–840.
[4] Maneecharoen J, Htwe W, Bergado D T and Baral P 2013 Ecological Erosion Control by Limited Life Geotextiles (LLGs) as Well as with Vetiver and Ruzi Grasses *Indian Geotechnical J* **43(4)** 388–406.
[5] Maffra C, Sousa R, Sutili F, and Pinheiro R 2019 The Effect of Roots on the Shear Strength of Texturally Distinct Soils *Floresta e Ambiente* **26(3)**.
[6] Hasan M M and Ankan C M 2020 Analysis of Vegetation Effects on Slope Stability of Embankment, *AJCE* **8(4)** 77-86.
[7] Dalton P A, Smith R J and Truong P N V 1996 Vetiver grass hedges for erosion control on a cropped flood plain: hedge hydraulics *Agricultural Water Management Elsevier* **31(1-2)** 91–104.
[8] Anaswara S and Shivashankar R 2015 Effect of Vegetation on Stability Of Slopes *50th Indian Geotechnical Conf.* Pune.
[9] Mickovski S B, Beek L P H V and Salin F 2005 Uprooting of Vetiver Uprooting Resistance of Vetiver Grass (Vetiveria zizanioides) *Plant and Soil Springer* **278(1-2)** 33–41.
[10] Wang S, Zhao M, Meng X, Chen G, Zeng R, Yang Q, Liu Y and Wang B 2020 Evaluation of the Effects of Forest on Slope Stability and Its Implications for Forest Management: A Case Study of Bailong River Basin, China *Sustainability* **12(16)** 6655.
[11] Chok Y H, Kaggwa W S, Jaksa M B and Griffiths D V 2004 Modelling the Effects of Vegetation on Stability of Slopes *Proc. 9th Australia New Zealand Conf. on Geomechanics Auckland.*

IOP Conf. Series: Earth and Environmental Science 1032 (2022) 012004 doi:10.1088/1755-1315/1032/1/012004

Impact of COVID-19 Lockdown on the Risk of Breast Cancer: A Case Study

Akriti Sharma[1]*, Nishtha Hooda[2], Nidhi Rani Gupta[3] and Renu Sharma[1]

[1] Chandigarh University, Mohali 140413 (Punjab), India

[2] Indian Institute of Information Technology, Una 177220 (Himachal Pradesh), India

[3] Multani Mal Modi College, Patiala 147001 (Punjab), India

* Corresponding author: akriti.7799@gmail.com

Abstract. Air pollution is among the world's major environmental concerns. It remains a major health threat in India and is the leading environmental cause of morbidity in the country. There is considerable evidence that heavy and prolonged exposure to several air contaminants increases the cancer risk. The prevalence of breast cancer in citified environments with high exposure to air pollution has been seen to be elevated. Among various Indian cities, the Delhi cancer registry is having a high breast cancer incidence (28.6%). Owing to the recent and unprecedented global outbreak of coronavirus infectious disease (COVID-19), India is exploring every possible way of controlling its vigorous human transmission. Work from home culture is adopted so as to maintain social distancing during the lockdown. This momentary stoppage is substantially reducing the level of air pollution in several city areas across India dramatically. This paper (i) Overviews the breast cancer and air pollution association; (ii) Compiles the air quality data of Delhi monitored by CPCB during the COVID-19 pandemic lockdown time and compares it with pre-lockdown air quality data; (iii) Explores the reduced threat of breast cancer in Delhi during the nationwide lockdown. This work concluded that Air pollution serves a significant part in breast cancer occurrence. The countrywide lockdown in an attempt to prevent Covid-19 transmission has greatly improved the air quality of various Indian cities like Delhi. Also, with an unprecedented drop in rates of air pollution over Delhi, breast cancer occurrence may also decrease.

Keywords: Breast cancer; Risk analysis; COVID-19; Air pollution

1. Introduction

Cancer incidence has continued to rise in India over the last 20 years, as it has in other developing nations. The pattern has also changed along with the incidence of cancer. Changes in food habits, lifestyle, and increased life expectancy have made a significant contribution to this transition. Among various types of cancer, breast cancer incidence is rising and is expected to be the most widely known cancer amongst women in India in the coming years [1]. Although the majority of features and underlying causes are uniform throughout the world there is uniqueness for particular cancer in every region. Most Indian cities have seen breast cancer as the most prevalent cancer. Every four minutes in India 1female is identified with breast cancer and 1 female dies every eight minutes from breast cancer. Approximately 70,218 women passed away from breast cancer in India, the highest in the world for the year 2012 [2].

As per the Globocan 2012, India collectively counts for approximately one-third of the global burden of breast cancer, along with the USA and China. India is confronting a challenging situation owing to a rise of 11.54 percent in incidence and a rise of 13.82 percent in breast cancer mortality during the period 2008–2012 [3]. As per the individual registries, breast cancer is ranked as highest (Dibrugarh, Bangalore, New Delhi, Mumbai, and Chennai) in the 2012-2014 periods. Growing westernization and urbanization coupled with dietary preferences and changing lifestyles have led to breast cancer hitting a top place in all major urban registries [4]. Based on all the PBCRs, Delhi held top place with an age-adjusted rate of 41.0/lakh proceeded by Chennai (37.9) in second, Bangalore (34.4) in third, and Thiruvananthapuram District (33.7) in the fourth position, as shown in Figure 1. The ratio of mortality/incidence (MIR) is a novel indicator for determining incidence-related cancer deaths. It is used to determine whether there is a higher death in an area than could be estimated on the basis of its incidence. Delhi registry had a minimal mortality/incidence ratio of 8.0 without being affected by elevated incidence (28.6%), likely due to awarded population, better medical facilities, and high literacy in metropolitan cities [4].

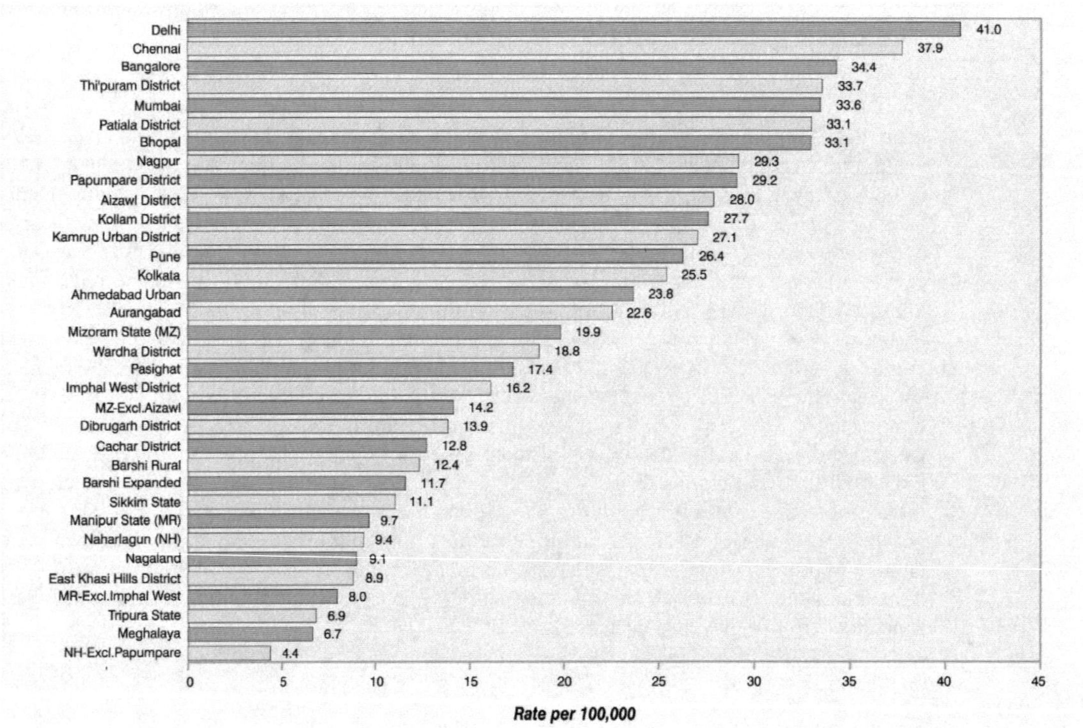

Figure 1. Comparison of Age-Adjusted Incidence Rates (AARs) of all PBCRs BREAST (ICD-10: C50) - Females [5].

In Delhi, the leading cancer sites among males were: lung-10.5 percent, mouth-6.9 percent, prostate-6.7 percent, tongue-6.5 percent, and larynx-5.7 percent. The corresponding AAR and CR per 100,000 for the sites were: lung (17.2 and 11.8), mouth (9.5 and 7.8), prostate (12.4 and 7.6), tongue (9.3 and 7.3), and larynx (8.9 and 6.4) [5]. The leading cancer sites among females were: breast-28.6 percent followed by cervix uteri-10.8 percent, gall bladder-7.9 percent, ovary-7.2 percent, and corpus uteri-3.5 percent. The respective AAR and CR per 100,000 population for the sites were: breast (41.0 and 34.8), cervix uteri (15.5 and 13.2), gall bladder (11.8 and 9.6), ovary (10.0 and 8.7), and corpus uteri (5.5 and 4.3) [5] as shown in Figure 2.

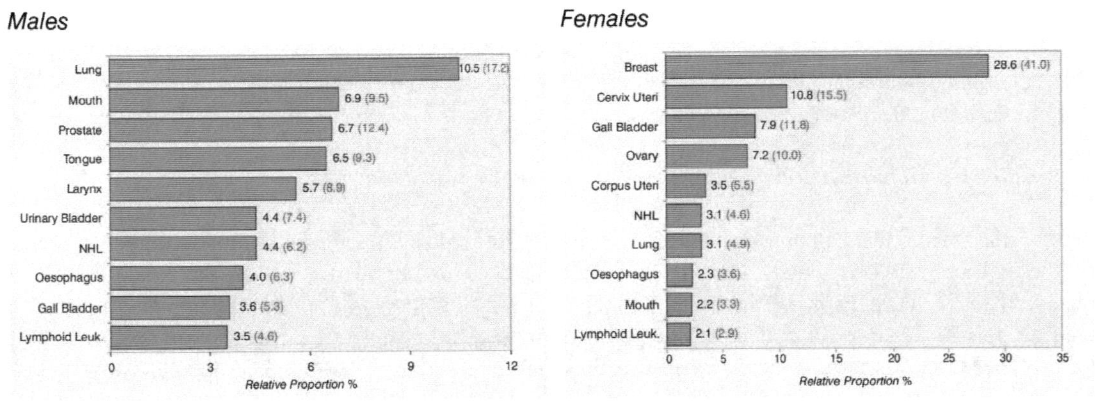

Figure 2. Ten leading sites of cancer (2012) (AARs given in parentheses) [5]

Breast cancer etiology is complex, and its occurrence is attributed to many risk factors. They contain elements linked to reproductive health, genetic mutations, family history, environmental, and occupational exposures to contaminants [6]. Environmental factors are considered to perform an important part in the pathophysiology of the disease. Findings have identified the possible risk of breast cancer with susceptibility to toxicants in ambient air [7].

Rapidly Increasing rates of air pollution are of significant concern to public health, especially in the cities, and have been directly linked to a variety of health hazards, including cardiovascular disease, childhood asthma, and lung cancer [8]. Outdoor air pollution has been listed as a carcinogen of category-1 by International Agency for Research on Cancer [9] and air pollution linked to vehicles covers multiple carcinogenic substances, such as polycyclic aromatic hydrocarbons (PAHs) [8]. PAHs generated through partial burning of organic content are suspected to be carcinogens for the lungs and can be of particular interest to breast cancer as they can adhere to DNA and form DNA adducts in breast tissue. Furthermore, animal-based research has also revealed that polycyclic aromatic hydrocarbons are capable of causing mammary tumors [10].

In a database monitored and presented by World Health Organization in April 2018, which includes 100 nations, India holds fourteen of the leading fifteen cities accounting for the worst PM2.5 levels. Delhi tops among the world's megacities for PM10 pollution [11]. Also, the Delhi cancer registry is having the highest breast cancer incidence (28.6%) [4]. The World Health Organization (WHO) has considered the recent and unprecedented worldwide emergence of a contagious coronavirus disease (COVID-19) as a global health emergency [12]. A nationwide lockdown of twenty-one days was imposed in India from March 25, 2020, to April 14, 2020, to prevent its spread. All institutes, factories, offices, and transport facilities were closed, except for essential services [13]. The nationwide halting of industrial activities and transportation systems has resulted in an apparent substantial decline in the level of a variety of gases released from the transport and energy sectors [12].

The remaining part of the work discusses the research related to the correlation of ambient air pollution with breast cancer and the COVID-19 lockdown influence on air quality. Then compiles AQI data of Delhi during the lockdown time and compares this with the AQI value before lockdown. The last section concludes the study while presenting further research concerns.

2. Literature Review

This section summarizes the related literature on the correlation of air pollution with breast cancer and COVID-19 lockout influence on air quality.

2.1 Related work on the association of air pollution with breast cancer

Nearly one out of three emerging instances of breast cancer are attributed to familiar factors and most of the pathophysiology persists to be uncertain [14]. The prevalence of breast cancer has been observed to be more in big cities with greater air pollution levels [15]. Local vehicle traffic is the leading source of air pollution in citified environments. Exhaust from the transport sector includes gases, VOCs, and particulate matter, most of these are deemed as potential/accepted carcinogens [16]. Studies indicate that elevated vehicle emissions in the US are linked to higher breast cancer risk [17].

Research findings have revealed that outdoor and indoor air contamination correlates with the possibility of breast cancer [7]. Toxic metals have demonstrated estrogenic and carcinogenic properties. Another research assessed the possibility of breast cancer occurrence in a large U.S.-wide cohort with respect to proximity to toxic metallic pollutants in the air. Increased levels of certain airborne metals, namely cadmium, lead, and mercury are linked to a greater probability of postmenopausal breast cancer [19].

Another research investigated data to ascertain if ambient air pollution has an influence on breast cancer prevalence. Geographic differences and trends referring to time in the rate of breast cancer have been examined with respect to the release of air pollutants [20]. There was a remarkable rise in the overall breast cancer prevalence in the US during the period 1986 to 2000, which could emerge as a consequence of the increased level of air pollution from automobiles and industrialization. The prevalence of breast cancer in metropolitan and heavy-emission areas has been noted to increase. This research finds a possible link across the United States between air pollution and breast cancer [20]. Table 1 presents the literature review showing association of air pollution with breast cancer.

Table 1. Literature review showing the link of air pollution with breast cancer.

TITLE OF PAPER	DESCRIPTION	FINDINGS	RISK FACTORS OF BREAST	REFERENCE
"Exposure to traffic emissions throughout life and risk of breast cancer: the Western New York Exposures and Breast Cancer (WEB) study"	This work Examined the possibility of breast cancer in reference to life-long exposure to air pollution from transport sector emissions.	Early-life proximity has an effect on breast cancer occurrence and shows a significant influence of traffic pollution in its potential risk.	Traffic emissions	Nie et al., 2007 [21]
"Exposure to traffic-related air pollution and the risk of developing breast	The research examines the correlation between historical NO2 exposure metrics	This work finds that air pollution from traffic emissions could be associated with	Susceptibility to air pollution through traffic	Hystad et al., 2015 [22]

cancer among women in eight Canadian provinces: A case-control study"	and incidences of post and pre-menopausal breast cancer.	breast cancer development, with greater correlations among premenopausal females.		
"Breast Cancer Risk and Exposure in Early Life to Polycyclic Aromatic Hydrocarbons using Total Suspended Particulates as a Proxy Measure"	This work conducted a study of exposure to polycyclic aromatic hydrocarbons (PAHs) in early life with respect to breast cancer risk.	The findings suggest that exposure to PAHs in formative years might increase the possibility of postmenopausal breast cancer	Polycyclic aromatic hydrocarbons (PAH)	Bonner et al., 2005 [23]
"Postmenopausal Breast Cancer Is Associated with Exposure to Traffic-Related Air Pollution in Montreal, Canada: A Case-Control Study"	This study assesses how well the occurrence of postmenopausal breast cancer is correlated with air pollution susceptibility in big urban sprawls.	This study shows that the occurrence of postmenopausal breast cancer is linked to exposure to nitrogen dioxide.	Air pollution from traffic	Crouse et al., 2010 [7]
"Hazardous air Pollutants and breast cancer risk in California teachers: a cohort study"	This research investigated the association between the emergence of breast cancer cases and air pollutants which are identified as mammary gland carcinogens (considering the modeled levels of air pollutants)	This work reported significant correlations between elevated chances of breast cancer with respect to Housing in locations where vinyl chloride and propylene levels are higher.	Hazardous air pollutants	Garcia et al., 2015[24]
"Air Pollution and Breast Cancer: a Review"	The purpose of this work is to examine epidemiological information concerning the involvement of air pollution with the	The study reveals the possible relationship between air pollution in breast cancer emergence	Air Pollution	White et al., 2018 [25]

	prevalence of breast cancer.	while considering NO2 as a marker for air contamination from the transportation sector.		
"Association between ambient air pollution and breast cancer risk: The multiethnic cohort study"	The research investigated the likelihood of breast cancer in Southern California Multiethnic Cohort by exposure to air contaminants over longer periods of time.	Women living near major roads were at a higher likelihood of breast cancer as they were more susceptible to air pollution.	Ambient air pollution	Cheng et al., 2020 [26]
"Residential Exposure to Estrogen Disrupting Hazardous Air Pollutants and Breast Cancer Risk: the California Teachers Study"	The link between the threat of breast cancer and the susceptibility of housing to ambient estrogen disruptors is investigated in this work.	There is a possibility of breast cancer with minimal dose exposure to inorganic arsenic or ambient cadmium for a longer duration.	Estrogen Disrupting Hazardous Air Pollutants	Liu et al.; 2015 [18]
"Metallic Air Pollutants and Breast Cancer Risk in a Nationwide Cohort Study"	The study assessed the breast cancer risk in a large U.S.-wide cohort with respect to proximity to toxic metallic pollutants in the air.	Elevated levels of certain airborne metals, namely Pb, Cd, and Hg are linked to an increased threat of postmenopausal breast cancer.	Metallic Air Pollutants	White et al., 2019 [19]
"PAHs and PM2.5 emissions and female breast cancer incidence in metro Atlanta and rural Georgia"	In this research, emission levels of PM2.5 and polycyclic aromatic hydrocarbons have been examined in reference to breast cancer emergence	This study concluded that air pollution, especially PM2.5 and PAHs, may significantly affect the likelihood of suffering from breast cancer	ambient air pollution, especially PAHs and PM2.5	Parikh et al., 2016 [27]

	in rural Georgia and Atlanta metro.	among females in urban settings.	

2.2 Influence of lockdown imposed during COVID-19 on air quality

The COVID-19 pandemic, which emerged in Wuhan, China, has been recognized by the World Health Organization as a public health emergency at a worldwide scale. Consequently, several countries-imposed restrictions on flights to China [28]. With the abrupt onset of the novel coronavirus (COVID-19), lockdown measures have been adopted in various countries across the world [29]. In India, a countrywide twenty-one days lockdown was implemented from March 25, 2020, to April 14, 2020, to limit the transmission of the coronavirus pandemic. Consequently, all educational establishments, offices, transport facilities, and factories are to be closed during the lockdown period excluding essential services [13]. The countrywide temporary suspension of major industrial units and public transport systems has contributed to a reduction in the emissions of gases from the transportation and energy sector [12] which resulted in improved air quality. Table 2 presents a review of the literature showing COVID-19 lockdown's influence on air quality.

Table 2. Literature review showing the effect of COVID-19 lockdown on the quality of air.

TITLE OF PAPER	DESCRIPTION	FINDINGS	REFERENCES
"A Study on Air Quality Index (AQI) of Bengaluru, Karnataka during Lockdown Period to Combat Coronavirus Disease (Covid-19): Air Quality Turns 'Better' from 'Hazardous'"	The focus of this research is to examine the trend in AQI during the lockdown time in Bengaluru.	Bengaluru's AQI turned 'better' from 'hazardous' within twenty-four hours after the lockdown was imposed.	Kambalagere et al., 2020 [13]
"COVID-19, City Lockdown, and Air Pollution: Evidence from China"	Using a comprehensive and timely dataset, the study examined the influence of city lockdown on China's air quality.	The PM2.5 concentrations and the Air Quality Index dropped significantly by 25 percent within weeks.	He et al., 2020 [30]
"Changes in air quality during the lockdown in Barcelona (Spain) one month into the SARS-CoV-2 epidemic"	This work attempts to demonstrate shifts in air pollution rates in the city of Barcelona (NE Spain) during lockdown measures.	Urban air pollution declined considerably after fourteen days of lockdown however with considerable variations between pollutants.	Tobías et al., [31]
"The dramatic impact of Coronavirus outbreak on air quality: Has it saved as much as it has killed so far?"	This work presents the first case study that compares the pre and post-crisis air quality.	The study concluded that through COVID-19 lockdown measures a greater number of people might have been saved by	Isaifan (2020) [32]

		reduced air pollution levels than by limiting the spread of COVID- 19.	
"Abrupt declines in tropospheric nitrogen dioxide over China after the outbreak of COVID-19"	This study provides observations into the unforeseen economic and environmental implications of reduced economic activities.	Satellite observations indicate a forty-eight percent decrease in vertical column densities of tropospheric NO2 from the average twenty days prior to the 2020 Lunar New Year to the average twenty days later.	Liu et al., 2020 [33]
"COVID-19 lockdowns cause global air pollution declines with 2 implications for public health risk"	The study tested that lockdown has reduced ground-level and tropospheric air pollution levels using a network of more than ten thousand air quality stations and satellite data.	It is observed that lockdown practices have decreased air pollution concentrations by around twenty percent across twenty-seven countries.	Venter et al., 2020 [29]
"Good in The Worst: COVID-19 Restrictions and Ease in Global Air Pollution"	This document analyzes and outlines valuable data that appeared in newspapers and media covering information on air pollution mitigation in major cities that were severely affected by COVID-19.	The countrywide lockout of large industrial units and mass transit systems has ended in noticeable reductions in the emission of gases from the transportation and energy sectors.	Anjum et at., 2020 [12]
"Lockdown caused by COVID-19 pandemic reduces air pollution in cities worldwide"	The study examines the variations in levels of six air pollutants namely NO2, O3, CO, PM2.5, SO2, PM10 between February, March 2019, and 2020, in forty cities.	In February and March 2020, the mean monthly levels of PM10 and PM2.5 were found significantly less than in most cities in the year 2019.	Shrestha et al., 2020 [34]

Following the lockdown, several air quality studies found a considerable drop in pollution levels. Pratima and Dogra compared the levels of particulate matter 2.5 and 10 as well as ground-level ozone, SO2, and NO2 before and after lockdown. As per the findings, the levels of SO2, NO2, PM2.5, and PM10 in Delhi were reduced by 19 percent, 60 percent, 49 percent, and 55 percent, respectively, in the post-lockdown phase [45]. Another study compared the AQI before and after lockdown considering seven air pollutants. During the lockdown, PM2.5, PM10, and CO levels in Delhi fell by 52 percent, 39 percent, and 13 percent, respectively. In the post-lockdown period, these pollutants showed the same decreasing pattern, with 4 percent, 4 percent, and 12 percent for Delhi [46].

3. Material and methods

This section discusses the study area and methods adopted during the air quality analysis n Delhi.

3.1 Study Area

The current work is conducted on Delhi, India's capital city, located on the western side of the Yamuna River, and captures a range of 1,490 square kilometers [38]. Delhi is among the large cities worldwide that are grappling with fast urban sprawl and enormous rates of emissions from household, transport, and industrial activities [39]. It has undergone a rapid pace of urbanization noticeable from the growing population. The sharply growing population has changed considerably the pattern of land utilization [40].

Delhi has been consistently listed among the world's severely polluted cities in recent years. Particulate Matter 2.5 pollution over Delhi poses a challenge to India's capital. The concentration of Particulate Matter 2.5 in Delhi surpasses the national air quality level by over 300 percent [41]. In addition, there are around 4.8 million registered vehicles, 125,000 industrial units (medium and small-sized), and 3 coal-fired thermal power stations in Delhi. The key pollution sources in Delhi include vehicular emissions, power plants, domestic cooking, and small-scale industries [42].

Delhi, once renowned for its diverse historical architectural attractions, is now a smog-smothered place which often compromises the attractiveness of these historical structures due to reduced visibility [43]. Delhi is undergoing health implications from various contaminants, notably respirable particulate matter. The findings in the year 2010 revealed that lung cancer has caused 223,000 fatalities worldwide attributable to air pollution. The IARC, which is a WHO specialized agency has categorized outdoor air pollution in Group-2 carcinogenic to mankind in October 2013 [44]. Based on all the PBCRs, Delhi held top place with AAR 41.0 (per 100,000). Delhi registry had a minimal mortality/incidence ratio of 8.0 without being affected by elevated incidence (28.6%) [4].

3.2 Sampling Stations

Various air quality stations have been established in various parts of Delhi with the goal of determining air quality trends so that suitable steps can be taken if necessary. Table 3 lists the various monitoring stations run by the CPCB in Delhi.

Table 3. Ambient Air Quality Monitoring Stations, Delhi

Ambient Air Quality Monitoring Stations, Delhi	
"Ambient Air Quality Monitoring Station 1	Alipur, Delhi- DPCC
Ambient Air Quality Monitoring Station 2	Anand Vihar, Delhi - DPCC
Ambient Air Quality Monitoring Station 3	Ashok Vihar, Delhi - DPCC
Ambient Air Quality Monitoring Station 4	Aya Nagar, Delhi - IMD
Ambient Air Quality Monitoring Station 5	Bawana, Delhi - DPCC
Ambient Air Quality Monitoring Station 6	Burari Crossing, Delhi - IMD
Ambient Air Quality Monitoring Station 7	CRRI Mathura Road, Delhi - IMD
Ambient Air Quality Monitoring Station 8	DTU, Delhi - CPCB
Ambient Air Quality Monitoring Station 9	Dr. Karni Singh Shooting Range, Delhi - DPCC
Ambient Air Quality Monitoring Station 10	Dwarka-Sector 8, Delhi - DPCC
Ambient Air Quality Monitoring Station 11	IGI Airport (T3), Delhi - IMD
Ambient Air Quality Monitoring Station 12	IHBAS, Dilshad Garden, Delhi - CPCB
Ambient Air Quality Monitoring Station 13	ITO, Delhi - CPCB
Ambient Air Quality Monitoring Station 14	Jahangirpuri, Delhi - DPCC
Ambient Air Quality Monitoring Station 15	Jawaharlal Nehru Stadium, Delhi - DPCC

Ambient Air Quality Monitoring Station 16	Lodhi Road, Delhi - IMD
Ambient Air Quality Monitoring Station 17	Major Dhyan Chand National Stadium, Delhi - DPCC
Ambient Air Quality Monitoring Station 18	Mandir Marg, Delhi - DPCC
Ambient Air Quality Monitoring Station 19	Mundka, Delhi - DPCC
Ambient Air Quality Monitoring Station 20	NSIT Dwarka, Delhi - CPCB
Ambient Air Quality Monitoring Station 21	Najafgarh, Delhi - DPCC
Ambient Air Quality Monitoring Station 22	Narela, Delhi - DPCC
Ambient Air Quality Monitoring Station 23	Nehru Nagar, Delhi - DPCC
Ambient Air Quality Monitoring Station 24	North Campus, DU, Delhi - IMD
Ambient Air Quality Monitoring Station 25	Okhla Phase-2, Delhi - DPCC
Ambient Air Quality Monitoring Station 26	Patparganj, Delhi - DPCC
Ambient Air Quality Monitoring Station 27	Punjabi Bagh, Delhi - DPCC
Ambient Air Quality Monitoring Station 28	Pusa, Delhi - DPCC
Ambient Air Quality Monitoring Station 29	Pusa, Delhi - IMD
Ambient Air Quality Monitoring Station 30	R K Puram, Delhi - DPCC
Ambient Air Quality Monitoring Station 31	Rohini, Delhi - DPCC
Ambient Air Quality Monitoring Station 32	Shadipur, Delhi - CPCB
Ambient Air Quality Monitoring Station 33	Sirifort, Delhi - CPCB
Ambient Air Quality Monitoring Station 34	Sonia Vihar, Delhi - DPCC
Ambient Air Quality Monitoring Station 35	Sri Aurobindo Marg, Delhi - DPCC
Ambient Air Quality Monitoring Station 36	Vivek Vihar, Delhi - DPCC
Ambient Air Quality Monitoring Station 37	Wazirpur, Delhi – DPCC”

4. Result and discussion

This section discusses the air quality in Delhi using the air quality index.

4.1 National Air Quality Index

The National Air Quality Index (NAQI), a measure initiated by the government of India as an effort to track air quality on a daily basis under the 'Swachh Bharat Programme' as 'One Number - One Colour-One Definition' for the layman to understand air quality [13].

The National Air Monitoring Program (NAMP), which includes 240 cities across the country, is administered by the Central Pollution Control Board in coordination with State Pollution Control Boards. In addition, in a few cities, continuous monitoring systems are also installed which provide data on a near-real-time basis. Air Quality Index (AQI) is a tool adopted to effectively portray information to people on air quality [35]. AQI can be measured by employing a formula based on a comprehensive evaluation of air pollutant levels, which can be utilized by government authorities to describe the air quality status at a specific location [36]. There are 6 categories of AQIs: Severe, Very Poor, Poor, Moderately Polluted, Satisfactory, and Good. The proposed Air Quality Index would include eight pollutants (NO2, PM10, CO, PM2.5, Pb, SO2, NH3, and O3) of which National Ambient Air Quality Levels are specified for an average duration of up to twenty-four hours [35].

The Air Quality Index reports daily for air quality. It informs on whether the air is polluted or clean, and what could be the possible health concerns. This emphasizes health consequences that might be encountered in about a very few hours/ days upon inhaling toxic air [37]. The AQI ranges from 0-500. A higher AQI value reflects an elevated degree of air pollution, suggesting a greater health concern. An Air Quality Index value of a hundred usually refers to the standard of air quality set at the national level regarding the pollutants and is also the threshold decided by Environment Protection Agency for health protection in India. Its values below a hundred are deemed satisfactory. As the values increase, it becomes dangerous initially for a sensitive group of individuals and gradually for all [37].

In reference to the measured ambient values along with their potential health impacts, a sub-index for all pollutants is calculated. Associated potential health implications for different pollutants and AQI categories have been indicated, with critical inputs from medical experts. The Air Quality Index values and respective ambient levels (health breakpoints) along with the associated potential health consequences for the eight pollutants are shown in Figure 3 and Table 4 [35].

AQI Category (Range)	PM_{10} 24-hr	$PM_{2.5}$ 24-hr	NO_2 24-hr	O_3 8-hr	CO 8-hr (mg/m^3)	SO_2 24-hr	NH_3 24-hr	Pb 24-hr
Good (0-50)	0-50	0-30	0-40	0-50	0-1.0	0-40	0-200	0-0.5
Satisfactory (51-100)	51-100	31-60	41-80	51-100	1.1-2.0	41-80	201-400	0.5 – 1.0
Moderately polluted (101-200)	101-250	61-90	81-180	101-168	2.1- 10	81-380	401-800	1.1-2.0
Poor (201-300)	251-350	91-120	181-280	169-208	10-17	381-800	801-1200	2.1-3.0
Very poor (301-400)	351-430	121-250	281-400	209-748*	17-34	801-1600	1200-1800	3.1-3.5
Severe (401-500)	430 +	250+	400+	748+*	34+	1600+	1800+	3.5+

Figure 3. AQI Category, Pollutants and Health Breakpoints [35].

Table 4. Air quality index categories and precautionary health advice [35].

AQI	Remark	Color Code	Possible Health Impacts
0 to 50	Good		Minimal Impacts
51 to 100	Satisfactory		Mild respiratory distress to sensitive individuals
101 to 200	Moderate		Breathing difficulty for individuals suffering from asthma, heart, and lungs diseases
201 to 300	Poor		Breathing distress to most individuals on long exposure
301 to 400	Very Poor		Breathing distress following prolonged exposure
401-500	Severe		Influences healthy individuals and has significant implications for those with existing illness

4.2 Air Quality Trends in Delhi

This study describes the pattern of the AQI during lockdown time in Delhi. With the COVID-19 outbreak, a lockdown of twenty-one days was imposed across the entire country including Delhi to curb the transmission of coronavirus. With no factories functioning and limited vehicles on the roads, Delhi's air quality has greatly improved amid the lockdown. This study has compiled the data from the Central Pollution Control Board website without modifications.

4.2.1. Air quality Analysis Oct 2019 to March 2020. Air pollution is of concern to health and the environment, especially in urban areas [35]. For several years, CPCB has been measuring the air quality of different regions across Delhi. Monitoring sites have been categorized based on the land utilization patterns i.e., traffic intersections, and residential and industrial areas [37]. In this work, the everyday data from all the air sampling sites in Delhi put by CPCB is procured. Six months are taken into account for examining the variability of the AQI values as follows: October, November, December, January and February, and March (including the nationwide lockout period beginning on March 25). The month-wise average of the Air Quality Index of these stations was then computed as shown in Figure 4. The highest AQI value is observed in December as per the analysis done, and the lowest value is observed in March, which includes the lockdown period. Delhi showed much lower levels of harmful contaminants emitted by power plants and vehicles during the lockdown.

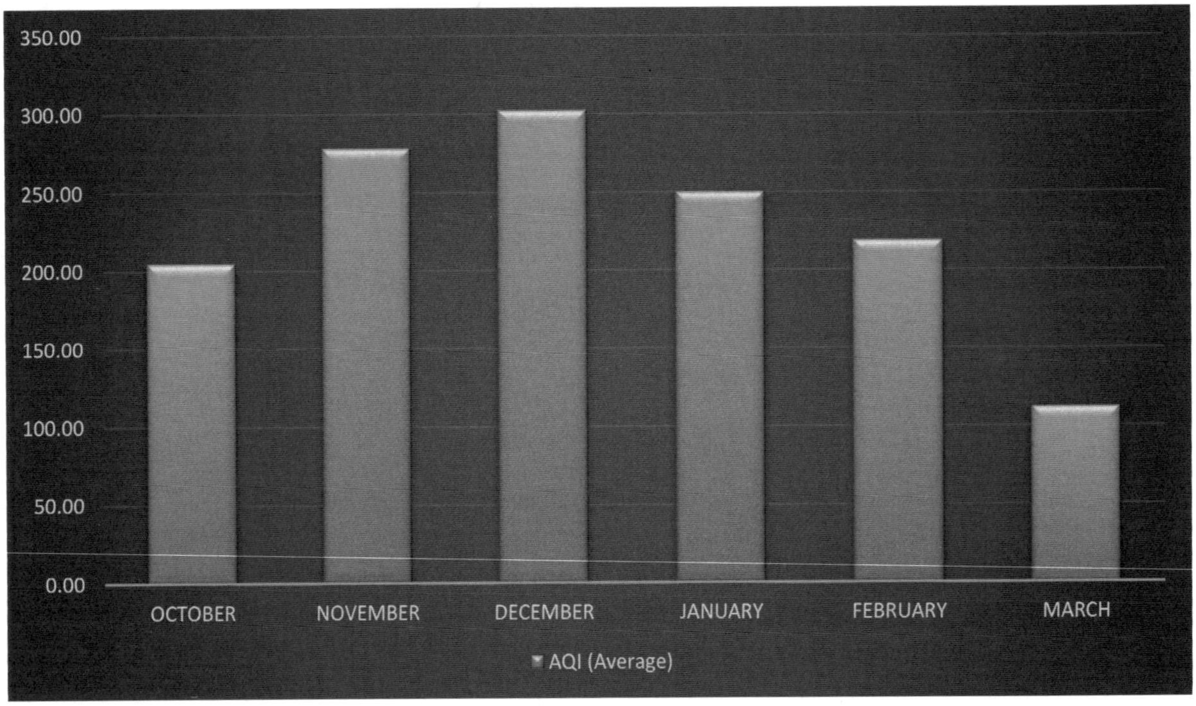

Figure 4. AQI Trends in Delhi (Oct 2019 to March 2020).

The patterns are compared on monthly basis for different monitoring stations in Delhi considering the observations for declining air quality standards from AQI Categories, which depict poor, moderate, and very poor categories. The Air Quality Index presented the following pattern in the declining order: December > November > January > February> October > March. Decreasing AQI order indicates that the air quality goes from the inferior to improved category e.g., this means that there is the worst air quality in December and the best in March. Delhi's Air Quality Index (AQI) has shown a sharp change from earlier very poor and poor levels to 'Moderate' levels in March as shown in Table 5. Air quality gradually improved by the end of March.

Table 5. Air quality index categories and precautionary health advice in Delhi (Oct 2019 to March 2020).

MONTH	AQI (Average) in Delhi	AQI	Remark	Color Code	Possible Health Impacts
OCTOBER	204.31	201 to 300	Poor		Breathing difficulty after long exposure to most individuals.
NOVEMBER	278.05	201 to 300	Poor		Breathing difficulty after long exposure to most individuals.
DECEMBER	302.11	301 to 400	Very Poor		Respiratory illness on prolonged exposure.
JANUARY	250.04	201to 300	Poor		Breathing difficulty after long exposure to most individuals.
FEBRUARY	218.56	201 to 300	Poor		Breathing difficulty after long exposure to most individuals.
MARCH	112.17	101 to 200	Moderate		Breathing difficulty for individuals with asthma, heart, and lungs disease.

4.2.2. Air quality Analysis 25th to 31st March 2018, 2019, 2020. In the study, the Average air quality of Delhi is computed for the year 2018-2020 from 25th to 31st March. Delhi's air quality index (AQI) showed a sharp fall to 'Satisfactory' levels amid the lockdown period in comparison to the last two years as shown in Table 6 and Figure 5. In the previous two years, AQI was Moderate and Poor during the same period with a possibility of causing breathing discomfort. The countrywide lockout adopted to fight Covid-19 positively affected the air quality. The air quality index procured from CPCB indicated relatively safer AQI levels. The analysis of the average AQI of the past two years (from 25th to 31st March) and its comparison with the 7 days of the lockdown period also showed a significant improvement in the air quality as shown in Figure 6.

Table 6. Air Quality Index of Delhi (25th to 31st March 2018, 2019, 2020).

LOCKDOWN DAYS (March)	2018	2019	2020
25	193	138	59
26	186	161	64
27	229	173	60
28	259	227	34
29	214	232	40
30	221	223	56
31	256	183	68
Average AQI	**223**	**191**	**54**

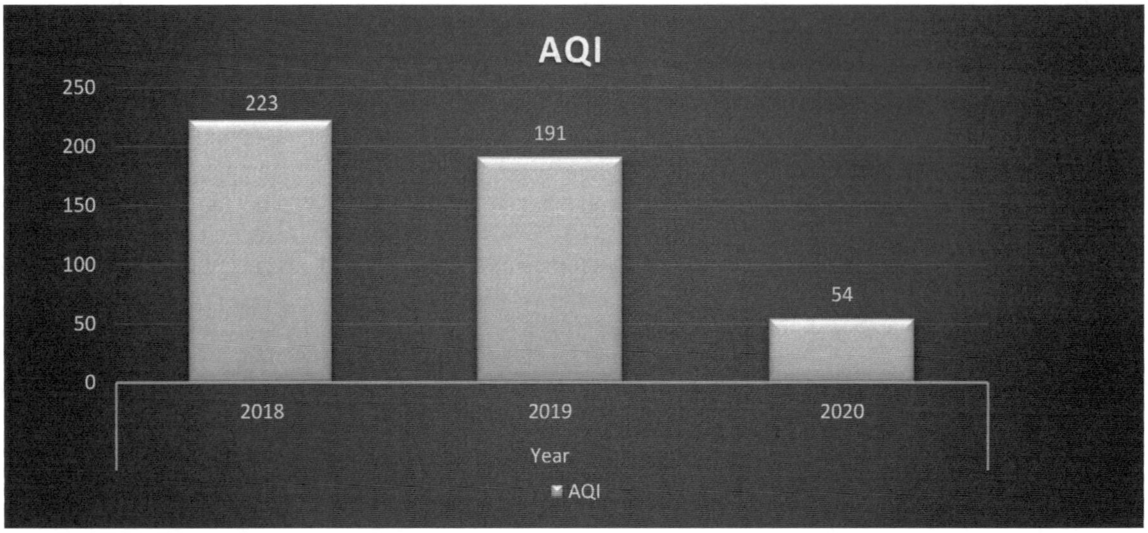

Figure 5. Average Air Quality Index of Delhi (25th to 31st March 2018, 2019, 2020).

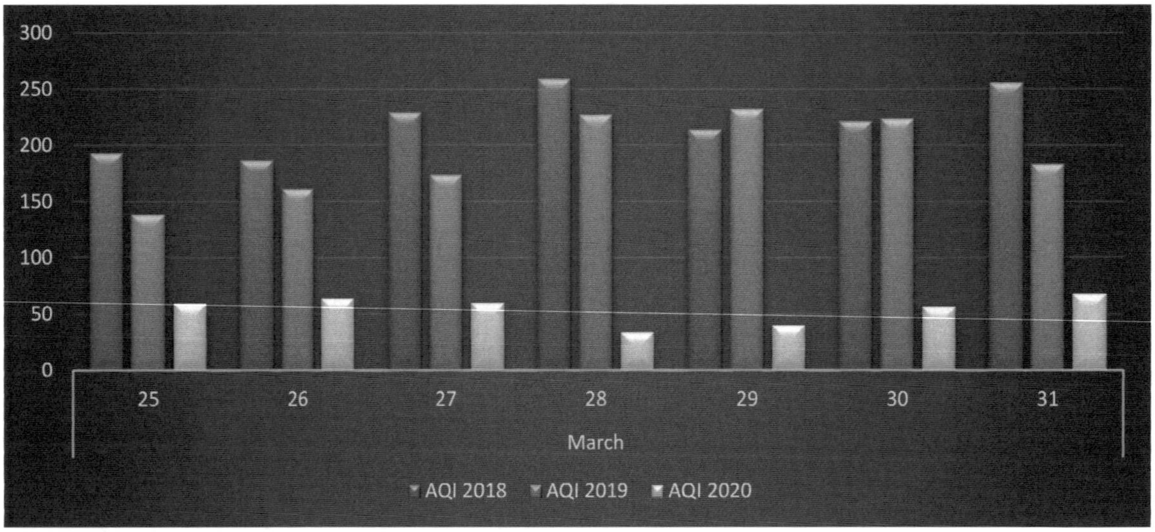

Figure 6. AQI on the days of lockdown in comparison with the same days of the previous two years.

5. Conclusions

Breast cancer is caused by the uncontrolled division of breast cells. There are several causes that have been linked to breast cancer occurrence including environmental causes. Among various environmental causes, air pollution has become a matter of concern for the environment and has been acknowledged for its adverse effects on the human body. There has been a variety of health effects attributable to air pollution including respiratory illness, asthma, cancer, etc. Several findings support the involvement of ambient air pollution in breast cancer occurrence. The nationwide temporary suspension of large industrial units and transportation facilities amid the COVID-19 lockdown has led to obvious cutbacks in the emission of transport and energy-related gases. The countrywide lockdown in an attempt to prevent Covid-19 transmission has greatly improved the air quality of various Indian cities like Delhi. Delhi has a high incidence of breast cancer (28.6 percent), and with an unprecedented drop in rates of air pollution over Delhi, breast cancer occurrence may also decrease. The current study concludes that air pollution serves a significant part in breast cancer occurrence, but more work is required to determine the potential association between breast cancer and different air pollutants.

6. References

[1] Mallick, S., Kumar, C., Benson, R., &Rath, G. K. (2020). Cancer in India. In *Practical Radiation Oncology* (pp. 299-300). Springer,Singapore.

[2] Snowprince. (2020). Breast Cancer India. Retrieved from https://www.breastcancerindia.net/statistics/stat_global.html

[3] Ferlay, J., Soerjomataram, I., Dikshit, R., Eser, S., Mathers, C., Rebelo, M., ...& Bray, F. (2015). Cancer incidence and mortality worldwide: sources, methods and major patterns in GLOBOCAN 2012. International journal of cancer, 136(5), E359-E386.

[4] Malvia, S., Bagadi, S. A., Dubey, U. S., & Saxena, S. (2017). Epidemiology of breast cancer in Indian women. Asia-Pacific Journal of Clinical Oncology, 13(4), 289-295.

[5] Monesh B.V. (2010). NCRP Annual Reports. Retrieved from http://ncdirindia.org/NCRP/Annual_Reports.aspx

[6] Gangemi, S., Miozzi, E., Teodoro, M., Briguglio, G., De Luca, A., Alibrando, C., ... &Libra, M. (2016).Occupational exposure to pesticides as a possible risk factor for the development of chronic diseases in humans. Molecular medicine reports, 14(5), 4475-4488.

[7] Crouse, D. L., Goldberg, M. S., Ross, N. A., Chen, H., &Labrèche, F. (2010). Postmenopausal breast cancer is associated with exposure to traffic-related air pollution in Montreal, Canada: a case–control study. Environmental health perspectives, 118(11), 1578- 1583.

[8] Shmuel, S., White, A. J., & Sandler, D. P. (2017). Residential exposure to vehicular traffic-related air pollution during childhood and breast cancer risk. Environmental research, 159, 257-263.

[9] Loomis, D., Grosse, Y., Lauby-Secretan, B., El Ghissassi, F., Bouvard, V., Benbrahim- Tallaa, L., ...&Straif, K. (2013). The carcinogenicity of outdoor air pollution. Lancet Oncology, 14(13),1262.

[10] International Agency for Research on Cancer. (2010). Some non-heterocyclic polycyclic aromatic hydrocarbons and some related exposures (Vol. 92). IARC Press, International Agency for Research on Cancer.

[11] World Health Organization. (2018). WHO global ambient air quality database (update 2018). World Health Organization: Geneva,Switzerland.

[12] Anjum, N. A. (2020). Good in The Worst: COVID-19 Restrictions and Ease in Global Air Pollution.

[13] Kambalagere, Y. (2020). A Study on Air Quality Index (AQI) of Bengaluru, Karnataka during Lockdown Period to Combat Coronavirus Disease (Covid-19): Air Quality Turns 'Better'from 'Hazardous'. Studies in Indian Place Names, 40(69),59-66.

[14] Coyle, Y. M. (2004). The effect of environment on breast cancer risk. Breast cancer research and treatment, 84(3),273-288.

[15] Binachon, B., Dossus, L., Danjou, A. M., Clavel-Chapelon, F., &Fervers, B. (2014). Life in urban areas and breast cancer risk in the French E3N cohort. European journal of epidemiology, 29(10),743-751.

[16] International Agency for Research on Cancer, & International Agency for Research on Cancer. (1987). *Overall evaluations of carcinogenicity: an updating of IARC monographs volumes 1 to 42* (p. 53). Lyon, France::IARC.

[17] Chen, F., &Bina, W. F. (2012). Correlation of white female breast cancer incidence trends with nitrogen dioxide emission levels and motor vehicle density patterns. *Breast cancer research and treatment*, *132*(1),327-333.

[18] Liu, R., Nelson, D., Hurley, S., Hertz, A., & Reynolds, P. (2015). Residential exposure to estrogen disrupting hazardous air pollutants and breast cancer risk: the California Teachers Study. *Epidemiology (Cambridge, Mass.)*, *26*(3),365.

[19] White, A. J., O'Brien, K. M., Niehoff, N. M., Carroll, R., & Sandler, D. P. (2019). Metallic air pollutants and breast cancer risk in a nationwide cohort study. *Epidemiology (Cambridge, Mass.)*, *30*(1),20.

[20] Wei, Y., Davis, J., &Bina, W. F. (2012). Ambient air pollution is associated with the increased incidence of breast cancer in US. *International journal of environmental health research*, *22*(1),12-21.

[21] Nie, J., Beyea, J., Bonner, M. R., Han, D., Vena, J. E., Rogerson, P., ...&Freudenheim, J. L. (2007). Exposure to traffic emissions throughout life and risk of breast cancer: the Western New York Exposures and Breast Cancer (WEB) study. *Cancer Causes & Control*, *18*(9), 947-955.

[22] Hystad, P., Villeneuve, P. J., Goldberg, M. S., Crouse, D. L., Johnson, K., & Canadian Cancer Registries Epidemiology Research Group. (2015). Exposure to traffic-related air pollution and the risk of developing breast cancer among women in eight Canadian provinces: a case–control study. *Environment international*, *74*,240-248.

[23] Bonner, M. R., Han, D., Nie, J., Rogerson, P., Vena, J. E., Muti, P., ...&Freudenheim, J. L. (2005). Breast cancer risk and exposure in early life to polycyclic aromatic hydrocarbons using total suspended particulates as a proxy measure. *Cancer Epidemiology and Prevention Biomarkers*, *14*(1), 53-60.

[24] Garcia, E., Hurley, S., Nelson, D. O., Hertz, A., & Reynolds, P. (2015). Hazardous air pollutants and breast cancer risk in California teachers: a cohort study. *Environmental Health*, *14*(1),14.

[25] White, A. J., Bradshaw, P. T., &Hamra, G. B. (2018). Air pollution and breast cancer: a review. *Current epidemiology reports*, *5*(2),92-100.

[26] Cheng,I.,Tseng,C.,Wu,J.,Yang,J.,Conroy,S.M.,Shariff-Marco,S.,...&Whittemore,A. S. (2020). Association between ambient air pollution and breast cancer risk: the multiethnic cohort study. *International journal of cancer*, *146*(3), 699-711.

[27] Parikh, P. V., & Wei, Y. (2016). PAHs and PM2. 5 emissions and female breast cancer incidence in metro Atlanta and rural Georgia. *International journal of environmental health research, 26*(4),458-466.

[28] Lau, H., Khosrawipour, V., Kocbach, P., Mikolajczyk, A., Schubert, J., Bania, J., &Khosrawipour, T. (2020). The positive impact of lockdown in Wuhan on containing the COVID-19 outbreak in China. Journal of Travel Medicine.

[29] Venter, Z. S., Aunan, K., Chowdhury, S., &Lelieveld, J. (2020). COVID-19 lockdowns cause global air pollution declines with implications for public health risk.medRxiv.

[30] He, G., Pan, Y., & Tanaka, T. (2020). COVID-19, City Lockdown, and Air Pollution: Evidence from China.medRxiv.

[31] Tobías,A.,Carnerero,C.,Reche,C.,Massagué,J.,Via,M.,Minguillón,M.C.,...&Querol, X. (2020).Changes in air quality during the lockdown in Barcelona (Spain) one month into the SARS-CoV-2 epidemic.*Science of The Total Environment*, 138540.

[32] Isaifan, R. J. (2020). The dramatic impact of Coronavirus outbreak on air quality: Has it saved as much as it has killed so far?.*Global Journal of Environmental Science and Management*, *6*(3),275-288.

[33] Liu, F., Page, A., Strode, S. A., Yoshida, Y., Choi, S., Zheng, B., ...& van der A, R. (2020). Abrupt declines in tropospheric nitrogen dioxide over China after the outbreak of COVID-19. *arXiv preprint arXiv:2004.06542*.

[34] Shrestha, A. M., Shrestha, U. B., Sharma, R., Bhattarai, S., Tran, H. N. T., &Rupakheti, M. (2020). Lockdown caused by COVID-19 pandemic reduces air pollution in citiesworldwide.

[35] Pib.gov.in. 2020. Retrieved from <https://pib.gov.in/newsite/PrintRelease.aspx?relid=110654> [Accessed 24 May 2020].

[36] Kumar, A., &Goyal, P. (2011). Forecasting of daily air quality index in Delhi. *Science of the Total Environment*, *409*(24),5517-5523.

[37] Mohan, M., &Kandya, A. (2007). An analysis of the annual and seasonal trends of air quality index of Delhi. *Environmental monitoring and assessment*, *131*(1-3),267-277.

[38] Pandey, P., Kumar, D., Prakash, A., Masih, J., Singh, M., Kumar, S., ...& Kumar, K. (2012). A study of urban heat island and its association with particulate matter during winter months over Delhi. *Science of the Total Environment*, *414*,494-507.

[39] Mohan, M., Dagar, L., &Gurjar, B. R. (2007). Preparation and validation of gridded emission inventory of criteria air pollutants and identification of emission hotspots for megacity Delhi. *Environmental monitoring and assessment*, *130*(1-3),323-339.

[40] Mohan, M., &Kandya, A. (2015). Impact of urbanization and land-use/land-cover change on diurnal temperature range: A case study of tropical urban airshed of India using remote sensing data. *Science of the Total Environment*, *506*,453-465.

[41] Chowdhury, S., Dey, S., Tripathi, S. N., Beig, G., Mishra, A. K., & Sharma, S. (2017). "Traffic intervention" policy fails to mitigate air pollution in megacity Delhi. *Environmental science & policy*, *74*,8-13.

[42] Tiwari, S., Pipal, A. S., Hopke, P. K., Bisht, D. S., Srivastava, A. K., Tiwari, S., ... &Pervez, S. (2015). Study of the carbonaceous aerosol and morphological analysis of fine particles along with their mixing state in Delhi, India: a case study. *Environmental Science and Pollution Research*, *22*(14), 10744-10757.

[43] Chetambath, R., &Jesin Kumar, C. (2019). Air Pollution in Delhi and the Health Emergency. The Indian Practitioner, 72(11), 8.

[44] Khanna, I. (2015). Health risks associated with heavy metals in fine particulate matter: a case study in Delhi city, India. Journal of Geoscience and Environment Protection, 3(02),72.

[45] Kumari, P., & Toshniwal, D. (2022). Impact of lockdown measures during COVID-19 on air quality–A case study of India. International Journal of Environmental Health Research, 32(3), 503-510.

[46] Mandal, J., Samanta, S., Chanda, A., & Halder, S. (2021). Effects of COVID-19 pandemic on the air quality of three megacities in India. Atmospheric Research, 259, 105659.

Acknowledgement

"This work did not receive any specific grant from funding agencies."

Effectiveness of PNL technique in disaster damage assessment: evidence from selective case studies

P Dave[1] and S Pasari[1]
[1] Mathematics Department, Birla Institute of Technology and Science, Pilani, Jhunjhunu – 333031, Rajasthan, India

Abstract. Natural disasters often cause large scale infrastructural damage and disruption of services. In such cases, a rapid damage assessment technique might prove beneficial to assist the disaster management efforts. Percent of normal light (PNL) can be one such technique. In this study, we test the effectiveness of PNL technique in rapid damage assessment from four selected case studies, namely the 2014 cyclone Hudhud, 2015 Gorkha (Nepal) earthquake, 2016 Central Italy earthquake, and the 2018 flood in Kerala. The dataset used has been taken from the Visible Infrared Imaging Radiometer Suite–Day/Night Band (VIIRS-DNB) scan data from the Joint Polar Satellite System (JPSS-1). The change in radiance values from the VIIRS-DNB dataset enables PNL computation to map the disaster affected regions. The results depict that PNL can be a viable alternative for rapid damage assessment.

1. Introduction

Remote sensing has been widely used in studying various features of the Earth surface (Chuvieco, 2016). Night-light time imagery, based on night-time capture, forms a useful data set in remote sensing. It is suitable to study human activities on Earth and their implications. The associated dataset primarily includes illumination due to various sources, such as city lights, fires, volcanoes, and gas flares. Night-time visible imaging was first initiated by the Defense Meteorological Satellite Program–Operational Linescan System (DMSP-OLS) in the 1960s, which was the only source of visible nighttime images until the launch of the Suomi National Polar-orbiting Partnership–Visible Infrared Imaging Radiometer Suite (SNPP-VIIRS). The nighttime light data set is available through various sources in the form of different data products. They vary in terms of the pre-processing step and associated information channels. A few major sources of night-time light data products are the NASA's Black Marble and the Earth Observations Group (EOG) at NOAA/NCEI. A detailed summary on the generation and usage of VIIRS-DNB satellite data is available in Doll (2008) and Elvidge et al. (2017).

2. Literature review

Several studies have explored the applicability of nighttime light data in disaster management. For instance, Gillespie et al. (2014) utilized nighttime light data from the DMSP-OLS to study the effect of the 2004 Indian Ocean tsunami in Indonesia. Extensive field surveys were carried out to validate the results and to establish a relationship between the aftermath economic expenditure and the night-time light data. Zhao et al. (2018) used the night-time PNL method from the National Polar-orbiting Partnership Visible Infrared Imaging Radiometer Suite Day/Night Band (NPP-VIIRS DNB) to assess damages from three selected case studies related to earthquakes, storms, and floods. One month before and ten days after a disaster was averaged to obtain pre-disaster and post-disaster values. A longer time-period for pre-disaster was chosen so as to account for the variations of the light intensity due to clouds and other effects. A larger period averaged out fluctuations of normal time and reflected the true light in a normal scenario. A shorter aftermath period was considered based on the assumption that it represents a critical period where post-disaster work is underway. Seasonal changes were not considered as the time periods were relatively small. The non-parametric Mann-Whitney U-test was employed to carry out significance test of the emanated results. In 2018, Wang et al. studied the power outages in USA as an aftermath effect of the 2012 Hurricane Sandy and the 2017 Hurricane María. Their study used the NASA Black Marble Product data that eliminates cloud-affected pixels and accounts for several disturbances due to atmospheric effects, terrain changes, vegetation cover, snow reflection, lunar effects, and stray light properties on the VIIRS-DNB radiances. The PNL technique was used to assess the damage. As a part of the study, the authors estimated power recovery metrics, comprising date of initiation, period, and proportion of recovery. The results were compared for different levels of urbanization, providing new insights to the disaster management efforts.

3. Methodology

In the present study, we have selected four natural disasters that occurred during 2013-2019, when the VIIRS-DNB data by NOAA/NCEI was available. A cyclone namely the 2014 cyclone Hudhud, two earthquakes namely the 2015 Gorkha (Nepal) earthquake and the 2016 Central Italy earthquake, and the recent 2018 flood in Kerala (India) have been considered here.

3.1. Dataset

The VIIRS-DNB sensor data has been used for this study. The data is available in monthly and annual packets for different regions of the world. It is divided into six tiles according to the geographical coordinates. The data-product is a processed version of the radiance data collected by the VIIRS instrument on the NOAA-20 satellite. The processing has been carried out to remove disturbances due to clouds and artifacts of stray lights. There are three types of data, such as steady lights, calibrated radiance, and mean digital number based on different parameters of detection frequency, radiance, and digital number, respectively. Such differences in data characteristics make the datasets usable in a number of applications such as economic activity, urbanization, light pollution, population growth, greenhouse gas emissions, and moreover disaster management.

3.2. Percent of normal light

The percent of normal light (PNL) is a technique used to estimate the affected areas after a disaster. It is based on the principle that an affected area will have structural damage thus affecting the urban lights from the area. As the nighttime light data captures these lights, they are used for applying the PNL method. The radiance values before and after the disaster are compared by taking their ratio to detect the change. The PNL method is susceptible to various sources of errors, such as erroneous data, insignificant change in radiance values, and other aspects of damage. Thus, it is necessary to test the efficacy of the PNL method prior to its implementation on practical applications such as discussed here.

3.3. Procedure

The VIIRS-DNB data for a month preceding and succeeding the disaster along with the month of the occurrence of the disaster is collected. As these datasets are for a whole tile, the data for the region of interest is extracted. The radiance values are threshold by 0.3 $nWcm^{-2}sr^{-1}$ to remove the effect of background noise and other disturbances. The disaster-affected months are assigned weights on the basis of the number of days affected by the disaster. The monthly dataset is then averaged to generate pre-disaster and post-disaster radiance values for the region. Using these values, the PNL image is generated by taking the ratio of pre-disaster radiance values (Rad_{pre}) and post-disaster radiance values (Rad_{post}). A PNL value less than 1 represents a drop in the radiance values and thus areas with PNL<1 are classified as affected. A summary of the entire methodology is shown in figure1 below.

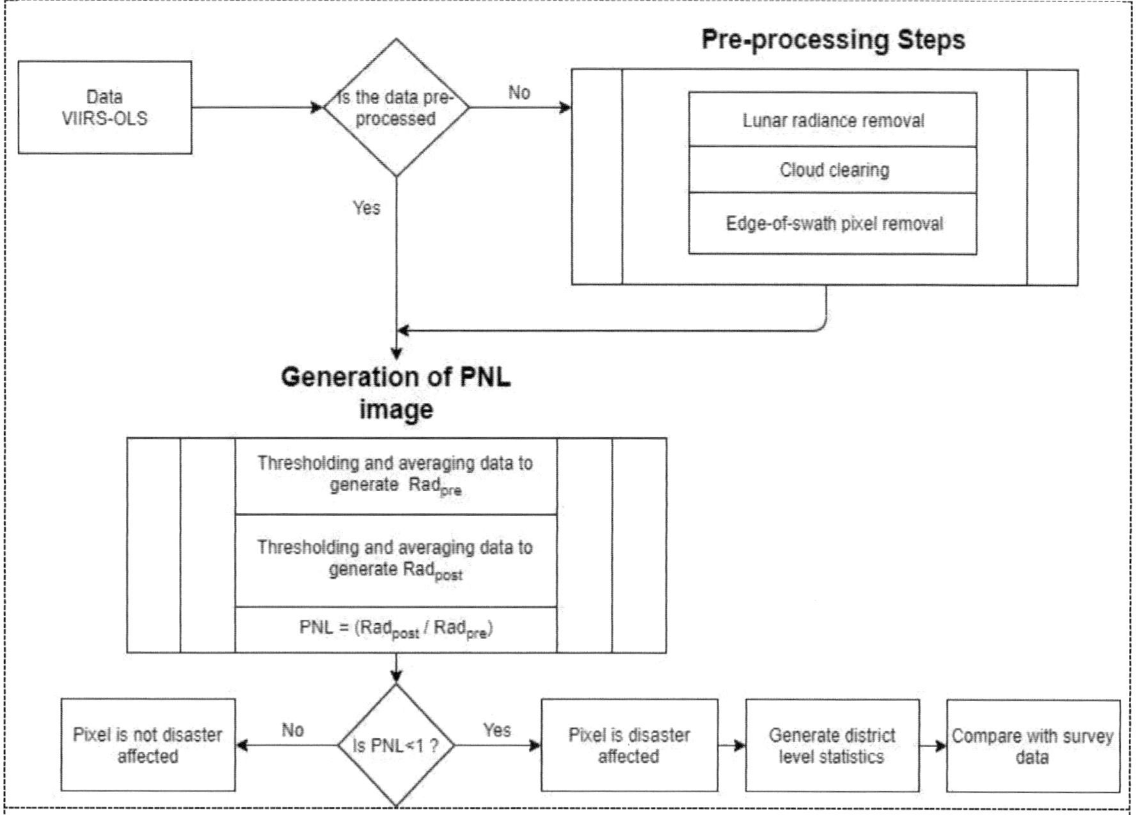

Figure 1. Flowchart of the proposed methodology.

4. Results and conclusions

In this study, disaster affected areas were identified (figure 2) for the four selected case studies using the PNL technique. The figure provides a colour coded map to depict the areas affected by the natural disaster. For example, the blue and yellow regions in figure 2(a) highlight the areas that had more devastation due to the 2014 Cyclone Hudhud, whereas the regions in blue in figure 2(b) show the most affected epicentral region of Kathmandu. In figure 2(c), the blue regions depict the regions that were most severely affected by the earthquake in Italy as identified by the PNL method. Similarly, the districts highlighted in blue in figure 2(d) are the worst affected ones by the 2018 Kerala flood, and it matches with the official reports of the state administration.

In conclusion, the present study implemented PNL technique for rapid damage assessment and found the results corroborating to the effectiveness of the technique. The PNL technique proves to be a promising alternative to traditional disaster damage assessment methods.

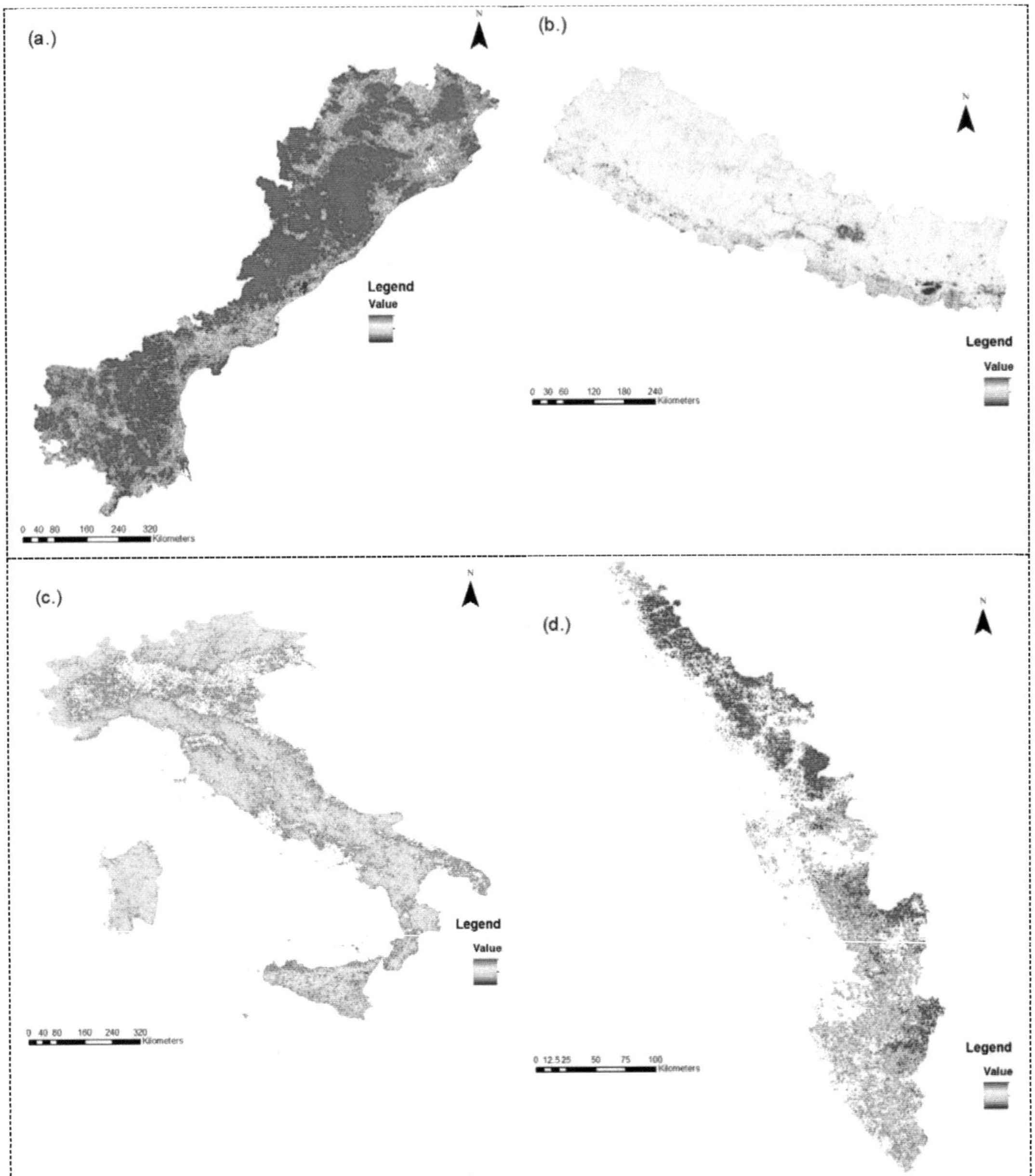

Figure 2. The PNL-based assessment of affected areas; (a) effect of the 2014 Cyclone Hudhud in eastern states of India such as Andhra Pradesh and Odisha, in which the yellow and blue lines highlight the areas of worst hit; (b) affected areas of the 2015 Gorkha (Nepal) earthquake; (c) affected areas of the 2016 Central Italy earthquake; and (d) the affected areas of the 2018 Kerala flood, in which the areas in blue represent the most devastated regions.

5. References

[1] Doll CNH 2008 *CIESIN thematic guide to night-time light remote sensing and its applications*. NY: Center for International Earth Science Information Network of Columbia University. Retrieved from http://sedac.ciesin.columbia.edu/tg/, last accessed on 30 April, 2021.

[2] Elvidge CD, Baugh K, Zhizhin M, Hsu FC and Ghosh T 2017 Supporting international efforts for detecting illegal fishing and GAS flaring using VIIRS. *IEEE International Geoscience Remote Sensing Symposium*. doi: 10.1109/igarss.2017.8127580.

[3] Gillespie TW, Frankenberg E, Chum KF and Thomas D 2014 Night-time lights time series of tsunami damage, recovery, and economic metrics in Sumatra, Indonesia. *Remote Sensing Letters* **5**(3), 286–294.
[4] Zhao X, Yu B, Liu Y, Yao S, Lian T, Chen L, …, Wu J 2018 NPP-VIIRS DNB daily data in natural disaster assessment: evidence from selected case studies *Remote Sensing* **10**(10), 1526.
[5] Wang Z, Román MO, Sun Q, Molthan AL, Schultz LA and Kalb VL 2018 Monitoring disaster-related power outages using NASA black marble nighttime light product. *ISPRS - International Archives of the Photogrammetry, Remote Sensing and Spatial Information Sciences*, **XLII**-3, 1853–1856.
[6] Chuvieco E 2016 Fundamentals of satellite remote sensing: an environmental approach. *Boca Raton: CRC Press*, Taylor & Francis Group.

Acknowledgments
The authors thankfully acknowledge anonymous reviewers for this useful comments and suggestions. Partial financial support was available from IRDR ICoE Taipei.

Life cycle cost analysis of wastewater treatment technologies

Purva Rathore[1], D J Killedar[2], Divyesh Parde[3] and Akansha Sahare[1]

[1] Post Graduate Student
Shri Govindram Seksaria Institute of Technology and Science, Indore, Madhya Pradesh, India
[2] Professor
Shri Govindram Seksaria Institute of Technology and Science, Indore, Madhya Pradesh, India
[3] Junior Research Fellow
IIT Bhubaneshwar, Odisha, India

E-mail – rathorepurva@ymail.com

Abstract. With the ever-increasing population, volumes of wastewater treatment are a major concern in our country. The Activated Sludge Process (ASP), Biological Filtration and Oxygenated Reactor (BIOFOR), Upflow Anaerobic Sludge Blanket (UASB), and Moving Bed Bio Reactor (MBBR) are all monetarily investigated in the present study using the Life Cycle Cost Assessment (LCCA) tool. In this study, life cycle costing is done using the present value method, which involves discounting the costs for a 20-year economic life. The costs of treating wastewater per million litres per day (MLD) of wastewater treatment technology are obtained from the literature. Moreover, this study takes into account the capital, annual operation, energy, salvage, and replacement costs to compare the life cycle costs of ASP, UASB, BIOFOR, and MBBR to make the best guess of an economical technology. The LCCA demonstrates that the MBBR has the highest costs of treatment, resulting in the highest Life Cycle Cost (LCC). BIOFOR has the largest energy requirement making LCC the second-highest among the technologies. In India, ASP is one of the most widely used technologies, whose LCC is the third most advanced of the four technologies. Because of its lower energy and operating costs, UASB has the lowest LCC.

Keywords. Life Cycle Cost Assessment (LCCA), Life Cycle Cost (LCC), Wastewater Treatment Technology, Activated Sludge Process (ASP), Upflow Anaerobic Sludge Blanket (UASB), Moving Bed Bio Reactor (MBBR).

1. Introduction

Water supply and sanitation are vital commodities. The wastewater generation will increase with an increase in population. From a sustainability point of view, wastewater should be treated and reused to meet future water demands. Apparently, there is a need for proper infrastructure for wastewater treatment to meet the non-potable water quality standards. There are numerous technologies available and being used for the treatment of wastewater [1]. The wastewater treatment plant is a costly affair requiring skilled personnel, expensive units, etc.

The choice of method to treat wastewater is dependent on many factors like process dependability, land area requirement, influent standards, process complexity, required grade of treated wastewater, cost of treatment, environmental concerns, etc. All these aspects need to be assessed and weighed against cost considerations [2]. To evaluate the best option considering the economic aspects of wastewater treatment Life Cycle Cost Analysis is done in the present study.

Life Cycle Cost Assessment (LCCA) analyses the treatment methodologies throughout its life taking into account the costs. The LCCA tool is used to determine the optimum solution in terms of the lowest life-cycle cost, which encompasses project capital expenditures, operations, maintenance, repair, replacement, and disposal costs [3]. It is often utilized to choose the best system, guaranteeing that perhaps the alternative preferred has a reduced average ownership cost while still ensuring adequate quality and functionality.

The current study compares four technologies, ASP, UASB, BIOFOR, and MBBR, to determine which is the most cost-effective option to choose in given conditions. Among the most widely used technologies in sewage treatment plants are ASP, UASB, and MBBR [4]. BIOFOR is a relatively new technology that is being implemented in a few places such as Delhi, India [5][6]. This research will provide a comprehensive analysis of BIOFOR to the other three methods. All costs and other data for treating 1 MLD of wastewater are considered to compare on a single platform. The technologies are classified based on their effluent quality after treatment performance. A comparison of their energy costs and area requirements per MLD is performed so that a correct decision can be made in the field. With the guidance of the present study, a judgment on an optimal wastewater treatment process can be reached.

2. Wastewater Treatment Technologies

While there are several treatment technologies adopted for sewage treatment. The two commonly known conventional technologies ASP and UASB are taken for analysis whereas MBBR and BIOFOR are advanced technologies. Following is a brief description of the wastewater treatment processes assessed in this study.

2.1. Activated Sludge Process (ASP)

Activated Sludge Process is an aerobic treatment method invented in 1914 by Arden and Lockett [7]. It is a suspended growth process in which oxygen is introduced into raw wastewater to form a biological floc. The floc minimizes the organic mass in influent thus reducing biological oxygen demand (BOD) and total suspended particles from wastewater. It was named so because it involved the creation of an active mass of microorganisms capable of aerobically stabilizing organic matter in wastewater. The reason for its wide applicability is because its performance is not affected by variation in season and takes up less area. It usually requires a continuous energy supply for aeration and sludge recirculation [8]. The performance of ASP in terms of percentage removal of BOD, COD, TSS, and Fecal coliform is given in Table1. The advantage of this method is even due to regular fluctuations in wastewater properties, the performance of wastewater treatment technologies is unaffected. The drawback of ASP has been that it needs a steady supply of electricity, which has a considerable impact on efficiency [8].

2.2. Upflow Anaerobic Sludge Blanket (UASB)

Upflow Anaerobic Sludge Blanket reactor, a suspended anaerobic bioreactor processes wastewater by moving it upward from the reactor's bottom. Bacteria in the suspended sludge blanket, filter and purify wastewater by converting organic particles to biogas. There is no need for an aeration device in this process. On top of the unit, baffles are provided to keep the sludge layer from leaving the reactor. UASB alone cannot meet the effluent standards. Table 1 describes the effluent quality from UASB in terms of BOD, COD, and TSS removal percentages [2][7]. UASB has the benefit of ensuring minimal sludge management and being able to bear hydraulic and organic shock loads. The process has the flaw of not being able to treat wastewater effectively on its own [8].

2.3. *Biological Filtration and Oxygenated Reactor (BIOFOR)*

Biological Filtration and Oxygenated Reactor is M/s Degremont Ltd.'s patented inventions. It is an aerobic attached growth process. BIOFOR uses a media known as Biolite to filter the unwanted particles. In this process, chemical dosing is needed in the form of coagulant and polyelectrolyte. Biolite is a layer of special clay granules on a bed of multiple media to filtrate the subsequent granules formed in flocculation. Diffusers induce air into the reactor thereby introducing an airy environment to oxidize the organic content in the raw influent. BIOFOR is a compact process requiring less space and more energy compared to ASP. Treatment efficiency (BOD, COD, and TSS % removal) of BIOFOR is stated in Table 1 [6][8]. Higher aeration efficiency thanks to a co-current diffused oxygenation device; less space required because secondary sedimentation is eliminated; and the capacity to endure volume flow and biological load fluctuations. The demerit is that it requires a high dosage of chemicals for treatment [8].

2.4. *Moving Bed Biofilm Reactor (MBBR)*

Moving Bed Biofilm Reactor is a hybrid of the activated sludge process (suspended growth) and the biofilter process (attached growth). The biofilm comprises plastic carriers that decompose organic waste thereby reducing wastewater pollution. Air bubbles introduced by the aerator shake the carriers, causing them to move. The biodegradable mass is then allowed to settle. MBBR requires skilled manpower and a high energy supply for its operation [2]. There are a variety of biofilms available for treatment [9]. BOD, COD, TSS, and Faecal Coliform removal by MBBR are stated in Table 1. MBBR requires skilled manpower and a high energy supply for its operation leading to higher expenses. Its primary characteristic is that it can achieve the precise removal of various metals by employing a specialized membrane [2].

Table 1. Data acquired from literature and sources.

	Description of WWTTs			
	ASP	UASB	BIOFOR	MBBR
Area of land required m^2/MLD	1000	2000	400	550
Capital cost Rs/MLD	₹ 50,00,000	₹ 30,00,000	₹ 11,00,000	₹ 1,14,19,000
Annual operation and maintenance cost Rs/MLD	₹ 9,00,000	₹ 12,50,000	₹ 12,00,000	₹ 1,10,00,000
Power required daily in kWh/MLD	225	150	335	280
% TSS removal	75-90	70-80	90-95	85-90
% BOD removal	84-95	75-78	75-90	85-90
% COD removal	75-94	70-80	90-95	85-90
Faecal Coliform removal	Log Scale unit 2-3		Log Scale unit 2	Log Scale unit 2-3

3. Life Cycle Cost Analysis (LCCA)

Life-cycle cost analysis (LCCA) method is used to calculate all of a project's, product's, or measure's expenses throughout time. It considers all expenses, which include initial costs like invested capital, acquisition, and setup, as well as future cash flows such as power, operating costs, maintenance costs, capital cost of repairs, and interest expense, as well as any resale, salvage, or disposal costs incurred during the system's or product's life cycle. The Life Cycle Cost Analysis (LCCA) method can be used to make long-term financial assessments of a system's lifespan.[3][10].

In this study, LCCA is done by the Present Value (PV) method according to IS 13174. The method is often referred to as the Net Worth or Present Value or Present Worth (PW) method. PW is an approach that requires all future cash flows to be converted to baseline, taking into account both inflation and the opportunity cost of capital [11][12]. PW is an economic approach for discounting future costs to their present equivalents and adding these up, taking into account the inflationary effects and depreciation and also the anticipated return on investment over time.

3.1. Components of LCCA

3.1.1. Capital costs - Capital cost comprises civil, mechanical, electrical and any other related items during the construction of the plant. Land purchase, equipment costs, building and infrastructure costs, contingency, materials, and planning are all elements of capital expenditures [5].

3.1.2. Operation_and maintenance costs (O&M) - O&M costs incurred usually account for labour requirement, energy, chemicals, repair and replacement of electro-mechanical materials [13][5].

3.1.3. Energy costs – The cost of total power (electricity) required by wastewater treatment technology to operate is energy costs [5].

3.1.4. Replacement costs - Replacement cost is the sum of all repair and equipment replacement costs anticipated over the life of the system [5].

3.1.5. Salvage value - The monetary sum expected from the disposal of an asset at the end of its economic life, or at the end of the study period. Also, termed as resale value or scrap value in certain cases depending upon how the asset may be treated after disposal [5].

3.2. Some important definitions used in LCCA

3.2.1. "Discounting - A technique for converting future cash flows to equivalent amounts at present or on an agreed prior date" [11].

3.2.2. "Discount Factor - A multiplication number for converting cost and benefits occurring at different times to a common baseline" [11].

3.2.3. "Discount Rate - The rate of interest reflecting the investor's time value of money, used to determine discount factors for converting benefits and costs occurring at different times to a baseline" [11].

$$\text{Discount rate} = \frac{(1+interest\ rate)}{(1-inflation\ rate)} - 1 \ [14] \tag{1}$$

3.2.4. "Life Cycle Cost - It covers all the costs from project conception to final scrapping and disposal and includes all costs of operation, repairs, maintenance, energy consumption, rentals, insurance. etc, in addition to the initial costs of development and/or acquisition, all discounted to the same point in time" [11].

3.2.5. "Present Worth/Present Value - The value of a benefit or cost found by discounting future cash flows to the baseline" [11].

3.2.6. "Present Value Factor/Present Worth Factor - The discount factor used to convert future benefits and costs to the present" [11].

4. Methodology

This study takes some assumptions in LCCA for baseline consideration. They are as follows

a) The study period (economic life) of each technology is considered as 20 years. (n)
b) The inflation rate is considered and assumed as 4.5%
c) The interest rate is assumed to be 10%
d) The discount rate is assumed to be 5%
e) The cost of land is considered at rate of Rs.100/m^2 and the rate of power is assumed to be Rs. 8/kWh unit.
f) The replacement cost is taken as 10% of capital cost. The replacement is considered at 10 years of its life.
g) The salvage value is taken as 10% of civil construction cost.

The capital costs, annual operation and maintenance costs of each technology are obtained from various literature & sources referred in the study and are noted in Table 1. Table 1 also contains the amount of energy required daily by the respective technology to treat 1 MLD of wastewater, as well as the amount of land required by the technology to treat 1 MLD of wastewater. Both of these values are derived from resources. The cost of required land is factored into the total capital cost [13]. The cost of land is estimated as shown in the equation below at a cost of Rs. 100 per m^2. During the life of the wastewater treatment process, the operation and maintenance costs reoccur on an annual basis. The energy costs are not included in operation and maintenance costs and are evaluated separately for LCCs. Because it is a recurring cost, the values of daily energy required by each technology are transformed annually. The salvage cost is assumed to be 10% of the construction cost, hence the construction cost is taken as 60% of the capital cost for ASP, 55% of the capital cost for UASB, 58% of the capital cost for BIOFOR, and 40% of the capital cost for MBBR. Table 2 below summarises the calculated capital costs, energy costs, replacement costs, and salvage costs.

Table 2. Costs and performance of wastewater treatment technologies (WWTT)

Row no	Description	WWTTs			
		ASP	UASB	BIOFOR	MBBR
(1)	Rate of land Rs/m^2	₹ 100	₹ 100	₹ 100	₹ 100
(2)	Area of land required m^2/MLD	1000	2000	400	550
(3)	Cost of land Rs/MLD (1)×(2)	₹ 1,00,000	₹ 2,00,000	₹ 40,000	₹ 55,000
(4)	Capital cost Rs/MLD	₹ 50,00,000	₹ 30,00,000	₹ 11,00,000	₹ 1,14,19,000
(5)	Total Capital cost Rs/MLD (3)+(4)	₹ 51,00,000	₹ 32,00,000	₹ 11,40,000	₹ 1,14,74,000
(6)	Annual Operation and Maintenance cost Rs/MLD	₹ 9,00,000	₹ 12,50,000	₹ 12,00,000	₹ 1,10,00,000
(7)	Rate of power Rs/kWh	₹ 8	₹ 8	₹ 8	₹ 8
(8)	Power required daily kWh/MLD	225	150	335	280
(9)	Energy Costs Rs/MLD (7)×(8)×365	₹ 6,57,000	₹ 4,38,000	₹ 9,78,200	₹ 8,17,600
(10)	[a]Salvage cost Rs/MLD	₹ 3,00,000	₹ 1,65,000	₹ 63,800	₹ 4,56,760
(11)	Replacement cost Rs/MLD {10% of (4)}	₹ 5,00,000	₹ 3,00,000	₹ 1,10,000	₹ 11,41,900

[a] Salvage Cost is 10% of construction cost. Construction cost is some percentage of capital cost.

The equations used to calculate the costs are as follows.

Cost of land = Rate of land per m^2 × Area required for treatment m^2

Total Capital cost = Capital cost + Cost of land.

Energy Cost = Rate of energy per kWh × Energy required kWh × 365

Salvage Cost = 10% × Given percentage for each technology to convert capital cost to construction cost (60 for ASP; 55 for UASB; 58 for BIOFOR; 40 for MBBR) × Capital cost

The LCC is sum of present values of Capital, O & M, Energy, Salvage and Replacement costs. These costs are converted to discounted present value by multiplying itself to present value factor. The present value factors are of two kinds

Uniform Present Value factor – The costs recurring annually are converted to present value cost by this factor. Denoted as UPV/UPWF

$$\text{UPWF} = \frac{(1+d)^{(n-1)}}{d(1+d)^n} \text{ [14]} \qquad (2)$$

where n – life of study in years; d- discount rate

Single Present Value factor – For converting the initial investments or one-time costs. Denoted as UPV/UPWF

$$\text{SPWF} = \frac{1}{(1+d)^n} \text{ [14]} \tag{3}$$

The values of SPWF and UPWF are given in Table 3. The present value factors for different costs are different because they vary with the number of years.

Table 3. Discount factors

Calculation of SPV and UPV				
	Discount rate (d)	Number of years (n)	Discounting factor	Round up value
SPV_1	0.05	1	0.95238095	0.95
SPV_{10}	0.05	10	0.61391325	0.61
SPV_{20}	0.05	20	0.37688948	0.38
UPV_{20}	0.05	20	12.4622103	12.46

Table 4. Life cycle cost calculation

Row no.	Costs	WWTT			
		ASP	UASB	BIOFOR	MBBR
(1)	Capital base cost in Rs/ MLD	₹ 51,00,000	₹ 32,00,000	₹ 11,40,000	₹ 1,14,74,000
(2)	Capital present value cost in Rs/ MLD (1)×SPV_1	₹ 48,45,000	₹ 30,40,000	₹ 10,83,000	₹ 1,09,00,300
(3)	Annual O&M base cost in Rs/MLD	₹ 9,00,000	₹ 12,50,000	₹ 12,00,000	₹ 1,10,00,000
(4)	Annual O&M present value cost in Rs/MLD (3)×UPV_{20}	₹ 1,12,14,000	₹ 1,55,75,000	₹ 1,49,52,000	₹ 88,01,99,320
(5)	Annual Energy base cost in Rs/MLD	₹ 6,57,000	₹ 4,38,000	₹ 9,78,200	₹ 8,17,600
(6)	Annual Energy present value cost in Rs/MLD (5)×UPV_{20}	₹ 81,86,220	₹ 54,57,480	₹ 1,21,88,372	₹ 1,01,87,296
(7)	Replacement base cost in Rs/MLD	₹ 5,00,000	₹ 3,00,000	₹ 1,10,000	₹ 11,41,900
(8)	Replacement present value cost in Rs/MLD (7)×SPV_{10}	₹ 3,05,000	₹ 1,83,000	₹ 67,100	₹ 6,96,559
(9)	Salvage base cost in Rs/MLD	₹ 3,00,000	₹ 1,65,000	₹ 63,800	₹ 4,56,760
(10)	Salvage present value cost in Rs/MLD (9)×SPV_{20}	₹ 1,14,000	₹ 62,700	₹ 24,244	₹ 1,73,569
(11)	Life Cycle Cost in Rs/MLD (2)+(4)+(6)+(8)-(10)	₹ 2,44,36,220	₹ 2,41,92,780	₹ 2,82,66,228	₹ 15,86,70,586
(12)	Life Cycle Cost in Lakh Rs/MLD	₹ 244	₹ 242	₹ 283	₹ 1,587

The Life Cycle Cost (LCC) takes into account the present value costs of each cost component. The Table 4 tabulates the LCC calculations, wherein the costs are first converted to present costs by multiplying its by respective discount factors and then summed up for total LCC. The capital cost is multiplied by single present value factor (SPV_1) to obtain its present cost because it is a one-time investment cost at the initial year of life cycle. The annual operation and maintenance as well as annual energy costs recur every year throughout the lifetime in same amount hence it is multiplied by UPV_{20} for present cost conversion. For purpose of obtaining present cost of replacement cost of WWTTs it is multiplied by SPV_{10} since the replacement period is assumed to be 10 years. Similarly, the salvage cost is procured at the end of life hence it is multiplied by SPV_{20} for further calculations. The last row represents the approximate total LCC in Lakhs required to treat 1 MLD of wastewater.

5. Results and Discussions

UASB is the technology with the highest area requirement of the four technologies. The limitation of UASB is that it cannot withstand discharge standards on its own; the effluent requires tertiary treatment before discharge or reuse. It is incapable of removing faecal coliform from wastewater. UASB generates biogas, which is a great source of revenue. UASB does not require any mechanical aeration device, which reduces its energy requirement and ensures that its performance is unaffected by power fluctuations [8]. UASB, despite having the lowest LCC, is not recommended for decentralised treatment because it may require tertiary treatment before reuse, and the generation of biogas puts it at a disadvantage due to its odour (for decentralised).

ASP is the technology with the second lowest LCC. Because of its high level of treatment and low cost, it is the most widely used technology. However, ASP necessitates a larger land area than BIOFOR and MBBR. It has operational issues during the winter season and is impacted by power fluctuations.[8]

BIOFOR is ranked third among LCCs. BIOFOR is a compact treatment system because it takes up less space and has the lowest capital cost. However, because of the continuous chemical and power requirements, the LCC is affected by O & M cost, despite the fact that it requires less manpower for supervision.[8] It has the highest energy requirement, resulting in a high LCC.

Because of its high capital and operating and maintenance costs, MBBR has a very high LCC. Such high costs are observed as a result of the membrane used for treatment, which provides the best effluent quality of the four technologies.

Wastewater treatment technologies are being used around the country to treat domestic and industrial wastewater. Recent cost data for wastewater treatment technology is not explicitly accessible. And the expenses for LCCA are acquired from literature; if the expenses are not as specified, the analysis will be inaccurate. The expenses computed may or may not be predicated on India's schedule of rates (SOR).

Figure 1 depicts a comparison of the area required by each technology. The highest area required is by UASB compared to other three and BIOFOR requires lowest space to treat 1 MLD of wastewater.

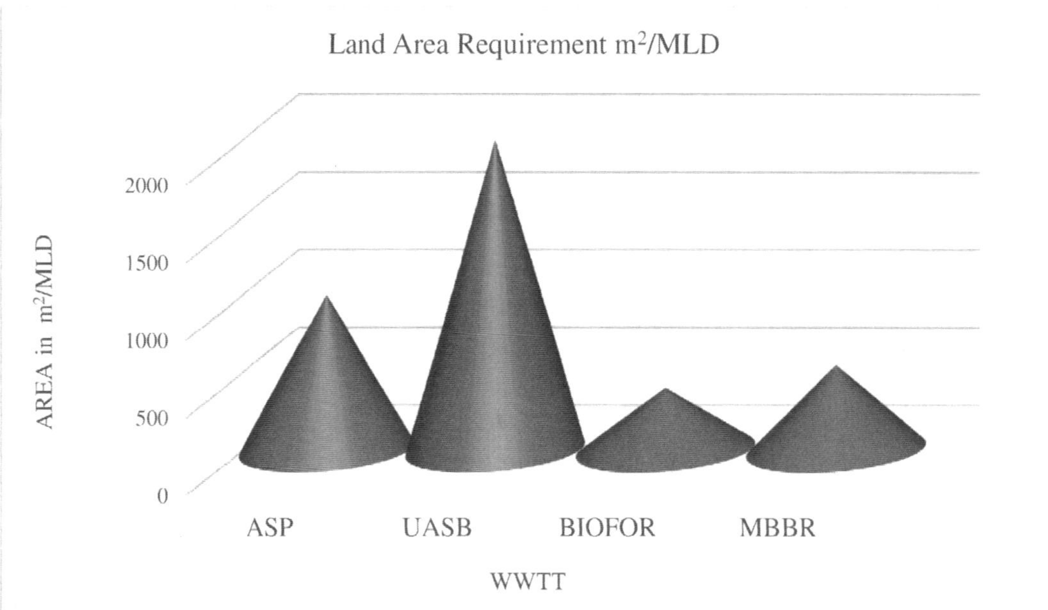

Figure 1. Area required for wastewater treatment technologies (WWTTs)

Figure 2 depicts the total capital cost and annual operation and maintenance (O & M) costs for treating 1 MLD of wastewater using all four technologies. MBBR being the technology with highest capital and O & M cost because of the use of membrane adopted for treatment. And BIOFOR has lowest capital cost as it requires minimum area when compared to others and ASP with lowest operation and maintenance as it does not involve any expensive material to maintain.

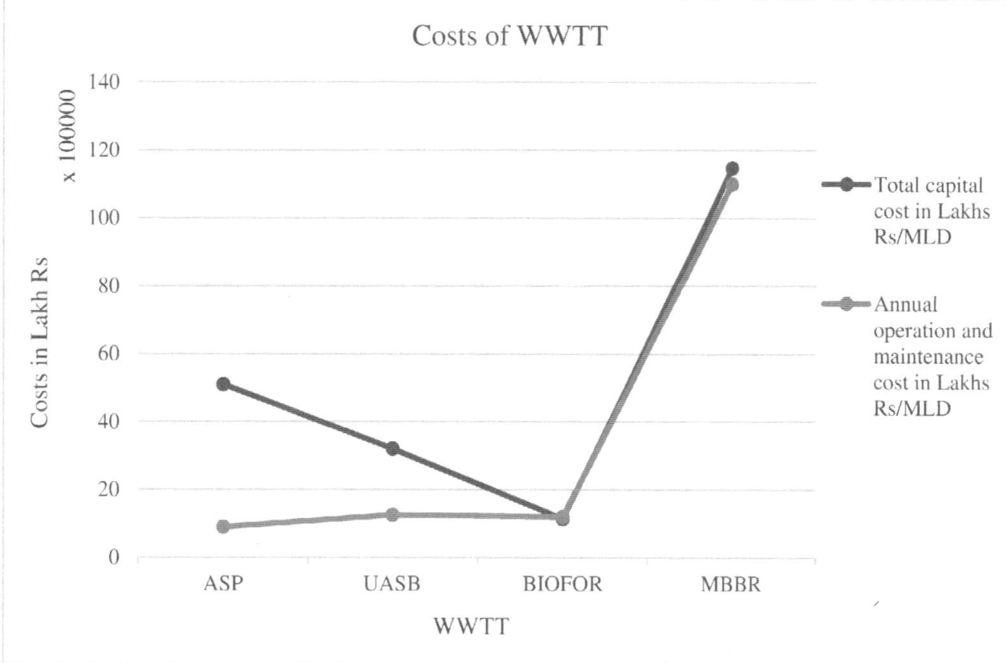

Figure 2. Capital and O & M cost of wastewater treatment technologies (WWTT)

Figure 3 depicts a comparison of area versus Life Cycle Cost (LCC) for each technology. The LCC is highly dependent on area requirement and its cost. Cost of land changes geographically and temporarily hence it is an uncertain factor. The cost of land influences the capital cost and thus affecting the LCC. In case of UASB, the land required is much higher than its LCC and hence the higher price of land will highly affect the LCC of UASB. Same is for ASP. While for MBBR, if the price of area is high it will increase the LCC, but comparatively it won't have much significant effect as in case of UASB.

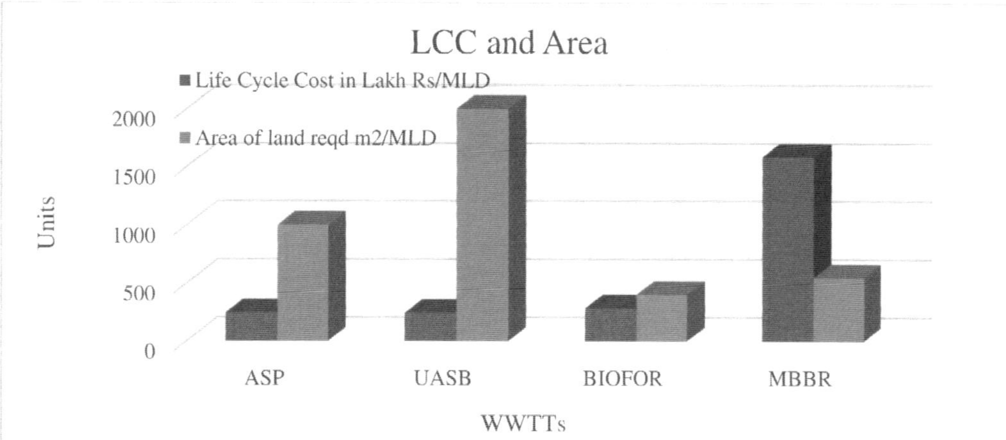

Figure 3. Area requirement and life cycle cost comparison

Figure 4 shows a graph of the cost components of each technology per MLD. There are two vertical axes shown for comparing the costs; the primary vertical axis is reference for life cycle costs of all four technologies while the secondary axis on right side is reference for values of capital costs, energy costs and annual operation and maintenance (O & M) costs. It can be seen through the graph that O & M costs highly affects the LCC of wastewater treatment technologies (WWTT). BIOFOR has highest energy cost which makes its LCC higher than the other two (ASP and UASB).

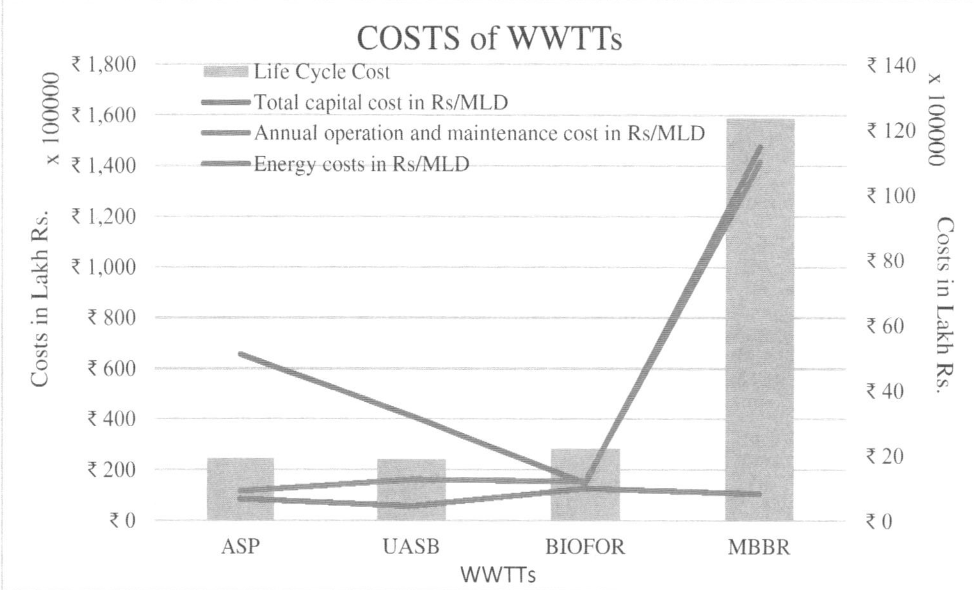

Figure 4. Cost comparison of wastewater treatment technologies (WWTTs)

6. Conclusions

i. The selection of wastewater treatment technology is a critical decision that is influenced by a variety of factors. All of these factors, as well as the costs of the technologies, should be presumed in relation to the purpose of installing the treatment technology.

ii. LCCA can be used to assess costs. The findings indicate that the order of LCC of the four technologies is UASB<ASP<BIOFOR<MBBR.

iii. The present LCCA study is a literature-based cost assessment of the four wastewater treatment technologies.

iv. UASB is a cost-effective option if there is no land limit and effluent quality is not a primary concern for treatment. It has the lowest treatment efficiency of the four technologies. It is being installed in sewage treatment plants with tertiary treatment.

v. If land is not a major consideration, ASP is a viable option. It can be installed to treat a larger volume of sewage or to provide centralised treatment. ASP outperforms UASB in terms of removal efficiency.

vi. Because it takes up less space, BIOFOR may be a good choice for decentralised treatment. BIOFOR has a higher percentage of COD and TSS removal than MBBR.

vii. MBBR can be used where cost is not an issue due to its high treatment efficiency. MBBR can be installed for decentralised treatment. Because of the medium, the filter used for treatment, MBBR has the highest capital and maintenance costs. MBBR is used for both industrial and domestic wastewater treatment.

viii. The LCCA is affected by change of any of the cost components and depends on the actual site conditions where it is adopted.

References

[1] Kamble S J, Chakravarthy Y, Singh A, Chubilleau C, Starkl M, and Bawa I 2017 A soil biotechnology system for wastewater treatment: technical, hygiene, environmental LCA and economic aspects *Environ Sci Pollut Res.*
[2] Tare V 2010 *Sewage Treatment in Class I Towns: Recommendations and Guidelines*, Report Code: 003_GBP_IIT_EQP_S&R_02_Ver 1_Dec 2010 (Kanpur: Professor and Coordinator Development of GRB EMP IIT Kanpur)
[3] Rawal N and Duggal S K 2016 Life Cycle Costing Assessment-Based Approach for Selection of Wastewater Treatment Units *Natl. Acad. Sci. Lett.* 103-107
[4] CPCB 2021 *National Inventory of Sewage Treatment Plants* (Delhi: Central Pollution Control Board, Ministry of Environment and Forests, Govt. of India)
[5] Chawda D S 2016 *Techno-Economic Evaluation of Sewage Treatment Technologies in India* Retrieved from http://www.dspace.dtu.ac.in:8080/jspui/bitstream/repository/14448/3/final_report.PDF
[6] Sharma C, and Singh S K 2013 Performance Evaluation Of Sewage Treatment Plant Based On Advanced Aerobic Biological Filtration And Oxygenated Reactor (BIOFOR) Technology- A Case Study Of Capital City -Delhi, India. *International Journal of Engineering Science and Innovative Technology* 435-442
[7] Metcalf & Eddy Inc., Tchobanoglous G, Burton F L, Tsuchihashi R and Stensel H D 2003 *Wastewater Engineering: Treatment and Resource Recovery* Fourth Edition (Boston: McGraw-Hill Companies Inc.)
[8] Tare V and Bose P 2009 *Compendium of Sewage Treatment Technologies* (Kanpur, Delhi: National River Conservation Directorate (NRCD))
[9] CPHEEO 2013 *Manual on Sewerage and Sewage Treatment Systems Part A Engineering* (New Delhi: Central Public Health And Environmental Engineering Organization, Ministry Of Urban Development)
[10] Chhipi-Shrestha G, Hewage K, and Sadiq R 2017 Fit-for-purpose wastewater treatment: Conceptualization to development of decision support tool (I) *Science of the Total Environment* 600-612
[11] Bureau of Indian Standards 1990 *IS 13174 - 1 :Life Cycle Costing, Part 1: Terminology* (New Delhi, India: BIS)
[12] Bureau of Indian Standards 1994 *IS 13174 (Part 2) : 1994 Life Cycle Costing, Part 2 : Methodology* (New Delhi, India: BIS)
[13] Kumar S, Bhargava R, and Singhal M K 2008 Optimization of Sewage Treatment Technologies on the basis of Life Cycle Cost Analysis (LCCA) *Proc. of the National Seminar on Conservation and Restoration of lakes (CAROL-08)* **2** (Nagpur: National Institute of Hydrology pp 807-816
[14] Fuller S K, and Peterson S R 1996 *Life- Cycle Costing Manual for the Federal Energy Management Program* (Washington, U.S.: National Institute of Standards & Technology(NIST) Handbook) **135**
[15] Bhoye Y V, Aher P D, and Saner A B 2016 Life Cycle Cost Analysis of Sewage Treatment Plants. *International Journal of Modern Trends in Engineering and Research* 426-429
[16] Kumar S D, and Sekran V 2013 FBBR and MBBR Bioreactors for Sewage Treatment. *International Journal of Environmental Engineering and Management* 73-88
[17] Reddy V R, Kurian M, and Ardakanian R 2015 Life-cycle Cost Approach (LCCA): Framework and Concepts *Life-cycle Cost Approach for Management of Environmental Resources* (New York City: Springer, Cham.) pp. 17-37
[18] Kumar M D, and Tortajada C 2020 Wastewater Treatment Technologies and Costs *Assessing Wastewater Management in India* (Singapore: Springer Briefs in Water Science and Technology) pp. 35-42

A concise air quality study on impact of COVID-19 lockdown for two industrial cities of Chhattisgarh

Akansha Sahare[1],D.J.Killedar[2],Ajay .V. Ahirwar[3] and Purva Rathore[1]

[1]Post Graduate Student
Shri.G.S.Intitute Technology & Science ,Indore, M.P ,India
[2] Professor
Shri.G.S.Intitute Technology & Science ,Indore, M.P ,India
[3]Assistant Professor
Civil Engineering Department, NIT Raipur

Email- akanshasahare18@gmail.com

Abstract.'Good Outcomes from evil situation' this phrase perfectly fits into the COVID-19 circumstances as several restrictions on anthropogenic activities provided an improvement in the ambient air quality status globally. The study deals with the consequences of COVID-19 lockdown on ambient air quality for 2 major industrial cities (Raipur and Bilaspur) in Central Indian state Chhattisgarh moreover a comparison of air quality data was made with non-lockdown year (2019) .The AQI and critical parameters (such as PM_{10}, $PM_{2.5}$, SO_2 and NO_x) were acquired form online available source and then analysed for the study period (2019 and 2020) .Noteworthy reduction in AQI and concentration of pollutants in Raipur was detected whereas there was reduction in Bilaspur but it was less than Raipur. Evident changes in the level of pollutants (NO_x and PM) were observed during the study. Meteorological parameters such as temperature and relative humidity were also examined for Raipur. Statistical analysis between data of meteorological parameters and AQI for capital city Raipur was also carried out.

Key words: COVID-19, AQI, Meteorological parameter and Statistical analysis

1.Introduction

COVID-19 rapid proliferation has turned into global public health emergency. In the end of December 2019, a contagious disease was detected at Wuhan city, China and named Corona virus disease (COVID 19) also known as SARS COV 2[1,2]. The virus began to spread swiftly across the globe in early 2020 infecting nearly many countries. On 30th January 2020 first case was reported in India [3].

The virus eventually expanded to 215 countries throughout the world and the WHO declared it as global pandemic on March 11,2020 [4].

To avoid the situation from getting more worse, as a prevention from this pandemic period lockdown were imposed at different places in different parts of the world. As a result of the increasing number of cases in various parts of globe because of the virus, the Government of India (GOI) has adopted an early call to eradicate the spread of the COVID by declaring a nationwide lockdown which started on 25th March 2020 to 14th April 2020 further which extended 31st May 2020[5].The Indian government imposed various restriction on several anthropogenic activities like commercial, industrial, education ,transportation sector ,tourism and public gathering .During this lockdown time only very essential services like hospital, ATMs, bank, grocery shops ,pharmacies and petrol pump were permitted[6,7,8].

The lockdown due to pandemic on the human activities proved as a gift for the environment which gave it much needed time to heal. A great improvement in air quality status was observed around worldwide and a lot of study is going on this[9,10].As a matter of fact ,in India there has been a significant improvement in air quality and reduction in critical pollutant in many metropolitan cities and other part of the country, numerous research papers are present to justify the fact[3,4, 11,12,13,14].Noteworthy reduction in the concentration of pollutants such as(NO_2 and particulate matter)due to various constraints in lockdown over transportation and industrial sector emission [8,15]. Satellite data and visual maps are also made used in variety of work to get a better understanding of the condition of air quality in the atmosphere [4,5,8,11].The main objective of the study to analyse consequences of COVID-19 lockdown on ambient air quality for 2 major industrial cities in Chhattisgarh and to compare the air quality data with non-lockdown year (2019) .

2.Study Area

Chhattisgarh was granted the status of separate and sovereign state of India in November, 2000.Soon after with the central government aid, the state experienced a boom in all sectors and is now the fastest growing state in the country [16].The state's rapid industrialization is due to the abundance of ores and minerals resources such as coal, limestone, granite, iron ore and bauxite [16,17].In the study major 2 industrial cities Raipur and Bilaspur have been analysed. Raipur is bordered by several industrial areas such as Urla, Siltara,Tilda, Bhanpuri and the district Durg which is also a large industrial city[18,19,20].Several cement industries, sponge iron units ,steel industry, ferro alloy plant ,rice mills and power plants are located in the various industrial area near Raipur. The next selected city, Bilaspur is also major industrial city[17,21].Number of industrial area such as Sigritti ,Tifra, Dagori and Silphari are surrounding the city Bilaspur(CSIDC). South Eastern Coalfields Limited (SECL) and National Thermal Power Corporation Limited (NTPC) are also located at the city.

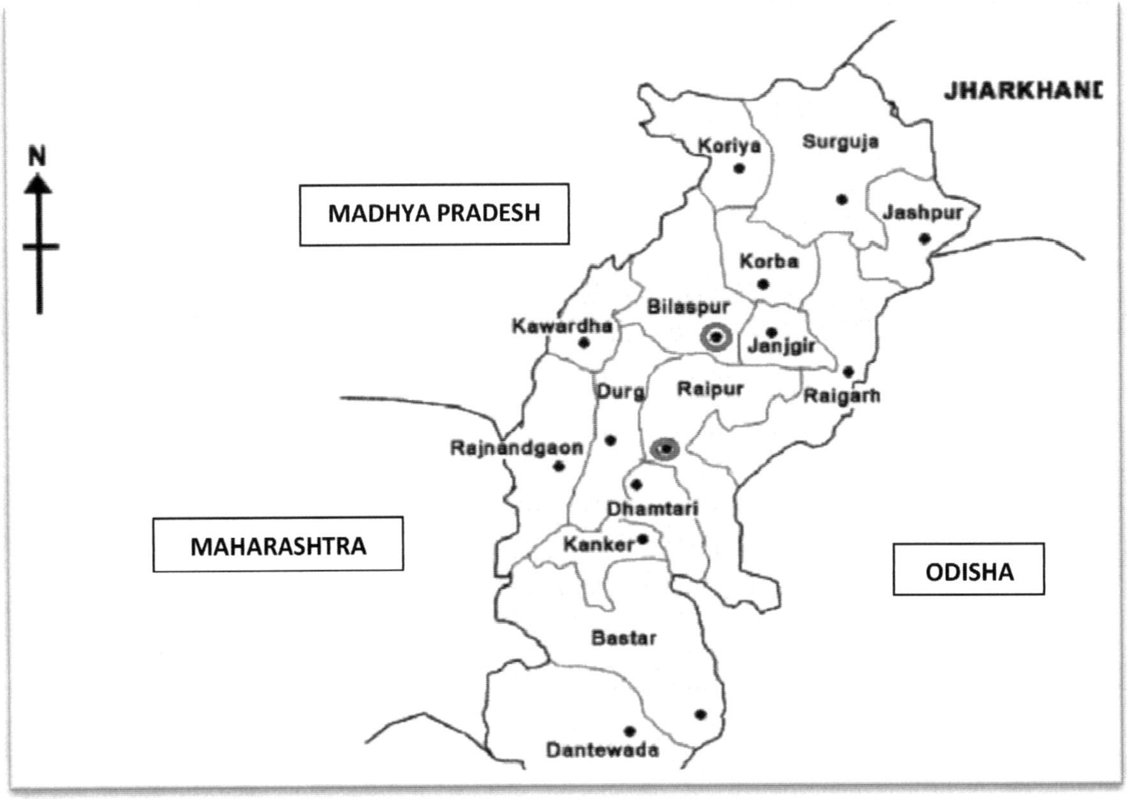

Figure 1: Map of Chhattisgarh, red circled shows the selected cities for the study.

3.Material and Methods

CECB (Chhattisgarh Environment Conservation Board) has been established as an authorised department of Chhattisgarh Government, with its headquarter in Raipur, state capital [17]. The required data for study were obtained from the official site of https://www.enviscecb.org/. For two years (2019 and 2020) the online data for AQI and concentration of pollutant (PM_{10}, $PM_{2.5}$, SO_2 and NO_x) were obtained for the months of March and April. Further the meteorological data were procured from the weather underground services (https://www.wunderground.com/) for Raipur city.

Assessment of the air data (AQI and concentration of pollutant) has been carried out using MS-Excel and various available research works. For the convenient analysis the study has been divided into 2 phase:

1) Before Lockdown Phase (BLP)[1st March -24th March]
2) During Lockdown Phase (DLP)[25th March- 25th April]

for 2 years to monitor the consequences of lockdown on air quality status in 2 industrial cities (Raipur and Bilaspur) of Chhattisgarh. Fixed interval of phase for both the years (2019 and 2020) has been taken so that the comparison and study could become easier.

4. Results and Discussion

4.1 AQI and concentration of pollutants variation for 2 cities

AQI and concentration of pollutants (such as PM_{10}, $PM_{2.5}$, SO_2 and NO_x) data were downloaded from the official site of CECB and after the processing of the data it was divided it into 2 phase further (Table1) and (Table 2) were prepared.

Table1 summaries the average AQI and % change over 2 different phases for non-lockdown year (2019) and lockdown year (2020).From Table 1 it was observed that Raipur experienced a greater decrease in AQI which was about 37.3% when the comparison was made between 2020 DLP and 2019 DLP. Moreover, when the comparison was made within the same year (i.e., 2020) between 2 phases (DLP and BLP) the result was also noticeable which about 26% decline in AQI. Presence of numerous of industries around city which were shut down during the lockdown could be one of the reasons for falling off in AQI.

On the other hand, for Bilaspur city the results which were observed were not as good as Raipur. Decrease of 10.9% and 12.5% in AQI were noticed when 2020 DLP was compared with 2019 DLP, and 2020 DLP was compared with 2020 BLP respectively for Bilaspur city. Presence of NTPC near the city could be considered as a factor that there was sight change in AQI. In addition to the fact, the power hub of the state, Korba which has several power stations which were operational during the lockdown also is located near to city [15,17].

(Figure2) describes graphs for study area with average AQI trend in both phases. Form (Figure 2) it was observed that for Raipur city decreasing trend was noted whereas sight change in AQI for Bilaspur was noticed. Table 2 give details of the average concentration for critical parameter such as (PM_{10}, $PM_{2.5}$, SO_2 and NO_x) and difference % for the lockdown year (2020) which were analysed for Raipur and Bilaspur. Prime source (NO_2 and PM) in Raipur is from several industrial sectors, high intensity vehicular population and re suspension of dust particulate [21,22,23]. As the lockdown was announced in the year 2020 hence due several restrictions on anthropogenic activities such as commercial work, construction activities, industrial and vehicular emission resulted in improvement in the level of concentration of several pollutants. From (Table 2) it was observed that exception decrease of 26.0% and 24.5% in PM_{10} and NO_x respectively for Raipur was found when comparison was made between 2020 DLP and 2020 BLP in the study.

For the second selected city Bilaspur a small decrease% was noticed which was about 11.1% and 8.7% when 2020 DLP was compared to 2020 BLP for PM_{10} and NO_x respectively. From (Table 2) it was observed that for Raipur the concentration level of $PM_{2.5}$ and SO_2 were not as remarkable as (NO_2 and PM_{10}) but reduction in level of concentrations were found about 15.9% and 13% when BLP was compared to DLP for 2020.Besides for Bilaspur city mere change in the concentration of 4% and 2.4% was recorded for $PM_{2.5}$ and SO_2 respectively during the analysis. The reason for this could be operation of various thermal power plant near the study area [15,24].

(Figure 3) depicts the graphical representation of parameters and respective average concentration in both phases of 2020 for selected cities. The result obtained for Raipur and Bilaspur were presented on the graphs.

Table 1. Average AQI and % changes over 2 phase for two major cities of Chhattisgarh state.

CITIES	2019		2020		ΔAQI % in DLP 2020 wrt DLP2019	ΔAQI % in DLP 2020 wrt BLP 2020
	BLP	DLP	BLP	DLP		
RAIPUR	71	59	50	37	-37.3	-26
BILASPUR	54	55	56	49	-10.9	-12.5

BLP-Before Lockdown Phase, DLP-During Lockdown Phase

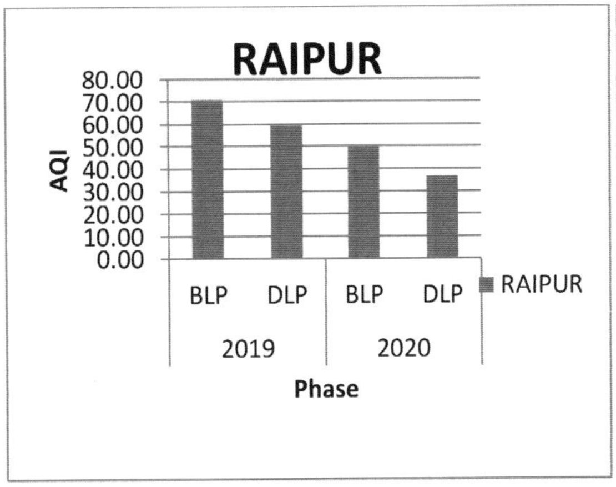

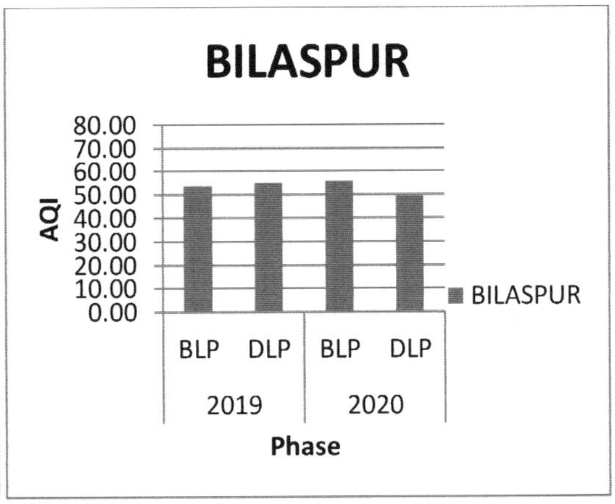

Figure 2. Average AQI in 2 years for selected cities in 2 phases.

Table 2. Average concentration of parameters and % difference for study areas for year 2020.

RAIPUR (2020)				
PARAMETERS	BLP	DLP	DIFFERENCE	DIFFERENCE %
PM_{10}	49.8	36.8	-13	-26.0
$PM_{2.5}$	27.4	23.1	-4.4	-15.9
SO_2	14.2	12.3	-1.8	-13.0
NO_X	13.7	10.4	-3.4	-24.5
BILASPUR (2020)				
PARAMETERS	BLP	DLP	Difference	Difference %
PM_{10}	55.6	49.4	-6.2	-11.1
$PM_{2.5}$	16.9	16.2	-0.7	-4.0
SO_2	4.9	4.8	-0.1	-2.4
NO_X	15.3	14.0	-1.3	-8.7

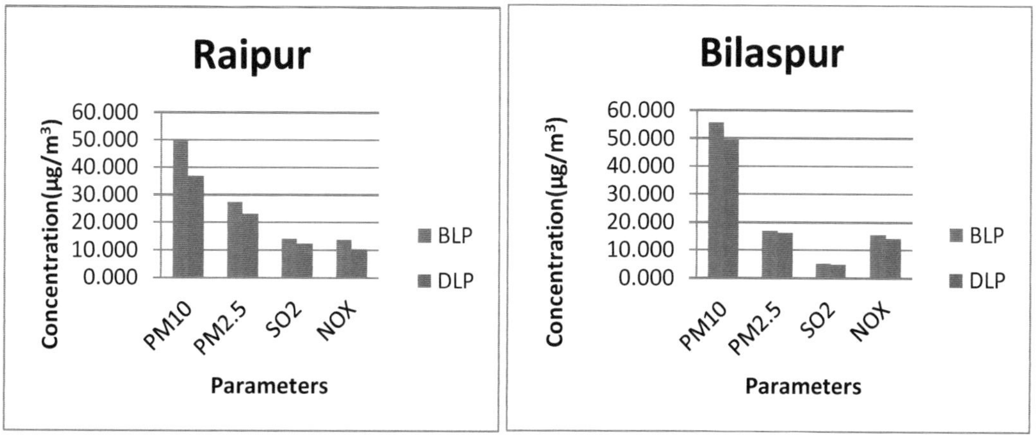

Figure3. Average concentration of parameters in the lockdown year 2020 for 2 cities.

4.2 Change in Meteorological parameters for Raipur city

Parameters such as temperature, relative humidity, wind speed, direction and precipitation greatly influence the air quality for any region. Pollutants in the environment can be easily dispersed, diluted and washed away with the help of meteorological parameters [25,26].

As the 2 phases which are selected for the study period lies in 2 different seasons. The BLP which lies at end of winter season and DLP come under the summer season. In winter due to prevailing meteorological conditions and low mixing height the pollutants are not easily dispersed whereas for summer the dispersion is better than winter season. Level of concentration is greatly influenced meteorological parameters [27].

Table 3 depicts the average temperature and relative humidity for year 2019 and 2020 for Raipur. The meteorological parameters such as temperature and relative humidity were obtained from the weather underground site for Raipur city and then these were analysed for 2 phases for year 2019 and 2020.

In the analysis of meteorological parameter (temperature and relative humidity) for the Raipur city it was observed that temperature variation for BLP in year 2019 and 2020 ranged between 24.7- 34.8°C and 23.1-28.7 ° C respectively. Further for DLP of both the year it found to be 30.4- 36.0° C and 26.5-32.9° C respectively. On the other hand, relative humidity variation was seen 26 - 69.4% and 55.87-82.3% in BLP period of 2019 and 2020, and for DLP it was found 24-53.6% and 38.2- 63.3%.

Rise in the temperature and drop in humidity when comparison was made between the 2 phases (BLP and DLP) of same year this was remarked from analysis and also from the Table 3.This result was observed for both the years. From various other studies it was justified that high temperature and low humidity were helpful in improving the air quality status [5,14,26]. Hence it was concluded that there was improvement in AQI during this period of pandemic and meteorological parameter also supported it.

Table 3. Meteorological parameters over 2 phases for Raipur.

Meteorological Parameter Analysis for Raipur				
Year	Temperature(˚C)		Relative Humidity (%)	
	BLP	DLP	BLP	DLP
2019	29.5	33.0	43.2	33.6
2020	25.9	30.7	69.0	47.1

4.2.1 Correlation between the meteorological parameters and AQI.

The correlation analysis was performed to get relation between meteorological parameter and AQI with the accessible data of the year 2021 for Raipur city. In the atmosphere, meteorological parameters play a vital role in carriage, dispersal and natural purification [25,26]. So, for the study statistical analysis to determine the influence of meteorological parameters on AQI for Raipur the analysis was carried out using MS Excel. Several correlation analyses for different parameters had been performed earlier for the study area in different works [27,28].The meteorological parameters such that temperature and relative humidity were worked out for this study. The parameters were procured form the Weather Underground site for the month March-April 2021. Figure4(a),(b),(c) and (d) show the graphs representing the correlation between parameters(temperature and relative humidity) and AQI. After carrying out the analysis for March 2021 it was observed that positive correlation of (0.188) between temperature and AQI Figure4(a), and correlation between relative humidity and AQI was (0.109) Figure4(b). Moreover, when the analysis was done for April 2021 a poor positive correlation of (0.001) between temperature and AQI Figure4(c) and a correlation of (0.006) were acquired between relative humidity and AQI Figure4(d).

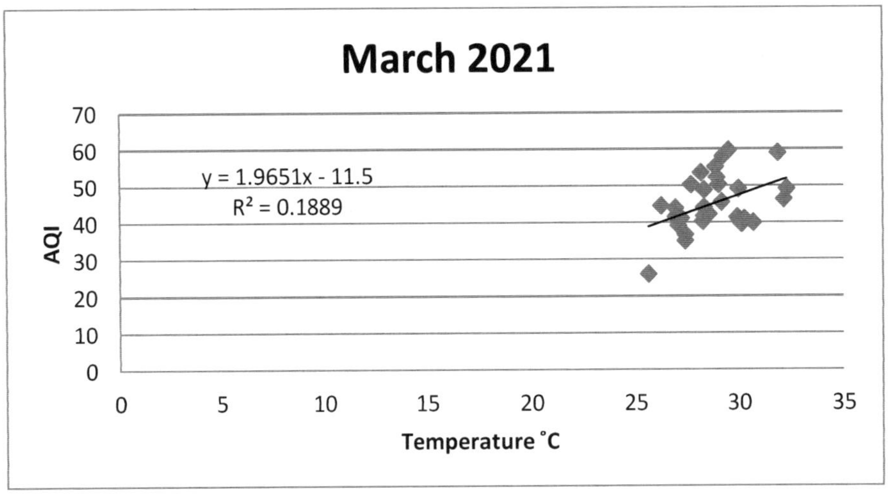

(a)

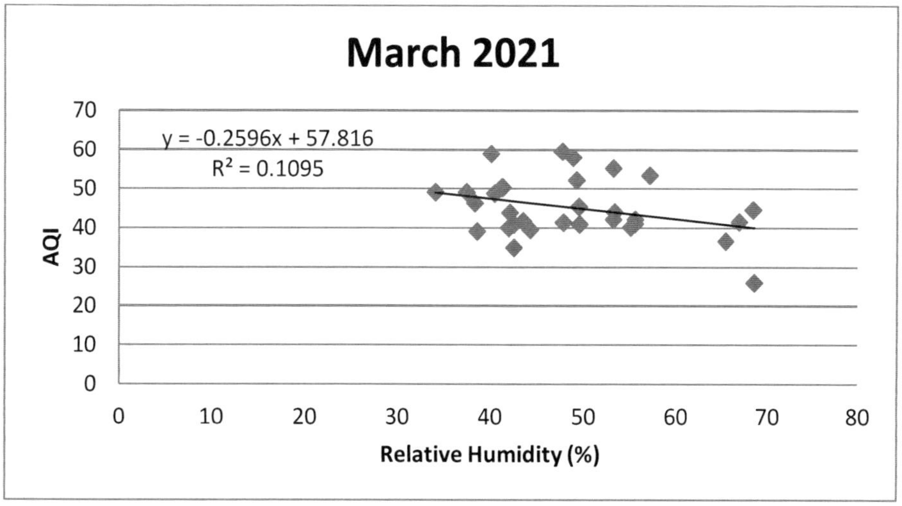
March 2021
y = -0.2596x + 57.816
R² = 0.1095
AQI
70
60
50
40
30
20
10
0
0
10
20
30
40
50
60
70
80
Relative Humidity (%)

(b)

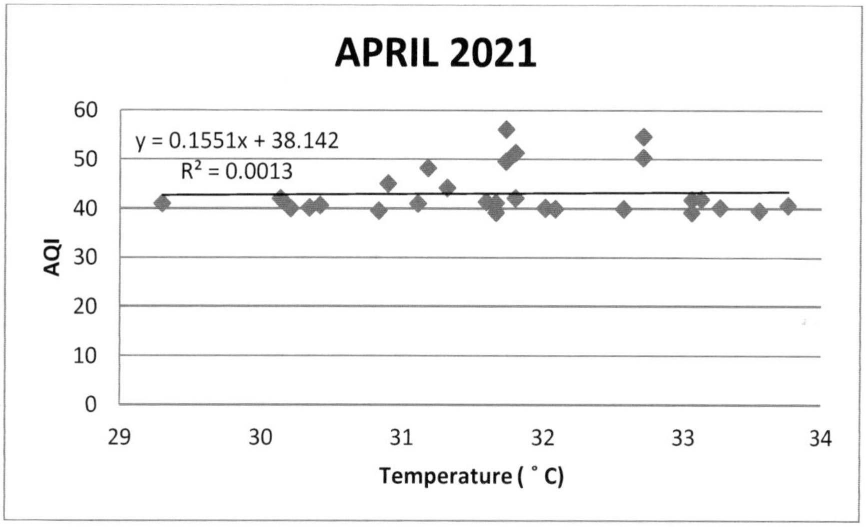
APRIL 2021
y = 0.1551x + 38.142
R² = 0.0013
AQI
60
50
40
30
20
10
0
29
30
31
32
33
34
Temperature (° C)

(c)

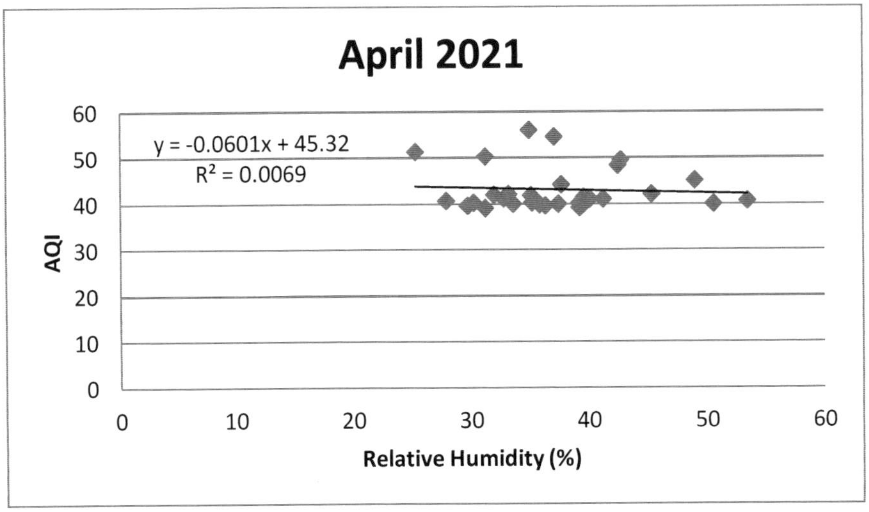

(d)

Figure 4.(a), (b), (c) and (d) Correlation between Meteorological Parameters and AQI for Raipur city for March –April 2021.

5. Conclusion

The study analysed the impact of COVID 19 on the Air Quality Index and concentration of pollutants (PM_{10}, $PM_{2.5}$, SO_2 and NO_x) for 2 selected cities of Chhattisgarh state, further the results were compared with the non-lockdown year 2019 with the available CECB data. A part of the study also comprised of correlation of meteorological parameter with air quality index which was found to be poorly correlated with the available data of year 2021 for Raipur city.

After analysis of air quality data during the study it was observed Raipur showed an appreciably result in improvement of AQI, about decrease of 37.99% when the comparison was made between 2020 DLP and 2019 DLP. Whereas for Bilaspur results of AQI reduction were not as good as Raipur only 10.04% of decrease when 2020 DLP was compared with 2019 DLP. For both the parameter (PM_{10} and NO_2) remarkable decline in level of concentration was observed, little less in Bilaspur. Meteorological parameters also supported in enhancing air quality. All these obtained results were possible due to the lockdown which was imposed and several limitations which were put on pollution producing activities. The pandemic turned out to be gift for the environment and helped a lot in improvement of air quality.

References

[1] Agarwal A , Kaushik A , Kumar S and Mishra R K 2020 Comparative study on air quality status in Indian and Chinese cities before and during the COVID 19 lockdown peroid Air Quality, *Atmosphere & Health*. 1-12

[2] He G , Pan Y and Tanaka T 2020 The short-term impacts of COVID-19 lockdown on urban air pollution in China *Nature Sustainability* .**3**.

[3] Vadrevu K P, Eaturu A , Biswas S , Lasko K , Sahu S , Garg J K and Justice C 2020 Spatial

and temporal variations of air pollution over 41 cities of India during the COVID19 lockdown peroid *scientific reports* .**10**

[4] Dumka U C , Kaskaoutis D G , Verma S , Ninogombam S S, Kumar S and Ghosh S 2021 Silver linings in the dark clouds of COVID-19: Improvement of air quality status in India and Delhi metropolitian area from measurement and WRF-CHIMERE model simulation. *Atmospheric Pollution Research* .**12** 225–242

[5] Chimurkar N, Patidar G and Phuleria H C 2020 Examining Effects of the COVID-19 National Lockdown on Ambient Air Quality across urban India *Aerosol and Air Quality Research.* 20 1759–1771

[6] Garg A , Kumar A and Gupta N C 2021 Impact of lockdown on ambient air quality in COVID19 affected Hotstop cities of India: Need to readdress Air pollution migitation policies *Environtmetal claims journal*. **33** 65–76

[7] Lokhandwala S and Gautam P 2020 Indirect impact of COVID-19 on environment: A brief study in Indian context. *Environmental Research*. **188**.

[8] Verma R L and Kamyotra J S 2021 Impacts of COVID-19 on air quality in India *Aerosol and Air Quality Research* . **21**

[9] Gautam S 2020 COVID-19: Air pollution remains low as people stay at home *Air Quality, Atmosphere & Health*.**13** 853–857

[10] Venter Z S , Aunan K , Chowdhury S and Lelieveld J 2020 COVID-19 lockdowns cause global air pollution declines. **117** 18984-18990.

[11] Bedi J S, Dhaka P , Vijay D, Aulakh R S , and Singh Gill J P 2020 Assessment of Air Quality Changes in the Four Metropolitan Cities of India during COVID-19 Pandemic Lockdow *Aerosol and Air Quality Research* . **20** 2062–2070

[12] Khaiwal R , Singh T , Biswal A , Singh V, and Mor S 2021 Impact of COVID-19 lockdown on ambient air quality in megacities of India and implication for all pollution control strategies *Environmental Science and Pollution Research*. **28** 21621–21632.

[13] Rahaman S , Jahangir S , Chen R , Kumar P and Thakur S 2021 COVID-19's lockdown effect on air quality in Indian cities using air quality zonal modelling *Urban climate*. **36**

[14] Sharma S, Zhang M , A , Gao J, Zhang H and Kota S H 2020 Effect of restricted emissions during COVID-19 on air quality in India *Science of the Total Environment.*

[15] Siddiqui A , Halder S , Chauhan P and Kumar P 2020 COVID-19 Pandemic and City-Level Nitrogen Dioxide (NO2) reduction for urban centres of India *Journal of the Indian Society of Remote Sensing*. **48** 999–1006

[16] Deshmukh D K , Deb M K , Tsa Y I and Mkoma S L 2011 Water Soluble Ions in $PM_{2.5}$ and PM1 aerosols in Durg City, Chhattisgarh, India *Aerosol and Air Quality Research* . **11** 696-708

[17] Verma M K , Patel A , Sahariah B P , and Choudhari J K 2016 Computation of ar quality index for major cities of Chhattisgarh state *Environmental Claims Journal* . **28** 1-11

[18] CSIDC(2021).Chhattisgarh State Industrial Development Corporation,(http://www.csidc.in/).(Jun.21,2021)

[19] Jaiswal N K , Ramteke S , Patel K S , Saathof H , Nava S , Lucarelli F , Yubero E and Viana M 2019 Winter Particulate Pollution over Raipur India *J. Hazard. Toxic Radioact. Waste* . **23**

[20] Deshmukh D K , Deb M K , Verma D and Nirmalkar, J 2013 Seasonal air quality profile of size-segregated aerosols *Bull Environ Contam Toxicol* . **91** 704-710

[21] Deshmukh D K , Deb M K and Verma S K 2010 Distribution pattern of Coarse ,Fineand ultra fine Atmospheric Aerosol particulate metter in major cities of Chhattisgarh. *IJEP* . **30** 184-197

[22] Deshmukh D K , Deb M K , Verma D , Verma S K and Nirmalkar J 2012 Aerosol size distribution and seasonal variation in an urban area of an Industrial city in Cental India *Bulletin of Environmental Contamination and Toxicology* . **89** 1098-1104

[23] Sinha D and Jaya D 2018 Seasonal variations in mass concentrations of PM10 and PM2.5 at traffic intersection and residential sites in Raipur city *Research Journal of Chemistry and Environment* .**22** 25-31

[24] Tiwari M K , Bajpai S and Dewangan U K 2019 Environmental Issues in Thermal Power Plants – Review in thermal powerplant -review in Chhattisgarh contex *Journal of Materials and environmental and sciences* , 1123-1134.

[25] Dubey N and Pervez S 2008 Investigation of variation in ambient PM10 levels within an urban-industrial environment *Aerosol and Air Quality Research* .**8** 54-64

[26] Sharma M , Jain S , and Yadav B Y 2020 Epigrammatic study on the effect of lockdown amid Covid-19 pandemic on air quality of most polluted cities of Rajasthan,India *Air Quality, Atmosphere & Health* .**13** 1157–1165

[27] Khobragade P P and Ahirwar A V 2019 Heavy metal in PM10 aerosols over an urban industrial city *International Journal of Engineering and Advanced Technology*. **9**

[28] Ahirwar A V and Bajpai S 2017 Seasonal Variability of TSPM, PM10 AND PM2.5 in Ambient Air at an Urban Industrial Area in Eastern Central Part of India *International Journal of Civil Engineering and Technology*. **8** 252-261

Machine learning algorithms in geostatistical data analysis: formulation and observation

S Pasari[1], P Asudani[1] and A Mehta[2]
[1] Mathematics Department, Birla Institute of Technology and Science, Pilani, Jhunjhunu – 333031, Rajasthan, India
[2] Civil Engineering Department, Institute of Infrastructure Technology, Research and Management, Ahmedabad – 380026, Gujarat, India

Abstract. Geostatistical analyses on spatiotemporal datasets have been increasingly popular on diverse disciplines including environmental sciences, soil and earth studies, meteorology, hydrology, and oceanography. They help in documenting and summarizing information to understand the variation of the process and other parameters. Due to high spatial variability and noise in the dataset, traditional geostatistical tools often fail to produce desirable results. In view of this, the present study proposes two machine learning methods, namely artificial neural network (ANN) and k-nearest neighbour (kNN), along with the kriging interpolation to derive temporal distribution of soil parameters. Results in terms of smooth porosity maps demonstrate the effectiveness of the proposed algorithm.

1. Introduction

Geostatistics deals with spatial or spatiotemporal datasets in describing spatial patterns, carrying out interpolation, and thereby analysing the underlying process. Large sized, unstructured and multivariate datasets are often associated with geostatistical analysis. Many areas of science and engineering disciplines directly or indirectly benefit from geostatistical techniques. For example, in mining industry, geostatistics is routinely used to characterize mineral resources and to examine its economic feasibility; in environmental studies, geostatistics is used to estimate and organize pollutant levels to ensure its proper remediation; in the field of soil science, mapping of subsurface topography in terms of acoustic impedance, carbon content, porosity, permeability, temperature and moisture, soil nutrients, soil electrical conductivity and related parameters is performed to prescribe proper fertilizer amount and agricultural practices by agri-consultants; geostatistics is inevitable in weather prediction in meteorology, crime mapping to social science analysis, seismological studies, medical statistics and several such domains. In all of these applications, spatial interpolation using traditional methods like variography and kriging are employed to estimate the value of a variable at new locations. In the present analysis, we implement ANN and kNN models along with kriging to demonstrate the suitability of machine learning models in the estimation of geo-parameters, such as soil maps.

As raw dataset often contains significant amount of measurement errors and associated uncertainty, variogram and other conventional geostatistical tools often fail to produce desirable results, leading to the formulation of machine learning models [1]. In geostatistics, kriging is the predominant method for spatial interpolation, though it assumes that predictions are linearly related to observations [1-2]. On the other hand, deep learning models can easily reveal patterns in high-dimensional and non-linear data related to any complex process. It is observed that neural networks are very useful for modelling patterns in data, though there is a risk of over fitting, whereas geostatistical techniques may overcome such problems using predictions or simulations [1-4].

2. Literature review

Some previous works have demonstrated the application of hybrid models as a mixture of traditional geostatistical tools (e.g., kriging) and machine learning techniques. For instance, Tadic et al. [3] proposed a hybrid model comprising kriging and ANN as spatial interpolator using real airborne carbon-di-oxide mixing-ratio and synthetic data. They compared the performance of the proposed hybrid model with ordinary kriging based on isotropic variogram. Foresti et al. [4] presented a hybrid approach for mapping of precipitation data using geostatistical methods and neural networks. Integrating geostatistics and machine learning algorithms, they obtained both spatial patterns and extreme values, simultaneously. Two case studies were undertaken to demonstrate the model. Wang et al. [5] proposed a kNN-based spatial prediction method that integrates deep learning neural network into a geostatistical model. Based on the implementation on a massive forestry dataset, the authors noted that it is a stochastic process with better performance than available geostatistical models in simulated non-Gaussian data. Kanevski et al. [6] proposed another hybrid approach, namely a feed-forward neural network known as neural network residual kriging (NNRK) for approximating nonlinear trends followed by the application of geostatistical estimator or simulator on residuals. In addition to the above, effects of spatial correlation on hyper-parameter tuning and performance checking have been studied by several machine-learning algorithms such as k-nearest neighbour, boosted regression trees, random forest and support vector machine along with conventional parametric algorithms, such as logistic regression and semi-parametric generalized additive models [1-6].

3. Dataset

An observed non-stationary (non-Gaussian) dataset of porosity and temperature for 294 geographic locations is considered. The porosity values lie between 0.0411 and 0.2069, with their mean and standard deviation 0.1274 and 0.0205, respectively. The temperature data ranges between -1.8790 and 31.0330, with their mean and standard deviation 14.4756 and 11.0929, respectively.

4. Methodology and results

For the studied dataset, first, kriging is implemented to understand the relationship between temperature and porosity (figure 1). It provides a good local estimation of the spatial distribution and porosity's dependency on temperature. After this, an ANN model is implemented to test its efficacy in modelling spatial changes of subsurface soil porosity. The network functions in two parts, namely training and testing data in about 8:2 ratios. The input parameters are fed to the network to provide porosity values as output. The architecture of the ANN is a feed forward network with four dense layers. The Adam's optimizer is used with a learning rate of 0.005 and MSE (mean squared error) as the loss function. Results are summarized in figure 2 and figure 3.

It is observed that ANN model is somewhat smooth but not as distributed and accurate as expected, probably due to the high dependency on the neighbouring data. This suggests us to implement a non-parametric kNN model that promises to account for the regression using neighbourhood data exclusively. The kNN captures the idea of similarity (e.g., distance, proximity, or closeness) in its working.

For the implementation of kNN, one needs to choose k number of neighbours, compute distances, and set the hyperparameters (e.g., weights, n-neighbours and distance metric, such as Euclidean and Manhattan distance) iteratively. As kNN uses a nearest training sample search in feature space (like k-means clustering), one may remove the impact feature range from the approach by standardizing predictor features with zero mean and unit variance. Here, the *scikit-learn* preprocessing library of Python is used to perform this step.

As an illustration of the above process, we provide the kNN results with similarity index as distance with $k = 5$ and $k = 15$ (Figure 4 and Figure 5, respectively). Small k results in a local specific prediction model, whereas large k results in a smoother, globally fit prediction model over the predictor features space.

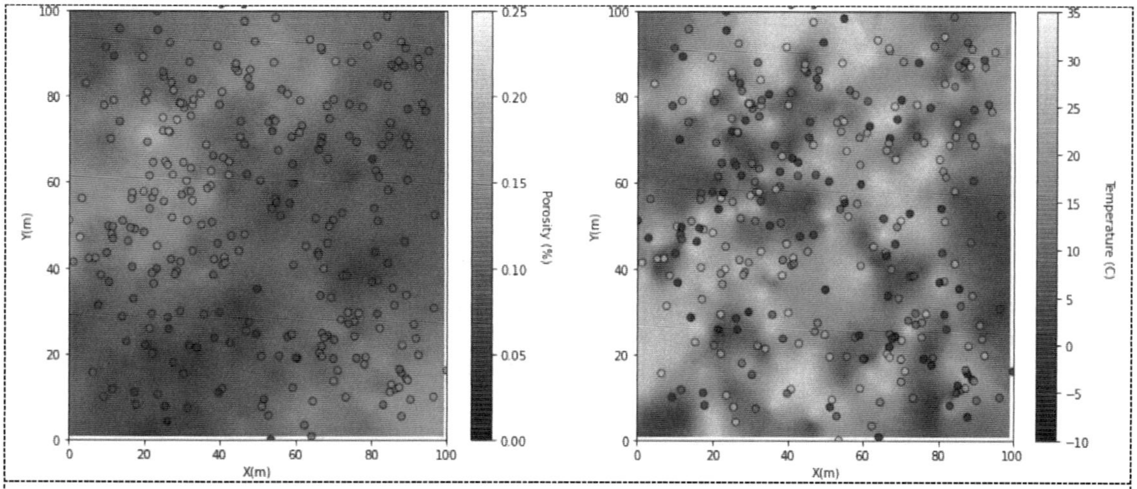

Figure 1. Results of kriging interpolation on porosity and temperature on the studied dataset.

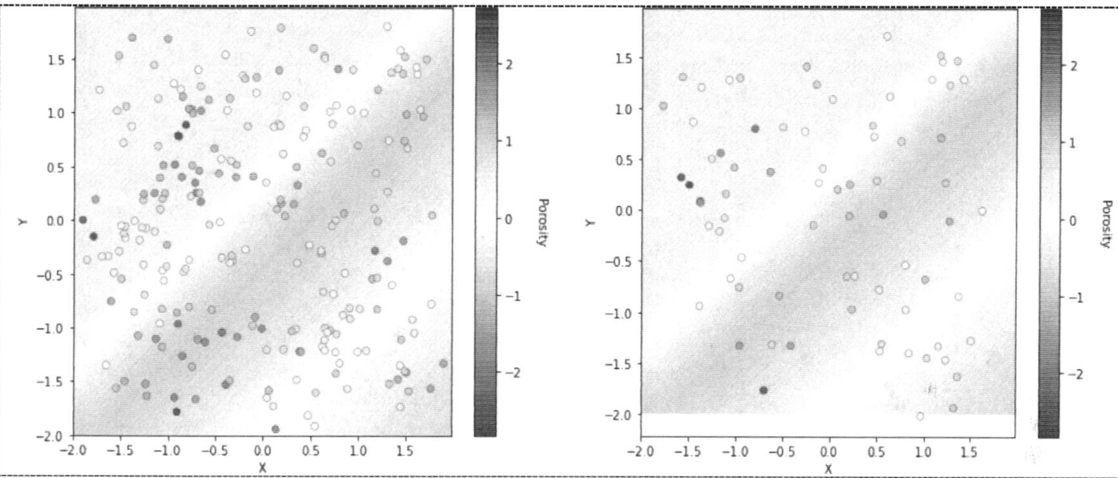

Figure 2. Results of ANN estimate (training and testing) on the selected dataset.

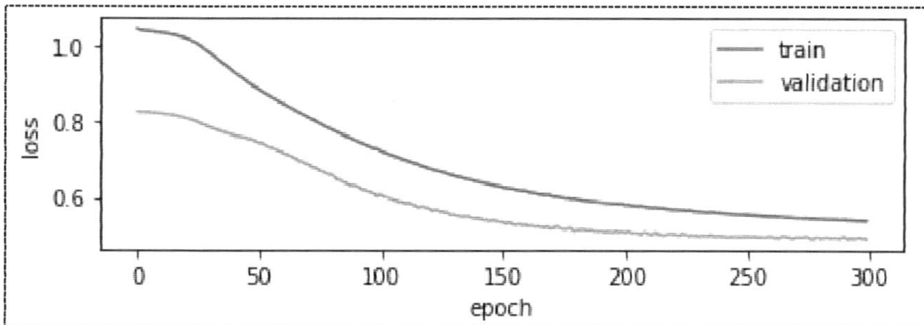

Figure 3. Loss function of the proposed ANN model.

5. Summary

In most of the datasets, the scatter plot of porosity suggests that a simple linear regression model may be inadequate to capture the intrinsic patterns of data. Therefore, nonparametric approaches are often employed to fit the nonlinear response patterns in the predictor feature space.

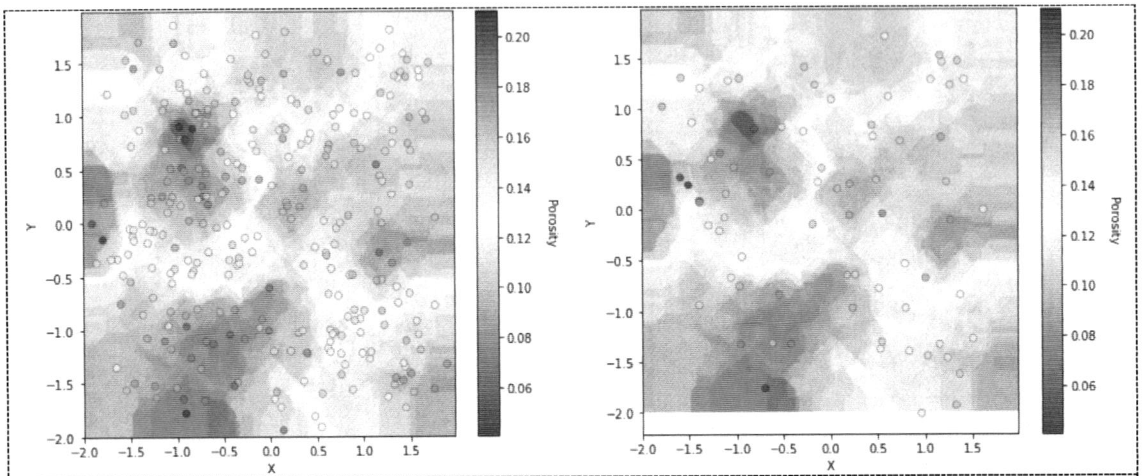

Figure 4. Results of kNN estimate (training and testing) on the studied dataset with $k = 5$.

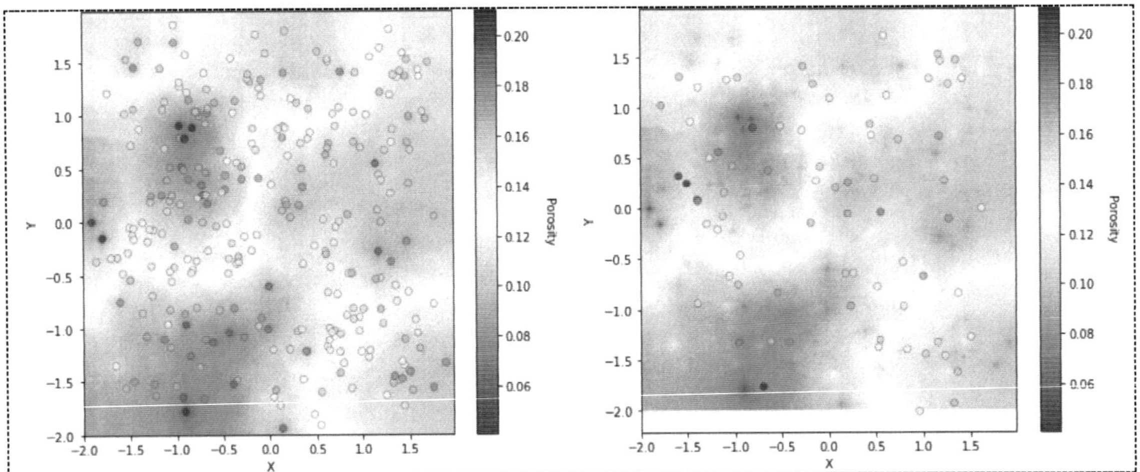

Figure 5. Results of kNN estimate (training and testing) on the studied dataset with $k = 15$.

In this study, we implemented a traditional geostatistical method (kriging) and two machine learning based models (ANN and kNN) to derive temporal distribution of soil parameters, such as porosity. We followed the standard practice of pre-processing (to observe, for example, outliers and data gaps), processing, post-processing (validation) of porosity values. Results in terms of smooth porosity maps demonstrate the effectiveness of the proposed kNN algorithm. The kNN model studied here is extendable to multivariate spatiotemporal datasets.

6. References

[1] Chiles JP and Delfiner P 2012 *Geostatistics: Modeling Spatial Uncertainty*, Wiley.

[2] Schratz P, Muenchow J, Iturritxa E, Richter J and Brenning A 2019 Hyperparameter tuning and performance assessment of statistical and machine-learning algorithms using spatial data, *Ecological Modelling* **406**, 109–120.

[3] Tadić JM, Ilić V and Biraud S 2015 Examination of geostatistical and machine-learning techniques as interpolators in anisotropic atmospheric environments, *Atmospheric Environment* 111, 28–38.

[4] Foresti L, Pozdnoukhov A, Tuia D and Kanevski M 2010 Extreme precipitation modelling using geostatistics and machine learning algorithms, In *Quantitative Geology and Geostatistics*, 41–52.

[5] Wang H, Guan Y and Reich B 2019 Nearest-neighbor neural networks for geostatistics, *International Conference on Data Mining Workshops*.

[6] Kanevski M, Parkin R, Pozdnukhov A, Timonin V, Maignan M, Demyanov V and Canu S 2004 Environmental data mining and modelling based on machine learning algorithms and geostatistics, *Environmental Modelling and Software* **19** (9), 845–855.

CHANNEL MIGRATION AND CONSEQUENTIAL LAND USE LAND COVER CHANGES OF SUBANSIRI RIVER, ASSAM, NORTH-EASTERN INDIA

Shukla Acharjee[1], Rajib Tarani Das*[2], Shubhendu Shekhar[3], Neelratan Singh[4], Uttam Goswami[5], M.M.G.S. Dilini[6], Y.K. Mawale[7], B.K. Iyengar[8] and R S Negi[9]

1. Centre for Studies in Geography, Dibrugarh University, Assam, India
2. Centre for Studies in Geography, Dibrugarh University, Assam, India
3. Department of Geology, Center for Advanced Studies, University of Delhi, Delhi, India.
4. CSIR - Central Institute of Mining & Fuel Research, Nagpur Research Centre Nagpur, Maharashtra, India
5. Department of Applied Geology, Dibrugarh University, Assam, India
6. Postgraduate Institute of Agriculture, University of Peradeniya, Sri Lanka
7. Department of Geology, SGB Amravati University, Amravati, India.
8. Department of Applied Geology, NIT Raipur, India
9. Department of Rural Technology, HNB Garhwal University Srinagar, Garhwal, Uttarakhand

*Corresponding author. e-mail: rajibtaranidas@gmail.com

Abstract. Subansiri River is the largest tributary of the Brahmaputra River running through the Indian states of Assam and Arunachal Pradesh, and Tibet, the Autonomous Region of China. The Subansiri River is 442 km long with a drainage basin of 32,640 km^2 and it contributes approximately 7.92% of the Brahmaputra's total flow. Sequential Channel shifting has been witnessed as the most important characteristic of the Subansiri River of Assam. The detailed study on channel migration of the present course of the Subansiri River through the upper floodplain of Brahmaputra valley indicates that the area is under active erosion for a long time. Therefore, an attempt has been made to understand the relationship between the rate of channel migration and successive land use/land cover changes in its surrounding floodplain area. The Support Vector Machine (SVM) and the Artificial Neural Network (ANN) algorithms are applied on Landsat images of the years 1973, 1988, 2001, and 2017 for generating land use/land cover maps through supervised classification technique. The overall accuracy of the land use/land cover classification ranges between 81% (for the year 1988) and 84% (for the year 2017). The land use/land cover maps show an increase in the built-up area and a decrease in the agricultural area. The change has been observed *vis-a-vis* channel migration indicating that the migration directly affects the floodplain habitats which in turn affects the land use, Findings of this study highlight geomorphological instabilities of the study area and the vulnerability of the habitations residing near the Subansiri river.

Keywords: Subansiri River, Channel migration, Land Use Land Cover, Sinuosity, Support Vector Machine (SVM), Artificial Neural Network (ANN)

1. Introduction

An alluvial floodplain area is always susceptible to active bank erosion due to river dynamics caused by natural and anthropogenic factors, hence it is considered as one of the most endangered areas continuously experiencing degradation by river regulations and enhanced land-use pressure (1, 2, 3). In the last few decades, the floodplain of the Subansiri River is subjected to erosion and deposition by the river to reach the equilibrium condition (4, 5, 6, 2). Therefore, mapping and monitoring of altered channel positions are crucial for assessing the erosional hazard and changes in land use/land cover (LULC) characteristics and also for understanding the cause of the alterations (2). While reviewing research paper on river dynamics and its effects on landforms it is found that many researchers have studied the morphological and erosional aspects of the Brahmaputra River and its tributaries and to name a few like Coleman, Goswami , Bristow, Klaassen and Masselink, Thorn, et al. and Kotoky and Sarma (7, 8, 9, 10, 11, 12). Bank erosion studies using topographic maps and satellite data have been carried out for the river Brahmaputra and of its tributaries by Sarma and Basumallick, Kotoky et al., Sarma, Sarma and Phukan, Sarma et al., (13, 14, 15, 16, 17,18).The present study has been carried out to understand the relationship between the channel migration and subsequent LULC changes in its surrounding floodplain area utilizing satellite remote sensing data of the period from the year 1973 to 2017. Historical satellite imageries provide an opportunity to go back in time and understand the changes in LULC categories that have taken place concerning the recent data.

The Subansiri River, the largest tributary of the Brahmaputra, originates from the Himalayas at an altitude of 5340m. In its upper reaches, the river takes an easterly course from its origin around 91°33''04"E and 28°29'38"N to 94°01'48"E and 28°21'32N and then takes south-easterly course up to the point at 94°20'49"E: 27°43'26N, from where it takes a sharp turn towards the southwest and continues up to 94°12''43" E: 27°36''18" N. It afterward travels to the position 94°17'02" E: 27°13'52" N, which the area of the study and then takes a curvilinear trend convexing southeastward and meets the Brahmaputra River at an acute angle. The river takes its southerly course emerging out of the Himalayas and enters the Brahmaputra River valley near Gerukamukh. In its upper reaches, the river is known as Tsari Chu. In mountainous terrain, the river length is 208 km, and from the dam site to the confluence with Brahmaputra (near Jamuguri, 5 km southeast of Hawajan) is 126 km. The Lower Subansiri dam site is located at latitude 27°33′15″ N and longitude 94°15′30″E. The present study is carried out to understand the relationship of river channel migration with changes in LULC (figure1).

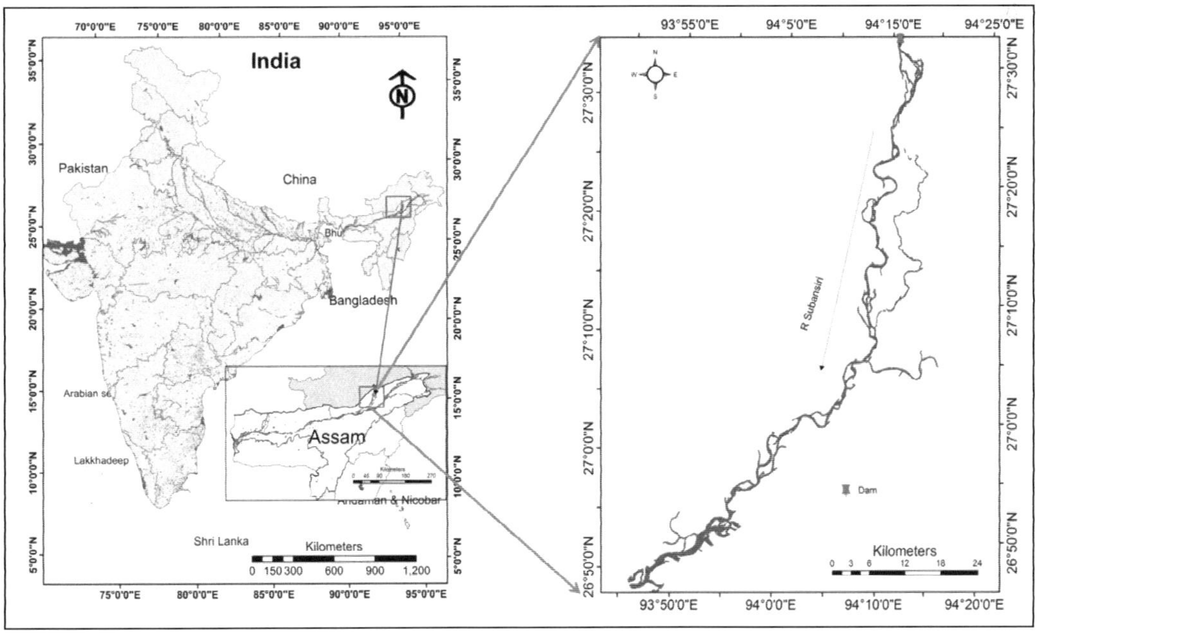

Figure 1. Location map showing the extension of the basin of the Subansiri River.

2. Database and Methodology

Historical channel change by using topographical maps and satellite images combined with the GIS technique has been used by many researchers in their river morphological changes (19, 20, 21, 7, 22, 23, 24, 25, 26, 6, 27, 28). In this current, work Remote Sensing and GIS techniques are used to delineate the channel dynamics and its impact on land use. Landsat 4 & 5 TM images of the years 1973, 1988, 2001, and Landsat 8 OLI image of the year 2017 are used in the study as they provide continuous multispectral data of 30m spatial resolution for landscape analysis and are available for download from USGS Earth Explorer data portal. For this study, the entire course of the river from Subansiri dam to confluence point was divided into 3 reaches and 25 sites were selected for cross-sections. For identifying the river and its characteristic parameters such as channel width, shift, sinuosity index, deposition, area of erosion, unchanged area (table-1), and historical migrated area were measured and studied along the cross-sections (3).

Table 1. Characteristics of reaches of Subansiri River.

Parameter	Reach-1	Reach-2	Reach-3
Areal extent	Subansiri Dam to Chaldhowa ghat	Chaldhowa ghat to ChenimoraKangkur No 1	ChenimoraKangkur No 1 to the confluence at Brahmaputra River
Length	12.33 km	79.56 km	27.57 km
Sinuosity index	1.13	1.36	1.11
Dominant bed material	Sand	Sand and silt	Sand
Morphological character	Sinuous and Anabranching	Sinuous and Anabranching	Sinuous and Anabranching

The cross section-wise study is applied for detection of widening of channel, whereas for calculating sinuosity index and capturing of the total area by the river, the reach-wise study is applied (29). The riverbank lines for both the left and right banks were identified and digitized from topographical maps and satellite images. Then the digitized vector data were overlaid and migration of the channel was identified and extracted from the overlaid maps. Deposition, Erosion, unchanged, and historical migrated areas were calculated (3).

Sinuosity index, which deals with the meandering nature of the river and indicates, meandering in quantitative terms, is calculated as the ratio between observed (actual) length and the expected straight distance of a meandering river between the endpoints. It is expressed as follows:

Sinuosity index (S) = OL / EL ……. (Schumm; 1963)
where S = sinuosity index,
OL = observed (actual) length of a river, and
EL = expected straight distance of a meandering river.

For mapping the LULC classes, two classification techniques were used. Support Vector Machine (SVM) is a biased classifier that is defined by a separable hyper plane, it is used in classification or regression for learning algorithms but is mostly used for classification (30).

Artificial Neural Network (ANN) is a computer technology-based method that is inspired by the human brain networking system. ANN has features to store experimental information and keep the information and able to use when it is required. Because of the ability in domains with non-linear relationships ANN is proposed as an alternative and valuable tool to evaluate.
The process of valid and factual comparison is made between SMV (31, 32) and ANN for the selection of better land use classification techniques; this was done by using ERDAS Imagine 9.2 and ENVI 5.2 by collecting different samples of land use categories. Kappa coefficient was used in the post-classification analysis for the accuracy assessment of LULC classified images (2).
For accuracy assessment, 155 points were taken randomly and superimposed on the classified images. The LULC class type for each point was compared with the LULC class as observed on high spatial resolution Google Earth data and ground truth data obtained through field survey (30).

3. Result and Discussion

Based on the techniques used to analyze channel plan form and land use classification, the results of the study are presented into two parts – Channel widening and migration, and subsequent adjustment of land use is associated with the change (3).

3.1 Channel widening and migration.

Lateral channel erosion is a significant characteristic that has been observed throughout the Subansiri River primarily leading to the widening of the channel (5, 33). Two maps of 1973 and 2017 have been used to determine the width of the channel and for the understanding of the vertical and lateral erosional processes and morphological characteristics of the bank. The continuous recession of the water level leads to disequilibrium conditions between the water level and river banks, resulting in massive erosion (3). Erosion is intensified by the non-cohesive sandy bank which ultimately leads to the widening of the channel (3).
For understanding river widening and migration, the Subansiri river has been measured at 25 stations (profiles) in the two maps to observe reach-wise channel migration (figure 2). It was observed that the Subansiri river has migrated vigorously to west, northwest, and southeast directions along with the stations (table 2). Maximum erosion (8.06km) has been observed in the left bank near cross-section KK' whereas in the right bank it was near cross-section LL' (7.2km) (table 2). The cross section-wise measurements of active width indicate some important aspects of the river. It is observed that the overall width of the river in the upper, middle and lower courses have increased (table 2, figure 3).

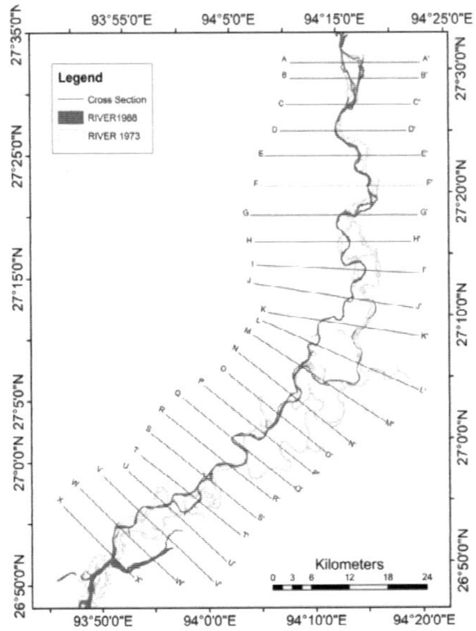

(a)

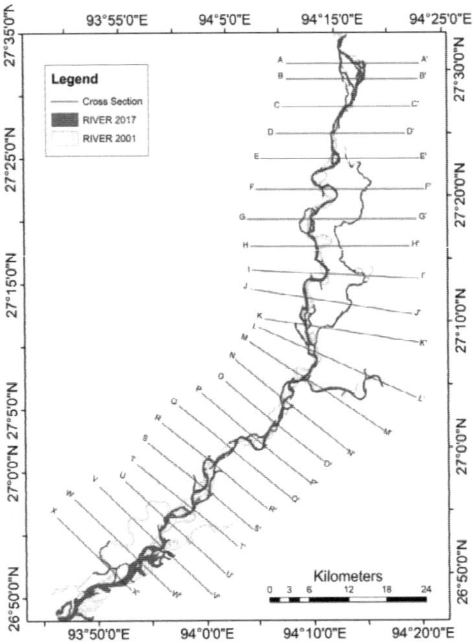

(b)

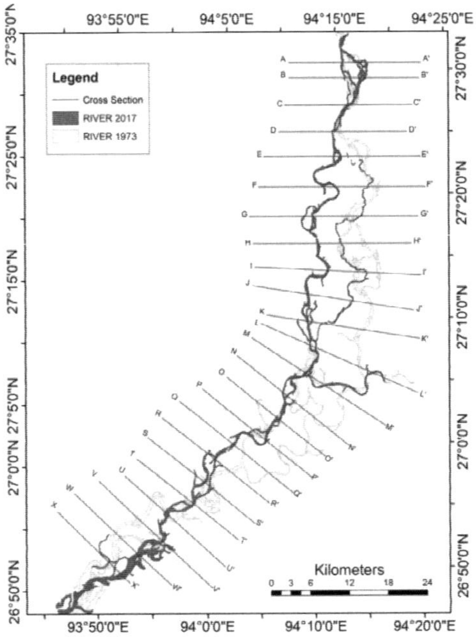

(c)

Figure 2. Channel migration of Subansiri River between (a) 1973 – 1988, (b) 2001-2017 and (c) 1973 - 2017.

Table 2. Channel migration and Variations in the active channel of Subansiri River between the years 1973 and 2017

Station	Direction of migration	Migration (m)	Channel width (in m)

			1973	2017
AA'	Westward	5	521.44	1058.56
BB'	Westward	16	618.23	1187.32
CC'	Westward	223.66	739.56	519.46
DD'	Westward	812.32	659.14	631.7
EE'	Westward	4146.29	1024.26	667.41
FF'	Westward	5921.11	601.85	375.17
GG'	Westward	6129.93	774.26	773.21
HH'	Westward	7278.4	113.58	699.81
II'	Westward	4510.54	561.28	502.19
JJ'	Westward	9705.4	215.19	636.63
KK'	Westward	8068.06	303.61	621.72
LL'	Noth-Westward	7229.82	396.22	494.46
MM'	South-Eastward	780.28	551.53	415.65
NN'	Noth-Westward	1943.09	280.78	663.55
OO'	South-Eastward	115.4	363.23	808.94
PP'	Noth-Westward	1220.33	370.5	661.88
QQ'	Noth-Westward	1928.82	448.11	488.58
RR'	South-Eastward	2629.31	606.81	850.07
SS'	South-Eastward	1746.16	606.43	666.04
TT'	South-Eastward	1243.59	538.8	412.65
UU'	South-Eastward	1685.14	356.79	970.03
VV'	South-Eastward	4319.83	614.95	444.96
WW'	South-Eastward	3921.29	485.58	1000.04
XX'	South-Eastward	1711.41	278.95	1116.09

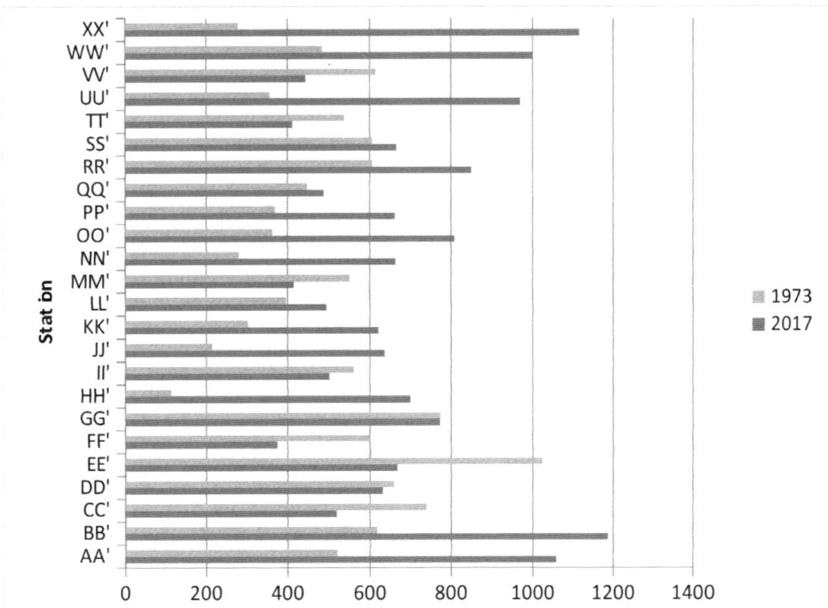

Figure 3. Cross section-wise width of Subansiri River during the years 1973 and 2017 (width in meters).

3.2 The area occupied by the river and associated changes in LULC:

Riverbank erosion is a complex phenomenon caused by the interplay of several factors including river discharge, hard rock, and/or riparian vegetation cover (3). Based on the remote sensing data derived observations, the projected area drained by the river Subansiri has been grouped into four classes - area under erosion, area remained unchanged, the area under deposition, and migrated area (3) (table 3). The total eroded area has been calculated from the active course in 2017 by subtracting from the unchanged area. It is observed that from 1988 to 2017, river Subansiri covers a total of 25842.19 hectares area under deposition (figure 4).

Table 3. Projected area drained by the Subansiri River

Category	Area in Hectare
Unchanged Area	2676.9
Erosion	5272.45
Deposition	25842.19
Historical migrated area	60167.65

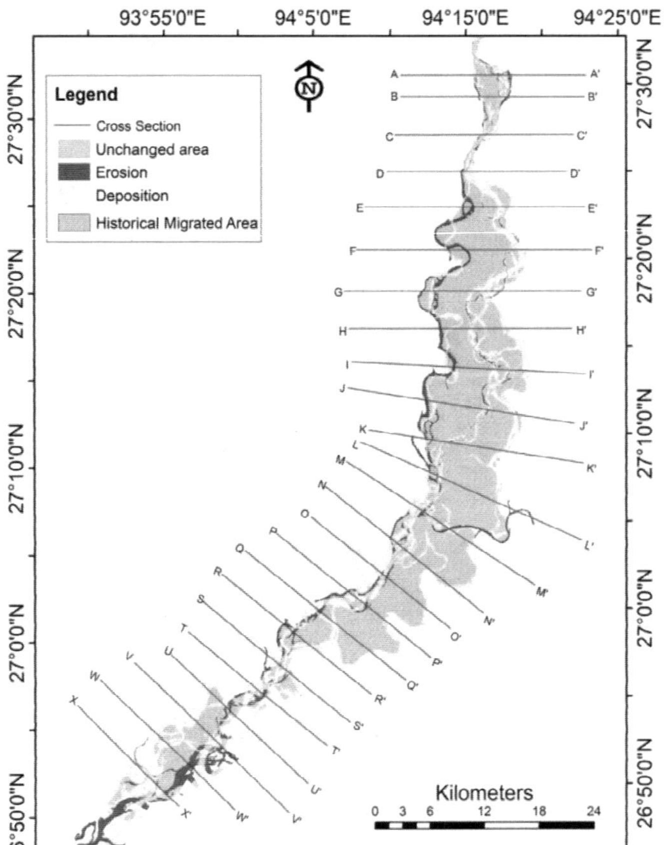

Figure 4. The projected area drained by the Subansiri River.

Table 4. Land use/Land cover classes of the study area in the years 1988 and 2017.

	1988		2017		Change
	Area (in hectare)	% of area	Area (in hectare)	% of area	Area (in hectare)
Waterbody	10583.55	4.797309	10812.78	4.901214	229.23
Vacant land	15703.47	7.118065	14236.92	6.453308	-1466.55
Vegetation	29583.18	13.40946	29455.29	13.35149	-127.89
Agriculture	148327.1	67.23368	138146.9	62.61917	-10180.26
Current Fallow land	3487.86	1.580976	7956	39.76716	4468.14
Built-up Area	12929.13	5.860513	20006.46	9.068524	7077.33

Subansiri River has witnessed a substantial and rapid change in land use configuration in its surroundings as observed in the LULC maps of the three years (figure 6). The rapid land-use changes particularly in the agricultural area can be attributed to urbanization in the last decade (34).

Landsat imageries of the three-time periods were classified using the supervised classification method employing the Support Vector Machine (SVM) algorithm to generate LULC maps and understand the land-use transition taken place during the time scale. Kappa coefficients of the classified images (1988 and 2017) are 0.81, 0.84 which indicates decent agreement between the actual land-use patterns and classified images (34).
Table 4 tabulates the land use statistics of the classified images of different temporal periods. Land use analysis shows that between 1988 and 2017, built-up area has increased from 5.86% to 9.06%. On the other hand, there is a decline of the vacant land from 7.11% in 1988 to 6.45% in 2017 and a decline of agricultural land from 67.23% to 62.61% from 1988 to 2017 respectively (34).

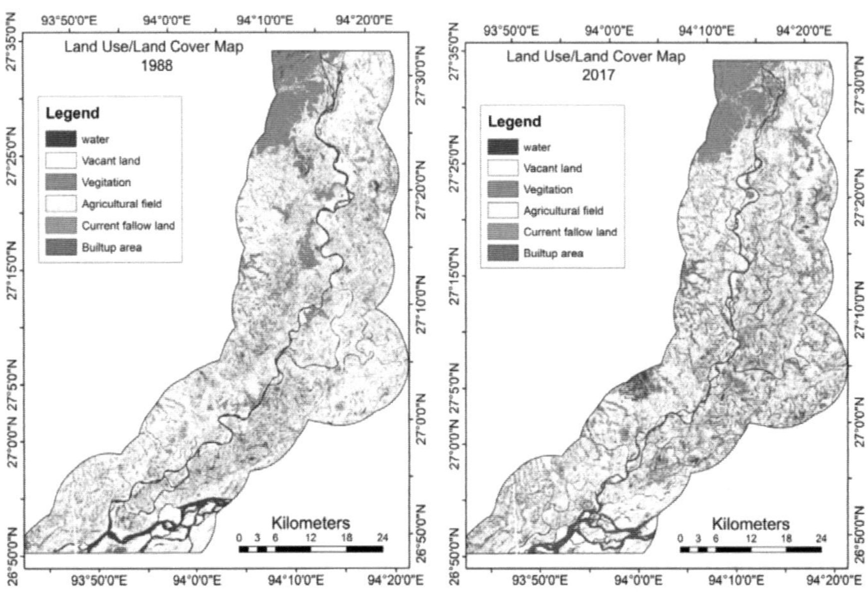

Figure 5. Land use/Land cover change of the study area between the years 1988 and 2017 adjoining the Subansiri River.

For analyzing the result of the land-use transition based on 'from' and 'to' changes, a matrix of land-use changes from the year 1988 to 2017 was generated (table 5).

Table 5. LULC change matrix of the Subansiri River

	Waterbody	Vegetation	Vacant land	Built-up Area	Agriculture	Current Fallow land	Total 1988
Waterbody	223.6	82.0	201.4	57.1	484.5	32.6	108.13
Vegetation	43.3	1618.3	29.7	137.7	1111.6	05.0	294.55
Vacant land	213.8	99.4	420.5	60.1	575.9	53.9	142.37
Built-up Area	68.0	274.6	77.4	211.4	1333.6	35.6	200.06
Agriculture	441.2	793.9	754.4	777.9	10855.7	191.5	1381.47
Current Fallow land	68.4	90.1	86.9	48.7	471.3	30.2	79.56
Total 2017	1058.4	2958.3	1570.3	1292.9	14832.7	348.8	

The transition matrix indicates that between 1988 and 2017, 575.9 hectares of vacant land were converted to agricultural land, while at the same period 777.9 hectares of agricultural land were converted to built-up land (34)(figure 7). From the other land use categories transformation, it is observed that nearly 1111.6 hectares of vegetation cover has been converted to agricultural land and 10.65% of it transformed into built-up land. The scrupulous observation over the land use classified maps makes it clear that from 1988 to 2017, vacant land, vegetation, and built-up areas have changed significantly. The built-up area has increased at the cost of the decrease in agricultural land, vegetation, and waterbody areas (table 6) (34).

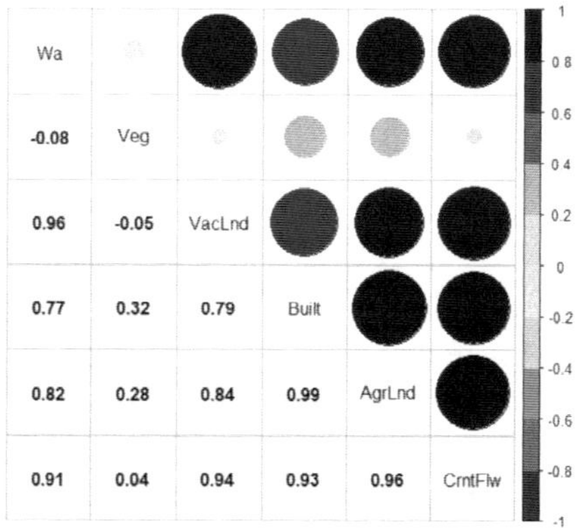

Figure 6. Correlation matrix between LULC classes 1988 and 2017 adjoining the Subansiri River.

Table 6. Losses and Gains between 1988 and 2017 in Subansiri River (Area in hectare).

Landuse class	Losses	Gains	Total change
Water	-8348	8577	229
Vacant land	-11498	10032	-1467
Vegetation	-13401	13273	-128
Agricultural field	-39770	29590	-10180
Current Fallow land	-3186	7654	4468
Built-up area	-10815	17892	7077

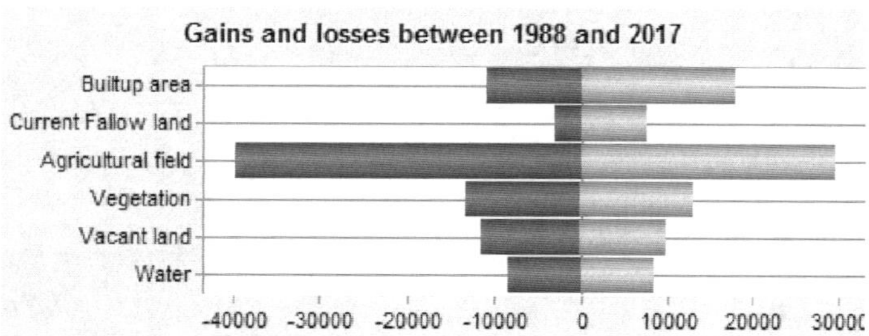

Figure 7. Losses and Gains between 1988 and 2017 in Subansiri River (Area in hectare).

Table 7. Net change between 1988 and 2017 in the Subansiri River (Area in hectare).

Water	229
Vacant land	-1467
Vegetation	-128
Agricultural field	-10180
Current Fallow land	4468
Built-up area	7077

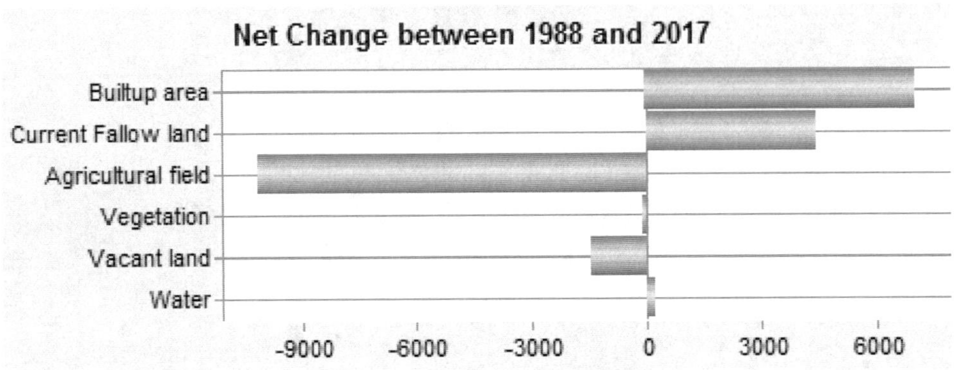

Figure 8. Net change between 1988 and 2017 in the Subansiri River (Area in hectare).

Table 8. Contribution to the net change in water between 1988 and 2017 in the Subansiri River (Area in hectare).

Water	0
Vacant land	-124
Vegetation	388
Agricultural field	433
Current Fallow land	-357
Built-up area	-109

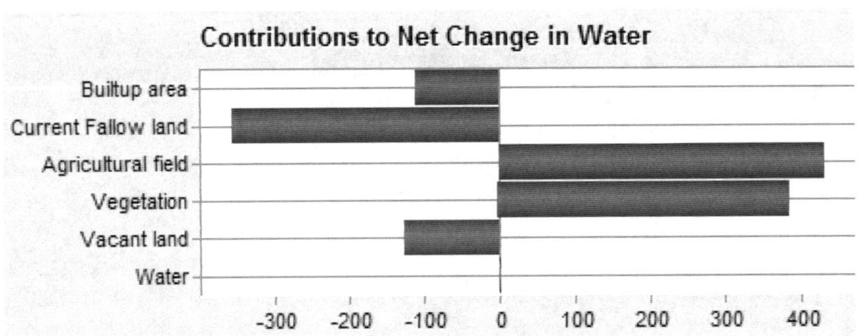

Figure 9. Contribution to the net change in water between 1988 and 2017 in the Subansiri River (Area in hectare).

Table 9. Contribution to the net change in vacant land between 1988 and 2017 in Subansiri River (Area in hectare).

Water	124

Vacant land	0
Vegetation	698
Agricultural field	-1785
Current Fallow land	-330
Built-up area	-173

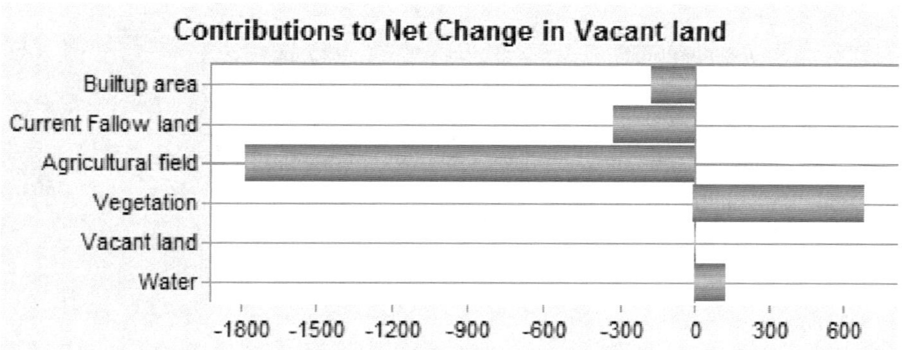

Figure 10. Contribution to the net change in vacant land between 1988 and 2017 in Subansiri River (Area in hectare).

Table 10. Contribution to the net change in the buildup area between 1988 and 2017 in the Subansiri River (Area in hectare).

Water	109
Vacant land	173
Vegetation	1369
Agricultural field	5557
Current Fallow land	-131
Built-up area	0

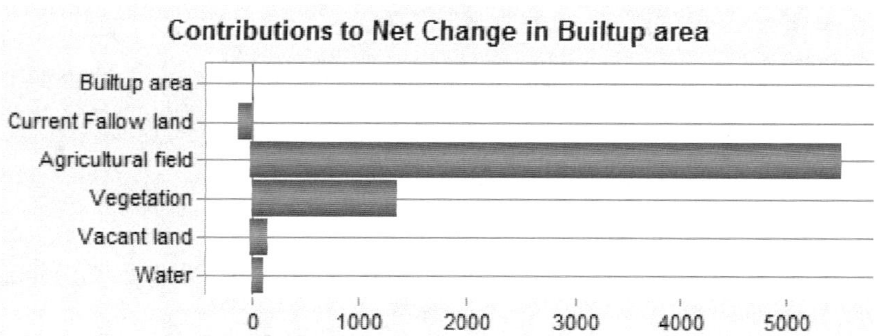

Table 11. Contribution to the net change in current fallow land between 1988 and 2017 in Subansiri River (Area in hectare).

Water	357
Vacant land	330
Vegetation	851
Agricultural field	2799
Current Fallow land	0
Built-up area	131

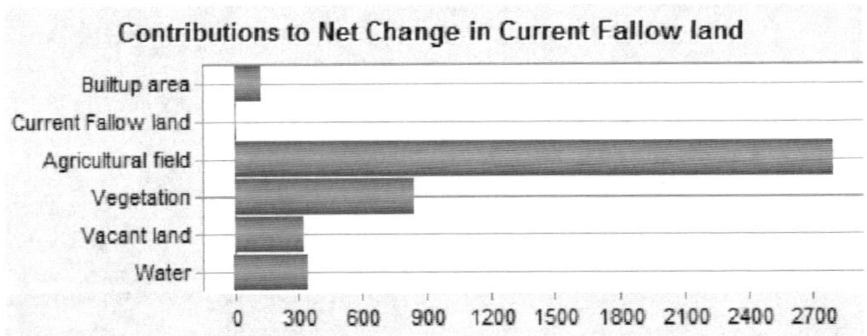

Figure 12. Contribution to the net change in current fallow land between 1988 and 2017 in Subansiri River (Area in hectare).

Table 12. Contribution to the net change in agriculture land between 1988 and 2017 in Subansiri River (Area in hectare).

Water	-433
Vacant land	1785
Vegetation	-3177
Agricultural field	0
Current Fallow land	-2799
Built-up area	-5557

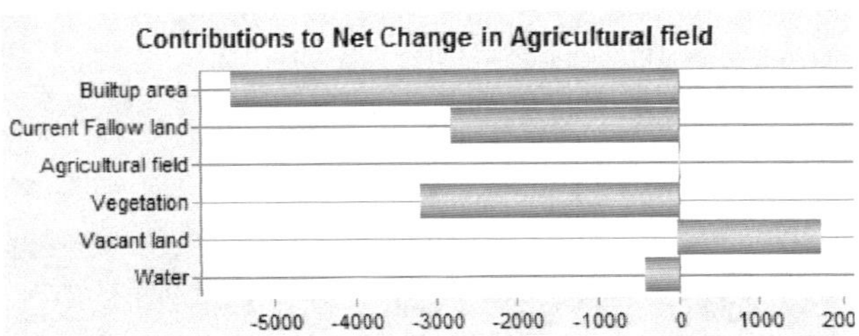

Figure 13. Contribution to the net change in agriculture land between 1988 and 2017 in Subansiri River (Area in hectare).

Table 13. Contribution to the net change in vegetation between 1988 and 2017 in the Subansiri River (Area in hectare).

Water	-388
Vacant land	-698
Vegetation	0

Agricultural field	3177
Current Fallow land	-851
Built-up area	-1369

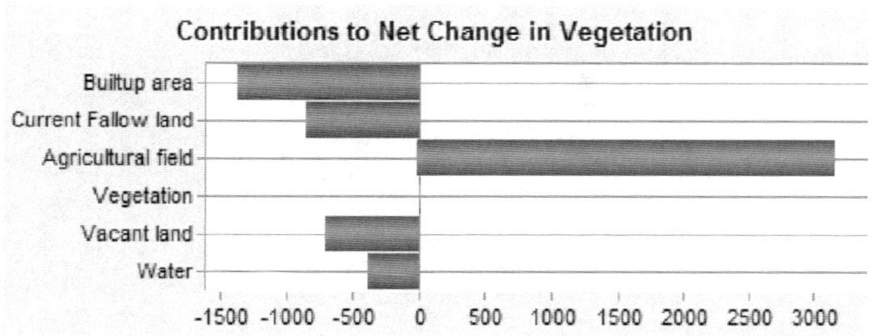

Figure 14. Contribution to the net change in vegetation between 1988 and 2017 in the Subansiri River (Area in hectare).

Table 14. Area on the file: Gain and loss in vacant land.

Category	Hectares	Legend
0	508764.06	
1	11498.22	Losses
2	4205.25	Persistence
3	10031.67	Gains

Table 15. Gain and loss in the area of water in Subansiri river (from 1988 to 2017).

Category	Hectares	Legend
0	515338.6	
1	8347.7	Losses
2	2235.7	Persistence
3	8577	Gains

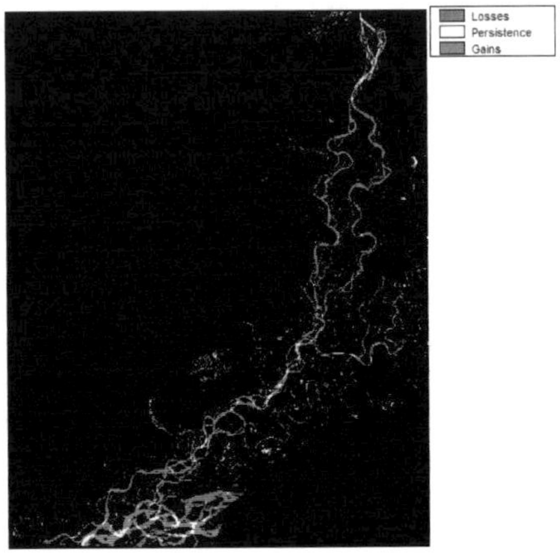

Figure 15. Gain and loss in the area of water in Subansiri river (from 1988 to 2017).

The gain and loss of land during 1988 to 2017 as shown in table 6 and figure 7 for all the landuse class shows that loss of fertile agricultural land is more than it gained. Likewise, open or vacant land loss is more compared to its gain. A negative change is observed for the agricultural land during the study period (table 7, figure 8). How different land use categories contributed to the net change between 1988 to 2017 is shown in table 8, 9,10, 11, 12, and 13 and graphical representation in figure 9, 10, 11, 12, 13, and 14). The gain and loss statistics indicate that 8347.7 hectares fall under loss and area gain is 8577 hectares whereas 2235.7-hectare area is unchanged (table 15). It is found that drastic land-use change has taken place in this part of the river course.

4. Conclusion

Sequential Channel shifting has been witnessed as the most important characteristic of the Subansiri river of Assam. The study on channel migration of the present course of the Subansiri river through the upper floodplain of Brahmaputra valley indicates that the area is under active erosion for a long time (3). These changes in the course are due to the changes in the hydrologic regime of the river, e.g., an increase or decrease of either water or sediment discharge provided to the channel reach. Since Lower Subansiri Hydro Electric Power Project (LSHEP) construction works are going on from the year 2005, the hydrologic regime may have been disturbed and it may have also influenced the channel morphological changes. The classified image generated using the Support Vector Machine (SVM) and the Artificial Neural Network (ANN) shows change in the course of the river and the resultant land use change. Both the SVM classification approach and ANN were found very promising for such analysis. The land use/land cover classification shows that there is an increase in the built-up area and a decrease in an agricultural area. The channel migration directly affects the land use and this change in land-use change thereby affects the floodplain inhabitants (3, 35). The present study of the geomorphological instabilities of the area highlights the vulnerability of the people residing near the Subansiri river. This study proves the importance and utility of the application of remote sensing and GIS technology in assessing river morphological changes as well as detecting the resultant Land use and land cover changes.

Acknowledgment

The authors are thankful to the Centre for Studies in Geography, Dibrugarh University for providing all the required amenities to carry out the present research work.

References

[1] Sarma J N and Acharjee S 2012 A GIS-based study on bank erosion by the river Brahmaputra around Kaziranga National Park *Earth Syst. Dynam.* Discuss. **3** 1085–1106

[2] Hazarika N, Das A K and Borah S B 2015 Assessing land-use changes driven by river dynamics in chronically flood-affected Upper Brahmaputra plains, India, using RS-GIS techniques, Egypt. *J. Remote Sensing Space Sci.* **18** 107-18

[3] Debnath J, Pan N D, Ahmed I and Bhowmik M 2017 Channel migration and its impact on land use/land cover using RS and GIS: A study on Khowai River of Tripura, North-East India. *The Egyptian Journal of Remote Sensing and Space Science* **20** 197-210.

[4] Das PK, Duarah BP and Goswami DC1996 Channel migration study of the Pagladiya River, Assam using remote sensing and field survey data. *J. Geosci.*138-45

[5] Goswami U, Sarma JN and Patgiri A D 1999 River channel changes of the Subansiri in Assam, India. *Geomorphology.* **30** 227-44

[6] Das A K, Sah R K and Hazarika N 2012 Bankline change and the facets of riverine hazards in the floodplain of Subansiri-Ranganadi Doab, Brahmaputra Valley, India *Nat. Hazards* **64** 1015–28

[7] Coleman J M 1969. Brahmaputra River: channel process and sedimentation *Sediment. Geol.* **3** 129–239.

[8] Goswami D C 1985 Brahmaputra river, Assam, India : physiography, basin denudation and channel aggracdation, *Water Resour. Res.* **21** 959–78

[9] Bristow C S 1987 Brahmaputra River: channel migration and deposition, recent development in fluvial sedimentology, edited by: Ethridge, F G Flores, R M and Harvey M D, *Soc. Eco. Paleo. Miner.*, Special Pub. **39** 63–74

[10] Klaassen G J and Masselink G 1992 Planform changes of a braided river with fine sand as bed and bank material *Proc. 5th Int. Symp. River Sedimentation* 459–71

[11] Thorne C R, Russell A P G and Alam M K 1993 Planform pattern and channel evolution of the Brahmaputra River, Bangladesh, in: Braided Rivers, edited by: Best, J. L. and Bristow, C. S., *Geol. Soc.*, London, 257–76

[12] Kotoky P and Sarma J N 2001 Hydrogeomorphological study of the Brahmaputra river from Majuli to Kaziranga, Unpub. *Report, RRL, Jorhat*, 100

[13] Sarma J N and Basumallick S 1984 Bankline migration of the Burhi dihing river, Assam, Ind. *J. Earth. Sci.***11** 199–206

[14] Kotoky P, Bezboruah, D and Sarma J. N 2005 Nature of bank erosion along the Brahmaputra river channel, Assam, India, *Curr. Sci.* **88** 634–39

[15] Sarma J N 2002 Pattern of erosion and bankline migration of the river Brahmaputra, Assam, *Report – National seminar on Disaster management*, Guwahati, 50–53

[16] Sarma J N and Phukan M K 2004 Origin and some geomorphological changes of the river island Majuli of the Brahmaputra in Assam, India *Geomorphology* **60** 1–19

[17] Sarma J N and Phukan M K 2006 Bank erosion and bank line migration of the river Brahmaputra in Assam, India, during the twentieth century, *J. Geol. Soc.* **68** 1023–36

[18] Sarma J N, Bora D and Goswami U 2007 Change of river channel and bank erosion of the Burhi Dihing river (Assam), assessed using remote sensing data and GIS *J. Ind. Soc. Remote Sens.* **35** 93–100

[19] Duncanson H B 1909 Observations on the shifting of the channel of the Missouri river since 1883. *Science* **29** 869–71

[20] Fisk H 1952 Mississipi river valley geology relation to river regime. *Trans. Am. Soc. Civ. Eng.* **117** 667–89

[21] Sundborg A 1956 The river Klaraven, a study of fluvial process. *Geog. Annaler* **38** 127–31

[22] Carey W C 1969 Formation of flood plain lands. *J. Hydraul.* D. Am. Soc. Civ. Eng. **95** 981–94

[23] Bluck B J 1971 Sedimentation in the meandering River Endrick *Scott. J. Geol.* **7** 93–138.

[24] Lewin J 1972 Late-stage meander growth. *Nature, Phys. Sci.* 240, 116 pp

[25] Mosley M P 1975 Channel changes in the River Bollin, Chesire, 1872–1973. *East Midland Geogr.* **6** 185–99

[26] Laczay I A 1977 Channel pattern changes of Hungarian rivers: The example of the Hernad River. In: Gregory, K.J. Ed. Ž . *River Channel Changes. Wiley*, Chichester, pp. 185–92

[27] Fuller I C, Large A R G and Milan D J 2003 Quantifying channel development and sediment transfer following chute cutoff in a wandering gravel-bed river. *Geomorphology* **54** 307–23

[28] Harmar O P and Clifford N J 2006 Plan form dynamics of the Lower Mississippi River. Earth. *Surf. Processes Landforms* **31** 825–43

[29] Ebisemiju F S 1994 The sinuosity of alluvial river channels in the seasonally wet environment: case study of river Elemi, Southwestern Nigeria. *Catena* **21** 13-25

[30] Tzotsos A and Argialas D P 2008 A support vector machine approach for object based image analysis chapter · January 2008 DOI: 10.1007/978-3-540-77058-9_36

[31] Negnevitsky M 2005 Artificial Intelligence, a Guide to Intelligent Systems. *Pearson Education*, p.440

[32] Theodoridis S and K Koutroumbas 2003 Pattern Recognition. Second Edition. *Elsevier Academic Press*

[33] Yao Z, Ta W, Jia X and Xiao J 2011 Bank erosion and accretion along the Ningxia-Inner Mongoloia reaches of the yellow river from 1958 to 2008. *Geomorphology* 127, 99–106.

[34] Dutta I and Das A 2019 Exploring the dynamics of urban sprawl using geo-spatial indices: A study of English Bazar Urban Agglomeration, West Bengal. Applied Geomatics, 11(3), pp.259-276.

[35] Debnath J, Pan N.D, Sharma R and Ahmed I 2019 Impact of confluence on hydrological and morphological characters of the trunk stream: a study on the Manu River of North-east India. *Environmental Earth Sciences*, 78(6), p.190.

A GIS-BASED PARAMETRIC ANALYSIS OF IRRIGATIONAL TANKS IN THE OLDEST CIVILIZED CITY IN INDIA-MADURAI

Praveenkumar J[1], Yazhini Malar Pa[2], Akash Kannan K[3], Sathya Bama B[4], Mohammed Mansoor Roomi S[5].

[1] Department of Civil Engineering, Thiagarajar College of Engineering, Madurai, India
[2,3,4,5] Department of Electronics and Communication Engineering, Thiagarajar College of Engineering, Madurai, India

E-mail: praveenkumar120580@gmail.com

Received xxxxxx
Accepted for publication xxxxxx
Published xxxxxx

Abstract

The irrigational tanks being traditional Water harvesting structures designed by rulers and chieftains dating back centuries. These are constructed in such a way that they act as a major water resource in drought-prone rural communities. The main aim of this paper is to develop a GIS-based Parametric Maps of Maximum Capacity(MC), Maximum Water Level(MWL), Ayacut Area(AA), Water Spread Area(WSA), Sub-basins and Basins of irrigational Tanks which form the basic irrigation information system to assist agriculturalists, farmers, and Government organization for the proper maintenance of irrigation tanks with suitable indexing parameters, to mitigate & forecast the water and food requirements by evaluating currently available water for the given population density and irrigation lands. The study area selected is the oldest civilized city in India which is in the state of Tamilnadu-Madurai district, which is served by the Vaigai River and consists of 1338 tanks. This GIS-based information system assists the decision-makers to perform surveillance, evaluate the efficiency of the tank at any space & time and take proper managerial methods to utilize the resources economically.

Keywords: Tank Irrigation; GIS; Ayacut area; Sub-basins; Maximum Water Level; Water Spread Area;

1. Introduction

Agriculture, being the lower backbone of our country's economy, relies absolutely on water resources. Water plays a major role in all living lives. It is known to be the Elixir of Life. Water resources are available as rainfall, ponds, lakes, rivers, and groundwaters. Water covers 71% of Earth's surface which accounts for nearly 326 million cubic miles of area. 97% is filled as the ocean which is nearly 320 million cubic miles of area. The remaining 3% is the usable freshwater. 2.5% is locked up as glaciers and ice caps, atmosphere, and soil. 0.5% of water is found utilizable which is continually accumulated, filtered, and circulated back again. Due to the rise in the Earth's temperature, drastic changes in rainfall patterns affect irrigation. Hence there is a vital need for proper maintenance and storage of irrigation tanks and rainwater. Also, water is taken into consideration to be the factor that's essential for improvement planning.

India receives total precipitation of about 4000 x $_{10}^{9}$ m^3. Out of this, water available on the surface accounts for 1780 x $_{10}^{9}$ m^3. But only half of the available surface water can be used due to other restraints. 420 x $_{10}^{9}$ m^3 of water is available as groundwater. Due to the erratic rainfall pattern, water availability is extremely unbalanced. Precipitation is limited to as it were almost 3 to 4 months with 20-45 critical blustery days within the year. Consequently, there is a crucial need for the proper storage, management, and required use of rainwater in places like Lakes, Tanks, Ponds, Aquifers, Reservoirs for conventional and communal resolves.

The irrigation tanks act as a natural and customary way of rainwater harvesting especially in places that are scarce of water. Tanks are low earthen bund that is built along with the tilt of terrain as well as valleys to store the precipitated water. Tamilnadu contains almost 39,202 tanks which are scattered all over the state. The system of a tank consists of a command area, sluice outlets, field distributaries, surplus weir, channel feeder, water spread area, catchment area, and tank bund. Over the years giant damage to the capable applicability of several tanks had been an acting reason of overwhelming siltation, immoderate infection, and encroachments. GIS plays an important part in enhancing the information system which is adapted in the rural areas to make decisions and sustain management functions of that particular area [10]. Thus, GIS will be an advancement in the field of agriculture in locating tanks, identifying their water level and other parameters for the prolonged functioning of agricultural activities.

Irrigation tanks are used recurrently for agriculture, humans, and cattle as drinking water. The possible outcomes from tank irrigation were assessed and concluded that this system may be a productive technology in financial, natural, and societal terms, but under current circumstances of administration, the conditions of these tanks are worsening [4]. Proper communication and knowledge of all these irrigation tanks would help society in many ways. Thus, the information system assists the management of resources. [17].

The proper management of runoff water leads to rainfall -surface water inside a catchment could be a beneficial resource of water to supply high demands [5]. The administration can receive societal endurance of tank rehabilitation projects if they could tackle the multifaceted issues and concede that irrigation tanks benefit the common people in addition to farmers [16]. The surveillance of irrigating crops & inaccessible areas like contours of reservoir water is made easier by Remote sensing techniques which involve reduced manpower, are cost-effective, precise [9]. The approach for economically unifying aerial photography and satellite data with surveys of ground to obtain current guideline information with respect to resources of tanks and irrigated areas under their command [23]. The strategy for the classification of water bodies based on various parameters is essential and effective [22].

Taking into consideration other productive uses (except domestic and cattle), the total production increases 13%, and the cumulative income received by tanks increases by 213% [14]. The research suggests that the correlation could be related between the potential storage and the diffusion area of the tanks. With the usage of remotely sensed images, Water spread areas can be mapped exactly [15]. The production of sustainable crops needs better performance than small training structures tanks, that require reconstruction of physical structure, effective distribution of water in agricultural fields, as well as in farmers, and correct maintenance of the reservoir system after modernization by the participation of farmers [1]. Urbanization was considered the main cause of the drastic runoff of catchment (40 to 60% of rainfall). For this reason, the availability of seasonal water in each reservoir in watershed waters was observed incoherent (0 to 15%), with the summer which comprises about the minimum water memory (0 to 8%) [21]. Geospatial technologies such as remote sensing and geographic information system (GIS) serve as effective aids to assess these impacts of sustainable development in a water management system [19]. The contribution of this paper includes developing a GIS-based tank information system for the study area, creation of parametric Maps of tanks with the satellite image as a base Map and the analysis of parametric maps to understand the present status of irrigation tanks in the study area for future planning and development.

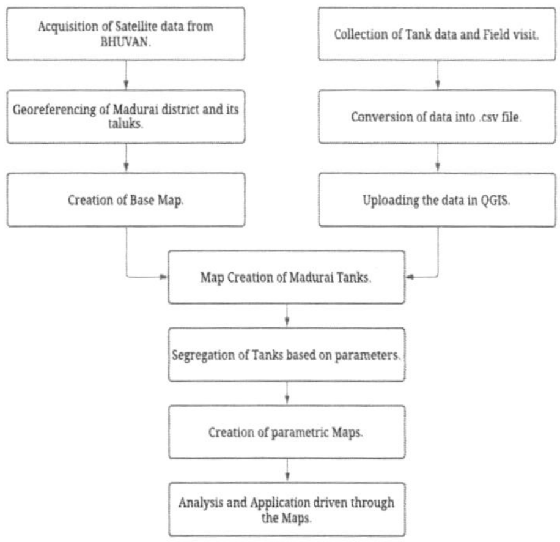

Figure 1: Methodology

2. Materials and Methods

2.1 Study area

The study area chosen is Madurai district in Tamil Nadu, India. Madurai District is in the Southcentral Region of Tamilnadu lying between longitudes E 77° 2' and E 78° 2' and between latitudes N 9° 30' and N 10° 27'. Madurai has been an important settlement for two millennia and is one of the oldest continuously inhabited cities in the world. The general slope of Madurai District is from West to East. The geographical area of the district is 3741.73 square kilometres, has a cultivating area of 150704 ha & consists of 7 Basins. All the required non-spatial and spatial data on the command area for all tanks in Madurai district (1338 tanks) were collected.

- The Central Part of the district is drained by the Gundar River and Vaigai River that forms parts of Vaigai.
- The Northern Part of the district is drained by the river Pambar.
- The Sedapatti and Kallupatti blocks (Southeastern part of the Madurai district) fall into the Vaippar basin and the area is drained by the river Arjuna Nadi.

It consists of 1 Municipal Corporation,2 Revenue divisions,669 villages,7 Taluks,13 Panchayat Unions (Blocks), and 431 Village Panchayats. It has a total population of 25,62,279. The Madurai district is continuously irrigated by the Periyar Vaigai Irrigation System which irrigates about 57,000 ha, Sathaiyar Reservoir Irrigating about 200ha, Minor Irrigation Tanks, and Dug wells & Tube wells.

Table 1 shows the population census of the Madurai district, India as per the 2011 census.

Table 1-Madurai Population

	Total Population	Male	Female	Area (Sq.km)	Density (/Sq.km)
Total	2578201	1303363	1274838	3451.5	746.98
Rural	1134025	573036	560989	244.49	4638.33
Urban	1444176	730327	713849	3696.0	390.74

2.2 Database

In this paper, all required non-spatial and spatial data on the command area for all tanks in Madurai district (**1338 tanks**) were collected (Fig. 2). The satellite data of the Madurai region has been obtained through BHUVAN (The national Geo-portal developed and hosted by ISRO comprising of Geo-Spatial Data, Services and Tools for Analysis - https://bhuvan.nrsc.gov.in/) portal. This Satellite data was captured by **Advanced Wide Field Sensor (AWiFS)** which is a part of the **Resources at-1 satellite**. AWiFS operates in 4 spectral bands, which have 56 meters of spatial resolution. The secondary data collected are Maximum Capacity (MC), Ayacut Area (AA), Maximum Water Level (MWL), Water Spread Area (WSA), Basins, and Sub Basins of Irrigational Tanks.

The Tank details & data of the Public Work Department (PWD) resources, Government of Tamil Nadu, have been used as a secondary dataset. The database contains 2 components (i) spatial data represents the distribution of features in the control area and (ii) attribute database represents Maximum Capacity (MC), Ayacut area (AA), Maximum Water Level (MWL), Water Spread Area (WSA), Basins, and Sub Basins of Irrigational Tanks.

The software which is used here is QGIS version 3.16.15 'Hannover'. The Coordinate Reference System (CRS) used here is WGS 84 (EPSG 4326).

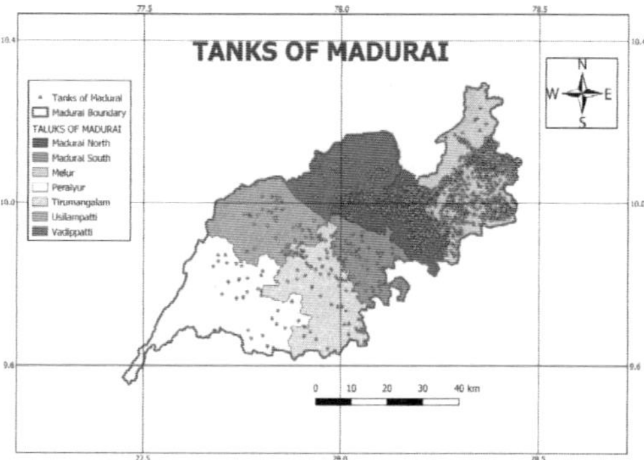

Figure 2: Tanks of Madurai

2.3 Methodology

The digital database of non-spatial and spatial data about the command area has been created in the GIS environment. As prime work for this project, the satellite data of the Madurai region were downloaded from the BHUVAN portal.

Furthermore, $(1/4)^{th}$ details of data have been verified for data accuracy through field visits. All data combined with the acquisition of new data related to technical aspects of Tanks were verified with the Latitude and Longitude coordinates of Tank location using a GPS device. The spatial database was created using QGIS (Quantum Geographic Information System).

2.3.1 Digitization of Boundaries

The Map of the Madurai district and its taluks have been downloaded from the Madurai district corporation website. Using BHUVAN data as a base map, this district and taluk map have been georeferenced using QGIS. Then we have digitized the boundaries of Madurai district and its seven taluks in QGIS using the above-georeferenced data (Fig. 3).

2.3.2 Addition of Data in QGIS

All the data have been recorded and updated into an excel sheet which consists of technical data such as Latitude, Longitude, Maximum Capacity (MC), Ayacut Area (AA), Maximum Water Level (MWL), Water Spread Area (WSA), Basins, and Sub Basins of Irrigational Tanks. After that, we have converted the excel sheet into comma-separated values (.csv) file. Then this .csv file has been imported into QGIS as point features. Using various attributes available in the tanks' data, five different maps have been created.

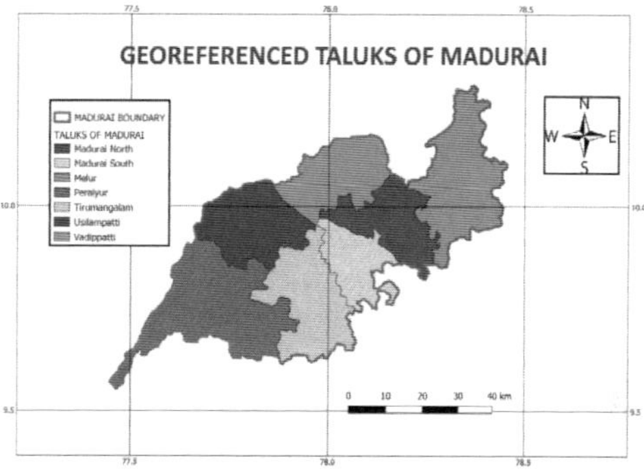

Figure 3: Georeferenced Taluks of Madurai

3. Parameters Taken

The following are the parameters taken related to Irrigational Tanks which come under the Irrigation Engineering of the Civil Engineering domain.

3.1 Maximum Capacity (MC) of Tank

Gross volume of water which can be stored in the Tank/storage structures.

3.2 Maximum Water Level (MWL) of Tank

The Water level is reached during the passage of the design flood. It depends upon the specified preliminary reservoir level and the spillway gate operation rule. It is also called the Highest Flood Level (HFL) or Highest Reservoir Level (HRL).

3.3 Ayacut Area (AA) of Tank

The area that is served by the Tanks for irrigation purposes is the Ayacut Area of the tank.

3.4 Water Spread Area (WSA) of Tank

It is an area covered by water, i.e., land of the Tank occupied by water (submerged area).

3.5 Sub-basin of Tank

Sub-basin is a structural geological feature in which a larger basin is divided into a series of smaller basins with intermediate intra-basin heights.

4. Results and Discussion

From the analysis of various geospatial data as well as non-geospatial data, the numerical decisions have been estimated using simple Mathematics which have been given below,

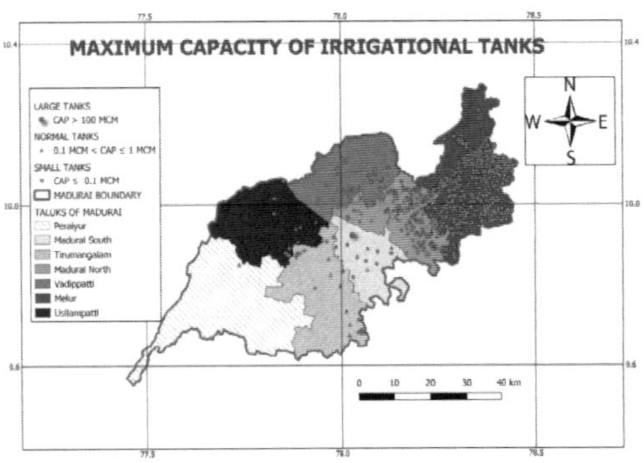

Figure 4: Maximum Capacity of Irrigational Tanks

Using the Map of the Maximum Capacity of Madurai tanks, **Madakulam Tank** (Thiruparankundram Village, Madurai South Block) is the Largest Tank in this District, having a Maximum Capacity of $166.90 \times 10^6 m^3$ serving Agricultural land of 1043.30ha. Also, there exist many small tanks of magnitude less than 1 MCM (Million Cubic Meter($10^6 m^3$)). It has been estimated that the Cumulative Capacity of 1338 Tanks which accounts for $2143.48 \times 10^6 m^3$ shown in Fig. 4.

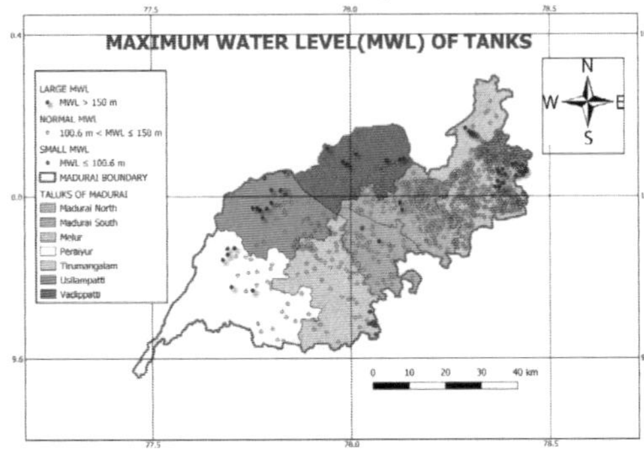

Figure 5: Maximum Water Level(MWL) of Tanks

With the application of the Maximum Water Level (MWL) Map of Madurai, **Veppankulam Tank** has the highest Maximum Water Level (MWL) of 497.41m, located in the Madurai North Taluk- Kulamagalam Village comes under the Sub-basin of Sathaiyar, Vaigai basin. In addition to, there exists a Tank possessing the smallest MWL less than 1m like Chandrankulam Tank, located in the Thirumangalam Taluk- Kallikudi Block which comes under the Sub-basin of Therkar, Gundar basin shown in Fig. 5.

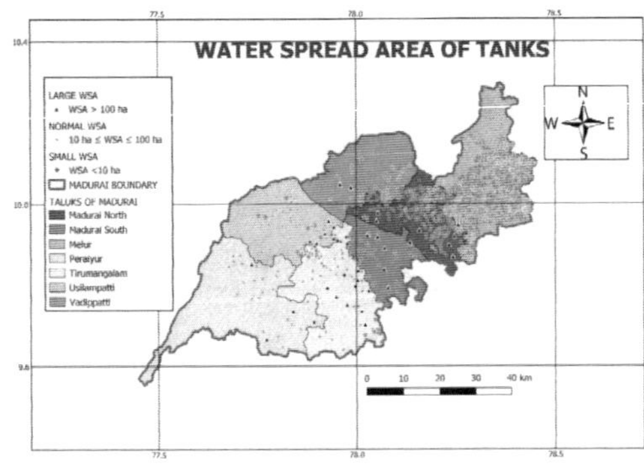

Figure 6: Water Spread Area of Tanks

The Water Spread Area (WSA) of Madurai was mapped and it is found that **Valandur Tank** has the largest Water Spread Area (WSA) of 1014.0ha, located in the Valandur Village-Usilampatti Taluk and Veeraputhukulam (Katchirayanpatti Village- Melur Taluk), Somakulam (Maruthankudi Village- Thirumangalam Taluk) tank has Water Spread Area (WSA) less than 1ha shown in Fig. 6.

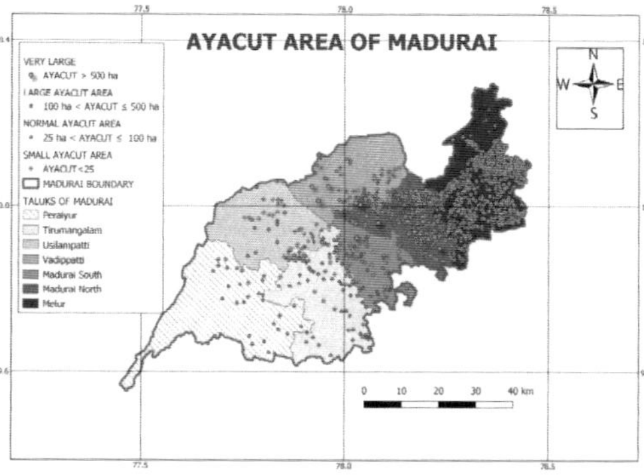

Figure 7: Ayacut Area of Madurai

It has been inferred from the Ayacut Area (AA) Map of Madurai that Madurai Tanks has a cumulative Ayacut Area (AA) of almost 52197.21ha(ha=hectare), spreading over 669 villages namely Elumalai, Mellapuram, Perungamanallur, Kallappanpatty, Poosalapuram, Sembarani, Kuppalnatham, Sedapatti, etc. **Madakulam Tank,** serving the largest Ayacut Area (AA) of about 1043.30ha to its neighboring villages with a capacity of 166.90 x $10^6 m^3$, comes under the Sub-basin of Gridhamaal Nadhi, the basin of Gundar. A small ayacut served by a tank of less than 1ha like **Chinnakulam Tank** is also available shown in Fig. 7.

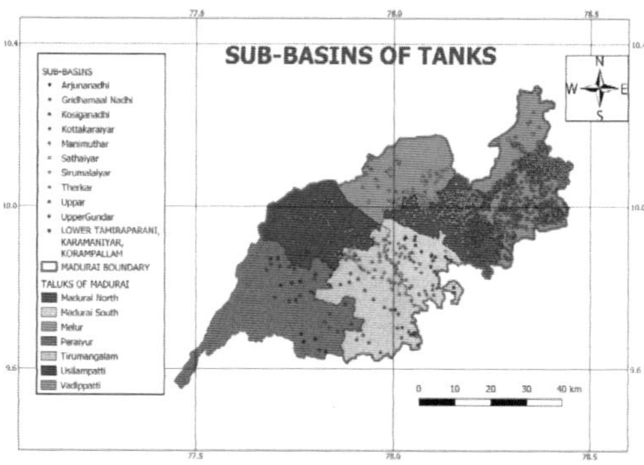

Figure 8: Sub-Basins of Tanks

Table 2 shows the data analyzed from the Sub-Basins map of Madurai district shown in Fig. 8.

Table 2: Sub-Basins-Tanks

S. NO	Name of the Sub-Basin	Number of Tanks that falls under these Sub-Basins
1	Arjunanadhi	5
2	Gridhamaal Nadhi	24
3	Karamaniyar	1
4	Korampallam	1
5	Kosiganadhi	11
6	Kottakaraiyar	31
7	Manimuthar	330
8	Sathaiyar	187
9	Sirumalaiyar	22
10	Lower Tamiraparani	5
11	Therkar	137
12	Uppar	540
13	Upper Gundar	44
	TOTAL	1338

Basin data that has been collected shows that there are almost 7 Basins in Madurai which in turn are divided into Sub-Basins (-No. of tanks) namely Pambar Kottakaraiyar-363, Vaippar-16, Vaigai-748, Nambiyar-1, Kallar-1, Gundar-204, Tamiraparani-5.

4.1 Application & Use of GIS Map

With the inference from each parametric GIS Map of Madurai tanks as well as generic irrigation tanks all over the world, we can conclude that applying Information System will help to reduce the Encroachments and siltation in the water spread areas of Tanks, supply channels, and surplus courses, which take precautions to mitigate Major flood loss. Using this data, the Government can allocate relief funds for farmers who have lost their grown crops in their field due to natural calamities like cyclones, floods, heavy rains, etc. These data directly contribute to the region of Natural wellness meant for potential & productive agricultural areas and give a clear vision to protect those particular areas with immense beneficiary data. With the cumulative capacities of tanks, currently, available water for the given population density and irrigation lands can be evaluated.

Proper rationing of water for both domestic purposes as well as agricultural utilities can be validated. Cost for Maintenance of Conduits and its new installation can be done in a very effective manner. Seasonal opening of sluices may be decided by knowing the Capacity of Tank. It also greatly helps to increase the depth of the Tank and to meet the future forecasted water requirements which would be found by Mathematical modelling. By monitoring the water level by taking these values as extreme values Major flood loss can be mitigated. An alarm for timely Desilting of Tank (Desiltation process) can also be given. Geological /Soil Analysis data combined with our obtained data can be combinedly used to suggest suitable agricultural crops that can be cultivated in those regions concerning uniform Sub-Basins.

With the help of this data, Renovation/Repair work in any particular Sub-Basin can be made since repair work simultaneously affects the series of Tanks associated with the particular Sub-Basin. These data enable us to directly measure the available surface water in the particular zone, which can be an index factor for balancing the eco-systems. Using these data, the Government can easily allocate funds for State Public Works Department (SPWD) and Central Public Works Department (CPWD) for the Civil repair Project.

5. Conclusion

Tank irrigation is an economical technology and proper management of tanks for agriculture and irrigation is the need of the hour. By providing an effective tank information system, current obstacles in agriculture and water resource management can be resolved. A study was conducted at Madurai district, Tamil Nadu, with the objective of developing a GIS-based irrigation tanks information system to assist the agriculturalists and governmental administrators for proper planning and management. Thus, irrigational tanks information was collected for various tank attributes including Maximum Capacity (MC), Maximum Water Level (MWL), Water Spread Area (WSA), Ayacut Area (AA), Sub-Basins, and Basins were uploaded in GIS to process the data. The study revealed that spatial, as well as non-spatial data, can be integrated and analyzed in different formats. Results in the form of maps, tables, and graphs were acquired with the help of GIS which could help farmers, irrigation engineers, and government officials to maintain the tanks' performance and resources effectively.

References

[1] "Sustaining Agriculture through Modernization of Irrigation Tanks: An Opportunity and Challenge for Tamilnadu, India V. Anbumozhi*, K Matsumoto** and E. Yamaji*," vol. III, pp. 1–12, 1995.

[2] J. S. Amarnath and P. Karthik Raja, "An Economic Analysis of Tank Rehabilitation in Madurai District of Tamil Nadu," Agric. Econ. Res. Rev., vol. 19, no. June, pp. 187–194, 2006.

[3] A. Anand et al., "Assessing the water spread area available for fish culture and fish production potential in inland lentic waterbodies using remote sensing: A case study from Chhattisgarh State, India," Remote Sens. Appl. Soc. Environ., vol. 17, p. 100273, 2020, doi: 10.1016/j.rsase.2019.100273.

[4] V. Anbumozhi, K. Matsumoto, and E. Yamaji, "Towards improved performance of irrigation tanks in semi-arid regions of India: Modernization opportunities and challenges," Irrig. Drain. Syst., vol. 15, no. 4, pp. 293–309, 2001, doi: 10.1023/A:1014420822465.

[5] G. de Winnaar, G. P. W. Jewitt, and M. Horan, "A GIS-based approach for identifying potential runoff harvesting sites in the Thukela River basin, South Africa," Phys. Chem. Earth, vol. 32, no. 15–18, pp. 1058–1067, 2007, doi: 10.1016/j.pce.2007.07.009.

[6] P. Fernández Moniz, J. S. Almeida, A. T. Pino, and J. P. Suárez Rivero, "A GIS-based solution for urban water management," Water Int., vol. 45, no.

6, pp. 660–677, 2020, doi: 10.1080/02508060.2020.1765130.

[7] M. Gireesh, N. Nagaraj, and M. G. Chandrakanth, "Rehabilitation of irrigation tanks in eastern zone of Karnataka - An economic analysis," Indian J. Agric. Econ., vol. 52, no. 2, pp. 231–240, 1997, doi: 10.22004/ag.econ.297533.

[8] Z. B. Irfan, V. Ling, and J. Shan, "Ecological health assessment of the Ousteri wetland in India through synthesizing remote sensing and inventory data," Lakes Reserv. Res. Manag., vol. 25, no. 1, pp. 84–92, 2020, doi: 10.1111/lre.12300.

[9] C. J. Jagadeesha and V. G. Palnitkar, "Satellite data AIDS in monitoring reservoir water and irrigated agriculture," Water Int., vol. 16, no. 1, pp. 27–37, 1991, doi: 10.1080/02508069108686096.

[10] M. Krishnaveni and A. Rajeswari, "GIS Technology for Agricultural Management of Tank Irrigation Systems in South India," vol. 3, no. 1, pp. 723–733, 2014.

[11] Latha, Rajendran, and Murugappan, "Comparison of GIS based SCS-CN and Strange table Method of Rainfall-Runoff Models for," Int. J. Sci. Eng. Res., vol. 3, no. 10, pp. 3–7, 2012.

[12] J. Li and S. Wang, "An automatic method for mapping inland surface waterbodies with Radarsat-2 imagery," Int. J. Remote Sens., vol. 36, no. 5, pp. 1367–1384, 2015, doi: 10.1080/01431161.2015.1009653.

[13] E. Molina-Navarro, A. Nielsen, and D. Trolle, "A QGIS plugin to tailor SWAT watershed delineations to lake and reservoir waterbodies," Environ. Model. Softw., vol. 108, pp. 67–71, 2018, doi: 10.1016/j.envsoft.2018.07.003.

[14] K. Palanisami and R. Meinzen-Dick, "Tank performance and multiple uses in Tamil Nadu, South India," Irrig. Drain. Syst., vol. 15, no. 2, pp. 173–195, 2001, doi: 10.1023/A:1012927722965.

[15] Nasir N., Selvakumar R. (2021) Estimating the Loss of Water Spread Area in Tanks Using Remote Sensing and GIS Techniques in Ambuliyar Sub-basin, Tamilnadu. In: Ramanagopal S., Gali M., Venkataraman K. (eds) Sustainable Practices and Innovations in Civil Engineering. Lecture Notes in Civil Engineering, vol 79. Springer, Singapore. https://doi.org/10.1007/978-981-15-5101-7_23

[16] V. Reyes-García, O. Aubriot, P. Ariza-Montobbio, E. Galán-Del-Castillo, T. Serrano-Tovar, and J. Martinez-Alier, "Local perception of the multifunctionality of water tanks in two villages of Tamil Nadu, South India," Soc. Nat. Resour., vol. 24, no. 5, pp. 485–499, 2011, doi: 10.1080/08941920802506240.

[17] C. Santhi and N. V. Pundarikanthan, "Application of management information system in improving irrigation system performance in a developing country," Water Int., vol. 24, no. 3, pp. 229–239, 1999, doi: 10.1080/02508069908692166.

[18] R. Seenivasan and J. Kanagavalli, "Dying Tanks in Urban Areas: What can be done with them?," Rev. Dev. Chang., vol. 19, no. 1, pp. 109–122, 2014, doi: 10.1177/0972266120140106.

[19] S. Shukla, M. V. Khire, and S. S. Gedam, "Remote sensing and GIS based impact assessment of land transformations on ground water level of a rapidly urbanizing watershed of India," Int. Geosci. Remote Sens. Symp., vol. 2016-Novem, pp. 787–790, 2016, doi: 10.1109/IGARSS.2016.7729198.

[20] R. Sivanpillai and S. N. Miller, "Benefits of pan-sharpened Landsat imagery for mapping small waterbodies in the powder River Basin, Wyoming, USA," Lakes Reserv. Res. Manag., vol. 13, no. 1, pp. 69–76, 2008, doi: 10.1111/j.1440-1770.2007.00356.x.

[21] A. Srivastava and P. Chinnasamy, "Water management using traditional tank cascade systems: a case study of semi-arid region of Southern India," SN Appl. Sci., vol. 3, no. 3, pp. 1–23, 2021, doi: 10.1007/s42452-021-04232-0.

[22] F. Sun, W. Sun, J. Chen, and P. Gong, "Comparison and improvement of methods for identifying waterbodies in remotely sensed imagery," Int. J. Remote Sens., vol. 33, no. 21, pp. 6854–6875, 2012, doi: 10.1080/01431161.2012.692829.

[23] S. Thiruvengadachari, "Remote sensing of tank irrigated areas in tamil nadu state, india," Int. J. Remote Sens., vol. 4, no. 3, pp. 545–554, 1983, doi: 10.1080/01431168308948572.

Petrology and geochemistry of Charnockite patches and host Leptynites from Digapahandi area, EGB: Evidence for incipient growth of Charnockites

Sagar Misra[1], Varun Kumar Mishra[2], Rabindra Nath Tiwari[3], Neeraj Vishwakarma[1]*
[1] Department of Applied Geology, National Institute of Technology Raipur, Chhattisgarh
[2] National Mineral Development Corporation Limited, Panna, Madhya Pradesh
[3] Govt. Model Science college, Rewa, Madhya Pradesh
*Corresponding author

E-mail: nvishwakarma.geo@nitrr.ac.in

ABSTRACT. Occurrences of the small, elongated, irregular shape of charnockite patches along the foliation planes of the leptynites in Digapahandi area, Ganjam district, Odisha of Eastern Ghats Belt (EGB) are reported. Petrography and mineral chemistry of charnockite patches and host leptynites suggest that both the rocks have analogues mineral assemblages, except the presence of ortho-pyroxene in charnockite patches. The mineral chemistry studies revealed that mineral phases common in both rocks are chemically identical. Based on whole rocks geochemistry it is evident that both charnockite patches and host leptynites are iso-chemical. The adamellite composition, per-aluminous nature with positive Europium anomaly of host leptynite and charnockite patches suggests their co-genetic relation. Higher values of $(La/Yb)_N$ and $(Gd/Lu)_N$ of leptynites indicate the highly fractionated HREE pattern of leptynites. High concentrations of HREE in charnockite patches are due to the influence of fluid phase in granulite facies metamorphism. The charnockite patches are characterized by higher K_2O content, higher HREE concentration, and less abundance of biotite indicating the in-situ growth of patchy charnockites.

Keywords: Incipient charnockite, Patchy charnockite, Incipient growth, Eastern Ghats Belt.

1. Introduction

The term *incipient charnockite* was first described by Pichamuthu (1960) [29] for the rocks from Dharwar Craton. Towards the end of the twentieth century, the presence of incipient charnockites in the Eastern Ghats Belt (EGB) as well as in the Southern Granulite terrain has been reported by several workers [1][5][17][41][42]. The term patchy charnockite, incipient charnockite, or arrested charnockites are generally used to designate the ortho-pyroxene bearing granitic assemblages, which occurs as patches and/or lenses within the quartzo-feldspathic assemblages. The processes of formation for patchy charnockite may be described in two major different ways. First, the development of charnockite within leucocratic gneiss through in-situ metamorphic transformation commonly referred to as 'charnockitization process' [1][9][15][17][24][27][42]. The second school of thought proposed the relict nature of pre-existing charnockite disseminated within the leptynite masses [5][6][18]. In the first case, the charnockitization process needs an open system environment [15][30][32][37][38] that allows foreign compounds to take part in the process. In the majority, these compounds are CO_2 and H_2O phases [25][30]. Hence it requires several channels within the pre-existing rock masses to migrate the above-mentioned compounds. Among these compounds, CO_2 is taken as an abundant phase by some authors [15][42] and some were preferred brine inclusion [3][44]. Migration of these phases might be taken as a prime responsible factor for charnockite development and localization. To resume the process, it needs to lower the water activity of the system, either through the presence of CO_2 [15][16][24][34][36] or the decrease in the fluid pressure [41]. As a major dehydrated phase, ortho-pyroxenes are hence formed from the breakdown of amphibole and/or biotite under dehydration conditions. This metamorphic change is generally described as prograde, and hence the charnockitization process has been described as a result of prograde metamorphism of upper amphibolites facies to granulite facies mineral assemblages.

The occurrences of charnockite patches within the host leptynites from EGB are either due to the late-stage development of charnockite [9] or as a remnant of previously formed charnockites [5][18]. Earlier studies on charnockite patches have proved the presence of both types of charnockite patches throughout the EGB. The present paper deals with the detailed mineralogical and petrological studies of charnockite patches along with their host leptynites exposed near the Digapahandi area, Ganjam district, Odisha. The mineralogical and petrological studies are correlated with the geochemical nature of the rocks which through light on their mode of origin.

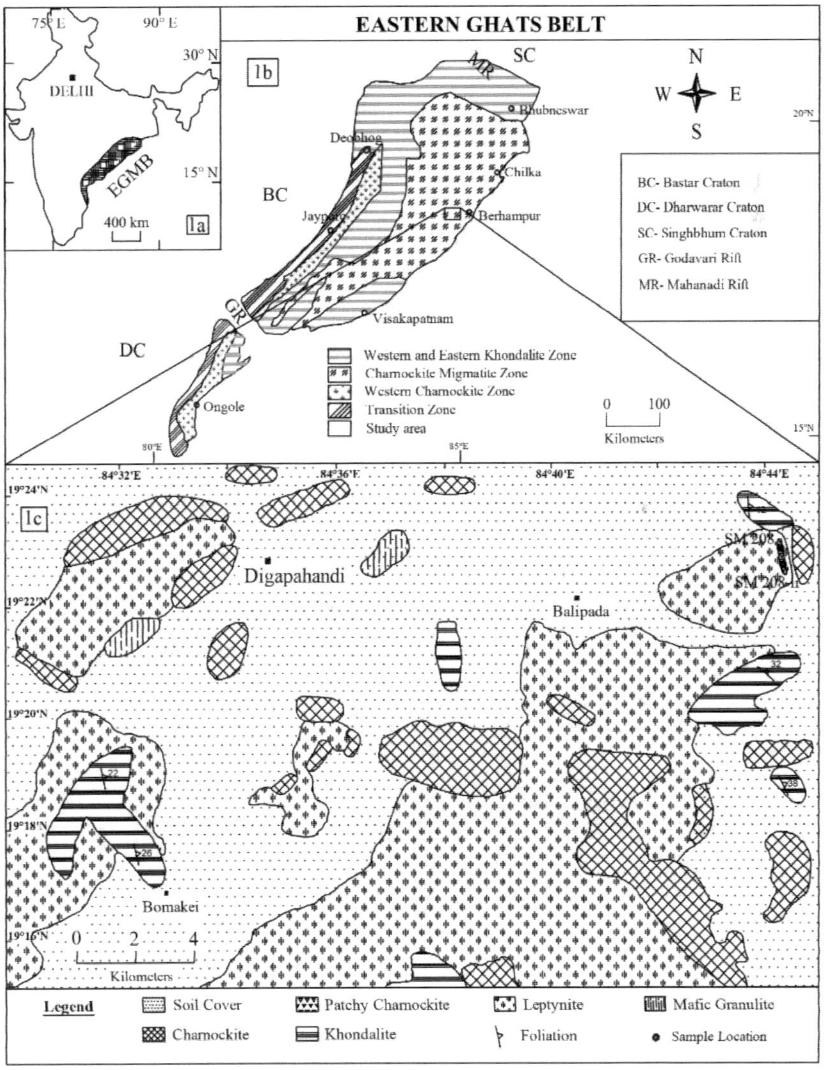

Figure 1. (a) Inset map of India representing Eastern Ghats Belt (EGB). (b) Zonation map of EGB [46] showing study area location along with the previously reported patchy charnockite. (c) Detailed Geological Map of the study area with representative sample locations.

2. Geological setup

The NE-SW trending Eastern Ghats Belt is a high-grade, poly-metamorphic terrain along the East Coast of the Indian sub-continent, where principal rock types are charnockites, khondalites, and leptynites. The EGB is bounded by three Archean Cratons, namely; Singhbhum Craton in the north,

Bastar Craton in the west, and Eastern Dharwar Craton in the southwest. The EGB is dissected by two Gondwana Rifts; i.e., Mahanadi Rift at the north and Godavari Rift at the south end of the belt (Fig. 1a). Presences of charnockite patches within the belt have been reported by several authors. The marked locations are Adaba region of Ganjam district, Odisha [23], Angul sector [14], Chilka Lake region [5][6][9], Deobhog region [13], Jenapur area [18], Rengali sector [8] and Titilagarh area [20]. Among these charnockites, relict growth has been reported from the Jenapur and Chilka area whereas other areas are marked as charnockite development in gneiss through the breakdown of biotite and hornblende.

Digapahandi and the adjacent area belong to the Central Migmatite zone of the EGB, situated at about 200 km south-west of Bhubaneswar, Odisha (Figure. 1a and 1b). Different lithologies exposed in and around the Digapahandi area are leptynites, khondalites, charnockites, and mafic granulites. The detailed geological map of the study area is present in figure.1c. Charnockites of the study area can be grouped into two categories; such as Massif charnockite and patchy Charnockite. Massif charnockites are compositional ranges from charnockite to enderbite concerning their feldspar content. The charnockite patches are present in host leptynites and gneisses. The development of charnockite patches in the host leptynites is found to restrict along the foliation planes of the leptynites (Figure. 2a). These charnockite patches have diffused margins with the host leptynites (Figure. 2b). Both charnockites and leptynites are separated through a thin layer of biotite matrix. The khondalites of the study area are present in association with the large leptynite bodies. The general trends of the khondalites are almost in the North-South direction. The mafic granulite bodies are present as dykes stretching along the NE-SW direction. These rocks occur in the vicinity of leptynites and charnockites. Mafic granulites are showing dark color, coarse to medium grain, porphyritic texture characterized by the presence of orthopyroxene, clinopyroxene, and hornblende.

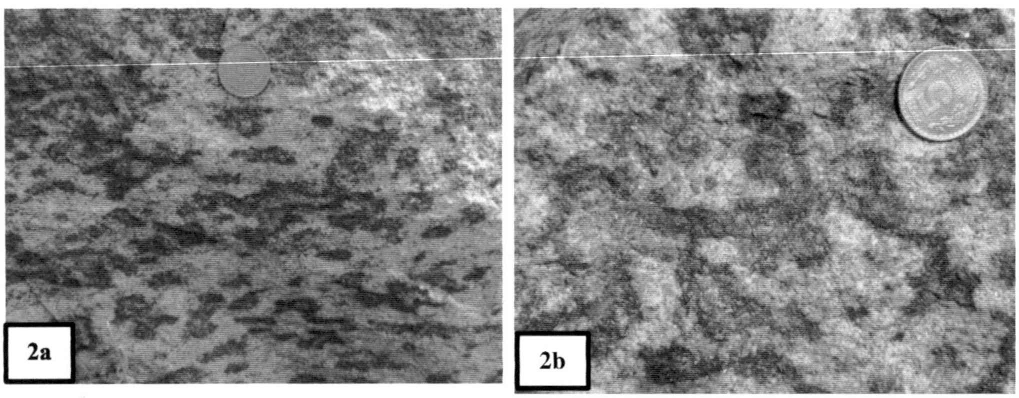

Figure 2. (a) Charnockite patches developed along the foliation plane of host leptynite. (b) Diffused margin around the charnockite patches where biotite present as rim in between host leptynites and charnockite patches.

3. Petrography

(a) Leptynite Host

Host leptynites are characterized by the presence of abundant plagioclase and alkali feldspar that to be of equal proportion. Myrmekitic intergrowth of quartz and feldspars are common. Other minerals like biotite, garnet, and ilmenite are also present in leptynite. Biotite is arbitrarily distributed throughout the rock. Small ilmenite inclusions are also common in biotite. The granoblastic texture is a common phenomenon in leptynites. At places, small patches of pelitic assemblages (sillimanite – biotite – garnet

– feldspar – quartz ± ilmenite) are present as xenoliths within the leptynite (Figure.3a). Sen and Bhattacharya [39] suggested that these leptynites are the product of partial melting of khondalites during granulite facies metamorphic conditions. The presences of relict pelitic assemblages within leptynites also advocate that the khondalites were the protolith for leptynites.

(b) Charnockite Patches

Overall, the charnockite patches show granoblastic texture. Mineral assemblage of charnockite patches is ortho-pyroxene, plagioclase, orthoclase, garnet, biotite and apatite. Orthoclase contents are moreover the plagioclase. Ilmenite is an important opaque mineral phase that is present in charnockite patches. Ortho-pyroxene grains are mostly euhedral to subhedral and have sharp contact with each other (Figure. 3b). Garnet and biotites proportion in charnockites are comparatively less than the host leptynites. Small inclusions of ilmenite and biotite are also present in some ortho-pyroxene grains of charnockite (Figure. 3c) suggesting the formation of ortho-pyroxene through biotite dehydration. Biotite inclusions in garnet are also common in charnockites which indicate the development of garnet by the biotite breakdown reaction. Biotite proportion in charnockite patches is quite low as compared to the host leptynite. It is probably due to the consumption of biotite to form ortho-pyroxene through dehydration melting. However, the proportions of mafic phases are comparatively higher in charnockite than leptynites. At a few places, ortho-pyroxene seems to be growing within the biotite grains (Figure. 3d), which also favours the generation of ortho-pyroxene through the biotite-dehydration process.

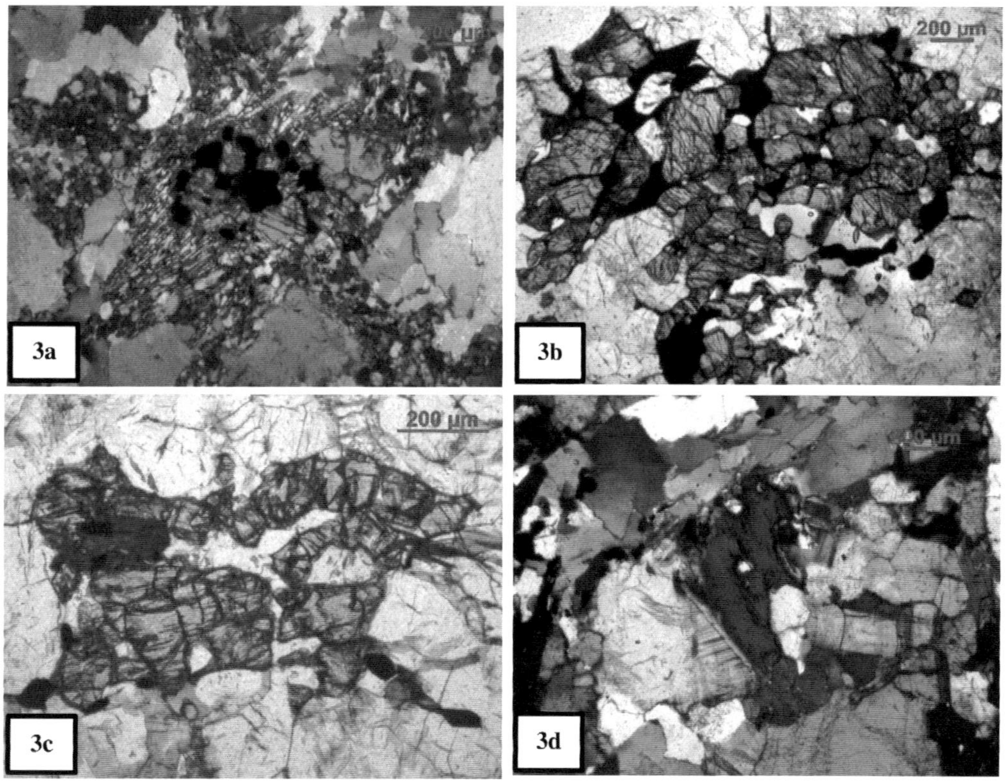

Figure 3. (a) Xenoliths of restite pelitic assemblages present within the leptynites. (b) Euhedral grains of orthopyroxene are present along with the ilmenite and biotite in charnockite. (c) Biotite and ilmenite are present as inclusion in pyroxene grains. (d) Growth of orthopyroxene within biotite. Mineral abbreviations are from Whitney and Evans [45].

4. Mineral chemistry

To determine mineral chemistry, the Electron Micro Probe Analyses (EPMA) was carried out at Electron Probe Laboratory in IIT Bombay (India) by using Cameca SX Five electron microprobe, operating at 15 keV accelerating voltage and 20na beam current. The beam size of the instrument is 1 µm and the dwell time for each analysis is around three minutes. EPMA data are represented by the weight percentage of major oxides. Ferric proportions were calculated by using the stoichiometric method [10] where ever necessary. The representative mineral phases from both leptynite (host) and charnockite (patches) were analyzed and presented in table 1.

Table 1A. Representative mineral chemistry of analysed phases from Patchy charnockites

			Biotite					Garnet				Plagioclase		
SiO_2	36.75	37.25	36.38	36.38	35.64	37.67	37.39	37.81	37.99	37.59	59.87	55.92	60.16	60.18
TiO_2	5.27	5.41	5.36	5.01	5.41	0	0	0	0	0.09	0	0.01	0	0
Al_2O_3	14.61	14.5	14.69	14.62	14.6	20.78	20.73	20.85	21.12	20.85	24.19	22.92	25.14	25.14
Cr_2O_3	0.11	0	0.11	0.13	0.12	0	0	0.01	0	0.03	0.02	0.47	0	0
FeO (T)	20.7	20.02	20.28	20.01	19.1	33.23	32.91	33.81	33.72	33.17	1.34	4.98	0.06	0.09
MgO	10.15	10.29	9.74	10.46	9.72	3.41	2.77	3.1	2.96	3.36	0.45	1.01	0	0
MnO	0.02	0.02	0.02	0.06	0	2.29	2.54	2.42	2.46	2.57	0	0.06	0	0
CaO	0	0.03	0.03	0.06	0.09	3.68	5.07	3.64	3.83	3.75	6.81	6.51	7.49	7.42
Na_2O	0.13	0.14	0.08	0.15	0.13	0.03	0.03	0.03	0.02	0.04	6.71	6.45	7.78	7.71
K_2O	9.69	9.36	9.65	9.58	9.15	0	0.02	0	0.03	0	0.24	0.14	0.2	0.13
F	0.24	0.3	0.3	0.28	0.2									
(O) Basis			24					12				8		
Si	5.79	5.85	5.79	5.78	5.78	2.99	2.96	2.99	2.99	2.97	2.69	2.60	2.67	2.67
Ti	0.62	0.64	0.64	0.60	0.66	0.00	0.00	0.00	0.00	0.01	0.00	0.00	0.00	0.00
Al	2.71	2.68	2.75	2.73	2.79	1.94	1.93	1.94	1.96	1.94	1.28	1.26	1.31	1.31
Cr	0.00	0.00	0.00	0.00	0.00	0.00	0.00	0.00	0.00	0.00	0.00	0.02	0.00	0.00
Fe2+	2.73	2.63	2.70	2.66	2.59	2.11	2.03	2.15	2.15	2.08	0.05	0.19	0.00	0.00
Fe3+	0.00	0.00	0.00	0.00	0.00	0.09	0.15	0.09	0.06	0.11	0.00	0.00	0.00	0.00
Mg	2.38	2.41	2.31	2.48	2.35	0.40	0.33	0.37	0.35	0.40	0.03	0.07	0.00	0.00
Mn	0.00	0.00	0.00	0.01	0.00	0.15	0.17	0.16	0.16	0.17	0.00	0.00	0.00	0.00
Ca	0.00	0.01	0.01	0.01	0.02	0.31	0.43	0.31	0.32	0.32	0.33	0.32	0.36	0.35
Na	0.04	0.04	0.03	0.05	0.04	0.01	0.01	0.01	0.00	0.01	0.58	0.58	0.67	0.66
K	1.95	1.87	1.96	1.94	1.89	0.00	0.00	0.00	0.00	0.00	0.01	0.01	0.01	0.01
∑ Cations	16.20	16.10	16.20	16.20	16.10	8.00	8.00	8.00	8.00	8.00	5.00	5.00	5.00	5.00
Almandine						70.90	68.60	72.00	72.10	70.20				
Grossular						10.50	14.60	10.30	10.80	10.70				
Pyrope						13.50	11.10	12.20	11.60	13.30				
Spessartine						5.20	5.80	5.40	5.50	5.80				
X An											35.40	35.50	34.40	34.50
X Ab											63.10	63.60	64.60	64.80
X Or											1.50	0.90	1.10	0.70
X Mg	0.33	0.34	0.32	0.34	0.34	0.09	0.08	0.08	0.08	0.09				
X Fe	0.67	0.66	0.68	0.66	0.66	0.91	0.92	0.92	0.92	0.91				

	Pyroxene							
	From Margin		From Core					
SiO_2	49.32	51.06	48.43	48.90	49.32	48.68	48.42	48.26
TiO_2	0.02	0.06	0.06	0.03	0.06	0.07	0.07	0.08
Al_2O_3	0.73	0.82	1.25	1.04	0.79	1.25	1.12	1.26
Cr_2O_3	0.03	0.03	0.05	0.04	0.02	0.00	0.00	0.01
FeO (T)	33.69	33.34	37.36	37.49	37.31	37.30	37.56	36.81
MgO	13.91	15.05	11.17	11.15	11.52	11.24	11.27	11.42
MnO	0.68	0.65	1.11	1.12	1.08	1.17	1.18	1.11
CaO	0.39	0.33	0.36	0.39	0.37	0.40	0.38	0.37
Na_2O	0.01	0.00	0.04	0.01	0.02	0.01	0.05	0.07
K_2O	0.00	0.00	0.03	0.00	0.00	0.00	0.00	0.00
(O) Basis	6				6			
Si	1.969	1.977	1.949	1.964	1.972	1.955	1.946	1.948
Ti	0.001	0.002	0.002	0.001	0.002	0.002	0.002	0.002
Al	0.035	0.037	0.059	0.049	0.037	0.059	0.053	0.060
Cr	0.001	0.001	0.002	0.001	0.001	0.000	0.000	0.000
Fe^{2+}	1.098	1.075	1.216	1.239	1.232	1.226	1.209	1.199
Fe^{3+}	0.027	0.005	0.042	0.021	0.016	0.027	0.053	0.044
Mg	0.828	0.869	0.670	0.667	0.687	0.673	0.675	0.687
Mn	0.023	0.021	0.038	0.038	0.037	0.040	0.040	0.038
Ca	0.017	0.014	0.016	0.017	0.016	0.017	0.016	0.016
Na	0.001	0.000	0.003	0.000	0.001	0.000	0.003	0.000
K	0.000	0.000	0.001	0.000	0.000	0.000	0.000	0.000
∑ Cations	4.0	4.0	4.0	4.0	4.0	4.0	4.0	4.0
Ferrosilite	56.5	54.9	63.9	64.4	64.6	65.2	65.3	64.6
Enstatite	42.6	44.4	35.2	34.7	34.6	33.9	33.9	34.6
Wollastonite	0.9	0.7	0.8	0.9	0.8	0.9	0.8	0.8
X_{Mg}	0.43	0.45	0.36	0.35	0.36	0.35	0.36	0.36
X_{Fe}	0.57	0.55	0.64	0.65	0.64	0.65	0.64	0.64

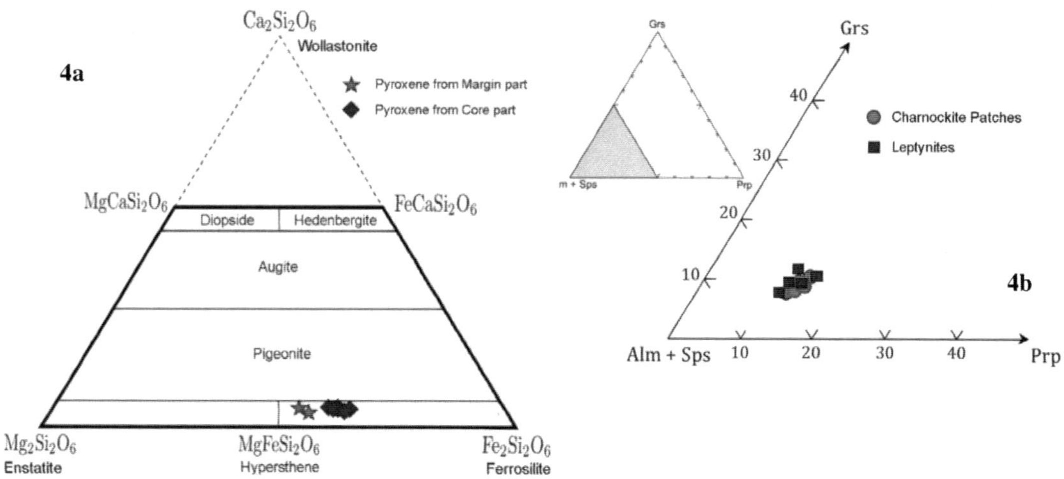

Figure 4. Compositional diagrams of minerals (a) Pyroxene composition range [22] from core and margin part of charnockite patches, (b) Garnet compositional diagram [7], (c) Biotite compositional diagram [19], (d) Plagioclase compositional diagram.

(a) Orthopyroxene

Orthopyroxene is an essential mafic phase present in the charnockite patches and compositionally marked as hypersthene (Figure. 4a) as per the orthopyroxene classification proposed by Morimoto [22]. In general, ortho-pyroxene grains (from the core part of the charnockite patch) have an X_{Mg} value of 0.35 to 0.36, whereas ortho-pyroxene grains are present near the margin of charnockite patches have X_{Mg} value 0.43 – 0.45. Comparative study of both pyroxenes indicates that the pyroxenes at the margin of charnockite patches have lower FeO and higher MgO. The compositional variations among ortho-pyroxene grains are due to the migration of the basic components (FeO, Al_2O_3, and MnO) from the marginal part towards the core during the metamorphic process under the influence of the fluid phase. The distance of migration depends upon the intensity of the fluid-rocks interaction [28].

(b) Garnet

Compositionally the garnets are Fe rich with a subordinate concentration of Mg, Mn, and Ca. The almandine composition of garnets in charnockite patches is higher as compared to leptynites whereas the other end members of garnet i.e., pyrope, spessartine, and grossularite proportion are almost similar. Compositions of garnet are plotted in the garnet mineral classification diagram (Figure. 4b) after Coleman et al. [7]. The core parts of garnet have higher MgO content than their rim counterpart. Such variations in garnets are common among the rocks which have suffered progressive regional metamorphism. The compositional variation of garnets, among both the rock types is mainly controlled by the bulk rock composition as well as the diversity in the mineral assemblage.

(c) Biotite

Biotite is a common phase in both charnockite patches as well as leptynite hosts. The compositions of biotite from both portions are quite similar. The TiO_2 and Na_2O content of biotite are slightly higher in the charnockite portion as compared to host leptynite. Similarly, the MnO content of biotite is higher in leptynite. The X_{Fe} values for all the biotites are almost constant and range from 0.65-0.68. The X_{Fe} values are plotted in X_{Fe}-Al_{IV} compositional diagram (Figure. 4c) after Lalonde and Bernard [19] to mark the compositional differences. The fluorine content of biotite is more in host leptynites than charnockite patches. Tropper et al. [44] observed that there is an increase in the modal abundance of the F-apatite in granulite facies rocks when compared to their amphibolite facies precursors. The lower fluorine content in biotite from charnockite patches might be due to the mobility of fluorine from biotite to apatite during the granulite facies metamorphism under the fluid dominant environment.

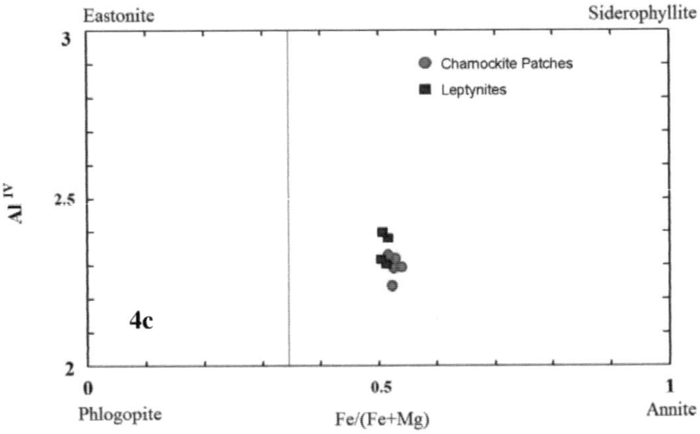

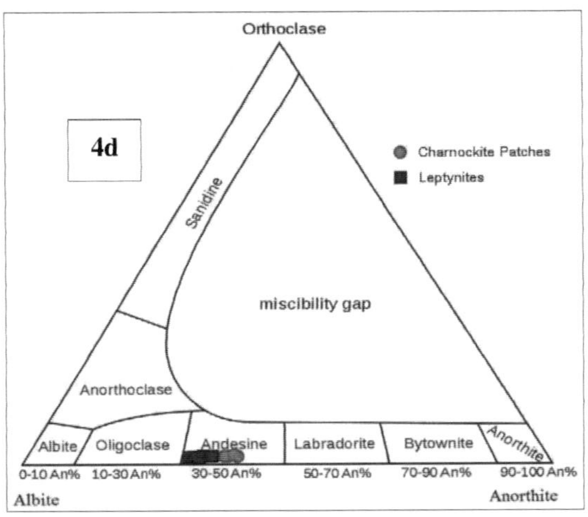

(d) Plagioclase

Another mineral phase that is commonly present in both charnockite patches and host leptynite is plagioclase with andesine composition (Figure. 4d). The albite content of plagioclase from charnockite patches is $Ab_{63.1-64.8}$ whereas in the leptynite portion it is $Ab_{65.2-67.7}$. Plagioclases from leptynites have higher CaO (6.53 – 7.31%) and Na_2O content (7.74 – 7.90%) than the plagioclases from charnockites, where the CaO and Na_2O content varies from 6.81 – 7.48% and 6.44 – 7.78% respectively.

Table 1B. Representative mineral chemistry of analysed phases from Leptynites.

	Biotite				Garnet					Plagioclase				
SiO_2	36.56	36.64	36.38	36.03	38.39	37.93	37.41	37.31	38.66	60.39	60.49	60.88	60.57	60.82
TiO_2	4.88	3.74	4.35	4.11	0.01	0.01	0.01	0.02	0.01	0.00	0.00	0.01	0.00	0.00
Al_2O_3	15.12	15.51	13.94	14.41	21.42	20.81	20.32	20.73	21.20	24.95	25.18	25.16	25.23	24.90
Cr_2O_3	0.00	0.00	0.00	0.01	0.00	0.00	0.00	0.00	0.00	0.00	0.01	0.00	0.01	0.00
FeO (T)	20.78	20.69	20.75	19.88	31.73	32.11	32.51	32.67	34.48	0.02	0.06	0.01	0.09	0.08
MgO	10.26	10.87	10.36	10.55	3.58	3.51	3.64	3.16	2.99	0.00	0.01	0.00	0.01	0.00
MnO	0.01	0.08	0.10	0.06	2.44	2.65	2.46	2.76	2.70	0.02	0.01	0.00	0.00	0.00
CaO	0.05	0.13	0.14	0.00	4.27	3.71	3.63	3.63	3.71	7.24	7.28	7.17	7.31	6.53
Na_2O	0.04	0.07	0.10	0.11	0.03	0.00	0.01	0.01	0.02	7.90	7.83	7.77	7.74	7.88
K_2O	9.52	9.44	8.82	9.58	0.00	0.00	0.01	0.01	0.00	0.22	0.24	0.19	0.15	0.24
F	0.21	0.34	0.34	0.41										
(O) Basis		24					12					8		
Si	5.76	5.77	5.86	5.82	3.01	3.01	2.99	2.98	3.00	2.68	2.67	2.68	2.67	2.70
Ti	0.58	0.44	0.53	0.50	0.00	0.00	0.00	0.00	0.00	0.00	0.00	0.00	0.00	0.00
Al	2.80	2.88	2.65	2.74	1.98	1.95	1.92	1.95	1.94	1.30	1.31	1.31	1.31	1.30
Cr	0.00	0.00	0.00	0.00	0.00	0.00	0.00	0.00	0.00	0.00	0.00	0.00	0.00	0.00
Fe^{2+}	2.74	2.72	2.80	2.69	2.07	2.11	2.09	2.11	2.17	0.00	0.00	0.00	0.00	0.00
Fe^{3+}	0.00	0.00	0.00	0.00	0.01	0.02	0.09	0.07	0.07	0.00	0.00	0.00	0.00	0.00
Mg	2.41	2.55	2.49	2.54	0.42	0.42	0.43	0.38	0.35	0.00	0.00	0.00	0.00	0.00
Mn	0.00	0.01	0.01	0.01	0.16	0.18	0.17	0.19	0.18	0.00	0.00	0.00	0.00	0.00
Ca	0.01	0.02	0.02	0.00	0.36	0.32	0.31	0.31	0.31	0.34	0.34	0.34	0.35	0.31
Na	0.01	0.02	0.03	0.03	0.01	0.00	0.00	0.00	0.00	0.68	0.67	0.66	0.66	0.68
K	1.91	1.90	1.81	1.98	0.00	0.00	0.00	0.00	0.00	0.01	0.01	0.01	0.01	0.01
$\sum$ Cations	16.20	16.30	16.20	16.30	8.00	8.00	8.00	8.00	8.00	5.00	5.00	5.00	5.00	5.00
Almandine					68.80	69.90	69.60	70.70	72.30					
Grossular					11.90	10.50	10.40	10.40	10.30					
Pyrope					13.90	13.80	14.50	12.60	11.50					
Spessartine					5.40	5.90	5.60	6.30	5.90					
X_{An}										33.20	33.50	33.40	34.00	31.00
X_{Ab}										65.60	65.20	65.50	65.20	67.70
X_{Or}										1.20	1.30	1.00	0.80	1.30
X_{Mg}	0.33	0.34	0.33	0.35	0.10	0.10	0.10	0.09	0.08					
X_{Fe}	0.67	0.66	0.67	0.65	0.90	0.90	0.90	0.91	0.92					

5. Geochemistry

The major oxide analyses of the samples were done by X-ray Fluorescence at National Centre for Earth Science Studies (NCESS), Trivandrum, India. The rare earth and trace element concentrations were analyzed by HR-ICP-MS available at National Geophysical Research Institute (NGRI), Hyderabad, India. Details of the ICP-MS analysis procedure and operating parameters are described by Balaram and Rao [4]. The major oxide geochemical data and their normative mineralogical calculations are presented in table 2. The trace and REE concentration data are presented in table 3.

As per major oxide analysis, the SiO_2 content in leptynites (76.19 – 76.30%) is more than charnockite patches (72.97 – 73.92%) whereas the MgO, FeO^T, K_2O, and TiO_2 contents are higher in charnockite patches. Overall, the major oxide data for both charnockite patches and host leptynites are analogous.

Feldspar composition diagram of O'Connor [26] shows that both charnockite patches and host leptynites belong to adamellite composition (Figure. 5). As per the normative mineral composition, orthoclase and hypersthene proportions in charnockite patches are higher than the host leptynites whereas quartz proportion is higher in leptynites. Except for hypersthene, there is no compositional difference of minerals existing between host leptynite and charnockite patches. Based on the alumina saturation index [40] both leptynite and charnockite patches belong to peraluminous (Figure. 6). The charnockite shows I-type affinity while leptynites fall within the S-type field of Maniar and Piccoli [21]. Similarly, as per the alkali lime index [12] host leptynites belong to calcic whereas charnockite patches represent calc-alkali nature (Figure. 7). It is mainly due to increasing K_2O content in the charnockite portion during the granulite facies metamorphism. The K_2O content of charnockite is slightly higher than leptynite. Endo et al. [11] mentioned that there is a slight increase in K_2O from garnet-biotite gneiss (leptynite) to charnockite, as ortho-pyroxene becomes more stable with increasing K_2O content.

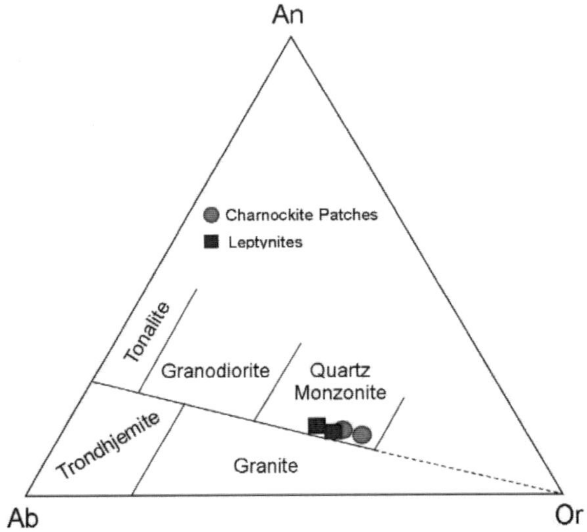

Figure 5. Or-Ab-An normative proportion [26] of charnockite patches and their host leptynites of Digapahandi area, EGB.

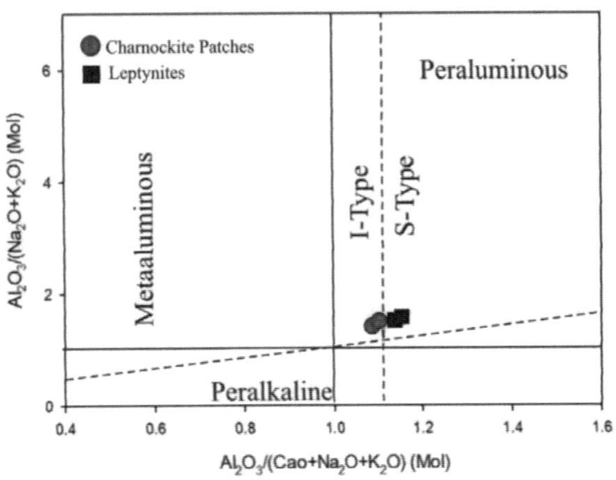

Figure 6. Discrimination diagram of Alumina saturation index [40] with the relative source of rocks [21] showing peraluminous nature for charnockite patches as well as host leptynites.

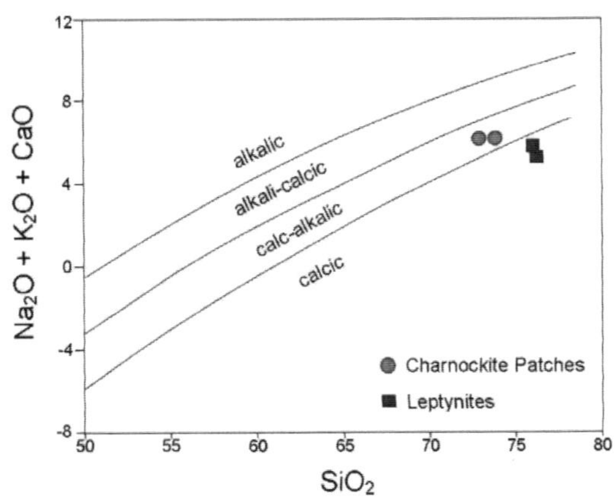

Figure 7. Alkali – lime index diagram [12] for charnockite patches and host leptynites.

Table 2. Major oxide data of Patchy charnockite and host leptynites along with their norm calculation

Rock type	Charnockite		Leptynite	
Sample no	SM 208	SM 208 ii	SM 208	SM 208 ii
SiO_2	72.97	73.92	76.19	76.3
TiO_2	0.21	0.2	0.1	0.17
Al_2O_3	14.72	13.67	14.03	13.99
MnO	0.02	0.03	0	0
Fe_2O_3	1.48	1.82	0.42	0.36
CaO	1.81	1.61	1.65	1.75
MgO	0.21	0.43	0.02	0.01
Na_2O	2.46	2.14	2.41	2.52
K_2O	5.48	5.58	4.97	4.49
P_2O_5	0.04	0.05	0.04	0.04
Total	99.4	99.45	99.83	99.63
Norm Calculation				
Quartz	33.16	35.52	39.62	40.91
Corundum	1.55	1.31	1.78	1.9
Orthoclase	32.62	33.21	29.43	26.64
Albite	20.97	18.23	20.43	21.41
Anorthite	8.78	7.71	7.94	8.45
Hypersthene	2.04	3.06	0.39	0.14
Ilmenite	0.4	0.38	0.19	0.32
Magnetite	0.38	0.45	0.11	0.12
Apatite	0.09	0.11	0.09	0.09
Total	99.99	99.98	99.98	99.98

Table 3. Trace and REE data of Charnockite patches and host leptynites.

Rock type	Charnockite		Leptynite	
Sample no	SM 208	SM 208 ii	SM 208	SM 208 ii
Sc	16.954	24.192	9.214	8.935
V	84.005	88.314	23.453	26.86
Cr	14.815	17.966	16.708	11.651
Co	7.754	9.27	2.803	2.287
Ni	1.094	1.165	0.975	0.902
Cu	60.42	58.585	53.38	44.657
Zn	15.272	19.38	9.935	11.261
Ga	13.516	13.322	12.334	10.674
Rb	8.647	10.731	9.454	6.849
Sr	76.812	110.683	104.357	57.808
Y	3.042	4.475	2.582	1.982
Zr	19.555	21.274	17.468	22.638
Nb	0.836	0.398	0.313	0.57
Cs	0.04	0.05	0.044	0.028
Ba	193.167	257.658	225.7	150.461
Hf	0.478	0.536	0.425	0.569
Ta	0.067	0.035	0.028	0.047
Pb	6.77	8.104	7.681	6.548
Th	0.142	0.216	0.189	0.319
U	0.014	0.019	0.014	0.016
La	4.826	5.669	5.614	4.619
Ce	9.811	10.668	10.727	9.628
Pr	1.111	1.241	1.236	1.28
Nd	4.449	4.943	4.846	5.187
Sm	1.059	1.165	1.111	1.26
Eu	1.312	1.553	1.436	1.036
Gd	0.967	1.144	1.004	1.114
Tb	0.126	0.154	0.118	0.126
Dy	0.633	0.779	0.484	0.482
Ho	0.138	0.176	0.095	0.088
Er	0.368	0.488	0.22	0.182
Tm	0.056	0.083	0.029	0.024
Yb	0.414	0.617	0.187	0.151
Lu	0.065	0.098	0.026	0.021
$(La/Yb)_N$	7.922	6.239	20.435	20.763
$(La/Sm)_N$	2.845	3.04	3.154	2.29
$(Gd/Lu)_N$	1.827	1.438	4.696	6.577
Eu/Eu*	3.952	4.1	4.145	2.666
Ba/Rb	22.338	24.011	23.874	21.967
Ba/Sr	2.515	2.328	2.163	2.603
Sr/Y	25.253	24.732	40.42	29.163
LREE	21.257	23.686	23.534	21.974
HREE	2.767	3.54	2.163	2.188
LREE/HREE	7.682	6.691	10.879	10.044
K/Rb	5260.54	4316.474	4363.92	5441.68
Rb/Sr	0.113	0.097	0.091	0.118

The trace element concentrations of both the rock types are almost comparable. The Sc, V, Co, Zn, Ga, and Y concentrations are higher in charnockite patches as compared to host leptynite. The Rb/Sr ratios for both the rock types are low and almost similar (0.09 – 0.12). Similarly, low to moderate Ba/Sr ratios (2.16 – 2.60) of both the rocks suggest the fractionation of plagioclase through the dehydration melting of the crustal source material [35]. The primitive mantle normalized spider plot [43] shows the depleted nature of Cr, Ni, Zn, Rb, and Y content whereas enrichment in Ba and Sr contents (Figure. 8a). Strongly depleted Rb in rocks is due to the increasing metamorphic grade [2]. When it reaches the anatectic point, the Rb gets partitioned into the melt phase.

The chondrite normalized REE plot for all investigated samples is represented in figure 8b. The REE pattern for both host leptynites and charnockite patches is quite similar with a strong positive Eu anomaly, due to the incorporation of plagioclase in the rocks. The Eu/Eu* ratio varies from 2.67 – 4.14 and is almost constant for both the rocks. This is also supported by the abundance of plagioclase in rocks and is evidenced through petrography. These samples show LREE enrichment with depleted HREE (Figure. 8b). Chondrite normalized [43] REE plot for both the rock types, show relatively steeper

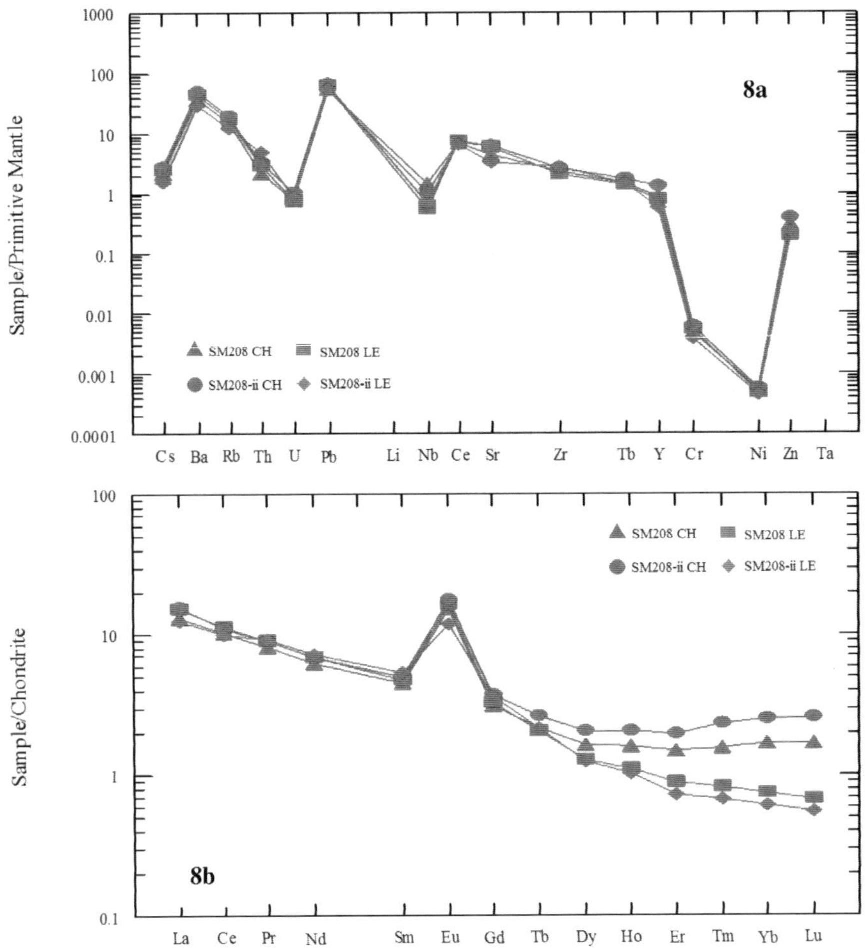

Figure 8. REE and Trace element concentration of charnockite patches and host leptynites from Digapahandi area, EGB. Normalization values are from Sun and Mc Donough [43]. (a) Primitive mantle normalized spider diagram of rocks (b) Chondrite normalized REE pattern of rocks.

LREE pattern as well as slightly steeper to nearly flat HREE pattern for both leptynite and charnockite respectively. The LREE/HREE ratio for charnockite patches (6.69 – 7.68) is lower than the leptynites (10.04 – 10.88). The charnockite patches show more HREE enrichment than host leptynite. The $(La/Yb)_N$ ratio for charnockite patches is low (6.24 – 7.92) as compared to host leptynites (20.43 – 20.76). This pattern indicates that the host leptynites are highly fractionated compared to charnockite patches. The higher $(Gd/Lu)_N$ ratio of leptynites is evident for HREE fractionation rather than LREE.

6. Discussion

The origin of patchy charnockites in different parts of EGB was proposed by different authors [5][8][9][13][18][23]. It is unanimously accepted that the charnockite patches are commonly present in host felsic gneisses and leptynites. The earlier views on the formation of charnockite patches are quite distinct. Bhattacharya et al. [5] and Kar [18] described the patchy charnockites as 'relict charnockites' based upon the diverse geochemical signatures as well as structural characteristics. Dobmeier and Raith [9] describe these charnockites as 'arrested type of charnockites', which probably originated through in-situ dehydration process within the leptynite and were presumably formed due to the localized syn-kinematic fluid migration through structurally controlled sites. About these two types of formation, nature of the studied patchy charnockites is described here.

As the major oxide and trace element concentrations are almost similar, it suggests that both the rocks are co-genetic in origin. The higher K/Rb ratio for both the rocks advocates the dehydration process. The Al_2O_3 versus SiO_2 (Figure. 9a) as well as Sr versus SiO_2 plots (Figure. 9b) after Rajesh et al. [32] for both the rocks of the study area also favours the dehydration process. The occurrence of small irregular-shaped elongated charnockite patches along the foliation planes within leucocratic leptynite host at Digapahandi area of EGB appears to be structurally controlled as manifested from field relationships which commonly exhibit diffuse margin. The presence of ortho-pyroxene in the charnockite portion makes it different from the host leptynite, otherwise, the rest of the mineral phases are common and compositionally analogues, as reflected by the mineral chemistry. As per the bulk rock geochemistry, both host leptynites and charnockite patches show iso-chemical nature. They are adamellite in composition with per-aluminous nature. The charnockite patches were developed within

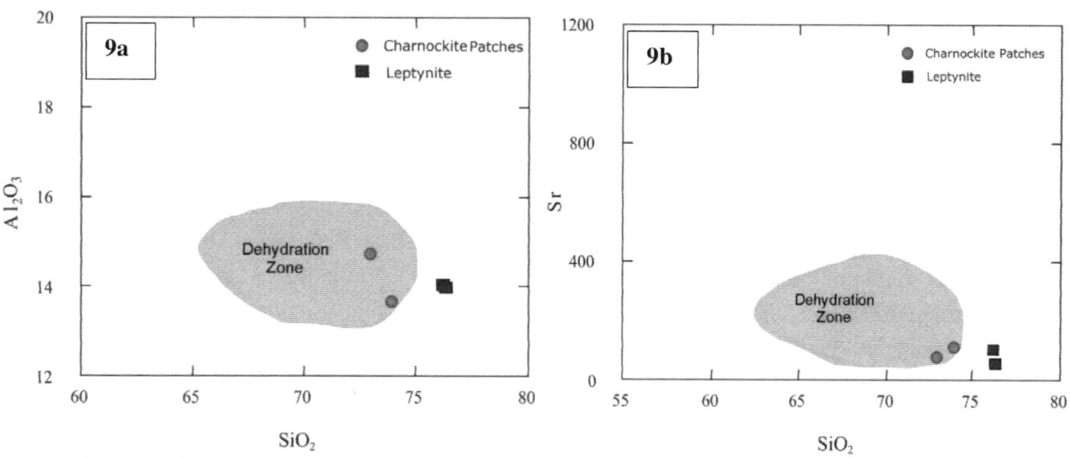

Figure 9. Compositional range of dehydration zone [15][16][32] applied for charnockite patches and host leptynites from Digapahandi area, EGB. (a) SiO_2-Al_2O_3 graph and (b) SiO_2-Sr graph showing the dehydration zone for charnockite patches.

the leptynite host through the biotite-dehydration melting process at structurally controlled sites, through the interaction of the fluid phase.

According to Tropper et al. [44], the solubility of HREE appears to be much more than LREE in the fluorine-bearing fluids during granulite facies metamorphism. The relatively higher HREE concentration within the charnockite patches as compared to the host leptynite seems to be due to the presence of a fluid phase. However, the fluid inclusion study was not performed in the present study but the biotites of charnockite patches having fluorine content compared to leptynite indicates the activity of CO_2 rich fluid phase in the formation of these charnockite patches. The detailed fluid inclusion studies on these rocks will certainly elucidate the role of fluid phases in their origin.

7. Conclusion

Charnockites occur as small patches within the host leptynites and granitic gneisses throughout the EGB. These charnockite patches have been described as 'relict charnockite' as well as 'arrested or incipient charnockite' based upon the structural and geochemical studies. Based on detailed petrography, minerals chemistry and geochemical studies carried out on the charnockite patches occurring within the leptynite host at Digapahandi area of EGB may be classified as incipient charnockite. The host leptynites are believed to be the product of anatectic melt generated through the partial melting of the khondalite during an early stage of evolution. The incipient charnockites were developed within the host leptynites, along the foliation planes, through biotite-dehydration melting under the influence of fluid phase, at a later stage of the evolutionary history. The mineral chemistry and whole-rock geochemistry data suggest both the charnockite patches and the host leptynites are iso-chemical. Hence the causative factor for the incipient growth of charnockite patches within-host leptynite is supposed to be the iso-chemical transformation of leptynites into charnockites under the high-grade granulite facies metamorphic condition.

Acknowledgment
Analytical works i.e., EPMA, XRF, and ICP-MS analyses of rock samples were carried out respectively at IIT Bombay, NCESS (Trivandrum), and NGRI (Hyderabad). We are thankful to the concerned authorities of the above-mentioned labs for their kind assistance.

8. References

[1]. Aftalion, M., Bowes, D.R., Dash, B. and Dempster, T.J. (1988) Late Proterozoic Charnockites in Orissa, India: A U-Pb and Rb-Sr Isotopic Study. *J Geol*, **96**, 663–676.
[2]. Aranovich, L.Y., Makhluf, A.R., Manning, C.E. and Newton, R.C. (2014) Dehydration melting and the relationship between granites and granulites. *Precambrian Res*, **253**, 26–37.
[3]. Aranovich, L.Y. and Newton, R.C. (1997) H2O activity in concentrated KCl and KCl-NaCl solutions at high temperatures and pressures measured by the brucite-periclase equilibrium. *Contrib to Mineral Petrol.*, **127**, 261–271.
[4]. Balaram, V. and Rao, T.G. (2003) Rapid determination of REEs and other trace elements in geological samples by microwave acid digestion and ICP-MS. *At. Spectrosc.*, **24**, 206-212.
[5]. Bhattacharya, S., Sen, S.K. and Acharyya, A. (1993) Structural evidence supporting a remnant origin of patchy charnockite in the Chilka Lake area, India. *Geol Mag.*, **130**, 363–368.
[6]. Bhattacharya, S., Sen, S.K. and Acharya, A. (1994) The structural setting of the Chilka Lake granulite-migmatite-anorthosite suite with emphasis on the time relation of charnockites. *Precambrian Res.*, **66**, 393–409.

[7]. Coleman, R.G., Lee, D.E., Beatty, L.B. and Brannock, W.W. (1965) Eclogites and eclogites: their differences and similarities. *Geol Soc Am Bull.*, **76**, 483-508.

[8]. Crowe, W.A., Nash, C.R., Harris, L.B., Leeming, P.M. and Rankin LR (2003) The geology of the Rengali Province: Implications for the tectonic development of northern Orissa, India. *J Asian Earth Sci.*, **21**, 697–710.

[9]. Dobmeier, C. and Raith, M.M. (2000) On the origin of "arrested" charnockitization in the Chilka Lake area, Eastern Ghats Belt, India: a reappraisal. *Geol Mag.*, **137**, 27–37.

[10]. Droop, T.R. (1987) A General Equation for Estimating Fe3+ Concentrations in Ferromagnesian Silicates and Oxides from Microprobe Analyses, Using Stoichiometric Criteria. *Mineral Mag.*, ***51***, 431–435.

[11]. Endo, T., Tsunogae, T. and Santosh, M. (2013) Granulite formation in a Gondwana fragment : petrology and mineral equilibrium modeling of incipient charnockite from Mavadi , southern India. *Mineral Petrol.*, **107**, 727–738.

[12]. Frost, B.R., Barnes, C.G., Collins, W.J., Arculus, R.J., Ellis, D.J. and Frost, C.D. (2001) A Geochemical Classification for Granitic Rocks. *J Petrol.*, **42(11)**, 2033–2048. doi:10.1093/petrology/42.11.2033

[13]. Gupta, S., Bhattacharya, A., Raith, M., and Nanda, J. (2008) Contrasting pressure-temperature-deformation history across a vestigial craton-mobile belt boundary: The western margin of the Eastern Ghats Belt at Deobhog, India. *J Metamorph Geol.*, **18**, 683-697.

[14]. Halden, N.R., Bowes, D. and Dash, B. (1982) Structural evolution of migmatites in granulite facies terrane: Precambrian Crystalline Complex of Angul, Orissa, India. *Transactions of the Royal Society of Edinburgh Earth Sciences*, **73**, 109-118.

[15]. Hansen, E.C., Janardhan, A.S., Newton, R.C., Prame, W.K.B.N. and Ravindra Kumar, G.R. (1987) Arrested eharnockite formation in southern India and Sri Lanka. *Contrib Mineral Petrol.*, **96**, 225–244.

[16]. Janardhan, A.S., Newton, R.C. and Hansen, E.C. (1982) The transformation of amphibolite facies gneiss to charnockite in southern Karnataka and northern Tamil Nadu, India. *Contrib Mineral Petrol.*, **79**, 130–149.

[17]. Janardhan, A.S., Newton, R.C. and Smith, J.V. (1979) Ancient crustal metamorphism at low pH2O:charnockite formation at Kabbaldurga, south India. *Nature*, **278**, 511.

[18]. Kar, R. (2001) Patchy charnockites from Jenapore, Eastern Ghats granulite belt, India: Structural and petrochemical evidences attesting to their relict nature. *Proc Indian Acad Sci - Earth Planet Sci.*, **110**, 337–350.

[19]. Lalonde, A. and Bernard, P. (1993) Composition and color of Biotite from Granites: Two useful Properties in the characterization of plutonic suites from the Hepburn internal zone of Wopmay orogen, Nortwest territories. *Can Mineral.*, **31**, 203–217.

[20]. Mahapatro, S.N., Nanda, J.K., Ghosh, S.P. and Tripathy, A.K. (2009) Patchy Charnockites from the Eastern Ghats Mobile Belt - Examples from Titilagarh and Khandpara areas of Orissa. *Vistas Geol Res.*, **8**, 190–197.

[21]. Maniar,P.D. and Piccoli, P.M. (1989) Tectonic discrimination of granitoids. *Geol Soc Am Bull.*, **101**, 635-643.

[22]. Morimoto, N. (1988) Nomenclature of Pyroxenes. *Min Mag.*, **52**, 535-550.

[23]. Nanda, J.K. and Pati, U.C. (1998) Mafic granulites and charnockites from Berhampur - Jeypore transect in Eastern Ghats of Orissa sector: a petrochemical study. *Geol Surv India Spl Pub.*, **44**, 276-285.

[24]. Newton, R.C. (1992) Charnockitic alteration: evidence for CO2 infiltration in granulite facies metamorphism. *J Metamorph Geol.*, **10**, 383–400.

[25]. Newton, R.C., Smith, J.V. and Windley, B.F. (1980) Carbonic metamorphism, granulites, and crustal growth. *Nature*, **188**, 45–50.

[26]. O'Connor JT (1965) A classification for quartz rich igneous rocks based on feldspar rations, *U.S. Geol Survey Prof Paper*, **525(B)**, 79-84.

[27]. Paul, D.K., Barman, T.R., McNaughton, N.J., Fletcher, I.R., Potts, P.J., Ramakrishnan, M. and Augustine, P.F. (1990) Archean-Proterozoic Evolution of Indian Charnockites: Isotopic and Geochemical Evidence from Granulites of the Eastern Ghats Belt. *J Geol.*, **98**, 253–263.

[28]. Perchuk, L.L. and Gerya, T.V. (1993) Fluid control of charnockitization. *Chem Geol.*, **108**, 175–186.

[29]. Pichamuthu, C.S. (1960) Charnockite in the Making. *Nature*, **188**, 135.

[30]. Raith, M., Hoernes, S., Klatt, E. and Stähle, H.J. (1989) Contrasting Mechanisms of Charnockite Formation in the Amphibolite to Granulite Grade Transition Zones of Southern India. In: Bridgwater, D. (Eds) Fluid Movements - Element Transport and the Composition of the Deep Crust. *NATO ASI Series (Series C: Mathematical and Physical Sciences)*, **281**. Springer, Dordrecht.

[31]. Rajesh, H.M. and Santosh, M. (2012) Charnockites and Charnockites. *Geoscience Feontiers*, **3(6)**, 737-744.

[32]. Rajesh, H.M., Santosh, M. and Yoshikura, S. (2011) The Nagercoil Charnockite : a Magnesian , Calcic to Calc-alkalic Granitoid Dehydrated during a Granulite-facies Metamorphic Event. *J Petrol.*, **52**, 375–400.

[33]. Ramakrishnan, M., Nanda, J.K. and Augustine, P.F. (1998) Geological evolution of the Proterozoic Eastern Ghats Mobile Belt. *Geol Surv India Spec Publ.*, **44**, 1-21.

[34]. Ravindra Kumar, G.R. and Chacko, T. (1986) Mechanisms of Charnockite Formation and Breakdown in Southern Kerala: Implications for the Origin of the Southern Indian Granulite Terrain. *J Geol Soc India*, **28(4)**, 277-288.

[35]. Ravindra Kumar, G.R. and Sreejith, C. (2016) Petrology and geochemistry of charnockites (felsic ortho-granulites) from the Kerala Khondalite Belt, Southern India: Evidence for intra-crustal melting, magmatic differentiation and episodic crustal growth. *Lithos*, **262**, 334–354.

[36]. Santosh, M. (1986) Carbonic metamorphism of charnockites in the southwestern Indian Shield: A fluid inclusion study. *Lithos*, **19**, 1–10.

[37]. Santosh, M., Harris, N.B.W., Jackson, D.H. and Mattey, D.P. (1990) Dehydration and Incipient Charnockite Formation: A Phase Equilibria and Fluid Inclusion Study from South India. *J Geol.,* **.98**, 915–926.

[38]. Santosh, M. and Omori, S. (2008) CO2 flushing: A plate tectonic perspective. *Gondwana Res.*, **13**, 86–102.

[39]. Sen, S.K. and Bhattacharya, S. (1997) Dehydration melting of micas in the chilka lake khondalites: the link between the metapelites and granitoids. *Proc Indian Acad Sci Earth Planet Sci.*, **106**, 277–297.

[40]. Shand, S.J. (1943) The Eruptive Rocks, 2nd edn. *John Wiley*, New York, 444 p.

[41]. Srikantappa, C., Raith, M. and Spiering, B. (1985) Progressive Charnockitization of a Leptynite-Khondalite Suite in Southern Kerala, India- Evidence for Formation of Charnockites Through Decrease in Fluid Pressure? *J Geol Soc India*, **26(12)**, 859-872.

[42]. Stähle, H.J., Raith, M., Hoernes, S. and Delfs, A. (1987) Element mobility during incipient

granulite formation at Kabbaldurga, Southern India. *J Petrol.*, **28**, 803–834.

[43]. Sun, S.S. and McDonough, W.F. (1989) Chemical and isotopic systematics of oceanic basalts: implications for mantle composition and processes. In: Saunders, A.D. and Norry, M.J. (Eds.), Magmatism in the Ocean Basins. *Geological Society Special Publication*, **42**, 313–345.

[44]. Tropper, P., Manning, C.E. and Harlov, D.E. (2013) Experimental determination of $CePO_4$ and YPO_4 solubilities in H_2O-NaF at 800??C and 1 GPa: Implications for rare earth element transport in high-grade metamorphic fluids. *Geofluids*, **13**, 372–380.

[45]. Whitney, D.L. and Evans, B.W. (2010) Abbreviations for names of rock-forming minerals. *Am Mineral.*, **95**, 185–187.

[46]. Ramakrishnan, M. and Vaidyanadhan, R. (2010) Geology of India. *Geological Society of India*, Vol. 1.

IOP Conf. Series: Earth and Environmental Science 1032 (2022) 012012 doi:10.1088/1755-1315/1032/1/012012

Water quality index assessment for drinking and irrigation purpose for Wardha district

Ashay Devidas Shende [1,a*], M.M.Sahoo [2,b] and Shrikant Tekade[3,c]

[1]Research Scholar, School of Civil Engineering, Lovely Professional University, Phagwara - 144411, India
[2]Assistant Professor, School of Civil Engineering, Lovely Professional University, Phagwara - 144411, India
[3] Assistant Professor, Department of Civil Engineering, Government Engineering College, Nagpur-441108, India

Email: [a*]ashayshende220@gmail.com; [b]mrunmayee.23405@lpu.co.in; [c]shrikant.tekade@yahoo.com

Abstract. It is always said that groundwater is safe for consumption i.e. free from micro-organisms. But in many places, the groundwater is subjected to contamination due to human and industrial activities. The unsafe disposal of domestic and industrial wastes is found to be the main source for contamination of groundwater with micro-organisms, nitrates, etc. Therefore, it is inevitable to assess the quality of groundwater in terms of the Water Quality Index (WQI) before it is supplied to the end-user. Water Quality Index (WQI) is an index that converts complex data about water quality into the simplest form. This manuscript discusses the level of contamination of groundwater in the villages situated in the vicinity of Wardha River in Wardha District. Groundwater samples were collected during ninety days starting from 1st May 2021 to 30th July 2021. Analysis was carried out for the assessment of 12 parameters and their WQI. The results reveal that 40% of the 12 parameters assessed are higher values whereas, 60% of the 12 parameters assessed are reported safe values as per Indian Standards for Drinking Water. On the other hand, WQI for all the groundwater samples (barring very few) has shown bad quality.

Keywords: Ground Water Quality, Water Quality Index, physio-chemical

1. Introduction

Water is the only source for the survival of all living beings. All human activities (Domestic, Agricultural, and Industrial) requires water. The water quality appears to be of utmost importance for domestic, especially from a potability viewpoint. The usefulness of water is determined by its physicochemical, metallic and bacteriological analysis.[1][2].

Water, being the priceless natural resource, is available for the end-user as surface and groundwater. A large amount of groundwater is extracted for domestic as well as industrial use. Due to human activities and industrial effluent, the quality of groundwater getting contaminated day by day. On the other hand, access to clean drinking water in the world is a matter of grave concern. An interesting fact about water is that 97% of water is available in oceans 2% is available in glaciers and the remaining 1% it is available for human purposes out of which only 0.2% is in form of freshwater, which is called pure water. Similarly, A large number of irrigation activities are usually dependent on groundwater more than 60% of the total area under irrigation depends upon groundwater sources. Ever since the rise of the industrial era, groundwater has gotten polluted besides human activities. It is indispensable to carry out the investigations of water quality to figure out this issue and hence, the concept of water quality index is chosen to mark the suitability of the source for its uses.

2. Study Area

Wardha district is part of Maharashtra state in western India having a population of 1,300,774 (Census 2011). Hinganghat, Seloo, Arvi, and Wardha are the major cities in the district. The district had a population of 1,300,774 as of 2011. The Wardha District lies between 20°44' north latitudes and 78° 36' east longitudes covering an area of 6310 sq. km. The present study deals with the analysis of water for one hundred and six villages of Wardha District.

3. Background

Wardha is a district place that boasts habitable as well as industrial units. Industries such as sponge iron, khadi, dairy, sugar, fertilizers, polyester, etc. are responsible for the effluents, which find their way to the groundwater source. The increase in population and the industrial activities in Wardha have led to water contamination of groundwater in the last decade (presence of faecal coliforms and nitrates) [3][4][5][6].

Determination of Water Quality Index is a numeric expression used to transfer a large quantity of water characterization data into a single number, representing the level of water quality[7][8][9]. For WQI, a total of nine different parameters viz. temperature, pH, dissolved oxygen, turbidity, faecal coliforms, biochemical oxygen demand, total phosphate, nitrates, and total solids are taken into consideration[1][10][11][12][13].

The 100 point Water Quality Index has been divided into the following ranges.

Table 1 WQI range and respective quality level [15]

Range	Quality
90 – 100	Excellent
70-90	Good
50-70	Medium
25-50	Bad
0-25	Very Bad

4. Material and Methodology

The villages situated in the vicinity of Wardha River in Wardha District were selected as the source for groundwater sample collection. Besides sampling bottles, testing kits, chemicals and other accessories were used to assess the groundwater quality parameters.

Groundwater samples from the designated villages for 90 days were collected in sampling bottles. The parameters like Temperature, pH, and dissolved oxygen, were noted at the place of sample collection using the respective testing kit following the process duly mentioned in the manual IS10500 - 2012. The samples were brought to the laboratory for testing their suitability for intended end-use. Other required parameters were tested in the laboratory following the standard equipment and methods for the examination of water and wastewater [14]. The results are tabulated in Tables 2 to 5.

5. Result and Discussion

The analysis result for 12 parameters for all the 106 samples of Wardha District are given in Tables 2 to 5. And the calculated WQI for all the 106 water samples is shown in figures 1 to 3. All the result values were cross-checked [16] and compared with Indian Standard for Drinking Water [15]

5.1 Temperature

The temperature of nearly all the samples was lying very close to 24°C.

5.2 pH

The pH value ranges between 7.0 and 8.4. The lowest pH is for water sample number 78, i.e., 7.0; whereas, a higher pH value is shown by samples 51 and 101; while other water samples showed good pH values.

5.3 Turbidity

All the water samples show turbidity less than 5 NTU, which is well within the prescribed limits.

5.4 Electrical Conductivity

Maximum electrical conductivity (EC) is recorded for sample number 51 (5440 μ mhos/cm) and the minimum for sample number 97 (449 μ mhos/cm). The EC for 16 samples is above the prescribed limit of 2000 μ mhos/cm.

5.5 Total Solids

Total Solids (TS) ranged between 292 mg / L to 3536 mg / L. TS value of 292 mg / L is for sample number 97, whereas sample number 51 showed the highest value of 3536 mg / L. Whereas all the water samples are well within the limits.

5.6 Dissolved Oxygen

The dissolved oxygen (DO) level for the water sample varied between 0 to 1.5 mg/L.

5.7 Alkalinity

The alkalinity values ranged between 120 - 740 mg / L. Nearly 90% of the water samples show alkalinity above 200 mg / L and 10% of the water samples are very close to the permissible limit of 600 mg / L. The highest value of 740 mg / L is for water sample number 80.

5.8 Nitrate

The nitrate ions concentration lies in the range of 1 to 371 mg/L. 36 Water samples out of 106 show values above 100 mg/L while 50% of the water samples show values greater than the desirable limit of 45 mg / L.

5.9 Total Phosphorus

The total phosphorus values for all the 106 water samples varied between 0.001 to 0.396 mg/L. The highest value is shown by sample no 77.

5.10 Faecal Coliforms

According to the Indian Standards for Drinking Water, there should not be any coliforms in the 100mL sample. The Most Probable Number (MPN count) for nearly all the water samples is undesirable. It ranges from 0 to 15 MPN/100 mL.

5.11 Water Quality Index

From the figures, it is seen that 41 groundwater samples out of 106 show the WQI less than 50, which is considered as bad water quality. On the other hand, only 10 groundwater samples show the WQI greater than 60. None of the samples shows good water quality i.e. above 70. While 20 samples falling in the 'Medium' quality range of WQI are very close to the lower limit of the Medium range, i.e. 50.

Table 2: Groundwater analysis data for samples 1 – 30

S N	Village	Temp	PH	Turbidity NTU	DO mg/l	EC	TS mg/l	Alkalinity mg/l	TH mg/l	Nitrate mg/l	T.Phosp. mg/l	BOD mg/l	Fecal Coliforms	WQI
1	BHUGAON	24.2	7.9	0.6	0.8	2138	1390	280	600	102	0.019	< 2	3.6	53.55
2	DHOTRA RAILWAY	24.0	8.0	0.80	0.7	2323	1510	400	520	151	0.127	< 2	7.3	47.86
3	WADADHA	24.1	7.6	0.2	1.2	1449	942	360	480	92	0.11	< 2	3.6	50.29
4	SELSURA	24.2	8.3	3.2	0.0	3878	2521	520	1040	362	0.312	< 2	15	42.84
5	SIRASGAON DHANADE	24.1	7.8	0.2	1.2	1214	789	520	400	51	0.037	< 2	3	53.84
6	DHANOLI (MEGHE)	26.0	8.1	1.3	0.5	2554	1660	400	680	204	0.208	< 2	7.3	46.31
7	MOZARI	24.3	7.1	1.9	1.3	1114	724	280	320	42	0.028	< 2	3	53.87
8	SEKAPUR	24.2	8.0	1.6	1.0	1800	1170	440	600	182	0.2	< 2	7.2	46.68
9	KAPSI	24.1	7.9	0.8	1.3	889.2	578	320	320	44	0.041	< 2	3	47.14
10	KATRI	24.2	7.6	1.2	1.2	815.4	530	160	120	2	0.001	< 2	0	62.71
11	KANHOLI	24.5	7.9	0.4	0.9	1172	762	320	400	122	0.146	< 2	3.6	49.14
12	DAWLAPUR	24.2	7.9	0.8	0.6	2418	1572	400	240	50	0.076	< 2	3	51.74
13	POTI	24.2	7.9	2.2	0.9	1754	1140	480	560	122	0.122	< 2	3.6	49.14
14	BHAYYAPUR	24.7	7.6	0.4	1.4	910.8	592	360	280	42	0.04	< 2	3	54.01
15	KHANGAON 1	24.3	7.6	0.8	1.5	640	416	320	240	38	0.021	< 2	3	54.35
16	KHANGAON 2	24.4	7.6	0.6	0.8	1357	882	320	280	66	0.104	< 2	3.6	50.25
17	SATI	24.2	7.2	0.1	0.9	1062	690	240	280	32	0.068	< 2	3	52.87
18	WARUD	24.2	7.4	0.6	1.0	892.3	580	280	240	42	0.071	< 2	3	52.69

19	KOSURLA(MOTHA)	24.1	7.4	0.9	0.8	2022	1314	600	200	90	0.113	< 2	3.6	50.13
20	PUJAI	24.5	8.0	0.2	0.7	1865	1212	480	480	242	0.261	< 2	11	45.53
21	KARANJI	24.5	8.1	0.9	0.7	1803	1172	320	360	212	0.223	< 2	7.3	46.25
22	BORGAON (MEGHE)	24.4	7.4	1.9	0.9	1040	676	260	320	40	0.016	< 2	3	54.2
23	ALODI	24.4	7.8	0.6	1.3	961.5	625	280	220	43	0.023	< 2	3	54.07
24	DATTAPUR	24.5	7.4	2.2	1.4	641.5	417	160	160	30	0.033	< 2	3	53.91
25	SATODA	24.6	8.0	3.0	1.2	1386	901	200	400	204	0.197	< 2	9.1	46.04
26	AJANSARA	24.5	8.1	0.4	1.4	1000	650	360	360	52	0.071	< 2	3	51.75
27	SONEGAON	24.7	7.2	2.2	1.4	1280	832	360	280	82	0.098	< 2	3.6	50.18
28	MANSAOLI	24.6	7.7	0.9	1.5	713.8	464	280	240	12	0.001	< 2	0	58.54
29	NIDHA	24.4	7.9	0.6	1.4	815.4	530	280	360	34	0.024	< 2	3	53.88
30	TAKLI	24.5	7.9	0.8	1.2	1117	726	360	400	84	0.099	< 2	3.6	49.89

Table 3: Groundwater analysis data for samples 31 – 60

S N	Village	Temp	PH	Turbidity NTU	DO mg/l	EC	TS mg/l	Alkalinity mg/l	TH mg/l	Nitrate mg/l	T.Phosp. mg/l	BOD mg/l	Fecal Colifo rms	WQI
31	SIRASGAO N	24.6	7.7	0.8	1.0	1988	1292	480	400	104	0.082	< 2	3.6	51.44
32	FUKTA	24.8	7.5	1.0	1.1	1309	851	200	240	44	0.051	< 2	3	53.53
33	FUKTA	24.7	7.9	0.7	0.8	2465	1602	480	840	51	0.056	< 2	3	52.82
34	NAGAPUR	24.5	7.9	0.3	0.9	1665	1082	360	480	54	0.071	< 2	3.6	51.86
35	KARANJI (BHOGE)	24.6	7.4	0.5	0.8	2111	1372	320	760	244	0.311	< 2	15	45.32
36	SHIRPUR	24.6	7.8	0.2	1.0	1754	1140	400	520	58	0.074	< 2	3.6	51.91
37	SONDLAPU R	24.7	7.9	0.1	1.0	929.2	604	200	360	41	0.038	< 2	3	53.62
38	WADGAON	24.7	7.4	0.6	1.5	723	340	240	220	12	0.001	< 2	0	58.83

39	GADEGAON	24.5	7.6	1.6	1.1	1034	672	360	440	42	0.031	< 2	3	53.94
40	DIGRAS	24.6	8.0	0.8	0.9	1006	654	280	240	39	0.026	< 2	3	53.51
41	ZADGAON	24.6	7.9	1.2	0.9	1357	882	280	480	104	0.217	< 2	7.3	46.74
42	DHANORA	24.6	7.6	2.0	1.1	1205	783	520	440	102	0.181	< 2	11	46.79
43	RAGHUNATHPUR	24.5	7.8	3.1	0.7	1742	1132	440	560	182	0.304	< 2	14	44.55
44	BHAGWA	24.7	7.2	0.2	1.3	1014	659	240	280	14	0.001	< 2	0	58.57
45	CHANKI	24.8	7.3	1.2	1.2	1012	658	280	360	2	0.001	< 2	0	62.79
46	NALWADI	24.8	7.4	0.6	1.1	1108	720	240	360	68	0.101	< 2	3.6	50.45
47	TALEGAON	24.6	7.4	0.9	1.2	956.9	622	320	360	20	0.011	< 2	0	57.86
48	JAULGAON	24.6	7.8	1.7	1.0	1615	1050	400	368	182	0.207	< 2	14	45.81
49	MANDAWA	24.7	7.9	1.5	1.0	1372	892	400	240	2	0.001	< 2	0	62.15
50	KUTKI	24.6	8.3	2.2	0.9	2203	1432	740	984	256	0.346	< 2	15	43.17
51	KHARANGANA	24.8	8.4	1.7	0.0	5440	3536	200	840	342	0.339	< 2	15	42.67
52	SONEGAON (STATION)	24.8	7.6	0.5	1.3	473.8	308	200	120	1	0.001	< 2	0	64.97
53	SONDI	24.8	7.4	0.2	1.5	452.3	294	200	120	1	0.001	< 2	0	65.17
54	GAIMUKH	24.7	7.9	0.2	1.5	621.5	404	280	240	32	0.046	< 2	3	53.41
55	JUNGAD	24.6	7.9	2.7	1.4	963.1	626	280	320	78	0.118	< 2	3.6	49.17
56	GHORAD	24.7	7.8	0.2	1.2	1511	982	440	416	152	0.214	< 2	11	46.51
57	SURGAON	24.7	7.8	1.7	1.2	1572	1022	480	496	104	0.119	< 2	9.1	47.89
58	SURGAON	24.7	7.6	2.4	1.4	863.1	561	280	240	16	0.001	< 2	0	57.48
59	NABABPUR	26.0	7.9	0.2	1.2	1445	939	240	404	1	0.001	< 2	0	64.56
60	REHAKI (KALA)	24.9	7.8	0.6	1.2	1360	884	160	280	1	0.001	< 2	0	64.66

Table 4: Groundwater analysis data for samples 61 – 90

S N	Village	Temp	PH	Turbidity NTU	DO mg/l	EC	TS mg/l	Alkalinity mg/l	TH mg/l	Nitrate mg/l	T.Phosp. mg/l	BOD mg/l	Fecal Colifo rms	WQI
61	PALASGAO N (BAI)	26.8	7.6	0.2	1.1	1720	1118	400	424	124	0.178	< 2	11	47.23
62	ASHTA	24.9	7.9	2.1	1.4	910.8	592	240	360	34	0.022	< 2	3	53.58
63	PAVNAR	25.0	8.2	4.2	0.2	3680	2392	720	552	284	0.296	< 2	15	43.27
64	ITALA	25.0	7.6	1.2	1.4	818.5	532	284	280	30	0.04	< 2	3	53.83
65	JANGLAPU R	25.2	7.9	2.1	1.1	1451	943	400	320	124	0.157	< 2	9.1	46.99
66	MAHABAL A	25.3	7.8	2.1	1.2	1372	892	400	300	104	0.146	< 2	7.3	47.75
67	SALAI (PEVATH)	26.0	7.6	1.2	1.5	800	520	240	320	39	0.019	< 2	3	54.29
68	GOHDA (KALA)	25.4	7.5	1.4	1.5	772.3	502	240	80	2	0.001	< 2	0	62.78
69	AKOLI	25.4	7.9	0.8	1.4	723.1	470	240	272	8	0.001	< 2	0	59.33
70	AMGAON (MADNI)	25.3	7.6	2.1	1.2	932.3	606	320	320	38	0.054	< 2	3	53.12
71	MADNI	26.0	7.8	0.2	1.3	723.1	470	280	280	8	0.001	< 2	0	59.64
72	MASALA	25.4	7.1	0.2	1.5	732.3	476	360	160	1	0.001	< 2	0	64.96
73	KANGAON 1	25.5	7.8	1.0	1.2	1155	751	320	200	29	0.046	< 2	3	53.39
74	KANGAON 2	25.3	7.4	0.2	1.3	1218	792	320	120	20	0.019	< 2	0	57.96
75	NANDGAO N	25.4	7.8	0.5	1.5	870.8	566	280	320	38	0.043	< 2	3	53.67
76	MADNI	25.4	7.9	1.7	1.3	812.3	528	240	200	24	0.054	< 2	3	52.75
77	MUDHAPU R	25.7	8.3	4.0	0.0	3849	2502	400	123 2	371	0.396	< 2	15	42.33
78	TIGAON	25.8	7.0	0.2	1.5	603.1	392	280	280	10	0.001	< 2	0	59
79	BARBADI	26.0	7.9	0.1	0.8	2443	1588	400	784	132	0.232	< 2	7.3	46.61

80	WADGAON KH.	25.6	7.4	0.5	1.4	1025	666	120	352	17	0.008	< 2	0	58.18
81	SALAI (KALA)	25.7	7.8	1.2	1.5	686.2	446	360	216	12	0.001	< 2	0	58.32
82	TAKALI (KITE)	26.0	7.6	0.1	1.0	1169	1769.2	480	432	84	0.143	< 2	7.2	48.5
83	KHAPRI	25.6	7.4	2.8	1.0	1354	880	360	428	131	0.167	< 2	9.1	47.25
84	KOLGAON	25.5	7.4	1.9	0.9	1335	868	380	352	44	0.082	< 2	3	51.73
85	JUNONA	25.4	7.4	0.2	0.8	1726	1122	360	352	222	0.281	< 2	14	45.73
86	WADGAON (KALA)	25.6	7.9	0.2	1.4	843.1	548	480	276	9	0.001	< 2	0	59.13
87	WAIFAD	25.7	7.6	0.9	1.5	716.9	466	280	248	32	0.073	< 2	3	52.47
88	PIMPALGAON	25.6	7.9	2.3	0.8	2169	1410	560	400	162	0.196	< 2	7.3	46.76
89	DIGRAJ	25.8	7.9	2.2	0.9	1954	1270	520	400	62	0.054	< 2	3.6	52.27
90	PARSODI	25.7	7.9	4.8	0.7	2129	1384	400	320	102	0.181	< 2	9.1	46.05

Table 5: Groundwater analysis data for samples 91 – 106

SN	Village	Temp	PH	Turbidity NTU	DO mg/l	EC	TS mg/l	Alkalinity mg/l	TH mg/l	Nitrate mg/l	T.Phosp. mg/l	BOD mg/l	Fecal Coliforms	WQI
91	PIPRA	25.8	8.3	3.4	1.1	1600	1040	400	320	58	0.136	< 2	3.6	47.66
92	DHONDGAON	26.0	7.6	4.8	1.2	1326	862	400	360	62	0.127	< 2	3.6	49.04
93	KHADKI (AMGAON)	25.7	8.2	0.9	1.0	1723	1120	480	640	168	0.204	< 2	11	45.52
94	BORI (KOKATE)	25.9	7.5	0.2	1.4	668	434	280	200	11	0.001	< 2	0	59.1
95	KELI	26.0	7.8	1.2	1.1	1126	732	400	440	65	0.104	< 2	3.6	49.89
96	DINDODA	25.7	7.9	0.2	1.2	1338	870	520	400	52	0.086	< 2	3	51.31

97	MHASALA	25.8	7.5	2.8	1.8	449.2	292	240	160	3	0.001	< 2	0	61.9
98	SAKHARA	26.0	7.2	1.3	1.3	698.5	454	280	240	34	0.044	< 2	3	53.69
99	SAWALI SAKHARA	25.6	7.9	4.0	1.4	646.2	420	320	400	21	0.017	< 2	0	56.53
100	BORKHED I	25.6	7.9	1.5	1.2	1148	746	360	360	105	0.136	< 2	9.1	47.46
101	BOTHALI	26.0	8.4	3.1	1.1	1218	792	320	400	161	0.211	< 2	11	44.36
102	ANTARGA ON	25.8	7.9	1.4	1.5	858.5	558	280	280	44	0.055	< 2	3	52.8
103	HIWARA	25.7	7.8	0.9	1.4	984.6	640	360	320	50	0.068	< 2	3	52.46
104	KAMTHI	25.8	7.8	1.4	1.2	1646	1070	440	640	188	0.261	< 2	15	45.2
105	ALAMDOH	25.6	7.6	0.2	0.9	2462	1600	520	320	182	0.224	< 2	14	46.19
106	EKURLI	25.6	7.6	3.0	1.1	1634	1062	440	440	174	0.198	< 2	15	45.81

6. Conclusion

The analyzed data for all one hundred and six places of collection in Wardha district show that a few parameters including nitrate, alkalinity and bacterial count are above the permissible limits. From the analysis, it is clear that the groundwater is contaminated and is not fit for drinking without treatment. The WQI calculated is below 50 for 41 water samples showing the deteriorated quality of water. While 20 samples falling in the 'Medium' quality range of WQI are very close to the lower limit of the range, i.e. 50.

References

[1] Malviya Niharika, Deo Sujata and Inam Farhin. 2011. Determination of Water Quality Index for drinking and agricultural purpose, *International journal of Basic and Applied Chemical Sciences (JCS)* **1(1)** 79-88.

[2] Tatawat, R K, Chandel, C P Singh, 2007, Quality of Groundwater of Jaipur–City, Rajasthan,(India)and its suitability for Domestic and Irrigation Purpose. *Applied Ecology and Environmental Research.* **6(2)** 79-88.

[3] Chavhan Sudarshan, Khan Arif, 2020. Analysis Of Total Dissolved Solid of Arvi (Wardha) Region. *International Journal Of Scientific & Technology Research* **9(3)** 156 – 159.

[4] Budhlani G N, Musaddiq M, 2014, Water Quality and Pollution Status of Wardha River (Kaundyanapur) from Amravati District, (MS) India. *International Journal of Researches in Biosciences, Agriculture and Technology* 75-81

[5] Rajankar P N, Wate S R, Tambekar D H, Gulhane S R, 2013, Assessment of Groundwater Quality using Water Quality Index (WQI) in Wardha District, Journal of Environmental Science and Sustainability **1(2)** 49 – 54

[6] Nagarnaik P B, Patil Pankaj N 2012, Analysis of Ground Water of Rural Areas of Wardha-City Using Physico – Chemical and Biological parameters. *International Journal of Engineering Research and Applications* **2(3)** 803-807

[7] Shivhare Niharika, Gour Anita, Pandya Anubha V and Trivedi Nair Rati, 2020, Water Quality Index Assessment of Ground Water of Indore City (MP) India, Ecology, Environment and Conservation, **26** 126-S130.

[8] Saanchez E, Colmenarejo M F, Vicente, J Rubio, Garci, M G Travieso, L Borja R 2006, Use of water quality index and dissolved oxygen deficit as simple indicators of water shed pollution. *Eco. Indic.***7** 315–328

[9] Bordalo, A A, Teixerra, R and Wiebe, W J 2006, A water quality index applied to an international shared river basin: the case of Douro river, *Environmental. Management* **38** 910-920

[10] Gaikwad Asha, Shivhare Niharika and J Shinde Dhira, 2018, Water Quality Index for Assessment of Raw Water at Different Cities in MP, India. *Global Journal of Engineering Science and Researches* 430-435.

[11] Krishan Gopal, Singh Surjeet, Kumar, C P, Gurjar Suman and Ghosh, N C 2016, Assessment of Water Quality Index (WQI) of Ground water in Rajkot District, Gujarat, India *Journal of Earth Science & Climatic Change.* **7(3)** 1-4.

[12] Bhardwaj Divya and Verma Neetu, 2017, Analysing impact of Various Parameters on Water Quality Index, *International Journal of Advanced Research in Computer Science.* **8(5)** 2496-2498.

[13] Phadatare Sneha S and Gawande Sagar 2016, Review Paper on Development of Water Quality Index, *International Journal of Engineering Research & Technology.* **5(5)** 765-767.

[14] APHA.2012. *Standard Methods for the Examination of Water and Waste Waters*, Washington, DC, American Public Health Association

[15] IS: 10500:2012, *Specifications for Drinking Water*, New Delhi, Bureau of Indian Standards

[16] Bassin, J K, 2007, An Automated Workbook for Checking Correctness of Water Analyses. *Journal of Indian Water Works Association* **39(4)** 259-264.

Footprints of Sedimentation on Loss of Reservoir Life using Satellite Remote Sensing Technique

Tulshidas M. Jibhakate[1, a], **Yashwant B. Katpatal**[2, b]

[1]Research Scholar, Civil Engineering Department, Visvesvaraya National Institute of Technology, Nagpur-440010, Maharashtra, India
[2]Professor, Civil Engineering Department, Visvesvaraya National Institute of Technology, Nagpur-440010, Maharashtra, India

Email: [a]tulshimj@gmail.com, [b]ybkatpatal@rediffmail.com

Abstract: Reservoir is the most important water storage systems especially in areas where groundwater potential is low which can be used for domestic, irrigation as well as industrial water supply. The sedimentation which is a natural phenomenon of settling sediment particles in a reservoir is observed to be enhanced with construction of dams due to change in flow dynamics. Sedimentation in a reservoir takes place in dead storage zone as well as in live storage zone causing reduction in useful capacity of reservoir over a period of time, thereby affecting the designed water supply planning. Sedimentation in live storage zone can be estimated using Remote Sensing Technique which will gives us synoptic view of a reservoir. Present study aimed at estimating a reduction in live storage capacity with the help of satellite imageries. For this, the Totladoh Reservoir which is situated on the boundary of Maharashtra and Madhya Pradesh, India has been selected for the study. Totladoh reservoir is situated in calc gneisses type of metamorphic rocks. Green and near-infrared bands were used to calculate the Normalized Difference Water Index (NDWI). The rate of siltation for Totladoh reservoir has been estimated for the period of 1982-2020 using LANDSAT 8 satellite imageries having 30 m spatial resolution. Percentage loss of live storage for year 2020 is found to be 17.63. Rate of Siltation in the reservoir is 1173 m^3/km^2/year which is more under prevailing conditions when compared to CWC values.

Keywords: Reservoir, Sedimentation, Live storage Capacity, LANDSAT 8, Satellite imageries, Totladoh Reservoir, Elevation Capacity Curve.

1. Introduction

Sedimentation is a natural phenomenon that takes place when high velocity of water is flowing through the catchment carrying the dissolved and suspended load which is ultimately deposited along its way to the reservoir; here maximum deposition occurs at the reservoir. Sedimentation in reservoirs takes place in both the dead and the live storage zone. It is essential to conduct the sedimentation surveys to determine the percentage of sediment deposition and the rate of sedimentation/siltation after a specific time interval, using the impounding year as a reference. Sedimentation in reservoir can be estimated and collate with the original designed data. Elevation capacity curve needs to be modified from time to time for the management of available water to users.

The life of the reservoir depends upon many factors but the one which is influencing more is Sedimentation. Designed life of a reservoir will reduce, if sedimentation rate is higher than the designed one. Proper management of sedimentation can make the reservoir renewable otherwise it will be exhaustible (Jibhakate et al. 2022). Earlier Sedimentation surveys were carried out by different methods i.e. hydrographic survey, inflow-Outflow method, mathematical models and empirical methods, etc. These conventional methods are cumbersome and time-consuming. Hydrographic surveys were often used for sediment deposition surveys, which were tedious, expensive, and less accurate when compared to remote sensing techniques (Foteh et al. 2018). The technology of satellite remote sensing (SRS) is now commonly employed for assessing sediment deposition (Jain et al. 2002; Katpatal et al. 2017).

The current study looks into the sedimentation of the Totladoh Reservoir in Maharashtra, and adjoining Madhya Pradesh, India using the spatial approach i.e. LANDSAT 8 satellite imageries have been used in satellite remote sensing technique for estimating the past 38 years of sedimentation. Totladoh reservoir shares its boundary with Maharashtra and Madhya Pradesh which makes it important for both states.

2. Study Area

Totladoh dam is a gravity dam at the Pench River close to Ramtek in Nagpur district within the state of Maharashtra and adjacent Madhya Pradesh in India. Figure 1 shows the index map of the research area.The reservoir's latitude and longitude are 21° 39′ 32.28″ N and 79° 13′ 55.59″ E. The reservoir was first impounded in 1982, Minimum drawdown level (M.D.D.L.) = 464 m, Full reservoir level (F.R.L.) = 490 m, Gross storage capacity (MCM-Million Cubic Metre) = 1241.11, Dead storage capacity (MCM) = 149.58, Live Storage Capacity (MCM) = 1091.53, Catchment Area= 4273 Km^2, Area at F.R.L. = 77.71 Km^2 (Source: India Water Resources Information System).

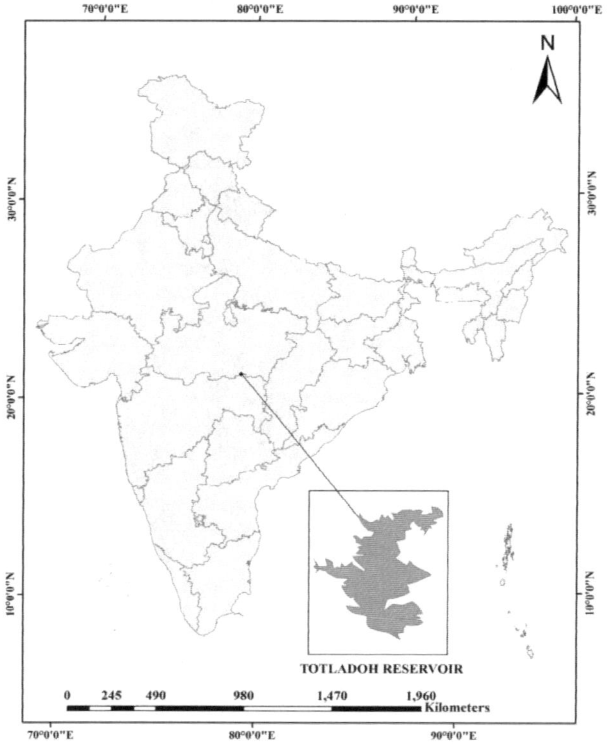

Figure 1. Index Map of Totladoh Reservoir, Maharashtra and Madhya Pradesh (INDIA)

3. Data Used

For assessment of sedimentation, variation of reservoir levels should cover the live storage zone which is from M.D.D.L. to F.R.L. (ideally) within a year. But due to non-availability of cloud free satellite imageries within a year, three years of satellite imageries (2018, 2019, and 2020) were considered for the present study to cover the maximum live storage zone.

Green and Near-Infrared bands are essential to quantify NDWI. To cover the maximum useful storage zone, cloud-free satellite imageries (LANDSAT 8) were acquired from earth explorer (USGS) for the years 2018, 2019, and 2020 by considering the minimum drawdown level as lowest level and full reservoir level as the highest level. Reservoir levels on the date of satellite passing were downloaded from India Water Resources Information System (IWRIS) for the same year mentioned above. All the specifications of the reservoir and dam were collected from the same source of IWRIS.

4. Methodology

The basic concept of sediment assessment is to find out the water spread area using spatial data for different water levels between M.D.D.L. to F.R.L. The difference between synoptic view of water spread area between current year and earlier years is the aerial extent of sedimentation or silting at these levels. While applying satellite remote sensing techniques to determine the storage capacity of any reservoir, a water spread area is necessary. The elevation capacity curves were drawn by obtaining the original reservoir elevation values w.r.t live storage values at the time of the first impoundment and revised one. NDWI was used for extracting the water spread area. Estimation of reservoir capacity at different elevations has been calculated using trapezoidal formula. Figure 2 illustrates a flowchart for assessing sediment deposition using satellite remote sensing.

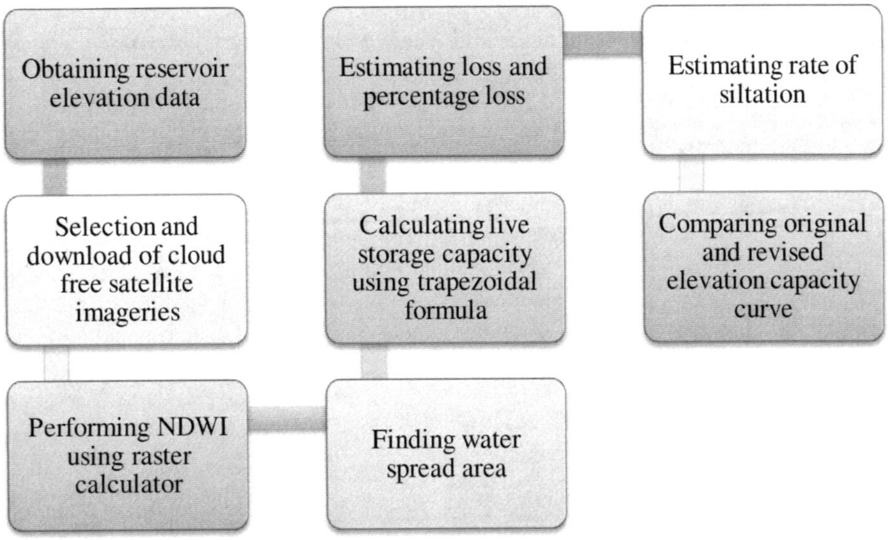

Figure 2. Flowchart of Assessment of Sedimentation using Satellite Remote Sensing Technique

Due to sedimentation, the water spread area shrinks at different altitudes inside the reservoir submergence region, reducing the reservoir's live and dead storage capacity. The live storage zone is the area between the minimum drawdown level (M.D.D.L.) and the full reservoir level (F.R.L.). Water spread area can be calculated using satellite imageries (cloud-free images) at various elevations by computing the pixel counts.

NDWI is widely used spectral index for water body mapping. NDWI value varies between -1 to 1. Zero is the threshold. As suggested by McFeeters (1996) if, NDWI values greater than zero represents water spread area and if NDWI is less than or equal to Zero, it covers Soil, vegetation, and other features (McFeeters 1996).

Water pixels that give water spread area are obtained using the normalized difference index (NDWI). The water spread area and elevation between two reservoir levels can be multiplied to get the live storage capacity for any height interval. NDWI formula is mentioned below.

$$\text{NDWI} = \frac{(Green\ band - NIR\ band)}{(Green\ band + NIR\ band)}$$

Water spread area (m^2) = No. of water pixels*each pixel area.
Spatial resolution of LANDSAT 8 satellite is 30 m. Hence, Each Pixel area is 30 m*30 m.
The following is the trapezoidal formula for calculating reservoir capacity between two successive water levels or reduced levels (R.L):

$$V = \frac{h}{3}\left[A1 + A2 + \sqrt{A1 * A2}\right]$$

Where, V is capacity/volume of a reservoir between corresponding intervals, $A1$ is water spread area at RL 1 and $A2$ is water spread area at RL 2, h is the difference between RL or height interval (Jain et al. 2002).

5. Results and Discussions

Water spread area can be determined using the raster calculator (i.e. NDWI) in ArcMap software by multiplying the number of water pixels with the spatial resolution, i.e. pixel size. By using the Trapezoidal formula, the revised live storage capacity of a reservoir was obtained between the maximum (489.85 m) and minimum (464.14 m) which are near to F.R.L. and M.D.D.L. respectively. All findings related to sediment assessment are shown in table 1.

Table 1. Observed data and Findings of sediment assessment

Date of satellite pass	Elevation of reservoir (m)	Pixel count (Water pixels)	Spatial Resolution (m)	Revised area (Km^2 or Mm^2)	Revised live capacity (Mm^3)	Cumulative revised live capacity (Mm^3)
16/05/2019	464.14	11072	30	9.96	0	0
14/04/2019	466.1	13947		12.55	22.01	22.01
24/01/2019	470	18642		16.78	56.99	79
20/10/2018	473.05	25809		23.23	60.75	139.75
04/10/2018	476.7	34186		30.77	98.23	237.98
18/05/2020	487.02	68346		61.51	467.1	705.08
25/10/2020	489.85	76852		69.17	184.8	889.88

The spreading of water area for different elevations is shown in Figure 3 from near Minimum drawdown level (16/05/2019) to near full reservoir level (25/10/2020). Figure 3 indicates a chronological increase in water spread area concerning reservoir elevations.

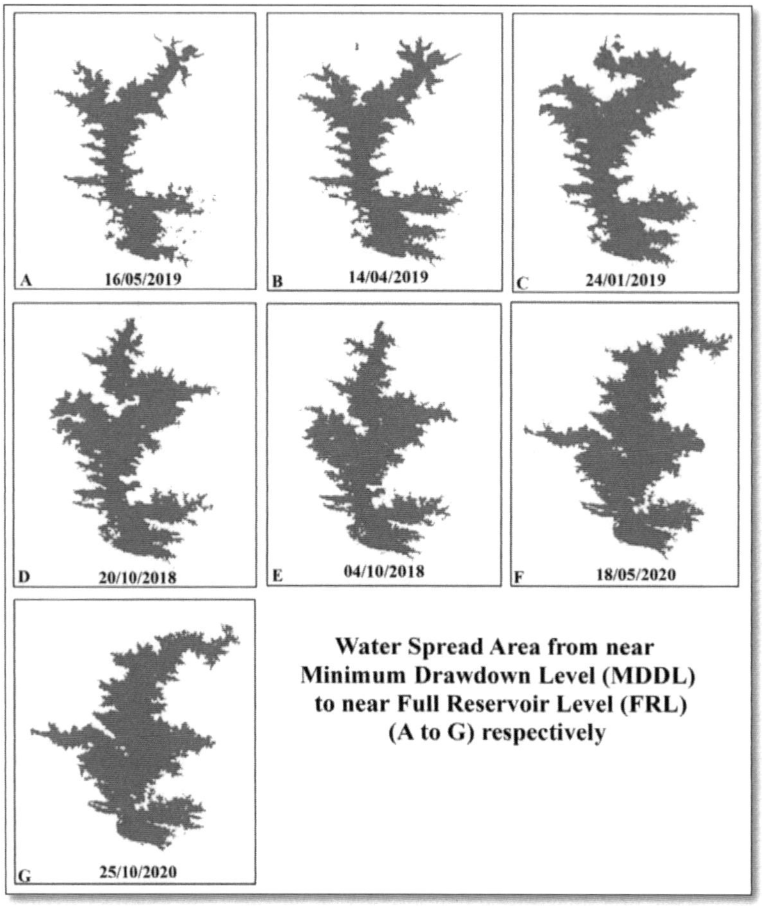

Figure 3. The water spread area at different reservoir elevations (i.e. From near M.D.D.L. to near F.R.L.)

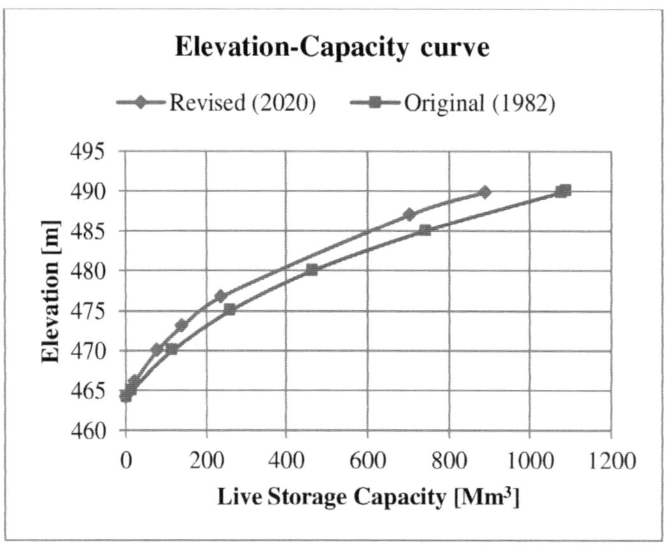

Figure 4. The Elevation Capacity Curve of Live Storage Capacity for Totladoh Reservoir

The graph indicates that the curve has shifted leftward at the same reservoir elevation as the original pattern of reservoir, which was first impounded in 1982, and the revised pattern of elevation capacity curve for the year 2020 (Figure 4).

Cumulative revised live storage capacity= 889.88 Mm^3
Designed original live storage capacity= 1080.32 Mm^3
Loss of live storage capacity= (1080.32 -889.88 /1080.32)*100 = 17.63 %

$$\text{Rate of siltation} = \frac{Cumulative\ loss\ in\ live\ storage\ capacity}{Catchment\ area * year\ of\ last\ survey}$$

$$= \frac{190.44}{4273*38} = 1.173\ \text{Th.m}^3/\text{km}^2/\text{year}$$

The revised live storage capacity of Totladoh Reservoir for the year 2020 is anticipated to be 889.88 Mm^3 between near F.R.L. at R.L. 489.85 m and near MDDL at R.L. 464.14 m, compared to a designed capacity of 1080.32 Mm^3 and a loss of 190.44 Mm^3 in 38 years. In the live storage zone, the rate of siltation is estimated to be 1.173 Th.m^3/km^2/year.

In the below table, the current study covers a live storage zone between 464.14 m (near M.D.D.L.) and 489.85 m (near F.R.L.). The current study has been compared with the CWC survey which indicates that the rate of siltation per year increases from 0.632 to 1.173 Th.m^3/km^2/year.

Table 2. Comparison of results of previous survey with present study

Reservoir Name	Totladoh	Totladoh
Year of Survey	**2009**	**2020**
Agency Name	CWC	Present Study
Gross Storage (MCM)	1241.11	1241.11
Live Storage-Original (MCM)	1091.53	1080.32
Dead Storage Capacity (MCM)	149.58	149.58
Present Live Capacity (MCM)	1018.58	889.88
Total Loss of Live storage up to last survey	72.95	190.44
% Loss of Live storage up to last survey	**6.68**	**17.63**
Average Observed rate of siltation (Th.m^3/km^2/year)	**0.632**	**1.173**

6. Conclusion

Till the completion of life of reservoir, the dead storage zone is supposed to be fully filled with sediment particles. Live storage capacity of a reservoir is reduced with time, due to improper management in handling the issue of sedimentation. As the live storage capacity of the reservoir is reduced, reservoir life is drastically reduced which ultimately raises the question of unbalanced water management for the command area. The percentage decline in live storage capacity of Totladoh reservoir in the current study region is significant when compared to the previous study. In a 38-year period, the percentage loss of live storage for the year 2020 is 17.63. The rate of siltation was found to be 1173 m^3/km^2/year, which is higher than CWC values under prevailing circumstances (CWC Report, 2020). The Totladoh reservoir shares its boundary with Maharashtra and Madhya Pradesh so it is necessary to estimate present live storage capacity to resolve the water dispute. Study proves that sedimentation studies using satellite data are effective.

7. References

[1] Ashish Pandey, U.C. Chaube, S.K. Mishra & Dheeraj Kumar 2016 Assessment of reservoir sedimentation using remote sensing and recommendations for desilting Patratu Reservoir, India, *Hydrological Sciences Journal*, 61:4, 711-718, DOI: 10.1080/02626667.2014.993988.

[2] CWC Report 2020 "Compendium on sedimentation of reservoirs in India 2020" by Watershed & Reservoir Sedimentation Directorate Environment Management Organisation Water Planning and Projects Wing Central Water Commission

[3] Foteh, R., Garg, V., Nikam, B.R. et al. 2018 Reservoir Sedimentation Assessment through Remote Sensing and Hydrological Modelling. *J Indian Soc Remote Sens* 46, 1893–1905 https://doi.org/10.1007/s12524-018-0843-6

[4] Jain, S.K., Singh, P., and Seth, S.M., 2002 Assessment of sedimentation in Bhakra Reservoir in the western Himalayan region using remotely sensed data. *Hydrological Sciences Journal*, 47 (2), 203–212. doi:10.1080/02626660209492924 [Taylor & Francis Online], [Web of Science ®], [Google Scholar]

[5] Jibhakate T., Mirajkar A. 2022 Reduction in Live Storage Capacity of a Reservoir Due to Sedimentation Using Satellite Imageries. In: Laishram B., Tawalare A. (eds) *Recent Advancements in Civil Engineering. Lecture Notes in Civil Engineering*, vol 172. Springer, Singapore. https://doi.org/10.1007/978-981-16-4396-5_76

[6] Katpatal Yashwant B, Patil Sanjay A, Singh Chandan Kumar 2017 Estimation of sediment yield within mining watershed to assess impact of mine dumps using satellite data: a modified approach, *J. Env Eng*. 10.1061/(ASCE)EE.1943-7870.0001261.

[7] McFeeters, S.K., 1996. The use of the Normalized Difference Water Index (NDWI) in the delineation of open water features. *International Journal of Remote Sensing,* 17 (7), 1425–1432. Doi:10.1080/01431169608948714

Estimation of shape factor for irregular particles using three-axial measurement approach

Mahendra Umare[1,a], **Swati Ambadkar**[2,b]
[1]Research Scholar, G H Raisoni University, Amravati, India
[2]Assistant Professor, G H Raisoni University, Amravati, India

Email: [a]mahendraumare72@gmail.com; [b]swati.ambadkar@ghru.edu.in

Abstract. Particle shape is an imperative term in civil engineering applications that play a significant role in the overall behavior of the particles; however, the shape factor becomes complex when it is related to the irregular-sized particle. The manuscript focuses on quantitative estimation of the shape factor of highly irregular metal particles of diameters ranging from 2.00 mm to 5.00 mm using a three-axial microscopic measurement. The measured data is used to compute the nominal diameter of a particle representing a circle of equivalent diameter and shape factor is computed. The result has been compared with the previously established studies and found to be corroborative.

Keywords: Shape factor, nominal diameter, irregularity.

1. Introduction

In geology, interest in particle shape emerged earlier than in geotechnical engineering. The particle shape is attained by its transportation from the original position to the deposits or during the machining process required in the metal industries.

There are also considerations for the process of particle genesis itself (rock structure, mineralogy, hardness, etc.). Many factors have been considered to define particle shape to classify and compare grains (axis lengths, perimeter, surface area, volume, etc.) to specify a particle shape and put forth empirical equations to substantiate the same. On that line [1] are endorsing form, roundness, and surface texture to describe the shape of a particle. Over the years, several attempts have been made to develop a methodology for measuring particle shape. Besides, other techniques of characterising a particle shape, a manual method including chart comparison [2] [3], also sieving [4], and, more recently, three-axial measurements using microscope were adopted in an industry with good results of characterizing a particle shape. Furthermore, using the computer-aided approach to measure particle shape saves a significant amount of effort [4].

1.1 The Objective

The shape of a particle can be described qualitatively or quantitatively. The qualitative description of the particle (e.g., elongated, spherical, flaky, etc.) is expressed in words, whereas the quantitative description relates to the measured dimensions; the quantitative description is more important in the engineering field due to reproducibility.

Particle quantitative geometrical measurements can be used to support qualitative classification. To describe the particle form, a few qualitative measures and several quantitative measures can be used.

Despite the abundance of qualitative descriptions, none were widely accepted. To analyse particle dimensions and shape factors, microscopic measurements are required.

In myriad applications, particle shape is crucial to its behaviour. This property is finding its place in the various fields of engineering such as hydraulics, cross-drainage works, transportation, mining, etc. due to its importance of changing behaviour in different applications.

The global form, major surface feature scale, and surface roughness scale are used to determine particle shape. Each scale reflects aspects of the particle's formation history and contributes to the overall behaviour of the particle, from particle movement to mechanical response.

2. Background

A particle's shape is captured using three independent relative scales [5] form, which describes differences in particle proportions

Roundness refers to the variations at the particle's corners that are superimposed on it. The characteristics that are superimposed on both corners and surfaces are referred to as roughness. Because these descriptors are distinct, one can differ significantly without affecting the others. The Fourier method, as well as fractal analysis if the shape is self-similar, can be used to characterise it.

2.1 Form, Roundness and Roughness

Form

The shape of a particle can be described using terms such as cubical, spherical, elliptical, elongated, flat, tubular, platy, lathlike, and needle. The form can be quantified using the length ratios of the three orthogonal axes. Barrett (1980) provides a list of at least 15 parameters that can be defined using these ratios. The aspect ratio is the ratio of the long axis L to the intermediate axis B. It is also known as the elongation. The ratio of the intermediate, B, and short axes is another term for flatness.

Two more mathematical descriptors of the form are sphericity and eccentricity. Sphericity is defined by [6] as the ratio of particle volume to circumscribing sphere volume. This definition is flawed because it includes a measure of roundness. The ratio of the particle's surface area to the surface area of an equal volume sphere is an improved definition of sphericity [7]. Eccentricity is defined as the 8p/Rp ratio of an elliptical particle, whose two-dimensional outline has been expressed as Rp = 8p.cos (20).

Roundness

The radius of curvature of each corner is averaged and compared to the radius of the particle Wadell's maximum inscribed circle to determine roundness [6]. The procedure is two-dimensional, but it can be made three-dimensional by replacing circles with spheres. Due to the difficulty in determining what constitutes a corner, subjectivity enters the test.

Roughness

Scale considerations are critical in the characterization of particle roughness. Because all surfaces are rough at some scale, roughness must be characterized at the scale deemed relevant to the problem at hand [7].

Krumbein's chart is used in this study to determine sphericity and roundness (1963). The procedure is as follows: A pinch of sand is placed on a Petri dish and examined under a Leica MZ6 stereomicroscope. Particles are indirectly illuminated by light reflected from a reflective shield. Approximately 30 grains are studied at various magnifications. The chart is then used to calculate representative grain sphericity and roundness finds a 10% variation in the roundness of particles from the same vial of sand when multiple students use the Powers chart.

The roughness of the particles subjected to shear wave velocity testing is also determined. The roughest sample particles are arbitrarily assigned a roughness value of 3, the smoothest sample particles a value of 1, and all others are assigned an intermediate value based on their relative roughness [6][3][2][5] depicts particle shape irregularity on three different scales:

Sphericity S is preferred over ellipticity or flatness, roundness R over angularity, and smoothness over roughness. The sphericity is computed based on a diametric ratio of the largest inscribing sphere and smallest circumscribing sphere. On the other hand, roundness is computed by comparing a curvature radius with a radius of the biggest sphere inscribing a particle under consideration. Surface features that are much smaller than particle diameter are referred to as roughness.

Sphericity and roundness can be estimated visually using charts like the one shown suggested by Folk 1955, and Barrett 1980. However, advanced techniques such as digital image analysis can also be applied to characterize the particle of irregular shape and size [8].

Because rough surfaces are fractal, they lack a characteristic scale, making the direct measurement of roughness difficult. As a result, the relevant roughness observation length is transformed into the inter-particle contact area: this is a particle's connection to its neighbor.

The shape parameters can also be derived from the soil mass's macro-scale behavior. The fall of particles or flow of particles in various applications is affected due to its shape, which results in complex drag force phenomena.

3. Methodology, Data Measurement and Estimation of Shape Factor

In hand measurement technique, sliding rod caliper, instruments were used to obtain the accurate data of particle geometry [3]. A particle under consideration needs to adjust on the sliding rod caliper to get the length accurately using a graduated scale attached to the instrument. Similarly, a convexity gauge is used to measure the curvature of the particle. To arrive at the shape of rock particles, an instrument was used and concluded that the tool gives good attribution [9]. The results so obtained was further reviewed and analyzed by many researchers.

3.1 Microscopic Measurements

Irregular mineral particles can be conveniently measured under a microscope. The accuracy further can be improved by attaching a video camera linked to a computer. The most obvious method is to compute the arithmetic or geometric mean of the number of measurements taken as well as the average distance between two cross-hairs. The third dimension (Z-axis) is obtained using a screw gauge. The extremities of the measured distance are standardized and used to calculate the mean cord length of the particle defined by Martine's or Ferret's diameter. Martine's diameter is the length of the line that divides the practical image in half. The dividing line is drawn in a parallel fixed direction regardless of the practical orientation. The mean distance between two tangents on opposite sides of the apparent outline of the particle is defined as Ferret's diameter. And the arithmetic or geometric mean value of length computed using both the methods is equal to the diameter of the reference circle; where arithmetic mean diameter is given as:

$$\left(\frac{D_{max}+D_{min}}{2}\right) = D_{am} \tag{1}$$

where,

Dmax and Dmin are the mean diameters of several Martine or Ferret measurements, and Dam is the arithmetic mean diameter.

3.2 Computation of Shape Factor

The coefficient of Drag and Reynolds Number can be very well related for particles of regular shape. However, it becomes difficult for irregularly shaped particles as the drag becomes complex. The empirical relationships have been developed for regularly shaped particles by many researchers to calculate the drag on it when it falls in a liquid media but to develop a correlation between Cd and Re for irregularly shaped particles, more precisely minerals between particles, shape factor is an essential parameter the can truly define an irregular particle in an equivalent sphere. S.F. is a shape factor that appears to be as satisfactory as any other.

$$S.F. = \frac{(D_3)}{\sqrt{(D_1 * D_2)}} \tag{2}$$

where D_1 is the longest axis, D_2 is the intermediate axis, and D_3 is the shortest axis of the three mutually perpendicular axes.

The shape factor considers three of its axial dimension of irregular particles and estimates the particle shape. Particles with the same shape factor can have rounded angular, rough, or smooth shape factors. This study is related to the particle of natural grain and some of the minerals having irregular shape and size. Shape factors based on particle roundness, sphericity, or other physical properties could be used, but they would not be adequate for hydraulic studies.

The set of irregular particles have been taken and three-axial measurements are noted using a microscope (major and intermediate axes) and screw gauge (thickness, as minor axis). Using equation 2, the shape factor is estimated as shown in Table 1. The plot of average diameter against the shape factor is shown in Figure 1

Table1. Microscopic measurement and shape factor

SN	Major axis(X) (D_1)	Minor axis(Y) (D_2)	Perpendicular axis(Z) (D_3)	Shape factor
1	4.039	1.941	1.512	0.540
2	3.014	3.821	1.999	0.589
3	3.387	2.953	2.063	0.652
4	4.050	1.995	1.302	0.458
5	3.958	2.653	1.401	0.432
6	2.899	1.941	1.512	0.432
7	2.885	3.821	1.999	0.720
8	2.667	2.953	2.066	0.435
9	3.041	1.995	1.353	0.341
10	1.933	2.653	1.398	0.521
11	2.913	0.863	1.533	0.966
12	2.978	1.922	1.403	0.586
13	2.948	2.968	1.050	0.354
14	2.933	1.907	0.386	0.163
15	2.957	1.907	1.013	0.426
16	2.968	2.040	1.050	0.426
17	3.972	1.927	0.090	0.020
18	3.910	2.000	0.090	0.032
19	3.095	1.957	1.060	0.430
20	3.040	2.923	1.030	0.345
21	1.316	1.497	1.030	0.733
22	2.649	1.018	0.866	0.527
23	1.360	1.525	1.020	0.708
24	2.637	1.057	0.866	0.518
25	2.948	0.998	0.733	0.427
26	2.004	2.000	1.026	0.512
27	2.974	1.800	0.866	0.357
28	2.005	1.961	1.033	0.520
29	2.036	2.010	1.026	0.506
30	2.009	1.970	1.013	0.512
31	2.003	2.000	1.026	0.512

32	2.001	1.977	0.866	0.372
33	2.331	1.962	1.033	0.436
34	1.092	2.014	1.026	0.966
35	1.919	1.943	1.013	0.511

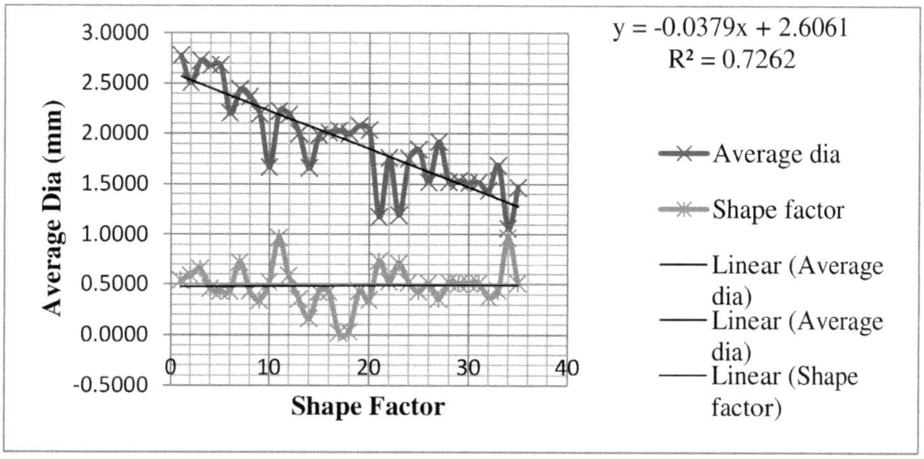

Figure1. Average diameter Vs. shape factor

4. Discussion and Conclusions

Based on the above microscopic measurements, experimental analysis, and graphical presentations, it is concluded that the shape factor's value varies with change in the perpendicular direction of the particle (Z-axis).

The graph depicts how the shape factor changes in the shorter direction. As the shape factor is directly proportional to the shortest axis.

The graph shows that the shape factor decreases as the minor axis value increases.

Microscopic analysis is objective, repeatable, produces quick results, and works with more data, but it still requires improvement to avoid the edition process being poorly- contrast particles.

Although there are numerous methods for defining shape factors and describing quantities, the measurement-based breakdown is quite practical and useful.

When performing microscopic analysis, the resolution must be considered because the effects can be significant. The resolution must be based on the requirements. The R-value reflects the effect of resolution on diameters as a perimeter.

References

[1] Seracettin Arasan, A.Samet Hasiloglu, Suat Akbulut, 2010 Shape Properties of Natural and Crushed Aggregate using Image Analysis, *International Journal of Civil and Structural Engineering,* **1**, 221-233

[2] Jacek Bogusław Szma´nda and KarolWitkowski, 2021, Morphometric Parameters of Krumbein Grain Shape Charts— A Critical Approach in Light of the Automatic Grain Shape Image Analysis, *Minerals, MDPI publisher*, **11**, 1-16

[3] W C Krumbein, 1941, Measurement and Geological Significance of Shape and Roundness of Sedimentary Particlesw. C. Krumbein', J. of Semmentary Petrology, 11, 64-72.

[4] Juan M. Rodriguez, Tommy Edeskär, and Sven Knutsson, 2013, Particle Shape Quantities and Measurement Techniques–A Review, *EJGE,* **18**, 169-198

[5] P. J. Barrett, 1980, The shape of rock particles, a criticaI review *Sedirnentolopy,* **27**, 291-303

[6] Hakon Wadell, 1932, Volume, Shape, and Roundness of Rock Particles , *Journal of Geology*,

University of Chicago Press, **40,** 443-451

[7] J. Carlos Santamarina, Katherinem Moheb A, 2001, *Soils and Waves: Particulate Materials Behavior, Characterization and Process Monitoring, Soil and Waves*, University of Toronto, Canada, Wiley.

[8] Gye-Chun Cho, Jake Dodds, and J. Carlos Santamarina, 2006, Particle Shape Effects on Packing Density, Stiffness, and Strength: Natural and Crushed Sands, *J. of Geotechnical and Geoenvironmental Engineering, ASCE*, **132**, 591-602

[9] Éva Farics, Dávid Farics, József Kovács, and János Haas, 2017, Interpretation of Sedimentological Processes of Coarse-grained Deposits applying a novel combined cluster and discriminant analysis, *Open Geosci, DeGruyter publication*, **9**, 525-538

Hydro-geochemical elucidation and its implications in the Wardha valley coalfields of central India

P S Ganvir[1*] **and Rajeeva Guhey**[2]
[1]Assistant Professor, Department of Geology, M. G. Arts, Science and Late N. P. Commerce College Armori, Maharashtra, India.

[2]Professor, Department of Geology, Govt. N. P. G. College of Science, Raipur, Chhattisgarh, India.

*priyadarshan.ganvir@gmail.com

Abstract. The performed endeavour is to elucidate the hydro-geochemistry of the groundwater in the Wardha valley coalfield of central India. Being a coalfield of well-appreciated coal exploiting region, many studies has also disclosed the prevailing contamination issues in the same. The study is motivated to understand the groundwater chemistry and its possible implications in contamination prevalence. The measured hydro-geochemical parameters from 45 groundwater samples have manifested the rock-water interaction as a controlling mechanism with the dominance of CaMgCl facies to the extent of 67 % to 81 % in pre- and post-monsoon respectively. The hydro-geochemical facies analysis specified the ascendency of strong acid (SO_4^{2-}-Cl^-- NO_3^-) over weak acids (HCO_3^-). Such domination not only lowers the pH of the groundwater but also provide an encouraging hydro-geochemical situation for contaminations. To mitigate the contaminations like of heavy metals in and around coalfields, the pre-deliberations of such hydro-geochemical studies should be much anticipated and the present study is in agreement to it.

1. Introduction

The key role of the water for the existence of the life is indisputable and its sufficient quality supply must be accessible to all [1]. The groundwater quality is one of the furthermost vital concerns for human habitations and the standards are established accordingly by various agencies depending upon situatedness [2]. The usual chemical makeup of the groundwater is driven by the geochemistry of the solid portion of the earth and the hydrochemistry of the hydrosphere [3]. In contemporary period the human interference has affected many groundwater regimes; especially in and around coalfields, where the groundwater quality and quantity alters simultaneously. The quality of groundwater has always been in limelight around the coalfields [4]. Numerous endeavours have ensured that the contaminants including heavy metals from the coal mines are potential to get released into the adjacent environment by processes like leaching [5]. Such contaminants may affect the groundwater regime for prolong period of time; say for example the heavy metals are stubborn pollutants in the groundwater and cannot be mitigated with ease. Though the pollutants like trace metals are largely immobile, their mobility under definite chemical setup exceeds the conventional rates and became a menace [6].

The endeavoured study in the present article deals with the elucidations of the hydro-geochemistry of the groundwater to comprehend the chemical setup augmenting the mobility of the contaminants like trace metals in the groundwater regime of the Wardha Valley coalfields. The interpretations made will not only descript the hydro-geochemistry but will also disclose the factors amplifying the long known groundwater contamination issues of the study area.

1.1. Study Area and Geology

The study area of 894.35 km^2 can be traced in the two district of Maharashtra viz., Chandrapur and Yavatmal, bounded within the latitude 19^0 50' 00" to 20^0 10' 00'' N and longitude 78^0 55' 00" to 79^0 25' 00" E and is traceable on toposheet no. 55L/16, 55 P/4, 55 P/8, 56 I/13, 56 M/1 and 56 M/5 (Figure 1). The study area is under uninterrupted quarrying for coal since many decades [7]. The landscape of the study area is somewhat uniform in elevation with very few raised topographies around mining region (especially anthropogenic) and along margin. The study area is a part of NW-SE trending Pranhita-Godavari sub-basin of the Gondwana basin located on the eastern limb of anticline plunging towards NNW [8]. The study area is fairly dominated by the rocks of Gondwana supergroup with Archean as the basement rocks (Table 1). The rocks of the Kamthi formation consisting ferruginous sandstone and variegated shale makes the top layer with the rocks of Talchir as the lowest layer sandwiching the grey-white sandstone of the Barakar formation. The only coal seam also referred as the Principal seam of the study area is hosted by the Barakar Formation [9]. Apart from the Gondwana rocks, few insignificant patches of neoproterozoic sediments are also observable in patches (Figure 1).

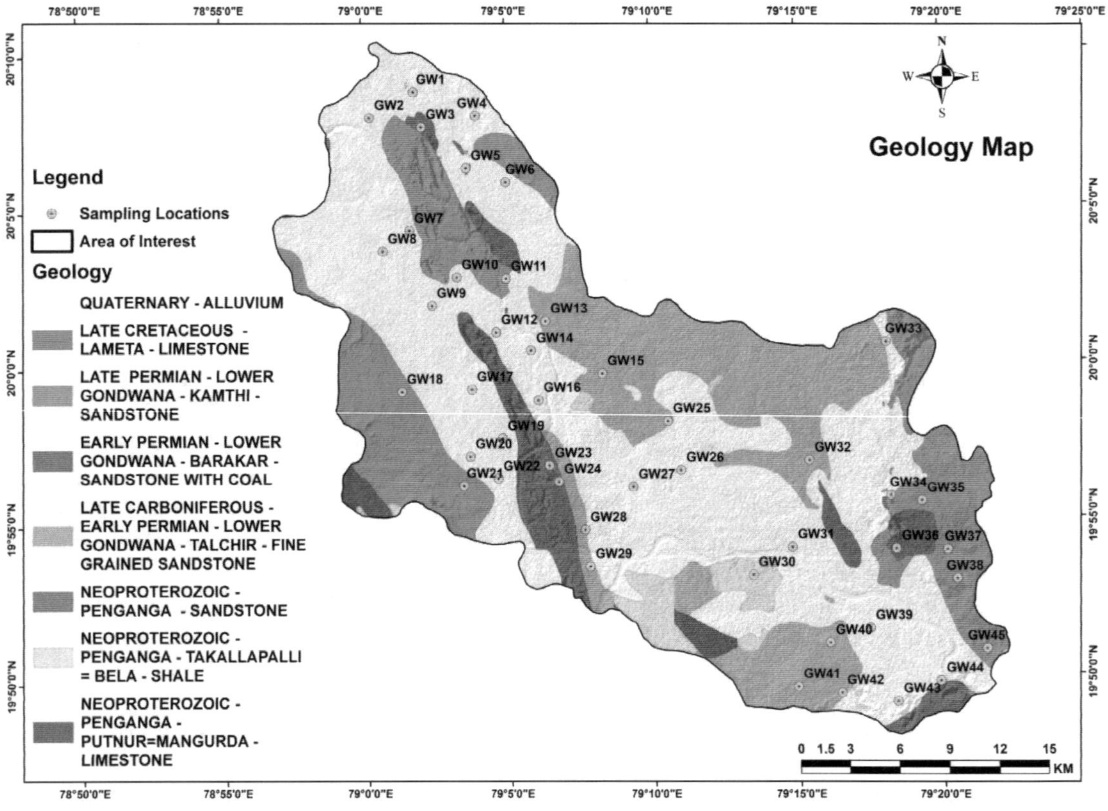

Figure 1. Geology map of the study area with groundwater sample locations [4].

Table 1. General Geological succession from Wardha Valley Coalfield, Maharashtra, [9].

Age	Group/Formation	Lithology
Recent	-	Alluvial gravel bed, black cotton soil
Eocene	Deccan Trap	Basalts
	-------------Unconformity-----------	
Cretaceous	Lameta Formation	Limestone, charts and silicified sandstone
	-------------Unconformity-----------	
Late Triassic	Maleri Formation (To South East)	Fine to med. Grained sandstone and red shale
Late Permian-Early Triassic	Kamthi Formation	Red, brown sandstone, reddish siltstone and variegated shale & sandstone
	-------------Unconformity-----------	
Early Permian	Barakar Formation	Light grey to white sandstone, shale & coal seam
Late Carboniferous-Early Permian	Talchir Formation	Tillites, turbidites, varves, needles shale & sandstone
	-------------Unconformity-----------	
Precambrian	Sullavai Sandstone	White and light brown quartzitic sandstone
	-------------Overlap-----------	
	Pakhal Limestone	Grey, bluish or pinkish limestone and chert
	-------------Unconformity-----------	
Archean		Quartzites, Granite gneisses, etc.

2. Methodology

In all, 45 groundwater samples were collected from pre-determined locations of the study area for both the seasons of 2020. The sampling locations were so fixed that to attain uniformity and appropriate representation to the manipulating factors like mining area, residential area and agricultural land. The samples were collected in agreement with the procedures suggested by APHA [10]. A pre-washed narrow mouth polyethylene container rinsed with the sample water was used to sample the groundwater. Physical suitability of the sampled groundwater was preferred to avoid any situation of filtration, which occasionally contributes contamination. To measure the fluctuating attributes like pH EC and TDS, spot analysis was performed by digital meter. The major ions in the sampled groundwater were analysed in the laboratory by the well-established procedures. The Ca^{2+} and Mg^{2+} are measured by respective titration methods; whereas, the Na^{+} and K^{+} are done by flame emission photometric method. Among anions, the HCO_3^- and Cl^- are measured by titrimetric method; while the SO_4^{2-} and NO_3^- are done by UV- spectrophotometric method. Once the results are deduced, the hydro-geochemistry was elucidated by the facies analysis through piper plots. The cause attributing the hydro-geochemistry was elucidated by Gibbs plots.

3. Result and Discussion

The assessment of the 45 groundwater samples collected from the study area has revealed the particulars of local hydro-geochemistry and has assisted in the corresponding elucidations. The data for the two succeeding seasons (pre-monsoon and post-monsoon) of 2020 was assessed attain the aspired precision. The attempt to understand the hydro-geochemistry not only revealed the chemical nature of the groundwater but also reflected the effect of the host rocks over it.

3.1. Hydro-geochemical characterization of the groundwater

On equating to the post-monsoon samples, the pre-monsoons are certainly intense in concern to pH, TDS and anionic-cationic concentrations. The pH values of the samples are reasonably evident. The EC measurements has characterized that the 49 % of pre-monsoon observations are intense and lies in the range of 1500 μS/cm to 3000 μS/cm (not permissible); whereas, the post-monsoon observations are mild and only 09 % lies in the cited range (Table 2 and 3). The enhanced EC values can be attributed to the dissolved salts, which are quite obvious in pre-monsoons. As the EC somehow reflects the TDS; hence, the bulk of the observations range between the prescribed acceptable (500 mg/l) and permissible limit (2000 mg/l) (Table 2 and 3) [11].

The Ca^{2+} was observed to be the foremost cation among both the seasons (Figure 2) and its abundance in the observations is surpassing the recommended permissible limit of 200 mg/l in pre-monsoons by 07 % of total sample share; while, around 80 % of the them were between the recommended acceptable limit of 75 mg/l and permissible limit of 200 mg/l (Table 2a) [11]. In post-monsoon, the value of 80 % is reduced to 60 % (Table 3). The Mg^{2+} concentration in around 02 % observations of the pre-monsoons was found to be exceeding the recommended permissible limit of 100 mg/l. Whereas, in an average 87 % of the observations among them were between the recommended acceptable limit (30 mg/l) and permissible limit (100 mg/l) (Table 2) [11]. Similarly, around 30% of the observations were between the acceptable and permissible limit in the post-monsoon with no sample surpassing the 100 mg/l value (Table 3). The Na^+ is the second abundant cation in the groundwater of the study area (Figure 2a and 3a). The average concentration of Na^+ is 104.16 mg/l and 76.95 mg/l in pre- and post-monsoon respectively (Table 2 and 3). The concentrations of Na^+ range widely within inter-seasons. The K^+ existence can be accounted to the mineral like orthoclase, microcline and biotite from the lower Gondwana sediments. The concentrations of K^+ were reasonably much lower than the Na^+ (Table 2 and 3).

The HCO_3^- is the foremost factor which controls the alkalinity in the groundwater regime, especially when the pH is low. The observation reveals the HCO_3^- as the only anion accountable for the alkalinity in the groundwater of the study area with an average concentration of 295.57 mg/l and 217.86 mg/l in pre-and post-monsoon respectively (Table 2 and 3). The limestone beds of the Lameta formation in the study area can certainly be the source of the HCO_3^- in the groundwater. The average concentration of the Cl^- ranges from 122.1 mg/l to 74.2 mg/l for the pre-monsoon and post-monsoon seasons (Table 2 and 3). The concentration of the Cl^- was found to be well within the permissible limit of 1000 mg/l for all observations; whereas, around 4 % of the observations in the pre-monsoon were above the recommended acceptable limit of 250 mg/l [11]. The SO_4^{2-} is the principal anion in the observed groundwater samples especially in pre-monsoon season (Figure 2b). Around, 44 % and 04 % of the samples were above the permissible limit of 400 mg/l in pre- and post-monsoon respectively (Table 2 and 3). The NO_3^- is the least anionic contributor in the observed groundwater samples of both the seasons with mere an average concentration of 15.36 mg/l and 15.11 mg/l in pre- and post-monsoon respectively (Table 2 and 3).

Table 2: The results obtained for hydro-geochemical parameters from pre-monsoon groundwater samples.

Sample	Locations		pH	EC	TDS	Ca^{2+}	Mg^{2+}	Na^{+}	K^{+}	HCO_3^-	Cl^-	SO_4^{2-}	NO_3^-
GW1	20°08'46"N	79°01'48"E	6.9	1071	685	104	25	72	7.91	255	102	98	24.22
GW2	20°07'58"N	79°00'16"E	7.2	1240	794	141	29	62	5.77	215	132	162	3.12
GW3	20°07'39"N	79°02'04"E	6.5	1681	1076	141	64	82	9.12	291	81	410	12.08
GW4	20°07'59"N	79°03'58"E	6.7	1276	817	92	46	91	10.22	301	88	156	46.29
GW5	20°06'20"N	79°03'38"E	7.1	998	639	64	41	59	7.59	212	82	155	20.12
GW6	20°05'52"N	79°05'00"E	6.9	1177	753	102	39	64	5.33	229	159	115	22.89
GW7	20°04'21"N	79°01'37"E	6.4	1498	959	132	51	70	6.99	231	24	489	4.12
GW8	20°03'42"N	79°00'40"E	6.7	1079	691	98	42	51	6.42	270	110	102	37.09
GW9	20°01'57"N	79°02'23"E	6.1	2278	1458	141	88	160	21	278	290	444	7.33
GW10	20°02'51"N	79°03'16"E	6.2	1655	1059	132	81	70	10.22	217	33	562	5.22
GW11	20°02'47"N	79°04'59"E	6.1	1777	1137	163	66	81	7.71	266	31	555	7.88
GW12	20°01'05"N	79°04'37"E	5.9	2280	1459	151	86	173	16.07	322	217	428	5.44
GW13	20°01'25"N	79°06'20"E	6.6	1604	1027	91	66	119	10.1	329	101	271	33.77
GW14	20°00'29"N	79°05'50"E	5.9	2317	1481	194	102	98	8.23	298	189	504	21.62
GW15	19°59'43"N	79°08'17"E	6.1	1232	788	125	38	61	5.56	241	132	200	12.32
GW16	19°58'54"N	79°06'04"E	5.9	2602	1665	228	96	142	12.47	335	262	492	12.03
GW17	19°59'16"N	79°03'45"E	6	2174	1391	173	66	158	13.8	294	177	527	3.03
GW18	19°59'13"N	79°01'17"E	7.2	1343	860	127	41	77	8.32	274	121	150	12.42
GW19	19°57'42"N	79°04'48"E	5.9	2110	1350	170	66	142	13.07	351	139	490	6.02
GW20	19°57'08"N	79°03'40"E	6.4	1305	835	93	37	113	8.3	277	144	162	14.12
GW21	19°56'12"N	79°03'26"E	6.9	1176	753	93	42	70	5.12	341	76	122	8.6
GW22	19°56'25"N	79°04'38"E	5.7	2503	1602	222	91	131	16.41	369	199	561	11
GW23	19°56'49"N	79° 06'25"E	5.9	2543	1628	234	79	142	14.23	402	182	524	24.55
GW24	19°56'18"N	79°06'44"E	6.1	2378	1522	179	88	141	12.04	323	211	490	9.09
GW25	19°58'11"N	79°10'33"E	6.4	1286	823	91	55	74	14.23	299	91	139	47.11
GW26	19°56'37"N	79°10'59"E	7.3	702	449	51	14	61	6.12	273	19	38	14.08
GW27	19°56'06"N	79°09'19"E	7.1	1136	727	77	34	101	7.67	351	77	133	8.22
GW28	19°54'45"N	79°07'37"E	6.2	2058	1317	172	55	152	15.89	388	162	389	5.43
GW29	19°53'35"N	79°07'47"E	6.3	1647	1054	129	78	73	9.02	276	41	421	10.9
GW30	19°53'15"N	79°13'25"E	6.8	1101	705	83	38	64	8	302	47	159	27.56
GW31	19°54'07"N	79°14'47"E	6.6	1095	701	62	48	81	6.06	298	48	142	15.02
GW32	19°56'53"N	79°15'24"E	7.2	831	532	69	27	52	6.77	326	31	55	4.12
GW33	20°00'38"N	79°18'07"E	6.3	1896	1213	106	60	171	21.88	211	73	561	18.9
GW34	19°55'44"N	79°18'14"E	5.9	2232	1428	136	77	196	20.2	280	235	471	13.08
GW35	19°55'33"N	79°19'18"E	6.6	822	526	54	22	72	4	304	42	66	4.2
GW36	19°54'02"N	79°18'24"E	5.9	1992	1275	117	91	143	12.7	223	212	414	31.88
GW37	19°53'59"N	79°20'10"E	6.3	1178	754	117	44	38	7.1	348	82	139	22
GW38	19°53'03"N	79°20'29"E	7.1	1133	725	104	42	53	4.89	342	67	106	9.7
GW39	19°51'31"N	79°17'28"E	6.2	1300	832	96	57	63	7.73	321	130	133	31.51
GW40	19°51'03"N	79°16'04"E	6.7	1376	881	73	51	113	12.76	389	90	157	8.7
GW41	19°49'40"N	79°14'56"E	7	976	625	82	30	61	4.78	293	72	90	5.53
GW42	19°49'27"N	79°16'27"E	5.8	1921	1229	131	52	169	9.77	219	178	434	21.12
GW43	19°49'10"N	79°18'24"E	5.8	2121	1357	127	76	183	18.67	221	154	591	8
GW44	19°49'48"N	79°19'53"E	5.7	2175	1392	139	84	162	14	342	170	436	8.66
GW45	19°50'48"N	79°21'29"E	6.1	2056	1316	139	61	176	22.07	374	190	374	11.13
	Minimum		**5.7**	**702**	**449**	**51**	**14**	**38**	**4**	**211**	**19**	**38**	**3.03**
	Maximum		**7.3**	**2602**	**1665**	**234**	**102**	**196**	**22.07**	**402**	**290**	**591**	**47.11**
	Average		**6.41**	**1607.36**	**1028.7**	**123.22**	**57.02**	**104.16**	**10.58**	**295.57**	**122.1**	**302.6**	**15.36**

Values in mg/l

Table 3: The results obtained for hydro-geochemical parameters from post-monsoon groundwater samples.

Sample	**Locations**		**pH**	**EC**	**TDS**	**Ca^{2+}**	**Mg^{2+}**	**Na^{+}**	**K^{+}**	**HCO_3^{-}**	**Cl^{-}**	**SO_4^{2-}**	**NO_3^{-}**
GW1	20°08'46"N	79°01'48"E	7.1	973	623	63	44	59	4.23	213	87	126	12.11
GW2	20°07'58"N	79°00'16"E	7.4	932	596	87	18	72	5.12	222	110	98	8.23
GW3	20°07'39"N	79°02'04"E	6.9	1299	831	104	47	63	7.88	231	79	247	12.22
GW4	20°07'59"N	79°03'58"E	7	813	520	74	22	58	6.1	190	89	102	27.85
GW5	20°06'20"N	79°03'38"E	7.6	730	467	61	22	52	5.14	232	71	62	8.74
GW6	20°05'52"N	79°05'00"E	7.1	1062	680	97	32	68	5.32	231	134	99	13.1
GW7	20°04'21"N	79°01'37"E	6.8	1242	795	123	37	69	4.3	214	88	280	23.21
GW8	20°03'42"N	79°00'40"E	7.2	973	623	68	39	61	4.56	256	89	102	16.21
GW9	20°01'57"N	79°02'23"E	6.4	1263	808	107	47	77	9.07	242	70	251	8.22
GW10	20°02'51"N	79°03'16"E	6.6	1380	883	131	33	97	3.9	223	72	362	13.22
GW11	20°02'47"N	79°04'59"E	6.3	1512	968	125	52	93	8.81	218	105	352	22.76
GW12	20°01'05"N	79°04'37"E	6.2	1214	777	94	44	74	5.5	173	84	262	22.71
GW13	20°01'25"N	79°06'20"E	7.1	982	628	103	19	62	7.34	241	79	121	10
GW14	20°00'29"N	79°05'50"E	6.3	1195	765	82	38	101	7.67	188	66	284	31.69
GW15	19°59'43"N	79°08'17"E	6.9	992	635	73	32	51	10.24	219	89	64	51.67
GW16	19°58'54"N	79°06'04"E	6.1	1373	879	110	41	95	11.32	212	65	303	31.14
GW17	19°59'16"N	79°03'45"E	6.9	1178	754	116	38	47	12.67	213	45	271	21.01
GW18	19°59'13"N	79°01'17"E	7.8	702	449	64	19	41	5.83	271	53	27	6
GW19	19°57'42"N	79°04'48"E	6.2	1322	846	117	43	69	23.21	238	60	302	14
GW20	19°57'08"N	79°03'40"E	6.8	806	516	67	32	42	6.5	180	83	91	7.67
GW21	19°56'12"N	79°03'26"E	7.6	779	499	67	29	42	7.9	222	65	83	14.32
GW22	19°56'25"N	79°04'38"E	6.2	1304	835	104	53	74	3.75	249	77	287	8.75
GW23	19°56'49"N	79° 06'25"E	6.3	1269	812	113	34	76	8	153	46	333	21.44
GW24	19°56'18"N	79°06'44"E	6.7	966	618	82	26	71	5.5	173	23	250	4.43
GW25	19°58'11"N	79°10'33"E	6.7	1020	653	63	24	111	8.02	220	124	102	33.6
GW26	19°56'37"N	79°10'59"E	7.5	883	565	51	29	75	7.3	295	60	73	7.38
GW27	19°56'06"N	79°09'19"E	7.1	826	529	85	21	43	4.1	240	69	92	11.28
GW28	19°54'45"N	79°07'37"E	6.5	1323	847	76	34	137	14.8	191	53	353	38.31
GW29	19°53'35"N	79°07'47"E	6.4	1379	883	88	46	109	10.04	183	44	391	27.84
GW30	19°53'15"N	79°13'25"E	7.1	871	557	58	26	69	5.87	271	59	82	8.66
GW31	19°54'07"N	79°14'47"E	7	799	511	59	25	73	8.34	283	71	42	12.7
GW32	19°56'53"N	79°15'24"E	7.4	660	422	43	12	71	8.33	284	31	41	4.55
GW33	20°00'38"N	79°18'07"E	6.4	1636	1047	97	61	132	14.99	236	97	414	18
GW34	19°55'44"N	79°18'14"E	6.2	1273	815	109	52	61	8.9	193	61	351	4.23
GW35	19°55'33"N	79°19'18"E	6.9	839	537	71	29	51	6.1	215	72	80	13.22
GW36	19°54'02"N	79°18'24"E	6.4	1230	787	84	31	121	7.17	232	31	295	23.54
GW37	19°53'59"N	79°20'10"E	6.9	832	532	74	21	53	7.67	181	112	77	6.2
GW38	19°53'03"N	79°20'29"E	7.4	727	465	46	21	71	3.12	185	83	51	7.16
GW39	19°51'31"N	79°17'28"E	7.1	821	525	56	22	72	6.16	193	111	51	7.91
GW40	19°51'03"N	79°16'04"E	6.9	833	533	52	18	81	9.14	194	104	71	9.5
GW41	19°49'40"N	79°14'56"E	7.1	831	532	88	30	54	4.3	170	88	79	6.43
GW42	19°49'27"N	79°16'27"E	6.1	1253	802	103	40	86	7.1	145	38	375	9.03
GW43	19°49'10"N	79°18'24"E	6.3	1699	1087	94	62	142	12.03	253	71	444	6.43
GW44	19°49'48"N	79°19'53"E	6	1591	1018	104	44	137	11	232	89	374	9
GW45	19°50'48"N	79°21'29"E	6.5	1278	818	91	38	100	13.7	204	42	348	4.54
	Minimum		**6**	**660**	**422**	**43**	**12**	**41**	**3.12**	**145**	**23**	**27**	**4.23**
	Maximum		**7.8**	**1699**	**1087**	**131**	**62**	**142**	**23.21**	**295**	**134**	**444**	**51.67**
	Average		**6.78**	**1085.88**	**694.93**	**84.97**	**33.93**	**76.95**	**7.95**	**217.86**	**74.2**	**198.77**	**15.11**

Values in mg/l

IOP Conf. Series: Earth and Environmental Science 1032 (2022) 012015 doi:10.1088/1755-1315/1032/1/012015

3.2. Interpretations

3.2.1. Percentage contribution. The percentage contributions of the cations and the anions for the pre-monsoon has disclosed the abundance in the order of Ca^{2+}>Na^{+}>Mg^{2+}>K^{+} and SO_4^{2}>HCO_3^{-}>Cl^{-}>NO_3^{-} respectively (Figure 2). Whereas the order for post-monsoon is Ca^{2+}>Na^{+}>Mg^{2+}>K^{+} and HCO_3^{-}>SO_4^{2-}>Cl^{-}>NO_3^{-} (Figure 3).

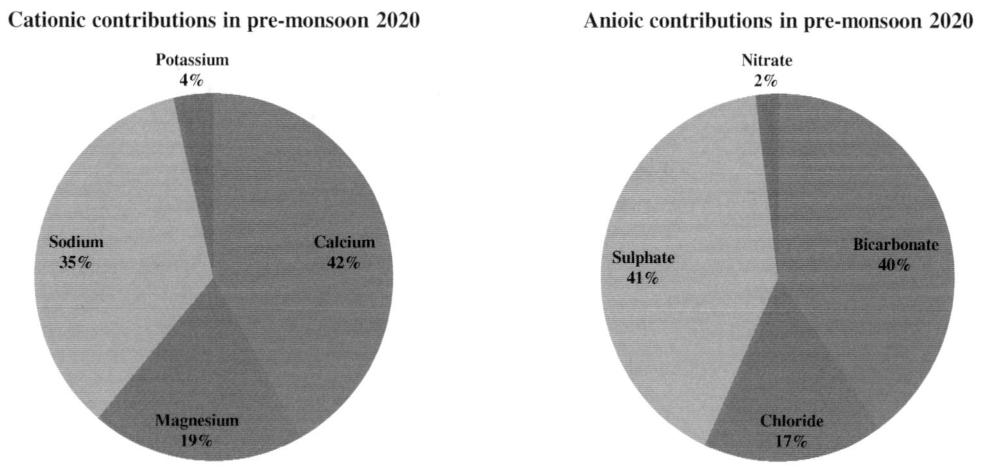

a. **b.**

Figure 2: Percentage contribution of the ions in the pre-monsoon; **a.** Cations and **b.** Anions.

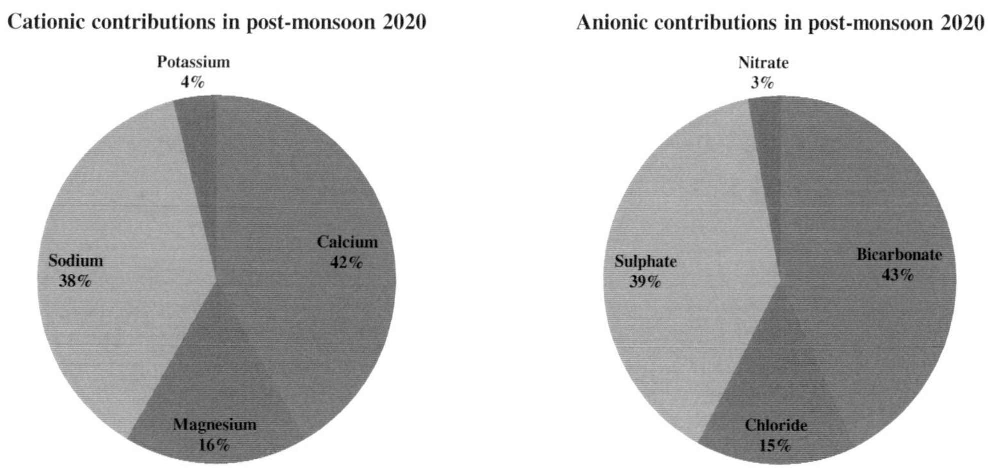

a. **b.**

Figure 3: Percentage contribution of the ions in the post-monsoon; **a.** Cations and **b.** Anions.

3.2.2. Piper Plots. In the pre-monsoon season, around majority (67 %) of the groundwater samples point to the ascendancy of mixed CaMgCl type (zone 4); While, the rest samples are distributed between $CaHCO_3$ type (zone 1) and CaCl (zone 5) (Figure 4). The majority of the groundwater falls in Ca^{2+}-Mg^{2+}-Cl^--SO_4^{2-} facies. In the cationic facies, about 93 % of the groundwater samples indicate no dominance. Whereas, in anionic facies, around 53 % of the samples are in the no dominance field and among the rest, 25 % are in the SO_4^{2-} type and 22 % are in the HCO_3^- type (Figure 4).

In post-monsoon, approximately majority (81 %) of the groundwater samples specifies the ascendency of mixed CaMgCl type (zone 4); while the rest are distributed among mixed $CaNaHCO_3$ type, $CaHCO_3$ type (zone 1) and CaCl (zone 5) (Figure 5). The majority of the groundwater falls in Ca^{2+}-Mg^{2+}-Cl^--SO_4^{2-} facies in post-monsoon too. In the cationic facies, about 89 % of the groundwater samples are in no dominance. While in anionic facies, around 43 % of the samples are in the no dominance field, 41 % is in SO_4^{2-} type and rest 16 % is in HCO_3^- type.

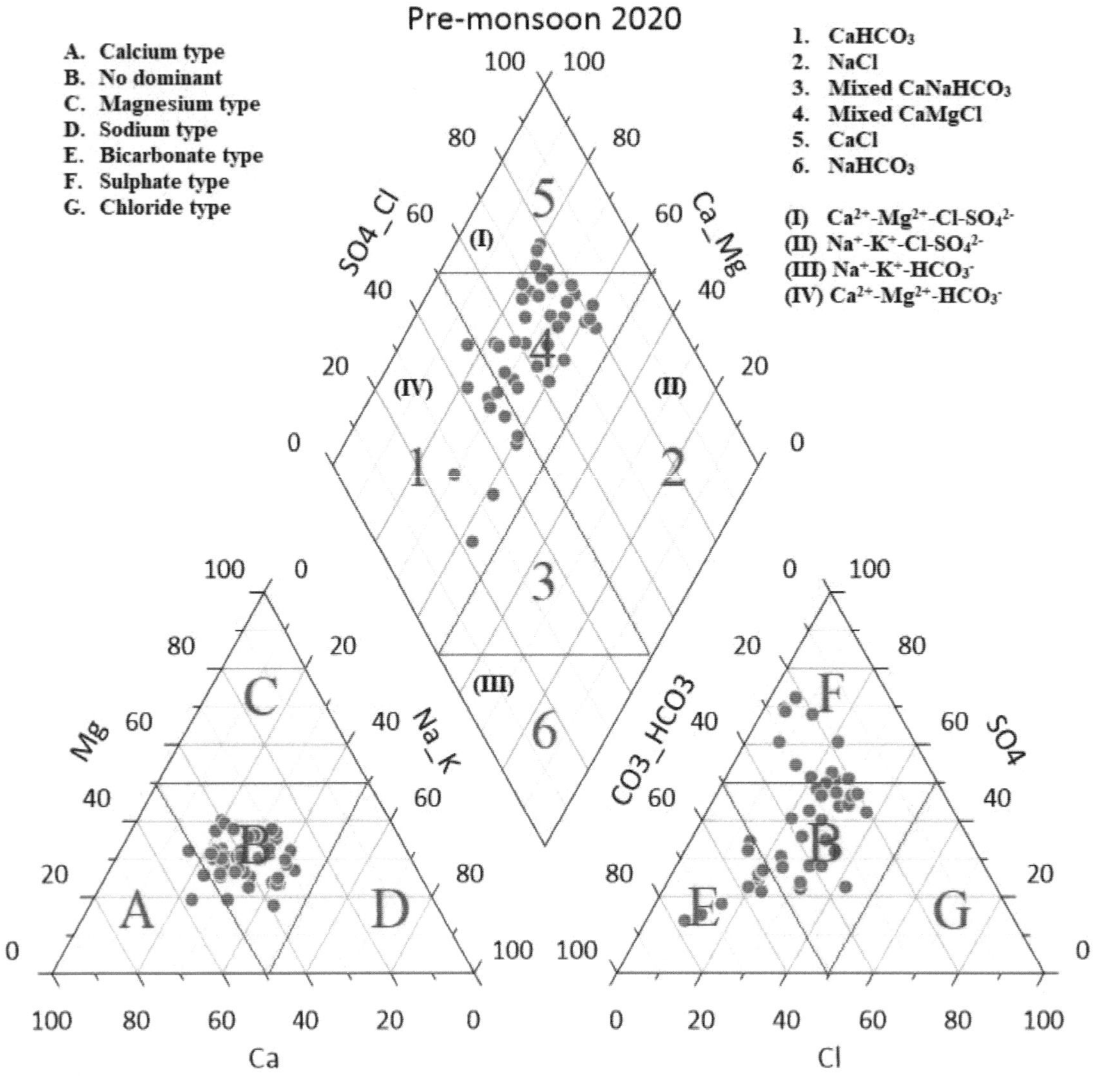

Figure 4: Piper diagram indicating the groundwater type in the pre-monsoon.

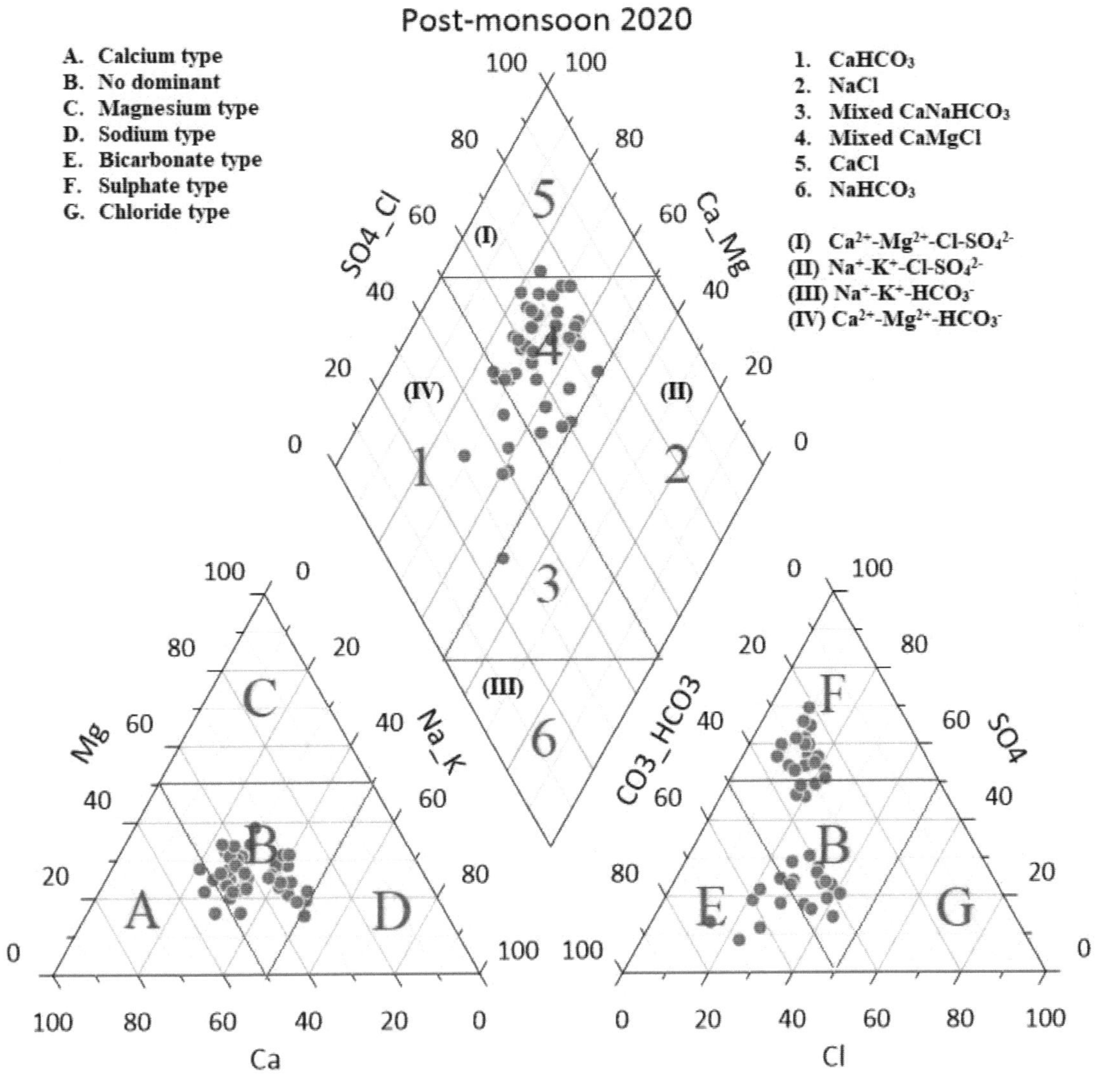

Figure 5: Piper diagram indicating the groundwater type in the post-monsoon.

3.2.3. Gibbs plot. The Gibbs plot indicates that the chemistry of the groundwater for the both the seasons. The groundwater chemistry is primarily governed by the rock-water interaction, with an insignificant influence of evaporation in pre-monsoon (Figure 6 and 7).

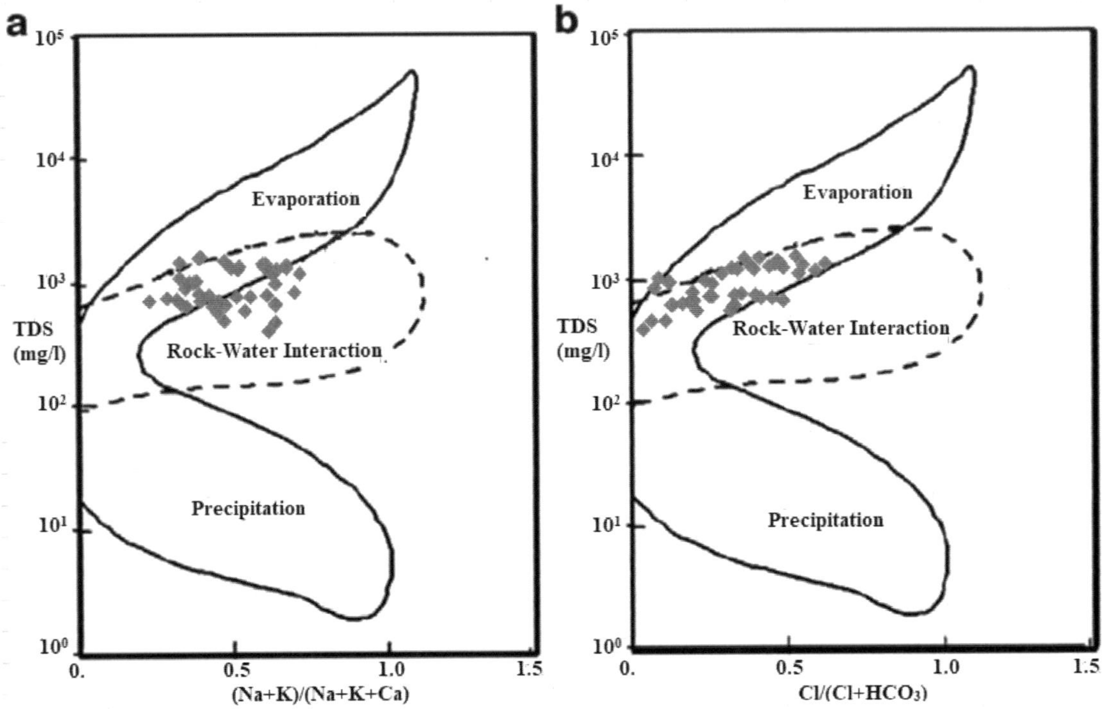

Figure 6: Gibbs diagram for the pre-monsoon; **a.** TDS vs. cations and **b.** TDS vs. anions.

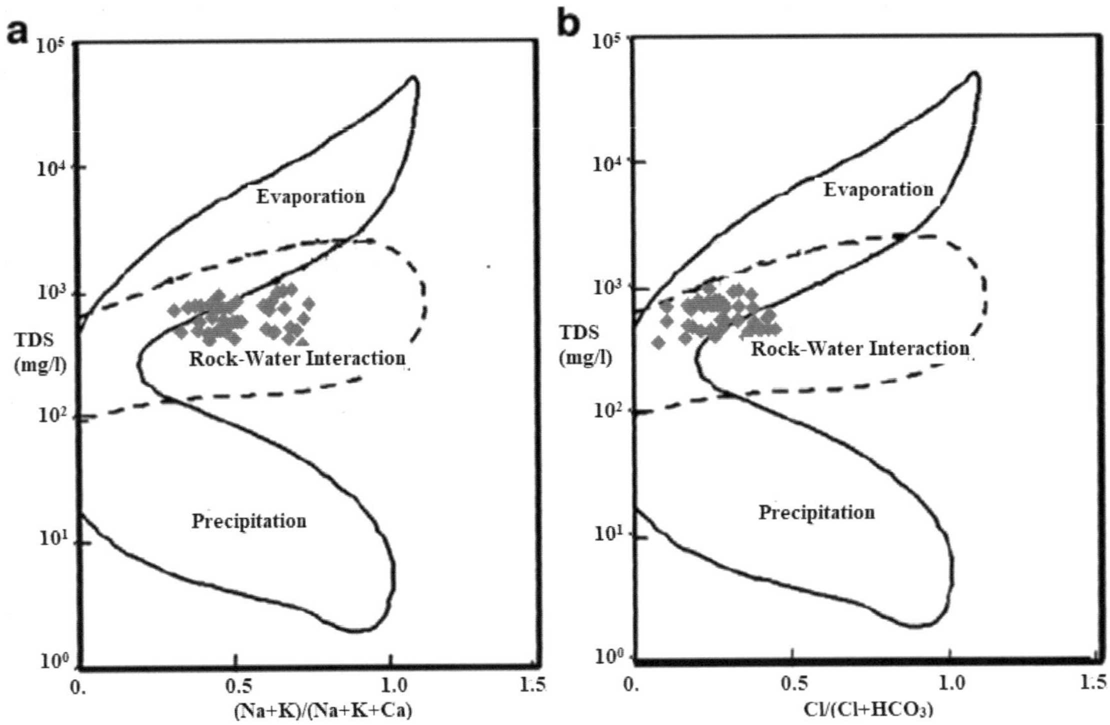

Figure 7: Gibbs diagram for the post-monsoon; **a.** TDS vs. cations and **b.** TDS vs. anions.

3.2.4. Implications of hydro-geochemical elucidations. The enhanced pH of the sampled groundwater can be attributed to the contributors like SO_4^{2-}, Cl^- and NO_3^- and its consequence in the hydro-geochemistry cannot be underrated. Water with low pH keeps potential of higher dissolving capacity as a result of which contaminants [12]. The acidic water ousted from coal mines are usually charged with the leachates from various sources like mine tailings, settling ponds, etc. resulting into severe pollutions [3]. As the groundwater from with lower pH is susceptible to to have high dissolvability, the higher TDS was quite expected. The enriched TDS can be attributed to the evaporative dominance conditions resulting into precipitation of the salts. The enriched concentration of the Ca^{2+} can be attributed to the plagioclase feldspar and the calcareous cementing material in the Barakar rocks of the Wardha valley coalfields is fairly attributable [13]. The Mg^{2+} concentration in the groundwater is certainly the result of water interaction with minerals like orthoclase, microcline and biotite of the lower Gondwana sediments. Similarly, the Na^+ can be guessed to be released from the minerals like feldspar, clay minerals, etc. of the lower Gondwana rocks of the study area. The limestone beds of the Lameta formation in the study area can certainly be the source of the enriched $HCO3^-$ concentration in the groundwater and is sole responsible for the alkalinity (predicted due to low pH). The dissolution of the NaCl from the rocks and soil is the prime process responsible for the higher Cl^- concentrations; whereas, the sewage effluents are also the considerable source for the same. The presence of the SO_4^{2-} in the groundwater regime lowers the pH through acid mine drainage [14]. The existence of the coal and sulphide minerals in the Barakar formation is accountable for the high concentration of the SO_4^{2-} in the groundwater [15]. Though the NO_3^- concentration is insignificant but could be responsible in lowering of pH of the groundwater. The most typical cause of in the groundwater is the decaying organic matter, sewage wastes and fertilizers.

The Gibbs plot has revealed the rock-water interaction as the controlling mechanism for groundwater chemistry. The same is also somewhat predictable by the ionic manifestation in the sampled groundwater. The hydro-geochemical facies analysis of the sampled groundwater has elucidated that the alkali earth (Ca + Mg) and the strong acids (SO_4^{2-}-Cl^-- NO_3^-) exceeds the alkalis (Na + K) and the weak acids (HCO_3^-) respectively. Such hydro-geochemical circumstances with dominance of strong acids are often favourable for the contaminations of heavy metals [16]. On correlating the sample locations and hydro-geochemical parameters, the intensities are found to be protruding around mining regions. Hence, observations are in somewhat agreement with the protruded contaminations in and around the mining locations are typical and go on decreasing away from the mining locations [17].

4. Conclusion

The major portion of the groundwater regime is dominated by the Ca^+-Mg^+ in cations and HCO_3^--SO_4^{2-} in anions. The hydro-geochemistry of the study area is majorly controlled by the rock-water interaction, where minerals from the host rocks are leached out. This process is augmented by the acidic nature of the groundwater (especially near mining regions) which can be attributed to the high SO_4^{2-} concentrations. The major groundwater is of CaMgCl type with no dominance of cations and distributed dominance of anions among no-dominance-HCO_3^--SO_4^{2-}. The hydro-geochemical facies analysis implies the ascendency of the strong acids (SO_4^{2-}-Cl^--NO_3^-) over weak acids (HCO_3^-). As a consequence of which, the groundwater conditions are quite favourable to the dissolution of contaminants like of heavy metals. The mitigation of such conditions is the most expected initiative to keep the groundwater potable in impending perspectives.

References

[1] WHO 1984 *Guidelines for drinking water qualities* (Washington DC: World Health Organisation) pp 35-43

[2] Ganvir P S and Guhey R 2020 Groundwater quality assessment with reference to some heavy metals toxicity and its probable remediation around Ballarpur area of Wardha valley coalfields, Maharashtra *IOP Conference Series: Earth and Environmental Science*, **597(1)** 012001

[3] Satapathy D R, Salve P R and Katpatal Y B 2009 Spatial Distribution of Metals in Ground/Surface Waters in the Chandrapur District (Central India) and heir Plausible Sources *Environmental Geology* **56** pp. 1323–52

[4] Ganvir P S and Guhey R 2021 Geochemical Studies of some Heavy Metals' Toxicity in Groundwater with their Plausible Sources around Gondwana Supergroup, Wardha valley Coalfields, Maharashtra *Journal Geological Society of India*, **97** pp. 1415-1421

[5] Filcheva E and Noustorova M 2000 Organic Accumulation and Microbial Action in Surface Coal-Mine Spoils, Pernik, Bulgaria *Ecological Engineering* **15** pp. 1–15

[6] Scokart P, Meeus-Verdinne K and DeBorger R 1983 Mobility of Heavy Metals in Polluted Soils near Zn Smelters *Water Air Soil Pollution* **20** pp. 451–63

[7] Singh R M and Chandra D 1983 Occurrence, Distribution and Probable Source of the Trace Elements in Ghugus Coals, Wardha Valley, District Chandrapur and Yeotmal, Maharashtra, India *International Journal Coal Geology* **24** pp. 371–81

[8] Sarate O P 2017 Petrography of Coal Seams from Wilson area, Chandrapur District, Maharashtra and their Depositional History *Geophytology* **47** pp. 193–201

[9] Raja Rao C S 1982 *Coal Resources of Tamil Nadu, Andhra Pradesh, Orissa and Maharashtra* (*Bulletin of the Geological Survey of India, coalfields of Maharashtra, Wardha Valley Coalfield* A vol 45) pp 85–87

[10] APHA 1998 *Standard Methods for the Examination of Water and Waste Water* 20th (Washington DC: Water Pollution Control Federation) pp 10 – 161

[11] BIS 2012 *Drinking Water - Specification (Second Revision) IS: 10500* (Bureau of Indian Standards) pp 2–3

[12] Zipper C, Skousen J and Jage C 2014 *Passive Treatment of Acid Mine Drainage* (Virginia: Virginia Polytechnic Institute and State University) pp 1–13

[13] Kale M G and Soman G R 2012 Petrography of Barakar Sediments from Ghugus Area, Pranhita-Godavari Basin, Maharashtra *Gondwana Geological Magazine* **27(1)** pp 67-78

[14] Singh G 1987 Mine Water Quality Deterioration due to Acid Mine Drainage *International Journal of Mine Water* **6** pp 49-61

[15] Karanth K R 1997 *Groundwater Assessment, Development and Management* Tata McGrraw Hill New Delhi p 720

[16] Ganvir P S and Guhey R 2020 Assessment of Some Heavy Metals Toxicity and Its Probable Remediation in Groundwater around Telwasa and Ghugus Area of Wardha Valley Coalfields, Maharashtra *Journal of Indian Association of Sedimentologists* **37(1)** pp 33-43

[17] Peplow D 1999 *Environmental Impacts of Mining in Eastern Washington* Centre for Water and Watershed Studies Fact Sheet, University of Washington, Seattle

Application of Hybrid Support Vector Machine model for Streamflow Prediction in Barak valley, India

A Sahoo[1a], and D K Ghose[2b*]

[1]Research Scholar, Department of Civil Engineering, National Institute of Technology Silchar, Fakiratilla, Silchar - 788010, Assam, India
[2]Assistant Professor, Department of Civil Engineering, National Institute of Technology Silchar, Fakiratilla, Silchar - 788010, Assam, India

Email: [a]bablusahoo1992.@gmail.com; [b]dillipghose_2002@yahoo.co.in

Abstract. Forecasting streamflow (Q_{flow}) is vital in flood and water management, determining potential of river water flow, agricultural practices, hydropower generation, and environmental flow study. This research aims to explore capability of hybrid support vector machines (SVM) with Whale Optimisation Algorithm (WOA) model for forecasting streamflow at Badarpur Ghat gauging station of Barak river basin and evaluate its enactment with the conventional SVM model. Root mean squared error (RMSE), mean absolute error (MAE), and Nash-Sutcliffe efficiency (NSE) statistical measures are considered as evaluating standards. Assessment of outcomes indicates that the optimization algorithm could enhance the accurateness of standalone SVM model in monthly streamflow forecasting. Compared to conventional artificial intelligence methods without a data pre-processing system, the comparatively good performance of applied hybrid model gives an effective alternate to achieve better precision in streamflow forecasting. Results confirm that enhanced SVM model can better process a multifaceted hydrogeological data set, have higher prediction accuracy, and possess better generalisation capability.

Keywords: Barak river basin, Streamflow, SVM, SVM-WOA

1. Introduction

Precise streamflow forecasting is crucial for decision-makers related to water supplies, navigation, hydropower generation, flood control, and drought mitigation [1, 2, 3]. Nonetheless, reliable and accurate streamflow modeling is still challenging to attain because of the complex behavior of streamflow itself driven by its frenzied features like stochasticity, non-stationarity, and non-linearity [4, 5, 6]. Incorporating only appropriate input variables into a prediction model is crucial for ensuring high precision of predicted streamflow data. Yet, intrinsic non-linear relationship amid input and output variables in Q_{flow} forecast is proving to be a continuing scientific test as traditional regression-based methods cannot accurately model streamflow data.

Models forecasting streamflow can be categorised into data-driven (DD) and physical-based models. Physical-based models give an idealsed demonstration of hydrogeological developments by simulating complex relations amid meteorological, subsurface, and land surface constituents of water cycle [7]. In contrast, DD models mathematically capture non-linear or linear interactions amid Q_{flow} and its descriptive variables [8]. DD methods like ANN (artificial neural network), SVM, and ANFIS (adaptive neuro-fuzzy inference system), have been successfully applied by many investigators in hydrological field of study; for example, sediment transport modelling [9, 10, 11, 12], runoff modeling [13, 14, 15], water table depth prediction [16, 17, 18], streamflow forecasting [19, 20, 21]; pan evaporation estimation [22, 23]; flood prediction [24, 25, 26, 27, 28].

Zakaria and Shabri [29] explored ability of SVM in forecasting streamflow at ungauged stations and assessed its performance against MLR (multiple linear regression) models. Results revealed that SVM outperformed prediction capability of conventional MLR. Shabri and Suhartono [30] investigated potential of least-squares SVM model for improving accuracy of streamflow forecasting using recorded data from two stations of Kinta River in Perak, Malaysia. Kisi et al. [31] evaluated accurateness of ANN, SVM, and ANFIS models for daily intermittent streamflow forecasting considering data from Uzunkopru and Babaeski stations located in Thrace region of Turkey. Their findings revealed that ANN performed better in Uzunkopru station whereas ANFIS performed better in Babaeski station. Alizadeh et al. [32] assessed performance of feed-forward NN, recurrent NN, radial basis function NN, time delay NN, K-nearest neighbors (KNN), and SVM models to predict monthly flow of Alavian and Dez river basins located in Iran.

Interest in developing hybrid streamflow forecasting models has grown in last two decades. Hybrid models combine different fitting techniques and decomposition approaches. Guo et al. [33] proposed an advanced SVM model using adaptive insensitive features to predict monthly streamflow and achieved outcomes were compared with ANN results and conventional SVM models. They found that improved SVM model processed complex hydrological data series better with high prediction accuracy. Yaseen et al. [34] treated non-stationary characteristics and uncertainty of streamflow data (Kelantan and Johor Rivers) utilising grey models (GM) and rolling mechanism (RM) pre-processing techniques before considering them as input. Again, Yaseen et al. [35] applied a hybrid ANFIS-FFA model to predict streamflow using recorded monthly Q_{flow} data of River Pahang, Malaysia. Comparative analysis of results revealed that FFA was capable of improving forecasting accurateness of conventional ANFIS model. Zuo et al. [36] projected a hybrid VMD-LSTM (variational mode decomposition with long short-term memory) to predict 1–7 days ahead of daily streamflow for Han and Jing Rivers, China. Results indicated that hybrid VMD-LSTM model was more efficient and robust than other applied models. Pan et al. [37] proposed a new oil layer recognition model utilising semi-supervised SVM and improved WOA. Results showed that improved SVM-WOA had better recognition precision in oil layer recognition.

In current study, potential effectiveness of SVM-WOA method was compared with standard SVM model for evaluating prediction accuracy. The novelty lies in the use of SVM-WOA for streamflow prediction in Barak river basin, as this river basin receives flood water almost every year. Hence streamflow prediction becomes of utmost importance for planners and decision-makers of this study location.

2. Study area

The Barak Valley is collectively formed by Karimganj, Cachar, and Hailakandi districts falling amid 24°8' and 25°8' N, and 92°15' and 93°15' E, and covers an area of 6,922 sq. km (Figure 1). A warm and humid climate is experienced in the valley with an average annual precipitation of 2,440 - 4,100 mm, most of which is during May to September (southwest monsoon season), and average monthly minimum and maximum temperature of 9.2°C and 33.9°C. Barak is the region's main river, with its distributaries and tributaries. The Barak river traverses the valley in a westerly direction upto Karimganj supporting agricultural activities and giving food source to inhabitants. The valley has a rising and falling topography contributed by grasslands, floodplains, wetlands, hillocks, mountains, etc. [38].

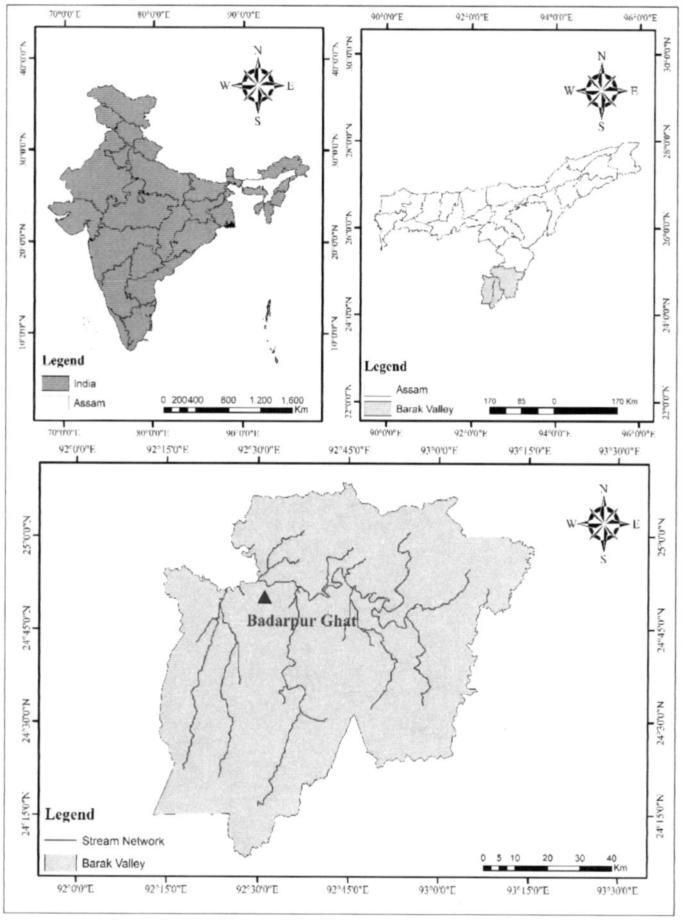

Figure 1. Location of Badarpur Ghat gauging station in Barak valley, Assam

3. Material and Methodology

3.1. SVM

Vapnik and Chervonenkis [39] first introduced SVM, on basis of statistical learning theory. SVM is a group of supervised learning techniques utilised for regression and classification works governed by binary classification in random property space and hence, an appropriate technique [40]. A proficient learning system based on efficient optimisation theory, SVM applies structural risk minimization principle for attaining a general optimum solution.

A training dataset is used for determining $f(x)$ by computing W (coefficient vector) and b (bias term). After that, coefficients obtained are applied for minimizing error function as given in following equations:

$$Min\ \frac{1}{2}\|W\|^2 + C\sum_{i=1}^{N}(\xi_i + \xi_I^*)$$

Subject to:

$$W\phi(x_i) + b - y_i \leq \varepsilon + \xi_I^*$$

$$y_i - W\phi(x_i) + b \leq \varepsilon + \xi_i$$

$$\xi_i, \xi_I^* \geq 0, i = 1, 2, \dots, N$$

Existence of training error is unavoidable in training process. Hence, there must be a penalty term that is shown by C. N and ϕ - number of samples, and kernel function, respectively. Furthermore, ξ_i, ξ_I^* - two slack variables indices. ε - loss function associated with training process accuracy.

3.2. WOA

An alternate swarm algorithm established on basis of natural behaviour of humpback whales is known as WOA [41]. In WOA, whales travel in a dimensional search space where number of parameters is represented by n. Every whale course signifies solution of a candidate. Preliminary step for determining a global solution is by modification of every solution's location. Following equation is used for updating location:

$$\vec{X}_{t+1} = \vec{X}^*(t) - \vec{A}\left|\vec{C}.\vec{X}^*(t) - \vec{X}(t)\right|$$

where$\vec{X}(t)$ - solution vector in tth iteration; $\vec{X}_{t+1}$- solution vector in $t+1$th iteration; $\vec{X}^*(t)$ - probable location of prey in ith iteration; $\vec{A}$ and $\vec{C}$ - coefficient vectors. In every iteration, $\vec{A}$and $\vec{C}$ are updated:

$$\vec{A} = 2\vec{a}.\vec{r_1} - \vec{a}$$

$$\vec{C} = 2.r_2$$

where $\vec{a}$ - reduces from 2 to 0; r_1 and r_2 - arbitrary vectors. By decreasing $\vec{a}$ value, shrinking Encircling Mechanism (SEM) is executed. New location of a search tool can be described anywhere between range of instrument's actual location and the present location of best agent by adjusting random values for $\vec{A}$. Following equation is applied for defining spiral-shaped movement of whales:

$$\vec{X}(t+1) = D'.e^{bl}.\cos(2\pi l) + \overrightarrow{W^*}(t)$$

where $D' = \overrightarrow{W^*}(t) - \vec{X}(t)$, b - a constant value to identify logarithmic spiral shape, and k - an arbitrary number. The working procedure of hybrid SVM-WOA model is showcased in Figure 2.

3.3. Evaluating Constraint

The authors collected 20 years (2001-2020) of discharge data (Q_t) from CWC, Shillong whereas precipitation (P_t); average temperature ($T_{avg(t)}$), and relative humidity ($R_{h(t)}$) are collected from IMD, Pune. 70% of collected data, i.e., 2001-2014, and rest 30%, i.e., 2015-2020 are considered to train and test proposed models. In present work, forecasting performance of established models (SVM and SVM-WOA) is assessed utilising different quantitative statistical evaluation indices. Statistical indices utilised here are RMSE, NSE, and MAE. In mathematical terms, these indicators are expressed as:

$$\text{RMSE} = \sqrt{\frac{1}{N}\sum_{i=1}^{N}[T_{i,o} - T_{i,f}]^2}$$

$$\text{NSE} = \frac{\sum_{k=1}^{N}[T_{i,o}-T_{i,f}]^2}{\sqrt{\sum_{k=1}^{N}[T_{i,o}-\overline{T_{i,o}}]^2}}$$

$$\text{MAE} = \frac{1}{N}\sum_{i=1}^{N}\left|T_{i,o} - T_{i,f}\right|$$

where $T_{i,f}$ and $T_{i,o}$= Predicted and collected discharge data; $\overline{T_{i,f}}$ and $\overline{T_{i,o}}$= Average of predicted and collected discharge data; N = data length.

Table 1. Modeling input combination structures

Input combination	Model description	
	SVM	SVM-WOA
Q_t	SVM1	SVM-WOA 1
Q_t, P_t	SVM 2	SVM-WOA 2
Q_t, P_t, $T_{avg(t)}$	SVM 3	SVM-WOA 3
Q_t, P_t, $T_{avg(t)}$, $R_{h(t)}$	SVM 4	SVM-WOA 4

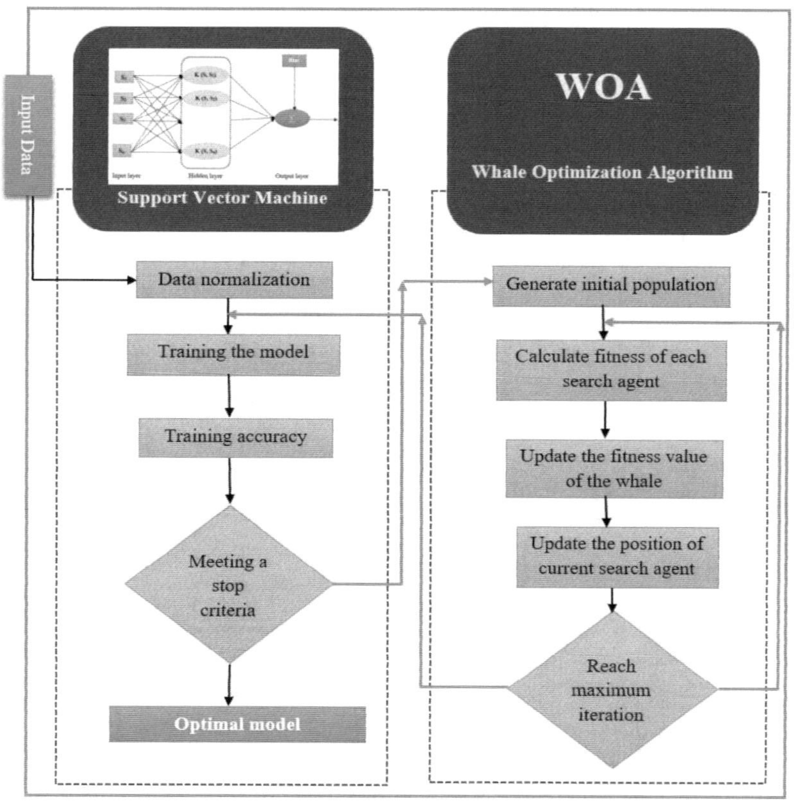

Figure 2. Flowchart of SVM-WOA algorithm

4. Result and Discussion

In present work, to forecast monthly Q_{flow} values, two models, a hybrid SVM-WOA, and standalone SVM, were used in Badarpur Ghat gauging station located in Barak valley. Classification of input data was done with four parameters as shown in Table 1 and in this way four input combinations were created. For easier comparison and briefness, outcomes of applied models are given in Table 2 and Figures 3, 4 and 5. As stated in Table 2, NSE coefficients using SVM-WOA range between 0.9764-0.9863 in training phase, and 0.961- 0.9748 in testing phase. Similarly for SVM, NSE values range between 0.9472-0.9536 in training and 0.9207-0.9302 in testing. Among all models, the best value is found for SVM4 (RMSE- 19.2213, NSE-0.9536, MAE-9.973) and SVM-WOA4 (RMSE-10.534, NSE-0.9863, MAE-5.028). From Figures 3 and 4 it is clear that the prediction values by the hybrid model are closer to the observed values, whereas for the basic model there is a considerable difference between the forecasted and observed time series, especially in peak values.

Table 2. Performance of model

Station Name	Model Name	RMSE	MAE	NSE	RMSE	MAE	NSE
		Training			Testing		
Badarpur Ghat	SVM1	22.367	15.835	0.9472	27.9436	18.829	0.9207
	SVM2	21.5218	14.949	0.9451	26.754	16.747	0.9235
	SVM3	21.0574	12.054	0.9494	25.862	15.442	0.9284
	SVM4	19.2213	9.973	0.9536	24.615	12.204	0.9302
	SVM-WOA1	13.205	9.429	0.9764	18.261	11.521	0.961
	SVM-WOA2	11.5349	8.362	0.9805	17.395	10.67	0.9682
	SVM-WOA3	11.087	6.175	0.9831	16.3489	8.136	0.9706
	SVM-WOA4	10.534	5.028	0.9863	15.573	7.384	0.9748

Figures 3 and 4 illustrate the scatter plot and time series plot between forecasted and observed Q_{flow} by SVM and SVM-WOA models for test period. The solid violet line in scatter plot exhibit regression line fitted to forecasted and observed Q_{flow}. The dotted black line is the 1:1 line. It is observed from Figure 3 that R^2 values generated by hybrid SVM-WOA4 model were better than values specified by standard SVM4. Considering time series plot (Fig. 4), it is evident that forecasted Q_{flow} values by SVM-WOA4 model give a good match with observed time series. An important point to note here is that extreme and low events are of significant concern to water resources planners and decision-makers. This is because decision-makers require accurate predictions of these two events as it is very crucial for avoiding drought and flooding conditions.

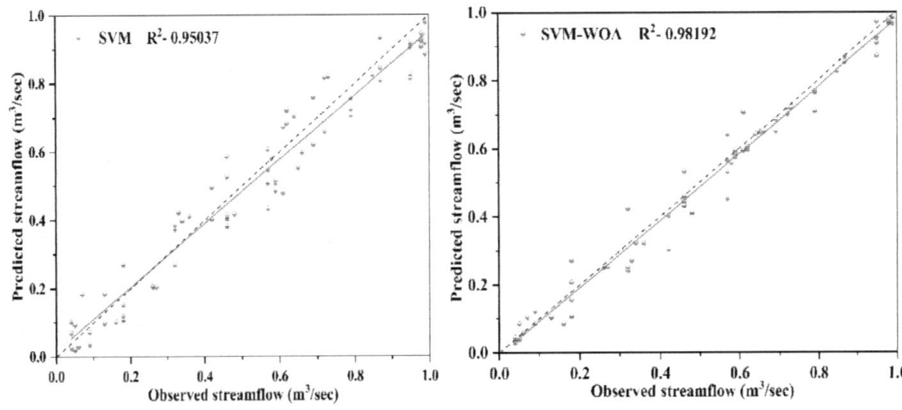

Figure 3. Scatter plot for observed vs predicted streamflow by proposed models during test phase

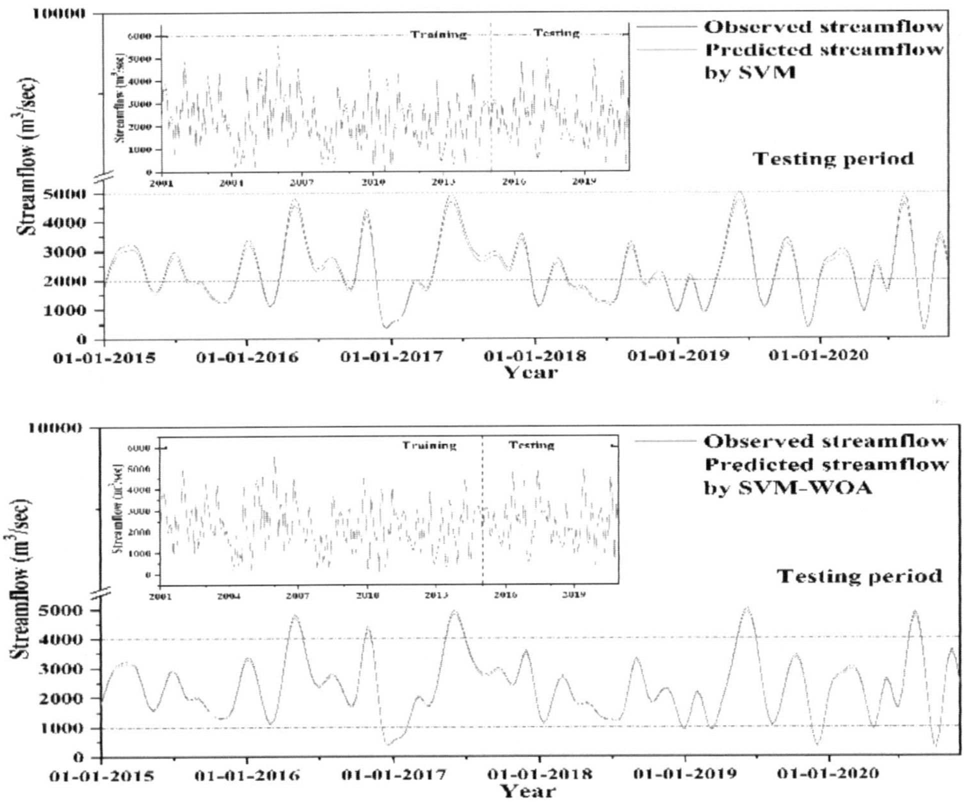

Figure 4. Time series showing comparison of observed predicted monthly streamflow using the (a) SVM and (b) SVM-WOA models

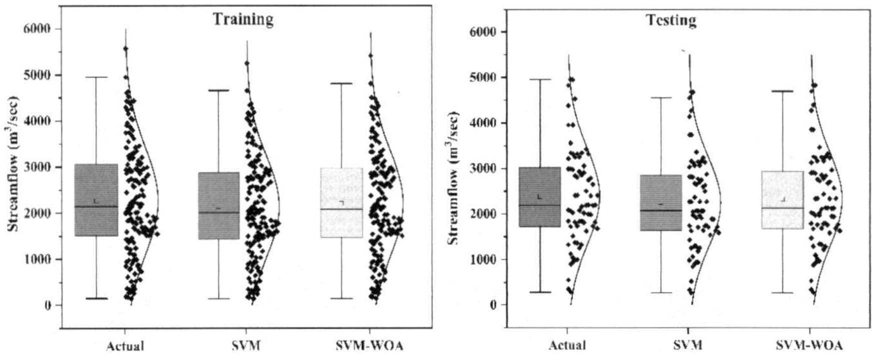

Figure 5. Box-plots of observed vs predicted streamflow from SVM and SVM-WOA models.

Box-plots with corresponding distributions were generated for demonstrating how closely SVM-WOA model forecasts against original data series and SVM approach. In Figure 4, boxplots are utilised for indicating degree of overall extent in observed and forecasted data according to corresponding quartile whiskers and values. In present study, application of proposed soft computing models is not restricted to Q_{flow} forecasting and can also be utilised for forecasting other hydrological variables such as temperature, precipitation, groundwater, etc.

5. Conclusion

Streamflow forecasting is significant for integrated water resources management, planning, and building river structures with other industrial processes. In this study, precision of classical SVM technique was enhanced by integrating SVM with WOA algorithm. Hybrid SVM model performance comparison with classical SVM model revealed the evolutionary algorithms' ability to optimize SVM membership function to minimize prediction error. Outputs suggested that the hybrid SVM-WOA model had more precision in predicting streamflow than the SVM-based model. Forecasting Q_{flow} achieved a comparatively good agreement with observed Q_{flow}, particularly the peak Q_{flow}. In future, other integrated intelligent systems may be explored for Q_{flow} forecasting in the Barak river basin that may increase forecasting performance.

References

[1] Fang W, Huang Q, Huang S, Yang J, Meng E and Li Y 2017 Optimal sizing of utility-scale photovoltaic power generation complementarily operating with hydropower: A case study of the world's largest hydro-photovoltaic plant Energy Convers. Manag. **136** 161–72

[2] Yu L and Li Y P 2019 A flexible-possibilistic stochastic programming method for planning municipal-scale energy system through introducing renewable energies and electric vehicles J. Clean. Prod. **207** 772–87

[3] Yaseen Z M, Kisi O and Demir V 2016 Enhancing long-term streamflow forecasting and predicting using periodicity data component: application of artificial intelligence Water Resour. Manag. **30** 4125–51

[4] Narsimlu B, Gosain A K, Chahar B R, Singh S K and Srivastava P K 2015 SWAT model calibration and uncertainty analysis for streamflow prediction in the Kunwari River Basin, India, using sequential uncertainty fitting Environ. Process. **2** 79–95

[5] Bayazit M 2015 Nonstationarity of hydrological records and recent trends in trend analysis: a state-of-the-art review Environ. Process. **2** 527–42

[6] Ismail S, Samsudin R and Shabri A (2010) River flow forecasting: a hybrid model of self organizing maps and least square support vector machine. Hydrol Earth Syst Sci Discuss **7** 8179–8212

[7] Rosenberg E A, Wood A W and Steinemann A C 2011 Statistical applications of physically based hydrologic models to seasonal streamflow forecasts Water Resour. Res. **47**
[8] Hadi S J and Tombul M 2018 Streamflow forecasting using four wavelet transformation combinations approaches with data-driven models: a comparative study Water Resour. Manag. **32** 4661–79
[9] Samantaray S and Sahoo A 2021 Prediction of suspended sediment concentration using hybrid SVM-WOA approaches Geocarto Int. 1–27
[10] Samantaray S, Sahoo A and Ghose D K 2020 Assessment of Sediment Load Concentration Using SVM, SVM-FFA and PSR-SVM-FFA in Arid Watershed, India: A Case Study KSCE J. Civ. Eng. **24** 1944-1957
[11] Afan H A, Keshtegar B, Mohtar W H M W and El-Shafie A 2017 Harmonize input selection for sediment transport prediction. Journal of Hydrology, **552** 366-375
[12] Rahman S A and Chakrabarty D 2020 Sediment transport modelling in an alluvial river with artificial neural network J. Hydrol. **588** 125056
[13] Chang F-J, Chen P-A, Lu Y-R, Huang E and Chang K-Y 2014 Real-time multi-step-ahead water level forecasting by recurrent neural networks for urban flood control J. Hydrol. **517** 836–46
[14] Samantaray S and Sahoo A 2020 Prediction of runoff using BPNN, FFBPNN, CFBPNN algorithm in arid watershed: A case study Int. J. Knowledge-Based Intell. Eng. Syst. **24** 243-251
[15] Van S P, Le H M, Thanh D V, Dang T D, Loc H H and Anh D T 2020 Deep learning convolutional neural network in rainfall–runoff modelling J. Hydroinformatics **22** 541–61
[16] Luo Q, Wu J, Yang Y, Qian J, and Wu J, 2016 Multi-objective optimization of long-term groundwater monitoring network design using a probabilistic Pareto genetic algorithm under uncertainty. J. Hydrol. **534** 352–363
[17] Samantaray S and Sahoo A 2021 Modelling response of infiltration loss toward water table depth using RBFN, RNN, ANFIS techniques. International Journal of Knowledge-based and Intelligent Engineering Systems, **25** 227-234
[18] Yadav B, Gupta P K, Patidar N and Himanshu S K 2020 Ensemble modelling framework for groundwater level prediction in urban areas of India Sci. Total Environ. **712** 135539
[19] Sahoo A, Samantaray S and Ghose D K 2019 Stream flow forecasting in mahanadi river basin using artificial neural networks Procedia Computer Science vol 157
[20] Samanataray S and Sahoo A 2021 A Comparative Study on Prediction of Monthly Streamflow Using Hybrid ANFIS-PSO Approaches KSCE J. Civ. Eng. **25** 4032–43
[21] Dalkiliç H Y and Hashimi S A 2020 Prediction of daily streamflow using artificial neural networks (ANNs), wavelet neural networks (WNNs), and adaptive neuro-fuzzy inference system (ANFIS) models Water Supply **20** 1396-1408
[22] Kisi O 2015 Pan evaporation modeling using least square support vector machine, multivariate adaptive regression splines and M5 model tree. J. Hydrol. **528** 312–320
[23] Wang L, Niu Z, Kisi O and Yu D 2017 Pan evaporation modeling using four different heuristic approaches. Computers and Electronics in Agriculture, **140** 203-213
[24] Sahoo A, Samantaray S and Ghose D K 2021 Prediction of Flood in Barak River using Hybrid Machine Learning Approaches: A Case Study J. Geol. Soc. India **97** 186–98
[25] Samantaray S, Sahoo A and Agnihotri A 2021 Assessment of Flood Frequency using Statistical and Hybrid Neural Network Method: Mahanadi River Basin, India J. Geol. Soc. India **97** 867–80
[26] Agnihotri, A., Sahoo, A. and Diwakar, M.K., 2022. Flood Prediction Using Hybrid ANFIS-ACO Model: A Case Study. In Inventive Computation and Information Technologies (pp. 169-180). Springer, Singapore
[27] Chang F-J and Tsai M-J 2016 A nonlinear spatio-temporal lumping of radar rainfall for modeling multi-step-ahead inflow forecasts by data-driven techniques J. Hydrol. **535** 256–69
[28] Xie S, Wu W, Mooser S, Wang Q J, Nathan R and Huang Y 2021 Artificial neural network based hybrid modeling approach for flood inundation modeling J. Hydrol. **592** 125605

[29] Zakaria Z A and Shabri A 2012 Streamflow forecasting at ungaged sites using support vector machines Appl. Math. Sci. **6** 3003–14
[30] Shabri A and Suhartono 2012 Streamflow forecasting using least-squares support vector machines Hydrol. Sci. J. **57** 1275–93
[31] Kisi O, Nia A M, Gosheh M G, Tajabadi M R J and Ahmadi A 2012 Intermittent streamflow forecasting by using several data driven techniques Water Resour. Manag. **26** 457–74
[32] Alizadeh Z, Yazdi J, Kim J H and Al-Shamiri A K 2018 Assessment of machine learning techniques for monthly flow prediction Water **10** 1676
[33] Guo J, Zhou J, Qin H, Zou Q and Li Q 2011 Monthly streamflow forecasting based on improved support vector machine model Expert Syst. Appl. **38** 13073–81
[34] Yaseen Z M, Fu M, Wang C, Mohtar W H M W, Deo R C and El-Shafie A 2018 Application of the hybrid artificial neural network coupled with rolling mechanism and grey model algorithms for streamflow forecasting over multiple time horizons Water Resour. Manag. **32** 1883–99
[35] Yaseen Z M, Ebtehaj I, Bonakdari H, Deo R C, Mehr A D, Mohtar W H M W, Diop L, El-Shafie A and Singh V P 2017 Novel approach for streamflow forecasting using a hybrid ANFIS-FFA model J. Hydrol. **554** 263–76
[36] Zuo G, Luo J, Wang N, Lian Y and He X 2020 Decomposition ensemble model based on variational mode decomposition and long short-term memory for streamflow forecasting J. Hydrol. **585** 124776
[37] Pan Y, Xia K, Wang L and He Z 2021 A Novel Approach to Oil Layer Recognition Model Using Whale Optimization Algorithm and Semi-Supervised SVM Symmetry (Basel). **13** 757
[38] Vapnik V 1995 The nature of statistical learning theory, Springer, New York.
[39] Pai P and Hong W 2007 A recurrent support vector regression model in rainfall forecasting Hydrol. Process. An Int. J. **21** 819–27
[40] Mirjalili S and Lewis A 2016 The whale optimization algorithm Adv. Eng. Softw. **95** 51–67
[41] Mazumder M K, Boro F, Barbhuiya B and Singha U 2014 A study of the winter congregation sites of the Gangetic River Dolphin in southern Assam, India, with reference to conservation Glob. Ecol. Conserv. **2** 359–66

RICE LEAF DISEASE CLASSIFICATION USING CNN

PallapothalaTejaswini[1], Priyanshi Singh[2], Monica Ramchandani[3], Yogesh Kumar Rathore[4], Rekh Ram Janghel[5]

DEPARTMENT OF INFORMATION TECHNOLOGY

NATIONAL INSTITUTE OF TECHNOLOGY, RAIPUR[1,2,3,4,5]

monicaramchandani4042@gmail.com[1], pallapothalatejaswini@gmail.com[2], priyanshisingh1099@gmail.com[3], yogeshgrathore23@gmail.com[4], rrjanghel.it@nitrr.ac.in[5]

***Abstract* -** Rice is amongst the majorly cultivated crops in India and its leaf diseases can have a substantial impact on output and quality. The most important component is identifying rice leaf diseases, which have a direct impact on the economy and food security. Brown spot, Leaf Blast, Hispa are the most frequently occurring rice leaf diseases. To resolve this issue, we have studied various machine learning and deep learning approaches for detecting the diseases on their leaves by calculating their accuracy, recall, and precision to measure the performance. This study helps the farmers by detecting the diseases in rice leaves in order to get a healthy crop yield. The deep learning models perform well when compared with the machine learning methods. After analyzing all of the deep learning models, we found that the 5-layer convolution model had the best accuracy of 78.2 %, while others, such as VGG16, had a lower accuracy of 58.4%.

Keywords: rice leaf diseases; deep learning; Convolutional neural networks; machine learning; transfer learning;

1. INTRODUCTION

India is a land of Agriculture [1] as it plays an essential role in our country because a lot of the people are dedicated to the agricultural industry. Crop production is amongst the major factors which are affecting domestic market conditions in our country. Agricultural firms began to search for new high-yield, cost-effective inventions as a consequence of expanding population, variable weather conditions, and political unpredictability. The health of the plant/crop is critical for achieving food security and sustainability in agriculture. However, the plants can quickly become infected with illnesses, which can cause major social and economic problems, due to a variety of factors. Crop diseases can affect its growth and development, and also crop yield and quality, and are one of the most common reasons for productivity loss. To avoid soil pollution, the illness should be detected and certain pesticides should be used from their beginning stage.

IOP Conf. Series: Earth and Environmental Science 1032 (2022) 012017 doi:10.1088/1755-1315/1032/1/012017

There are a variety of methods for detecting plant diseases in their early stages. The traditional method of plant disease detection is naked eye monitoring, which is ineffective and inaccurate for large crops[2].The major goal of this paper is to research and diagnose rice leaf illnesses in advance, as well as to identify the disease's name so that appropriate precautions can be followed. Rice is a standout amongst the most vital food plantations [4] in our country, as well as one of the crops having a variety of purposes and high nutritional worth, with a production volume of 104.80 million tons coming from various Indian states. Because our country is the largest producer of rice at the second position in the world, the country's rice-growing region is constantly expanding. It contains a high amount of carbohydrates and protein, as well as a significant amount of dietary fiber and minerals. Plant illnesses are caused by pathogens, fungus, bacteria, viruses, and other microbes in the majority of cases. Rice leaves are sensitive to diseases that are caused by fungus, viruses, and the varied field environment makes it simple for pathogens to infect the leaves[5]. Figure 1 shown below is an example of different diseases of rice leaf.

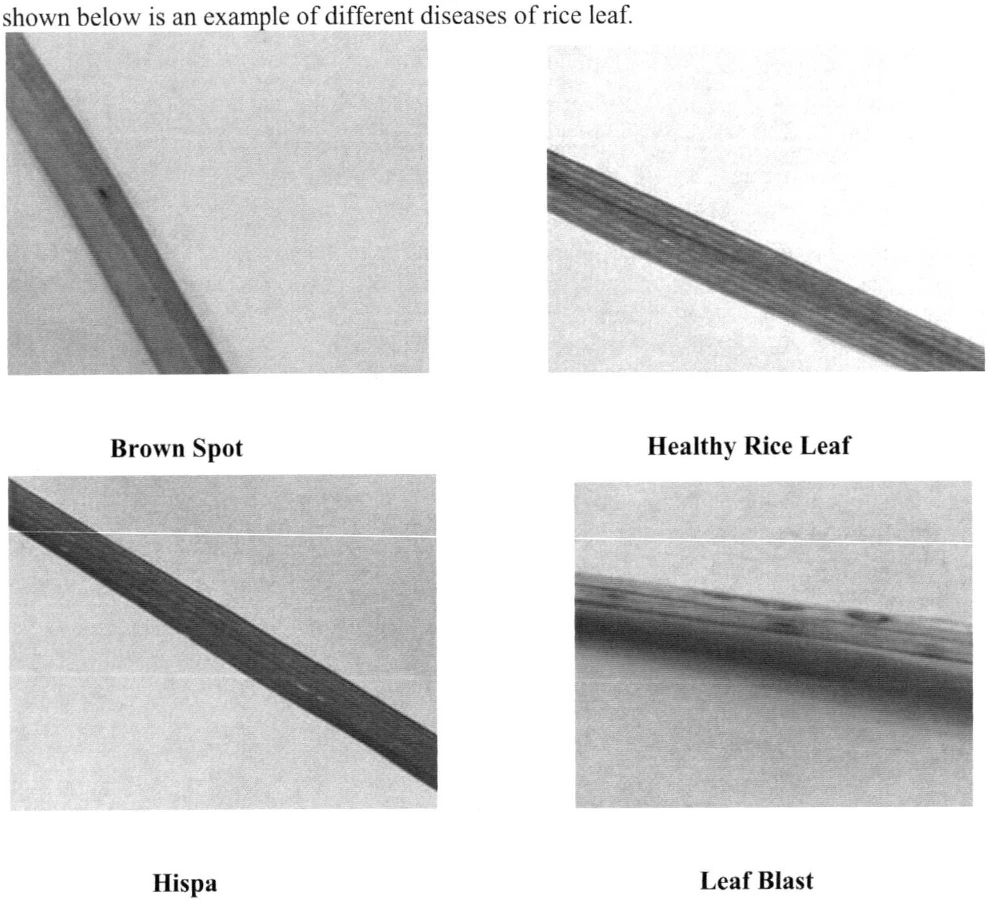

Fig.1. Types of Rice Leaf Diseases[6]

Climate changes [7] will create an ideal environment for those pathogens to thrive. The growth of crops is hampered in their initial stages due to fungi-caused illnesses. If illness strikes while the crop is still growing, it might reduce the crop's yield. Manually determining the presence of illnesses in large agriculture regions is quite challenging. Diseases, particularly in rice plants, have become a problem since farmers are unable to identify leaf disease with the naked eye, and they must consult the expert in order to discover that specific disease, which takes more time and requires much expense. The most frequently occurring diseases in rice leaf are Brown spot, Hispa, LeafBlast, Healthy.

Thus, disease detection in leaves is an important topic that provides many benefits in monitoring large fields of crops. Rice leaf disease can affect yield and quality by damaging the green layer from the leaves. The way to control these rice diseases is to rapidly and precisely detect the disease type and then implement appropriate corrective actions in a timely manner[8]. Using digital image processing techniques and deep learning networks, the detection of disease is efficient, consumes less time, and is accurate. Advances in Computer vision offer an opportunity to extend and increase plant protection[9].

The content of the paper is organized in the following manner: In Section 2, we'll go through a few previous kinds of research that have been done using image processing techniques, ML methods, and DL models for recognizing diseases in the leaves of rice plants. Section 3 is about the different deep learning methods studied for rice leaf disease detection, and the final section concludes the experimental results and future scope.

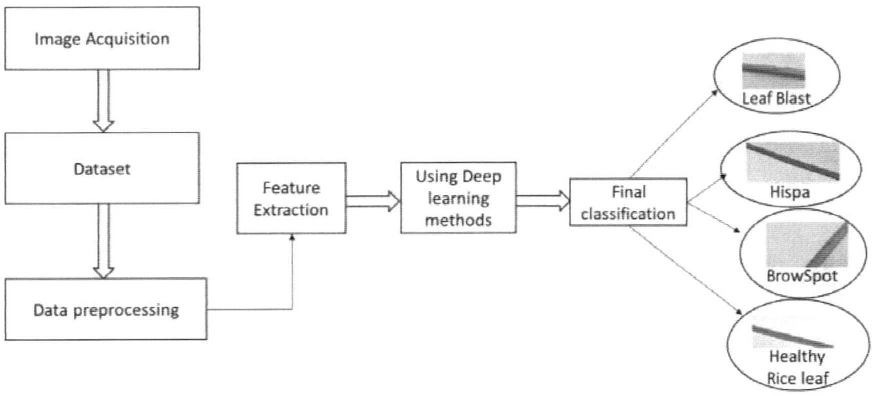

Fig.2. Overview of Proposed Method

In this paper we are working on an automated system to detect the fungal disease of rice plants which is a major cause for a loss of a rice plant. This type of disease may occur and spread due to climate change and moisture on the leaves. The above-mentioned figure (Fig1) represents the entire process of the method which we have proposed.The first step is acquiring the image. The act of obtaining a picture from a source is known as image acquisition. The input can be taken from different resources especially hardware such as sensors or cameras etc. This is a crucial step in the entire process because processing is impossible without an image. This step is always the initial phase in the process. The second step is dataset collection. The dataset in our method consists of a number of images of the four diseases of rice leaf which we are predicting. We train the dataset using a wide variety of images to get better accuracy. The next step is preprocessing of data. Collected data should be preprocessed involving data cleaning and removing all the inconsistencies from the data. We do this process to obtain a clean dataset for achieving better accuracy. Then we perform feature extraction which involves extraction of only those features which are important and which are most required. Then we use various DL methods which we are employing in our method and train and test the dataset accordingly. Finally, the images are classified according to their respective diseases. In our method, we have tested four diseases that affect the rice plant.

2. LITERATURE REVIEW

(HaixiaQi*et al.*,2021) [1], this paper dealt with the automatic identification of groundnut leaf diseases using the stack ensemble technique. The proposed research was conducted on diseased groundnut leaves to identify four groundnut leaf diseases, in this study, they merged deep learning models with traditional machine learning approaches. Deep layer networks, such as ResNet50 and DenseNet121, performed the best when it came to dataset prediction. The maximum accuracy for data augmentation was 97.59 percent.ResNet50 had the best identification performance when integrated with the LR model. (GowriShankar*et al.*,2020) [2], in this paper, they discussed the automatic identification of groundnut leaf diseases. A DL model was employed to increase the network's speed and accuracy in finding and classifying different disease-infected patches on groundnut leaves. To grow the efficiency of previous algorithms, they replaced the typical SVM classifier with KNN for distinguishing four different pathologies (Leaf Blight, Leaf Spot, Stem Rot, and Bud Necrosis).(Ramakrishna *et al.*,2015) [3], this paper discusses one of the most common illnesses that affect ground leaves in its early stages. The suggested scheme incorporates four leading phases for the detection and categorization of groundnut leaf disease. The initial process is to do a color renovation on the images that will be used as input. The plane separation would be the next step. The extraction of features is the next phase. The backpropagation algorithm is employed to detect the leaf disease as a last step. (NilamBhise*et al.*,2020) [4] In the suggested research, disease detection is carried out in two stages. The type of crop and the type of the disease are determined in the first step using a CNN. Tensorflow lite is used to categorize the uploaded picture numerical value to the dataset values, and Keras frameworks are used to classify the dataset values. In terms of disease diagnosis performance, the findings show that the Mobile Net Model outperforms other models.

(Salini *et al.*,2021) [5], The focus of this research is to reduce pesticide use in agriculture while making better quality and quantity of output. For feature extraction, they use image processing techniques, and for classification, they use SVM. To improve performance and provide a better outcome, the model was combined with data augmentation. This research aims to detect three major rice plant diseases: Bacterial Leaf Blight, Brown Spot, and Leaf Blight. The input to this model is the full image for processing, and the output will be the disease that has affected the plant, as well as the model's accuracy. (Mahalakshmi *et al.*,2021) [6], To identify the existence of disease as well as to detect specific types of disease, the author extracted color and texture features of corn leaves then the collected features are categorized using Binary SVM and multi-class SVM. The proposed system's accuracy is 85 percent, which is its best performance. (Saleem *et al.*,2019) [7], this paper reviewed the detection of several plant diseases and their classification by deep learning. Alternative deep learning models and machine learning methods for visualizing plant diseases are examined in this study with the conclusion that deep learning models are more accurate than conventional machine learning techniques. (Shruthi *et al.,*2019) [8], ML techniques were used to describe the steps involved in the detection of general plant disease. They have used a CNN to detect the diseases with high accuracy.

(Yang Lu *et al.,*2018) [9], this article mainly focuses on the identification of diseases on rice crops. To detect diseases in the rice crop, it employs deep Convolutional neural networks.. One of the backdrops of this study is that they used less data for training. (Azath*et al.,*2021) [10], this research paper presents the work on the cotton leaf for the detection of diseases and pest diagnosis using image processing and Deep Learning. The researchers used CNN to detect diseases in cotton leaves.In terms of recognizing specific diseases, the model has a 96.4 percent accuracy rate.Recently, different deep learning methods [11-17] have been applied with Convolutional neural networks and computer vision to detect plant disease, recognition of plant leaves for medicine purposes, pest detection, counting of the wheat head,*etc.* In real-time robots need to detect the plant leaf disease and provide proper pesticide over it in large farms and weed detection with a bounding box helps to remove the weed with herbicides.

The following section includes a tabular column that gives a quick overview of the recent works related to the different plant leaf disease detection. The various machine learning and deep learning algorithms used for identifying the diseases, which we will look at in the next section.

Table.1. – A quick summary of studied research works on the detection of leaf diseases using different classifiers.

Article	Crop	Dataset	Classifier	Accuracy	Advantages	Drawbacks
[1]	Peanut	6029 Lab images	ResNet50 DenseNet121	97.59%	Best prediction performance	It is complicated in field environment
[5]	Rice	Kaggle dataset with 1000 images	Image processing techniques and SVM	80%	Easy to train and flexible model	Low performance
[6]	Corn	1292 lab images	Binary SVM and Multi-Class SVM	85%	Exhibits best performance on the dataset created	Only suitable to cornfield and it is not suitable for other fields
[3]	Groundnut	100 lab images	Back Propagation	97.41%	Best Performance	Works with only one groundnut disease i.e.,

[9]	Rice	500 lab images	Convolutional Neural Network	95.48%	The proposed The solution is both feasible and effective	Used small dataset for training
[11]	Groundnut	250 lab images	KNN	92.14%	Best prediction performance	Need to reduce the probability of false classification
[4]	Papaya	160 lab images	Random Forest	70.14%	Easy to train	Low accuracy
[10]	Cotton	2400 lab images	CNN	96.4%	High accuracy	Doesn't suggest remedies

Findings from the Literature: From the above literature, we found that plant diseases are region-specific due to different environmental conditions and geographic locations. Here many authors proposed a deep learning-based model to detect plant diseases more accurately. We also observed that the deep learning models are suggested only when a large dataset is available. Here, in our paper, we are working on the fungal disease of rice plants and having a dataset of 1600 images. So, we are proposing an automated deep learning model in the next section for the prediction of such diseases.

3. PROPOSED METHODOLOGY

3.1 DATASET DESCRIPTION

In the present study, the leaf dataset consists of four types of diseased rice leaf images; these are Hispa, Brownspot, LeafBlast, and Healthy. This dataset consists of 1600 images of rice leaves with the various symptoms of the diseases, which consists of 400 images of Brownspot, 500 images of Hispa, 300 images of LeafBlast, and 400 images of Healthy. These images are all in JPEG format and have good resolution with a width and height of 1449*1449. The image background is also white so there is no need to apply any background subtraction method [32]. This leaf dataset contains a combination of Hispa, Brownspot, LeafBlast and Healthy. The sample images are shown in Fig.1.

Table 2. Information about the division of dataset

Disease Type	**Train Images**	**Test Images**	**Validation Images**
Hispa	200	150	150
Brownspot	200	100	100
LeafBlast	150	60	90

Healthy	250	100	50

3.2 Data Preprocessing:

In our present world scenario[11], input data involves a lot of noise, has missing values, outliers, and is inconsistent. Data preprocessing involves removal of the noise, missing data, and organizing data in a proper format so that accuracy is increased. It enhances the quality of the data.

This step involves data cleaning, data transformation, and data reduction (data compression)

Data cleaning involves cleaning the data. It removes the noise present in the data. Data transformation is transforming high-level data into low-level data for easier calculations.Data reduction involves reducing the data dimensions so that the data is not high dimensional but the quality of data remains the same. Data cube aggregation which is summarizing the data.

3.3 DEEP LEARNING METHODS FOR CLASSIFICATION

3.3.1 CNN:

Here, we applied a Convolutional neural network (CNN) based approach which is a method of DL that takes input as an image and gives importance to many other objects in the image, as well as differentiates between them. The amount of pre-processing needed by a CNN is substantially less than that required by other classification methods. While simple techniques need the hand-engineering of filters, with enough training, CNN learns these filters/characteristics[15].

Our architecture mainly contains the following layers:

- Convolution layer
- Pooling layer
- Fully connected layer

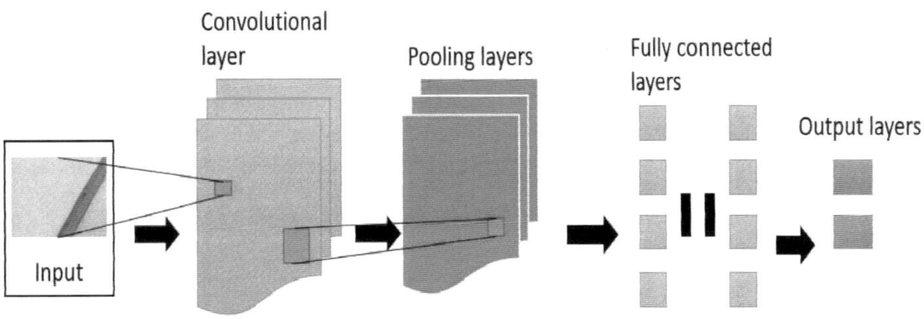

Fig.3. Layers of CNN

The above figure represents the working of CNN. The input in the form of the image after preprocessing the data and extracting the required features when passed through CNN passes through 3 layers of CNN and it is precisely represented. The final output is then displayed.

- **Input Layer:** The input layer of CNN consists of the dataset. The input data will be represented as a 3X3 matrix.
- **Convolution Layer:** A layer that uses filters to learn from smaller sections of input data to obtain features from an image.
- **Pooling Layer:** This layer is used to shrink the image's dimensionality, lowering the processing power required for subsequent layers. There are two variations of pooling. They are:

- ***Max pooling:*** The pixel with the maximum value as input is selected and transferred to the output while parsing input. It is the most used approach compared to average pooling.
- **Fully Connected Layer (Dense):** This is one of CNN's last layers, and it can recognize features that are significantly linked with the output class. The result is a one-dimensional vector created by flattening the pooling layer results.
- **Dropout Layer:** Used to reduce model overfitting problem by removing a random set of neurons in that layer. It is connected with the FC layer.
- **SoftMax Layer:** This is the network's last layer that assists in classifying individual input images of the dataset into several classes depending on the learned properties from the network.
- **Output Layer:** The output layer holds the final classification result.

3.3.2 VGG-19:

VGG-19 is a 19-layer deep CNN. You may use the ImageNet database [17], that has been trained on over a million photographs, to import a pre-trained version of the network. The network can sort photographs into 1000 distinct object categories.

A fixed-size (224 * 224) RGB image was supplied as input to this network, suggesting that matrix was of form (224,224,3). The one and only preprocessing was to eliminate the mean RGB value obtained across the whole training set from each pixel. To cover the complete visual notion, they employed kernels with the size of (3 * 3) as well as a stride size of 1 pixel. Spatial padding was used to maintain the image's spatial resolution. Maximum pooling was obtained with stride 2 across a 2×2-pixel frame. Since prior models used tanh or sigmoid functions, the Rectified linear unit (ReLu) was developed to add non-linearity to improve model classification and processing time, and it proved to be considerably better to them.

3.3.3 VGG-16:

We have used Visual Geometry Group (VGG) 16 layers model. The input image size has set to $224 \times 224 \times 3$, and it is connected to next convolution layer Conv1. Then combination of convolutional and max pooling layers is applied for feature extraction and feature reduction respectively. In a few settings, it includes 11 convolutional filters which are considered to extract the features from input image and send these features to subsequent layers for further processing[18].To extract the features from corners of the image we used "similar padding" of one row and column each side and stride is set to "1". Spatial pooling is done via five max-pooling layers that follow part of the Conv layers.

3.3.4 RESNET:

An Artificial Neural Network Based model ResNet (Residual Network) has been used here to solve the problem of the vanishing/exploding gradient. In this network, we use a technique called skip connections. The skip connection skips a few training steps and connects directly to the output. The shortcut link is introduced to a VGG-19-inspired 34-layer basic network architecture. As a result of these shortcut connections, the architecture becomes a residual network.

A residual neural network (ResNet) is a form of artificial neural network (ANN) based on cerebral cortex pyramidal cell structures. Residual neural networks employ skip connections, also known as shortcuts, to go around some layers. Double- or triple-layer skips with nonlinearities (ReLU) and batch normalization are used in the bulk of ResNet models. We have used the above network in its standard form[1].

3.3.5 Xception:

Xception is a deep CNN with 71 layers. You may use ImageNet to import a pre-trained network that is already trained on a number of input images. The network arranges all the input images into 1000 different categories, such as pencils, pens, books, and many more. The network has a massive library of various representations of features for a long-range of input datasets. The input data size for this pre-trained network is 299 * 299 pixels.

Xception is a convolutional neural network with just convolution layers according to their depth. The Xception architecture's feature extraction building is constructed up of 36 convolutional layers. In our experimental evaluation, we will just look at picture classification, therefore our results will be constrained. Just after the convolutional base layer, regression will be utilized. Fully-connected layers, as illustrated in the example, can be introduced before the logistic regression layer. The experimental assessment portion. Six categories are used to categorize the 36 convolutional layers. There are a total of 14 modules, each having its own linear residual connection. Everything else is constructed around them, with the exception of the first and last modules.

3.3.6 5-Layer CNN:

A five-layerCNN[19] is similar to a three-layer CNN in the exception that it has two additional layers apart from the five Layer CNN at the end. As usual, the input image went through three layers of the CNN and the output is passed through 2 additional layers namely:

- Dropout
- Activation Functions

Dropout: This is done when the output after the FC layer causes overfitting of the model. Overfitting generally occurs when the model works very well such that it gives a negative impact on the performance of the model.

In dropout, due to overfitting of the model, a few neurons are removed randomly from the

IOP Conf. Series: Earth and Environmental Science 1032 (2022) 012017 doi:10.1088/1755-1315/1032/1/012017

NN of the training process with reduced model size.

Activation Functions: It is amongst the most necessary parameters of the CNN Model. It defines the complex relationship between the variables in the model[20]. It decides the start of the network and the end of the network in the NL. The common activation functions used areReLu, tanH etc.

4. EXPERIMENTAL RESULTS

We have studied various deep learning algorithms such as VGG 16, VGG 19, ResNet, Xception, and 5D CNN on the same rice dataset to measure the accuracy of each method.The below screenshot (Fig 3) presented the training process. Whereas training and validation accuracy were shown in Fig 4, where the y-axis shows the accuracy obtained after each iteration represented in the x-axis. Similarly, Fig 5 contains the training and validation loss, where the y-axis shows the loss (in percentage) when training started and thereafter increase or decrease in loss after each iteration represented in the x-axis.

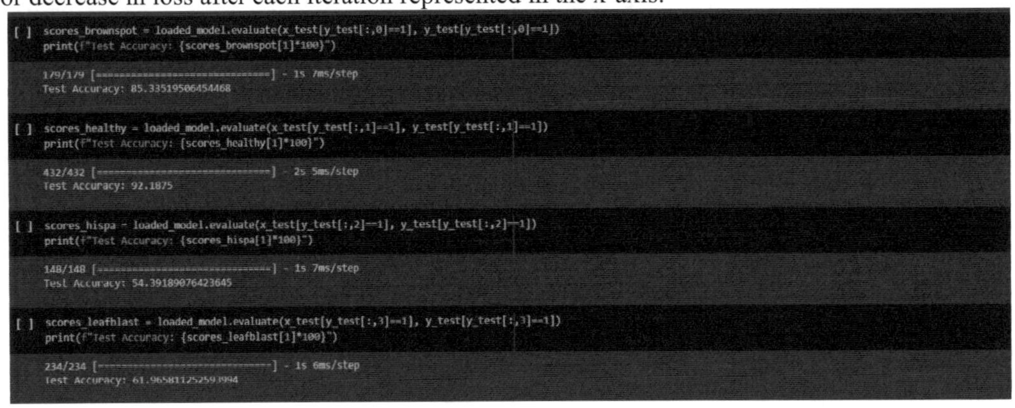

Fig.4. Test accuracy of Brownspot, Healthy, Hispa and Leafblast

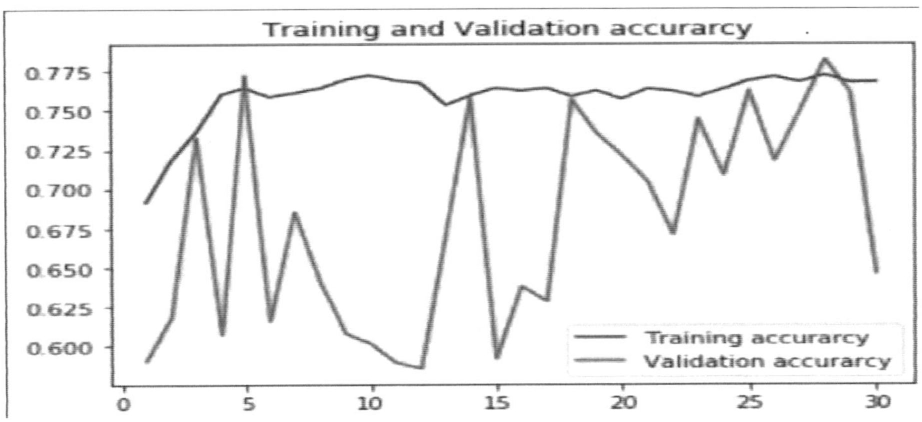

Fig.5. Graph of training and validation accuracy

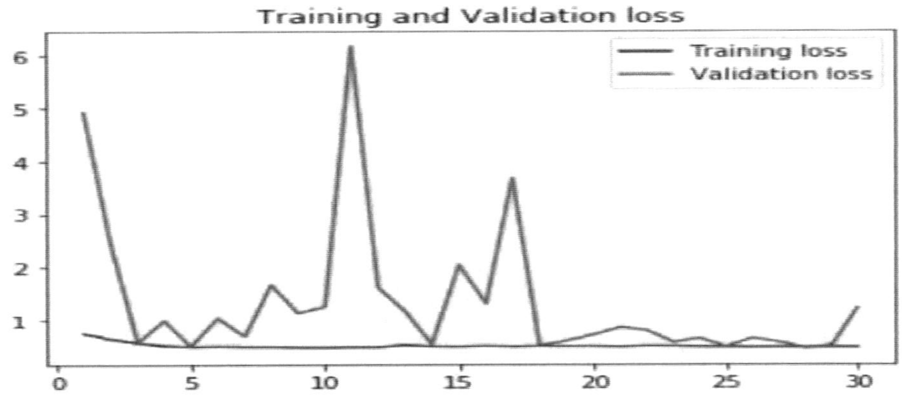

Fig.6. Graph of training and validation loss

Table.3. Accuracy of different DL Methods

Based Model	Accuracy
VGG 16	58.4%
VGG 16 (with first three block of frozen)	72.2%
VGG 19	72.4%
XCeption	72.2%
ResNet50	72.2%
5-layer convolution	78.2%

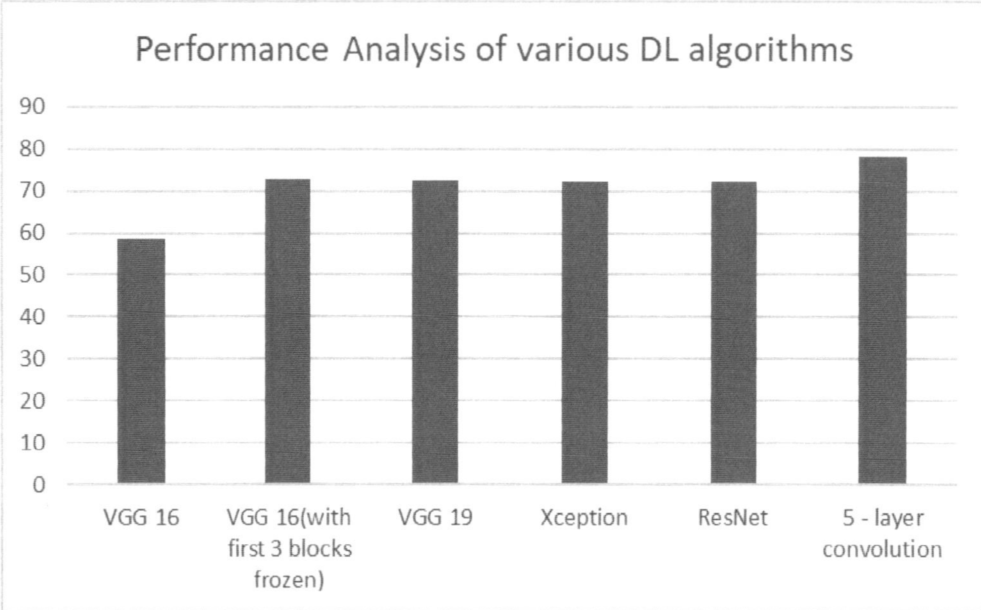

Fig.7. Performance analysis of various DL Methods based on the accuracy

5. CONCLUSION AND FUTURE SCOPE

In this study, we have performed the classification of various rice leaf diseases using a few DL methods for four rice leaf diseases. We used a dataset of diverse rice leaves with illnesses, which we subsequently processed using several standard deep learning methods like VGG19, VGG16, Xception, Resnet, along with a handmade 5-layer convolutional network. We discovered that the 5-layer convolutional network performs the best at identifying rice leaves out of all of them. From Fig 6 we can conclude that the accuracy of our proposed 5 layer CNN model is approximately 6 percent more than the other standard deep learning models. Also, we observed that by adjusting the training parameters like learning rate, number of epochs, and optimizer methods, we can get significant accuracy with a handmade model having less number of layers than the other traditional models. The better we can detect infections, the simpler it will be for farmers to protect their crops. In the future, we will broaden the scope to include more diseases and algorithms, making disease detection vast, easier and faster.

REFERENCES

[1] Qi, H., Liang, Y., Ding, Q., & Zou, J. (2021). Automatic Identification of Peanut-Leaf Diseases Based on Stack Ensemble. Applied Sciences, 11(4), 1950.

[2] Rahman, C. R., Arko, P. S., Ali, M. E., Khan, M. A. I., Apon, S. H., Nowrin, F., & Wasif, A. (2020). Identification and recognition of rice diseases and pests using convolutional neural networks. Biosystems Engineering, 194, 112-120.

[3] Ramakrishnan, M. (2015, April). Groundnut leaf disease detection and classification by using back propagation algorithms. In 2015 international conference on communications and signal processing (ICCSP) (pp. 0964-0968). IEEE.

[4] Ramesh, S., Hebbar, R., Niveditha, M., Pooja, R., Shashank, N., & Vinod, P. V. (2018, April). Plant disease detection using machine learning. In 2018 International conference on design innovations for 3Cs compute communicate control (ICDI3C) (pp. 41-45). IEEE.

[5] R Salini, A.J Farzana, B Yamini, (2021). Pesticide Suggestion and Crop Disease classification using Machine Learning , 63(5), 9015-9023.

[6] Mahalakshmi, S. D., & Vijayalakshmi, K. (2021). Agro Suraksha: pest and disease detection for corn field using image analysis. Journal of Ambient Intelligence and Humanized Computing, 12(7), 7375-7389.

[7] Sony, A. (2019). Prediction of Rice Diseases Using Convolutional Neural Network (in Rstudio). *I*nt. J. Innov. Sci. Res. Technol, 4(12), 595-602.

[8] Lu, Y., Yi, S., Zeng, N., Liu, Y., & Zhang, Y. (2017). Identification of rice diseases using deep convolutional neural networks. Neurocomputing, 267, 378-384.

[9] Lu, Y., Yi, S., Zeng, N., Liu, Y., & Zhang, Y. (2017). Identification of rice diseases using deep convolutional neural networks. Neurocomputing, 267, 378-384.

[10] Zekiwos, M., & Bruck, A. (2021). Deep Learning-Based Image Processing for Cotton Leaf Disease and Pest Diagnosis. Journal of Electrical and Computer Engineering, 2021.

[11] A. Jayakumar, "Detection and Classification of Leaf Diseases in Maize Plant using Machine Learning MSc Research Project Data Analytics." [Online]. Available: https://agriculturegoods.com/why-is-agriculture-important/

[12] Vaishnnave, M. P., Devi, K. S., Srinivasan, P., & Jothi, G. A. P. (2019, March). Detection and classification of groundnut leaf diseases using KNN classifier. In 2019 IEEE International Conference on System, Computation, Automation and Networking (ICSCAN) (pp. 1-5). IEEE.

[13] Shrivastava, V. K., Pradhan, M. K., Minz, S., & Thakur, M. P. (2019). Rice plant disease classification using transfer learning of deep convolution neural network. International Archives of the Photogrammetry, Remote Sensing & Spatial Information Sciences.

[14] Hasan, M. J., Mahbub, S., Alom, M. S., & Nasim, M. A. (2019, May). Rice disease identification and classification by integrating support vector machine with deep convolutional neural network. In *2019 1st international conference on advances in science, engineering and robotics technology (ICASERT)* (pp. 1-6). IEEE.

[15] Shrivastava, V. K., & Pradhan, M. K. (2021). Rice plant disease classification using color features: a machine learning paradigm. *Journal of Plant Pathology*, *103*(1), 17-26.

[16] Ahmed, K., Shahidi, T. R., Alam, S. M. I., &Momen, S. (2019, December). Rice leaf disease detection using machine learning techniques. In 2019 International Conference on Sustainable Technologies for Industry 4.0 (STI) (pp. 1-5). IEEE.

[17] Ramesh, S., &Vydeki, D. (2018, September). Rice blast disease detection and classification using machine learning algorithm. In 2018 2nd International Conference on Micro-Electronics and Telecommunication Engineering (ICMETE) (pp. 255-259). IEEE.

[18] Matin, M. M. H., Khatun, A., Moazzam, M. G., & Uddin, M. S. (2020). An Efficient Disease Detection Technique of Rice Leaf Using AlexNet. Journal of Computer and Communications, 8(12), 49.

[19] Liang, W. J., Zhang, H., Zhang, G. F., & Cao, H. X. (2019). Rice blast disease recognition using a deep convolutional neural network. Scientific reports, 9(1), 1-10.

[20] Sethy, P. K., Barpanda, N. K., Rath, A. K., & Behera, S. K. (2020). Deep feature based rice leaf disease identification using support vector machine. *Computers and Electronics in Agriculture*, *175*, 105527.

[21] Dataset available at: https://www.kaggle.com/minhhuy2810/rice-diseases-image-dataset, seen on 12 January.

IOP Conf. Series: Earth and Environmental Science 1032 (2022) 012018 doi:10.1088/1755-1315/1032/1/012018

Developmental Strategies for a Water Quality Assessment Model with Limited Datasets – A Case Study from River Bhavani, India

A Ravinashree[1], C Sivapragasam[2,*], M Vasudevan[3,*]

1Post-Graduate Student, Department of Civil Engineering, Kalasalingam Academy of Research and Education, Krishnankoil- 626126, Tamil Nadu, India.

2Senior Professor, Department of Civil Engineering, Kalasalingam Academy of Research and Education, Krishnankoil- 626126, Tamil Nadu, India.

3Assistant Professor, Department of Civil Engineering, Bannari Amman Institute of Technology, Sathyamangalam – 638401, Tamil Nadu, India.

*Corresponding authors Email: vasudevan@bitsathy.ac.in; sivapragasam@klu.ac.in

Abstract. Understanding the significance of rational attributes for a water quality model depends largely on the availability and reliability of virgin datasets pertaining to their spatial-temporal variations and interactions. Present study aims to identify certain crucial attributes for deriving a water quality model for a selected reach of River Bhavani in Tamil Nadu, India using the results of the QUAL2K model based on a limited dataset. The spatial variations in the concentrations of inorganic suspended solids (ISS) and carbonaceous biochemical oxygen demand (cBOD) were studied in detail corresponding to the possible variations in other model parameters (pertaining to the hydraulic, environmental and climatic conditions). The results indicate that the cBOD has a significant dependency on temperature, dissolved oxygen and pH, while the ISS has shown more dependency on alkalinity, conductivity and bottom algae. It is further inferred that variations in pH and dissolved oxygen can impose triggering effects when external pollution loads (point and nonpoint sources) are introduced while generating the corresponding profiles of ISS and cBOD. Based on the results, the conceptual strategies for developing a data-limited model is presented, which can be further developed to simulate water quality profiles for similar data-limited scenario.
Key Words: Biochemical oxygen demand; Data-limited Modelling; Inorganic suspended solid; QUAL2K; Water quality model

1. Introduction

The global scenario of accessibility for clean and safe drinking water is certainly challenging as driven by the exploitative captivation of surface and ground water resources. The key challenges presently existing and are likely to emerge in the near future pertaining to the water sector can be broadly understood under the classification of accessibility, availability, quality, ecosystem impacts, crisis preparedness and shared governance. As evidenced from many earlier studies, river pollution is one of the most serious challenges faced by the water sector of the present age. Rivers have become invariably the immediate recipients and conduits of different types of wastewaters as imposed by the growing population adding with higher industrial production requirements. Unfortunately, most of the perennial rivers are on the verge of imminent bereavement and eventual transition to mere effluent drains as they are practically supplemented by the incessant sewage flows. In order to address some of the existing conflicts in water resources management activities, usage of surface water quality models can play a major role by simulating the spatial-temporal variations in the pollutant concentrations thereby depicting appropriate strategies for effective waste management and ecosystem restoration.

As the source and nature of pollutants entering into a water body vary from place to place, it is quite difficult to ascertain their exact distribution within a given water resource profile without having sufficient data collection and monitoring network. Various attempts were made earlier to devise reliable systems for collection, monitoring and assessment of real-time water quality data from the rivers for the purpose of developing water quality models [1-4]. However, many of such models were intrinsically limited in their applications in terms of site-specific characterization of model parameters, requirement of huge dataset and incomplete derivation platform for developing futuristic decisions for pollution abatement/restoration of resources. Hence selection of suitable water quality assessment tools is very crucial to understand the present situation, to simulate futuristic scenarios and to derive a meaningful decision-support tool.

Though many sophisticated simulation tools are being developed for surface water quality assessment by various research groups, the open source tool QUAL2K (and its previous version QUAL2E) has been quite extensively used in simulating the water quality profile for medium and large-scale river basins [1]. Cho and Ha [5] had performed parametric optimization for QUAL2K model using calibration with influence coefficient algorithm. They reported that the majority of the errors were found for dissolved oxygen and chlorophyll-A concentrations in a multi-reach river simulation. Zhang et al. [6] developed a water quality improvement model using QUAL2K for Taihu Lake Basin, China to select an optimal management program. Fan et al. [7] reported a combined modeling approach using QUAL2K and HEC-RAS to assess the impact of tidal effect on river water quality. They have concluded that such calibration essentially requires enormous corrections on reaction rate constants especially when the location is suffering from data-insufficient condition. Chowdhury et al. [8] attempted scenario-based modeling for Sungai Selangor Basin using QUAL2K for remediation management and planning. They have used the water quality index based on a minimum number of parameters (DO, BOD, COD, SS, NH3-N and pH) and validated QUAL2K results with the observed readings.

In general, QUAL2K is mainly employed for simulating various waste loading scenarios in polluted streams as part of basin-based environmental management plan [6-7]. In order to integrate various water quality models based on a conceptual interpretation, Bui et al. [9] successfully attempted to utilize the inputs of the SWAT model (in terms of pollutant concentrations and loading rates) as a feed for the QUAL2K model. Vasudevan et al. [10] studied the waste loading scenario in River Yamuna for a certain reach near Delhi using QUAL2K modeling with detailed analysis. However, one of the prominent research topics related to the water quality assessment is the DO-BOD interactions which were modeled extensively using QUAL2K [7, 11-12].

Since most of the water quality models are influenced by the parametric sensitivity and scalability of the datasets, the data-driven models using soft computing techniques are emerging as reasonable solutions to predict the trend of common water quality features. One specific advantage of these models is that the emphasis on obvious input parameters can be isolated and a rational relationship can be brought out which is purely acceptable based on the statistical evidence. There are many studies related to the prediction of water quality (in terms of dissolved oxygen, biochemical oxygen demand, electrical conductivity etc.) using soft computing methods such as genetic programming and artificial neural networks [13-15]. Such models are useful to predict the discharge variations and eutrophication (in terms of algal blooming) in water bodies [16-17]. Such models provide avenues to address the constraints such as restricted parameters with limited dataset. In some cases, a non-parametric approach is considered to be better that can eventually address the issues pertaining to limited data sets and constrained repeatability.

Though the QUAL2K model is quite user-friendly in developing the simulation scenario and capable of providing free-formatted results, a realistic representation of the field requires a huge number of input data owing to account for most of the possible complex interactions happening in a selected river section. This is particularly difficult to apply under situations of limited measured data in order to justify the prevailing assumptions and to verify the results. In an attempt to address this issue, the present study investigates the identification of crucially influencing parameters and their parametric sensitivity for two categorically representative pollutants (inorganic suspended solids – ISS and carbonaceous biochemical oxygen demand – cBOD). The simulation is particularly carried out for a selected stretch of River Bhavani located in Sathyamangalam, Tamil Nadu, India. Considering the limitations on acquiring all the required dataset for the QUAL2K model, the present study aims to provide a sensible characterization of various developmental strategies for simulating futuristic water quality under limited datasets.

2. Materials and Methods

2.1. *Study Area*

The study area consists of a selected stretch of River Bhavani at Sathyamangalam comprising of 5km longitudinal distance (from $11^0$30’18.30” N & $77^0$13’53.50” E to $11^0$28’50” N & $77^0$15’37.54” E) covering the key aspects of the localities as mentioned in Figure 1. The Bhavani River originates from the Western Ghats' Nilgiri Hills, runs through Kerala's Silent Valley National Park, and returns to Tamil Nadu. It is a perennial river having a length of 217 km, draining from a watershed area of 0.62 million hectares which is scattered throughout the states of Tamil Nadu (87%), Kerala (9%) and Karnataka (7%). The Bhavani sub-basin is primarily characterized as agriculture-based with supportive forest and plantation area in the upper regions. It is provided with an intensely irrigated agricultural system with remarkable groundwater depletion. In order to ascertain the intricate aspects of water pollution in River Bhavani, a few point and nonpoint sources were identified for sampling as shown in Figure 1. The selected points are representative sources of unconnected drainage outlets of municipal effluents which are directly discharged to the river without any treatment.

Figure 1: Location of study area with identification of sampling points (PS indicates point sources and NPS indicates nonpoint sources)

2.2. *Data Collection*

The closed-type layout of the entire river sub-basin has the advantage of effective utilization of surface water if sufficiently harvested. Therefore, the biggest disadvantage of this layout is the persistence of pollutants emerging from the neighboring industrial belt. The primary data collection includes the monitoring and analysis of basic water quality parameters such as pH, alkalinity, electrical conductivity, inorganic suspended solids (ISS), dissolved oxygen (DO) and biochemical oxygen demand (cBOD) from the selected points of reference within a reach of 5km as shown in the Figure 1. The spatial sampling method was consistently followed to study the variations in the distribution of pollutants over the selected stretch of the river. Moreover, the selected QUAL2K model requires steady state conditions which has been safely considered to be a typical representation of the extreme conditions of persisting pollution loading to the river. Hence, temporal variations in the pollutant concentrations are omitted in the present study although a seasonal variation can be of significance to study the futuristic impacts of the modeling strategies. Apart from the water quality data, other required information about the hydraulics and weather conditions of the study area were gathered from the literature about the study area with suitable assumptions which are explained in the upcoming sections.

2.3. *Model Description*

The selected QUAL2K model is a one-dimensional, steady state, non-uniform flow and transport model where the channel is assumed to be well-mixed both vertically and laterally. It is a very simple spreadsheet based model which can be downloaded (https://www.qual2k.com/) and installed on any medium-sized processor-system. In addition to the flow simulation, the heat budget and temperature profiles can also be modeled as a function of meteorological parameters on a diurnal time scale. Another salient feature of this model is the inclusion of extensive modeling parameters pertaining to the kinetics of water quality which enables one to simulate the selected features on a daily/hourly time scale. As the focus of the present study is to identify the significant factors for simulating the water quality profile under limited dataset conditions, a reference condition is initially defined with all available data in order to serve as a means for the comparison of the results. The selected parameters and their values are described in Table 1.

Table 1: Sample database prepared for calibrating the QUAL2K model for the given scenario

Parameter type	Parameter description	Value	Unit	Remarks/ Significance
Hydraulic	Headwater flow rate	1.5	m^3/s	Normal range (1-10); selected for calibration
	Flow measuring method	Manning's formula	-	To compare with a weir, rating curve or Manning's formula
Water quality	Temperature	27	C	Can vary diurnally
	Electrical Conductivity	800	umhos	Chance of many dissolved ions
	Inorganic Solids	30	mg D/L	Presence of sediment, mineral soils; highly variant
	Dissolved Oxygen	5	mg/L	Vary between 4-8 depending on pollution
	Carbonaceous Biochemical Oxygen Demand (fast degrading type)	5	mg O_2/L	Readily degradable organic matter
	Organic Nitrogen	1800	ug N/L	Mainly from nonpoint source pollution
	NH_4-Nitrogen	500	ug N/L	Depends on undigested organic matter from point source
	NO_3-Nitrogen	300	ug N/L	Depends on degradation of organic matter from point source
	Organic Phosphorus	10	ug P/L	Mainly from nonpoint source pollution
	Pathogen	30	cfu/100 mL	Based on total coliforms
	Alkalinity	150	Mg $CaCO_3$/L	Total value
	pH	8	-	Depends on water flow rate and mixing rate

2.4. *Model Calibration*

The QUAL2K model is fed with the necessary initial dataset to run for the hydraulic, geometric and climatic conditions for River Bhavani. Prior to further trails, the model is first calibrated by adjusting the hydraulic parameters based on the observed discharge values. For this, three conditions were considered as possible means of accessing realistic discharge values – Manning's formula, Weir formula and Rating curve. For each case, the model is run to fine-tune the corresponding flow model parameters (velocity coefficient, depth coefficient and roughness coefficient) while other parameters were kept unaltered during each run in order to check the satisfactory match with the reported results.

2.5. *Sensitivity Analysis*

The model is further analyzed for the sensitivity of various model parameters such as discretization limit, headwater quality parameters, lateral loading of pollutants, diurnal variations in temperature and dissolved oxygen as well as the variability in pollutant loading rates. A series of case studies were framed for simulation experiments by varying the limiting values under low, medium and high categories. This approach will help not only to identify the sensitivity towards individual parameters, but also to estimate the extent to which the results (i.e. downstream water quality profile) can vary. The details of the case study conditions are represented in Table 2.

Table 2: Description of the case studies used in the QUAL2K model

S. No.	Parameter	Value	Remarks
		Basic database maintained	
1.	River length	5.0 km	Representative of the study area with identified pollution sources
2.	No. of reaches	5	Selected after trails on the optimum number of flow elements
3.	Headwater flow rate	1.5 m^3/s	Representative of the average flow for the last 15 years
		Influent water quality parameters	
4.	ISS	10, 100 and 500 (mg/L)	Typical range limits selected based on extremely cases of low flow and flood flow
5.	DO	4, 7 and 10 (mg/L)	Typical range limits selected based on literature values for varying oxygenation potential
6.	cBOD	5, 10 and 20 (mg/L)	Typical range limits selected based on observed and reported values representing the background concentration
		Diurnal variations	
7.	ISS	10, 100 and 500 (mg/L)	Similar range selected based on the diurnal peak factor and flow variability due to the influence of change in effluents
8.	DO	4, 7 and 10 (mg/L)	
9.	cBOD	5, 10 and 20 (mg/L)	
10.	pH	4, 7, 10 (-)	

3. Results and Discussion

3.1. *Calibration Results*

The simulation results of the QUAL2K model for the selected reach of River Bhavani has been analyzed on a pragmatic conceptual platform in order to draw some plausible conclusions for developing a simplified data-limited model. Based on the calibration tests with flow parameters, the model is found to be quite stable and consistent with the hydraulic components of simulation. The simulated values of discharge rate using the three conditions were found to be consistent for the selected reach of the river (Figure 2).

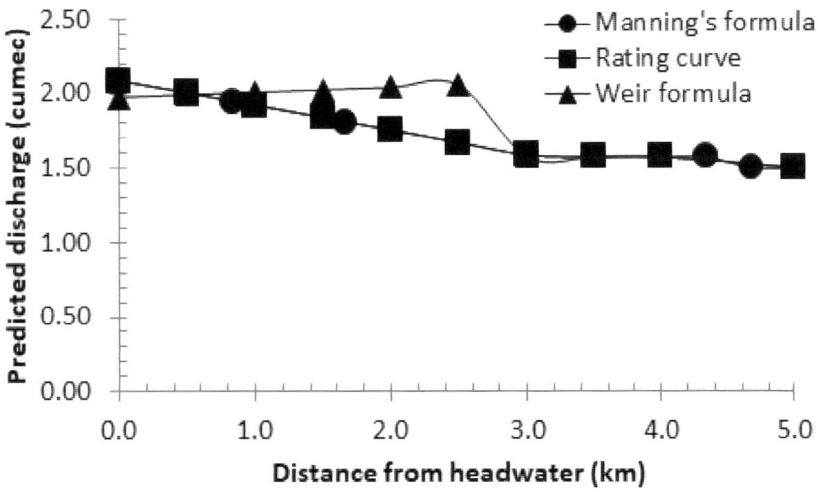

Figure 2: Calibration results for flow simulation using QUAL2K

3.2. *Simulated Contamination Profile for Inorganic Pollutants*

The spatial variations of organic and inorganic pollutants in the river are simulated for the selected reach of the river. It is observed that the inorganic pollutants have reached high concentrations towards the outer end when considered as a constant point source at the selected entry point. However, the concentration values are found to be lower after assuming a settling rate of 0.5 m/day for the ISS (Figure 3). This is in accordance with the assumption that the pollutants carried by the influent can be quite turbid in nature and may get deposited to the river bed as they travel with the given low discharge value ($1.5 m^3/s$). The same trend can be extended to analyze the expected profile of suspended solids within the vicinity of any point source of pollution. However, it is to be understood that the resuspension of sediments due to the turbulence can be a limiting criteria within the mixing reach of such loading.

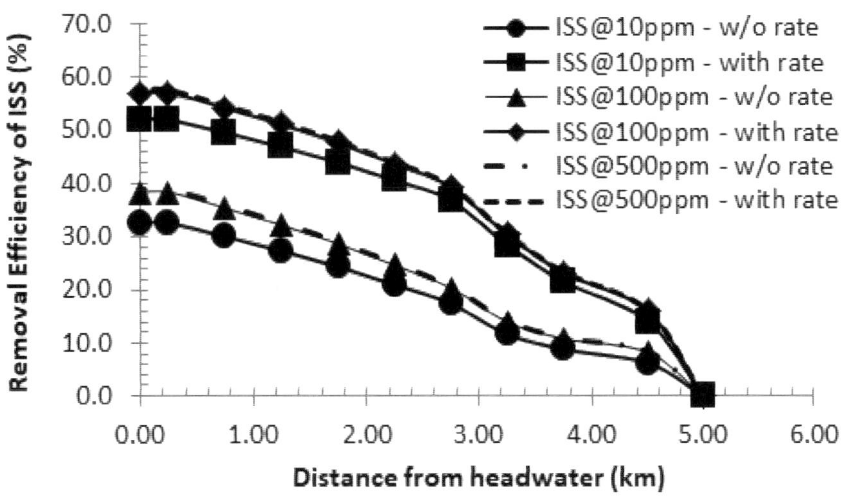

Figure 3: Comparison of removal efficiency of inorganic suspended solids (ISS) within the reach for various input and loading conditions

3.3. *Simulated Contamination Profile for Organic Pollutants*

The extent of organic pollution in the river is studied in terms of the variations in DO and cBOD using various simulation case studies as mentioned in Table 2. Based on the close analysis, the concentration of DO is found to be directly dependent on the temperature profile in the river in consideration with its diurnal as well as seasonal variations. It is observed that the concentration profile of cBOD increased drastically within the selected regime despite the fact that DO is sufficiently supplied through reaeration throughout the simulation. When the initial cBOD is low (5mg/L), the depletion in DO was higher when compared to the increased input of cBOD (20mg/L). In addition, the increase in cBOD profile towards the outer end of the selected reach shows the significant impact of waste loading situations. This indicates the limited influence of dilution in absence of turbulence in bringing back the DO levels. However, it is observed that there is no significant increase in the ISS concentrations for the corresponding increase in cBOD indicating the predominant organic nature of the point sources emanating from the Sathyamangalam Township. The higher values of cBOD are of concern in order to provide sufficient degradation period as they are assumed to be fast degrading species.

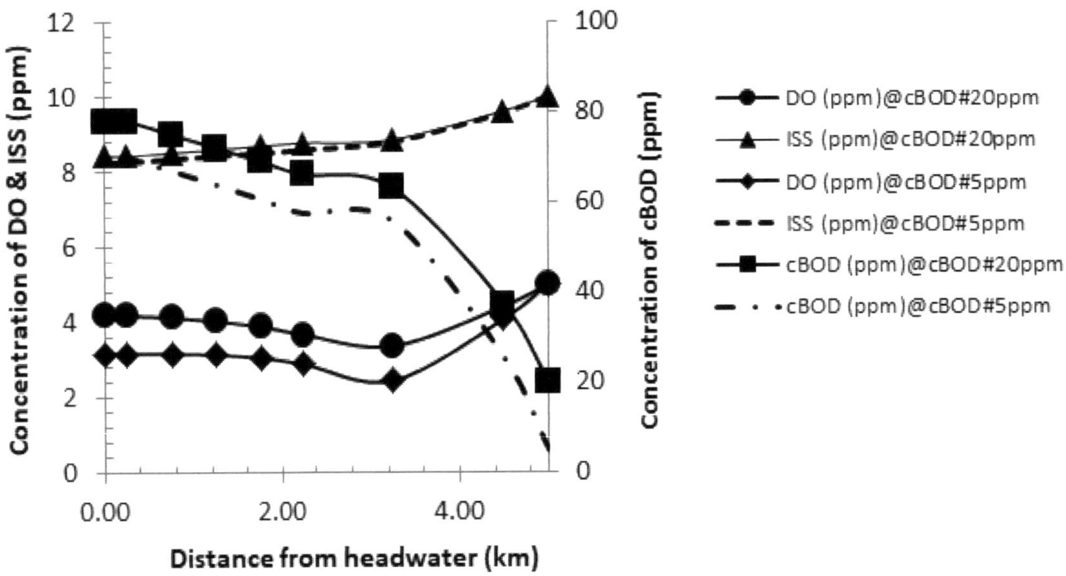

Figure 4: Comparison of simulated concentrations of dissolved oxygen (DO), inorganic suspended solids (ISS) and carbonaceous biochemical oxygen demand (cBOD) for various initial conditions of cBOD within the reach

Based on the above mentioned results, the influence of organic pollutants on the overall environmental condition of the river can be understood in terms of the three simulated parameters, viz. DO, ISS and cBOD. However, in order to justify the selection of these parameters, the influence of other most prominent parameters was also analyzed. It is found that there is a drastic decrease in the electrical conductivity (EC) values under low DO conditions (4 mg/L) whereas the decrease in EC seems to be marginal for higher DO values (7 and 10 mg/L) (Figure 5). Since the river is observed to receive large amounts of organic-dominated effluent, the apparent change in the concentration of inorganic species seems to be limited mainly towards gravitational settling. Based on the results, the overall biodegradation potential of the river can be expected to be still viable under the selected ranges of the organic pollutants in order to abate the pollution loads within a certain reach limit. However, this is not the case when the model is to be used for an extreme contamination scenario.

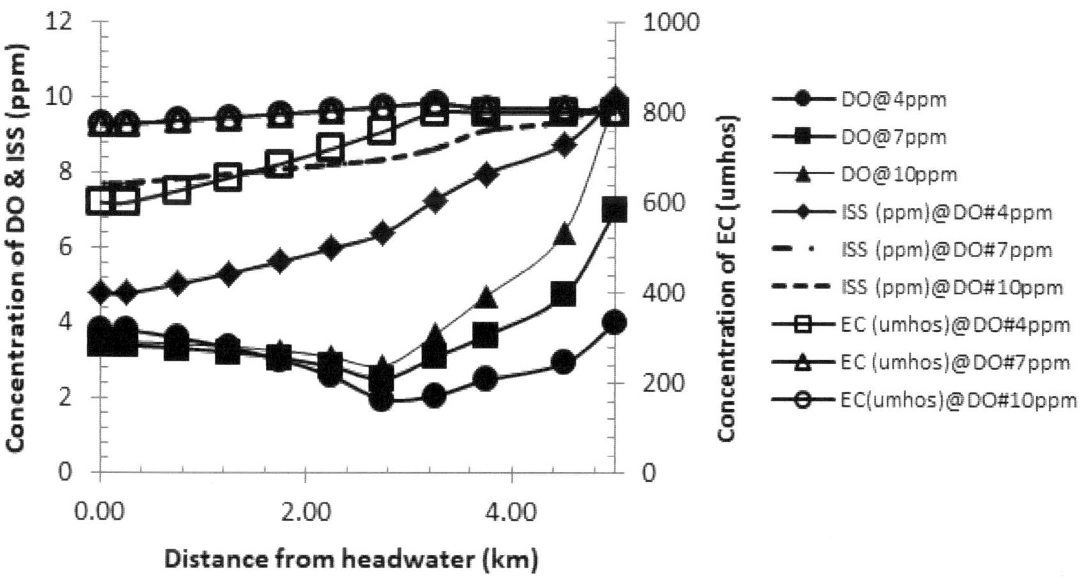

Figure 5: Comparison of simulated concentrations of dissolved oxygen (DO), inorganic suspended solids (ISS) and electrical conductivity (EC) for various initial conditions of DO within the reach

3.4. *Strategies for Modelling River Water Quality*

The success of management operations in the water resources depends largely on the effective monitoring and protection of water quality based on the available data [18-19]. As observed with the above results, the organic pollutants can be understood to have an unsurpassable role on the overall environmental condition of the river despite the incremental changes in the distribution of inorganic species. Quite apart from performing a rigorous model developmental approach, it is important to adopt and articulate a comprehensive strategy for identifying the important factors and necessary steps to derive a data-limited model for futuristic prediction. As highlighted in the introduction as well as based on the experience from the study, an outline for a data-driven modeling framework is proposed here (Table 3) with a view to extend the modeling exercises towards successful prediction of water quality under practical constraints on data availability.

Table 3: Proposed strategies for selecting modeling constraints for water quality assessment under data-limited conditions

Strategy type	Suggested steps	Suggested factors	Remarks from experience
Selection of inputs with limited records	Calibration and sensitivity analysis	Statistical error limits	Dependency on relevant literature is unavoidable. Model performance was satisfactory based on three-level sensitivity studies.
	Correlation of grouped parameters	Correlation and regression factors	
	Adaptation of design of experiments procedure	Dimensionality based on principal component analysis or clustering	
Data split for training	Optimization-based supervised methods	Statistical error limits	Not required in the selected model

Selection of model structure and architecture	Selection between linear and non-linear input-output functions	Computational speed and efficiency	Over-dependency on large number of model coefficients can be minimized by adopting simplified rule-based selections
	Selectivity of data-based models for hybrid approach		
Scope of validation	Trials on extended domains of same study area (replicability)	Statistical error limits	Selected model is quite extendable for higher reaches. Scenario-based simulations resulted in good predictability
	Application to similar scenario (or futuristic) (predictability)		
	Reduction of uncertainties (structural validity)		

4. Conclusions

The present study focuses on the water quality assessment using the QUAL2K model for a perennial stream under data limited conditions. The QUAL2K model is calibrated using the available hydraulic and environmental data for a limited stretch of River Bhavani in Tamil Nadu and further analyzed for the sensitivity of various model parameters. A number of case studies were developed to study the response of the model for the selected range of parameters. The model results suggest that ISS, DO and cBOD can frame the basic structure of the model if no other details of pollutants are available. The influence of DO on both organic and inorganic pollution was found to be critical in this study. The biodegradation potential of the river can be expected to be consistent under the selected ranges of the organic pollutants. However, this is not the case when the model is to be used for an extreme contamination scenario. The scope of developing a data-limited model can be evaluated based on the strategies for selection of input parameters, data splitting, selection of model structure and validation plan.

References

[1] Chapra SC 2008 *Surface water-quality modeling* (Waveland Press, Inc.) p 235-257
[2] Wu W, Dandy GC and Maier HR 2014 *Environ Model Soft* **54** 108-127
[3] Moriasi DN, Gitau MW, Pai N and Daggupati P 2015 *Trans ASABE* **58** 1763-85
[4] Ji ZG 2017 *Hydrodynamics and water quality: modeling rivers, lakes, and estuaries* (John Wiley & Sons) 473-503
[5] Cho JH and Ha SR 2010 *Sci Total Environ* **408** 1985-91
[6] Zhang R, Qian X, Yuan X, Ye R, Xia B and Wang Y 2012 *Int J Environ Res Public Health* **9** 4504-21
[7] Fan C, Ko CH and Wang WS 2009 *J Environ Manage* **90** 1824-32
[8] Chaudhary S, Dhanya CT and Kumar A 2018 *Hydrol Res* **49** 1042-55
[9] Bui HH, Ha NH, Nguyen TN, Nguyen AT, Pham TT, Kandasamy J and Nguyen TV 2019 *Ecohydrol Hydrobiol* **19** 210-23
[10] Mangottiri V, Nambi IM and Suresh Kumar G 2011 *Int J Adv Engg Technol* **2** 336-344
[11] H Ismail A and A Abed G 2013 *J Kerbala Univ* **9** 257-73
[12] Kalburgi PB, Shareefa RN and Deshannavar UB 2015 *Int J Engg Manufa* **5** 15-25
[13] Vanitha S, Sivapragasam C and Nampoothiri NV 2016 *Int Conf Theor Comput Sci Discr Math* (Springer, Cham) p 445-452
[14] Mahalakshmi R, Sivapragasam C and Vanitha S 2019 *Inform Commu Technol Intelligent Sys* (Springer, Singapore) p 401-408

[15] Sivapragasam C, Jegatheesan V, Arun VM and Vanitha S 2010 *Int J Eng Sci Technol* **2** 3128-36
[16] Sivapragasam C, Muttil N, Muthukumar S and Arun VM 2010 *Marine Pollu Bullet* **60** 1849-55
[17] Sivapragasam C and Muttil N 2005 *Water Resour Manage* **19** 505-20
[18] Singh RP, Mangottiri V and Pandiyan B 2020 *IOP Conference Series: Materials Science and Engineering* Ed Vasudevan M, Soundara B, Karthiga N and Preetha V (IOP Publishing) p 012085
[19] Sivakkumar SN, Mangottiri V, Narayanan N and Saragur SK 2020 *IOP Conference Series: Materials Science and Engineering* Ed Vasudevan M, Soundara B, Karthiga N and Preetha V (IOP Publishing) p 012103

Evaluation of a Preliminary Support Design of Railway Tunnel Adit in Inner Lesser Himalaya India: An Empirical Analysis

Md Alquamar Azad[1*], Neelratan Singh[2], S K Singh[1], Apurva Alok[1]

[1]Department of Geology (Centre for Advanced Studies) University of Delhi, Delhi-11007, India
[2]CSIR -Nagpur Research Centre (Ministry of Science & Technology, Govt. of India) 17 C, Telankhedi Area, Civil Lines, Nagpur - 440001, Maharashtra, INDIA

*e-mail Corresponding author: md.azaddu@gmail.com

Abstract. This article addresses the excavation method and support design for the adit tunnel in the Rudraprayag District, Lesser Himalayas of India, using Rock Mass Rating (RMR), Tunneling Quality Index (Q), and New Austrian Tunneling Method (NATM). Based on ONORM B 2203 correlations with RMR and Q systems, the New Austrian Tunneling Method rock structure classes were developed. Because the geology was constantly changing, NATM concepts were applied. The RMR-based rock mass estimates were overestimated, but the qualitative investigation was correct. The NATM method is more appropriate for a Garhwal Himalayan rock with varying rock mass uncertainty. The present adit research reveals several outstanding questions about rock mass quality, tunnel behavior during construction, and use. The analysis results might be used to build new tunnels in comparable terrain in other parts of the world.

Keywords: NATM; RMR; Tunneling Quality Index (Q); Excavation Method; Support Analysis; Lesser Himalaya

1. Introduction

In recent years, several highways and railway tunnels providing quick and efficient routes over mountainous terrain have been created in Uttarakhand Himalaya. The railway track between Rishikesh and Karnaprayag is being built in Uttarakhand's Garhwal region, which is prone to landslides and earthquakes. According to the Seismic Zone Map of India SEISAT [1], this location falls within zone V. This railway track is 125 kilometers long, and the entire tunneling distance will be 105.47 kilometers. The 9.5-kilometer-long tunnel-13 will travel through Rudraprayag and will be constructed using the 672-meter-long adit (Figure 1). The geological and geotechnical parameters of the adit, as well as the construction challenges, are examined in this section. Following extensive analysis, a suitable excavation process and support system were developed. It is horseshoe-shaped and measures 8.7 m width by 7.05 m height, when excavated of overall 672 m length of the adit tunnel, 426.5 m was driven utilizing the drill and blast technique using the Sandvik DT922i, Jumbo machine. The overburden thickness begins at 28 meters at the portal and reaches a maximum of 443 meters at the adir-tunnel's junction.

The techniques for classifying rock masses are very useful for selecting excavation methods and supporting design systems. Numerous approaches for classifying rock masses are employed to generate a picture of the rock mass composition and properties to offer early estimates of support needs [2]. The adit tunnel's rock mass was classified using an internationally known rock mass classification method, specifically the Q [3], RMR [4], and the NATM, based on ONORM B 2203 [5].

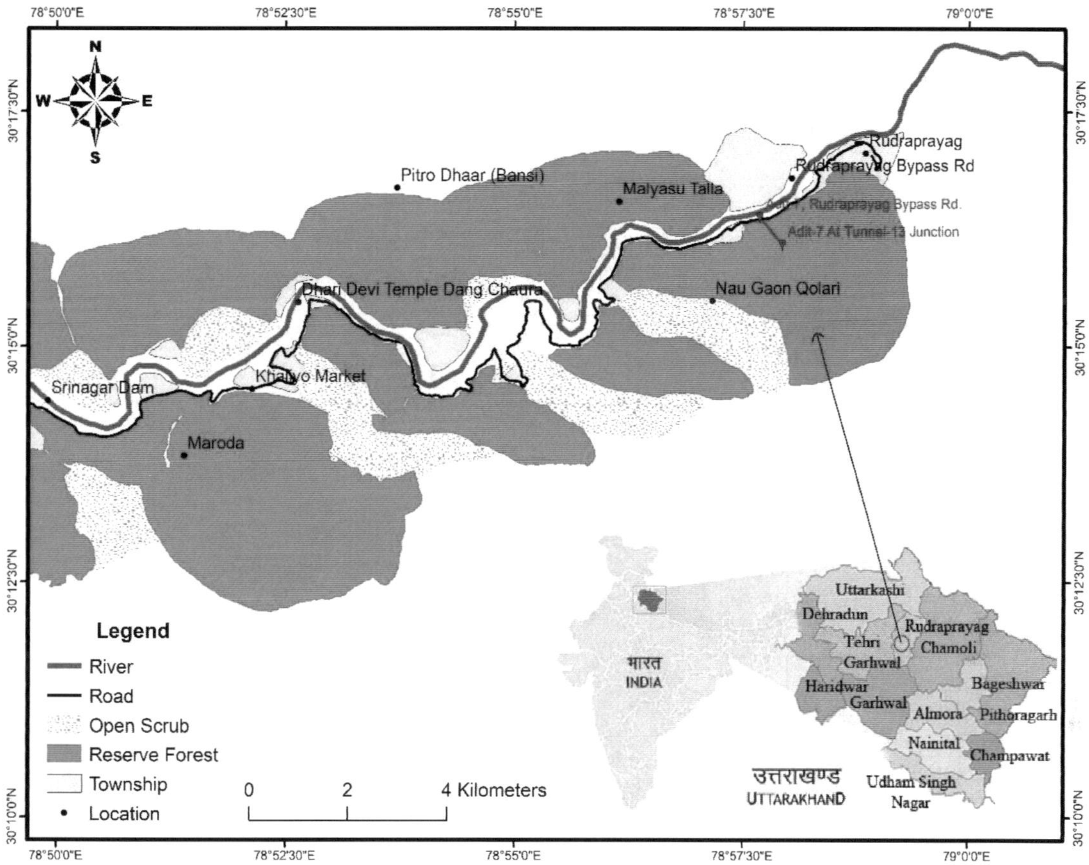

Figure 1. Location map of the study area showing the location of the adit

2. Geology of the adit site

The Himalayas are a textbook example of an orogenic system that was formed as a result of the continent and continent collision [6-7]. The basic layout of the Himalayan chain is about E-W. The orogenic belt has captured the interest of geoscientists worldwide because it denotes the southern limit of the Earth's biggest active crustal deformation zone, which resulted from the 55-million-year-old collision of India and Asia. According to [8-10], and others, it is separated longitudinally into five main geological divisions: Sub-Himalayan Sequence (SH), Lesser Himalayan Sequence (LH), Greater Himalayan Sequence Complex (GH), Tethyan Himalayan Sequence (TH), and Indus-Tsangpo Melange Zone (ITSZ).

Numerous scientists, including [11] and most recently [8], have made substantial contributions to the existing structural and stratigraphic understanding of the Garhwal Himalaya, which includes the study region. The region is mostly defined by the inner lower Himalayas. The North Almora Thrust (NAT), which marks the northern edge of Almora klippe, it acts as a barrier between the Outer lesser Himalaya and Inner Lesser Himalayas. It distinguishes the phyllites of Chandpur from the quartzites of the Garhwal Group. The region is geologically composed of quartzite with subordinate purple phyllite and basic metavolcanics from Berinag formation (Figure 2).

The structure is a key tectonic factor in the Himalayan orogeny. The Main Boundary Thrust (MBT) and the Main Central Thrust (MCT) in (Figure 3) define the Lesser Himalayan, which is influenced by a series of thrusts, some of which are closely folded. The Tons Thrust, Ramgarh Thrust, and Berinag Thrust are three of the region's most significant faults. Between the North Almora push and the Alaknanda fault lies the research region. The NAT is exposed in the Project area as a multiple fault zone with five major and several minor faults.

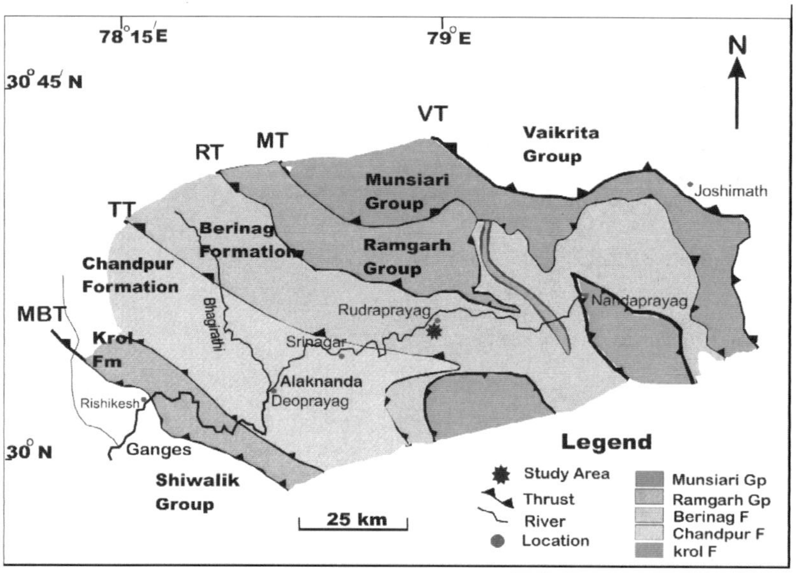

Figure 2. Study area geological map [12].

3. Engineering Geological Condition

This area is always full of surprises when it comes to underground construction because the geology changes so often [14]. Engineering geological parameters have been used to look at adit and figure out how good the rock is, how long it will take to stand up, and how much support it will need. [15] and his correlation were utilized to determine the RQD of the rock mass units in question by using a joint volumetric count. Ranges from 25% to 60%. III, IV, and V are the rock mass classes found from chainage 25.2 to 424 meters of the adit-7. At the portal, up to chainage 21.20 meters, there is an "adverse geological occurrence" (AGO) due to river-borne material (RBM). In the excavated face, there are three sets of joints that go from vertical to sub-vertical and horizontal to sub-horizontal. The detail of engineering geological property along the alignment is given in (Table 1).

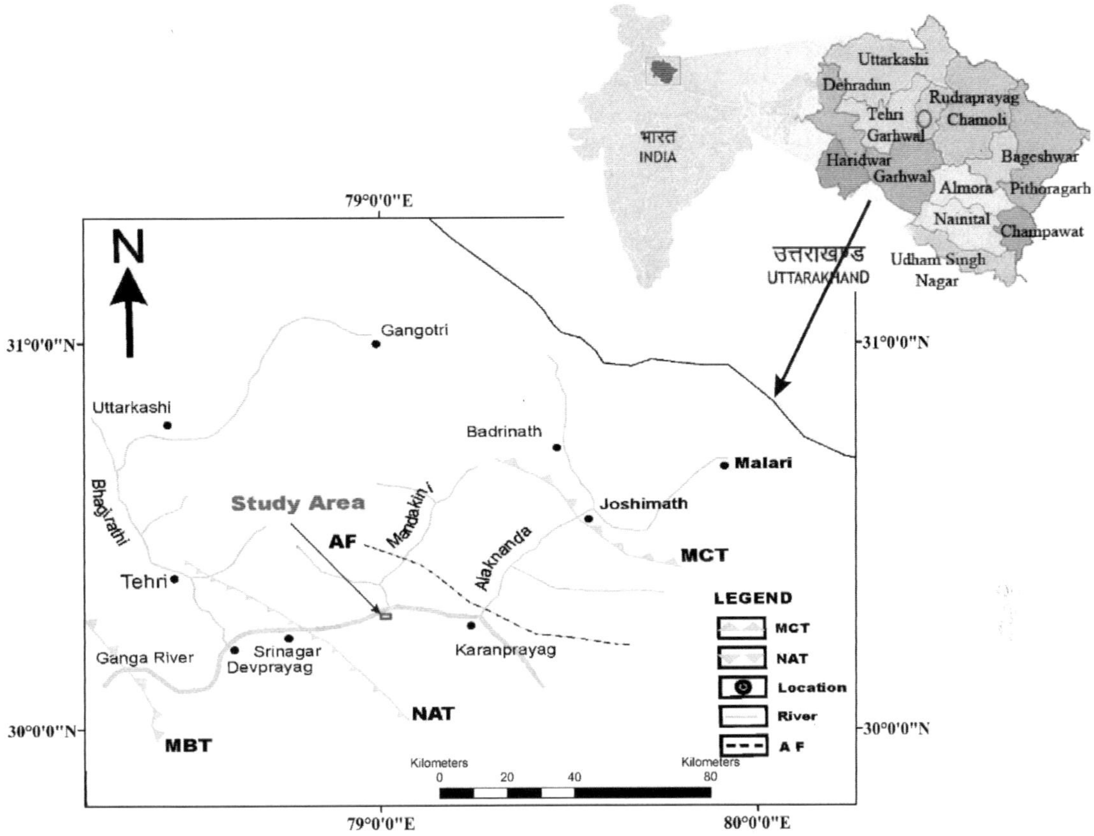

Figure 3. Main Tectonic structure of the study area [13]

Table 1. Anticipated engineering geological property along the adit alignment

			0.0 - 100	100 - 500	500 - 672
	Chainage (m)		0.0 - 100	100 - 500	500 - 672
Geology	Structural Domain		**Inner Lesser Himalaya**		
Geology	Group		**Garhwal Group**		
Geology	Formation		**Berinag Formation**		
Geology	Lithotype		Quartzite	Quartzite	Quartzite
Discontinuity Properties	Foliation	Persistence (m)	5-10m	5-10m	5-10m
Discontinuity Properties	Foliation	Roughness	Rough	Rough	Rough
Discontinuity Properties	Foliation	Filling	Crushed rock	Crushed rock	None
Discontinuity Properties	Foliation	Aperture (mm)	5-100m	5-100mm	Tight
Discontinuity Properties	Foliation	Spacing (mm)	5-200cm	6-200cm	15-80cm
Discontinuity Properties	Joint	Persistence (m)	0.5-1.50m	0.5-1.50m	0.5-1.50 m
Discontinuity Properties	Joint	roughness	Rough	Rough	Rough
Discontinuity Properties	Joint	Filling	Crushed rock	Crushed rock	Crushed rock
Discontinuity Properties	Joint	Aperture (mm)	Tight	Tight	Tight
Discontinuity Properties	Joint	Spacing (mm)	5-80 cm	5-80cm	5-80 cm
Rock Mass Properties	Rockmass strength		Strong to medi-strong	Strong - very strong	St. - V. strong
Rock Mass Properties	Weathering		Fresh - Slightly W.	fresh	Fresh
Rock Mass Properties	Anticipat. Q-values		0.38 - 5.4	3.3 - 3.9	3.3 - 5
Rock Mass Properties	Anticipated rock class		V	IV	III
Rock Mass Properties	Anticipate. G.W. condition		Dry - wet dry	dripping	
Rock Mass Behavio	Anticipat. wedge failure		Failure of from crown & wall,	wedge failure from the crown	
Rock Mass Behavio	Anticipated squeezing Co.		There will be no squeezing		mid squeezing
Rock Mass Behavio	Rock cover (m)		28 m	216 m	443m

4. Rock mass rating characterization system

[16] Developed the method for determining the rock mass rating. Determined from 49 case history based on his experiences in the shallow tunnel. Subsequently, classification has undergone several significant changes. [17] Regarded the RMR system as a very efficient method for examining rock mass conditions and establishing appropriate support, and enabling effective communication on engineering projects. The rock mass rating is computed by [18] from the total of ratings (weighting) of six input parameters: 0 to 15 for UCS, 3 to 20 for RQD, 5 to 20 for joint spacing, 0 to 30 for joint conditions, and 0 to 15 for groundwater conditions. The rating values range from 0 to 100, and the aggregate of the individual ratings for each parameter is used. Adjustments for discontinuity orientation in excavation are performed by adjusting the rating from 0 to -12. The RMR method was used to classify the rock mass quality over the entire excavated length of 426.5m adit-7. The entire excavated area was classified into six zones i.e. chainages 00.00m (at the entry portal up to 21m), Chainage 25.20m, chainages 41.9m, chainages 218.80m, chainages 401m, chainages 426.5m near the adit-tunnel junction (Table 2). The entry portal shows very poor rock mass due to adverse geological occurrence (AGO) it is a river-borne material. The RMR varies from 39 to 69 and rock description in different chainages is shown in (Table 2).

Table 2. Rock mass rating parameters and their ratings along adit-7 alignments

Parameters	Chainage 00.00m	Chainage 25.20m	Chainage 41.9m	Chainage 218.80m	Chainage 401m	Chainage 426.5m
UCS	AGO	7	12	12	7	12
RQD	AGO	4	8	13	8	13
Discontinuity spacing	AGO	10	10	15	10	10
Discontinuity condition	AGO	8	14	19	20	19
Ground-water	AGO	15	15	15	4	4
Orientation of joint	AGO	-5	-5	-5	-5	-2
Rating	-	39	54	69	44	56
Class number	-	IV	III	II	III	III
Description	-	Poor rock	Fair rock	Good rock	Fair rock	Fair rock

5. Tunneling quality index (Q) system

At the Norwegian Geotechnical Institute (NGI), [3] created the Q method of rock mass categorization for assessing tunnel quality of rock. Following a review of 200 tunnel case histories, the Q-system was suggested. Geological mapping in subterranean excavations may be used to calculate Q – values. It may be required to split the subsurface excavation into many pieces during mapping so that the fluctuation of Q-value within each segment is moderate. It is based on a numerical evaluation of the quality of the rock mass utilising six separate parameters and when the 6-parameters are combined, they define the three key elements that explain the stability in subsurface openings: block size, active stress factor, and inter-block shear strength. The block size is a ratio of the rock quality index (RQD) and multiple joints (Jn), while the shear strength of joints is defined by the ratio of joint roughness (Jr) to joint alteration (Ja). By dividing the joint water condition (Jw) by the stress reduction factor, the stress factor was established (SRF). At the adit-7 location, all characteristics were employed for various rock mass groups. Individual ratings were assigned using the tabular instructions, and final ratings were calculated by multiplying the three quotients together. The field study results were used to provide stress reduction factor (SRF) ratings, and the ground condition was appraised using empirical techniques.

$$Q = \frac{RQD}{Jn} \times \frac{Jr}{Ja} \times \frac{Jw}{SRF}$$

Q- Values were calculated during the excavation along the adit chainages (Table 3). The Q-Values at entry portal up to chainage 25m is varies from AGO to 0.62 showing very poor rock mass condition, visual observations describe the rock class V, while at chainage 41.9m and chainage 401m show poor rock. The rest of the entire chainage Q- values vary from 4.1 to 5 it corresponds to fair rock mass condition. In accordance with RMR and the Q-system, an excavation methodology and support system for adit has been proposed (Table 4). In (Figure 4), the cross-section of 672 m with horseshoe shape and dimensions of the excavated area are 8.7 m wide by 7.05 m high.

Table 3. Q-Rating along adit Alignment

S. No.	Chainage (in m)	Q- Values	Rock Class	Tunneling quality (Q)
1	00.00	A.G.O (RBM)	V	Very poor
2	3.00	A.G.O (RBM)	V	Very poor
3	3.4	A.G.O (RBM)	V	Very poor
4	5.3	A.G.O (RBM)	V	Very poor
5	15.60	A.G.O (RBM)	V	Very poor
6	21.20	A.G.O (RBM)	V	Very poor
7	25.2	0.62	V	Very poor
8	41.9	1.3	IV	Poor
9	218.80	5.00	III	Fair
10	230	4.58	III	Fair
11	238.6	4.58	III	Fair
12	291.60	4.68	III	Fair
13	401.21	2.83	IV	Poor
14	404.10	4.1	III	Fair
15	406.60	4.1	III	Fair
16	409.30	4.37	III	Fair
17	412.30	4.37	III	Fair
18	415.30	4.37	III	Fair
19	416.30	4.37	III	Fair
20	421.4	4.37	III	Fair
21	424	4.37	III	Fair
22	426.5	4.37	III	Fair

6. Excavation methodology and support system

Tunnel construction in Himalayan terrain is usually problematic due to the shifting geology, groundwater conditions, the rock mass characteristics, as well as in-situ stress conditions. According to geological and geotechnical findings, an adit tunnel was driven through a heterogeneous rock mass body that is jointed and of very poor, poor, and fair quality. Based on the above knowledge and experience in similar conditions in Himalayan rock, five basic support systems were suggested: lattice girders, rock bolts, steel ribs, shotcrete, and wire mesh. The support structure was designed by the New Austrian Tunneling Method (NATM). As part of the NATM strategy, a qualitative ground classification system must be taken into account [4]. It is a "Build as you go" attitude with the following limitations. The NATM Drill & Blast technique has the benefit of allowing for excavation flexibility without altering the excavation operation [19].

"Not too stiff, Nor too flexible
Not too early, Nor too late"

The adit tunnel excavation was carried out in the following sequence of activities:

- Drilling for underground excavation with Sandvic DT922i jumbos utilizing a suitable blast pattern to keep the excavation profile consistent with the support system design drawings.
- Scaling down/mucking out the spoils.
- Application of the first layer of shotcrete for the present blast and the second layer of shotcrete based on the strata's nature.
- Rock bolts were installed according to the rock profile.
- Drilling for the next blast (advancement)

The excavation method and support system for very poor, poor, and fair rock mass categories of the adit tunnel is given in (Table 4) and Figure 4, and a summary is given below. Before the excavation, forepolling was done in the face for portal and rock class-V excavation.

- Shotcrete- fiber shotcrete thickness of 300 mm wet and dry cast
- Wiremesh- welded wire mesh in crow 150 x150 x 6mm
- Rock bolts- ø25mm, 4m long grouted rock bolt staggered 1500 c/c applied
- Steel rib-ISHB 150 where required.
- Forepolling- forepolling required length=6m, ø32mm
- Lattice girder- 32/25/10 mm accordingly – girders were used to secure the roof and side walls in every quality class of the rock mass.

Table 4. Excavation and support quantities/m for a different type of rock class

Portal			
FULL FACE EXCAVATION	Round Length		0.75m^3
	Excavation Volume Up to Pay Line (B-Line)		59.83m^3
	Minimum Excavation Volume		56.84m^3
	Shotcrete	M35 / A10, Thick=300mm	3.93m^3
	Wiremesh	150 x150 x 6mm	0.019 MT
	Rock Bolts	Fy 200KN, 4m	69 R M
	Steel Rib	ISHB 150 (30.6k9/m) or equivalent section	0.576 MT
	Drainage Holes	Ø75mm, Length=5m	-
	Forepolling	Length=6m, Ø32mm	84 RM
Class-V (0.1 ≤ Q< 1)			
TOP HEADING	Round Length		1.0m
	Excavation Volume Up to Pay Line (B-Line)		39.73m^3
	Minimum Excavation Volume		37.31m^3
	Sfrs/Fibre Shotcrete	ds=300mm	2.23m^3
	Wiremesh	150 x150 x 6mm	0.019MT
	Sn Bolts	Fy 200KN, 4m	48 RM
	Steel Rib	ISHB 150 (30.6k9/m)	0.447 MT
	Drainage Holes	Ø 45mm, Length=5m	-
	Forepolling	Length=6m, Ø32mm	84RM
BENCHING	Round Length		2.0m^3
	Excavation Volume Up to Pay Line (B-Line)		21.12m^3
	Minimum Excavation Volume		20.52m^3
	Sfrs/Fibre Shotcrete	ds=300mm	0.9m^3
	Sn Bolts	Fy 200KN, 4m	16 RM
	Steel Rib	ISHB 150	0.13 MT
Class-IV (1 ≤ Q < 4)			
FULL FACE	Round Length		1.5m^3
	Excavation Volume Up to Pay Line (B-Line)		57.83m^3
	Minimum Excavation Volume		54.89m^3

	Sfrs / Fibre Shotcrete	M35 / A10, ds= 150mm	1.78m^3
	Lattice Girder	32/25/10	0.285 MT
	Rock Bolts	Fy 200KN, 4m	34.68 RM
	Drainage Holes	Ø 45mm, Length=5m	-
Class-III ($4 \leq Q < 10$)			
FULL FACE EXCAVATION	Round Length		2.0m^3
	Excavation Volume Up to Pay Line (B-Line)		56.84m^3
	Minimum Excavation Volume		53.93m^3
	Sfrs / Fibre Shotcrete	ds=100mm	1.90m^3
	Wiremesh	150 x150 x 6mm	0.019 MT
	Rock Bolts	Fy 200KN, 4m	22 RM
	Steel Rib	ISHB 150 (30.6k9/m) or equivalent section	0.576 MT
	Drainage Holes	Ø 45mm, Length=5m	-

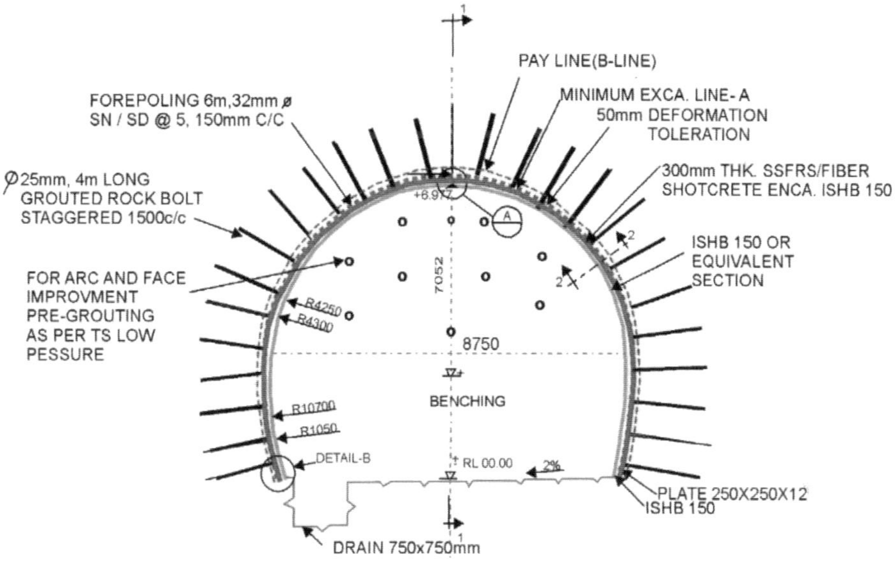

a.

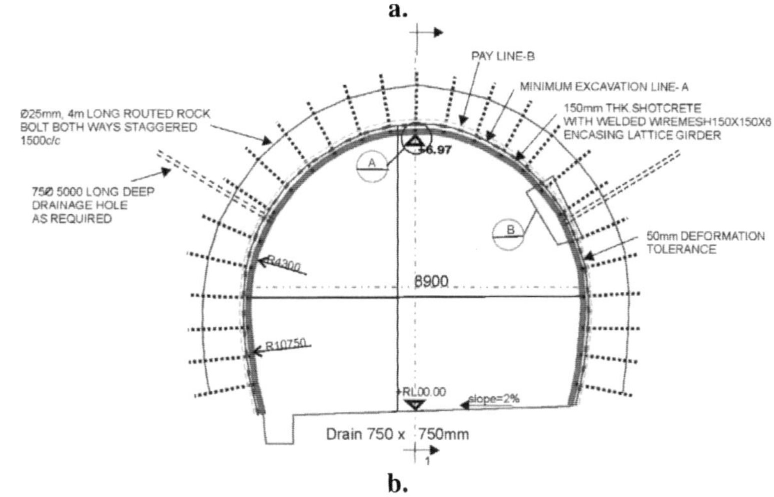

b.

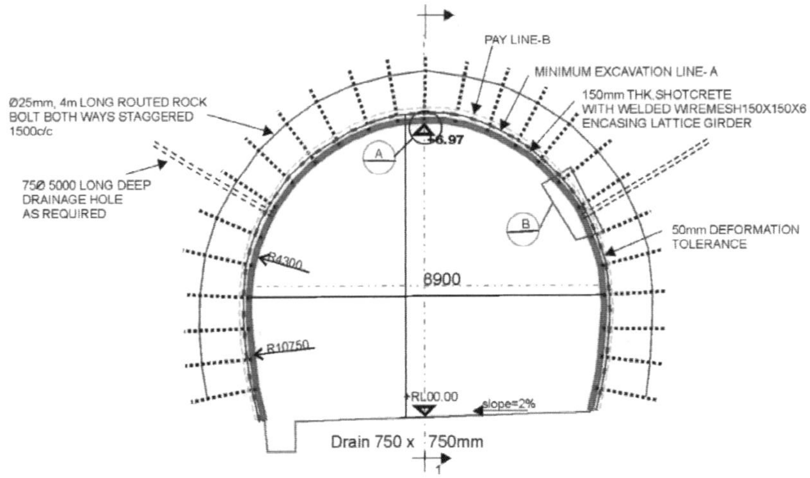

c.

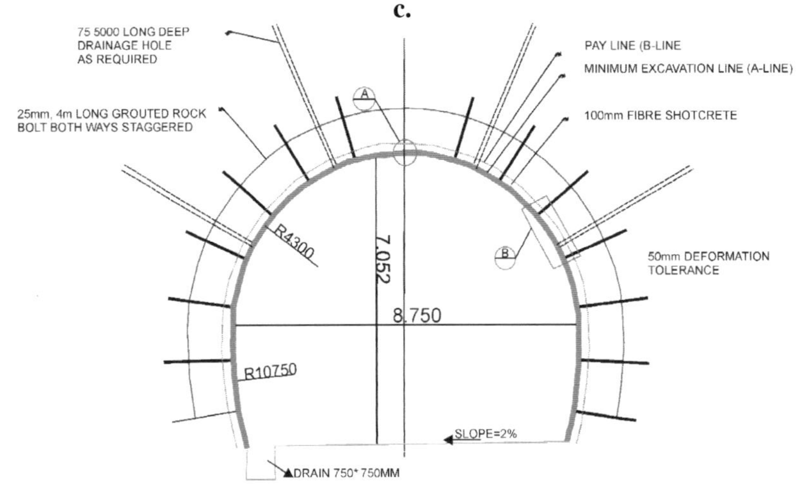

d.

Figure 4. Support system for **a.** Adit tunnel portal. **b.** class-V. **c.** class-IV and **d.** class-III

7. Discussion and conclusions

A total of 426.5 meters of the 672-meter-long adit tunnel of the ongoing rail project between Rishikesh and Karanprayag has been driven. The adit tunnel's excavated diameter is 8.7 meters. The rock mass conditions varied throughout the tunnel alignment. Quartzite was discovered to contain a significant percentage of tectonically disturbed rocks of fair grade. The strata were medium-grained, jointed in places, and had small shear seams. There are three pairs of joints in the slightly to moderately worn quartzite. Crushed rock fills the joints, which are spaced anywhere from 5 to 80 cm apart. Due to difficult geological conditions, the NATM approach was used to suggest excavation sequences and support systems.

In different tunnel sections, various engineering properties resulted in varying rock mass quality evaluations. Furthermore, the quantitative quality rating based on the RMR and Q classification systems has varied. The key problems for the Lesser Himalayan rock were the method for identifying rock mass quality and the application of the examined classes. Is the difference due to various approaches or a distinct feature of the rock mass around the adit? The RMR and Q classification methods, both based on quantitative features of the rock mass, are used to classify the rock mass. A support system design was developed using NATM, a qualitative categorization approach.

According to RMR, the rock mass classes in the research region were IV, III, and II (poor rock, fair, and good rock categories). The grade of rock mass up to 21.2 m in the entrance portal (section 1) was classed as extremely bad rock-borne material (RBM). For quartzite rock of the whole rest alignment, the RMR ranged from 39 to 69. Jointed to fractured rock mass had the lowest value, whereas fresh rock mass had the highest. The support system was adjusted as a consequence of the constant assessment of rock mass quality, which indicated that actual geotechnical conditions were far better than predicted.

Empirical categorization techniques were also utilized in the tunnel design. The results showed that the rock mass estimation using RMR was overstated, while the qualitative assessment was accurate. For this sort of terrain and rock mass, the NATM approach, which is more adaptable, may be employed, which may help to guarantee that a tunnel is completed on schedule.

References

[1] SEISAT, 2000. Seismotectonic atlas of India and its environs. Geological Survey of India, India

[2] Hoek, E., Kaiser, P.K. and Bawden, W.F., 2000. *Support of underground excavations in hard rock*. CRC Press.

[3] Barton, N., Lien, R. and Lunde, J., 1974. Engineering classification of rock masses for the design of tunnel support. *Rock mechanics*, *6*(4), pp.189-236.

[4] Bieniawski, Z.T., 1989. *Engineering rock mass classifications: a complete manual for engineers and geologists in mining, civil, and petroleum engineering*. John Wiley & Sons.

[5] Azad, M.A., Singh, S.K., Alok, A., Shekhar, S. and Kumar, P., 2022. Geotechnical and geological studies of Adit-6 of the railway tunnel between Rishikesh and Karnprayag in India focusing on the excavation methods and design of support analysis: a case study. *Arabian Journal of Geosciences*, *15*(1), pp.1-15.

[6] Dewey, J.F. and Bird, J.M., 1970. Mountain belts and the new global tectonics. *Journal of geophysical Research*, *75*(14), pp.2625-2647.

[7] Dewey, J.F. and Burke, K.C., 1973. Tibetan, Variscan, and Precambrian basement reactivation: products of continental collision. *The Journal of Geology*, *81*(6), pp.683-692.

[8] Yin, A., 2006. Cenozoic tectonic evolution of the Himalayan orogen as constrained by a long-strike variation of structural geometry, exhumation history, and foreland sedimentation. *Earth-Science Reviews*, *76*(1-2), pp.1-131.

[9] Gansser, A., 1964. Geology of the Himalayas.

[10] Célérier, J., Harrison, T.M., Beyssac, O., Herman, F., Dunlap, W.J. and Webb, A.A.G., 2009. The Kumaun and Garwhal lesser Himalaya, India: part 2. Thermal and deformation histories. *Geological Society of America Bulletin*, *121*(9-10), pp.1281-1297.

[11] Valdiya, K.S., 1980. Geology of Kumaun Lesser Himalaya. Wadia Institute of Himalayan Geology, Dehradun. *Google Scholar*, pp.290-291.

[12] Ahmad, T., Harris, N., Bickle, M., Chapman, H., Bunbury, J. and Prince, C., 2000. Isotopic constraints on the structural relationships between the lesser Himalayan series and the high Himalayan crystalline series, Garhwal Himalaya. *Geological Society of America Bulletin*, *112*(3), pp.467-477.

[13] Sati, S.P., Sundriyal, Y.P. and Rawat, G.S., 2007. Geomorphic indicators of neotectonic activity around Srinagar (Alaknanda basin), Uttarakhand. *Current Science*, pp.824-829.

[14] Azad, M.A., Singh, S.K. and Nair, A.S., 2020. Application of rock mass classification systems for future excavation and support design of proposed road tunnel from Khand Birkot to Sarot Doman in Tehri Garhwal District, Uttarakhand. *HIMALAYAN GEOLOGY*, *41*(2), pp.234-240.

[15] Palmstrom, A., 2005. Measurements of and correlations between block size and rock quality designation (RQD). *Tunnelling and Underground Space Technology*, *20*(4), pp.362-377.

[16] Bieniawski, Z.T., 1973. Engineering classification of jointed rock masses. *Civil Engineering= Siviele Ingenieurswese*, *1973*(12), pp.335-343.

[17] Bieniawski, Z.T., 1975. The point-load test in geotechnical practice. *Engineering geology*, *9*(1), pp.1-11.

[18] Bieniawski, Z.T., 1976. Rock mass classification in rock engineering applications. In *Proceedings of a Symposium on Exploration for Rock Engineering, 1976* (Vol. 12, pp. 97-106).

[19] Singh, B. and Goel, R.K., 1999. *Rock mass classification: a practical approach in civil engineering* (Vol. 46). Elsevier.

A study of Environmental and Ecological impacts due to Construction and Operation of Tehri-Polavaram Dam

S K Verma[1*], R T. Sahu[1], H Singh[1], A.D. Prasad[1] and M K Verma[1]
[1]Civil Engineering Department, National Institute of Technology, Raipur, Raipur, Chhattisgarh, India.

Abstract. Any water resources project has both a positive and a negative impact on the environment. Sometimes these environmental losses are unavoidable, so it is most important to minimize the environmental losses and their social-ecological and geopolitical impact. Environmental impact assessment plays an important role in identifying, predicting, and mitigating the various impacts on the environment. This paper presents the details of the construction of Tehri and Polavaram dam projects and their various impacts on the environment and ecological aspects. Any water resource expansion that fulfills minimum needs without endangering future generations' ability to fulfils their own will be considered sustainable. There are a number of factors to consider when it comes to the long-term effects of dams on the surrounding environment, including redevelopment and relocation of people and businesses, socioeconomic influence on the surrounding area, and environmental concerns such as sedimentation, among others. Sustainable development also means that, in addition to minimizing the dam's negative effects on the environment, making sure that the benefits of the dam outweigh the negative ones. As a result, it is recognized that water resource projects must be intended, implemented, and managed to meet future population demands while reducing disturbance to the existing eco-system and incorporating effective control measures at various phases to ameliorate any adverse effects.

Keywords: Tehri and Polavaram Dam, environmental impact, ecological aspects, future scenario

1. Introduction

Dams play a critical role in the efficient use of water. Many years have passed since they were built before the current state of knowledge about hydraulics and hydromechanics could be gleaned. There is nothing typical about these structures. When it comes to managing stream flows and satisfying water needs, dams have played a significant role in the evolution of civilization. Since ancient times, dams have been built to keep rivers dry, provide fresh water for irrigation, generate electricity, and all of the above. Additionally, dams have a wide range of good and bad impacts on the environment, such as regulating stream regimes, preventing floods, getting water from the dam's reservoir, and producing energy. In addition to their benefits, including providing society's fundamental needs and raising living standards, dams contain the potential for significant harm to living organisms [1]. Up to the 1950s, over 700 dams were constructed every ten years. After the 1950s, this number climbed significantly. As the dams were being constructed, it became apparent that something was missing, and this was damaging. Despite the fact that water has a profound impact on human life and civilization, it is claimed that the economic benefits anticipated from projects designed to utilize water resources will not be realized, and that necessary precautions to reduce environmental, economic, and social losses will not be taken. Some international institutions are conducting studies aimed at thwarting these water supply schemes in developing countries. In light of these findings, water management, considering economic, social, and cultural development, as well as environmental implications, has become increasingly important in the sustainable water management [12]. Because of this, water resource studies must have a legal foundation to assure long-term growth and sustainability. Water resources planning studies often conflict with laws concerning the responsibilities and powers of government entities and establishments. These relate directly to environmental issues throughout these water resources planning studies. Planning studies should not overlook these laws. The rules of the land are always evolving as technology and advancements progress. Scientific, technological, and educational investigations should be used to achieve the necessary modifications [6], [14]. A thorough review will be necessary if we are to minimize the inconsistencies in the legislation. It is imperative that the investigations continue.

Tehri Dam is the world's tallest dam. Initially, in 1994, the government approved this proposal for implementation. The Tehri Hydro Development Authority, a joint venture between the Indian and Uttar Pradesh governments, took over in 1998.On the Bhagirathi River, the Tehri dam is located, which is approximately 200 miles north-east of Delhi. It is the fifth-tallest dam in the world. With a total height of 840 ft (256 m), the dam can generate 244 million kilowatt-hours of electricity and provide water for an extent of 270,000 hectares. After a lot of thought, the Tehri dam was thought up in 1949. The location was chosen in 1961, the Planning Commission approved it in 1972, and work began on the dam in 1978 under a lot of security. For the Tehri Development Project, apreliminaryinvestigationwascompletedin1961, and after completion of the design of the dam it was approved in 1972. Construction of this dam startedin1978, and Coffer dam wasfinishedin1996. The last two tunnels were closed in December 2002, and the project is expected to be completed in August 2003. Whereas, Polavaram Project is one of the major projects in India approved by the central government. The proposed project is a multipurpose project mainly for irrigation, public water supply, and hydropower generation. The site is in Andhra Pradesh state in Polavaram village. Site selected for construction is in Godavari River in the east and west Godavari districts of Andhra Pradesh, which is under construction. The Godavari River is a perennial river in India which is the second largest in the country. The Polavaram project was initially planned in 1941. In 1870, Lt-Col. Haig, Superintending Engineer, after observation, prepared a plan and sent a report to construct a reservoir across the Sokileru River, which is a tributary of the Sabri River, for irrigation purposes of about 7,300 hectares. In the year 1896, Mr. Walch wrote in his book, "The Engineering Works of the Godavari Delta," that storage could be achieved at a cost that was not prohibitive, although a dam across the main Godavari River was not considered. The Polavaram reservoir was first proposed in 1941 by Diwan Bahadur. Diwan Bahadur L. Venkata Krishna Iyer, head engineer of the Presidency's irrigation department, after a detailed investigation, the project was finalized and named Rampada Sagar. The project name was given, keeping the mythic in view, as the back waters of the reservoir would touch the Lord Rama temple at Bhadrachalam. For this project, the total cost was estimated at 129 crores. But at the dam site, due to the absence of hard rock strata at a shallow depth, the proposed dam project was postponed. Therefore, due to the deeper depth of excavation for the foundation and construction difficulties, the project was not found to be economical.

Due to a flood in the Godavari River in 1953, the Visakhapatnam Steel Plant saw the need for a barrage on the Godavari River. Different agreements have been put up by the neighboring states to share the water of the Godavari River and for the completion of the project. After a long period of inactivity, detailed investigations had to be carried out in 1976, and the project was presented in 1978.once Again, after detailed investigations and on the basis of comments by the Central Water Commission in the 1978 and 1982 projects, that report was finalized for the Polavaram project. As per the rates in 1985-1986, the project cost is estimated at 2665 crores. Once again, due to political issues, project work could not be executed. The project was taken on by YSR Reddy, the chief minister of Andhra Pradesh, in 2004 to complete, but work on the project was halted to obtain clearance from the Ministry of Forests and Environment.

2. Study area

The Polavaram Project is an irrigation project on the Godavari River in Andhra Pradesh's West Godavari and East Godavari districts. The Indian government has designated the project as a national initiative. Its backwater goes up to Dummugudem Anicut, which is about 150 km downstream from the Polavaram Dam, and 115 km downstream on the Sabari River. It also goes into Chhattisgarh and Odisha. The reservoir covers the popular Papikonda National Park, boosting tourism in the Godavari Districts. On the left bank of the river are the Polavaram HEP and National Waterway 4. It is 40 kilometers upstream of the Sir Arthur Cotton Barrage and 25 km from Rajahmundry Airport. Whereas, Tehri Dam is India's tallest dam. It is a multi-purpose rock and earth-fill embankment dam on the Bhagirathi River in Uttarakhand. It is the main dam of the Tehri hydropower complex. Phase 1 ended in 2006. The Tehri Dam contains a reservoir that is used for agricultural, municipal water supply, and hydroelectric power generation. The dam's 1,000 MW variable-speed pumped-storage project is anticipated to be completed in 2025. 'Figure 1' shows the study area map which contains Polavaram dam and Tehri dam, respectively. Apart from that table 1 represents the silent features of the same.

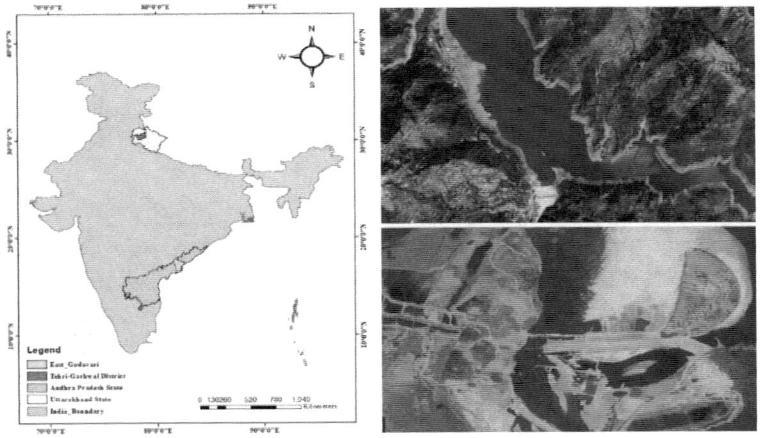

Figure 1. Study area map.

Table 1. Silent features of study area.

Technical details	Polavaram Dam	Tehri Dam
Type of dam	Rock fill type	Rock and earth fill type
River Impoundment	Godavari river	Bhagirathi river
Maximum height above ground level	48.32 m	260.5 m
Project completion period	1941-2018	1978-2003
Catchment Area	306643 sq. km	7293 sq km
Design flood discharge	1.02 lakh cumecs	15540 cumecs
District	Tehri garhwal	East Godavari
Length of dam	2454 m	575 m
Annual rainfall	1082.95 mm	1000 mm
Power generation	960 MW	1000 MW
Land area under submergence	47403 Hectare	5200 Hectare
Number of villages under	239	80
Rehabilitated population	1,88,370	24,000
Location	81° 39' 46" E to 17°16' 53" N	78° 28' 46.074 E to 30° 22'38.7984 N
Capacity of spillway	3600000 cusecs	15540 cusecs
Spillway types	Chute	Combination of Shaft and Chute spillway
Type of Gates	Radial	Controlled
Type of Turbine	Kaplan	Francis
Number of gates	42	2

3. Positive and Negative impacts of dam

3.1 Positive impacts of dam

3.1.1 Irrigation

The majority of people in Andhra Pradesh rely on agriculture, and agriculture accounts for 51.58% of total income. The Godavari River is the primary source of water for high areas in West Godavari and Krishna districts, including the region being cultivated by the Nagarjuna Sagar left bank canal. In the case of inadequate rainfall, there may be a possibility of a shortage of water for irrigation. In that case, the Polavaram dam can meet all of the needs and remove backwardness from these areas.

3.1.2 Water supply

The dam can supply 270 million gallons of water per day to Uttaranchal, Uttar Pradesh, and Delhi. Vishakhapatnam is a port city in Hyderabad and there are so many industries, steel plants, and proposed to be setup nearby, such as the harbor, naval complex, shipyard, refinery, etc., which require an adequate and continuous supply of water. People in Visakhapatnam, West Godavari, Krishna, and East Godavari also want drinking water during the summer.

3.1.3 Power generation

'Figure 2' shows the overview of hydroelectric power plant. In the Polavaram Project, power 12×80 MV capacity must be installed, which means that in so many areas, electricity can reach about 3,100 million units annually. Whereas, the Tehri dam can generate 244 million kilowatt-hours (MKWH) of electricity.

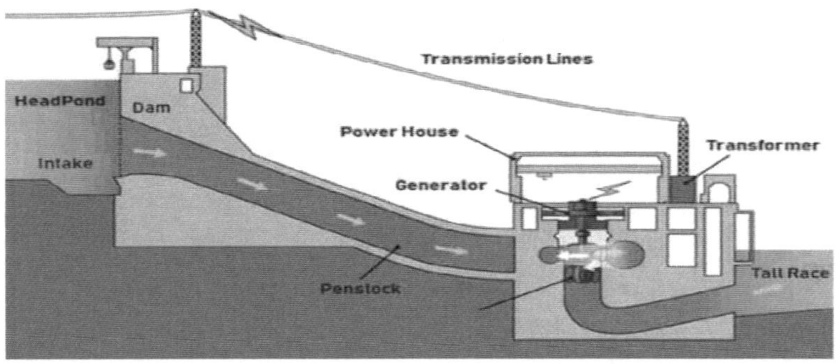

Figure 2. Hydroelectric power plant (Source: https://www.electricalfundablog.com/).

3.1.4 Flood control

As we know, in 1953, 1959, and 1986, when floods came into the Godavari River, it damaged and flashed away some agricultural crops, which was the reason for the huge loss in property and livestock across the plains. The Godavari River flood can be controlled via the Polavaram project. However, a total of 80 villages would be entirely or partially flooded due to the construction of Tehri dam.

3.1.5 Navigations

Due to the lack of water in the river (insufficient depth), navigation is not possible. In the Polavaram Project, there is a target for the formation of a foreshore lake, and the left canal of this dam is specially designed for navigation. Once the Polavaram project is completed, it will be helpful for cheap and quick transportation of products produced by forest and food grains in the market industry and for the supply of different coals and ores in the industries. However, the Tehri dam cannot be used for navigational purposes.

3.2 Negative impacts of dam

3.2.1 Displacement and Rehabilitation

The Polavaram Irrigation Project is slated to uproot 276 communities and 44,574 people, primarily in the state of Andhra Pradesh, where archaeological sites, coal deposits, a wildlife sanctuary, and many hectares of arable land will be submerged. After investigation, it is understood that 56% of agriculture, 15% of villages, 13% of vegetation, 9% of barren land and 8% of forest land are coming under the submergence area of the dam due to the construction of the same on the upstream of the river. Therefore, the results cause severe disturbance of natural, wildlife, and other species. However, in the case of the Tehri Dam, approximately 1, 000,000 individuals were affected by the disaster, but there was no plan to relocate them: some received compensation, others were given houses, and still others received nothing. Even though a reported sum of Rs. 1117.15 million was spent on rehabilitation, the results were dismal. Land acquisition started in the year 1979 without preparing any master plan or giving any compensation to affected people. As per the 2002 Status Report by the Public Works Department of Tehri, many people lost their land, but it was not officially recognized as project affected [13]. Furthermore, 'Figure 3' shows the really against Polavaram and Tehri dam, respectively.

Figure 3. Rally against the Polavaram and Tehri dam project (Source: https://www.sandrp.in/).

3.2.2 Earth crust
Even though it is claimed that the dam reservoirs have some seismic effects, it must be stated that this has not been proven scientifically [15].

3.2.3 Aquatic entity
The dam prevents animals from swimming from one mouth of the river to the other. Dams harm fish that reside in the springtime or in high-bodied dams' sluice gates, rotors, and pumps. This dam will disrupt the life cycle of these organisms and cause mass deaths. Bypass flow is designed for this purpose [7].

3.2.4 Disappearance and wane of historical places
After a reservoir is built, archaeological and historically esteemed sites, as well as geological and topographical wonders, are buried beneath it.

3.2.5 Scouring of river
Water flowing from reservoirs, even those with turbines, typically contains very little suspended sediments, which can result in river bed scouring and river bank erosion [3], [9].

3.2.6 Generation of new species
The temperature, salt content, and oxygen distribution of the water may alter vertically as a result of reservoir development. As a result, new living species may evolve, and the old environment may become extinct. (According to the International commission of large dam) [2], [8], [10], [11].

3.2.7 Human heath
According to 'Figure 4' it is possible that the dam's accumulation of wood and waste will lead to a rise in water-borne illnesses like fever, dysentery, and cholera that impair human health in the immediate vicinity of the dam and reservoir [5].

Figure 4. Accumulation of wood and garbage because of dam(Source: https://www.sandrp.in/).

3.3 Environmental and ecological impacts on dam

The Polavaram Project includes a large area consisting of agricultural, forest, and barren land. This area is becoming a submergence area due to dam construction. Natural, wildlife, and other species will suffer badly due to submergence of forest land area. Sloth bears and tigers are being degraded due to habitat loss and over-hunting of wild creatures such as tigers and gaurs. Therefore, the loss of wildlife will disturb the diversity of the region, and the famous tourist spot 'Papikondalu' is completely getting submerged due to its construction. The Polavaram dam is affecting the state of Andhra Pradesh but also has minor effects on the adjacent states of Orissa and Chhattisgarh. In the proposed area, there are so many agricultural areas from which various types of crops and food grains are generated. But after the dam is constructed, there may be a shortage of food security for people. Thousands of crores of rupees are expended to obtain the clearance for this project and the re-habitation of people. This project has gotten clearance after certain modifications by the government, and its design procedure is outdated. So, it's design procedure cannot be taken into fully consideration. It needs to consider the present status, otherwise it may cause damage to the ecology and people. According to the NCAER study, some of the submergence areas contain deposits of minerals, e.g., iron ore, chromite, graphite, etc. So, it is essential to collect information about this and make an assessment of the various impacts of submergence on this area. A coal field is coming under submergence area that contains a large amount of coal, which is part of the Sigaroni coal fields, one of the important coal fields in the south of India. Because these detailed investigations are not being conducted, it is impossible to predict what the true impact of these submerged areas will be. However, in the case of the Tehri dam, the protests have been led by environmentalists and residents of the area. Environmentalist Sunder Lal Bahuguna was the driving force behind the anti-Tehri Dam campaign from the 1980s and extending up to 2004. The anti-displacement movement sparked the demonstration, as did the environmental consequences of a degraded ecosystem, which sparked the anti-displacement movement. It has also sparked questions about how a dam of this size and scope will affect the delicate ecosystem of the Himalayan foothills, and Concerns have also been raised regarding the dam's geological durability and viability. A large fault zone and the Tehri dam are located near the Central Himalayan Seismic Gap. At the time of the earthquake, the dam was located 53 kilometers (33 miles) away from the epicenter. For this project, which will be in the Himalayas, the dam's proponents contend that the design is intricate enough to endure an earthquake with a magnitude of 8.4 on the Richter scale. Some seismologists believe that if an earthquake of magnitude 8.5 or greater occurs in the immediate area, a dam break might affect half a million people. Lalitpur and Jhansi Districts have been adversely affected by compensatory afforestation to the extent of 4586,07 hectares. To prevent soil erosion and protect 52.204 hectares of land drains were built to decrease water logging and salinity. The creation of a "greenbelt" (14.28 hectares) for the conservation of key plant species, a fish hatchery, and a fishery for Mahseer Siltation along the reservoir perimeter has resulted in the creation of an area known as a "greenbelt." Urban areas have a higher standard of living due to the availability of more services and the potential for future growth [4]. It is also important for rural areas to have a sufficient water supply and road network system, as well as improved infrastructure. Treatment and land reclamation work at the THDC were preferentially given to those who had been affected by the project schemes for earning money through training facilities for the khadi industry by the UP-state department of horticulture and food preservation.

4. Present scenario of the projects

The present Polavaram Project is also known as the Rampada Sagar Project, and it is getting late to complete the project due to political issues and some other reasons. Also, due to the modification of the project, the project cost has increased. The project has gotten all the necessary approvals and the government of India has designated the Polavaram Project as a National Project (section 90 of the AP Reorganization Act 2014). The project is expected to be completed by the end of March 2018 (excluding the installation of dam gates). The anticipated cost of this project is Rs. 16010.45 crores after the Planning Commission have given its final approval (price level based on 2010-11 rate). However, in the case of Tehri dam, a benefit-cost study by the Indian National Trust for Art and Culture Heritage (INTACH) found that the project costs would be at least twice the anticipated benefits, and more than US $2.5 billion was spent on the first phase in July 2006, compared to an estimated US $12 million in July 2006.

5. Conclusions

Water has always been a vital plus critical resource for the survival of civilizations around the planet. In some regions around the world, an imbalance between demand and accessible water, food, and energy threatens life's survival. Dams and reservoirs can and should cohabit with the social and natural environment of the region. The next challenge will be to employ dams and reservoirs to prudently manage the world's water supplies as part of each nation's social and economic development goals. The dam's negative environmental consequences can be reduced or eliminated with thorough planning and design that includes public participation and input early in the process. When relevant mitigation measures are discovered early in the dam and reservoir planning and design phase, they can be included effectively and efficiently into the project design, construction, and operation.

These projects have undeniably substantial and significant benefits. With increased electrical generation, irrigation canals and food production have improved. Dams protect downstream residents from floods. Dams can be decided after weighing the costs and benefits for a long time, and the benefits of dams may outweigh the drawbacks in the future. But these massive engineering feats should warn us that we cannot modify the entire environment. Because ecology connects entire chains A single broken chain link or cog piece will destroy the entire system. So, when planning, the environment should be thoroughly evaluated. Precautions should be taken to avoid major risks caused by minor sensitive responses. Aside from their significant social and environmental benefits, dams must be minimized in order to achieve sustainable development. The environmental impact assessment concept takes into account the above impacts and solutions. In addition, the overall summary, the environmental changes caused by dams vary in magnitude and impact. Hence, it is impossible to predict the positive and negative impacts of these interactions in advance. Each dam and reservoir should be estimated separately. However, it is incorrect to interpret the consequences negatively.

References

[1] Chen, S., Chen, B. and Su, M 2011 The cumulative effects of dam project on river ecosystem based on multi-scale ecological network analysis. Procedia Environmental Sciences, **5**, pp.12-17
[2] Altinbilek, D 2002 The role of dams in development. Water Science and Technology, **45(8),** pp.169-80
[3] Jeon, J., Lee, J., Shin, D. and Park, H 2009 Development of dam safety management system. Advances in Engineering Software, **40(8),** pp.554-63
[4] Bohlen, C. and Lewis, L.Y 2009 Examining the economic impacts of hydropower dams on property values using GIS. Journal of Environmental Management, **90,** pp.S258-S69
[5] Tetteh, I.K., Frempong, E. and Awuah, E 2004 An analysis of the environmental health impact of the Barekese Dam in Kumasi, Ghana. Journal of environmental management, **72(3),** pp.189-94
[6] Graf, W.L 2006 Downstream hydrologic and geomorphic effects of large dams on American rivers. Geomorphology, **79(3-4),** pp.336-60
[7] Adams, W.M. and Hughes, F.M 1986 The environmental effects of dam construction in tropical Africa: impacts and planning procedures. Geoforum, **17(3-4),** pp.403-10
[8] Schilt, C.R 2007 Developing fish passage and protection at hydropower dams. Applied Animal Behaviour Science, **104(3-4),** pp.295-25
[9] Giers, A., Freistühler, E. and Schultz, G.A 1998 Methodology for assessment of ecohydrological effects of dam construction in a headwater region. IAHS Publications-Series of Proceedings and Reports-Intern Assoc Hydrological Sciences, **248,** pp.509-14

[10] Sayadi, A. and Partani, S 2009 Environmental Impact Assessment of Gotv and Hydro-Electric Dam on the Karoon River Using ICOLD Technique. International Journal of Environmental and Ecological Engineering, **3(6),** pp.118-25

[11] Walker, K 2008 Environmental impact statement for Traveston Crossing dam (Mary River, Queensland): a review with regard for species of concern under the EPBC Act 1999. Environmental Defenders Office, Canberra, ACT, Australia

[12] Küçükarslan, S. and Coşkun, S.B 2003 Fifth National Conf. on Earthquake EngineeringDam-reservoir interaction including the reservoir bottom effects under earthquake ground motion.I.T.U., İstanbul, Turkey pp 26-30

[13] Duvail, S. and Hamerlynck, O 2003 Mitigation of negative ecological and socio-economic impacts of the Diama dam on the Senegal River Delta wetland (Mauritania), using a model-based decision support system. Hydrology and Earth System Sciences, **7(1),** pp.133-46

[14] Pytharouli, S. and Stiros, S 2008 Proc. on 13th FIG symposium on deformation measurement and analysis and 4th IAG symposium on Geodesy for Geotechnical and Structural Engineering Dam crest settlement, reservoir level fluctuations and rainfall: evidence for a causative relationship for the kremasta dam greece. LNEC Lisbon Portugal pp. 12-15

Groundwater quality assessment and its insinuations in some parts of Aheri taluka dist. - Gadchiroli, Maharashtra.

S T Peddiwar[1], D K Ingole[1*] and G R Nimbarte[1]
[1]Mohsinbhai Zaweri Mahavidyalaya, Desaiganj (Wadsa) Dist. Gadchiroli, Maharashtra, India 441207.
[*]Corresponding author e-mail: diptiingole25@gmail.com

Abstract

Aheri tehsil of Gadchiroli district in the Indian state of Maharashtra is located on the bank of the Pranhita River, a side stream of the Godavari River. In the present study, total five villages from Aheri taluka were selected to collect groundwater samples. A total number of 25 samples were collected, where five samples represent each village. This study is concerned with physicochemical parameters mainly electrical conductivity, total dissolved solids, temperature, chloride, fluoride, total hardness, pH, nitrate, alkalinity, iron and turbidity to analyse groundwater quality. Out of 25 samples, eight samples of water were detected to be unfit for consumption. Hence, some remedial measures are suggestible. The study revealed the associations among all physicochemical parameters by correlation coefficient analysis.

1. Introduction

Water is the most basic and essential requirement of life. As per human health, potable water plays a vital role in all health-related issues. In developing countries like India, the availability of pure and safe drinking water is still out of reach for people in some places. In rural areas of India, where water treatment is still unavailable and transportation facility does not exist, about 88% of the consumable and safe drinking water is from groundwater source [1]. Quality of water is responsible for causing and healing health-related problems. Generally, people are not that aware of the quality and purity of water they are drinking. Therefore, it is necessary that water quality testing be done accordingly to avoid health problems. Overall, the water for drinking purpose is mainly the groundwater at most of the places [2]. Groundwater is present beneath the surface in numerous small aquifers and different compartments [3]. Hydrosphere's hydrochemistry, earth's solid portion and the geochemistry of lithosphere's affects quality of groundwater and all these factors mainly control the whole quality of water [4]. Groundwater chemical composition chenges because of local lithology, the groundwater flow velocity, residence time and interactions between rock and water [5].

The study area lies in the villages of Aheri tehsil of Gadchiroli district in the Indian state of Maharashtra. The selected sites were Welgur (Latitude 19°31′18″N and Longitude 80° 03′ 38″E), Navegaon (Latitude 19°30′05″N and Longitude 80° 04′ 56″E), Develmari (Latitude 19°18′20″N and Longitude 79° 58′ 27″E), Venkatraopetha (Latitude 19°21′33″N and Longitude 79° 59′ 28″E), Allapali (Latitude 19°26′52″N and Longitude 80° 03′ 19″E). In the present study, total number of twenty-five Groundwater samples had been collected and tested analytically for different parameters mainly electrical conductivity, temperature, total dissolved solids, chloride, fluoride, pH, total hardness, nitrate, alkalinity, iron and turbidity.

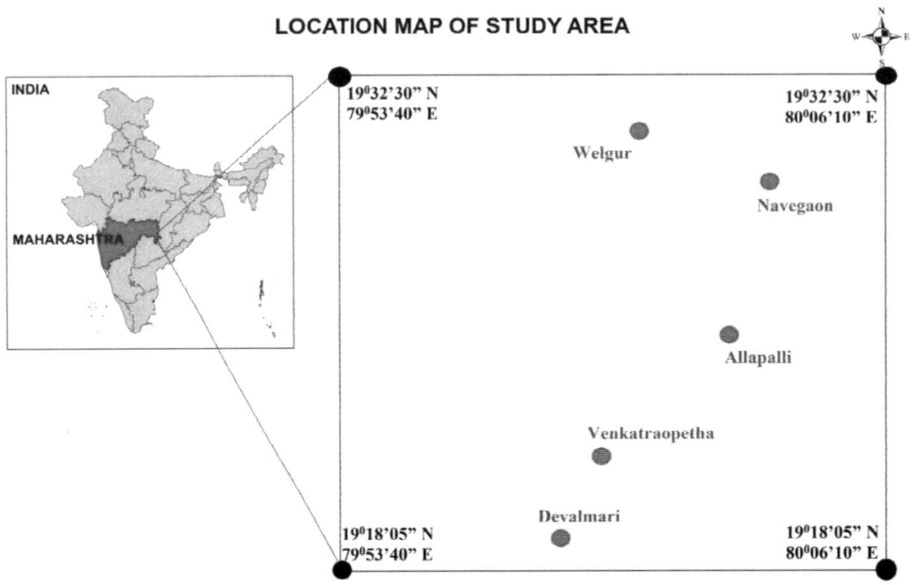

Figure 1: Map of study area indicating sampling locations

2. Material and methods

2.1 Collection of samples
From five villages of Aheri Taluka, namely Welgur, Navegaon, Develmari, Venkatraopetha, Allapali, a total number of 25 samples of groundwater had been collected in April 2021 at morning time. The temperature and pH were observed and recorded at the water collection sites and then brought to the laboratory to investigate various parameters.

2.2. Methodology
After collecting samples, temperature and pH were observed and recorded immediately using a digital thermometer and pH meter. The turbidity of the water sample was noted by using a turbidity meter. With the help of a conductometer and TDS meter, electrical conductivity (EC) and total dissolved solids (TDS) were determined, respectively. EDTA titration method was used to calculate the total hardness of the water sample. Quantitative estimation of nitrate was done by the phenol disulphonic acid method [6,7,8]. Standard laboratory methods estimated chloride, fluoride and iron.

3. Result and Discussion

The data of groundwater samples collected from the five different villages of Aheri taluka are summarized in table no.1.1 and 1.2, which includes the parameters such as temperature, pH, electrical conductivity, total dissolved solids, chloride, fluoride, total hardness, nitrate, alkalinity, iron and turbidity.

3.1 Depiction of Groundwater

3.1.1 Temperature. At higher temperatures, groundwater can dissolve more minerals from the rocks in the surrounding. This results in the higher values of some related parameters like EC and TDS. In the present study, all the parameters show a weak correlation with the temperature.

3.1.2 Turbidity. Estimation of the degree of muddiness or clarity of the water reveals the turbidity. It is also used to measure the optical property, which causes the light scattered and also absorbed by water. In the present work, the correlation value between turbidity and fluoride is **0.5209** which is in the moderate range. All the remaining parameters shows a weak correlation with turbidity. There is no permanent increase or decrease in turbidity levels due to environment brightness changes rather and only temporary changes could take place [9].

3.1.3 pH. The hydrogen ion concentration of the solution is the measure of pH. According to WHO, the pH range for drinking water between 6.5 to 8.5 is permissible. The water below 6.5 pH value is acidic and corrosive to the hand pumps. Water with a pH value of 8.5 is considered alkaline and may have a bitter taste [10]. In this study, 25 water samples were tested for pH and it was to be mentioned that all the values were under the acceptable limit and the range was between 7.2 to 8.4. In the present work, pH was found to be positively weak correlated with TDS, EC and Nitrate.

3.1.4 Electrical Conductivity (EC). EC values increase due to the high amount of salts in groundwater. EC values also depend upon temperature, types, and ions' concentration in the water [11]. The correlation value of EC with TDS was found to be 0.9258, which was near about the maximum range. This suggests that the amount of dissolved salts in the water samples was quite more. The moderate range of correlation of EC was observed with alkalinity and TH, which also confirmed the more dissolution of salts in the water. As per WHO, the maximum permissible value for EC is 1000μS/cm [12]. In the present work, all the samples showed higher values of EC except GW8, GW9, GW10, GW15.

3.1.5 Total dissolved solids (TDS). It is considered that the most important parameter of water to count its quality is TDS. It is affected and directly related to the increased alkalinity, hardness, conductivity and turbidity of water. The acceptable range is 500mg/L and it is suggested that the water with TDS value more than 300 mg/L should not be consumed for drinking purposes. The consumption of high TDS value water for drinking may induce gastrointestinal irritation and some unfavorable physiological reaction [13]. In the present work, it was to mention that maximum samples were not in the consumable range and it is due to more amount of dissolved salts, so it is highly recommended that the water should be well treated before consumption. Here it was noted that TDS was correlated in moderate range with Alkalinity and weakly correlated with TH and Cl^-.

3.1.6 Alkalinity. The high values of alkalinity of groundwater are mainly because of carbonates and bicarbonates. pH can be stabilized by alkalinity. Many substances in the water can be affected by alkalinity, pH and total hardness. The range between 200 to 600 mg/L for alkalinity is acceptable to consume. All the samples except GW11 were found to be safe for consumption. It was found that alkalinity is negatively correlated with pH (r = -0.1471) F^- (r = -0.2190), Nitrate (r = -0.1958) and Fe (r = -0.5243)

3.1.7 Chloride ions (Cl^-). Human-made problems are mainly responsible for the increased concentration of chloride content in water. Sewage water generally gets mixed with the water source, which ultimately leads to increase in the chloride concentration of water. Human-made pollution, improper sewage disposal, solid waste, septic tanks linkages, dumping of animal wastes, etc., are the leading causes of this high chloride concentration in groundwater. The chloride concentration permissible limit for drinking water is between 250 and 1000 mg/L [14]. In the present work, chloride concentration was weakly correlated with all the other parameters.

3.1.8 Fluoride ions (F^-). The fluoride content of water in proper concentration prevents bone damages and dental cavities. As per the Bureau of Indian Standards, the permissible range for fluoride concentration is between 1-1.5 mg/L. The increased level of fluoride may cause teeth yellowing, bone weakness, dental fluorosis and some other diseases. In the present study, it was observed that fluoride is moderately correlated with turbidity. It was also observed that fluoride weakly correlated with the other parameters.

3.1.9 Nitrate. As per WHO, the permissible nitrate content for drinking water is 45 mg/L. This study observed that the water samples GW3, GW4, GW6, GW21, GW24 exceeded the permissible range. These water samples were found to be unfit for consumption. In this work, nitrate was weakly correlated with the other parameters. The increase in nitrate level in groundwater is mainly due to natural processes, improper sewage disposal, excessive use of fertilizers, and animal and farm wastes.

3.1.10 Total Hardness (TH). Sulphates, chlorides, carbonates, and bicarbonates are mainly responsible for the high degree of hardness of the drinking water. The permissible range of the total hardness is in between 300 – 600 mg/L as per WHO and ICMR. It was observed that out of twenty-five water samples, only one sample (GW17) was found to be unfit for consumption due to 782 mg/L of total hardness. In the present study, total hardness (TH) was negatively correlated with pH, $F^-$and Fe; also, it was positively correlated with all other parameters. TH was found in moderate correlation with EC (r = 0.5148).

3.1.11 Iron (Fe). Due to the rain falling on the surface of the land, water drips through iron-bearing rock and soil, and iron gets dissolved in the water. The high concentration of Fe needs to the attribution to pyrite's oxidation which is present in rocks exposed because of mining [15, 16]. The and desirable and permissible value of iron in the water sample is 0.3 mg/L. The high concentration of iron present in water causes a bitter like and astringent taste with inky flavor and turbid nature. The soluble iron in water gives a clear appearance when pumped out and on exposure to air causes the precipitation of iron, resulting in rust like color and appear turbid in nature [17]. The dissolved oxygen is removed by some organic matter, which ultimately produces reduced conditions, and this results in increasing the solubility of Fe-bearing minerals [18]. The reddish-brown precipitate that appeared in the water sample and the vicinity of the tap and hand pumps specifies the presence of iron concentration in the high amount [19]. In the present study, all the samples showed the concentration of Fe in the acceptable range except sample numbers GW8 and GW10, with Fe concentration values as 1.28 and 1.16, respectively.

Table 1: Physicochemical Parameters of Groundwater samples.

Sample	Name of Village	Temp.	Tur (NTU)	pH	EC	TDS (mg/l)	Alkalinity (mg/l)
GW1	Welgur	27	0.9	7.62	1015	564	404
GW2	Welgur	24	1.02	7.8	1298	722	402
GW3	Welgur	22	0.38	7.72	1192	662	456
GW4	Welgur	23	0.45	7.78	1412	784	418
GW5	Welgur	21	1.11	7.75	1184	658	370
GW6	Navegaon	23	0.9	7.26	1124	652	482
GW7	Navegaon	22	0.54	8.27	1354	724	328
GW8	Navegaon	22	0.76	7.22	548	304	136
GW9	Navegaon	23	0.7	7.36	598	332	166
GW10	Navegaon	25.2	0.42	8.22	656	364	132
GW11	Devalmari	22	0.42	7.36	1010	622	**632**
GW12	Devalmari	24	1.02	8.24	1326	702	538
GW13	Devalmari	25	0.78	8.22	1444	792	488
GW14	Devalmari	22	1.32	8.12	1558	826	442
GW15	Devalmari	24	0.98	7.46	854	402	436
GW16	Venkatraopetha	24.7	0.88	7.38	1029	839	360
GW17	Venkatraopetha	24	0.64	7.72	1916	918	402
GW18	Venkatraopetha	24	0.63	7.76	1467	860	362
GW19	Venkatraopetha	24	0.33	7.62	1582	882	412
GW20	Venkatraopetha	24.2	0.29	7.62	1527	852	448
GW21	Allapalli	23	1.11	8.44	1231	684	296
GW22	Allapalli	24	0.42	8.2	1166	648	292
GW23	Allapalli	21	0.5	8.1	1181	656	312
GW24	Allapalli	24	0.29	8.29	1210	672	274
GW25	Allapalli	22	0.57	8.3	1206	670	302

Table 2: Physicochemical Parameters of Groundwater samples.

Sample	Name of Village	Temp.	Cl^- (mg/L)	F (mg/L)	Nitrate (mg/L)	Fe	TH
GW1	Welgur	27	168	0.32	41	0.01	492
GW2	Welgur	24	142	0.52	42	0.05	350
GW3	Welgur	22	196	0.02	**49**	0.04	346
GW4	Welgur	23	192	0	**47**	0.01	296
GW5	Welgur	21	132	0.39	38	0.02	396
GW6	Navegaon	23	83	0.1	**48**	0.22	478
GW7	Navegaon	22	87	0.1	42	0.24	464
GW8	Navegaon	22	37	0.3	39	**1.28**	224
GW9	Navegaon	23	60	0.4	21	0.02	198
GW10	Navegaon	25.2	103	0.3	38	**1.16**	214
GW11	Devalmari	24	75	0.1	36	0.06	298
GW12	Devalmari	25	52	0.2	19	0.16	228
GW13	Devalmari	22	48	0.1	15	0.2	288
GW14	Devalmari	24	52	0.3	14	0.2	274
GW15	Devalmari	24.7	96	0.1	17	0.1	282
GW16	Venkatraopetha	24	85	0.02	19	0.07	228
GW17	Venkatraopetha	24	125	0.1	43	0.05	**782**
GW18	Venkatraopetha	24	141	0.03	21	0.06	366
GW19	Venkatraopetha	24.2	141	0.05	25	0.30	408
GW20	Venkatraopetha	23	131	0.01	15	0.06	184
GW21	Allapalli	24	64	0.02	**49**	0.05	324
GW22	Allapalli	21	68	0.02	35	0.03	288
GW23	Allapalli	24	90	0.02	36	0.09	254
GW24	Allapalli	22	78	0.02	**48**	0.06	320
GW25	Allapalli	22	92	0.02	38	0.05	276

3.2 Graphical representation for various parameters are as follows:

Graph 1: Turbidity

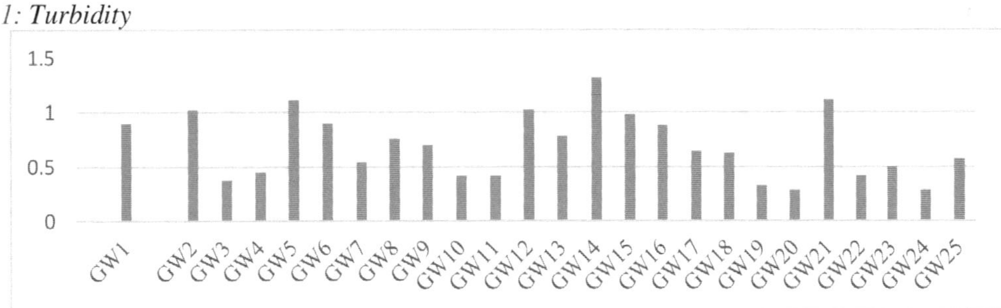

Graph 2: pH

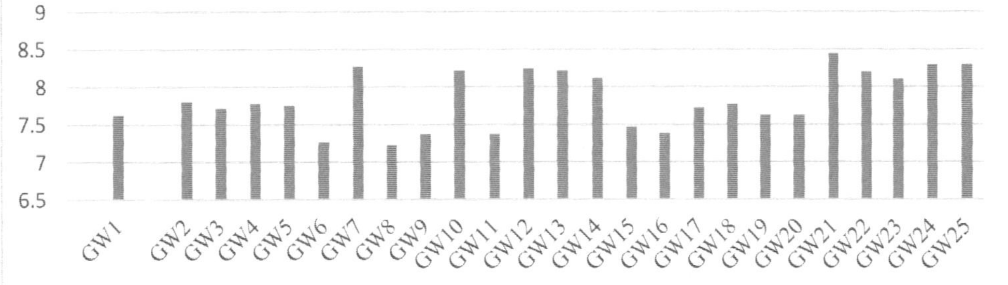

Graph 3: *Electrical Conductivity*

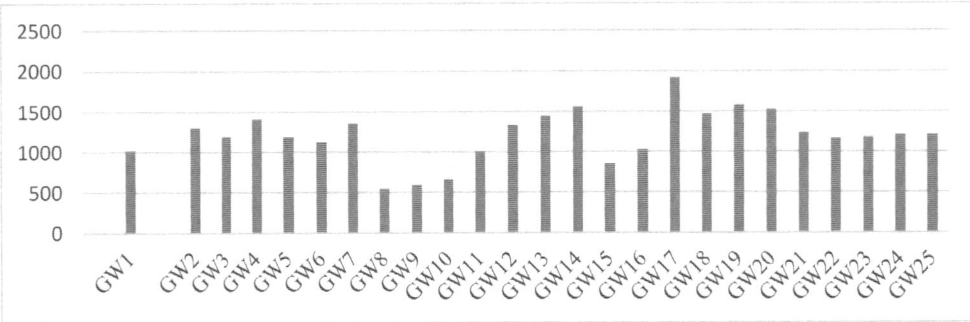

Graph 4: Total Dissolved Solids.

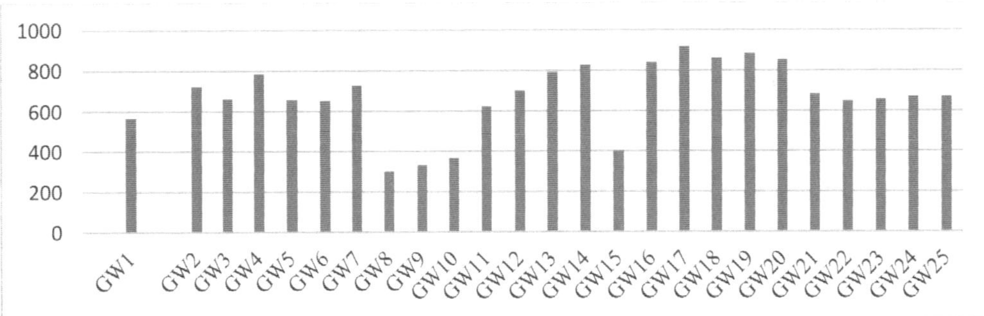

Graph 5: Alkalinity

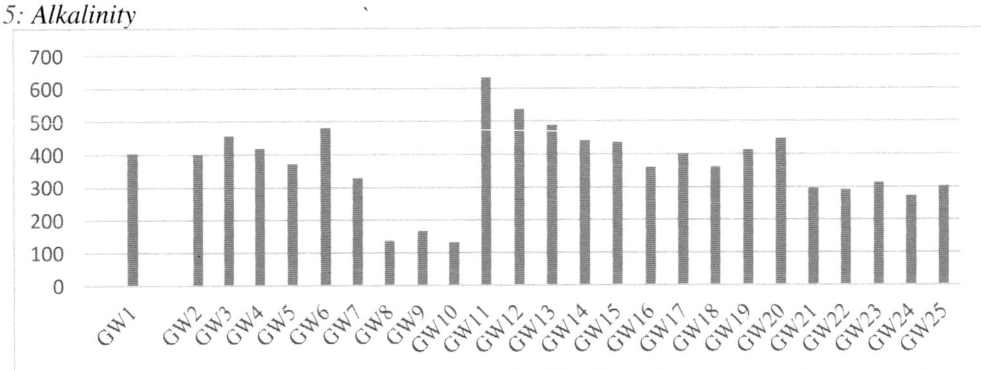

Graph 6: Chloride

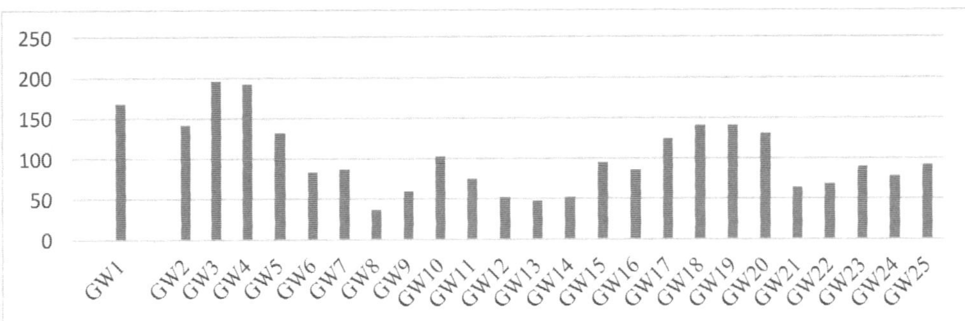

Graph 7: Fluoride

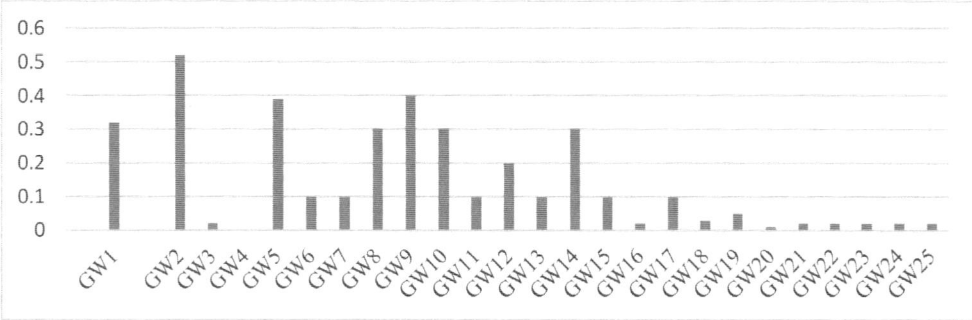

Graph 8: Nitrate

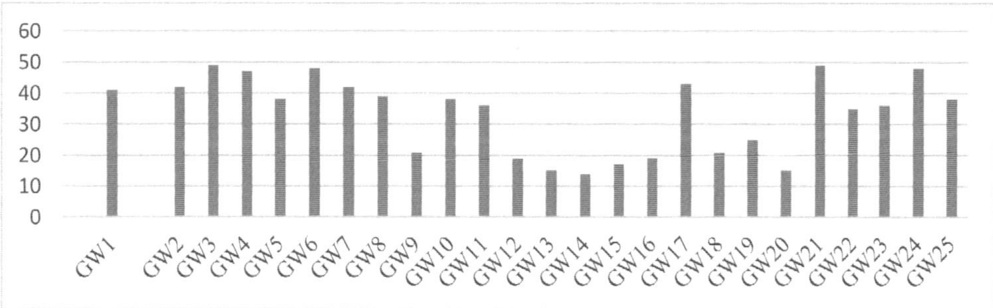

Graph 9: Iron

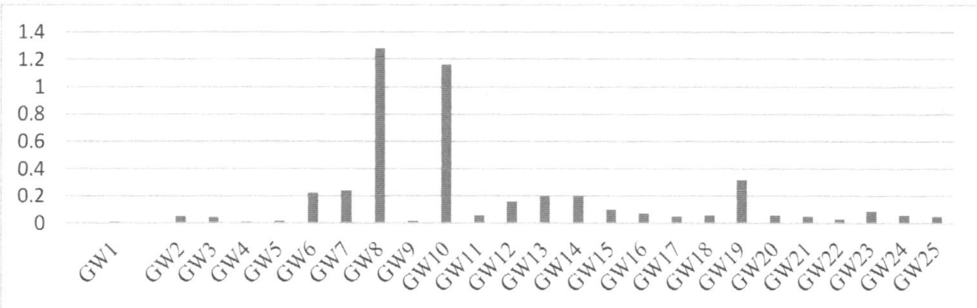

Graph 10: Total Hardness

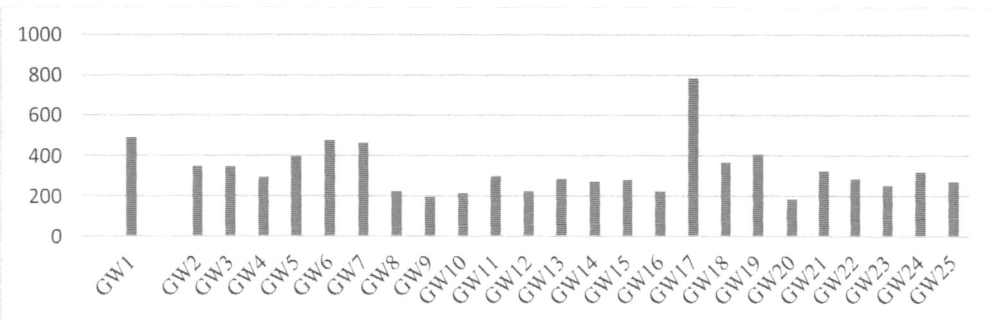

3.3 Correlation matrix of various physicochemical parameters

The quality of groundwater is mainly influenced by various physicochemical parameters and their associations can be efficiently sorted out by the correlation analysis [20]. The correlation matrix is shown in table no.3. This mainly involves statistical calculations and provides information about how any variable is influenced by the value of some other variable [21]. If the correlation coefficient (r) is less than 0.3, it is considered a weak correlation. If the correlation coefficient value is between 0.5 – 0.7, then it is assumed to be a moderate correlation. If the correlation coefficient value is greater than 0.7, it is considered a strong correlation. Weak values of correlation coefficient indicate the independency of the parameters from each other and the positive values of correlation infer the parameters are influenced by each other [22].

The relation between TDS with EC is found to be (r=0.9), which suggests that more salts are dissolved in water, leading to the increase in the electrical conductivity value. The correlation between alkalinity with EC (r=0.5118) and TDS (r=0.5296) suggests more dissolution of salts in the water sample.

Table 3: Correlation matrix of various physicochemical parameters.

	Tem.	Tur.	pH	EC	TDS	Alkal.	Cl^-	F^-	Nitrate	Fe	TH
Tem.	1										
Tur.	-0.02	1									
pH	-0.01	-0.02	1								
EC	0.02	-0.04	0.31	1							
TDS	0.06	-0.05	0.22	**0.92**	1						
Alkal.	0.02	0.13	-0.14	**0.51**	**0.52**	1					
Cl^-	0.15	-0.29	-0.18	0.29	0.29	0.20	1				
F^-	0.05	**0.52**	-0.19	-0.35	-0.43	-0.21	-0.07	1			
Nitrate	-0.22	-0.20	0.11	-0.07	-0.13	-0.1	0.32	-0.04	1		
Fe	0.02	-0.07	-0.08	-0.48	-0.51	-0.52	-0.29	0.27	0.04	1	
TH	0.10	0.05	-0.07	**0.51**	0.37	0.21	0.34	-0.02	0.42	-0.22	1

4. Conclusions and Recommendations

Out of 25 groundwater samples tested, nine samples were found unfit for consumption. The high level of nitrate ions in the five water samples was observed due to the animal wastes, poor drainage system, and some other sources. This could be minimized by selecting proper sites for the hand pumps and septic tanks that should be kept far from the hand pumps and also, the drainage system should be managed appropriately. It was observed that GW8 and GW10 samples were found to be unfit due to excess iron (Fe) concentration. To avoid health-related issues, it is suggested to use a water purification device. The sample number GW11 was found to be highly alkaline. Highly mineralized alkaline drinking water cause drying of the skin, as that could remove normal oils of the skin. It is suggested to use reverse osmosis process to remove alkalinity. Boiled and then filtered water could also be used to minimize the alkalinity. It was observed that the sample GW17 unfit for consumption due to the high value of total hardness. Temporary hardness could be removed by boiling and then filtering the water after cooling. Also, it was suggested not to use the water before proper filtration as it can ultimately affect health.

References

[1] Ravikumar P, Somashekar R K and Angami M 2011 Hydrochemistry and evaluation of groundwater suitability for irrigation and drinking purposes in the markandeya river basin, Belgaum District, Karnataka State, India *Environmental Monitoring and Assessment* **173** pp 459-87

[2] Bear J 1979 *Hydraulics of groundwater* McGraw- Hill International Book (New York)

[3] Vasanthavigar M, Srinivasamoorthy K, Vijayaragavan K, Ganthi R R, Chidambaram S and Anandhan, P 2010 Application of water quality index for groundwater quality assessment: Thirumanimuttar sub-basin, Tamil- nadu, India *Environmental Monitoring and Assessment* **171** pp 595–609

[4] Satapathy D R, Salve P R and Katpatal Y B 2009 Spatial Distribution of Metals in Ground/Surface Waters in the Chandrapur District (Central India) and heir Plausible Sources *Environ. Geol.* **56** pp 1323–52

[5] Todd D K 1980 *Groundwater hydrology* 2nd ed (New York: Wiley)

[6] Sharma S, Bhattacharya A 2017 Drinking Water contamination and treatment techniques *Applied Water Science* **7** pp 1043-67

[7] Jayalaxmi V L and Singaracharya M A 2011 Assessment of Physicochemical parameters of water and waste waters in and around.Vijayawada. *International Journal of Research Pharmaceutical and Biomedical Sciences* **3** pp 1040-46

[8] Bloomfield R A, Guyon J C, and Murmann R K 1965 Spectrophotometric Determination of Nitrate Ion Using Rhenium and α-Furildioxime *J. Analytical Chemistry* **2** pp 248-52

[9] Tempori J, Iskokangas A, Tuuttila T and Paavola M 2020 Functionality of turbidity measurement under changing water quality and environmental conditions *Environmental Technology* pp 1-9

[10] Goon A M, Gupta M K and Dasgupta B 1986 *Fundamental of Statistics* vol. I (Calcutta: The World Press Pv Ltd)

[11] Hem J D 1985 *Study and interpretation of the chemical characteristics of natural waters* 3rd ed USGS Water Supply Paper **2254** pp 117- 20

[12] World Health Organization 1993 *Guidelines for drinking water quality- I Recommendations* 2nd ed (Geneva WHO)

[13] Shankar B S, Balasubramanya N and Reddy M T 2008 Impact of industrialization on groundwater quality a case study of Peenya industrial area, Bangalore, India. *Environmental Monitoring and Assessment* **142** pp 263–68

[14] Asitkumar B and Surajit C 2015 Hydrogeochemistry and water quality index in the ssessment of groundwater quality for drinking Water *Environ.* **87** (7) pp 607-17

[15] Ganvir P S and Guhey R 2020 Groundwater quality assessment with reference to some heavy metals toxicity and its probable remediation around Ballarpur area of Wardha valley coalfields, Maharashtra, *IOP Conf. Series: Earth and Environmental Science* **597** -012001

[16] Ganvir P S and Guhey R 2021 Geochemical Studies of some Heavy Metals' Toxicity in Groundwater with their Plausible Sources around Gondwana Supergroup, Wardha valley Coalfields, Maharashtra *Journal Geological Society of India* **97** pp 1415-21.

[17] Jain C K, Bandyopadhyay A and Bhadra A 2010 Assessment of Groundwater quality for drinking purpose, District Nainital, Uttarakhand, India. *Environmental Monitoring and Assessment* **166** pp 663–76

[18] Mondal, N C, Singh V S, Puranik S C and Singh V P 2010 Trace element concentration in groundwater of Pesarlanka Island, Krishna Delta India *Environmental Monitoring and Assessment* **163** pp 215–27

[19] Ansari K and Hemke N M 2013 Water quality index for assessment of water samples of

different zones in chandrapur city *International Journal of Engineering Research* pp 233-37
[20] Ganvir P S and Guhey R 2020 Assessment of Some Heavy Metals Toxicity and Its Probable Remediation in Groundwater around Telwasa and Ghugus Area of Wardha Valley Coalfields, Maharashtra *Journal of Indian Association of Sedimentologists* **37(1)** pp 33-43
[21] Pearson K. 1896 Mathematical contributions to the theory of evolution III Regression heredity and panmixia *Philosophical Transactions of the Royal Society of London Series A* **187** pp 253–18
[22] Islam A T, Shen S, Bodrud-Doza M D and Rahman M S 2017 Assessing irrigation water quality in Faridpur district of Bangladesh using several indices and statistical approaches *Arabian Journal of Geosciences* **19** pp 418-36

Characterization of coke plant effluent by lysimeter test to evaluate safe disposal to land

Nikhil Kumar[1], H K Naik[2], Lipsa Mishra[3] and S Jena[4]

[1,2] Department of Mining Engineering,
[3,4] Department of Civil Engineering,
[1,2,3,4] National Institute of Technology, Rourkela, Odisha, India,

Email: hknaik@nitrkl.ac.in

Abstract. The importance of coal carbonization industries is evolving with great interest from the national point of view. In coke ovens, naturally found coal is converted to coke suitable for metallurgical industries. Enormous quantities of liquid effluents produced contain suspended solids high in COD, BOD, phenols, ammonia, and other toxic substances disposed into the water bodies without proper treatment contamination of the aquatic ecosystem. A large coke plant of Steel Authority of India Limited (SAIL) at Rourkela Steel Plant (RSP) has been surveyed for this purpose, and the impact of its effluent on surface water quality is discussed in this paper. The study found that concentrations of BOD, COD, total suspended, and dissolved solids have exceeded the tolerance limit as per I.S.: 2490 effluent water standard (inland surface water). Column lysimeters were prepared by collecting soil from the National Institute of Technology (NIT) campus, Jhirpani riverside, Mandira Dam side, putting it layer-wise into a column, and compacting it carefully to put on the original in-situ conditions. Raw effluents, diluted effluents, and water (as a blank) were applied to the three lysimeters twice a week with a 5cm inundating. The impact of raw and diluted effluent on groundwater quality and soil is also studied and discussed in this paper. The study found that the RSP soil sample and NIT campus soil sample were found more efficient for the disposal of effluents, and it is estimated that 2.73 hectares of land can be irrigated with effluents of the coke plant without deteriorating groundwater and soil quality.

1. Introduction

Coke is produced in coke plants for metallurgical purposes, and in the process of conversion, gas is produced as a by-product. These are responsible for the entire supplies of toluene, benzene, naphthalene, anthracene, and other coal tar products in India, constituting raw materials for manufacturing synthetic dyes, drugs, and high explosives [1]. A large amount of water is used for exhilarating hot coke and washing off gas produced from the oven. Effluents generated from the excitement of coke usually contain suspended matter of coke breeze (broken pieces of coke). Water used for washing gas comes out as ammoniac solid liquor, having a high concentration of phenol, cyanide, ammonia, and other toxic substances. These effluents cause severe surface water pollution in the area [2].

The iron and steel industry is one of the most energy-intensive sectors within the Indian economy. Steel production in India has increased by a compounded annual growth rate (CAGR) of 8% over the last ten years [3]. Each ton of steel production in India consumes 25–60 m^3 of water on average, and 5-9 tons of coal is fed as primary raw material. It may be noted that, in developed countries, the water consumption for each ton of steel production varies from 3 to 6 m^3, i.e., 8 to 10 times less than that of the water consumption in India (CPCB 2010) [4]. In an integrated iron and steel industry, wastewater generated from coke oven by-product plants is usually the most polluting.

Coke is produced by pyrolysis of coal, and it is also the primary component in the manufacture of steel. During pyrolysis, the gases produced from the coke oven are cooled further for steel-making purposes [5]. Tar, ammonia, phenols, naphthalene, light oil, and sulfuric compounds are some of the major pollutants found in the condensed water of coke oven gases; some of them, like tar and ammonia, can be recovered in a decanter for other purposes, the liquid left after the recovery is called coke-oven wastewater [6].

The agricultural potential of coke oven wastewater is being neglected in India. Wastewaters are substantial in nutrients, and they can be used in irrigating agrarian lands [7]. Disposal on land could recede pollution on surface water occurring due to removal, and it can also successfully remove the nutrients from the wastewater, thereby providing multistage water treatment and at the same time boosting agricultural growth. Disposal on land will help check the ecological balance of groundwater and control surface water pollution. The main challenge is to utilize all three aspects: chemical, physical, and biological properties of water with minimum unwanted effects on the crops, the characteristics of soil, and the quality of groundwater and surface water runoff. Marinating the aerobic condition under the ground is also possible by limiting the disposal amount of wastewater [8].

Rourkela Steel Plant (RSP), which has six coke oven batteries, effluents from it are considered for treatment. Rourkela does not have varied soil texture, but the various sites were chosen for soil sample collection based on the location and distance from the nearby water body, and a study was performed on it for disposal of effluents. In the process of production of coke by anaerobic carbonization of coal, gas is produced in a considerable quantity. Clean fuel gas is produced by treating this gas in the by-product plant after removing condensable, corrosive, or economically valuable components. The coal chemicals generally have reasonably good selling value. Coke wastewater generated during the washing of coke gas contains major chemicals and hazardous contaminants like ammonia, phenol, cyanide, and sulfide, whereas materials like oil, grease, and tar form an insoluble layer that hinders the access of oxygen from the air.

1.1 Objectives
The aim of the present study was:

- To determine the number of pollutants retained by soil so that the quality of groundwater remains unharmed.
- To construct a lysimeter column for the laboratory scale test.
- To perform tests on leachate by considering parameters such as pH, Temperature, D.O., COD, BOD, Dissolved solids, suspended solids, Total hardness, Ca hardness, and Mg hardness.

1.2 Scope of the study:
The present study focused on the retaining of pollutants by the soil, and the effluent received must be under the limits of disposal of effluent in sewage disposal of facilities.

2. Literature review
The treatment of coke oven wastewater, as one of the most complex and problematic industrial waste streams, is often discussed in the literature and widely investigated by many scientific and industrial R&D centers worldwide. The topics listed below include the previous research works that have been conducted on the treatment of wastewater:

- Industrial wastewater treatment
- Coke oven wastewater treatment

2.1 Industrial wastewater treatment
Biological methods have long treated industrial wastewater. Several studies indicate that physical treatments can remove organic pollutants from the enormous quantity of sewage at a relatively low cost

[10-14]. However, a long residence time is required due to the prolonged process of biodegradation. Autotrophic bacteria that oxidize ammonia during the biological nitrification process and other heterotrophic oxidizing bacteria are normally sensitive to the presence of toxic compounds such as phenol, cyanide, and high concentration of ammonia.

2.2 Coke oven wastewater treatment
In recent publications, electrochemical processes are proposed to decompose and destroy a range of contaminants (cyanides, sulfides, and thiocyanates) present in coke oven wastewater.

Ozyonar and Karagozogly [15] compared treating of pre-treated coke wastewater by electrocoagulation process (E.P.) and electrochemical peroxidation process (ECP) using direct pulse current. They used the air stripping process of ammonia as a physicochemical process for this purpose. A direct pulse current was used to prevent the passivity or polarization of electrodes and increase removal efficiency. They found that ECP was more efficient than E.P. in removing COD, total organic carbon, phenol, cyanide, and thiocyanate, but higher operating costs.

Pillai and Gupta [16] applied anodic oxidation of industrial wastewater from a coke oven plant having cyanide including thiocyanate (280 mg/l), chemical oxygen demand (COD 1520 mg/l), and phenol (900 mg/l) using a novel PbO_2 anode. Optimization was performed for maximizing the removal efficiencies of these three parameters simultaneously; those were 99.6%, 86.7%, and 99.7% for cyanide, including thiocyanate, COD, and phenol, respectively.

Water reclamation from coke oven wastewater effluent or introduction of membrane processes arranged in a different model to the coke oven wastewater treatment is also a subject of much research. Fascinating and promising results are obtained.

Kumar and Pal [17] designed an experiment and investigated a new system using forward osmosis–nanofiltration in a flat-sheet crossflow module. The aim was to separate reusable water from coke-oven wastewater with reduced concentration polarization and high flux using low energy. Removal of about 96–98% of cyanide, phenols, NH_4^+–N, and chemical oxygen demand from real coke-oven could be achieved along with pure water flux of 46 l/(m^{2}h) in a forward osmosis system under optimized conditions.

Jin et al. [18] investigated a full-scale plant using anaerobic, anoxic, and oxic processes (A1/A2/O), along with a pilot-scale membrane bioreactor (MBR), nanofiltration (N.F.), and reverse osmosis (R.O.) integrated system to treat coking wastewater for industrial reuse over a period of one year. The removal of pollutants (TCN, COD, BOD, ammonium nitrogen, SCN, fluoride) efficiency reached very high values during the A1/A2/O biological treatment stage, and all parameters were further reduced by over 96.0%.

Chang et al. [19] investigated the biological and chemical characteristics of coke-oven wastewater after ozonation treatment through the examination of selected parameters in a bench-scale bubble column reactor. Ghose [20] analyzed the physicochemical treatment of wastewater from the coking plant as a suitable option for the treatment of coke plant effluents.

Vazquez et al. [21] analyzed laboratory-scale activated sludge plants to study the biodegradation of coke wastewater.

3. Land treatment and details of the coke plant understudy
Compared to all other methods of treatment, land treatment is the cheapest in some old coke plants, availability of space might be a problem for biological treatment. Disposal to land does not require an expensive wastewater treatment plant. It acts as a primary, secondary, and tertiary treatment for wastewater and helps in controlling surface water pollution. This can help in minimizing the cost of fertilizers and help to grow the important crops in the region. This proves to be economical as well; the cost of irrigation can be checked. Also, groundwater gets restored by using this method.

3.1 The methodology adopted for the study:

3.1.1 Sampling and analysis of coke plant effluent: Wastewater from RSP was collected twice a week on Monday and Friday. Each day at 17:55 hours, they were poured into the Lysimeter; two irrigation per week with 5 cm flooding each. This was continued for a period of one month. The grab sampling method was used for sampling. The leachate and the raw effluent were analyzed by following the technique given in "standard methods for the examination of wastewaters" [11].
3.1.2 Collection of soil samples: Different areas of Rourkela were selected for the Collection of soil samples. Though Rourkela did not have varied soil texture (mostly sandy, loamy soil is found), different places were selected based on the location. The samples were collected from the following areas:

- *NIT Campus*- The soil on the NIT campus was sandy loamy, and the ground is stable and in a sparsely populated area.
- *Jhirpani (Riverside)* - This site was chosen because it was saturated and moving water is always there.
- *Mandira dam*- The soil sample is the same as Jhirpani, but here, the water is stable, and the soil is also saturated.
- *RSP waste disposal site*- RSP was chosen because the waste from industries is disposed of here, which contains a heavy amount of chemicals.

4. Preparation of Column Lysimeters:

The PVC pipes were used for the construction of the column lysimeter, which had a dimension of 2m in length and 0.2m in diameter. It consists of three ports with a diameter of 0.05m, at 0.3, 0.9, and 1.6m depth, which were driven inside lysimeters. One end of each port was connected to the plastic cup fixed tightly used for collecting leachate from each port, and the other end was closed to retain leachate from leaving the Lysimeter. To collect leachate, the top of the pipes in the ports were perforated, and a fine screen, sand, and gravel were placed so that leachate could be collected smoothly, as shown in figure 1 and figure 2(a & b). The soil collected was carefully placed and compacted in the Lysimeter to simulate the original in-situ condition pertaining to stratification and density.

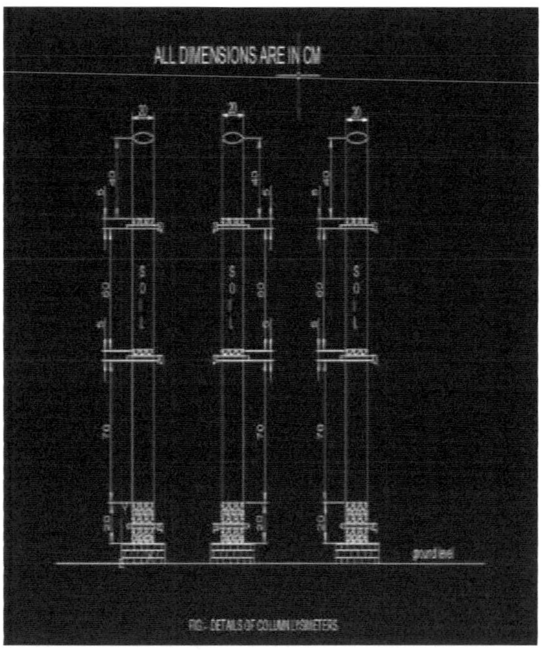

Figure 1. Details of Column Lysimeter

Figure 2.a Constructed Lysimeter

Figure 2.b Constructed Lysimeter.

4.1 Disposal of effluent and inspection:

The experiment was conducted in two phases:

Phase 1: All three lysimeters were introduced with different types of liquids keeping the soil (NIT Campus) in the Lysimeter the same. Distilled water was introduced in lysimeter No.1, and in lysimeter No.2, diluted effluent of the coke oven plant was introduced. In lysimeter No.3, raw effluent was introduced. There were two irrigations conducted per week in each Lysimeter for a period of one month.

Phase 2: In this phase, the raw effluent was used as the liquid, and the soil in each of the lysimeters was changed. Soil from Mandira dam was placed in lysimeter No.4, Jhirpani soil was placed in lysimeter No.5, and soil from the RSP dumping site was placed in lysimeter No.6. The soil was compacted to attain in-situ conditions, and the same procedure for irrigation was followed for the next month.

The leachate from lysimeters 2, 3, 4, 5, and 6 were collected to assess the impact of coke plant effluent on groundwater quality.

4.2 Impact on soil quality:

Soil from different lysimeters after completion of sampling was taken and put into separate pots. For the comparison, the untreated soil sample was also put into the pot alongside, and different seeds were sown into them in equal amounts.

5. Result and discussion

The leachate temperature from all lysimeter tests was found slightly higher than the temperature specified in the effluent water standard. The temperature in the first phase of the experiment was found to be 41°C for all the three lysimeters taken, and in the second phase, it was 46.1°C for all the three lysimeters. The limit for disposal in sewerage for effluent is 40°C, which is being slightly exceeded by effluents from the Lysimeter. However, the pH was found to be 7.5 for Lysimeter with distilled water, 7.48 for Lysimeter with the diluted sample, 7.4 for Lysimeter with raw effluent, and 7.48 for Lysimeter with Mandira dam soil, 8.98 for Lysimeter with Jhirpani soil, and 7.91 for Lysimeter with RSP soil sample. The pH range of all the soil and effluent samples was within the standard range of 5.5 – 9.0 (figure 4). Therefore, considering pH it is safe to dispose of it on land, and considering temperature, it is not suitable.

There is no standard limit for the initial D.O. content of the leachate. The BOD of the samples taken in the first phase was 510 mg/l for distilled water, 520 mg/l for the diluted sample, and 490 mg/l for raw effluent. In the second phase of the experiment, the BOD was found to be 340 mg/l for the RSP sample, 590 mg/l for Jhirpani soil, and 640 mg/l for the Mandira dam. The BOD is found to be very high as compared to the tolerance limit of effluent water standard (figure 5). This shows that survival of aquatic life is not possible due to the lesser amount of dissolved oxygen in the water and the bacteria and other micro-organisms are present in very high concentrations. The COD found was also much higher than the limits in the effluent water standard. The COD of distilled water was 1350 mg/l, 1220 mg/l for diluted sample, and 1270 mg/l for Lysimeter with raw effluent (figure 6). In the second phase of the experiment, the COD concentration found was much higher at 1650 mg/l for Lysimeter with Jhirpani soil, and 1650 mg/l for Lysimeter with Mandira dam soil, and 1670 mg/l for Lysimeter with RSP soil sample.

The number of dissolved solids in the first phase of the experiment was found to be below the tolerance limit of the effluent water standard. The number of dissolved solids in the distilled water sample and the diluted sample was 1085 mg/l and 870 mg/l, respectively. But the number of dissolved solids in the raw solid was 985 mg/l, which is lesser in comparison to the two others. However, the standard limit is 2100 mg/l. In the second phase of the experiment, TDS for RSP soil was 1585 mg/l, 8730 mg/l for the Jhirpani soil sample, and 2840 mg/l for the Mandira dam soil sample (figure 3). It is inferred that soil from RSP is suitable for disposal and within limits. TSP found was beyond the tolerance limit in both phases. The Mandira dam soil sample had 10880 mg/l, 21855 mg/l in Jhirpani, 5890 mg/l in RSP,1565 mg/l in distilled water, 1570 mg/l in the diluted sample, and 805 mg/l in raw effluent. TSP limit for effluent water standard is 100 mg/l.

For measuring the quality of soil, the original soil sample of Mandira dam soil and RSP soil was taken, and the soils from the Lysimeter after treatment were also taken (figure 7 and figure 8). Green seeds were planted in the soil, and the soil was irrigated twice a day because of the hot weather. The growth of the plant was observed for the next 15 days. The plant's growth in the soils after treatment was healthier than compared to the original soil (figure 9). This shows that it is suitable for plant growth, and the soil is enriched with nutrients also.

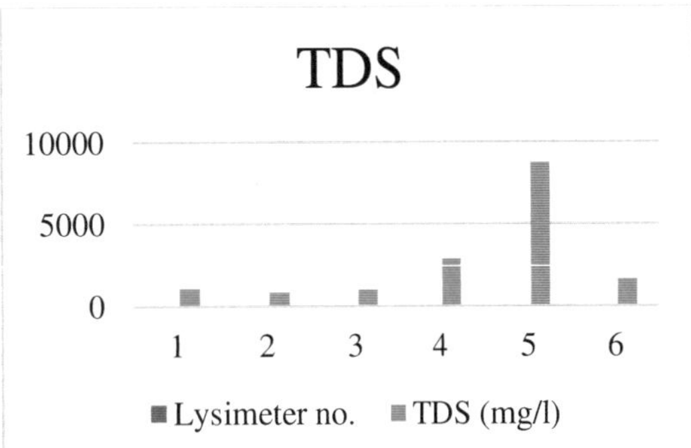

Figure 3. TDS vs. Concentration in mg/l

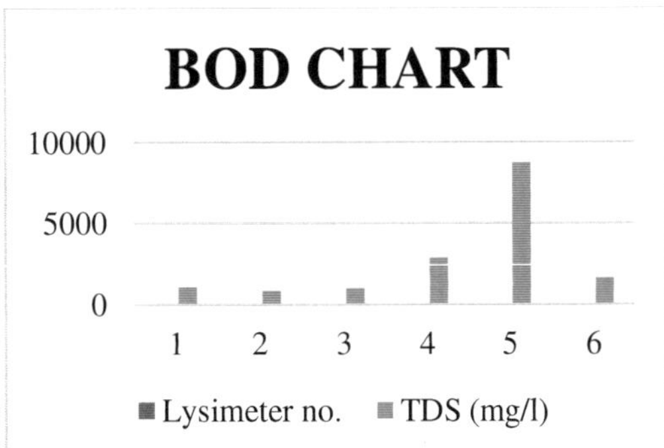

Figure 5. BOD vs. Concentration in mg/l

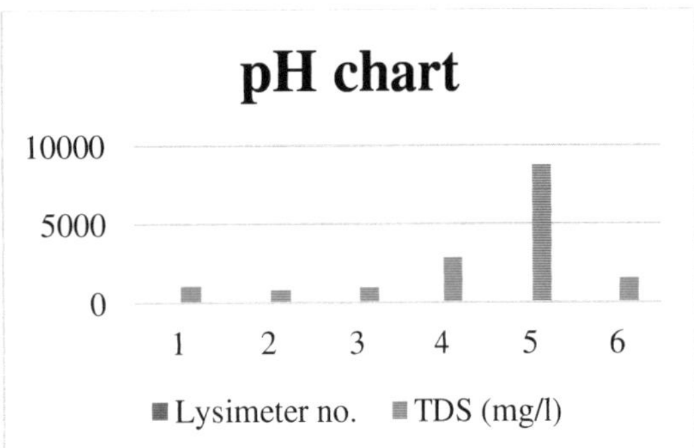

Figure 4. pH Chart

Figure 6. COD testing setup

Figure 7. Leachate collected from Lysimeter

Figure 8. Soil testing before sowing of seed

Figure 9. Soil testing after sowing of seed

Table 1. Measured parameters

Parameters	Distilled water sample (1)	Diluted water sample (2)	Raw Effluent (3)	Mandira Dam Soil Sample (4)	Jhirpani Soil Sample (5)	RSP Soil Sample (6)
pH	7.5	7.48	7.4	7.48	8.98	7.91
Temperature (^{0}C)	41	41	41	46.1	46.1	46.1
BOD (mg/L)	510	520	490	340	590	640
COD (mg/L)	1350	1220	1270	1650	1650	1670
Dissolved Solids (mg/L)	1085	870	985	1585	8730	2840
TSP (mg/L)	1565	1570	805	10880	21855	5890

6. Conclusion

From the above study, the following conclusions are drawn. With the use of wastewater, the nutrient in the soil was restored, and healthier plants could be grown. However, for direct disposal, it should be treated first for high BOD and COD, and then it can be disposed of on land. It will help in reducing surface water pollution and restoration of nutrients in the soil. For the initial treatment, much lesser space and capital are required, which will reduce the cost of setting up large treatment plants on the plant sites.

References

[1] Bone W A 1980 *Coal and Its Scientific Uses,* FRS
[2] Ghose M K and Roy S 1996 *Status of water pollution due to coke oven effluents- a case study,* Journal of IPHE **3** pp 1-9
[3] Spyridakis D E and Welon E B 1976 *Treatment Process and environmental impacts of waste effluent disposal on land. In Land treatment and disposal of Municipal and Industrial wastewater,* Sanks, R.; Asano, T.; Eds.; (Ann Arbor Science: Ann Arbor, MI)
[4] EIFAC 1973 *Wastewater Treatment and Reuse by Land Application;* U.S. Environmental Protection Agency: (Washington, D.C.) **1** and **2**
[5] Ghose M K 1994 *Control of pollution by recycling and reusing wastewater* Ind. Journal of Env. Prot. **12** pp 884-867
[6] Ferguson A H 1976 *Acceptability of Wastewater Effluents by Soils. In Land Treatment and Disposal of Municipal and Industrial Wastewater* Sanks, R.; Asano, T., Eds.; (Ann Arbor Science: Ann Arbor, MI)
[7] Ghose M K 2002 *Complete Physico-chemical treatment for coke plant effluents* Water Research 36 **5** pp 1127-1134
[8] Pal P and Kumar R 2014 *Treatment of Coke wastewater: A critical Review for Developing Sustainable Management Strategies* Separation and Purification Reviews, 43 **2** pp 89-123
[9] Davie T 2003 *Fundamentals of Hydrology* (Psychology Press)
[10] Rana G and Katerji N 2000 *Measurement and estimation of actual evapotranspiration in the field of Mediterranean climate: a review* European Journal of Agronomy, **13** pp 125-153
[11] Landscape Evolution observatory 2015| (Biosphere 2, biosphere2.org)
[12] Ghose M K and Roy S 1999 *Lysimeter test of coal plant effluent disposal to land in India* Air Waste Manage Assoc. **49** pp1245–9
[13] Bone W A 1980 *Coal and its scientific uses* FRS
[14] Dartnell J 1978 *Formed coke from steel writers in wings* Iron Steel Int **15** p155
[15] Ozyonar F and Karagozogly B 2015 *Treatment of pre-treated coke wastewater by electrocoagulation and electrochemical peroxidation processes* Sep. Purif. Technol., **150** pp 268–277
[16] Pillai I M and Gupta A K 2016 *Anodic oxidation of coke oven wastewater: multiparameter optimization for simultaneous removal of cyanide, COD, and phenol* Journal of Environmental Management Jul 1 **176** pp 45-53
[17] Kumar R and Pal P 2015 *A novel forward osmosis-nano filtration integrated system for coke-oven wastewater reclamation* Chemical Engineering Research and Design Aug 1; **100** pp 542-53
[18] Jin X, Li E, Lu S, Qiu Z and Sui Q 2013 *Coking wastewater treatment for industrial reuse purpose: combining biological processes with ultrafiltration, nanofiltration and reverse osmosis* Journal of Environmental Sciences Aug 1;25 **8** pp1565-74
[19] Chang E E, Hsing H J, Chiang P C, Chen M Y and Shyng J Y 2008 *The chemical and biological characteristics of coke-oven wastewater by ozonation* Journal of Hazardous Materials Aug 15;156(1-3): pp 560-7
[20] Ghose M K, 2004 *Physico-chemical treatment as a suitable option for treatment of coke plant effluent*, Journal of Institution of Engineers, Chemical Engineering Division, Mar 1
[21] Vázquez I, Rodríguez J, Maranon E, Castrillon L and Fernandez Y 2006 *Study of the aerobic biodegradation of coke wastewater in a two and three-step activated sludge process* Journal of hazardous materials. 2006 Oct 11;137 **3** pp1681-8

Study of effects of some metals and hydraulic retention time on performance of AMD treatment by SAPS

M. D. Patel[1]
[1]Associate Professor, Department of Mining Engineering, National Institute of Technology, Raipur, India
Email- mdpatel.min@nitrr.ac.in

Abstract: Acid mine drainage (AMD) is presently observed as one of the major water pollution problems in mining area throughout the world. Proper scientific management of AMD is essential for achieving the national goals for environmental norms. In this studyperformance of AMD treatment by laboratory successive alkalinity producing system (SAPS) by observing the effects of some metals and hydraulic retention time. In this study the experiments were conducted in laboratory successive alkalinity producing system (SAPS) with different synthetic AMDs for five Hydraulic retention times(HRTs).The successive alkalinity producing system shown the promising results during the AMD treatment. The effects of iron, aluminium and manganese were jointly investigated, whichshowed that Net alkalinity generation (NAG) increases with increase in influent metal contents.It was also observed thatNAG also showed increasing tendency with HRT during AMD treatment by SAPS. The obtained findings from this study can be applied for efficient design and operation of successive alkalinity producing system for AMD treatment process in mining industry.

1. Introduction

AMD is an environmental problem that deteriorates the water resources in mining areas. In India, the coal of Barakar series of Lower Gondwana system in Madhya Pradesh, Chhattisgarh and Maharashtra and tertiary system of Assam and Jammu and Kashmir contains sulfide, which is responsible for AMD generation. The occurrence of sulfide minerals dictates the generation of AMD besides other factors like hydrogeology of area, nature of adjoining rocks, presence of water, temperature, and oxygen supply, etc. [7, 18]. The AMD generation reaction can be expressed as

$$4\,FeS_2 + 15\,O_2 + 14\,H_2O \text{ -------}\rightarrow 4\,Fe(OH)_3\downarrow + 8\,H_2SO_4$$

The surrounding aquatic environment is adversely affected byAMD discharge and as a result aquatic livings are killed.AMD also increases the probability of corrosion in mining machine and fittings and reduces the mining machinery'slife. The adverse effects of AMD can be minimized by the treatment of AMD either by active treatment or passive treatment methods. SAPS is latest modified passive treatment system. SAPS eliminate the armoring of limestone by oxidation of iron, therefore life of SAPS passive treatment system increases. The treatment of AMD by SAPS was first observed in late 1980s [9]. Many SAPS systems were installed throughout the world for treatment of AMD[17, 5, 10, 2].

SAPSis a bioreactor in which biological and chemical processes takes place during treatment of AMD at certain HRT. The reduction in acidity, generation of alkalinity and removal of dissolved metals is achieved by utilizing organic matter and limestone in the SAPS units during treatment of AMD [11].The AMD contains commonly iron, aluminium and manganese there for these metals are selected for investigation. Presently very limited informationis available on the effects of composition of AMD, especially of total iron, ferric ferrous ratio, aluminium, manganese and HRT on NAG by SAPS. Further the ferric iron and ferrous iron in the AMD exhibit influence on the chemical activity inside the SAPS cell [15].In this investigation ferric/ ferrousratiowith the total iron is selectedfor more comprehensive understanding ofperformance of the SAPS. Finally, a correlation between the NAG and the ferric/ferrous ratio along with total iron, manganese, aluminum, iron and HRT has been established using SPSS software.

2. Methodology

Synthetic AMDs were prepared in the laboratory with variation in metal contents.The coal mine drainageis generally contaminated with dissolved iron, aluminum and manganese [4].The research work were performed on the four identical laboratory SAPS units, wherein 28 varieties of artificial AMDs of a known composition were treated in seven phases (a-b-c-d-e-f-g).The experiments with five HRT from 1 day to 10days were carried out. Hence, each phase of experiment involving of four varieties of AMD was conducted for the stipulated HRT periods. The next phases of experiments commenced on another four variety of AMDs on the identical SAPS units [12].

Four PVC vessels were selected for SAPS column units(80L each). Limestone of 15cmthickness is filled up at bottom, then saw dust and cow compost are filled up with 5 cm and 22 cm thickness respectively. Lastly, 2.5 cm gravel is packed at the top and then artificial AMD is allowed to fill up above the top gravel pack [14, 16]. The estimation of quantity of limestone (27 kg) and organic substrate (29 liters) was done on the basis of chemical calculations [12]. A perforated pipe is fixed at lowermost portion of SAPS column for the discharge of processed AMD to oxidation cell and flushing (Fig.1).

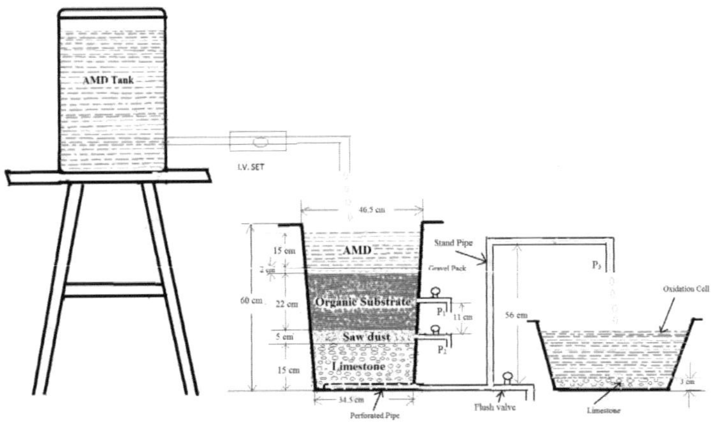

Fig.1. Laboratory arrangements for SAPS column study(Patel et al., 2019)

A polyethylene net is placed between different layers of filled materials for separation purpose. The cow compost is taken with sufficient number of microbial counts ranging from 2.96×10^5 to 2.9×10^6cfu/ml. The studies were conducted from 13.6°C to 43.3°C for 273 days in similar conditions. Limestone and organic substrate having similar compositionwere processed simultaneously in each phase (24 days), hence 28 synthetic AMDs of different compositions were processed in seven phases.

2.1 Composition of AMDs

All the 28 Artificial AMDs were prepared just before the commencement of each experiment (Fig. 2). The 28 synthetic AMDs were prepared according to ranges of composition given in Table 1.

Fig. 2. Four laboratory SAPS column operation [15]

Table 1. Summary of ranges of composition of 28 artificial AMDs

Parameter	**Range**
pH	2.44- 4.50
DO	5.52- 7.13 mg/L
ORP	104.30 -113.50 mV
Iron (Fe^{2+}, Fe^{3+})	85.70 – 196.80 mg/L
Aluminum	20-80 mg/L
Manganese	15-60 mg/L
Calcium	125 mg/L
Magnesium	100 mg/L
Sulphate	1020-1498 mg/L
Electrical conductivity	1939-2217 µS/cm

In phase a, the aluminum + manganese concentrations in AMD were uniformly 20+15 mg/Lfor AMD Aa,AMD Ba, AMD Ca and AMD Da. Similarly in phase b,c,d,e,f and g, the aluminum and manganese concentrationin AMDs were 40+15, 60+15, 80+15, 20+30, 20+45 and 20+60 respectively for respective group's AMDs. The FFR×TI for all the phase's AMDs is given in Table 3.

Table 2. Phase wise distribution of AMDs for experiments

Phase a	**Phase b**	**Phase c**	**Phase d**	**Phase e**	**Phase f**	**Phase g**
AMD Aa	AMD Ab	AMD Ac	AMD Ad	AMD Ae	AMD Af	AMD Ag
AMD Ba	AMD Bb	AMD Bc	AMD Bd	AMD Be	AMD Bf	AMD Bg
AMD Ca	AMD Cb	AMD Cc	AMD Cd	AMD Ce	AMD Cf	AMD Cg
AMD Da	AMD Db	AMD Dc	AMD Dd	AMD De	AMD Df	AMD Dg

2.2 Sample Testing

WTW multi 3620 IDS digital meter was used to measure pH, DO, ORP, temperature and electrical conductivity.The American Public Health association (APHA) standards were followed during sample testing and analysis for alkalinity, acidity, total iron,aluminum, manganese, ferric ferrous ratio, sulphate and for other elements [1]. The Lasany UV-VIS spectrophotometer was used in above mentioned tests [12].

3. Results and Discussion

The results obtained after seven phases SAPS experiments are summarized in Table 3.The observations are discussed firstly forNAG with variation in influent metal loading with fixed HRT then for NAG with variation in HRT with fixed influent metal loading.

Table 3. AMD's details and Incremental NAG value for all experiments for all five HRTs

AMD Sample	Total metal loading (Fe+Al+Mn) (mg/L)	FFR × T I	Incremental NAG value (From base value 430 of AMD Aa for 1d HRT) mg/L				
			HRT=1d	HRT=2d	HRT=4d	HRT=7d	HRT=10d
AMD Aa	120.70	4.627	00	90	260	415	415
AMD Ab	150.63	10.61	30	120	295	470	600
AMD Ac	183.60	14.01	120	190	470	700	820
AMD Ad	205.70	15.50	270	460	670	850	975
AMD Ae	141.70	10.45	05	110	270	475	600
AMD Af	157.50	12.67	10	115	280	505	645
AMD Ag	173.20	15.66	10	120	295	525	745
AMD Ba	153.70	30.62	20	150	355	565	710
AMD Bb	176.50	40.10	65	195	470	625	775
AMD Bc	209.00	43.01	190	330	515	740	850
AMD Bd	236.10	53.48	360	590	830	1060	1225
AMD Be	179.20	44.10	35	185	365	640	780
AMD Bf	193.90	49.50	45	250	380	690	880
AMD Bg	208.50	52.43	50	250	420	780	920
AMD Ca	206.60	147.75	185	345	500	715	875
AMD Cb	227.30	153.35	335	440	665	790	960
AMD Cc	249.00	160.08	420	645	855	1075	1275

AMD Cd	269.00	174.70	585	905	1235	1570	1780
AMD Ce	224.00	196.62	230	385	570	820	945
AMD Cf	237.80	207.01	250	390	575	850	950
AMD Cg	251.50	218.49	250	395	580	865	990
AMD Da	230.50	223.46	360	400	580	790	970
AMD Db	249.20	229.93	375	490	690	835	1015
AMD Dc	269.30	235.10	550	790	980	1120	1330
AMD Dd	291.80	261.94	670	1000	1290	1630	1840
AMD De	244.35	239.05	285	495	720	970	1080
AMD Df	258.00	250.90	310	545	760	1020	1160
AMD Dg	270.80	60.00	340	570	810	1065	1210

3.1. NAG with variation in influent metal loading with fixed HRT

In Phase a, For AMD Aa, the metal loading is (85.70mg/L iron +20 mg/L aluminum +15 mg/Lmanganese =120.70 mg/L) andmetal loading were gradually increased in AMD Ba, AMD Ca and AMD Da to 153.70, 206.60 and 230.50 (all in mg/L) respectively (Table 3). The NAGreported after SAPS treatment for AMD Aa was 430 mg/L. The AMD Aa was taken as base AMD because it was prepared first then metal loadings were increased gradually. Thereafter for Phase a, for fixed 1d HRT 20,185 and 360(all in mg/L) incremental value from base value of NAG (i.e. 430 mg/L) were reportedfor AMD Ba, AMD Ca and AMD Da respectively.Similarly for fixed 2d, 4d, 7d and 10d HRT and with above variations in metal loading, the incremental NAG range was 90-400, 260-580, 415-790 and 415-970 mg/L respectively.It was reported that NAG was showing increasing trend with increased metal loading in influent AMDs for a particular HRT. It is clear from above observationsthat the net alkalinity generation range were increasing gradually with fixed higher HRTs. The reason for above increase in NAG range is availability of more time for microbial reaction in organic substrate and chemical reaction of AMD with limestone at the bottom level, which results as more metal precipitation and alkalinity generation.

In Phase b, For AMD Ab, AMD Bb, AMD Cb and AMD Db, the metal loadings were 150.63, 176.50, 227.30 and 259.20 (all in mg/L) respectively. The incremental NAGfrom base value of NAG, after SAPS treatment for AMD Ab, AMD Bb, AMD Cb and AMD Dbfor Phase b, for fixed 1d HRT, were also reported in increasing order i.e. 30, 35, 335 and 375(all in mg/L) respectively. Similarly for fixed 2d, 4d, 7d and 10d HRT and with above variations in metal loading, the incremental NAG range were120-490, 295-690, 470-835 and 600-1015 mg/L respectively. The pattern of the incremental NAG range was higher as compared to phase a. This can be justified that metal loading is higher as compared to phase a.

In Phase c, For AMD Ac, AMD Bc, AMD Cc and AMD Dc, the metal loadings were 183.60, 209.00, 249.00 and 269.30 (all in mg/L) respectively. The incremental NAG from base value of NAG, after SAPS treatment for AMD Ac, AMD Bc, AMD Cc and AMD Dc for Phase c, for fixed 1d HRT, were also reported in increasing order i.e. 120, 190, 420 and 550 (all in mg/L) respectively.Similarly for fixed 2d, 4d, 7d and 10d HRT and with above variations in metal loading, the incremental NAG range were 190-

790, 470-980, 700-1120 and 820-1330 mg/L respectively. The incremental NAG range in Phase c, again indicating similar trends as discussed. The higher NAG range was obtained because of increase in metal loading in phase c.

In Phase d, For AMD Ad, AMD Bd, AMD Cd and AMD Dd, the metal loadings were 205.70, 236.10, 269.00 and 291.80 (all in mg/L) respectively. The incremental NAG from base value of NAG, after SAPS treatment for AMD Ad, AMD Bd, AMD Cd and AMD Ddfor Phase d, for 1d HRT, were also reported in increasing order i.e., 270, 360, 585 and 670 (all in mg/L) respectively.Similarly for fixed 2d, 4d, 7d and 10d HRT and with above variations in metal loading, the incremental NAG range were reported as 460-1000, 670-1290,850-1630 and 975-1840 mg/L respectively. The incremental NAG range is gradually increasing due to increase in metal loading.

In Phase e, For AMD Ae, AMD Be, AMD Ce and AMD De, the metal loadings were 141.70, 179.20, 224.00 and 244.35 (all in mg/L) respectively. The incremental NAG from base value of NAG, after SAPS treatment for AMD Ae, AMD Be, AMD Ce and AMD Defor Phase e, for 1d HRT, were also reported in increasing order i.e. 05, 35, 230 and 285 (all in mg/L) respectively. Similarly for fixed 2d, 4d, 7d and 10d HRT and with above variations in metal loading, the incremental NAG range were 110-495, 270-720, 475-970 and 600-1080 mg/L respectively.

In Phase f, For AMD Af, AMD Bf, AMD Cf and AMD Df, the metal loadings were 157.50, 193.90, 237.80and 250.90 (all in mg/L) respectively. The incremental NAG from base value of NAG, after SAPS treatment for AMD Af, AMD Bf, AMD Cf and AMD Df for Phase f, for 1d HRT, were also reported in increasing order i.e. 10, 45, 250 and 310 (all in mg/L) respectively. Similarly for fixed 2d, 4d, 7d and 10d HRT and with above variations in metal loading, the incremental NAG range were115-545, 280-760, 505-1020 and 645-1160 mg/L respectively.

In Phase g, For AMD Ag, AMD Bg, AMD Cg and AMD Dg, the metal loadings were 173.20, 208.50, 251.50 and 270.80 (all in mg/L) respectively. The incremental NAG from base value of NAG, after SAPS treatment for AMD Ag, AMD Bg, AMD Cg and AMD Dg for Phase g, for 1d HRT, were also reported in increasing order i.e., 10, 50, 250 and 340 (all in mg/L) respectively. Similarly for fixed 2d, 4d, 7d and 10d HRT and with above variations in metal loading, the incremental NAG range were 120-570, 295-810, 525-1065 and 745-1210 mg/L respectively.

Based on above findings it was indicated that metal loading affects the NAG in SAPS. The AMD with high metal loading has higher mineral acidity and their rate of reaction become faster and higher NAG was observed. In phase a, continuous increase in alkalinity generation reported because there was a continuous increase in metal loading along with iron component, FFR × T I (i.e.,4.627, 30.62, 147.75, 223.46). Similar trends were reported for all the phases.The reason for more NAG after metal removal is reduction in mineral acidity. Asper the mechanism SAPS AMD is passes through a layer of an organic material in environment favorable for reduction, where chemical and biological reactions take places. The SAPS processes generate alkalinity, increase the pH, reduce sulfates content, and encourage the metalsremoval. These findings are showing an agreement with research work of Johnson and Hallberg [8]. The above phenomena can be better explained by following equations:

$$2CH_2O + SO^{2-}_4 \text{-------}\rightarrow S^{2-} + 2CO_2 + 2H_2O$$

$$S^{2-} + 2CO_2 + 2H_2O \text{-------}\rightarrow 2HCO^-_3 + H_2S$$

where CH_2O represents a simple organic carbon source.The influent AMD high metal loading contains high concentration of sulphate. The high sulphate removal produces high bicarbonate alkalinity in the form HCO^-_3 and high amount of H_2S. The H_2S react with metals to form metal sulphide precipitates.Therefore, the increase in metal loading in influent AMD increases the alkalinity generation during treatment by SAPS, is in line with principle chemistry. Similar findings were also reported by Zipper and Skousen [20].

In all the experiments initially, alkalinity was generated with faster rate, then it became slower after passes of time due to reaching towards saturation level. This confirmed the slower dissolution rate of limestone at higher pH level which agrees with Jageet works [6]. In other words, more acidic AMD showed higher alkalinity generation trends in SAPS as compared to less acidic AMD.From trends of NAG, it was evident that NAG increases with increase in metal loading in AMD, the reason for more NAG after metal removal is reduction in mineral acidity .

3.2. NAG with variation in HRT with fixed influentmetal loading

In phase a, the NAG observations with variation inHRT with constant metal loadingi.e.,120.70 mg/L, for AMD Aa,was carried out to understand the behavior of NAG during AMD treatment by SAPS. The incremental NAG from base value of NAG (=430), for 2d HRT was reported as90 mg/L(Table 3). Further the incremental NAG for AMD Aa, for 4d,7d and 10d HRTs were reported as 260,415 and 415 (all in mg/L) respectively (Figure 3).Similarly the incremental NAG from base value, for AMD Ba withconstant metal loading of 153.70 mg/L for 1d, 2d, 4d,7d and 10d HRTs were reported as 20,150,355,565 and 710 (all in mg/L) respectively.The incremental NAG from base value, for AMD Ca with constant metal loading of 147.75 mg/L, for 1d, 2d, 4d,7d and 10d HRTs were reported as 185,345,500, 715 and 875 (all in mg/L) respectively. The incremental NAG from base value, for AMD Da with constant metal loading of 230.50 mg/L, for 1d, 2d, 4d,7d and 10d HRTs were reported as 360,400,580,790 and 970 (all in mg/L) respectively. Further it is clear that substantial increase in NAG were observed in all the four AMDs of phase a experiment with increasing HRT with constant metal loading.

In phase b, the metal loading for AMD Ab was kept constant at150.63mg/L whereas the incremental NAG from base value, for AMD Ab, for 1d,2d 4d,7d and 10d HRTs were reported as 30, 120, 295, 470 and 600 (all in mg/L) respectively (Figure 4).Similarly the incremental NAG from base value, for AMD Bb with constant metal loading of 176.50 mg/L, for 1d, 2d, 4d,7d and 10d HRTs were reported as 65,195,470,625 and 775(all in mg/L) respectively. The incremental NAG from base value, for AMD Cb with constant metal loading of 227.30 mg/L, for 1d, 2d, 4d,7d and 10d HRTs were reported as 335,440,665,790 and 960 (all in mg/L) respectively. The incremental NAG from base value, for AMD Db with constant metal loading of 249.20 mg/L, for 1d, 2d, 4d,7d and 10d HRTs were reported as375,490,690,835and 1015 (all in mg/L) respectively. It is again indicating that increase in NAG in all the four AMDs of phase b experiment with increasing HRT with constant metal loadings.

In phase c, the metal loading for AMD Ac was kept constant at183.60 mg/L whereas the incremental NAG from base value, for AMD Ac, for 1d,2d, 4d,,7d and 10d HRTs were reported as 120, 190, 470,700 and 820 (all in mg/L) respectively (Figure 5).Similarly the incremental NAG from base value, for AMD Bc with constant metal loading of 209.00 mg/L, for 1d, 2d, 4d,7d and 10d HRTs were reported as 190,330,515,740 and 850 (all in mg/L) respectively. The incremental NAG from base value, for AMD Cc with constant metal loading of 249.00 mg/L, for 1d, 2d, 4d,7d and 10d HRTs were reported as 420,645,855,1075 and 1275 (all in mg/L) respectively. The incremental NAG from base value, for AMD Dc with constant metal loading of 269.30 mg/L, for 1d, 2d, 4d,7d and 10d HRTs were reported as 550,790,980,1120 and 1330 (all in mg/L) respectively.

It was again confirmed thatthe increase in NAG in all the four AMDs of phase c experiment with increasing HRT with constant metal loadings.

NAG vs HRT plots for phase d, phase e, phase f and phase g are not shown for avoiding similarity and repetition. Similar trends for NAG for remaining 16 AMDs were also observed from phase d, phase e, phase f and phase g experiments for different HRTs(i.e., 1d,2d,4d,7d and 10d) with constant metal loading.(Table 3). Thus, based on the above experimental observations, it was found that NAG increases with increase in HRT with constant metal loadings.The NAG by the SAPS unit ranged between 430 mg/L to 2270 mg/L.

It was found that the pH level was increased with increase in HRT for all the seven phases' experiments. (Fig. 3, Fig. 4 and Fig. 5). Vasquez has also found similar findings during treatment of AMD for three different HRT by biochemical passive reactor [19]. The justification for above is availability of more time for microbial activities and limestone dissolution.This can be explained due to following reasons (i) When low pH acidic AMD solution passes through organic layer, the microbe responds quickly, and bicarbonate alkalinity generated with faster rate (ii) After some time saturation level of alkalinity generation by microbe and limestone reached and the rate of pH increase became slower. The pH values in the 28 influent synthetic AMDs ranged from 2.44 to 4.50 and they were found to rise to a range from 8.30 to 8.85 at the discharge for 10d HRT. The pH levels increased in all the seven phases of experiments which confirmed the alkalinity generation by SAPS units. The strong reducing environment is observed to inside the SAPS in organic substrate thereby resulting in the precipitation of metal sulfide and at the same time generation of bicarbonate alkalinity. This is an agreement with Choudhary [3].

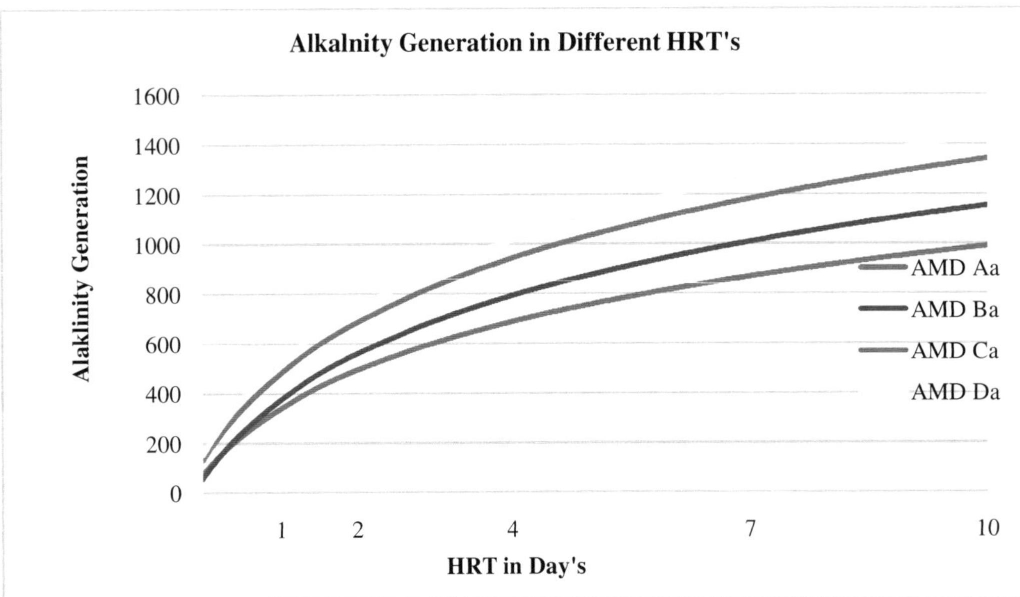

Fig. 3.NAG in different HRT: Phase a

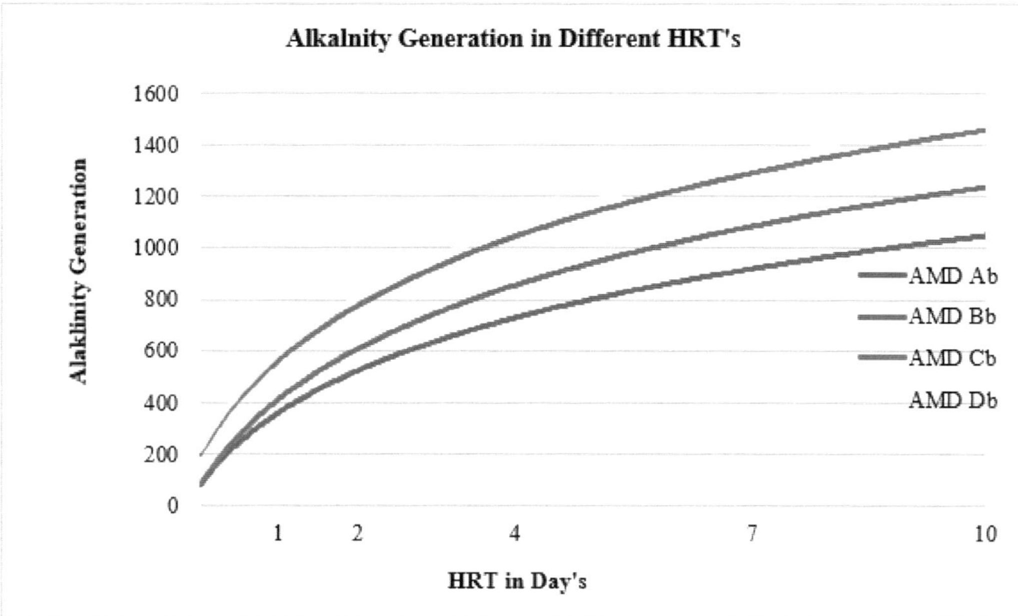

Fig.4.NAG in different HRT Phase b

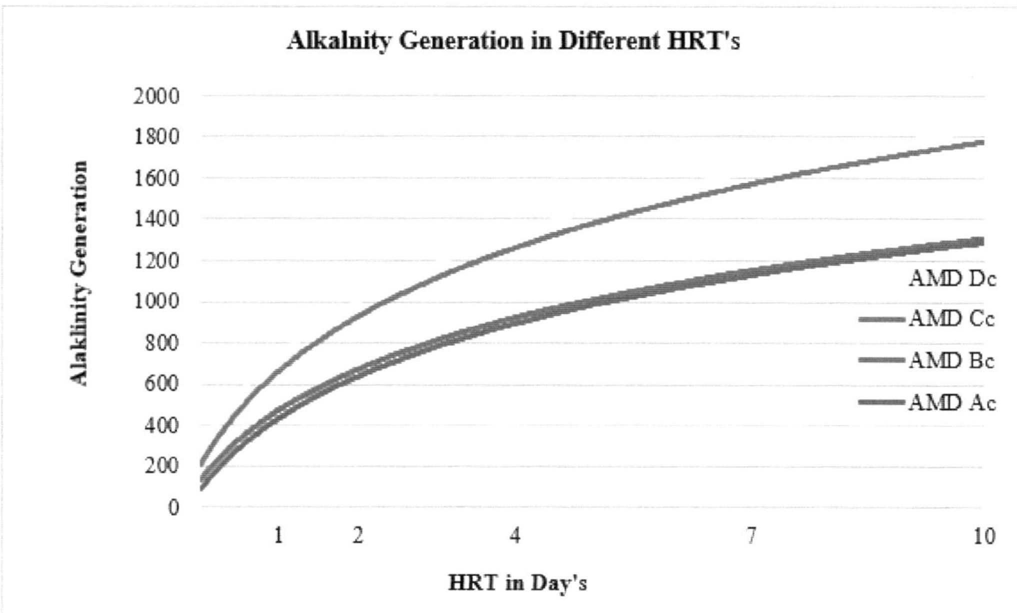

Fig. 5 NAG in different HRT Phase c

Based on plots between HRT and NAG for AMD Aa,AMD Ba, AMD Ca and AMD Da (Figure 3) at Port P_3, the following best fit logarithmic curve equations were obtained (Table 4):

Table 4. Logarithmic curve equations for HRT Vs NAG for phase I

AMD	Logarithmic curve equations	R^2 Value
Aa	380.9ln(x)+75.62	0.974
Ba	456.3ln(x)+59.06	0.988
Ca	504.5 ln(x)+132.5	0.956
Da	535.9ln(x)+154.6	0.948

Similar trends in logarithmic curve equations were also found for all the remaining 24 AMDs used in experiments. Therefore, it is clear that hydraulic retention time (HRT) has logarithmic relationship with NAG, hence natural log of HRT in hours is taken as input parameter for SPSS software analysis. The significant increase in the pH level in all the experiments was reported, which confirmed the NAG by SAPS units with maximum pH level of 8.85 (for AMD Dd).Thesimilar trends in pH increase were found for remaining 27 AMDs.The strong reducing environment was prevailing to inside the SAPS unit thereby resulting in the precipitation of metal sulfide and at the same time generation of bicarbonate alkalinity. This is an agreement with Choudhary [3].

3.3 Selection of iron component (FFR×TI)

The effects of iron on NAG by SAPS can be only understand better if its composition in the form of ferric iron and ferrous iron is considered alongwith total iron (TI). The ferric iron/ferrous iron ratio (FFR) gives an idea about redox status of the AMD solution. Therefore, total iron must be considered along with ferric iron/ferrous iron ratio. Generally, in very acidic AMD the ferric iron concentration is higher than ferrous iron and during treatment by SAPS, the most of ferric iron is converted to ferrous iron form. The representation of ferric/ferrous ratio is always associated with corresponding total iron. The expression of ferric/ferrous ratio alone does not give any meaning unless it is represented together with corresponding total iron [13]. Therefore, in this study we have selected an iron component factor as FFR×TI, a product of FFR and TI, so that the effects of ferric/ferrous ratio along with total iron on NAG can be evaluated in more appropriate manner. Patel [15] has also selected (FFR×TI) as important evaluating parameter for NAG by SAPS. Therefore, iron component (FFR×TI),aluminum and manganese contents are used in software analysis, for obtaining the more accurate relationship for NAG.

4. Analysis of SAPS Results by SPSS Software

The analysis of results obtained from laboratory SAPS experiments processed by SPSS statistical software. The analysis of SPSS software provides a relationship between NAG by the SAPS unit and the influencing parameters such as HRT, Iron component i.e.,total iron along with ferric iron/ferrous iron ratio (FFR×TI), aluminum and manganese contents in influent AMD solution. A total of 140 samples were collected from SAPS unit effluent at Port P_3 for all 28 synthetic AMDs for 1d, 2d, 4d, 7d and 10d HRTs. Total 140 AMD samples data are considered as qualified for SPSS software analysis by ANOVA multivariate analysis. Theinput data of SAPS treatment for NAG were processed by SPSS software and the following output data were obtained as given below

Descriptive Statistics

	Mean	Std. Deviation	N
NAG P3	1032.68	381.370	140
FFR*TI	120.012857	96.7195501	140
Al (mg/L)	37.14	22.578	140
Mn (mg/L)	27.86	16.933	140
HRT in natural log	4.443641	0.8375231	140

Correlations

		NAG P3	FFR*TI	Al (mg/L)	Mn (mg/L)	HRT in natural log
Pearson Correlation	NAG P3	1.000	0.478	0.422	-0.126	0.719
	FFR*TI	0.478	1.000	-0.008	0.101	0.000
	Al (mg/L)	0.422	-0.008	1.000	-0.581	0.000
	Mn (mg/L)	-0.126	0.101	-0.581	1.000	0.000
	HRT in natural log	0.719	0.000	0.000	0.000	1.000
Sig. (1-tailed)	NAG P3		0.000	0.000	0.068	0.000
	FFR*TI	0.000		0.461	0.118	0.500
	Al (mg/L)	0.000	0.461		0.000	0.500
	Mn (mg/L)	0.068	0.118	0.000		0.500
	HRT in natural log	0.000	0.500	0.500	0.500	
N	NAG P3	140	140	140	140	140
	FFR*TI	140	140	140	140	140
	Al (mg/L)	140	140	140	140	140
	Mn (mg/L)	140	140	140	140	140
	HRT in natural log	140	140	140	140	140

Model Summary

Model	R	R Square	Adjusted R Square	Std. Error of the Estimate	Change Statistics				
					R Square Change	F Change	df1	df2	Sig. F Change
1	0.967[a]	0.935	0.933	98.923	0.935	482.729	4	135	0.000

*a. Predictors: (Constant), HRT in natural log, Mn (mg/L), FFR*TI, Al (mg/L)*

ANOVA[a]

Model		Sum of Squares	df	Mean Square	F	Sig.
1	Regression	18895537.95	4	4723884.487	482.729	0.000[b]
	Residual	1321082.587	135	9785.797		
	Total	20216620.54	139			

a. Dependent Variable: NAG P3
*b. Predictors: (Constant), HRT in natural log, Mn (mg/L), FFR*TI, Al (mg/L)*

Coefficients[a]

Model		*Unstandardized Coefficients*		*Standardized Coefficients*	*t*	*Sig.*
		B	*Std. Error*	*Beta*		
1	*(Constant)*	*-1021.530*	*54.914*		*-18.602*	*0.000*
	*FFR*TI*	*1.858*	*0.087*	*0.471*	*21.264*	*0.000*
	Al (mg/L)	*8.280*	*0.457*	*0.490*	*18.106*	*0.000*
	Mn (mg/L)	*2.495*	*0.613*	*0.111*	*4.072*	*0.000*
	HRT in natural log	*327.253*	*10.018*	*0.719*	*32.666*	*0.000*

a. Dependent Variable: NAG P3

b. Predictors: (Constant), HRT in natural log, mn (mg/L), FFR*TI, Al (mg/L)

4.1. Interpretations of SPSS Software output

The following relationship is obtained for NAG and iron component i.e. ferric/ferrous iron ratio x total iron (FFR× TI), aluminum, manganese and HRT,

$$NAG = 1.86\ (FFR \times TI) + 8.28\ Al + 2.49\ Mn + 327.25\ \ln(HRT) - 1021.53$$

Where, NAG in mg/L (in terms of $CaCO_3$ equivalence), HRT in hours, TI (total iron), aluminum and manganese in mg/L.In the SPSS analysis 140 dataset were processed with ANOVA model for 95% confidence level and for ln(HRT) , FFR×TI , aluminum and manganese. Therefore, all these four parameters had significant impact on net alkalinity generation. The coefficient of multiple determinations was found as R^2=0.935, which means that above relationship for given dataset is better fit and highly acceptable. It is evident that from above obtained relationship that net alkalinity generation increases with metal loading and HRT. It is also very clear that the FFR×TI gives more clarity on behavior of iron. Jage [6] reported SAPS performance by incorporating HRT, total iron and non-manganese acidity and given following relationship

Net alkalinity generation = 59.33 ln(tr) +1.03 Fe+0.32 non-Mn acidity-80.43

Where tr = residence time in hours, Fe = total iron in mg/L, Non-manganese acidity in mg/L

Therefore, the relationship obtained by this research work (equation 5.2) is more comprehensive and acceptable because it had incorporated aluminum and manganese along with iron component (ferric/ferrous ratio×total iron).

5. Conclusions

- The experimental investigations, observations and analysis conducted in the research work led to the following important conclusions.The relationship between iron component,aluminum, manganese,HRT and NAG by SAPS is obtained as

$$NAG = 1.86\ (FFR \times TI) + 8.28\ Al + 2.49\ Mn + 327.25\ \ln(HRT) - 1021.53$$

- On the basis of above relationship, it can be concluded that the NAGincrease with increase in total iron, ferric/ferrous iron ratio, aluminum, manganese contents and hydraulic retention time in influent AMD.
- Based on the experiments in all cases of AMDs, the performance of SAPS as indicated by the NAG has been found to improve with the increase in HRT. The NAG by the SAPS unit ranged between 430 mg/L to 2270 mg/L.

- The above obtained relationship after SPSS software analysis shows the importance of composition of the influent AMD on the treatment process by SAPS. Large scale SAPS unit in the mine can be designed for better performance, cost effective and for longer life by taking into consideration the quality of the influent AMD and HRT.
- It is recommended that SAPS should be operated with appropriate influent metal load for longer life of SAPS. If influent AMD contains high metal concentration, then it will exhaust the carbon source and limestone rapidly. Therefore, a number of SAPS shall have to be used in series where treatment of AMD with higher metal content and acidic load is proposed.

References

1. Baird RB, Eaton AD, Rice EW (2017) Standard methods for the examination of water and wastewater. APHA -AWWA -WEF, Washington, D C.
2. Bhattacharya J, Yim GJ, Ji SW, Lee HS, Cheong YW (2007) SAPS influent and effluent geochemical parameters related to AMD pollution control: A case study- IMWA Symposium: Water in Mining Environments. Cagliari, Italy.
3. Choudhary RP, Sheoran AS (2012) Performance of single substrate in sulphate reducing bioreactor for the treatment of acid mine drainage. Miner. eng.39: 29-35.
4. Hedin RS, Watzlaf GR, Nairn RW (1994) Passive treatment of acid mine drainage with limestone.J. Environ. Qual. 23:1338-1345
5. Hedin R, Weaver T, Wolfe N (2010) Passive Treatment of Acidic coal mine drainage:The Anna S Mine Passive Treatment Complex. Mine Water Environ 29:165-175.
6. Jage, C.R. (2000). Water Quality Based Design Guidelines for Successive Alkalinity-Producing System Used in the Treatment of Acidic Mine Drainage (M.S. thesis). Virginia Polytechnic Institute and State University, Blacksburg Virginia. 14-80.
7. Jamal A, Dhar BB, Siddharth S, Tiwari RK (2003) Genesis of Pyrites of Coal and their relation with acid drainage from mine. J. Inst. Eng. (India) 83: 47-57
8. Johnson, D .B., & Hallberg, K. B. (2005). Acid mine drainage remediation options: a review. Science of the Total Environment, 338, 3-14.
9. Kepler DA, McCleary EC (1994) Successive alkalinity producing systems (SAPS) for the treatment of acidic mine drainage. American society of mining and reclamation195-204. https:// doi:10.21000/JASMR94010195
10. Lee JY, Khim J, Woo K, Ji WH (2013) A full-scale successive alkalinity producing passive system (SAPPS) for the treatment of acid mine drainage. Water Air Soil Pollut.224:1656. https:// doi:10.1007/s11270-013-1665-4
11. Ordonez A, Loredo J, Pendas F (1999) A successive Alkalinity Producing System (SAPS) as Operational unitin a Hybrid passive Treatment system for Acid Mine Drainage. Mine Water Environ IMWA Congress, 575-580, Sevilla, Spain
12. Patel MD (2020b) Investigation on The Effects of Ferric/Ferrous Ratio, Aluminum, Manganese And HRT on NAG by Successive Alkalinity Producing System for Synthetic AMD. Ph D Thesis, National Institute of Technology, Raipur, India.

13. Patel MD (2020c)Study of the effects of ferric iron/ferrous iron ratio in influent AMD and hydraulic retention time on alkalinity generation, iron, aluminum and manganese removals by SAPS. Desalination Water Treat.195: 256–266

14. Patel MD, Jade R, Dewangan P (2019) Study of performance of laboratory successive alkalinity producing system (SAPS) with reference to metal and NAG during acid mine drainage treatment. Res. J. Chem. Environ. 23(8): 76-84

15. Patel MD, Jade R, Dewangan P, Dash AK (2020a) Effects of ferric/ferrous ratio on performance of successive alkalinity producing system (SAPS) during acid mine drainage treatment. J. Indian Chem. Soc. 97:1152-1156

16. Patel MD (2021) Study of microbial contribution in NAG during treatment of acid mine drainage by laboratory successive alkalinity producing system . Indian J. Biochem. Biophys.58: 162-170

17. Rose AW (2004) Vertical flow systems - effects of time and acidity relations, Proc American society of Mining and Reclamation and the 25th West Virginia Surface Mine Task Force, Lexington, U.S.,1595-1615. https:// doi.org/10.21000/jasmr04011595

18. Seervi V, Yadav HL, Srivastava S K, Jamal A (2017) Overview of Active and Passive systems for Treating Acid Mine Drainage. International Advanced Research Journal in Science, Engineering and Technology4: 5

19. Vasquez, Y., Escobar, M.C., Neculita, C.M., Arbeli, Z., & Roldan, F. (2016). Biochemical passive reactor for treatment of acid mine drainage: Effect of hydraulic retention time on changes in efficiency, composition of reactive mixture and microbial activity: A review. Chemosphere, 153, 1-11. doi.org/10.1016/j.chemosphere

20. Zipper, C. E., & Skousen, J.G. (2010). Influent water quality affects performance of passive treatment systems for acid mine drainage, Mine Water Environ, 29, 135-143. doi:10.1007/s10230-010-0101-9.

GIS-based Industrial Land Suitability Analysis for locating Industrial Parks in Raipur and Nava Raipur

Ankur Baghel
Assistant Professor, Department of Planning and Architecture
National Institute of Technology Rourkela, Odisha, India
baghela@nitrkl.ac.in

Abstract This paper offers an integrated approach towards the preparation of a mechanism for allocation of land for the industrial parks assessing the land potential for different uses to regulate and manage the use of land and improve the surrounding environment. An Industrial land Suitability Analysis has been carried out to allocate suitable land parcels for industrial development. An Analytical Hierarchy Process (AHP) with the integration of Geographic Information System based Multi-criterion decision-making (MCDM) was adopted for weighing the parameters. With the pairwise comparison matrix, the establishment of the weights was carried out and based on obtained weights, overlaying was carried out in the GIS interface to generate outputs for the most suitable and least suitable areas for industrial land use with a synthesized suitability map. The final results showed that the prominent locations for future industrial development came mostly near roads and railway routes. Since the utilization of vacant and wastelands with preservation of agricultural fields were prime concerns while deciding on the ranking of the criteria, the suitability results showed lesser agricultural areas than in other areas.

1. Introduction

As per the Chhattisgarh Industrial Policy 2019-2024, the investment climate plays an essential role in the region's industrial development. Raipur district is widely known for its rice mills, cement, and steel plants. Raipur district has a rare distinction of having the most significant number of large and small cement plants. Raipur and Nava Raipur have emerged as prominent and essential destinations for industrial and commercial activities (Mohan et al., 2012). This development of a favourable industrial climate has led to rapid industrialization in areas near Raipur and Nava Raipur (Bhojwani & Bagga, 2016).

Alternatively, this has made Raipur one of India's critically polluted cities, as per a study from the Centre of Science and Environment (CSE, 2011). There exists a severe lack of the development of industrial estates, which is required for better growth (Verma, 2014). Being the new Capital and acting as one of India's largest steel manufacturing hubs lying in the industrial corridor due to expressway connectivity; The PM2.5 show a rise of 50% between 1998 and 2014 (Guttikunda, Pant, Nishadh, & Jawahar, 2019). Thus, it is essential to provide suitable land parcels for future industrial development in this region.

Suitability analysis is a valuable tool for assessing land suitability for specific purposes. It aims to utilize the land for the best possible use appropriate to the needs and preferences. Since this is a decision-making process involving the land's capacity for supporting a specific service and considers the other vital aspects like socio-economic and environmental features thus, the weighing of the criteria under consideration becomes difficult. A multicriteria decision-making process has been adopted to assist in the weight calculation to overcome this issue. This method can analyze problems, producing alternatives.

Integrating AHP used for multicriteria analysis and GIS in land suitability analysis is suitable for this study. Both these tools were used along with the mapping of existing land uses, and appropriate land parcels were found for future industrial development (Cengiz & Akbulak, 2009).

2. Methodology

2.1 Study area

The study area is situated in the fertile plains of the Chhattisgarh Region. This region is between 22° 33' N to 21°14'N latitude and 82°6' to 81°38'E longitude (figure 1). District Bilaspur surrounds the North, District Bastar and part of Odisha state in South, District Raigarh and part of Odisha state in East, and district Durg in West.

Road connectivity is an essential development determinant of any upcoming economy of the region. The Raipur Region is strategically positioned and crosses numerous critical national highways, including the Great Eastern Road (Bombay – Kolkata) and Raipur-Dhamtari-Vishakhapatnam. Aside from that, it has air connections to the country's major cities. National highways (NH-6, NH-43, and NH-200) and three state highways connect the region (SH-2, 7, and 9). In Bilaspur, the South Eastern Railway Headquarters manages the region's railway network. All of the lines run practically parallel to National Highway and State Highway. Raipur is located on the Bilaspur-Durg portion of the South-Eastern Railways' Mumbai-Howrah broad gauge line. Over the region, the railway lines and stations offer mass communication and goods movement.

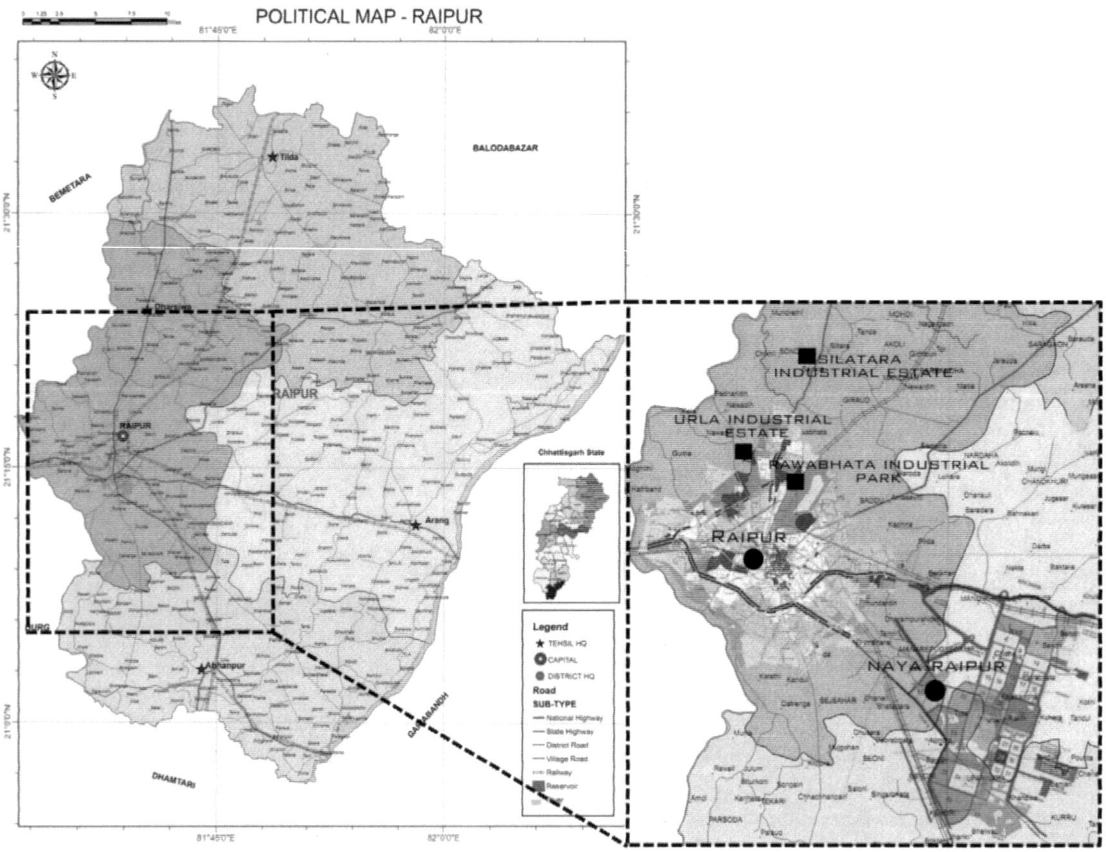

Figure 1 Location map of Study area

2.2 Criteria to be considered for the Land Suitability Analysis

An Industrial land Suitability Analysis has been carried out to allocate suitable land parcels for industrial parks. The areas with maximum susceptibility of action were considered in terms of industrial suitability, which were thus generated with the help of the land-use suitability method and included specific criteria. The criteria provided with respective scales based on expert opinions and reviews are as follows:

Table 1. Importance to different criterion

The scale of importance given to Different Criteria	
Criteria	*Scale*
Utilization of Wastelands	*10*
Utilization of Vacant land	*9*
Accessibility to Road	*8*
Accessibility to Railways	*7*
Preservation of Agricultural Field	*6*
Distance from Residential Areas	*5*
Distance from Water bodies	*4*
Distance from Core City area	*3*

2.3 Weighted Index Model

Saaty's 'Analytical Hierarchy Process' is the most widely accepted method for scaling the weights of different criteria (Jie, Jing, Wang, & Shu-Xia, 2010). A pairwise comparison matrix was made based on a scale of essential intensities. Ranks were given to criteria Factors after considering the expert's opinions.

Table 2. Saaty's intensity of importance on an Absolute Scale

Saaty's power of significance on an Absolute Scale		
Intensity of importance	Definition	Explanation
1	Equal importance	Two activities contribute equally to the objective
3	Moderate importance of one over another	Experience and judgment moderately favour one activity over another
5	Strong importance	Experience and judgment Strongly favour one activity over another
7	Extreme importance	An activity strongly favoured

9	Extreme importance	The evidence favouring one activity over another is of the highest possible order of affirmation
2,4,6,8	Intermediate values between the two adjacent judgment	When compromise is needed

2.4 Computation of criterion weight:

After forming the *Pairwise Comparison Matrix*, *criterion weight* has been calculated. The steps followed are as follows:

- The sum of each column of the pairwise comparison matrix has been found.
- Each element of the pairwise comparison matrix is divided by the column sum, which results in the formation of a normalized matrix.
- The average of all elements in each row of the normalized matrix is calculated and divided by some criteria. The averages that come out are the relative weights of the corresponding criteria.

Table 3. depicts the Pairwise comparison matrix

PAIRWISE COMPARISON MATRIX [C]								
	W	*V.L.*	*A.R*	*A.Ry*	*Ag*	*R.A*	*C.C*	*W.B*
W	*1.00*	*2.00*	*3.00*	*3.00*	*4.00*	*7.00*	*9.00*	*4.00*
V.L	*0.50*	*1.00*	*2.00*	*3.00*	*5.00*	*7.00*	*8.00*	*5.00*
A.R	*0.33*	*0.50*	*1.00*	*2.00*	*4.00*	*5.00*	*6.00*	*4.00*
A.Ry	*0.33*	*0.33*	*0.50*	*1.00*	*4.00*	*5.00*	*4.00*	*4.00*
Ag	*0.25*	*0.20*	*0.25*	*0.25*	*1.00*	*4.00*	*6.00*	*2.00*
R.A	*0.14*	*0.14*	*0.20*	*0.20*	*0.25*	*1.00*	*2.00*	*2.00*
C.C	*0.11*	*0.13*	*0.17*	*0.25*	*0.17*	*0.50*	*1.00*	*2.00*
W.B	*0.25*	*0.20*	*0.25*	*0.25*	*0.50*	*0.50*	*0.50*	*1.00*
	2.92	*4.50*	*7.37*	*9.95*	*18.92*	*30.00*	*36.50*	*24.00*

Table 4. shows the Normalized matrix for criteria weights

NORMALIZED MATRIX									
	W	*V.L.*	*A.R*	*A.Ry*	*Ag*	*R.A*	*C.C*	*W.B*	*Average Criteria Weight (W)*
W	*0.34*	*0.44*	*0.41*	*0.30*	*0.21*	*0.23*	*0.25*	*0.17*	*0.294*

V.L	*0.17*	*0.22*	*0.27*	*0.30*	*0.26*	*0.23*	*0.22*	*0.21*	*0.236*
A.R	*0.11*	*0.11*	*0.14*	*0.20*	*0.21*	*0.17*	*0.16*	*0.17*	*0.158*
A.Ry	*0.11*	*0.07*	*0.07*	*0.10*	*0.21*	*0.17*	*0.11*	*0.17*	*0.126*
Ag	*0.09*	*0.04*	*0.03*	*0.03*	*0.05*	*0.13*	*0.16*	*0.08*	*0.077*
R.A	*0.05*	*0.03*	*0.03*	*0.02*	*0.01*	*0.03*	*0.05*	*0.08*	*0.039*
C.C	*0.04*	*0.03*	*0.02*	*0.03*	*0.01*	*0.02*	*0.03*	*0.08*	*0.031*
W.B	*0.09*	*0.04*	*0.03*	*0.03*	*0.03*	*0.02*	*0.01*	*0.04*	*0.035*
	1.00	*1.00*	*1.00*	*1.00*	*1.00*	*1.00*	*1.00*	*1.00*	

2.5 Calculation of the Consistency Ratio

Calculating a consistency ratio to assess the consistency of the judgments is the next critical step. An identity matrix is multiplied by the normalised matrix. The consistency vectors are created by multiplying the reciprocal values of the final matrix by a matching relative weight.

The sum of all consistency vectors is calculated to get λmax. Consistency Index is measured from the following formula:

Equation 1. Consistency Index calculation

$$C.I = (\lambda max - n)/(n-1)$$

Table 5. Calculation for Consistency vector

[C] x W=Ws		*CONSISTENCY VECTOR*
		Ws x 1/W
2.63		*8.95*
2.17		*9.19*
1.47		*9.23*
1.16		*9.17*
0.69		*8.80*
0.33		*8.34*
0.26		*8.20*
0.30		*8.41*
	λmax	8.78

Consistency Index provides a measure of departure from consistency. For the determination of the goodness of C.I, the analytical hierarchy process compares it with Random Index, and the result we get is consistency ratio (C.R)

Equation 2. Consistency ratio

$$CR = CI/RI$$

Here C.R. is Consistency Ratio, CI is consistency index, and R.I. is Random Index.

The C.R is designed so that if C.R< 0.1, the ratio indicates a reasonable level of consistency in a pairwise comparison. If C.R>0.1, then the ratio is indicative of inconsistent judgments. In such cases, one should reconsider and revise the original values in the pairwise comparison matrix (Franek & Kresta, 2014).

Table 6. Random Index Table (Wedley, 1993)

Random Index Table (R.I.)										
N	*1*	*2*	*3*	*4*	*5*	*6*	*7*	*8*	*9*	*10*
RANDOM INDEX	*0*	*0*	*0.58*	*0.9*	*0.12*	*1.24*	*1.32*	*1.41*	*1.45*	*1.49*

Table 7. depicts the obtained CI and CR

CONSISTENCY INDEX (CI)	*CI = (λ-n)/(n-1)*
CI	*0.11*
CONSISTENCY RATIO (C.R.)	*CR=CI/RI*
CR	*0.0795*

Table 8 shows the Obtained Criteria weights

Criteria	*Obtained Criteria Weight*
Utilization of Wastelands	*0.294*
Utilization of Vacant lands	*0.236*
Nearness to road	*0.158*
Nearness to Railways	*0.126*
Preservation of Agricultural fields	*0.077*
Distance from Residential Areas	*0.039*
Distance from water bodies	*0.031*
Distance from Core City area	*0.035*

Once the weights were found out, the "Overlaying" method was employed in the GIS interface; with more than one parameter affecting the region's future development, this technique is most appropriate.

The generated output depicts the most suitable and least suitable areas in the Raipur region. Utilization of vacant and wasteland with preservation of agricultural fields were prime concerns while deciding on criteria ranking. Hence the suitability shown by agricultural areas was comparatively lesser than in other areas.

Table 9 Physical Parameters considered for land suitability analysis

Parameter	*Constraint*	*Development Constraint*	*Buffer/ Factors Considered*	*Ranks*
Accessibility to Roads	Infrastructure, Transport mode	Areas nearer to the transportation network have a higher potential for industrial development. Growth of infrastructure near the road saves cost and time for transportation	< 300 m	10
			300-500 m	8
			500-800 m	6
			800-1000 m	4
			> 1000 m	2
Accessibility to Railways	Transportation	Areas near to railway have a high potential for industrial development	< 500 m	10
			500-800 m	8
			800-1500 m	6
			1500-1800 m	4
			> 1800 m	2
Distance from water bodies	Environmental Pollution	Needs conservation for future	Other Areas	10
			Waterbody buffer of 500 m	5
Distance from Residential areas	Environmental Pollution	Areas away from residential areas are suitable for hazardous industrial development	Areas outside 1200 m Buffer	10
			800-1200 m	8
			600-800 m	6
			400-600 m	4
			< 400 m	2
Distance from Core City area	Infrastructure, Environmental Pollution	Areas away from City core areas are suitable for hazardous industrial development	13000-17000 m	10
			10000- 13000 m	8
			7000-10000 m	6
			3500-7000 m	4
			< 3500 m	2
Preservation of Agricultural fields	Agricultural Productivity	Productive agricultural and reserved green areas should not be considered for development	Other Areas	10
			Areas under Agriculture	5
Utilization of Wastelands	Developmental cost, Environment protection	Wastelands are the most suitable areas for industrial development	Areas under Wasteland	10
			Other Areas	5

Distance from vacant lands	Developmental cost, Environment protection	Vacant lands are more suitable for industrial development	Areas under vacant land	10
			Other Areas	5

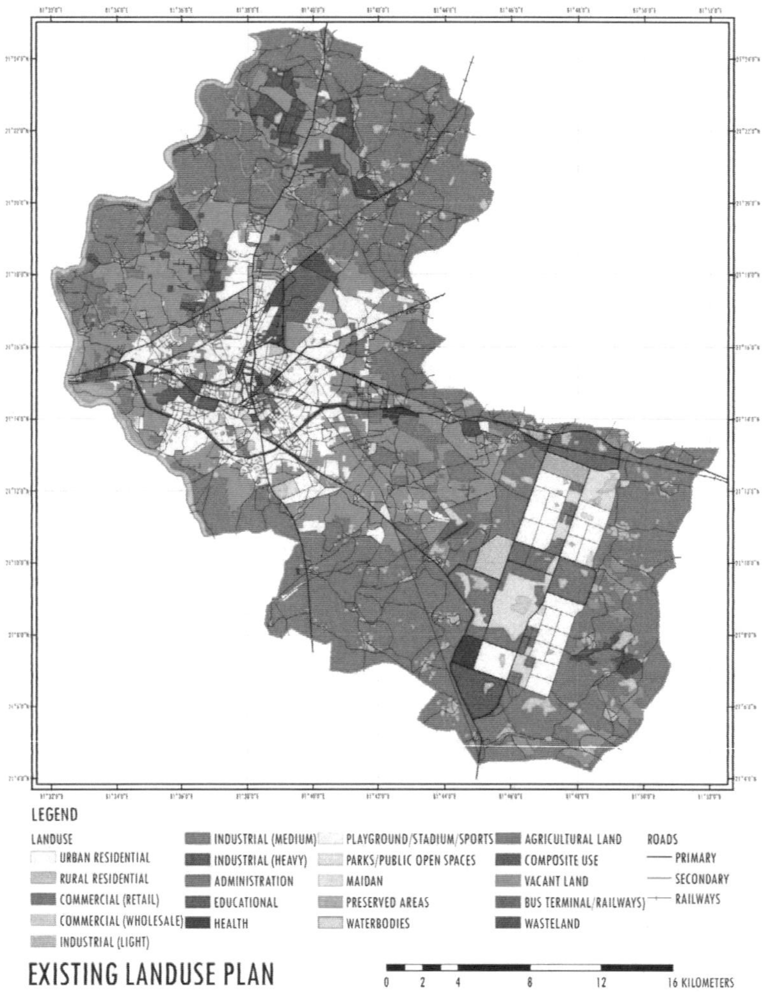

Figure 2. Existing Landuse map for the study area

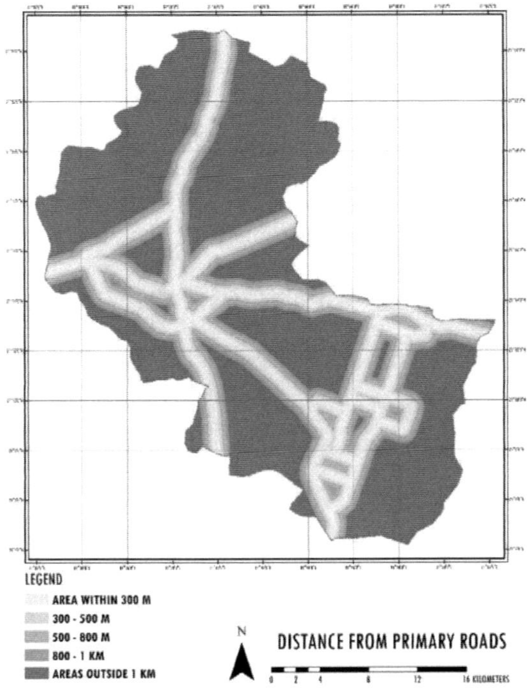

Figure 3. Generated map for distance from primary roads

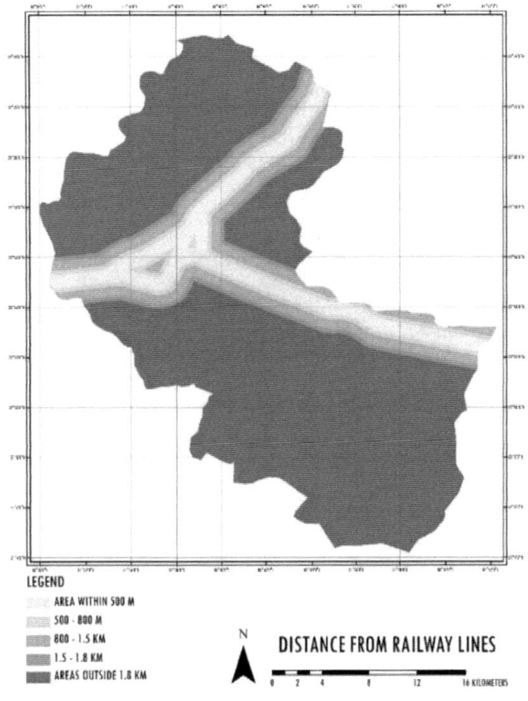

Figure 4. Generated map for distance from Railway lines

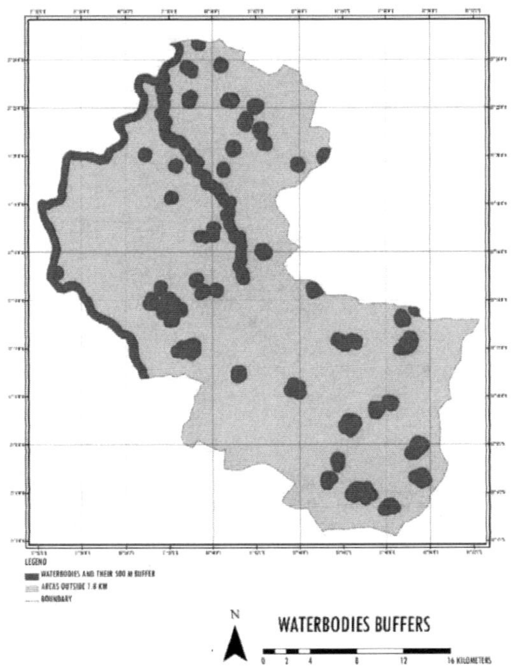

Figure 5. Generated map for waterbodies with buffers

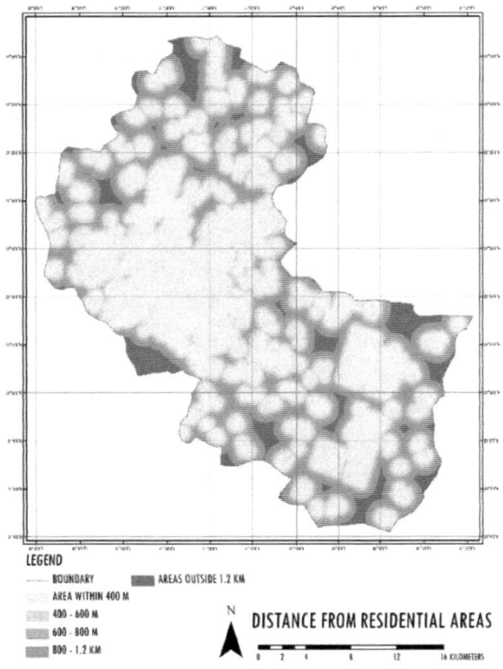

Figure 6. Generated map for distance from residential areas

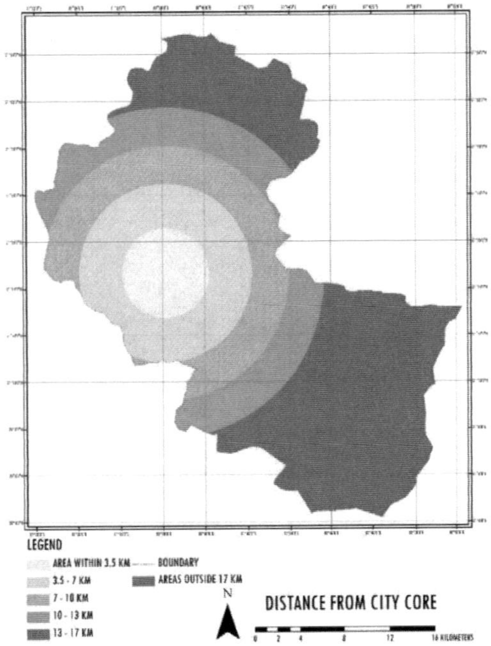

Figure 7. Generated map for distance from the city core

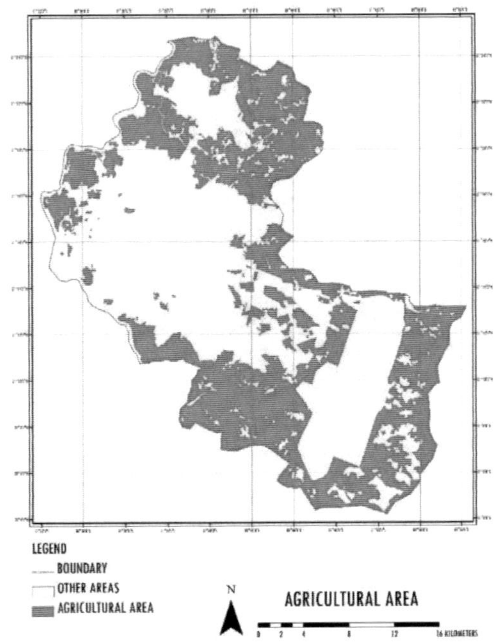

Figure 8. Generated map for agricultural areas

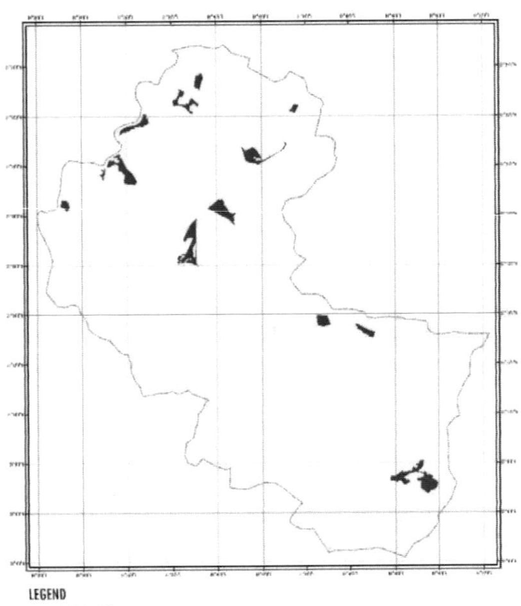

Figure 9. Generated map for Wastelands

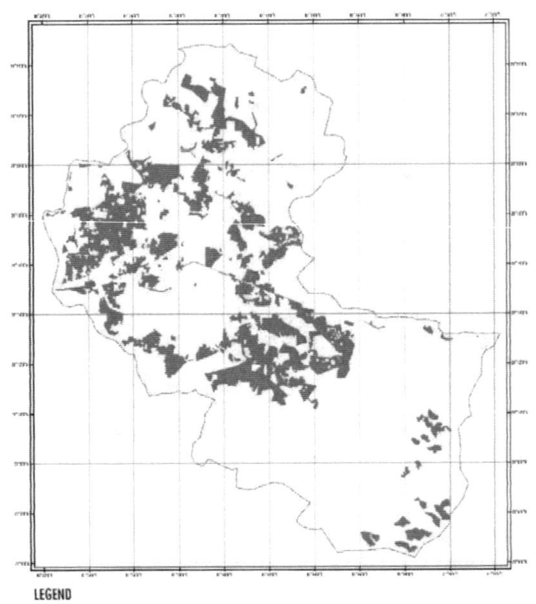

Figure 10. Generated map for Vacant land

3. Results

The generated output depicts the most suitable and least suitable areas for industrial development in the Raipur and Nava Raipur region. In the entire region, the most prominent areas or the most suitable areas with a higher probability of future industrial Estates or parks are mostly near roads and railways.

IOP Conf. Series: Earth and Environmental Science 1032 (2022) 012024 doi:10.1088/1755-1315/1032/1/012024

Utilization of vacant and wasteland with preservation of agricultural fields were prime concerns while deciding on criteria ranking. Hence the suitability shown by agricultural areas was comparatively lesser than in other areas.

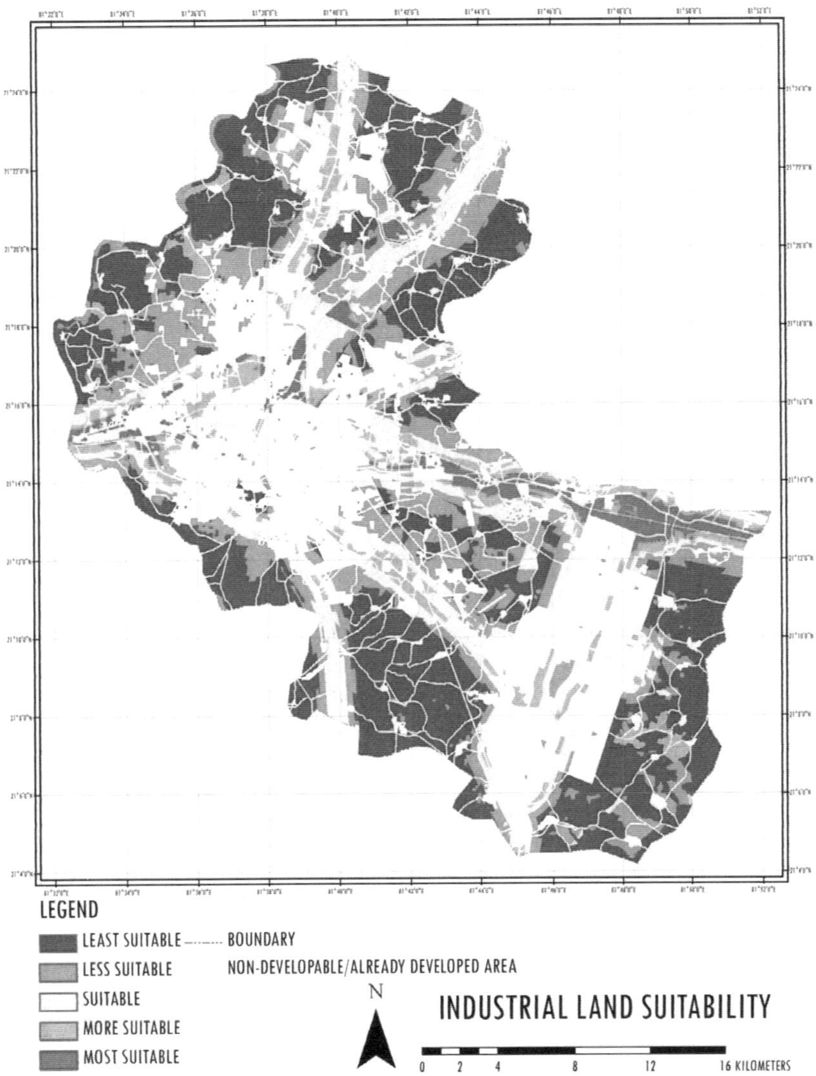

Figure 11. Map depicting Land Suitability for Industrial Development

4. Conclusion:

The present study attempts to improve the prevailing environmental condition in the Raipur and Nava Raipur region, keeping in mind that industrial growth cannot be restricted. Still, it can be carried out in a planned and systematic manner that creates a win-win situation for the industries and residents and healthily improves the ecosystem, enhancing the standard of living for the people under constant health risks. A solution for any urban issue, locating industries or minimizing pollution, must be convincing and follow logical steps, meaning more scientific tools and techniques.

References:

[1] Bhojwani, M., & Bagga, G. D. S. (2016). "Sponge Iron Industry in Chhattisgarh : a Perspective ". *Indian Journal of Applied Research*, *6*(2), 615–617.

[2] Cengiz, T., & Akbulak, C. (2009). Application of analytical hierarchy process and geographic information systems in land-use suitability evaluation: A case study of Dümrek village (Çanakkale, Turkey). *International Journal of Sustainable Development and World Ecology*, *16*(4), 286–294. https://doi.org/10.1080/13504500903106634

[3] CSE. (2011). *Sponge Iron Industry: The regulatory challenge*. 78. Retrieved from https://cdn.cseindia.org/attachments/0.25223700_1499927364_sponge_iron_layout.pdf

[4] Franek, J., & Kresta, A. (2014). Judgment Scales and Consistency Measure in AHP. *Procedia Economics and Finance*, *12*(March), 164–173. https://doi.org/10.1016/s2212-5671(14)00332-3

[5] Guttikunda, S. K., Pant, P., Nishadh, K. A., & Jawahar, P. (2019). Particulate matter source contributions for Raipur-Durg-Bhilai region of Chhattisgarh, India. *Aerosol and Air Quality Research*, *19*(3), 528–540. https://doi.org/10.4209/aaqr.2018.06.0237

[6] Jie, L., Jing, Y., Wang, Y., & Shu-Xia, Y. (2010). Environmental impact assessment of land use planning in Wuhan city based on ecological suitability analysis. *Procedia Environmental Sciences*, *2*(5), 185–191. https://doi.org/10.1016/j.proenv.2010.10.022

[7] Mohan, H., Gazzola, H., Bougnoux, F., Karthaus, R., Plane, A., Vigé-Hélie, C., … Panigrahi, L. . (2012). *Context Document Naya Raipur India: Shaping a new capital city to fit Indian life*. 1–102.

[8] Verma, T. K. (2014). *Physico-chemical studies of steel industries effluent of Urla, Raipur (Chhattisgarh) in relation to iron contamination.* (September). https://doi.org/10.13140/2.1.2709.0883

[9] Wedley, W. C. (1993). Consistency prediction for incomplete AHP matrices. *Mathematical and Computer Modelling*, *17*(4–5), 151–161. https://doi.org/10.1016/0895-7177(93)90183-Y

InSAR Data Analysis using Deep Neural Networks

Ayush Kumar Agrawal[1], Himanshu Verma[2,*], Sumanta Pasari[3]

[123]Department of Mathematics, Birla Institute of Technology and Science Pilani, Pilani Campus, Jhunjhunu - 333031, Rajasthan, India

[1]f20180938@pilani.bits-pilani.ac.in

[2]vermahimanshu0008@gmail.com

[3]sumanta.pasari@pilani.bits-pilani.ac.in

Abstract. Among all natural disasters, earthquakes cause the most severe devastation to both infrastructure as well as human lives. The pattern of surface deformation from satellite-based measures, such as interferometric synthetic aperture radar (InSAR), helps understand crustal dynamics and associated seismic hazards in tectonically active areas. In the last few years, the exponentially rising volume of InSAR data has led to the formulation of automatic crustal deformation algorithms through data-intensive machine learning models. In this study, we propose the long-short-term memory (LSTM) architecture of deep neural networks to forecast InSAR-based line-of-sight displacement. The model suitably captures the inherent temporal variations of SAR interferograms. The method implementation on a region near Central and East Java, Indonesia shows satisfactory performance (mean absolute percentage error around 2%) of the proposed LSTM model comprising four layers, each consisting of 100 nodes. We therefore conclude that deep neural network is a promising avenue for InSAR time series analysis.

Keywords: InSAR, Machine learning, LSTM-RNN, Crustal deformation.

1. Introduction

Occurrence of earthquakes is commonly explained by the theory of plate tectonics [12]. The spatio-temporal dataset, that are used to understand earthquake dynamics and consequent seismic hazards, primarily include tabular seismicity data and Earth's surface displacement from space geodesy [9, 11]. In recent years, the dependency on the dense InSAR data has increased manifold due to the inadequate resolution of global navigation satellite system (GNSS) stations [2]. As a consequence, several methods have been proposed to deal with InSAR data processing in order to infer the underlying crustal structure, fault kinematics, and earthquake potential in a defined region [4, 7]. Moreover, due to inherent diversity and exponentially rising volume of SAR interferograms, machine learning models may be employed to extract meaningful information in terms of "hidden-variables" [11]. Such models use feature vectors to simultaneously process millions of data points and they are less sensitive to human biases [1]. In this paper, we propose a long-short-term memory (LSTM) architecture of recurrent neural networks (RNNs) to capture inherent temporal variations of InSAR dataset.
A SAR data is a 2-D representation of the amplitude of target reflections inside the imaging region [12, 9, 11, 2, 4]. The amplitude represents target reflectance, while the phase represents surface variations. A SAR interferogram is formed by combining the phase of two SAR pictures obtained concurrently or at different times from satellite with slightly varying look angles [1, 3, 10, 13, 6]. The

phase difference of two pictures, called the master image and slave image, collected on the same region from two orbits separated by a given baseline distance is used in radar interferometry [2].
The InSAR data is useful in studying ocean currents, landscape studies, vegetation characteristics, terrain categorization, glacial motions, and landslide surveillance. It can be also used to study the ground surface displacement in the urban, rural or hilly regions [10]. Multiple SAR images can be combined to generate a radar interferogram. The interference pattern induced by the phase shift between these pictures can be used to quantify topography or small changes in terrain of order of a few millimetres along the spacecraft look path [1, 3, 10, 13, 6, 2, 14]. The changes in the returned phase can be accurately calculated using the phase difference between two radar picture obtained with the same point of look but at separate times. As a consequence, if the upper mantle moves closer or far from the satellite between two picture passes, phase difference may be computed with an efficiency equivalent to millimeter-level deformation [14]. Ding et al. [5] estimated crustal deformation of the eastern segment of the Altyn Tagh Fault (EATF) by using the D-InSAR technique. They used ERS-1/ERS-2 SAR data and Space Shuttle Radar Topography Mission (SRTM) digital elevation model (DEM) of the study area. Recently, Brengman and Barnhart [1] proposed a machine learning based SarNet method to identify surface deformation through InSAR data. Using convolutional neural network (CNN) architecture, they achieved an accuracy of 99.74% within synthetic interferograms. In this direction, the present study deploys an LSTM-RNN based neural network model to forecast InSAR-based line-of-sight (LoS) displacement.

2. Study Area & dataset

For illustration of the proposed method, a 100 km by 100 km region between central and east Java, Indonesia is selected (Fig. 1). Several unwrapped and coherent GeoTIFF files of Sentinel-1 (ascending) images from 08/12/2015 to 27/06/2021 with 1175 interferograms have been considered from the "COMET-LiCS Sentinel-1 InSAR" portal (https://comet.nerc.ac.uk/COMET- LiCS-portal/) to produce velocity map and time- series of velocity [8]. These velocity data will be used to forecast InSAR-based line-of-sight displacement using LSTM model.

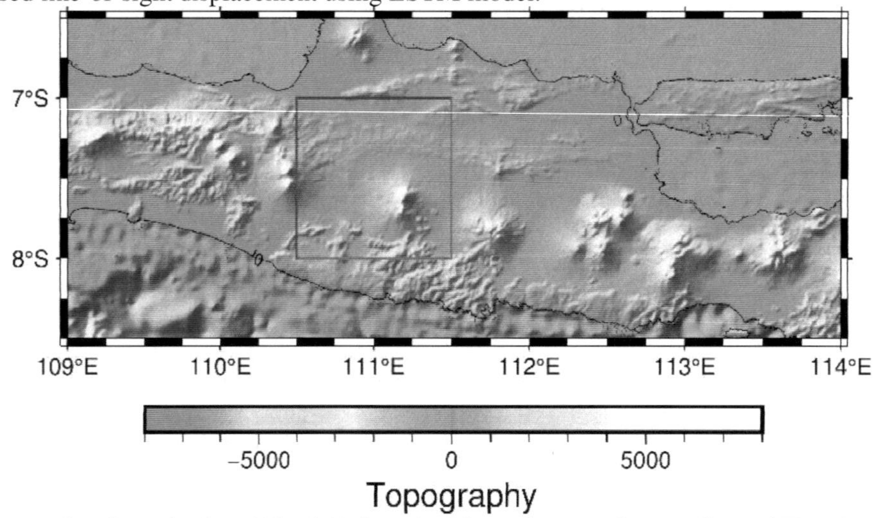

Fig 1. Topography along the Java Island, Indonesia. The red rectangle near Central-East Java represents the region of interest.

3. Methodology and results

The methodology comprises three steps: pre-processing of data, processing (LSTM implementation), and post-processing analysis. A summary of the proposed methodology is provided in Fig. 2.
As mentioned earlier, we use COMET-LiCS InSAR (LiCSAR) processing toolbox to extract the LoS velocity map of the study area. Using the LiCSAR toolbox, we downloaded the SAR data for the study

region in the first phase. Then to improve the accuracy of the velocity map, data from the Generic Atmospheric Correction Online Service (GACOS) has been used to account for the atmospheric correction [15]. Later, for time series analysis, erroneously unwrapped data that degrades the findings is discovered and deleted based on the unwrapped data's coherence and coverage, as well as loop closure [8]. To extract the displacement time series and velocity, the improved stacks of unwrapped data are inverted. The standard deviation of the velocity map (Fig. 3) shows millimeter level accuracy. We choose a pixel from the velocity map at location (7.08278°S, 110.60667°E) for observation and forecasting of InSAR time series.

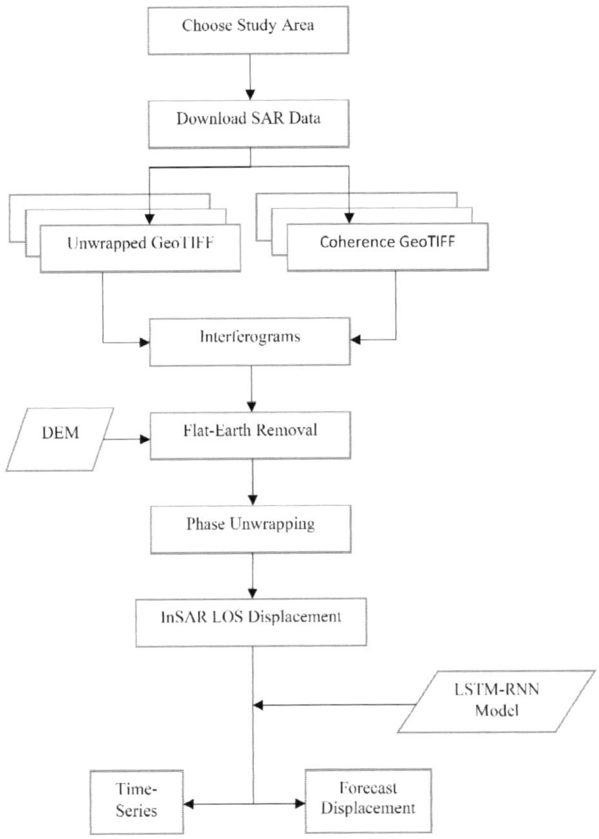

Fig 2. Flowchart of the proposed methodology.

To analyze the trend pattern in displacement (e.g., upliftment or subsidence), a time series data for six years (2016-2021) is plotted in Fig. 4. The nature of the dataset (negative values) reveals subsidence of the area, with a mean displacement of 23.69 mm/yr. Two statistical tests, namely the ADF and KPSS tests, with their corresponding p-values 0.155 and 0.100, suggest that the trend is non-stationary. However, the dataset exhibits a seasonal behavior. Therefore, a time series decomposition would be useful to understand the trend, seasonality, and residual components. The dataset also exhibits a significant auto-correlation, indicating that the future value is correlated (or depends) on the occurrence of the past values. In fact, by inspection, it appears from Fig. 4 that a time lag of 20 would be appropriate to consider in forecasting models. The time-series decomposition plot is shown in Fig. 5. From Fig. 5, we can visualize that the dataset has a strong downtrend behavior (subsidence), seasonality, and random variation that may be related to localized geospatial effects and observational noise due to wrapping and unwrapping, vegetation cover, and atmospheric disturbances [2]. Having pre-processed the dataset and analyzing several components of the time-series, we construct LSTM architecture for suitable forecasting of the LoS displacement. First, a single layered LSTM-RNN model with four nodes is considered. We consider a variable epoch size of 25, 50, and 100, split size as 0.8 (80% training data) and 0.9 (90% training data), and "Relu" as activation function. We consider two accuracy indicators, namely the root mean squared error (RMSE) and mean absolute percentage error (MAPE). Based on their values for several combination of models, it is found that a split size of

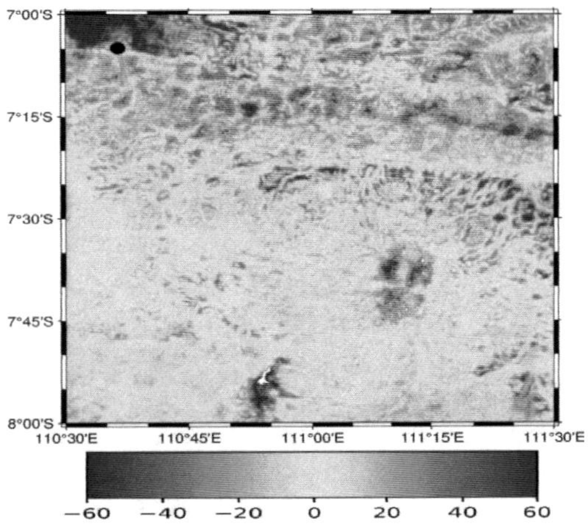

Fig 3. InSAR-based LoS velocity map of the study area in mm/yr. Black dot at the upper left corner represents the pixel that was chosen for observation and forecasting of InSAR time series.

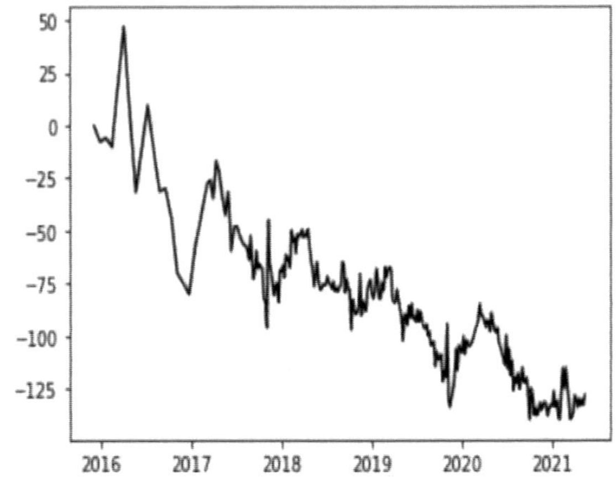

Fig 4. Displacement (in mm) versus time plot.

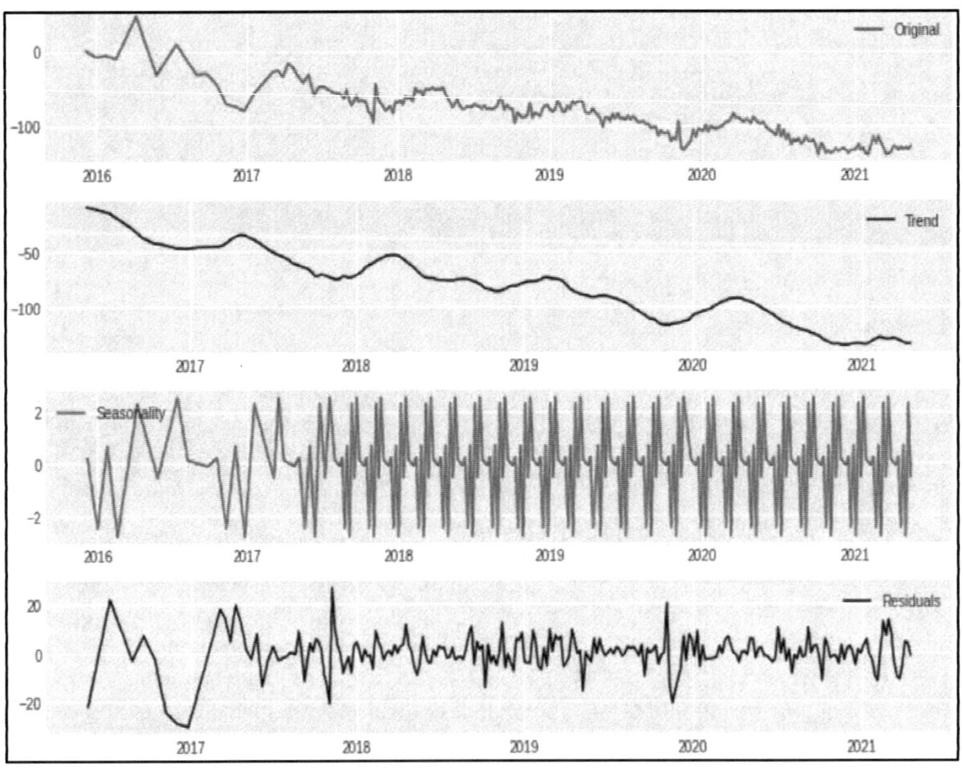

Fig 5. Time series decomposition of LoS displacement in mm/yr.

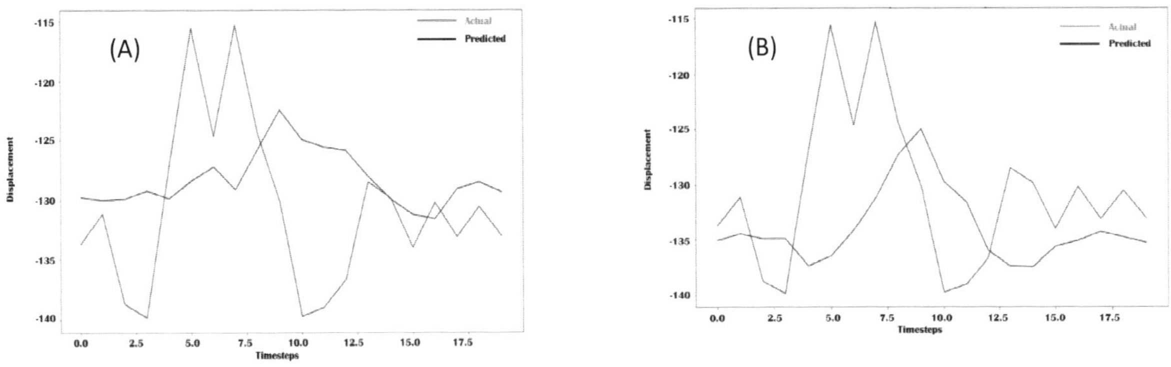

Fig. 6. Displacement versus time graph corresponding to the two types of LSTM architecture; (A) single-layered LSTM model with epoch size 100 and split size 0.9 provides an MAPE 4%, training RMSE score 8.27 and testing RMSE score 7.75; (B) Four-layered LSTM model with epoch size 100 and split size 0.9 provides MAPE 2%, training RMSE score 7.91 and testing RMSE score 8.10.

0.9 and epoch size of 100 provide better results (Fig. 6a). In fact, the model with epoch size 25 or 50 suffers heavy under-fitting. However, to further improve modeling accuracy, we have included three more LSTM layers to the previous model architecture. Epoch size 100 and split size 0.9 are chosen. This yields an MAPE score of 2%, with the training and testing RMSE values 7.91 and 8.10, respectively (Fig. 6b). The actual versus predicted values are plotted in Fig. 6.

4. Conclusions

This paper has provided a step-by-step implementation of the LSTM-RNN based machine learning technique to forecast InSAR-based LoS displacement of a pixel. It was observed that a multilayer LSTM model with four layers, each consisting of 100 nodes, provides desirable accuracy. In addition, the time series plots reveal land subsidence with a rate of 23.69 mm/yr. Although this study provides basic formulation of deep neural networks for InSAR data analysis in a selected pixel, the efficacy of the proposed approach may be assessed based on the results at several study regions. The deep-learning based LSTM model may also be implemented on a combined InSAR and GPS dataset to reveal better insights of crustal dynamics in a geographic area.

5. References

[1] Brengman, C. M., & Barnhart, W. D. (2021). Identification of Surface Deformation in InSAR Using Machine Learning. Geochemistry, Geophysics, Geosystems, 22(3), e2020GC009204.

[2] Bürgmann, R., Rosen, P. A., & Fielding, E. J. (2000). Synthetic aperture radar interferometry to measure Earth's surface topography and its deformation. Annual Review of Earth and Planetary Sciences, 28(1), 169-209.

[3] Chelbi, S., Khireddine, A., & Charles, J. P. (2011, December). Interferometry process for satellite images SAR. In 2011 7th International Conference on Electrical and Electronics Engineering (ELECO) (pp. II-200). IEEE.

[4] Diao, F., Walter, T. R., Solaro, G., Wang, R., Bonano, M., Manzo, M., Ergintav, S., Zheng, Y., Xiong, X., Lanari, R. (2016). Fault locking near Istanbul: Indication of earthquake potential from InSAR and GPS observations. Geophysical Supplements to the Monthly Notices of the Royal Astronomical Society, 205(1), 490-498.

[5] Ding, X., & Huang, W. (2011). D-InSAR monitoring of crustal deformation in the eastern segment of the Altyn Tagh Fault. International Journal of Remote Sensing, 32(7), 1797-1806.

[6] Ferretti, A., Monti Guarnieri, A., Prati, C., Rocca, F., & Massonnet, D. (2007). InSAR Principles-Guidelines for SAR Interferometry Processing and Interpretation, Part A. Netherlands: European Space Agency.
[7] Liu, F., Elliott, J. R., Craig, T. J., Hooper, A., & Wright, T. J. (2021). Improving the resolving power of InSAR for earthquakes using time series: a case study in Iran. Geophysical Research Letters, 48(14), e2021GL093043.
[8] Morishita, Y., Lazecky, M., Wright, T. J., Weiss, J. R., Elliott, J. R., & Hooper, A. (2020). LiCSBAS: an open-source InSAR time series analysis package integrated with the LiCSAR automated Sentinel-1 InSAR processor. Remote Sensing, 12(3), 424.
[9] Pasari, S. (2015). Understanding Himalayan tectonics from geodetic and stochastic modeling. PhD Thesis, Indian Institute of Technology Kanpur, 376.
[10] Radutu, A., Nedelcu, I., & Gogu, C. R. (2017). An overview of ground surface displacements generated by groundwater dynamics, revealed by InSAR techniques. Procedia Engineering, 209, 119-126.
[11] Rundle, J., Stein, S., Donnellan, A., Turcotte, D. L., Klein, W., & Saylor, C. (2021). The complex dynamics of earthquake fault systems: New approaches to forecasting and nowcasting of earthquakes. Reports on Progress in Physics, 84, 076801.
[12] Scholz, C. H. (2019). The mechanics of earthquakes and faulting. Cambridge university press.
[13] Sreejith, K. M., Agrawal, R., & Rajawat, A. S. (2016). Crustal deformation studies using synthetic aperture interferometry. Proceedings of the Indian National Science Academy, 82(3), 737-746.
[14] Verma, H., Sharma, Y., & Pasari, S. (2022). Synthetic Aperture Radar Interferometry to Measure Earthquake-Related Deformation: A Case Study from Nepal. In Disaster Management in the Complex Himalayan Terrains (pp. 133-140). Springer, Cham.
[15] Yu, C., Li, Z., Penna, N. T., & Crippa, P. (2018). Generic atmospheric correction model for Interferometric Synthetic Aperture Radar observations. Journal of Geophysical Research: Solid Earth, 123(10), 9202-9222.

Various Types of Constructed Wetland for Wastewater Treatment- A Review

Arvind Kumar Swarnakar[1], Dr. Samir Bajpai[2], Dr. Ishtiyaq Ahmad[3]

1. Research Scholar, Department of Civil Engineering, National Institute of Technology Raipur, Chhattisgarh, India akswarnakar@nitrr.ac.in
2. Professor, Department of Civil Engineering, National Institute of Technology Raipur, Chhattisgarh, India
3. Assistant Professor, Department of Civil Engineering, National Institute of Technology Raipur, Chhattisgarh, India

Abstract. As per the World Health Organization 80% of wastewater is released to the environment without satisfactory treatment. Constructed Wetlands (CWs) are one of the natural wastewater (WW) treatment methods. CWs have been recommended as a low technology, low maintenance, low operation cost, and green technology wastewater treatment system. Many types of CWs are currently in use. This paper studies these different types based on the climate, area, base materials, temperature, contaminant removal efficiency, removal mechanism and physicochemical analysis of various wastewater parameters. It is found that wetlands are successful in removing organic matter–Biochemical Oxygen Demand (BOD_5), Chemical Oxygen Demand (COD) and suspended solids and nutrients–total Nitrogen (TN) and total Phosphorus (TP). CWs need to be studied as a promising solution not only for effective treatment of wastewater but also as an economical method to improve the fertility of soil. Further, the paper discusses the scope of future research in CW to further improve the wastewater treatment technology.

Keywords: Constructed wetland, soil, gravel, wastewater, treatment,

1. Introduction

Natural wetland comprises of vegetation, soil and either water or wastewater. There are different conventional methods for wastewater treatment such as sand traps (grit chamber), septic tanks, Imhoff tank, baffled reactor (Anaerobic Baffled Reactor, ABR), anaerobic filter, green filters, Soil BioTechnology (SBT) or Constructed Soil Biofilter (CSB) or Constructed Soil Filter (CSF), anaerobic stabilization ponds, aerobic stabilization pond (Maturation Ponds/Oxidation Pond), Rotating Biological Contactor (RBC), Active Sludge Process (ASP), Upflow Anaerobic Sludge Blanket (UASB), Trickling Filter (TF), and micro-algae techniques etc. CWs are the manmade system with combination of base material, vegetation, and organic matter to provide wastewater treatment [1]. CWs require less infrastructure, investments, raw materials, energy consumption, operation, staff during operation, maintenance, odors, insects, flow variations, toxic substances, and by-products [2]. Categorization of CWs depends on flow direction, macrophytic growth and hydrology [3]. CWs are typically used as a secondary treatment unit. The CWs are used for treatment of domestic wastewater, animal wastewater, mine water, leachate remediation, industrial wastewater, urban stormwater and field runoff [4]. The main part of the CWs are aggregate, soil and gravel used as filter media and macrophytes (vegetation) such as Typha, Canna indica, Flax Lily, Banksias, Bottlebrush, P. australis, common rush, tapered rosette grass, Phragmites australis, common reed, Club-rush, Cattail, Common water plantain, Reed canary grass, Meadowsweet, Yellow flag, Compact rush used as vegetation [5]. According to water flow in substrate UN- HABITAT has divided CWs in three categories – 1. Vertical Flow Constructed Wetland (VFCW) 2. Horizontal Flow Constructed Wetland (HFCW) 3.

Hybrid Flow Constructed Wetlands (HyFCW). Most of the natural wetlands are Free Water Surface (FWS). Further subdivision is shown in fig.1. This paper briefly describes all types of CWs.

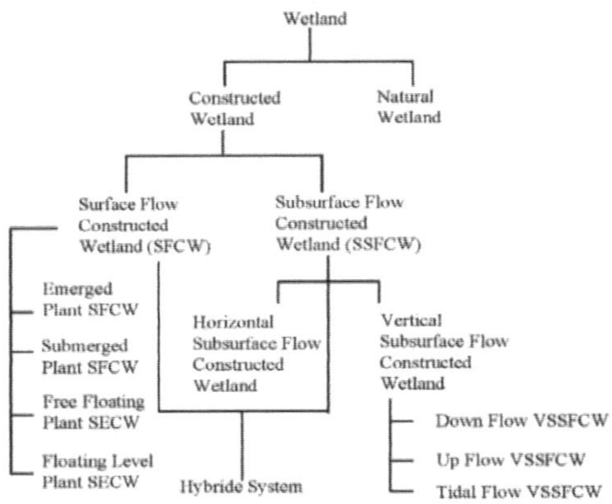

Figure 1. Various types of wetlands

2. Removal Mechanisms in Constructed Wetland

The main mechanism of treatment through CWs comprises biogeochemical transformation with solid and liquid separations. Macrophytes plays a very important role in removing various types of contaminants. The removal mechanism by macrophytes is shown in fig. 2. Contaminants uptake by root zone involves rhizodegradation process by microbial activity. The heavy metals are reduced through phytostablization. Plant enzymes break down the contaminants through phytodegradation. Plants and algae remove the contaminants from soil and sediments through phytoextraction. Plants release the contaminates from soil and sediments in atmosphere through phytovolatalization process [6, 7, 8, 9]. Removal of various contaminants involves various operations and process in CWs. Organic material (BOD) is removed by biological degradation, sedimentation, and microbial uptake – organic contamination including pesticides are removed by adsorption, volatilization, photolysis, and biotic degradation, suspended solids are removed by sedimentation and filtration, Nitrogen is removed by plant uptake, sedimentation, nitrification/ denitrification, microbial uptake, volatilization, Phosphorus is removed by sedimentation, filtration, adsorption, plant and microbial uptake, pathogens are removed by natural die-off, sedimentation, filtration, UV degradation, adsorption, heavy metals are removed by sedimentation, adsorption by vegetation and substrate and plant uptake process.

Figure 2. Macrophytes mechanism for contaminant removal

2.1 Role of plants in CWs

Different types of CWs plants are – free floating, rooted floating, emergent, submerged, shrub and trees. Most commonly, emergent herbaceous plants are used in CWs. Phragmites, phragmites australis, cattail, reeds, vetiver grass, typha latifolia, water grass and canna indica are some wetland plants. Selection of plants depends on type of constructed wetland systems, substrates, type of wastewater, viability of local plants and temperature [11]. Soil organic matter plays important role in plant growth in HFCWs [12]. The vegetation density (numbers in per sq-m area) depends on available filter media area [5-11]. Due to easy availability, Typha latifolia is a common plant for CWs and it also generates the higher biomass.

2.2 Selection of wetland

Selection of the wetland depends on type of pollutants, material of substrates, local available vegetations, temperature, Hydraulic Retention Time (HRT= Volume (Area x Water Depth x Porosity) / Discharge) and Hydraulic Loading Rate (HLR) [10]. As per the United State Environmental Protection Agency (USEPA) performance of constructed wetland depends on depth of water, slope of ground, and flow velocity. Table 1, 2 and 3 shows the efficiency of the HFCWs with various vegetation, HRT and HLR. Geo physicochemical parameters of filter media are necessary parameters for selection of CWs [7-49].

3. Vertical Flow Constructed Wetland (VFCWs)

In VFCWs the wastewater in flow is above the substrate. The wastewater is treated through percolates in root zone. Higher concentration of polluted water can be treated by VFCWs. The major parts of the VFCWs are shown in figure 2. The inlet of influents flows either by pump or by gravity. Oxidizing of ammonia is done quite well by VFCWs hence used in food proceeding and landfill leachates as shown in previous studies by Kadlec R.H. & Wallace S.D. (2009) [4]. The removal efficiency of vertical flow CWs is shown in table 1. It is used for continuous flow and good n utrient removal as compare to organic matter. It is less capable of removing total dissolve solids. Water stagnation, mosquitoes and aquatic animals are the problem with VFCWs. It has studied the textile waste with sugar bagasse and sylhet sand with plantation P.australis, D. sanderina and observed 79.2% BOD_5, 69.0-89.0% COD, 67.6-89.0% turbidity removal efficiency in VFCW [13]. It has been observed in pharmaceutical compounds in wastewater removal by pilot-scale VFCW by using scirpus grossus plants and sand, gravel as media. They practiced with different HRT (3, 4, and 5 days) and has been observed that 99% ibuprofen, 88% COD, 99% ammonia and 83% orthophosphate are effectively removed (HRT 5 days and 2 L/min aeration) [14]. Soil base constructed wetlands can also use for VFCWs.

Figure 3. Vertical Type Constructed Wetland

Table 1. Removal efficiencies (%) for VFCWs

S. No.	Type of WW	TSS	BOD_5	COD	NH_4-N	NO_3-N	TN	TP	Reference
1.	Household	97.0	96.0	-	88.4	-			[16]
2.	STP	-	-	85.0	77-100	-	64-80	90.0	[17]
3.	Municipal	89.4	41.5	55.8	73.0	34.6	-	67.8	[18]
4.	Domestic	90.0	-	84.0	92.0	-	-	-	[19]
5.	Landfill Leachate	-	-	78.0	87.0	-	-	-	[20]

4. Horizontal Flow Constructed Wetland (HFCWs)

HFCWs comprise of soil or gravel bed with vegetation. The wastewater flows below the top surface around the root of vegetation. Such wetlands can be called Sub Surface Flow Constructed Wetland system (SSFCWs). HSFCWs are used as secondary system for small community system [4].
Pretreated wastewater flows by gravity, horizontally or vertically, through the bed substrate where it contacts a mixture of facultative microbes living in association with the substrate and plant roots resulting in effective removal of BOD_5 and suspended solids (USEPA, 2000). Small scale HFCWs are shown in fig.3. The aerobic zones occur around roots and rhizomes that leak oxygen into the substrate. Organic compounds are very effectively degraded aerobically around roots and rhizome bacteria. As per the table 2, good efficiency for removing total suspended solid (TSS), organic matter (BOD) and COD as compared to nutrients. Mosquitoes and aquatic animal issues are limited in this CWs. HFCWs are used for all types of wastewaters. The various materials can be used for substrates like shale, soil sediments, zeolite, limestone, volcanic mineral, alum sludge, oyster shell, wood chip, plant waste, fly ash, slag, construction waste, ceramsite and activated carbon. Sewage with high *E. Coli* bacteria and suspended solids pit soil can be used [15]. Fecal coliform (FC) can effectively remove (82%) by HFCWs under 4.5 mg/L Dissolved Oxygen (DO) and temperatures 25 °C [21]. Soil base subsurface constructed wetland enhanced the potential for removing wastewater pollution [11-49].

Table -2. Removal efficiencies (%) for HFCWs

S.No.	Type of WW	TSS	BOD_5	COD	NH_4-N	NO_3-N	TN	TP	Reference
1.	Domestic	95.0	85.5	80.0	29.5	-	42.0	19.5	[22]
2.	Municipal	-	84.0	69.0	-	-	41.0	96.0	[23]
3.	Urban Sewage	82.7	91.6	-	55.3	-	-	58.2	[24]
4.	Sewage Treatment Plant	-	77.0	60	67.0	69.0	-	85.0	[25]
5.	Graywater	86.0	35.0	61.9	-	-	87.0	95.0	[26]

Fig 4 shows the layout of the HFCWs. Influents flow under gravity. Table 2 shows the comparative performance for wastewater treatment in HFCWs.

Figure 4. Horizontal Flow Constructed Wetland (HFCWs)

5. Hybrid Flow Constructed Wetlands (HCWs)

HFCWs are the combination of VFCWs and HFCWs. HFWS is effective for reduction of pathogens, ammonia and TN [9]. The various types of CWs can be arranged together to create a combined system, which is called hybrid constructed wetlands. Such mixed CWs are utilized to accomplish higher effectiveness of wastewater treatment as opposed to single CW. Investigation of HCWs was initiated by Seidel in Germany in 1980 [27] – ammonia was oxidized to nitrate, and in the following anoxic HF beds nitrate was reduced via denitrification [3]. Using a similar setup for the municipal wastewater, using the gravel, sand and soil with typha plant, the removal efficiency for $BOD_5$78%, TKN 66%, TSS 74% and P 86% while pilstia stratiotes removed BOD_5 84%, TKN 76%, TSS 82% and P 83% [28]. Fig. 5 shows the layout of HCWs.

Figure 5. Hybrid Constructed Wetland (HCWs)

HFCWs has some disadvantages like space consuming, capital cost, pretreatment, high supervision and not extremely lenient to cold atmospheres.

Table 3. Removal efficiencies (%) for HCWs

S.No.	Type of WW	TSS	BOD_5	COD	NH_4-N	NO_3-N	TN	TP	Reference
1	Domestic	89.0	-	85.0	83.0	-	83.0	64.0	[29]
2.	Domestic	-	-	59.0	79.2	-	79.7	-	[30]
3.	Dormitory wastewater	92.2	88.9	86.0	66.5	65.7	-	63.5	[31]
4.	Urban Wastewater	96.0	86.0	80.0	88.0	-	-	24.0	[1][32]
5.	Anaerobic baffled reactor(ABR)-HSSF-VSSF	95.9	90.1	90.0	-	69.5	-	26.1	[33]

[1]99.7% for fecal enterococci

6. Current scenario in Constructed wetland research

Various researchers have studied performance improvement of CWs by manipulating various parts and parameters such as, substrate, plants, Hydraulic Retention Time (HRT), Loading Rate (LR), aeration system, flow pattern, recirculation pattern, various combination for wetlands and plant density [USEPA]. For example, the locally available broken brick used for substratum in SSFCW for removal of nutrients, plants such as the typha dominensis , cyperus papyrus and dark green bulrush used for treatment for hospital wastewater resulting in 93% TSS, BOD_5 90%, COD 83%, TKN 64%, and 56% phosphate removal [35], different aeration methods like tide flow (TF), effluent recirculation (ER) and artificial aeration for all type of CW (VFCW, HFCW and HCW) improving the removal efficiency 89% TSS, 84% COD, 63% total Nitrogen (TN) and 81% ammonia nitrogen [35]. Recently, steel slag size 10 to 20 mm used as filer media in multistage pond CWs for removal of contamination in river water resulting in increase in total phosphorus (TP) reduction for first level HFCW by 151% [36].

Researchers have also studied different types of wastewaters such as removal of petrochemical waste through HFCW by using gravel and sand media, plants typha and phragmites, removing the BOD_5 (85%), COD (89%) and turbidity (66%). Similarly study on different plant species, like acacia nilotica, cotton, brassica napus carried out for decontaminating oilfield wastewater. BTEX (benzene, toluene, ethylbenzene and xylene), PAH (Polycyclic Aromatic Hydrocarbons) and aliphatic/ aromatic petroleum hydrocarbon can be removed by CW for refinery wastewater [37].
Heavy metals like Cu, Zn, Ni, Fe, Pb, Cd, B, Cr, Al, Co, As and Se from industrial wastewater, mine tailing waste, tannery wastewater, oil refinery, domestic wastewater can be removed from SFCW. Media used for removal of heavy metals from CW are sediments, compost, gravel, soil, ash with plants macrophyte, typha latifolia, macrophyte myriophyllum spicatum and algal species [38,39]. The removal efficiency of heavy metal from CW are reported as 69%- 96% [38,39,40].
Pharmaceutical, hospital and personal care products (PPCB), medical waste and cosmetic waste has been seen effectively removed by CWs. Phytoremidiation in various types of CW is a well stablished treatment method now [39,40, 41]. The removal of various PPCB was observed to be in the range of 70% to 90% by various CWs [42,43].
Agro based synthetic fertilizer, pesticides, fungicides, insecticides and herbicides leach with soil and contaminate the river, lakes, ponds, and ground water. CW removal efficiency for agricultural wastewater for COD (92%), BOD_5 (98%), TN (91%), TP (96%), TSS (96%), 94-99% E.Coli, Fecal Coliform (FC), and Total Coliform (TC) from single stage CWs in various conditions has been reported [44]. The researchers attributed the contribution of CW for treating agro-base runoff and industrial wastewater [45].
Group of researchers [46] recently observed that wastewater from textile dyeing process can be effectively treated by CW resulting in removal efficiencies–COD (87%), BOD_5 (86%), TSS (49%), TDS (46%), Orange3R (84%), alcohol oxidase (85%), lignin peroxidase (150%) and color (75%).
Compost and landfill leachates have a high amount of organic matter and it can be treated by CW [47]. A pilot-scale study was done in Iran where HSSFCW planted with vetiver grass and fine gravel is used as filter media. It is observed that 75% of BOD5, 53% of COD and 74% of TN were removed successfully [47]. Microbial fuel cell (MFC)-based horizontal flow (HF) constructed wetlands are novel systems that use interconnected electrodes (cathode and anode), saturated media, plant, and electrochemically active microbial population for wastewater polishing [48,49]. One study reported that the Microbial Fuel Cell (MFC) HFCW with aluminum plates anode and cathode in HFCW with stone aggregate as filter media and vetiver grass planted in CW, has 93% phosphorus, 83% coliform, 55-92% nitrogen and 80-100% organic impurities removal [50].

Microplastics (MP) are the emerging as one of the major environmental pollutants on the earth. In MP Synthetic Organic Polymer Plastics (SOPP) have been used majorly in households, clothing, agriculture, personal care products [51]. The increased dependency on plastic material has skyrocketed the demand for plastic products which eventually end up as a waste in the environment. The research shows that MPs travel from municipal solid waste landfill to aquatic systems [52]. Natural wetlands are also proving to be effective in removing MPs, a pollutant of emerging concern. Mangrove wetlands are proving to be effective in removing MPs from aquatic streams [53]. CWs can be a suitable treatment strategy in removing MPs before they reach natural water sources.

7. Conclusion and further directions

Due to a small pollution load in domestic and municipal wastewater, CWs can be an effectively part of treatment. CW provides high removal efficiency for organic contaminants, nutrients–nitrogen and phosphorus, and pathogenic microorganisms. Highly toxic contaminants and heavy metals from industrial wastewaters can also be effectively treated by CWs. Toxic heavy metals removal like lead, cadmium, iron, mercury, arsenic, copper, chromium, zinc, nickels, silver and manganese through CWs have been reported by many researchers. CWs provide many advantages over conventional wastewater treatment systems. CWs provide an economical, low technology, less expensive and high energy-saving treatment technology. CWs also show promise in providing effective treatment not only for the domestic, municipal wastewater but also for difficult to treat industrial wastewaters. Waterborne pathogens can also be reduced by wetland systems. CWs are environmentally friendly and natural resource treatment systems that contribute to the public health. CWs also show potential in removal of emerging contaminants of concerns that are difficult to remove by conventional treatment methods. Recent research shows that some endocrine disrupting chemicals were reduced by 48 to 99% by CWs [54]. More research is also required for removal of MPs by CWs. A long term monitoring of geo-physicochemical properties of CW under various operating conditions will give further research for better understanding of the key factors responsible for the optimization of CW [55]. CW has great potential efficiency for removal of all types of water pollution and wetland technology has made great strides in the last few years.

REFERENCE

[1] Kirby, J.A. (2000). Constructed wetland: Tertiary Treatment of Wastewater. Wastewater treatment symposium, A ssociation of Professional Engineer and Geoscientists of newfoundland, Corner Book, NF.

[2] Stefanakis, A.I., Akratos, C.S., & Tsihrintzis, V.A. (2014). Vertical flow constructed wetlands: Ecoengineering systems for wastewater and sludge treatment (1st Ed.). Amsterdam, The Netherlands: Elsevier Publishing.

[3] Vymazal, J. (2011). Plants used in constructed wetlands with horizontal subsurface flow: A review. *Hydrobiologia*, *674*, 133–156.

[4] Kadlec R.H, Wallace S.D. (2009). Treatment wetlands, Second edition. CRC Press, Page 10.

[5] Jethva K., Bajpai S., Chaudhari P.K., Swarnakar A.K. (2017). Wastewater nutrient removal through phytoremidiation: a review. 49th Annual Convention of IWWA on "Smart Water Management" January 19-21.

[6] United Nations Human Settlements Programme (UN-HABITAT), (2008). Constructed Wetland Manual. P-8.

[7] USEPA, Design Manual, Constucted Wetlands and Aquatic Plant system for Municipal Wastewater Treatment. EPA/625/1-88/022;1988

[8] Srivastava J, Gupta A, Chandra H. Managing water quality with aquatic macrophytes. Rev Environ Sci Biotechnol. 2008;7:255–266.

[9] Odinga C.A., Swalaha, F. M., Otieno F. A.O., Kumar R. Ranjith, and Bux. F., (2013). Investigating the efficiency of constructed wetlands in the removal of heavy metals and enteric pathogens from wastewater. Environmental Technology Reviews. DOI:https://doi.org/10.1080/21622515.2013.865086.

[10] Cooper,P.F., Job G.D., Green M.B. and Shutes R.B.E.(1996). Reed Beds and constructed wetland for wastewater treatment.WRc Swindon, UK.

[11] Jethwa K., Baipai S., (2016). Role of plants in constructed wetlands (CWS): a review. Journal of Chemical and Pharmaceutical Sciences. Issue 2: August 2016, P;1-10.

[12] Bajpai, S. & Swarnakar, A., (2021). Role of soil organic matter in plant growth for horizontal flow subsurface constructed wetlands for waste water treatment. EM International. https://doi.org/10.1080/09593330.2016.1273396.

[13] Saeed, T., Sun G., (2012). A review on nitrogen and organic removal mechanism in subsurface flow constructed wetland: depend on environmental parameters, operating condition and supporting media. J. Environ.Manag. 112, 429-448.

[14] Falahi, O. A. A., Abdullah, S. R. S., Hasan, H. A., Othman, A. R., Ewadh, H. M., Al-Baldawi, I. A., & Imron, M. F. (2021). Simultaneous removal of ibuprofen, organic material, and nutrients from domestic wastewater through a pilot-scale vertical sub-surface flow constructed wetland with aeration system. *Journal of Water Process Engineering*, *43*, 102214.

[15] Wang Y., Cai Z., Sheng S., Pan F., Chen F., Fu J., (2020). Comprehensive evaluation of substrate materials for contaminants removal in constructed wetlands. Sci Total Environ.701:134736. doi: 10.1016/j.scitotenv.2019.134736. Epub 2019 Oct 31. PMID: 31715485.

[16] Wu, S., Austin, D., Liu, L., Dong, R., (2011). Performance of integrated household constructed wetland for domestic wastewater treatment in rural areas. Ecol. Eng. 948-954.

[17] Cui, Y.B., Jiang, T.L., Guo, Z.Q., (2008). Nutrient removal mechanism and performance of constructed wetland in cold climate.J. Xi'an univ. Archit.Techno. 40, 121-125.

[18] Sani, A., Scholz, M., Bouillon, L., (2013), Seasonal assessment of experimental vertical- flow constructed wetlands treating domestic wastewater. Bioresour. Technol. 147, 585-596.

[19] Sklarz, M.Y., Gross, A., Yakirevich, A., Soares, M.I.M., (2009). A recirculating vertical flow

constructed wetland for the treatment of domestic wastewater. Desalination. 246 (1e3), 617e624.

[20] Lavrova, S., Koumanova, B., (2010). Influence of recirculation in a lab-scale vertical flow constructed wetland on the treatment efficiency of landfill leachate. Bioresour. Technol. 101, 1756e1761.

[21] Fahim. R., Xiwu L., Hussain I., Jilani G., Hussain J.,(2019). Inactivation Of Faecal Coli Form Indicator Microorganisms In A Constructed Wetland Under Diverse Environmental Conditions. International Journal of Advances in Science Engineering and Technology, Vol. 7, Issue -2 ISSN(p): 2321 –8991

[22] Hayward, J., Jamiesoon, R., Boutilier, L., Goulden, T., Lam, B., (2014). Treatment performance assessment and hydrological characterization of an arctic tundra wetland receiving primary treated municipal wastewater. Ecol. Eng. 73, 786-797.

[23] Jenssen, P.D., Maehlum, T., Krogstad, T., vrale. L. 2005. High performance constructed wetlands for cold climate. environ. Sci. Health 40, 1343-1353.

[24] Rai, U.N., Upadhyay, A.K. Singh, N.K., Dwivedi, S., Tripathi, R.D., 2015. Seasonal applicability of horizontal sub surface flow constructed wetland for trace elements and nutrient removal from urban wastes to conserve Ganga river quality at Haridwar, India. Ecol. Eng. 81, 115-112.

[25] Thalla A.K., Devatha C.P., Anagh K., Sony E., (2019), Performance evaluation of horizontal and vertical fow constructed wetlands as tertiary treatment option for secondary effluents. Applied Water Science 9:147.

[26] Raphael. O. D., Ojo S. I. A., Ogedengbe K., Eghobamien C., and Morakinyo A. O. (2018). Comparison of the performance of horizontal and vertical flow constructed wetland planted with Rhynchospora corymbosa. International Journal Of Phytoremediation. Https://Doi.Org/10.1080/15226514.2018.1488809.

[27] Sayadi, M. H., Kargar, R., Doosti, M. R., & Salehi, H. (2012). Hybrid constructed wetlands for wastewater treatment: A worldwide review. Proceedings of the international academy of ecology and environmental sciences, 2(4), 204

[28] Haydar, S., Anis, M., & Afaq, M. (2020). Performance evaluation of hybrid constructed wetlands for the treatment of municipal wastewater in developing countries. Chinese Journal of Chemical Engg. 28(60), 1717-1724.

[29] Mander U, Tooming A, Mauring T, et al. (2007). Performance dynamics of a LWA-filled hybrid constructed wetland in Estonia. Ecohydrol. Hydrobiol, 7: 297-302.

[30] He, S.,Wang, Y., Li, C., Li, Y., Zhou, J., (2018). The nitrogen removal performance and microbial communities in a two-stage deep sequencing constructed wetland for advanced treatment of secondary effluent. Bioresour. Technol. 248, 82–88.

[31] Gholipour, A., & Stefanakis, A. I. (2021). A full-scale anaerobic baffled reactor and hybrid constructed wetland for university dormitory wastewater treatment and reuse in an arid and warm climate. Ecological Engineering, 170, 106360.

[32] Melian JA, Rodríguez AJ, Arana J, et al. (2010). Hybrid constructed wetlands for wastewater treatment and reuse in the Canary Islands. Ecological Engineering, 36: 891-899.

[33] Singh SH, Haberl R, Moog O, (2009). Performance of an anaerobic baffled reactor and hybrid constructed wetland treating high-strength wastewater in Nepal—A model for DEWATS. Ecological Engineering, 35:654-660.

[34] Dires, S., Birhanu, T., & Ambelu, A. (2019). Use of broken brick to enhance the removal of nutrients in subsurface flow constructed wetlands receiving hospital wastewater. Water Science and Technology, 79(1), 156-164.

[35] Ilyas, H., & Masih, I. (2017). The performance of the intensified constructed wetlands for organic matter and nitrogen removal: A review. Journal of environmental management, 198, 372-383.

[36] Lu, H., Xiao, L., Wang, T., Lu, S., Wang, H., Guo, X., & Li, J. (2021). The application of steel slag in a multistage pond constructed wetland to purify low-phosphorus polluted river water. Journal of Environmental Management, 292, 112578.

[37] Agarry, S. E., Oghenejoboh, K. M., Latinwo, G. K., & Owabor, C. N. (2018). Biotreatment of petroleum refinery wastewater in vertical surface-flow constructed wetland vegetated with Eichhornia crassipes: lab-scale experimental and kinetic modelling. Environmental technology.

[38] Hadad, H. R., Maine, M. A., & Bonetto, C. A. (2006). Macrophyte growth in a pilot-scale constructed wetland for industrial wastewater treatment. Chemosphere, 63(10), 1744-1753.

[39] Lim, P. E., Tay, M. G., Mak, K. Y., & Mohamed, N. (2003). The effect of heavy metals on nitrogen and oxygen demand removal in constructed wetlands. Science of the total environment, 301(1-3), 13-21.

[40] Liu, M., Li, X., He, Y., & Li, H. (2020). Aquatic toxicity of heavy metal-containing wastewater effluent T reated using vertical flow constructed wetlands. Science of The Total Environment, 727, 138616.

[41] S. Dires, T. Birhanu, A. Ambelu, Use of broken brick to enhance the removal of nutrients in subsurface flow constructed wetlands receiving hospital wastewater, Water Sci. Technol. 79 (2019) 156–164, https://doi.org/10.2166/WST.2019.037.

[42] Kumari A., Mourya N.S., Tiwari B., Hospital wastewater treatment scenario around the globe. Current Developments in Biotechnology and Bioengineering. DOI: https://doi.org/10.1016/B978-0-12-819722-6.00015-8.

[43] N.A. Khan, R. El Morabet, R.A. Khan, S. Ahmed, A. Dhingra, M. Alsubih, A.R. Khan, Horizontal sub surface flow Constructed Wetlands coupled with tuben settler for hospital wastewater treatment, J. Environ. Manag. 267 (2020), 110627, https://doi.org/10.1016/j.jenvman.2020.110627.

[44] Nan, X., Lavrnić, S., & Toscano, A. (2020). Potential of constructed wetland treatment systems for agricultural wastewater reuse under the EU framework. Journal of Environmental Management, 275, 111219.

[45] Wang, M., Zhang, D., Dong, J., & Tan, S. K. (2018). Application of constructed wetlands for treating agricultural runoff and agro-industrial wastewater: a review. Hydrobiologia, 805(1), 1-31.

[46] Kadam, S. K., Tamboli, A. S., Chandanshive, V. V., Govindwar, S. P., Choo, Y. S., & hong Pak, J. (2022). Construction and implementation of floating wetpark as effective constructed wetland for industrial textile wastewater treatment. Journal of Hazardous Materials, 424, 127710.

[47] Bakhshoodeh, R., Alavi, N., Majlesi, M., & Paydary, P. (2017). Compost leachate treatment by a pilot-scale subsurface horizontal flow constructed wetland. Ecological Engineering, 105, 7-14.

[48] Corbella, C., Garfí, M., & Puigagut, J. (2016). Long-term assessment of best cathode position to maximise microbial fuel cell performance in horizontal subsurface flow constructed wetlands. Science of the total environment, 563, 448-455.

[49] Villasenor, J., Capilla, P., Rodrigo, M. A., Canizares, P., & Fernandez, F. J. (2013). Operation of a horizontal subsurface flow constructed wetland–microbial fuel cell treating wastewater under different organic loading rates. Water research, 47(17), 6731-6738.

[50] Saeed, T., Yadav, A. K., & Miah, M. J. (2022). Influence of electrodes and media saturation in horizontal flow wetlands employed for municipal sewage treatment: A comparative study. Environmental Technology & Innovation, 25, 102160.

[51] Hassan, I., Chowdhury, S. R., Prihartato, P. K., & Razzak, S. A. (2021). Wastewater Treatment Using Constructed Wetland: Current Trends and Future Potential. Processes, 9(11), 1917.

[52] Upadhyay, K., & Bajpai, S. (2021). Transport of Microplastics from Municipal Solid Waste Landfills to Aquatic system: An Overview. AIJR Proceedings, 222-227.

[53] Meera, S. P., Bhattacharyya, M., Nizam, A., & Kumar, A. (2021). A review on microplastic pollution in the mangrove wetlands and microbial strategies for its remediation. Environmental Science and Pollution Research, 1-15.

[54] Hsieh, C. Y., Yang, L., Kuo, W. C., & Zen, Y. P. (2013). Efficiencies of freshwater and estuarine constructed wetlands for phenolic endocrine disruptor removal in Taiwan. Science of the total environment, 463, 182-191.

[55] Swarnakar, A. K., Bajpai, S., & Ahmad, I. (2021). Geo Physicochemical Properties for Soil Base Subsurface Constructed Wetland System. AIJR Proceedings, 228-233.

Kimberlitic olivine – a proxy to kimberlite petrogenesis and ascent process

V. K. Mishra[1], R. N. Tiwari[2], C. B. Verma[1], A. Mukherjee[1], R. K. Dixit[3], G. Prabhakar[4]

[1] NMDC Limited, Khanij Bhavan, Castle Hills, Masab Tank, Hyderabad-500028.

[2] Department of Geology, Govt. Model Science College, Rewa-486001.

[3] Directorate of Geology and Mining, Bhopal, Madhya Pradesh.

[4] Department of Geology, Osmania University, Hyderabad-500007.

E-mail: varunmishra1975@gmail.com

ABSTRACT. In kimberlite, olivine is an important constituent and it is found as mineral grains of different sizes, which can be of xenocrystic and magmatic origin. To know the processes, that are involved and controls the compositional variations of olivine, can provide distinct understanding into the genesis and evolution of kimberlites. In addition to this the textural features of kimberlitic olivine, which are recorded as textural events during magma assent constrains the ascent of kimberlite and the processes involved.
Unambiguous identification of kimberlitic olivine and its textural features require careful petrographic examination combined with mineral compositional analysis and use of high magnification images (BSE-SEM), all this is integrated to know the origin of olivine grain which in turn constrain the process involved in ascent of kimberlite and its petrogenesis. In this study we review the use of kimberlitic olivine in deducing the upper mantle conditions and process for kimberlite petrogenesis.

1.Introduction:

Olivine, the most profuse mineral phase in the upper mantle and, a major mineral constituent of mantle derived rocks such as Kimberlite. In kimberlites olivine forms as 50% of total volume and sometime exceeds up to 60%, in case of coherent kimberlite [1, 2, 3, 4, 5]. Therefore, can be used as proxy for kimberlite petrogenesis, ascent process and also for diamond quality or grade of a kimberlite pipe. Kimberlitic olivine, includes three parageneses based on crystal morphology and size [6, 7] and accordingly are termed as megacryst (coarse grain, rounded crystal of > 10mm size); macrocryst (medium grain, rounded to sub-rounded crystal of 5-10mm); and phenocryst (medium to fine grain less then 5mm size) (Figure 1).

Various, theories and processes has been suggested for the origin of kimberlitic olivine, as expressed by their paragenesis: the macrocryst & xenocryst are of foreign origin to that of kimberlite magma and the phenocryst is considered to have crystallized from the kimberlite melt. Olivine in Kimberlite shows zoning between core and rim, and crystallise in three distinct phases (1) in the mantle from earlier kimberlite melts; (2) in the ascending kimberlite; (3) after emplacement [8]. There is a considerable debate regarding the nature of the mantle source of the olivine cores and the proposed origins ranges from kimberlite related dunites, megacryst and sheared peridotite, and an early magmatic origin [8, 9]. Hence, olivine contains rich repository of information on petrogenesis, metasomatism and process related to kimberlite evolution. Here, we review the implication and utility of kimberlitic olivine as a proxy to the evolution of kimberlite melt, upper mantle condition, metasomatism, kimberlite ascent process, which is mostly based on the extract of the recent studies, publications and research articles available on the internet on the subject.

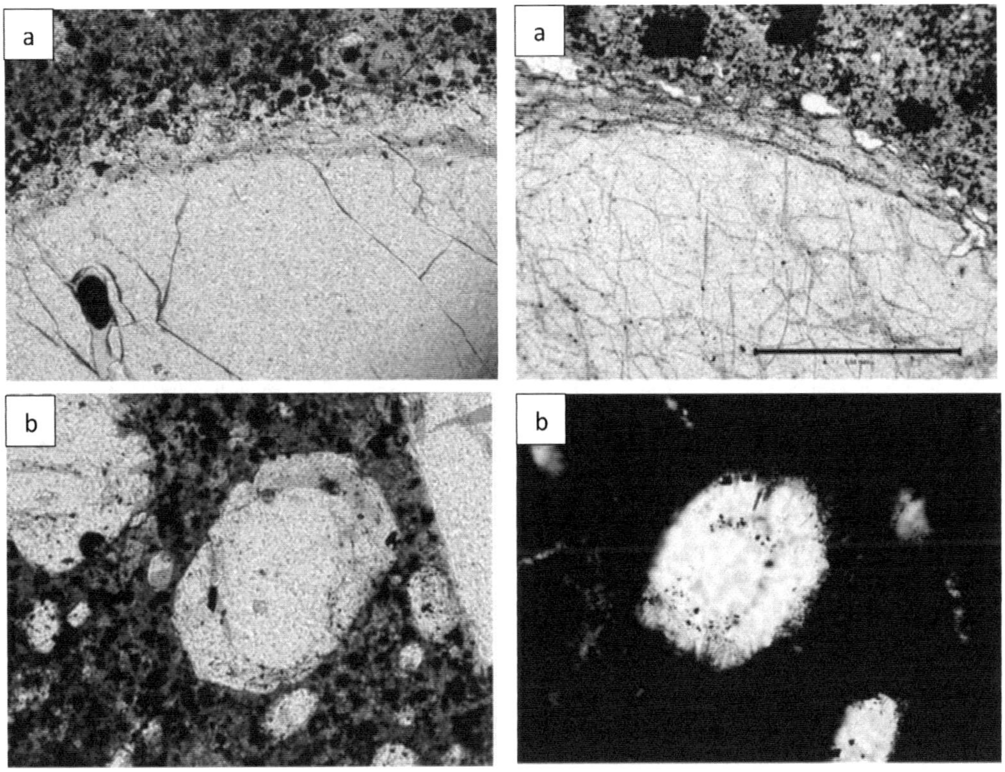

Figure 1. (a) Photomicrographs of olivine hosted by kimberlite; xenocrystic olivine (macrocrysts) with distinct rims formed by overgrowth of olivine [after 31], (b) Phenocryst-like, olivines comprising core of rounded olivine xenocryst and euhedral rim of olivine overgrowth

2. Kimberlitic Olivine – a monitor of upper mantle condition:

Olivines from kimberlite provides important information's, on the nature and evolution of kimberlitic melt, as its formation can persist from earliest stage to the latest stages of crystallisation [9] and witnesses the complete processes of kimberlite evolution. Olivine contain mantle lithologies and phases as inclusion, which provide clues to kimberlitic metasomatism during the early stages, zoning and textures of olivine grains contain information on the effects of mantle metasomatism at mantle depths. In addition to this zones and rims of kimberlitic olivine also testify the various processes involves in ascent and emplacement of kimberlite magmas.

2.1. Origin of Olivine:

Magmatic origin has been suggested for kimberlitic olivine, recently studies based on high precision analytical techniques such as BSE, EPMA, LA-ICPMS etc. indicated that the cores of olivine have inclusions of lithospheric mantle phases i.e., orthopyroxene, garnet, clinopyroxene, Cr-spinel [10, 11, 12] and challenges the magmatic origin. Based on the experimental and petrographic studies, it is evident that these minerals are unstable at low pressure within a melt [13, 14, 15, 16, 17] and, advocates the xenocrystic origin of olivine grain in kimberlite.

The core composition of olivine plays vital and important role in deducing information on kimberlite evolution and upper mantle conditions. Most of the core compositions from kimberlitic olivine plots in the area designated for peridotite (i.e.,89-94 of Mg# value; 0.30-0.45wt%NiO value and CaO < 0.1% wt%).

Olivine grains showing decreasing Mg#; Ni and Ca trends and high Mn, low Ca cores with Mg# 84-88 are believed to be formed from metasomatism of the lithospheric mantle by precursor kimberlite (or kimberlite-related) melts not long before entrainment in the ascending kimberlite [18, 19, 20, 21, 22]. Early metasomatism at depths may also result the, olivine cores with high Mg# > 89 and CaO more than 0.1wt% is attributed to the process of early metasomatism by Ca rich kimberlitic melt within the mantle [12, 23]. Similarly, Fe rich olivine cores are considered to be formed by deep metasomatic products of early kimberlitic melt [24]. Hence, the composition of olivine core provides an estimate of the degree of metasomatism along the magmatic conduit sampled by kimberlite magmas.

2.2. Kimberlitic olivine-its implication to melt evolution:

Proportion of magmatic and xenocrystic olivine in kimberlite is essential in calculating kimberlite melt composition. Relative proportion of both the olivines (magmatic and xenocrystic) within a kimberlite is variable, this variation can be due to paragenesis of olivine within a particular kimberlite or may be also due to different methods used for estimation. Higher and lower concentration of xenocrystic and magmatic olivine respectively, within bulk kimberlite is attributed to low Mg and Si concentration of parent kimberlite melt [1,13].

For its, antecrystic and xenocrystic nature, cores of kimberlitic olivine furnish significant insights into the conditions ranging from mantle to kimberlite magma conduits.

The information when integrated with thermometry data, based on olivine, garnet, diopside etc. can be applied for constraining the depth of olivine entrainment and offer a complementary approach for mantle compositional variation with depth [25,26].

Olivine Rim Composition: Rims of olivine in kimberlite, exhibits a trend of homogenous Mg# values and variable concentration of miner and trace element. It shows constant values Mg# in rims are constant, Ni values decreases and Mn and Ca values are variable. Fractional crystallisation of olivine and chromite during early stage of kimberlite crystallisation has been suggested for the above trend. In addition to this, different theories such as higher distribution coefficient of Fe-Mg between olivine and melt [2]; combination of fractionation and assimilation of olivine & orthopyroxene respectively in Ca-Mg carbonate melt has also been suggested for the olivine rim composition. (Figure 1 & 2).

However, to explain the composition pattern and constant Mg# of olivine rims, needs contribution of other additional processes. The process involving high Si and low Co_2 contents through assimilation of orthopyroxene [27, 2, 13, 14] and other mantle phase; and increasing oxygen fugacity during kimberlite crystallisation has been proposed for the composition of olivine rims worldwide. The study of olivine rim composition and associated process during kimberlite crystallisation can provide a profound in depth sights into the petrogenesis of kimberlite melts.

3. Kimberlite Ascent- as olivine reflects:

The rapid movement of kimberlite from the depths of mantle to the upper surfaces of lithosphere have been studied by various workers in details [28, 29]. Various models like, the propagation of a volatile rich magma-filled cracks upward through the mantle lithosphere [29]; deep seated exsolution of Co_2-rich fluids [30] have been suggested for this rapid phenomenon. Textural features within the olivine

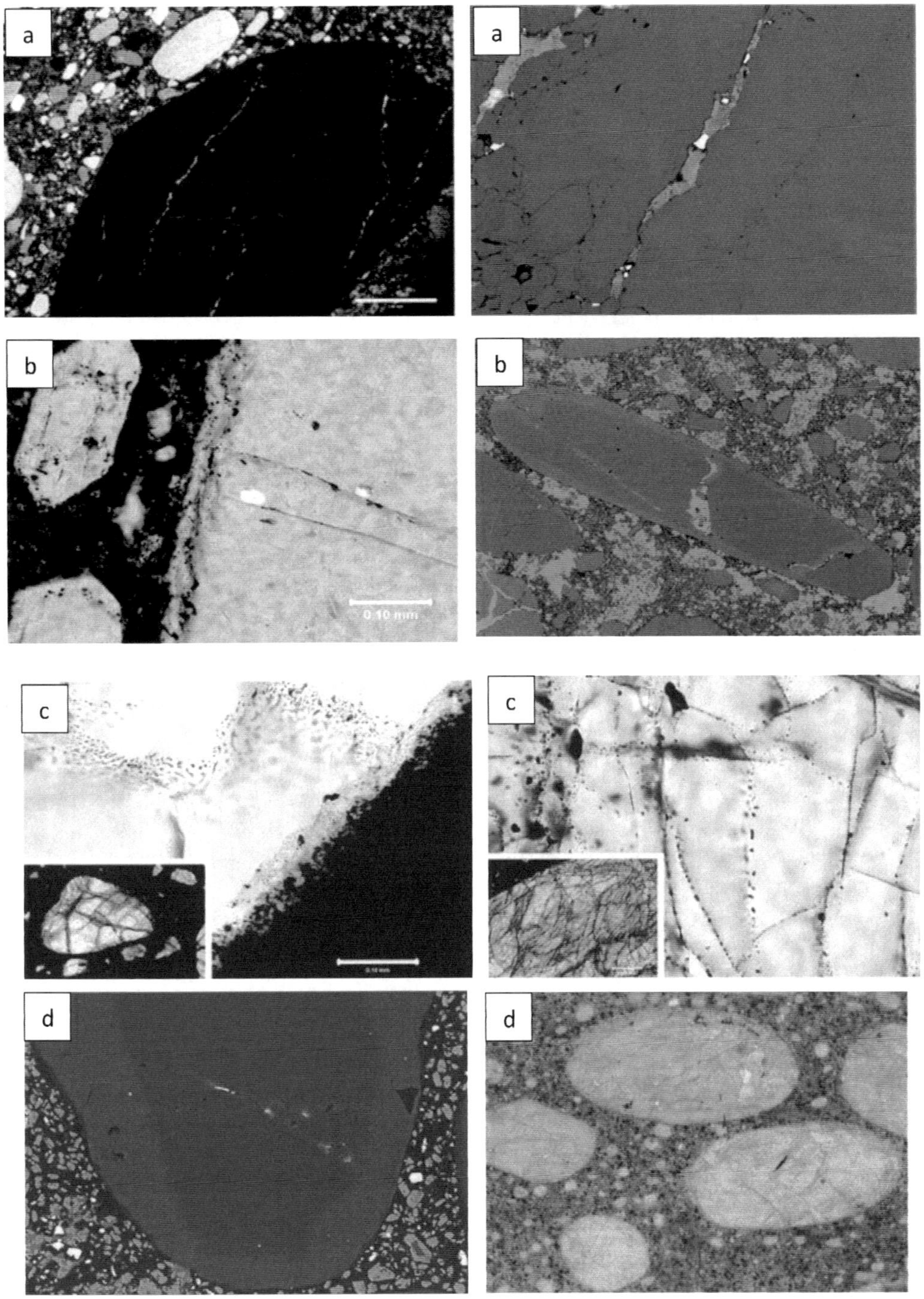

Figure 2. Photomicrographs showing different types of texture and their relationship in a Kimberlitic Olivine [after, 31]. (a) network of sub-parallel sealed cracks carbonate filled sealed crack, (b) Network of sub-parallel Healed cracks, fractures, overgrowth and carbonate filled crack (c) healed cracks defined by discontinuous trails of fluid inclusions, (d) Olivine overgrowth and shapes of olivine (rounded and ellipsoidal).

record and experiences physical and chemical changes of the ascending magma. These textural features can constrain the process operative during the kimberlite ascent and eruption (Figure 2).

3.1 Texture features within olivine

-Olivine Overgrowths: -development of growth rims on a larger olivine grain and the interface of this trapped fluids, melt inclusion, mineral phase can be seen in form of discontinuous trails, which are parallel to margins of core.

-Fractures: -It host a network of sub parallel to micro cracks, these are of different types.

-*Sealed cracks*- sets of fractures, cracks filled with inclusions, mostly of carbonate and oxide.

-*Healed cracks*: They are in form of discontinuous fractures or trails of fluids and mineral inclusion, representing later event of fracturing. These fractures are filled melt and fluid inclusion.

-*Sealed cracks*: They are continuous and discontinues segments and patches as elongated carbonate.

-*Shapes and surfaces*: - Olivine grains are rounded to elliptical and the surfaces are texturally distinct, they appear penetrative flaky and curving smooth when seen in high magnification.

The implication of above textural relationship can be utilised and a sequence of events occurring within the rising magma can be established. Within, the rising magma xenocrystic / xenolith minerals are not in equilibrium and react with the melt at different rates depending on their composition [3]. The textures observed in olivine are the results of the process involved during the ascent and interaction of the magma. The fractures sealed with carbonate rich fluids can develop during development of partially digested olivine. A later set of fractures filled with more sialic and hydrous magma can also develop which may be attributed to olivine saturation event, which is followed by overgrowths and capping of the sealed cracks. Rounded to ellipsoidal olivine grains may represent the last textural event. Rough and pitted textures in olivine grains may be due to the abrasion and attrition during the transport of kimberlitic magmas of a turbulent nature [31, 32].

4. Conclusion

Olivine grains in a Kimberlitic is a depository of details on the composition, metasomatism and thermal conditions of the mantle sampled by kimberlites. Important constraints on the evolution of kimberlitic melts are made as it is present from the earliest stage up to last stage of crystallisation. It requires a careful point selection and detailed textural identification under the microscope and high magnifying images (BSE-SEM). To address the petrogenesis of kimberlitic melt, it is essential to include and integrate the information based on textural relationship and analytical data. Studies suggest that olivine cores are the product of different types of disaggregate mantle rocks and large portions of the core is derived from megacryst, sheared peridotite and other lithologies. Textural information recorded by olivine, coupled with the intrinsic properties of olivine such as bulk modulus, viscosity and tensile strength can constrain magma ascent rates.

Overall, Olivine from kimberlite provide a complete information on the evolution of kimberlitic magma and its systems. It provides clues to the early metasomatic process operating within the upper mantle. Zoning and rims of kimberlitic olivine indicate the evidences of early crystallisation at mantle depths and also the process involved in ascent and emplacement of kimberlite.

5. References

[1] Mitchell R H 1986 Kimberlites: Mineralogy, Geochemistry, and Petrology. Plenum Press, New York

[2] Kamenetsky V S, Kamenetsky M B, Sobolev A V, Golovin A V, Demouchy S, Faure K, Sharygin V V and Kuzmin, D V 2008 Olivine in the Udachnaya-East Kimberlite (Yakutia, Russia): types, compositions and origins Journal of Petrology 49, 823–839

[3] Brett R C, Russell J K and Moss S 2009 Origin of olivine in kimberlite: phenocryst or impostor? Lithos 112S, 201–12

[4] Moss S, Russell J K, Smith B H S and Brett R C 2010 Olivine crystal size distributions in kimberlite American Mineralogist 95, 527

[5] Scott Smith B H, Nowicki T E, Russell J K, Webb K J, Mitchell R H, Hetman C M, Harder M, Skinner E M W and Robey J A 2013 Kimberlite terminology and classification. In: Pearson D G, Grütter S H, Harris W J, Kjarsgaard A B, O'Brien H, Rao C N V, Sparks S (Eds.), Proceedings of 10th International Kimberlite Conference: Volume 2. Springer India, New Delhi, 1–17Central Ground Water Board 2017 Ground water year book of Chhattisgarh 2016-2017 Ministry of Water Resources, River Development & Ganga Rejuvenation Government of India

[6] Clement C R, Skinner E M W and Scott B H 1977 Kimberlite redefined. 2nd International Kimberlite Conference, Santa Fe, NM, Expanded Abstracts

[7] Skinner E M W and Clement C R 1979 Mineralogical classification of southern African kimberlites. In: Boyd F R, Meyer H O A (Eds.), Proceedings of 2nd International Kimberlite Conference, Washington D.C. AGU, 129–139

[8] Giuliani A 2018 Insights into kimberlite petrogenesis and mantle metasomatism from a review of the compositional zoning of olivine in kimberlites worldwide. Lithos 312–313, pp. 322–342

[9] Mitchell R H, Giuliani A and O'Brien H 2019 What is kimberlite? Petrology and mineralogy of hypabyssal kimberlites Elements 15, 381-386

[10] Bussweiler Y, Foley S F, Prelević D and Jacob D E 2015 The olivine macrocryst problem: new insights from minor and trace element compositions of olivine from Lac de Gras kimberlites, Canada Lithos 220–223, 238–252

[11] Sazonova L V, Nosova A A, Kargin A V, Borisovskiy S E, Tretyachenko V V, Abazova Z M and Griban' Y G 2015 Olivine from the Pionerskaya and V. Grib kimberlite pipes, Arkhangelsk diamond province, Russia: types, composition, and origin. Petrology 23, 227–258

[12] Sobolev N V, Sobolev A V, Tomilenko A A, Kovyazin S V, Batanova V G and Kuzmin D V 2015 Paragenesis and complex zoning of olivine macrocrysts from unaltered kimberlite of the Udachnaya-East pipe, Yakutia: relationship with the kimberlite formation conditions and evolution. Russian Geology and Geophysics 56, 260–279.

[13] Mitchell R H 2008 Petrology of hypabyssal kimberlites: relevance to primary magma compositions Journal of Volcanology and Geothermal Research 174, 1–8

[14] Russell J K, Porritt L A, Lavallee Y and Dingwell D B 2012 Kimberlite ascent by assimilation-fuelled buoyancy. Nature 481, 352–356.

[15] Soltys A, Giuliani A, Phillips D, Kamenetsky V S, Maas R, Woodhead J and Rodemann T 2016 In-situ assimilation of mantle minerals by kimberlitic magmas — direct evidence from a garnet wehrlite xenolith entrained in the Bultfontein kimberlite (Kimberley, South Africa). Lithos 256–257, 182–196

[16] Stone R S and Luth R W 2016 Orthopyroxene survival in deep carbonatite melts: implications for kimberlites. Contributions to Mineralogy and Petrology 171, 1–9

[17] Sharygin I S, Litasov K D, Shatskiy A, Safonov O G, Golovin A V, Ohtani E and Pokhilenko N P 2017 Experimental constraints on orthopyroxene dissolution in alkali-carbonate melts in the lithospheric mantle: implications for kimberlite melt composition and magma ascent. Chemical Geology 455, 44–56

[18] Tappe S, Pearson D G, Nowell G, Nielsen T, Milstead P and Muehlenbachs K 2011 A fresh isotopic look at Greenland kimberlites: Cratonic mantle lithosphere imprint on deep source signal. Earth and Planetary Science Letters 305, 235–248

[19] Giuliani A, Kamenetsky V S, Kendrick M A, Phillips D, Wyatt B A and Maas R 2013 Oxide, sulphide and carbonate minerals in a mantle polymict breccia: Metasomatism by proto-kimberlite magmas, and relationship to the kimberlite megacrystic suite. Chemical Geology 353, 4 –18.

[20] Woodhead J, Hergt J, Giuliani A, Phillips D and Maas R 2017 Tracking continental-scale modification of the Earth's mantle using zircon megacrysts. Geochemical Perspectives Letters 4 1–6

[21] Ionov D A, Doucet L S, Pogge von Strandmann P A E, Golovin A V and Korsakov A V 2017 Links between deformation, chemical enrichments and Li-isotope compositions in the lithospheric mantle of the central Siberian craton. Chemical Geology 475 105–121

[22] Kargin A V, Sazonova L V, Nosova A A, Pervov V A, Minevrina E V, Khvostikov V A and Burmii Z P 2017 Sheared peridotite xenolith from the V. Grib kimberlite pipe, Arkhangelsk Diamond Province, Russia: texture, composition, and origin. Geoscience Frontiers 8 653–669

[23] Howarth G H and Taylor L A 2016 Multi-stage kimberlite evolution tracked in zoned olivine from the Benfontein sill, South Africa. Lithos 262 384–397

[24] Arndt N T, Guitreau M, Boullier A M, Le Roex A, Tommasi A, Cordier P and Sobolev A 2010 Olivine and the origin of kimberlite. J. Petrol. 51 573–602

[25] Griffin W L, O'Reilly S Y, Natapov L M and Ryan C G 2003 The evolution of lithospheric mantle beneath the Kalahari Craton and its margins. Lithos 71 215–241

[26] Griffin W L, O'Reilly S Y, Doyle B J, Pearson N J, Coopersmith H, Kivi K, Malkovets V and Pokhilenko N 2004 Lithosphere mapping beneath the North American plate. Lithos 77 873–922

[27] Bussweiler Y, Stone RS, Pearson D G, Luth R W, Stachel T, Kjarsgaard B A and Menzies A 2016 The evolution of calcite-bearing kimberlites by melt-rock reaction: evidence from polymineralic inclusions within clinopyroxene and garnet megacrysts from Lac de Gras kimberlites, Canada. Contributions to Mineralogy and Petrology 171 1–25

[28] Kavanagh J L and Sparks R S J 2009. Temperature changes in ascending kimberlite magma. Earth Planet. Sci. Lett. 286 (3) 404–413

[29] Sparks R S J 2013 Kimberlite volcanism. Annu. Rev. Earth Planet. Sci. 41 497–528

[30] Moussallam Y, Morizet Y, Massuyeau M, Laumonier M and Gaillard F 2014 CO2 solubility in kimberlite melts. Chem. Geol

[31] Brett R C, Russell J K, Andrews G D M and Jones T J 2015 The ascent of kimberlite: insights from olivine. Earth and Planetary Science Letters 424 119–131

[32] Jones T J, Russell J K, Porritt L A and Brown R J 2014 Morphology and surface features of olivine in kimberlite: implications for ascent processes. Solid Earth 5 313–326

Landuse/landcover monitoring and spatiotemporal modelling using multilayer perceptron and 'multilayer perceptron'-Markov Chain ensemble models: A case study of Dausa City, Rajasthan

Sangeeta Soni[1], Harvir Singh[1], Jameel Qurashi[1], Mirza Shuja[2], Manish Pandey[3], Aman Arora[3,4*]

[1]Jaipur National University, Jaipur-Agra Bypass, near New RTO office, Jagatpura, Jaipur, Rajasthan 302017, India.
[2]Lovely Professional University, Jalandhar, Punjab
[3]University Center for Research & Development (UCRD), Chandigarh University, Mohali-140413, Punjab, India
[4]Bihar Mausam Sewa Kendra, Planning and Development Department, Govt. of Bihar, Patna-800015

Email id: aman.jmi01@gmail.com

Abstract. The present work is an attempt to the LULC classification, monitoring, and spatiotemporal prediction using Artificial Neural Network - Multi-Layer Perceptron (MLP) and MLP-Markov Chain (MC) models. Dausa city and its surroundings of Rajasthan, India has been selected for this study for several reasons including arid climatic setting being a sensitive precursor to the climate change scenarios and the huge population pressure experienced by the area. The MLP based supervised classification for two periods 2001 and 2018 have been analyzed using Landsat 7 Thermal Mapper (TM) and Landsat 8 OLI satellite images. The images were classified into six LULC categories viz. Built-up (Settlements), Cultivated Lands (Agricultural/Cropland), Water Body, Uncultivated/Fallow Lands, Barren Lands, and Forest/Vegetation cover. The accuracy assessment for both classified images was performed using confusion matrix led Kappa Coefficient (K) technique. Reasonable accuracies, K=0.82 (2001) & K = 0.91 (2018), have been achieved for datasets selected for both periods of time. The MLP-MC model based spatiotemporal LULC prediction for the year 2045, using the trends in the classified LULC results for the period 2001-2018, prophecies that the 'built-up land' would increase to reach 76.10 km2 (67.60% increase) in 2045 with the reference year 2001 whereas the increase in this class of LULC would only be 39.34% during the period 2018-2045. The 'cultivated land' (2001-2045: -83.86%; 2018-2045: -65.20%), 'barren land', (2001-2045: -54.70%; 2018-2045: -4.86%), 'water body' (2001-2045: -96.43%; 2018-2045: -84.42%), and 'forest/vegetation' (2001-2045: -81.94%; 2018-2045: -20.59%), categories would experience continuous areal decline over this period, though some at faster pace and other at comparatively lower rate. The projected unprecedented exponential increase in 'follow land/uncultivated land' (2001-2045: +372.45%; 2018-2045: +6.39%) presents worrisome future picture of this ecologically sensitive and fragile region. The results of this study indicate and warrant intensive management and policy, and local level participation of communities to help maintain the deteriorating ecological balance in this ecologically sensitive arid ecosystem with fragile agricultural and natural vegetation traits.

Keywords: Land use land cover, Artificial Neural Network, Markov Chain, Remote Sensing & GIS, Dausa City, Rajasthan

1. Introduction

The relationship of landuse/landcover (LULC) change with climate change, ecological function, soil degradation, biodiversity, etc. is well established by now. Still ongoing researches and sizeable amount of research articles on this theme prove the need of a better refinement of understanding of LULC and other components of the physical and human spheres. The International Geosphere-Biosphere Programme (IGBP) run under the aegis of International Council for Science (ICS) recommended the monitoring and evaluating the LULC change and its pattern at a global, national, as well as local scales with the help of remote sensed (RS) satellite imageries and geographic information system (GIS) tools. Land-use change affects natural biogeochemical cycles at global, regional, and national scales [1]–[4]. It impacts are vividly noticeable in the domains of water resources, green-house processes, atmospheric processes, biodiversity, hydrological cycle, carbon cycle, and ecology [5]–[9]. Also, at regional scale, the LULC changes affects micro-climatic processes, water catchments/watersheds, surface water runoff, and groundwater levels [10]. The land-degradation and land subsidence processes have also been directly related tied to, as one of the important controlling factors, LULC. The study of landuse dynamics, therefore, has been found to be very crucial for protecting the natural way of functioning of different cycles of earth's system by monitoring and continuous evaluation of LULC pattern, both in spatial and temporal contexts.

The studies of LULC trend and pattern related issues, at local and/or regional, have increasingly been appearing in the research sphere from the last two to three decades [11]–[16]. These research, covering different aspects, in conjunction with its relationship with other domains of four spheres of our planet, at country/global scale have also been carried out by many international agencies i.e. International Geosphere–Biosphere Programme (IGBP), European Space Agency (ESA) Climate Change Initiative (CCI) Land Cover (LC) project, Global Human Settlement Layer (GHSL), United States Geological Survey (USGS), National Remote Sensing Centre (NRSC), etc. But the need to quantify LULC monitoring and their spatiotemporal prediction at finer scale have been emphasized by time to time by these agencies and similar such studies.

To provide more accurate LULC classification maps, different statistical models, in terms of structure and application, have been developed during last 2-3 decades to understand LUCC dynamics. These models range in characteristics from simple statistical models, Maximum likelihood classifier (MLC) based classifications [17], multi-criteria decision-making analysis algorithms i.e. Analytical Hierarchy Processes (AHP) based analysis [18], Fuzzy-AHP based study [19] to advanced machine learning models i.e. Support Vector Machine (SVM) based LULC studies [20]–[24], Bayesian algorithms based researchers [25], [26], applications of Artificial Neural Network (ANN) in LULC classifications [15], [27], [28], etc. All these models have their own set of advantages and disadvantages [29], [30]

Apart of using different models for LULC classification and comparing the accuracy several studies recently found in the interest of projecting the LULC based on earlier maps of two or more than two time periods. The LULC forecast is performed and analysed using the historical remotely sensed satellite images and in-field survey collected samples in Geographic information system (GIS) environment. The LULC patterns are the final mosaic of different landscape elements pertaining to hydrological, climatological and man-made features, which, to a certain extent, determine the forecast of the LULC dependency [30]–[32]. Many studies have been carried out to test the model accuracy performances using different set of simple as well as advanced simulation models [33], [34]. A range of simulation models-based studies have been published for LULC forecast. Some of them are Markov Chain model (MCM) [35], Cellular Automata (CA)- Markov Chain (CA-Markov) model [36], Multi-Layer Perceptron Markov Chain (MLP-MC) model [37], [38], multi-actor-based land use modelling (MAS) model [39], multi-criteria evaluation (MCE)-CA model [40], GeoCA-Urban model [41], and SLEUTH model [42]. Among these advanced simulation models, the MLP-MC model is highly recommended for its better performance and accuracy [38].

Inspiring from these studies and after a thorough literature review it has been found that very limited studies have been studied on the selected study area for present work. The present study uses MLP model for LULC classification of historical satellite images of Dausa city and its surroundings of Rajasthan state, India for two periods, 2001 and 2018, to forecast the LULC for year 2045 by using MLP-MC model. There has been found lack of any important study on this aspect of LULC classification using machine learning and future LULC scenario forecasting assessment for Dausa city/district. This research aims to fill research gap by providing accurate and reproducible LULC classified maps for said periods and based on the computed trend of LULC, the prediction for 2045

LULC map of Dausa city and its surroundings has been attempted to have information of LULC transition needed for sustainable development planning and habitat conservation practices as well as ecosystem service assessments. There was one limitation found in the study that the due to arid/semi-arid climate characteristics the

2. Study Area

The Dausa city, satellite city, is located about 55 km from Jaipur metropolitan city which is the capital of Rajasthan state. For present research, a highlighted portion covering Dausa city and its surroundings, of Dausa district has been undertaken to analyse the urban growth and its expansion in future. The study area lies between 26° 38' 41.3" N to 27° 8' 17.1" N latitude and 76° 10' 41.7" E to 76° 32' 55.2" E longitude, and covers an area of approximately 1277 km^2 (Figure 1).

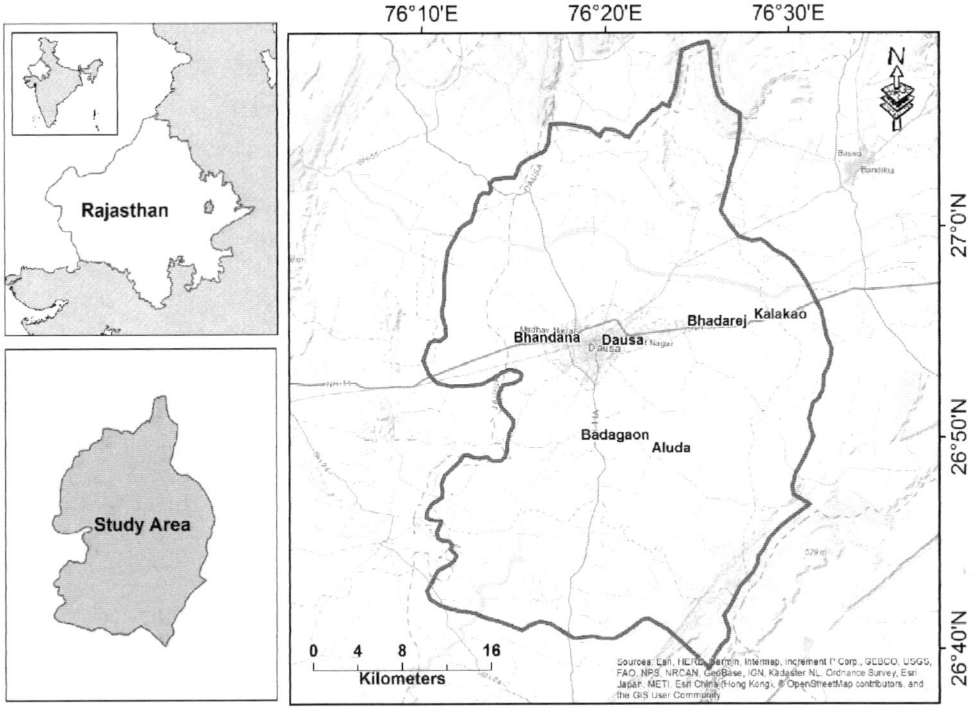

Figure 1. Location map of Dausa city and its surroundings

The population of Dausa city, as per 2011 census of India, was recorded as 85,960. Two national highways, NH-21 & 148, connect the city with major urban hubs like Jaipur, Agra, Mathura, etc. The easy and fast road connectivity, apart from connectivity of the nearby cities with air and train routes, attracts many tourism and other activities leading to inflow of people in the area which finally promotes population expansion and urbanization. The climatic conditions of the study area are characteristically hot semi-arid (Köppen climate classification *Bsh*) with monsoon rainfall during June-July-August months. The arid region records extremely hot summer days, for a very long period, ranges from March to October. The maximum temperature recorded in the region reaches up to ~48°C in the month of May and the average temperature is 32.2°C for a calendar year. The average annual rainfall varies from 50-70 mm. The existing deserts found approximately adjacent to the study area causes the temperature falls drastically during nights, and during winter period (November to February), the average temperature records attain the values in the range from 3°C to 5°C. The scant (and sometimes heavy but for low duration high) rainfall, and high temperature casus very low water availability in river basin throughout the year.

3. Materials and methods

In present study, the remotely sensed satellite Landsat satellite data of two different times has been utilized to prepare the LU/LC maps for the Dausa City and its surroundings. For two different years, 2001 and 2018, the L1TP level (level-1, pre-processed) satellite tiles for Landsat-5 Thematic Mapper (TM) and Landsat-8 Operational Land Imager (OLI) were downloaded from the earth explorer portal, a platform provided by the United States Geological Survey (USGS) to their users to order archived or latest satellite images, cartographic products, and aerial photographs for different purposes of studies and research. The L1TP products are already pre-processed meaning thereby they are radiometrically calibrated, orthorectified with optimum GCPs (ground control points) and correction of relief displacement computed has also been performed during pre-processing, using digital elevation model (DEM) [43]–[46]. The details of satellite data are provided in table 1. The complete methodology flowchart is given in figure 2.

Table 1. Characteristics of the Landsat satellite data in detail used in the study

S. No.	Landsat Image scene ID	Cloud coverage (%)	WRS Path/Row	Acquisition date	Spatial reference
1.	LE07_L1TP_147041_20011103_20170202_01_T1	Scene cloud cover: 0.0 Land cloud cover: 0.0	147/41	03-11-2001	Projection: UTM Zone: 43N Datum: WGS84 Ellipsoid: WGS84
2.	LC08_L1TP_147041_20181126_20181210_01_T1	Scene cloud cover 1.01 Land cloud cover 1.01		26-11-2018	

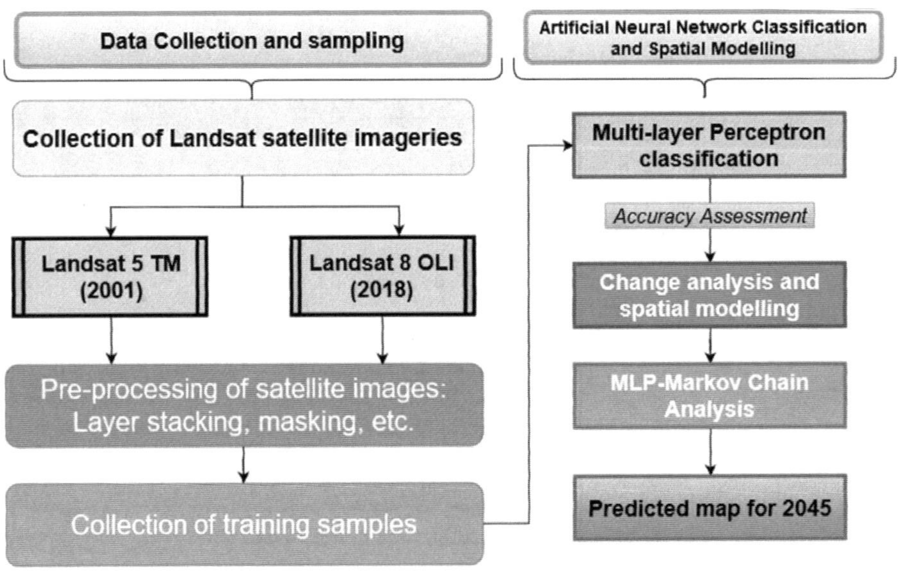

Figure 2. Flowchart of methodology

3.1. Pre-processing of satellite images

The collected imageries were projected in Universal Transverse Mercator (UTM), Zone-43 projection with datum of World Geodetic System (WGS) 1984, developed by Defence Mapping Agency (DMA) as global reference [47] for coherence of spatial data being produced at global scale. Further, the Region of Interest (ROI) has been extracted and false colour composite (FCC) was generated for period, 2001 and 2018, datasets using the procedural exercise of band combination.

3.2. Training Samples

In order to collect consistent training samples from the images, they should be cloud and shadow free or very low share of cloud clusters encourage better quality of sample collections. For both periods, 2001 & 2018, the selection of cloud coverage less than 2% (table no. 1) has been achieved by removing the low-quality sample collection constraint. In addition, before sample collection the two constraints have been followed: absence of cloud or shadow pixel(s) and avoidance of shared boundary classes collection [48]. For collection of samples the following criteria have been followed after comparing both datasets (2001 & 2018 images) and identifications of change pixels from one class of earlier period maps to any other classes in the later period maps [13]: (1) selection of maximum possible classes/places in ROI, (2) inclusion of change pixels in training samples, and (3) minimum-distance rule for avoiding overlap or duplication in collection. Hence, for training of the datasets (images) in the model, the samples were collected in six categories viz. Built-up (Settlements), Cultivated Lands (Agricultural/Cropland), Water Body, Uncultivated/Fallow Lands, Barren Lands, and Forest/Vegetation cover.

3.3. Multi-Layer Perceptron neural network

An artificial neural network (ANN) is an inter-connected network of several processing units which are prepared on basis of human brain's neurons to solve a complex problem. Neural networks, non-linear complex models, help to convert a data, remotely sensed imagery in this case, into a desired output (LULC classified image). Generally, there are six ANN models, popularly used for studies of pattern recognition in images; Carpenter/Grossberg classifier, Hamming network, Kohonen's self-organising feature maps, Hopefield network, single layer perceptron, and the last one is multi-layer perceptron [49]. The most popular among them, the multi-layer perceptron (MLP), described by Rumelhart *et al.*, (1986), is popularly being used in studies involving remote sensing data. As described in figure 3, it consists of minimum three layers which are interconnected to the preceding and subsequent layer and there is no connectivity of nodes, processing elements, with each other in same layer. In the structure of this model, the first layer is called input layer, holds the input or assigned values by user, distributes the values to the next layer for processing. The processed values are saved in output or last layer of the structure and between input & output layers, the hidden layer receives the feeds (values) forwarded by input layer for finding the better output to be transferred to the final output layer. A user can define the number of hidden layers; generally, one hidden layer is sufficient to solve a complex problem but for deep learning exercise or to solving more complex problems, the number may be increased accordingly [51]–[53].The hidden layer nodes are a critical part of MLP structure for functioning; in absence of it, the MLP is unable to learn from the previous records as well as it cannot perform further iterations to solve any problem.

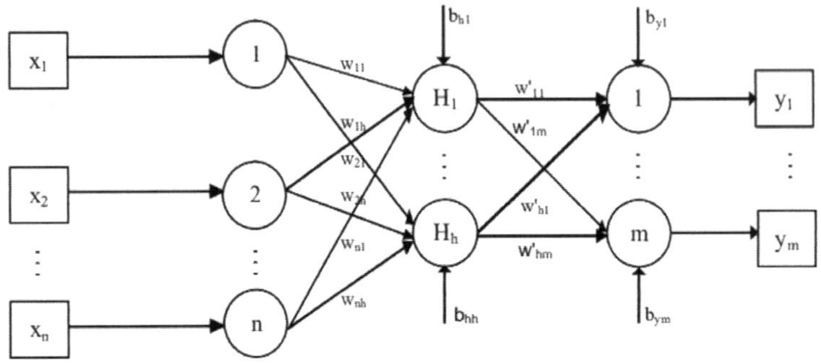

Figure 3. Schematic of MLP Neural Network. Source: (Shafiullah and Abido, 2018)

In order to accomplish the modifications of neuron weights (calculated from assigned values), the MLP functionalities involve forward and backward propagation. During training of the dataset, each sample is fed into the input layer and the receiving node calculates the weighted signals from all nodes to which it is connected in the preceding layer. Formally, the input layer that a single node receives is weighted according to the equation:

$$net_j = \sum_{i=1}^{m} w_{ij}\, O_i \tag{1}$$

Where, w_{ij} represents the weight between nodes i & j. O_i refers output from node i. The output from a given node j is computed by:

$$O_j = f(net_j) \tag{2}$$

In the function of MLP, the number of hidden layer nodes can be calculated as follows:

$$N_h = INT(\sqrt{N_i \times N_o}) \tag{3}$$

Where, N_h, N_i and N_o are the number of hidden, input and output layer nodes respectively.

3.4. LULC classification and accuracy assessment

The LULC classification for both periods 2001 & 2018 were carried out using supervised-MLP learning method by using collected training samples. Out of total collected samples, 70% were apportioned for training and the rest 30% was used for testing of the MLP model. For each class, 1000 samples were chosen with 10,000 iterations for better performance of the model. The complete parameter and performance of both period classifications is provided in the table.

Accuracy assessment was also performed for both classified maps through confusion matrix table derived kappa coefficient. The model performance has been assessed from two perspectives viz. 1) how well the model has performed in actual and how well it could performed by chance can be calculated through Cohen's Kappa [54]. The Kappa coefficient is calculated through following formula;

$$K = \frac{N\sum_{i=1}^{n} m_{i,i} - \sum_{i=1}^{n}(G_i C_i)}{N^2 - \sum_{i=1}^{n}(G_i C_i)} \tag{4}$$

Where, i denotes the class number, N refers the total number of obtained values through classification in comparison to the true values, $m_{i,i}$ is the obtained values belong to the truth class i, also been classified as class i, C_i & G_i refer total number of predicted and observed values in class i.

In addition, the overall accuracy, producer's accuracy and consumer's accuracy is also calculated from confusion matrices by using following methods;

$$Overall\ Accuracy = \frac{Total\ number\ of\ correctly\ classified\ pixels}{Total\ number\ of\ reference\ pixels} \quad (5)$$

Where, total number of reference pixels refers the sum of the obtained diagonal pixel counts in confusion matrix for each class.

$$Producer's\ Accuracy = \frac{Correct\ classified\ pixel\ counts\ within\ a\ class}{Total\ number\ of\ reference\ pixels\ of\ that\ class} \quad (6)$$

Where, the total number of reference pixels of that class refers the sum of the column for particular category in the confusion matrix table.

$$Consumer's\ Accuracy = \frac{Correct\ classified\ pixel\ counts\ within\ a\ class}{Total\ number\ of\ classified\ pixels\ in\ that\ class} \quad (7)$$

Here, the total number of classified pixels in that class refers the sum of the row for particular category in the confusion matrix table.

The error assessment of MLP model has been computed by the root mean square error (RMSE) test. The RMSE method is considered as standard statistical metric to assess a model's performance in various fields of Science, Technology, Applied sciences, etc. [55]. The RMSE used in this study was aimed to exercise a comparison between the predicted and observed output values. The equation of RMSE can represented mathematically as follows:

$$RMSE = \sqrt{\frac{\sum_{i=1}^{n}(P_i - O_i)^2}{n}} \quad (8)$$

Where P and O refer to predicted and observed values respectively. 'n' is the total number of input cases.

Further, the LULC change map was prepared and the derived LULC change map used to calculate selected LULC transitions and then the LULC transitions were utilized transition potential modelling for prediction of LULC by using the Markov Chain analysis.

3.5. Markov Chain Analysis for Prediction of LULC

Markov chain (MC) analysis, a stochastic process, is being widely used for spatial modelling and for prediction of LULC based on trends in the change of historical LULC pattern [14], [35], [49], [56]. The MC analysis for predication of LULC for a region uses different sets of matrices, represent multi-directional LULC changes between all LULC categories [35] and provides the probability matrix of transitions from one land category to another category. The transition probability matrix defines change trends from past to present and into the future probabilities of for a certain class type of LULC for each grid [49]. Correspond to the potential driver(s) and taken into consideration of different sets of variables, the probability matrix can be defined as a set of conditional probabilities for the transition of LULC from one state to a new state. The MC equation, adopted from Muller and Middleton (1994) for predictability of LULC distributions at the beginning (M) and at the end (M_{t+1}) can be expressed mathematically as follows:

$$MC_{LC} * M_t = M_{t+1} \quad (9)$$

Where, MC_{LC} = transition matrix which represents the LULC changes occurred during the period. M_t = the probability of any given point being classified as any category of LULC at time *t*.

The MC transformation matrices are used to predict the future LULC, but the MC based analysis only provides inadequate changes of spatial distribution whereas, the ensemble of MLP-MC based model provides higher accuracy level than MC based analysis [49]. Hence, the predicted map was prepared by using the ensemble of MLP-MC based model.

Parameters and performances of classified images are provided in Table 2.

Table 2. Parameters and Performance of classified images

Parameters	2001	2018
Input layer neurons	3 (bands)	3 (bands)
Hidden layers	1	1
Hidden layer 1 neurons	4	4
Output layer neurons	6	6
Samples per class	1000	1000
Final learning rate	0.0100	0.0100
Momentum factor	0.5	0.5
Iterations	10000	10000
Training RMS	0.2161	0.1421
Testing RMS	0.2243	0.1315

4. Results & Discussion

The result section has been described in three parts: LULC change analysis & accuracy assessment; LULC transitions, selection of LULC transitions and modelling, and prediction of future map using MLP-MC based model.

4.1. LULC Change Analysis & Accuracy Assessment

From the LULC change analysis, it has been observed that the water body in the area has reduced from 3.37 km^2 to 0.06 km^2. The built-up lands increased 22.84 km^2 from 112.57 km^2 in 2001 to 135.41 km^2 in 2018. Moreover, the forest cover has been drastically decreased from 330.58 km^2 to 75.19 km^2 which may have occurred due to the expansion of urbanization led deforestation in the area. The cultivated or croplands have also shrank by -53.62% from 2001 to 2018 but the uncultivated or fallow lands have observed exponential increase. It has increased from 169.89 km^2 to 754.44 km^2 in 2018. This indicates shift in the practice of agriculture during the 18 years and possibly because the cultivation of the lands could not take place in during the later years. The complete LULC statistics for both periods are given in table 3.

The LULC map for 2001 and 2018 are represented in the figure 4 & 5 respectively. By comparing both figures, it can be understood that the urbanization has been spread more in the central part of Dausa city than in its surrounding places.

Table 3. Area and area change statistics for 2001 & 2018 LU/LC

LU/LC Classes	2001		2018		2001-2018	
	Area (km^2)	Area (%)	Area (km^2)	Area (%)	Area (km^2)	Area Change (%)
Water Body	3.37	0.26	0.77	0.06	-2.60	-77.12
Cultivated Lands	250.78	19.63	116.31	9.10	-134.47	-53.62
Built-up Lands	112.57	8.81	135.41	10.60	22.84	20.29
Uncultivated/Fallow Lands	169.89	13.30	754.44	59.06	584.55	344.08
Barren Lands	410.29	32.12	195.35	15.29	-214.94	-52.39
Forest/Vegetation Cover	330.58	25.88	75.19	5.89	-255.38	-77.25

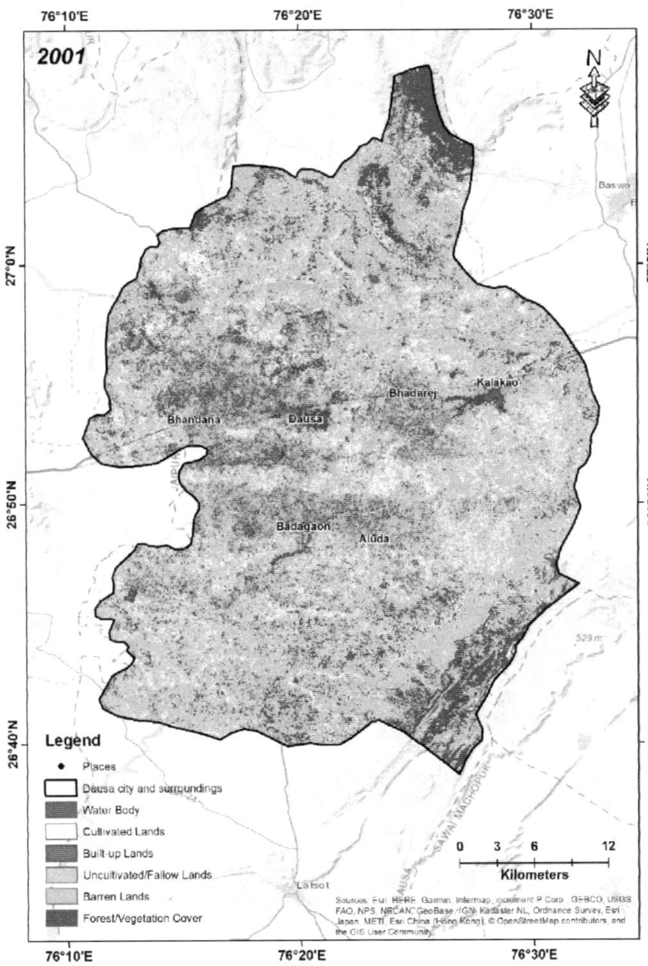

Figure 4. Land use land cover for 2001

Total gains and losses for each landuse class are demonstrated in figure 6. The exponential gain, as compared to meagre loss, is clearly visible in uncultivated/fallow lands and built-up lands during the study period. On the other hand, the maximum loss is recorded in barren lands, more than 30%, and less than 10% gain is recoded. The cultivated lands gained about 10% but lost ~15% in 2018. Water body is reduced to negligible amount from 2001 to 2018.

In the figure 7, the LULC change from one class to another has been presented. Total forty-two types of changes from one class to another were recorded. Among them, the more important changes from any of the classes to built-up land are displayed along with figure 7. The major change has been observed from uncultivated lands to built-up lands wherein ~ 38 km^2 area has been converted into built-up land. Another important change of approximately 20 km^2 and 19 km^2 was recorded from barren lands and cultivated lands into built-up respectively.

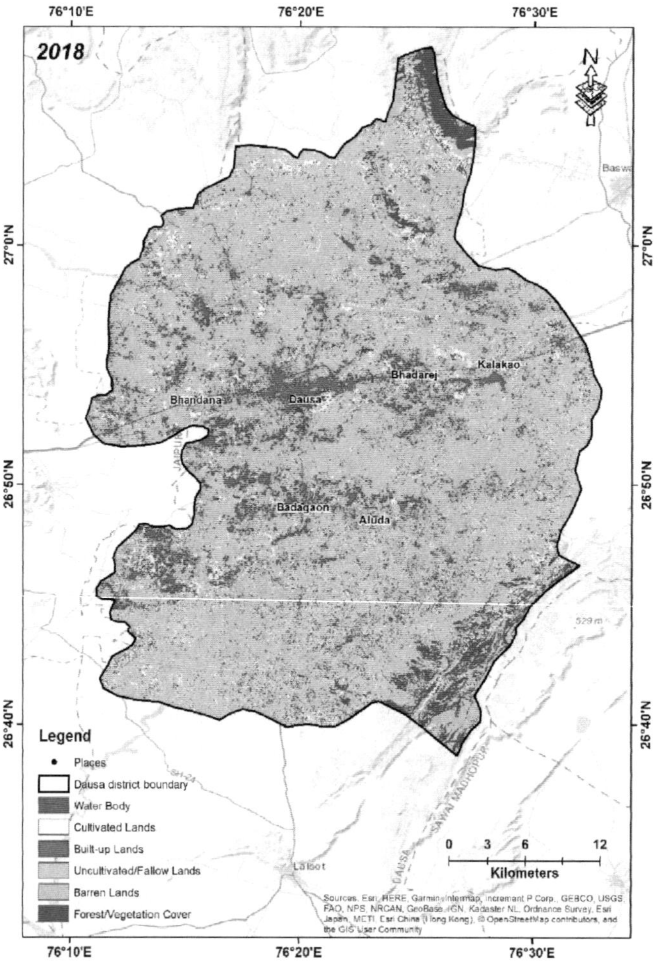

Figure 5. Land use land cover for 2018

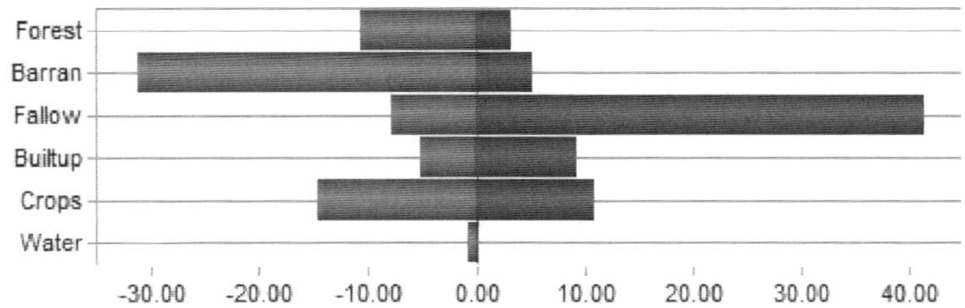

Figure 6. Gain and losses between 2001 and 2018 (in %)

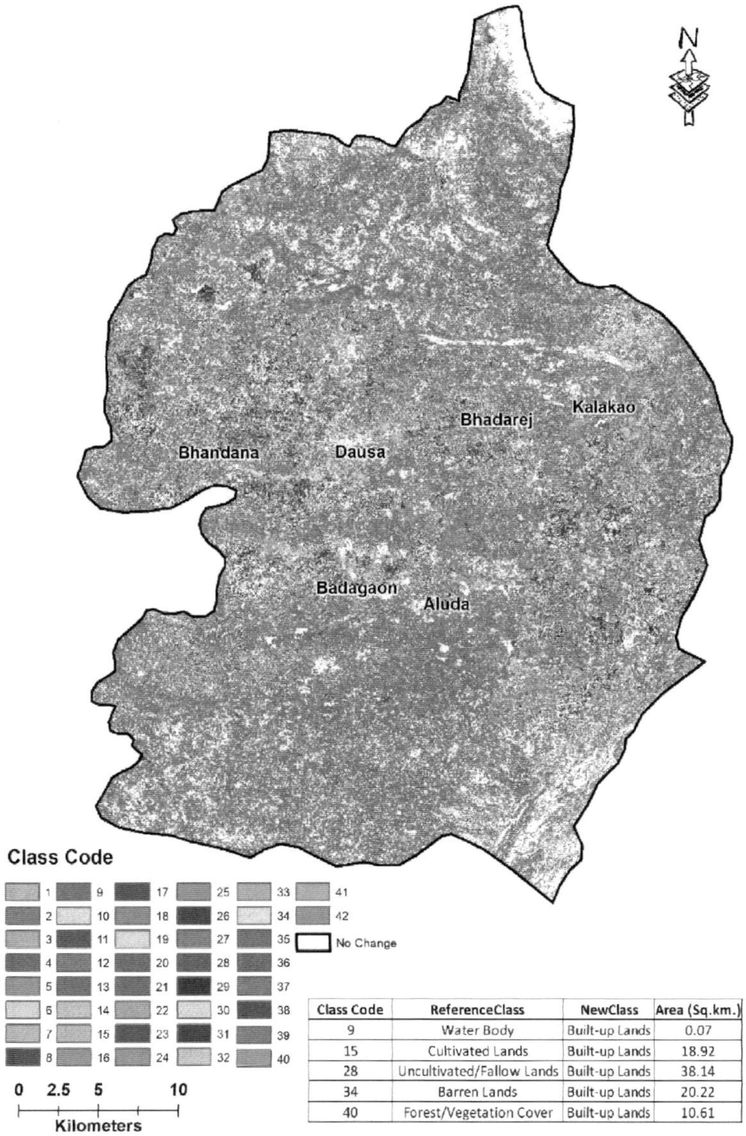

Class Code	ReferenceClass	NewClass	Area (Sq.km.)
9	Water Body	Built-up Lands	0.07
15	Cultivated Lands	Built-up Lands	18.92
28	Uncultivated/Fallow Lands	Built-up Lands	38.14
34	Barren Lands	Built-up Lands	20.22
40	Forest/Vegetation Cover	Built-up Lands	10.61

Figure 7. Land use land cover change from 2001 to 2018

The calculated results from confusion matrices for both periods, 2001 & 2018, are presented in the table 4. In the table of accuracy assessment for classified images, the following calculated parameters are presented: Consumer's Accuracy (CA), Producer's Accuracy (CA), Overall Accuracy and Kappa Statistics (KIA).

Table 4. Accuracy assessment for LULC classification

LU/LC Classes	2001		2018	
	CA (%)	PA (%)	CA (%)	PA (%)
Water Body	93.33	84.85	100.00	80.00
Cultivated Lands	81.03	62.67	88.89	80.00
Built-up Lands	88.24	85.71	93.18	100.00
Uncultivated/Fallow Lands	74.72	85.81	90.54	97.81
Barren Lands	80.71	86.92	86.49	84.21
Forest/Vegetation Cover	97.30	84.71	99.31	92.86
Overall Accuracy	83.83		91.11	
Kappa Statistics (KIA)	0.82		0.91	

The maximum CA is recorded for Forest/Vegetation Cover (97.30%) followed by Water Body (93.33%), Built-up lands (88.24%) and so on for the 2001 LULC image. The maximum PA is observed for Barren lands (86.92%) followed by Uncultivated/Fallow Lands (85.81%), Built-up lands (85.71%), etc. On the other hand, due to better quality of image and sample collection, the higher accuracy is recorded for 2018 dataset. The highest CA is recorded to be 100% for water body followed by Forest/Vegetation Cover (99.31%) and Built-up Lands (93.18%). In terms of PA, the 100% accuracy is recorded for built-up lands followed by Uncultivated/Fallow lands (97.81%) and Forest/Vegetation Cover (92.86%). The overall accuracy is found 83.83% and 91.11% for 2001 and 2018 images respectively. The reasonable KIA = 0.82 is also recorded for 2001 data and K=0.91 found for 2018 image.

4.2. Selection of LULC transitions and modelling

LULC transitions from all the six classes to other classes, between 2001 and 2018, were analysed in this study. Through the MLP method, the transition potential modelling has been conducted with considerable accuracy [37]. The evidential likelihood dataset, as a driving force, has been taken into consideration while calculating the conducting the transition potential modelling. Finally, the major and important transformations from one class to another have been taken into consideration for further sub-modelling for prediction. The following seven class-transitions have been recorded: Barren to Built-up lands, Cultivated to Barren lands, Cultivated to Built-up lands, Uncultivated/Fallow to Built-up lands, Forest/Vegetation Cover to Barren lands, Forest/Vegetation Cover to Built-up lands, and transition of water body to Built-up lands. The computed transition probability matrix, from 2001 to 2018, to predict the LULC for 2045, through MLP-MC analysis, is given in table 5.

Table 5. Transition probabilities matrix from 2001 to 2018

	Water	Crops	Built-up	Fallow	Barren	Forest
Water	0.017	0.132	0.167	0.440	0.073	0.172
Crops	0.026	0.120	0.120	0.612	0.075	0.048
Built-up	0.016	0.151	0.163	0.511	0.103	0.056
Fallow	0.026	0.111	0.121	0.608	0.090	0.045
Barren	0.014	0.111	0.118	0.616	0.098	0.044
Forest	0.014	0.129	0.118	0.499	0.055	0.185

4.3. Prediction using MLP-MC based model

The MLP-MC model based predicted LULC map for 2045 of the Dausa city and its surroundings is presented in figure 8.

Table 6. Predicted LU/LC area and area change statistics

LU/LC Classes	2045		2018-2045		2001-2045	
	Area (km^2)	Area (%)	Area (km^2)	Area Change (%)	Area (km^2)	% Change
Water Body	0.12	0.01	-0.65	-84.42	-3.25	-96.43
Cultivated Lands	40.48	3.17	-75.84	-65.20	-210.31	-83.86
Built-up Lands	188.67	14.77	53.26	39.34	76.10	67.60
Uncultivated/Fallow Lands	802.64	62.83	48.20	6.39	632.75	372.45
Barren Lands	185.85	14.55	-9.50	-4.86	-224.44	-54.70
Forest/Vegetation Cover	59.71	4.67	-15.48	-20.59	-270.86	-81.94

From the map, it is observed that the urban sprawl is spreading more towards northern part than southern counterpart. Also, the forest cover and cultivated lands are visible in patches. From the pixel counts, it has been calculated that, in future, the built-up lands would increase by approximately 53.26 km^2 from 2018-2045 and would cover ~188.67 km^2 of land in 2045. The cultivated lands will diminish to 40.48 km^2 in 2045 with a loss of 75.84 km^2 from 2018. Barren lands will also reduce to 185.85 km^2 with loss of 9.50 km^2. The complete change matrix is given in table 6.

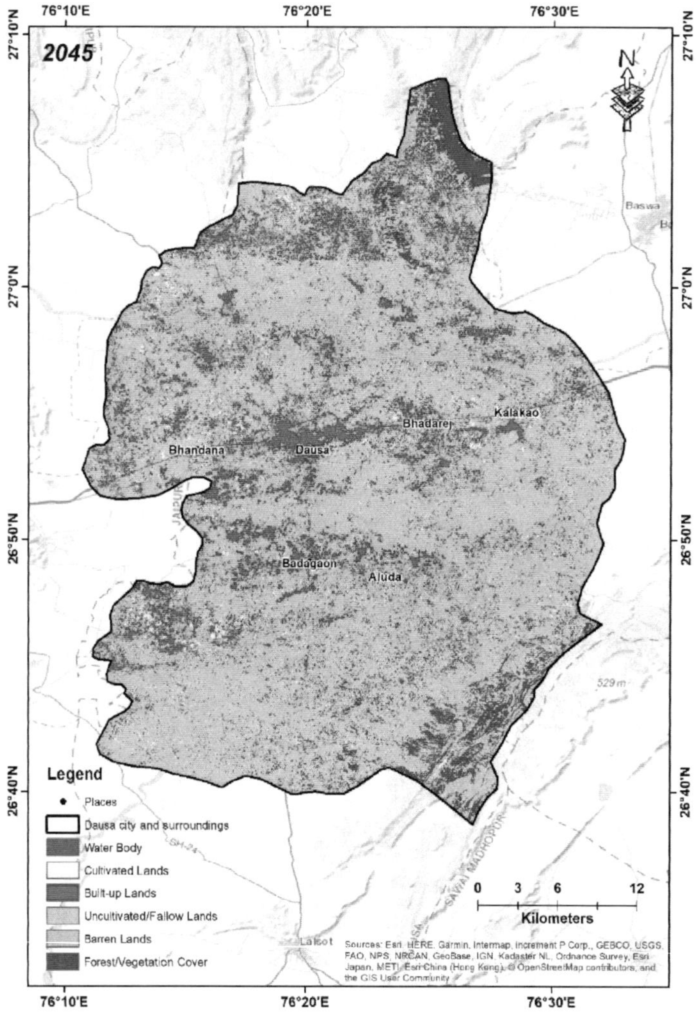

Figure 8. Predicted land use land cover for 2045

5. Conclusion

The LULC classification accuracy depends on the resolution and image quality of the input satellite images [57]–[59]. More advanced satellites images having greater quality than earlier satellite images, therefore, provide better accuracy in LULC classification. The higher resolution, of course, of a satellite or aerial photograph would be helpful for a user to collect more accurate samples and prepare LULC map with considerable accuracy. Moreover, the prediction is directly dependent on the quality of classification of earlier set of images [60]. The higher resolution and high accuracy of images will help to increase the prediction and hence more accurate future maps. Also, the success of a model is mostly dependent on the accurate LULC classification [49] and thereafter the transition classes could be used for prediction.

This study utilized MLP model to produce LULC maps for 2001 and 2018 and simulated LULC for 2045 using MLP-MC analysis. For the purpose, Landsat 5 TM & 8 OLI satellite images with 30-meter resolution for two different periods 2001 & 2018 were used for supervised MLP classifications. Sufficient numbers of samples were collected in each LULC class for both images and then the MLP model was applied for LULC classification. Due to low quality images and problem to clearly identify an entity in the image at the time of sample collection, the accuracy for the 2001 LULC map was found to be less than 90%; the same technique was applied on 2018 Landsat image but the identification of LULC classes at the time of sample collection was found to be more easy due to high quality of images, although the resolution for both datasets is the 30-meter, the accuracy for 2018 image was found to be more than 90%. Using both images, the LULC change map was also prepared. Afterwards, seven major transitions derived from LULC change analysis between 2001 and 2018 maps were taken as ingredient for prediction of LULC of 2045. The MLP-MC method was used to simulate the LULC for 2045. From the results, a major transformation in settlement (built-up) and cultivated lands is observed. The maximum changes are observed in the central and northern parts of the city. Due to high rate of urbanization the settlement would be also expanding in the outer fringes of the city. The present work can be helpful for the urban planners at local and national level to prepare future layout of the city towards sustainable development.

6. References

[1] R. A. Houghton and A. A. Nassikas, "Global and regional fluxes of carbon from land use and land cover change 1850-2015," *Global Biogeochem. Cycles*, vol. 31, no. 3, pp. 456–472, Mar. 2017, doi: 10.1002/2016GB005546.

[2] X.-P. Song *et al.*, "Global land change from 1982 to 2016," *Nature*, vol. 560, no. 7720, pp. 639–643, Aug. 2018, doi: 10.1038/s41586-018-0411-9.

[3] R. V. Pouyat, D. E. Pataki, K. T. Belt, P. M. Groffman, J. Hom, and L. E. Band, "Effects of Urban Land-Use Change on Biogeochemical Cycles," in *Terrestrial Ecosystems in a Changing World*, Berlin, Heidelberg: Springer Berlin Heidelberg, 2007, pp. 45–58.

[4] R. S. DeFries, C. B. Field, I. Fung, G. J. Collatz, and L. Bounoua, "Combining satellite data and biogeochemical models to estimate global effects of human-induced land cover change on carbon emissions and primary productivity," *Global Biogeochem. Cycles*, vol. 13, no. 3, pp. 803–815, Sep. 1999, doi: 10.1029/1999GB900037.

[5] H. Memarian, S. Kumar Balasundram, J. Bin Talib, C. Teh Boon Sung, A. Mohd Sood, and K. Abbaspour, "Validation of CA-Markov for Simulation of Land Use and Cover Change in the Langat Basin, Malaysia," *J. Geogr. Inf. Syst.*, vol. 04, no. 06, pp. 542–554, 2012, doi: 10.4236/jgis.2012.46059.

[6] P. H. VERBURG, W. SOEPBOER, A. VELDKAMP, R. LIMPIADA, V. ESPALDON, and S. S. A. MASTURA, "Modeling the Spatial Dynamics of Regional Land Use: The CLUE-S Model," *Environ. Manage.*, vol. 30, no. 3, pp. 391–405, 2002, doi: 10.1007/s00267-002-2630-x.

[7] C. Kamusoko, M. Aniya, B. Adi, and M. Manjoro, "Rural sustainability under threat in Zimbabwe – Simulation of future land use/cover changes in the Bindura district based on the Markov-cellular automata model," *Appl. Geogr.*, vol. 29, no. 3, pp. 435–447, 2009, doi: https://doi.org/10.1016/j.apgeog.2008.10.002.

[8] A. J. Hansen *et al.*, "Global change in forests: responses of species, communities, and biomes: interactions between climate change and land use are projected to cause large shifts in biodiversity," *Bioscience*, vol. 51, no. 9, pp. 765–779, 2001.

[9] V. Mishra, P. Rai, and K. Mohan, "Prediction of land use changes based on land change modeler (LCM) using remote sensing: A case study of Muzaffarpur (Bihar), India," *J. Geogr. Inst. Jovan Cvijic, SASA*, vol. 64, no. 1, pp. 111–127, 2014, doi: 10.2298/IJGI1401111M.

[10] W. Bewket and S. Abebe, "Land-use and land-cover change and its environmental implications in a tropical highland watershed, Ethiopia," *Int. J. Environ. Stud.*, vol. 70, no. 1, pp. 126–139, 2013, doi: 10.1080/00207233.2012.755765.

[11] Z. Dafang, "Research Activities on L a N D - U S E / C O V E R Change in the Past Ten Years in China Using Space Technology < >," vol. 9, no. 4, pp. 330–334, 1999.

[12] H. Fathizad, N. Rostami, and M. Faramarzi, "Detection and prediction of land cover changes using Markov chain model in semi-arid rangeland in western Iran," *Environ. Monit. Assess.*, vol. 187, no. 10, 2015, doi: 10.1007/s10661-015-4805-y.

[13] J. Chen, P. Gong, C. He, R. Pu, and P. Shi, "Land-use/land-cover change detection using improved change-vector analysis," *Photogramm. Eng. Remote Sensing*, vol. 69, no. 4, pp. 369–379, 2003, doi: 10.14358/PERS.69.4.369.

[14] D. G. Brown, B. C. Pijanowski, and J. Duh, "Modeling the relationships between land use and land cover on private lands in the Upper Midwest, USA," *J. Environ. Manage.*, vol. 59, no. 4, pp. 247–263, 2000.

[15] H. Lyu, H. Lu, and L. Mou, "Learning a transferable change rule from a recurrent neural network for land cover change detection," *Remote Sens.*, vol. 8, no. 6, pp. 1–22, 2016, doi: 10.3390/rs8060506.

[16] G. J. Scott, M. R. England, W. A. Starms, R. A. Marcum, and C. H. Davis, "Training Deep Convolutional Neural Networks for Land-Cover Classification of High-Resolution Imagery," *IEEE Geosci. Remote Sens. Lett.*, vol. 14, no. 4, pp. 549–553, 2017, doi: 10.1109/LGRS.2017.2657778.

[17] J. R. Otukei and T. Blaschke, "Land cover change assessment using decision trees, support vector machines and maximum likelihood classification algorithms," *Int. J. Appl. Earth Obs. Geoinf.*, vol. 12, no. 1, pp. S27–S31, Feb. 2010, doi: 10.1016/j.jag.2009.11.002.

[18] S. Kundu, D. Khare, and A. Mondal, "Landuse change impact on sub-watersheds prioritization by analytical hierarchy process (AHP)," *Ecol. Inform.*, vol. 42, pp. 100–113, 2017.

[19] R. Mosadeghi, J. Warnken, R. Tomlinson, and H. Mirfenderesk, "Comparison of Fuzzy-AHP and AHP in a spatial multi-criteria decision making model for urban land-use planning," *Comput. Environ. Urban Syst.*, vol. 49, pp. 54–65, 2015.

[20] T. Kavzoglu and I. Colkesen, "A kernel functions analysis for support vector machines for land cover classification," *Int. J. Appl. Earth Obs. Geoinf.*, vol. 11, no. 5, pp. 352–359, 2009, doi: 10.1016/j.jag.2009.06.002.

[21] L. S. Davis, "2002-An assessment of support vector machines for land cover classi cation.pdf," vol. 23, no. 4, pp. 725–749, 2002.

[22] R. Devadas, R. J. Denham, M. Pringle, R. S. Centre, and E. Percinct, "SUPPORT VECTOR MACHINE CLASSIFICATION OF OBJECT-BASED DATA FOR CROP MAPPING , USING MULTI-TEMPORAL LANDSAT IMAGERY," vol. XXXIX, no. September, pp. 185–190, 2012.

[23] X. Niu and Y. Ban, "Multi-temporal RADARSAT-2 polarimetric SAR data for urban land-cover classification using an object-based support vector machine and a rule-based approach," *Int. J. Remote Sens.*, vol. 34, no. 1, pp. 1–26, 2013.

[24] V. N. Mishra, R. Prasad, P. K. Rai, A. K. Vishwakarma, and A. Arora, "Performance evaluation of textural features in improving land use/land cover classification accuracy of heterogeneous landscape using multi-sensor remote sensing data," *Earth Sci. Informatics*, vol. 12, no. 1, pp. 71–86, Mar. 2019, doi: 10.1007/s12145-018-0369-z.

[25] M. J. Aitkenhead and I. H. Aalders, "Predicting land cover using GIS, Bayesian and evolutionary algorithm methods," *J. Environ. Manage.*, vol. 90, no. 1, pp. 236–250, 2009.

[26] I. Aalders, "Modeling land-use decision behavior with Bayesian belief networks," *Ecol. Soc.*, vol. 13, no. 1, 2008.

[27] H. Yuan, C. F. Van Der Wiele, and S. Khorram, "An automated artificial neural network system for land use/land cover classification from landsat TM imagery," *Remote Sens.*, vol. 1, no. 3, pp. 243–265, 2009, doi: 10.3390/rs1030243.

[28] M. Azari, A. Tayyebi, M. Helbich, and M. A. Reveshty, "Integrating cellular automata, artificial neural network, and fuzzy set theory to simulate threatened orchards: Application to Maragheh, Iran," *GIScience Remote Sens.*, vol. 53, no. 2, pp. 183–205, 2016, doi: 10.1080/15481603.2015.1137111.

[29] D. Anguita, A. Ghio, N. Greco, L. Oneto, and S. Ridella, "Model selection for support vector machines: Advantages and disadvantages of the Machine Learning Theory," *Proc. Int. Jt. Conf. Neural Networks*, 2010, doi: 10.1109/IJCNN.2010.5596450.

[30] R. Marrec *et al.*, "Spatiotemporal dynamics of the agricultural landscape mosaic drives distribution and abundance of dominant carabid beetles," *Landsc. Ecol.*, vol. 32, no. 12, pp. 2383–2398, Dec. 2017, doi: 10.1007/s10980-017-0576-x.

[31] J. Wegner and G. Merriam, "Use of spatial elements in a farmland mosaic by a woodland rodent," *Biol. Conserv.*, vol. 54, no. 3, pp. 263–276, 1990.

[32] Y.-C. Lee, K.-M. Lee, and J. Chon, "Landscape mosaic pattern analysis system using land cover map for micro-spatial analysis of regional planning," *J. Digit. Contents Soc.*, vol. 18, no. 7, pp. 1367–1375, 2017.

[33] W. Nie, Y. Yuan, W. Kepner, M. S. Nash, M. Jackson, and C. Erickson, "Assessing impacts of Landuse and Landcover changes on hydrology for the upper San Pedro watershed," *J. Hydrol.*, vol. 407, no. 1–4, pp. 105–114, 2011, doi: 10.1016/j.jhydrol.2011.07.012.

[34] M. D. Behera, S. N. Borate, S. N. Panda, P. R. Behera, and P. S. Roy, "Modelling and analyzing the watershed dynamics using Cellular Automata (CA)-Markov model - A geo-information based approach," *J. Earth Syst. Sci.*, vol. 121, no. 4, pp. 1011–1024, 2012, doi: 10.1007/s12040-012-0207-5.

[35] M. R. Muller and J. Middleton, "A Markov model of land-use change dynamics in the Niagara Region, Ontario, Canada," *Landsc. Ecol.*, vol. 9, no. 2, pp. 151–157, 1994, doi: 10.1007/BF00124382.

[36] E. Gidey, O. Dikinya, R. Sebego, E. Segosebe, and A. Zenebe, "Cellular automata and Markov Chain (CA_Markov) model-based predictions of future land use and land cover scenarios (2015–2033) in Raya, northern Ethiopia," *Model. Earth Syst. Environ.*, vol. 3, no. 4, pp. 1245–1262, 2017.

[37] V. N. Mishra and P. K. Rai, "A remote sensing aided multi-layer perceptron-Markov chain analysis for land use and land cover change prediction in Patna district (Bihar), India," *Arab. J. Geosci.*, vol. 9, no. 4, 2016, doi: 10.1007/s12517-015-2138-3.

[38] B. Mondal *et al.*, "Comparison of spatial modelling approaches to simulate urban growth: a case study on Udaipur city, India," *Geocarto Int.*, vol. 35, no. 4, pp. 411–433, 2020, doi: 10.1080/10106049.2018.1520922.

[39] A. Ligtenberg, M. Wachowicz, A. K. Bregt, A. Beulens, and D. L. Kettenis, "A design and application of a multi-agent system for simulation of multi-actor spatial planning," *J. Environ. Manage.*, vol. 72, no. 1–2, pp. 43–55, 2004, doi: 10.1016/j.jenvman.2004.02.007.

[40] S. Maithani, "Cellular Automata Based Model of Urban Spatial Growth," *J. Indian Soc. Remote*

Sens., vol. 38, no. 4, pp. 604–610, 2010, doi: 10.1007/s12524-010-0053-3.

[41] Y. Cao and N. Cao, "Dynamic Process of Urban Landscape Pattern Based on GeoCA-Urban Model-Taking the Neijiang City as Example," in *Advanced Materials Research*, 2014, vol. 869, pp. 205–208.

[42] K. C. Clarke, "Mapping and modelling land use change: an application of the SLEUTH model," *Lect. Notes Geoinf. Cartogr.*, no. 9783540691679, pp. 353–366, 2008, doi: 10.1007/978-3-540-69168-6_17.

[43] P. Potapov, S. Turubanova, and M. C. Hansen, "Regional-scale boreal forest cover and change mapping using Landsat data composites for European Russia," *Remote Sens. Environ.*, vol. 115, no. 2, pp. 548–561, 2011, doi: 10.1016/j.rse.2010.10.001.

[44] S. H. Winsvold, L. M. Andreassen, and C. Kienholz, "Glacier area and length changes in Norway from repeat inventories," *Cryosphere*, vol. 8, no. 5, pp. 1885–1903, 2014, doi: 10.5194/tc-8-1885-2014.

[45] D. Poursanidis, N. Chrysoulakis, and Z. Mitraka, "Landsat 8 vs. Landsat 5: A comparison based on urban and peri-urban land cover mapping," *Int. J. Appl. Earth Obs. Geoinf.*, vol. 35, pp. 259–269, 2015, doi: 10.1016/j.jag.2014.09.010.

[46] A. Arora, M. Pandey, M. A. Siddiqui, H. Hong, and V. N. Mishra, "Spatial flood susceptibility prediction in Middle Ganga Plain: comparison of frequency ratio and Shannon's entropy models," *Geocarto Int.*, vol. 0, no. 0, pp. 1–32, Nov. 2019, doi: 10.1080/10106049.2019.1687594.

[47] M. Kumar, "World geodetic system 1984: A modern and accurate global reference frame," *Mar. Geod.*, vol. 12, no. 2, pp. 117–126, Jan. 1988, doi: 10.1080/15210608809379580.

[48] A. Sharma, X. Liu, and X. Yang, "Land Cover Classification from Multi-temporal, Multi-spectral Remotely Sensed Imagery using Patch-Based Recurrent Neural Networks," pp. 1–18, 2017, [Online]. Available: http://arxiv.org/abs/1708.00813.

[49] M. E. MIRICI, "LAND USE/COVER CHANGE MODELLING IN A MEDITERRANEAN RURAL LANDSCAPE USING MULTI-LAYER PERCEPTRON AND MARKOV CHAIN (MLP-MC)," *Appl. Ecol. Environ. Res.*, vol. 16, no. 1, pp. 467–486, 2018, doi: 10.15666/aeer/1601_467486.

[50] D. E. Rumelhart, G. E. Hinton, and R. J. Williams, "Learning representations by back-propagating errors," *Nature*, vol. 323, no. 6088, pp. 533–536, 1986, doi: 10.1038/323533a0.

[51] S. Karsoliya, "Approximating Number of Hidden layer neurons in Multiple Hidden Layer BPNN Architecture," *Int. J. Eng. Trends Technol.*, vol. 3, no. 6, pp. 714–717, 2012.

[52] G. Panchal, A. Ganatra, Y. P. Kosta, and D. Panchal, "Behaviour Analysis of Multilayer Perceptronswith Multiple Hidden Neurons and Hidden Layers," *Int. J. Comput. Theory Eng.*, vol. 3, no. 2, pp. 332–337, 2011, doi: 10.7763/ijcte.2011.v3.328.

[53] F. Girosi and T. Poggio, "Networks and the best approximation property," *Biol. Cybern.*, vol. 63, no. 3, pp. 169–176, 1990, doi: 10.1007/BF00195855.

[54] J. Cohen, "A Coefficient of Agreement for Nominal Scales," *Educ. Psychol. Meas.*, vol. 20, no. 1, pp. 37–46, Apr. 1960, doi: 10.1177/001316446002000104.

[55] T. Chai and R. R. Draxler, "Root mean square error (RMSE) or mean absolute error (MAE)? - Arguments against avoiding RMSE in the literature," *Geosci. Model Dev.*, vol. 7, no. 3, pp. 1247–1250, 2014, doi: 10.5194/gmd-7-1247-2014.

[56] J. J. Arsanjani, W. Kainz, and A. J. Mousivand, "Tracking dynamic land-use change using spatially explicit Markov Chain based on cellular automata: the case of Tehran," *Int. J. Image Data Fusion*, vol. 2, no. 4, pp. 329–345, 2011.

[57] A. Marsetič, K. Oštir, and M. K. Fras, "Automatic orthorectification of high-resolution optical satellite images using vector roads," *IEEE Trans. Geosci. Remote Sens.*, vol. 53, no. 11, pp. 6035–6047, 2015, doi: 10.1109/TGRS.2015.2431434.

[58] B. Demir and S. Ertürk, "Improving Svm Classification Accuracy Using," pp. 2849–2852, 2009.

[59] P. L. N. Raju, H. Chaudhary, and A. K. Jha, "Shadow analysis technique for extraction of building height using high resolution satellite single image and accuracy assessment," *Int. Arch. Photogramm. Remote Sens. Spat. Inf. Sci. - ISPRS Arch.*, vol. 40, no. 8, pp. 1185–1192, 2014, doi: 10.5194/isprsarchives-XL-8-1185-2014.

[60] C. F. Chen *et al.*, "Multi-decadal mangrove forest change detection and prediction in honduras, central america, with landsat imagery and a markov chain model," *Remote Sens.*, vol. 5, no. 12, pp. 6408–6426, 2013, doi: 10.3390/rs5126408.

[61] M. Shafiullah and M. A. Abido, "S-Transform Based FFNN Approach for Distribution Grids Fault Detection and Classification," *IEEE Access*, vol. 6, no. February, pp. 8080–8088, 2018, doi: 10.1109/ACCESS.2018.2809045.

IOP Conf. Series: Earth and Environmental Science 1032 (2022) 012029 doi:10.1088/1755-1315/1032/1/012029

Challenges and solutions in COVID-19 related pandemic solid waste management (PSWM) - A detailed analysis with special focus on plastic waste.

Kshitij Upadhyay[1] & Samir Bajpai[2]

[1]Research Scholar, Department of Civil Engineering, National Institute of Technology Raipur, G.E.Road, Raipur - 492010, Chhattisgarh, India
[2]Professor, Department of Civil Engineering, National Institute of Technology Raipur, G.E.Road, Raipur - 492 010, Chhattisgarh, India

Email: **kshitij.upadhyay111@gmail.com**

Abstract. A pandemic like novel coronavirus 2' (SARS-CoV-2) not only poses serious public health repercussions but also affects the socio-economic and environmental conditions of the affected countries. The increased consumption of material resources in conjunction with 'containment and preventive measures' is generating an unprecedented amount of potentially infectious solid waste, especially that of plastic origin, which if mismanaged, is bound to affect the ecosystem and public health, as the virus can survive on fomites for longer duration. COVID-19 related pandemic waste, such as Personal protective equipment (PPEs), sanitizer and water bottles, disinfection wipes, and Single use Plastics (SUPs) products has already found its way to the aquatic and terrestrial environment. Even before the start of the COVID-19 pandemic, the management of plastic waste, an environmental stressor with trans-boundary migration capabilities, was a major environmental issue for every stake-holder.

In this paper, we propose a separate domain in the waste management framework for the effective management of pandemic related solid waste. Factors and sources contributing to increased plastic waste generation are discussed in detail. A concise picture of global plastic demand through sectors and polymer types is presented and speculations are made on how COVID-19 is going to affect the plastic demand. Current solid waste handling and management practices in developed and developing countries are critically examined from the perspective of this pandemic. We identified various challenges that waste management sectors are facing currently and offered possible solutions.

Concerns of transmission through fomites is bringing a change in public behavior and consumption pattern which affects 3R practices, while fear of secondary transmission from occupational infections is interfering with 3R practices at end-of-life plastic waste management. The legislative and restrictive frameworks on plastic use being currently put-on hold at the governmental level to ensure public safety are being used by the plastic industry to lobby for increased plastic consumption. The inability of the governments to win public confidence is further escalating unsustainable practices and slowing the shift towards sustainable economy. It is imperative to enforce sustainable practices without putting public safety at risk and to ensure that an unsustainable societal attitude wouldn't be reinstated in the post-pandemic world. Lastly, eight research and policy points suggested here may guide future studies and governmental frameworks in the domain of COVID-19 pandemic related solid waste handling and management.

Keywords - COVID-19, SOLID WASTE MANAGEMENT, PLASTIC WASTE, 3R, PSWM

1. Introduction

During an outbreak of COVID 19, "preventive and precautionary measures" have been shown to be effective. Accordingly, social distancing, cluster and national lockdowns, quarantine, isolation, and halting economic and transportation activities are being implemented regionally, nationally, and globally, to contain the spread of the virus. These preventive measures are showing a direct impact on socio-economic and environmental conditions at the local and global level **[1-9]**.

Nationwide lockdowns disrupted the regional, national and global economy due to the shutdown of transportation, business, industries, and recreational activities. Such measures showed a positive short-term positive on the environment such as decrease in atmospheric pollutants, reduction in carbon footprint, and decreased demand for fossil and petroleum fuels (due to restriction on traveling and industrial productions). Substantial reductions in environmental emission indices such as carbon, NO_2, ozone (O_3), SO_2, particulate matter $(PM)_{2.5}$ and $(PM)_{10}$ were recorded in major cities, around the globe **[10-15]**.

Besides the positive impact of 'preventive and restrictive measures' on the ecosystem, the negative consequences of COVID-19 on the environment are a major growing concern worldwide. One of the major environmental issues of COVID-19 related pandemic is a sudden global surge in liquid and solid wastes. Relevant global and national agencies tasked to deal with COVID-19 have issued advisories, guidelines and standard operating protocols to be used by healthcare professionals, first responders, and common public. To minimize human to human transmission of the virus, social distancing, good hygiene practices, frequent hand-washing, and respiratory etiquettes such as face-covering, using face masks, and coughing in elbows has been recommended **[16-18]**.

These life-saving recommendations are generating an enormous amount of waste. The global campaign to adopt better hygiene practices and frequent washing of hands resulted in acute water demand and subsequent increase in wastewater generation **[19-21]**. Similarly, the volume of solid waste is also increasing, as medical solid waste keeps increasing, especially the plastic waste composition in medical waste and municipal solid waste, as demand for plastic-based materials is an all-time high in the healthcare sector and society. With no effective treatment protocols available, healthcare workers rely on personal protective equipment (PPE) for their protection. Similarly, demand for medical plastic equipment like ventilators, and disposal plastic consumables like syringes, blood bags, drip sets, etc. have drastically increased with an increase in number of cases. Additionally, the consumption pattern of the public is changing, as visible from increased consumption of single-use plastics (SUPs), which is generating more solid waste. The public fear of contacting with the virus from a contaminated surface, as the virus possess capability to adhere to the surface of various materials up to 11 days, is swaying sustainable 3R practices, which is further contributing to the municipal solid waste generation **[21-33]**.

The burden of handling and managing solid waste, especially plastic waste has long been an important challenge for waste management facilities, and any significant change in volume or composition is likely to overwhelm waste handling and management infrastructure, as these systems are designed to cater to fixed volume of waste and have little margin for surplus volume **[34]**. The major composition of COVID-19 related solid waste is made up of plastic, an environmental stressor that the world was already struggling to manage. The COVID-19 related plastic waste will further add to the problem at an unimaginable scale. The extent of the problem can be understood by the projected global monthly consumption rate of facemasks and gloves (129 billion and 65 billion respectively), a figure obtained through extrapolation of monthly demand in Italy **[35-36]**.

The unsustainable plastics use, in combination with poor waste management practices, and governmental apathy towards the environment has created a new type of anthropogenic pollution, which is persistent in nature. The eco-toxicological impact of plastic pollution on biotic and abiotic environment has drawn significant attention from academia, public, and government resulting in accelerated research and governmental intervention **[37-39].**

Medical waste requires special waste handling and disposal practices, due to possible toxicity and the infectiousness of the waste. The municipal solid waste handling and management, apart from seeing a significant change in volume and composition, have to deal with logistical and operational challenges as well, such as low workforce, fear of secondary contamination, and feasibility. Clinical waste originating from temporary shelter houses and quarantine facilities can be potentially infectious. The direct impact (change in volume and composition) and indirect impact of the pandemic such as increased advocacy for plastic consumption, change in consumption pattern, plummeting oil price and increasing popularity of unsustainable lifestyle changes, in conjunction call for immediate actions from all stakeholders, if we are to keep moving towards a sustainable circular economy. Long term surge and potential toxicity of pandemic related solid waste require scientific and sustainable waste handling and management approach accommodating changing variables amid pandemic. Given the severity and nature of pandemic, pandemic related safety and precautionary measures are likely to remain in place till mass vaccination and a significant reduction in cases.

The central focus of this manuscript is to discuss the handling and management of COVID-19 related pandemic solid waste (PSW) with special emphasis on the plastic composition of the waste. Sources of pandemic related plastic waste have been identified and discussed in detail. The polymeric composition of plastics and demand trend by various sectors is reviewed and the effect of the pandemic on plastic demand in near future is speculated. The factors affecting the generation of pandemic related plastic waste and challenges in its handling and management have been discussed in brief, and the possible solutions for pandemic waste handling and management have been proposed. Moreover, the impact of the pandemic on sustainable circular economy from 'reduce, recycle, and reuse' perspective is reviewed and discussed. We also summarized the status of solid waste management practices in developing and developed countries from the pandemic perspective. Lastly, we suggest policy points and give a future perspective on which further studies can be conducted.

2. Sources of plastic waste amid COVID-19 pandemic

Any disaster event creates a stream of solid waste that requires efficient end-of-life management as the safety of public health and quality of the environment is directly correlated with these wastes **[40-42]**. The COVID-19 related solid waste generating from various sectors (medical facilities, quarantine centers, municipal solid waste from affected sectors, etc.) if not handled properly, creates a risk of secondary infection by exposing the public with the virus. In particular, under the COVID-19 pandemic, it is of paramount importance to not only ensure public safety but also minimize the environmental damage from pandemic related solid waste **[36,43].**

Given the extent at which the habitats are infected, sectors that are involved in containment of the virus, and material resources being used directly (PPE) or indirectly (SUPs), it is imperative to define COVID-19 related solid wastes in waste management framework, and to do so, here we propose using the term ***'Pandemic solid waste (PSW)'*** to club all the pandemic related solid wastes under one umbrella which also covers epidemic events. Pandemic solid waste (PSW) can be defined as ***"any solid waste, generated during a pandemic or epidemic event, with the potential of impacting human and animal health, regardless of whether they have potential to cause direct offense and infection or not, but not limited to healthcare facilities, households, quarantine centers, isolation centers, testing centers and so forth.***

To simplify, and for effective waste management, PSW can be classified, based on point of origin, as medical solid waste (MLSW), clinical solid waste (CSW), and municipal solid waste (MSW). Medical waste incorporates all types of waste originating from healthcare facilities. Clinical solid waste includes *"any waste that has the potential to cause injury, infection, or offense, arising but not limited to medical, dental, podiatry, healthcare services* and *so forth*" **[44]**. Municipal solid waste includes all types of solid waste generating from municipalities excluding medical, clinical, electronic, and nuclear waste.

The exponential growth in consumption of plastic, a versatile, cheap, and adaptable material and subsequent unsustainable mismanagement of plastic waste is now considered as global environmental menace. Due to adaptive properties, the plastics form the backbone of the healthcare industry. The monetary value of plastic in the global healthcare sector was valued at USD 23.5 billion in 2019 and is expected to grow at the rate of 8.6% from 2020 to 2027, with 7,667.6 kilo-tons of plastic being expected to be consumed in 2020. This figure will keep increasing as demand for the plastic is surging both for medical and public consumption [**27**]. The urgency fueled global demand of PPE has disrupted the overall life cycle and end-of life waste management as supply chain is disrupted at upstream, while discarded PPE is creating waste management problems at downstream. The demand for medical plastic products (PPE, disposable plastic components for ventilators, and general plastic consumables) varies with the number of people infected by the virus **[37].**

The potentially infectious COVID-19 related plastic waste is observed to be originating from 3 sectors, discussed below:

a. *COVID-19 related plastic waste from hospitals*

Initially, the human-to-human transmission of COVID-19 was conceptualized to be droplet transmission, but new shreds of evidence suggest that aerosol transmission is also possible **[46]**. Past experiences in dealing with infectious outbreaks, mandate the preventive measures as crucial in minimizing the transmission and containment of infectious diseases. The preventive measures depend upon the level of contact and surroundings **[47]**. Standard precaution in form of PPE equipment, proper hygiene including regular sanitization, safe disposal of medical waste, and respiratory etiquettes by healthcare workers, first responders, and patients, to prevent contact transmission is recommended by concerned agencies. Face shields, respiratory masks, hazmat suits, and social distancing are used to avoid droplet and aerosol transmission of the virus **[48-50]**.

This medical wisdom came in very handy during the early days of COVID-19 breakout resulting in a surge in demand of these items putting healthcare systems under unprecedented strain. Shortage of PPE including facemasks, for health professionals, first responders, and the common public, has been a matter of criticism on traditional and social media. Media reports of using common household items such as plastic bags by healthcare workers drew global criticism **[51-53].**

Plastic is the most widely used material in the healthcare industry and commonly used for the manufacturing of costly medical instruments to cheap protective gear. The most commonly used protective items (such as gloves, facemasks, face shields, respiratory masks, head caps, hazmat suits, etc) and consumables (like parts of ventilators, blood bag, drip set, pill casing, and a body bag, etc.) from medical industry end up in waste after single use due to its infectious nature. The urgency of plastic-based PPE and medical consumables is without a doubt, crucial for fighting this pandemic, though this may further aggravate an already existing epidemic, the world is facing, i.e. plastic pollution.

b. *COVID-19 related plastic waste from quarantine centers, testing facilities, research laboratories, and shelter homes*

The ongoing pandemic requires efforts on all fronts, from treating infected patients in hospitals, putting people (asymptomatic people, people with frequent travel history, migrants and people planning to migrate) in isolation and quarantine centers, and advances in scientific research. These efforts utilize ample resources, especially that of plastic origin, which is subsequently discarded as waste. People in isolation centers, quarantine centers, and shelter homes, may or may not be infected by the virus. The testing and research facilities are making huge efforts in testing for the virus for confirming the positive cases, finding treatments, and development of vaccines. The waste from these sectors knows as clinical solid waste, is infectious in nature and contains an ample amount of plastic products (PPE, medical consumables, consumables used in virus sequencing, testing swabs, swab containment bags, plastic consumables) used in testing and research and SUPs products used for general consumptions **[54-55]**.

c. COVID-19 related plastic waste from households, workplaces, educational institutions, recreational centers, community centers, leisure industry, etc.

The plastic composition in 2.01 billion tonnes of municipal solid waste generated annually was around 12% in 2019. COVID-19 pandemic stopped the mobility of people as authorities are implementing nationwide or clustered lockdown. The closure of public facilities, workplaces, educational centers, recreational facilities, and leisure industry affects the generation of municipal solid waste, with initial reports suggesting a lesser quantity of municipal solid waste being generated during the lockdown. The unsustainable lifestyle change, which changes the consumption pattern of the public, is observable from the increased consumption of SUPs, plastic bags, food containers, and plastic packaging materials. Mandatory use of PPE in public settings is further adding to the demand for facemasks and gloves. The fear of contamination from material surface (fomite) puts hold on sustainable 3R (reduce, recycle, and reuse) practices. All these factors combined to increase the generation of municipal plastic waste amid ongoing pandemic **[36,43].**

The plastic waste from the aforementioned sectors is of diverse nature which plays a crucial role in its handling and management. Identifying the polymeric composition, the plastic consumption behavior, and sector-wise demand for plastic will help us in assessing the sectors likely to generate more plastic waste further helping in drafting od suitable mitigation policies.

3. Types of plastic waste, polymeric composition, and demand by various sectors

The variation of polymeric composition and distribution across various sectors has grave implications for plastic waste management. The annual plastic waste generated globally is growing exponentially, in line with exponential growth in global plastic production and consumption. Plastic waste generated from all the sectors discussed in the previous section is likely to increase amid ongoing pandemic. It's too early to speculate whether this increasing trend returns to normalcy, once the pandemic subsides.

PPE (masks, gloves, and protective shields for eyes/face, head and shoe cover, apron, etc.) is mostly made from plastics due to its impervious properties which obstructs contaminated droplets to come in contact with skin by trapping them on the outer surface of the PPE. It's a common practice among healthcare professionals to change their PPE every eight hours but some PPE may require disposal after a single use **[22]**. Therefore, it's very challenging to estimate the quantity of PPE based on the consumer end. The dramatic increase in demand for PPE (made up of various plastic and rubber items) by healthcare industries, first responders, and the common public have been reported globally **[22,24,29,30]**. Similarly, the consumption of single plastic products and packaging products has also increased significantly which is made of various polymers. It's imperative to understand the polymeric composition of these items; their origin and nature to assess and devise effective waste management practices.

Table 1. Origin, category, type, and polymeric composition of COVID-19 related plastic waste.

Origin	Category	Type	Polymer Composition	References
Medical and clinical solid waste	PPE	Gloves	Nitrile and Vinyl	Avacare Medical, 2020
		Masks (Respiratory)	Polypropylene, Polyurethane, Polyacrylonitrile	Czigány and Ronkay, 2020
		Masks (Surgical)	Polypropylene Polystyrene, Polycarbonate, Polyethylene, and Polyester	Henneberry, 2020
		Gowns and Body Suits	Polyethylene, Polypropylene and Polyethylene Terephthalate	Ajmeri and Joshi, 2020
		Head Caps	Polyvinyl Chloride, Polypropylene, Polyethylene, Polystyrene, Nylon, Polyethylene Terephthalate, Polyimide, Polycarbonate,	McKeen, 2014
		Face Shields		
		safety glasses		
		Body Suits		
		Shoe covers		

	Consumables	Disposal syringes	Acrylonitrile Butadiene, Polyetheretherketone, and Polyurethane	
		Blood bags		
		Catheters		
		Pill casings		
Municipal solid waste	PPE	Disposable Gloves	Nitrile and Vinyl	Avacare Medical, 2020
	PPE	Disposable Masks	Polyethelene	WHO, 2020h
	SUPs	Bags, trays, containers, food packaging film	Light-density polyethylene	UNEP, 2018
		Milk bottles, freezer bags, shampoo bottles, ice cream containers	Light-density polyethylene	
		Bottles for water and other drinks, dispensing containers for cleaning fluids, biscuit trays	Polyethylene Terephthalate	
		Cutlery, plates, and cups	Polystyrene	
		Hot drink cups, insulated food packaging, protective packaging for fragile items	Expanded polystyrene	
		Microwave dishes, ice cream tubs, potato chip bags, bottle caps	Polypropylene	

Besides PPE, due to high mortality rates, body bags are needed in huge quantities to dispose of the deceased body. Additionally, medical consumables such as pill casings, disposal syringes, catheters, and blood bags are essential in the functioning of hospitals. These materials are often composed of plastic material and subsequently end up as medical waste. The staggering amount of plastic waste originating from medical facilities amid COVID-19 is going to be a challenge for waste management facilities for safe disposal **[57-58]**.

Apart from the medical sector, plastic waste composition in municipal solid waste is also affected due to pandemic. Due to social distancing, better hygiene practices, and concern for personal and public safety, disposable gloves and face masks are being used by the public in social and public settings to avoid contact with the outer surfaces and prevent droplet transmission. Local governments in countries like India have made it mandatory for the public to use face masks (gloves in few places) whenever they leave their homes **[59-60]**.

The pandemic is causing a shift in public lifestyle as people are being more cautious to avoid contacts with potentially exposed surfaces in fear of coming in contact with the virus. The lifestyle changes include avoiding going out, eating out, online shopping, avoid outside leisure activity, and avoiding reuse practices. To avoid human contact and due to lockdown, people are ordering more home deliveries and opting for takeaway, which means increasing amounts of SUPs (packing materials, disposable meal boxes, carry bags, spoons, straws, and plates, etc.) being used. These items are generally made up of polyethylene, polypropylene, and polystyrene as highlighted in [**table 1**]. Thus, it's safe to say that the fear of contracting COVID-19 virus has pushed people to use more plastic than ever before.

The distribution and demand for plastics on basis of polymers and sectors are depicted in **fig 1 & 2**. The plastic consumption in the medical sector, packaging sector, and leisure industry are expected to rise due to pandemic. "The global packaging sector which accounted for 40% of total plastic consumed in 2018 is forecasted to grow from USD 909.2 billion in 2019 to USD 1,012.6 billion by 2021 at a rate of 5.5% amid COVID-19 pandemic. Similarly, the global medical sector which accounted for 17% of total plastic consumed in 2018 is forecasted to grow from USD 22.8 billion in 2019 to USD 31.7 billion by 2024 at a rate of 6.8% amid COVID-19 pandemic". Polyethylene and polypropylene are the most dominant polymer being produced and used worldwide in both the healthcare industry and SUPs manufacturing industry. Polypropylene, a non-woven single-use polymer, is largely used in making PPE. Demand for all the polymer is expected to increase in 2020 and 2021 due to the looming uncertainty of COVID-19 pandemic **[61-62]**.

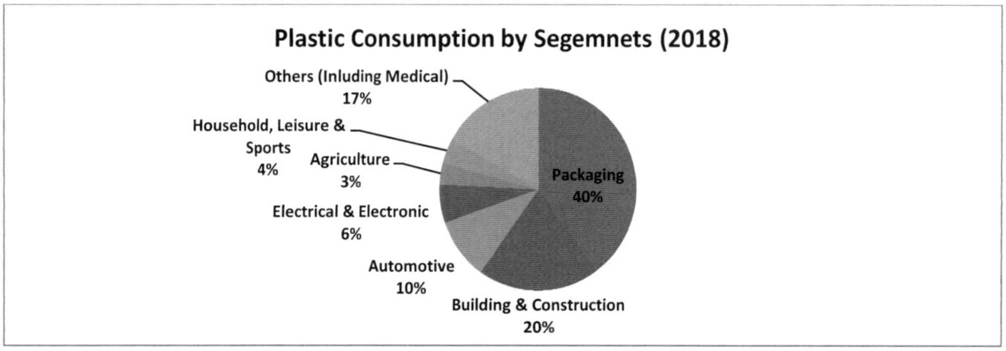

Fig 1. Global plastic consumption by segments in 2018 (adopted from Plastic Europe, 2018).

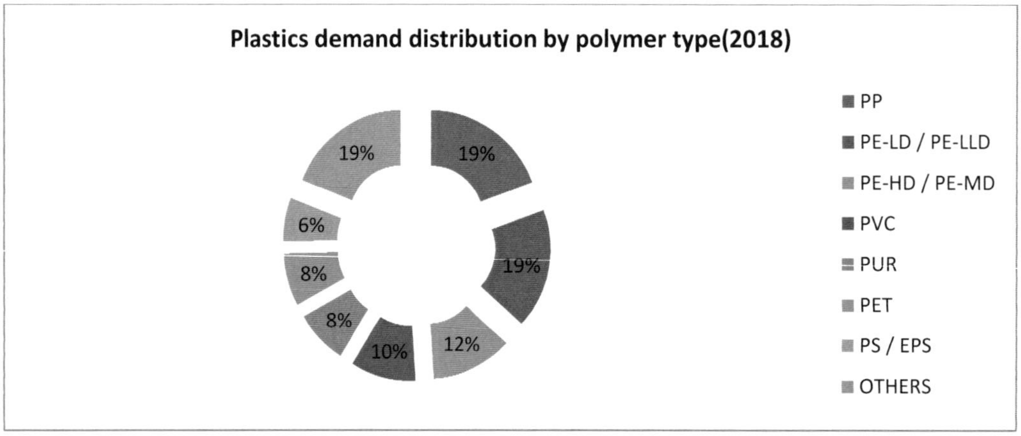

Fig 2. Global plastic demand by polymer type in 2018 (adopted from Plastic Europe, 2018)

4. Factors affecting the quantity of COVID-19 related plastic waste

The COVID-19 pandemic has highlighted the importance of plastics as a resource (material) in the healthcare industry and public utility. The direct and indirect socio-economic impact of pandemic governs the quantity of plastic waste generated, which interferes with its handling and management. In a sustainable circular economy, the solid waste handling and management (SWHM) are based on material balance and waste hierarchy, which advocate for lower consumption of resources at production end and prioritize 3R practices at consumer and disposal end. This is accomplished by the active participation of all stake-holders through legislative frameworks and awareness campaigns **[63-65].** The irony here is that pandemic is not only affecting the public health but also creating indirect drivers that hamper all the efforts made towards the circular economy. The key factors affecting the quantity of COVID-19 related plastic waste are discussed below:

a. *Plummeting oil price:* Immobility of humans and stoppage or reduction in industrial activities leads to a decrease in demand for fossil fuels like crude oil and natural gases **[66]**. While demand decreased, the global oil and gas production maintained a steady rate, resulting in a surplus of stock with no demand **[67]**. The economic interdependence of the plastics and fossil fuel industry, where fuel economy governs the economy of plastic production, consumption, and disposal, saves each other at the time of economic crisis.

Plummeting oil and gas prices indirectly affect plastic production, consumption, and recycling. Plastic is produced from byproducts of fossil fuels **(oil and gases)**. With oil and gases being cheaper than ever before, to avert an economic crisis, the fossil fuel industry is over-producing the plastics. The upstream production of cheaper virgin polymers and downstream increase in demand creates a material imbalance, which may have drastic environmental consequences. This type of economic phenomenon has also been observed in past [68-70]. With a decrease in monetary value and increased production, the entire value chain of plastics is now imbalanced. At one hand demand for plastic is skyrocketing due to increased consumption in the healthcare industry and SUPs by society; industries that produce plastic products are facing a dilemma, whether plastic recycling is still an economically viable option to curb plastic pollution which puts end-of-life plastic management at risk **[71].**

The overproduced cheap plastic already has demand in the market. The PPE of single-use nature, originating from households and society will end up being part of MSW. Similarly, though SUPs products are being discouraged, the infectious nature of COVID-19 is driving take-away culture and online shopping upwards, which drastically increases the demand for single-use and packaging plastics. Reusable plastic products are now treated as SUPs as the virus can survive on the surface of plastic products. This change in societal attitude, exponentially increases the plastic composition in municipal solid waste, while medical and clinical plastic waste is being generated at an unimaginable scale.

b. *Legislative setbacks*: The ongoing pandemic kick-started rapid production of PPE on the war front, globally, to meet the demand of the healthcare industry and public. The governments are trying to support healthcare systems while common masses are hoarding these products for personal use. As mismanaged discarded PPEs and SUPs are ending in the environment, due to ongoing pandemic, national and international environmental regulatory agencies are either scrapping, roll-backing, or delaying the laws and regulations, on the use of plastic products **[72-74, 36].**While these decisions seem temporary, given the nature of pandemic and increased concerns among the public, how long these regulatory policies remain dormant will be a matter of heated scientific discussion. Suddenly, environmental issues such as circular economy and pollution reduction have taken a backseat in governmental and societal attitude and the pandemic is of priority concern at the moment, rightly so, but the environmental consequences, especially in form of increased plastic pollution of an already polluted ecosystem are long-term phenomena, and are going to affect our fight with pollution.

This pandemic is changing the attitude of the common public to think that SUPs remain the safest choice. The plastic industries are set to lobby against any laws in the future that limit the use of SUPs. Understandably so, the latest research highlights that the COVID-19 virus survives the longest on plastic when compared with both cardboard and meta **[33].** This finding is bound to be used by advocates of the plastic industry to push further SUPs. Earlier, the restrictive regulations saw opposition from plastic manufacturing and consumer industries. Public information campaigns were held globally to discourage the use of SUPs and encouraging a sustainable lifestyle. Now, the plastics industry is trying to capitalize on public fear by promoting the use of SUPs products. Plastic industries will lobby against any future regulatory frameworks that will address post COVID-19 plastic pollution by using public fear and pandemic associated uncertainties.

c. *Duration of pandemic and mass vaccination*: Given the trajectory of outbreak and progress in finding a vaccine, experts suggest that the effect of COVID-19 will be seen until 2022 **[75]**. The increased stream of COVID-19 related plastic waste is directly related to the duration and severity of the pandemic. Increased duration of pandemic means increased hospitalization leading to an increase in medical waste generation. Also prolonged pandemic means prolonged social distancing and precautionary measures, resulting in increased PPE consumption by the public. Additionally, the unsustainable lifestyle practices might be observed till the end of pandemic, thereby increased consumption of single used plastics, increasing the volume of plastic waste.

5. Challenges in handling and management of COVID-19 pandemic related solid waste

The unprecedented consequences of waste generated due to environmental disasters (such as earthquakes, cyclones, tsunamis) on waste management facilities have been well researched by the scientific community and waste researchers **[43].** However, such a body of knowledge is not available for the infectious wastes generated during pandemic or epidemic events. The increased generation of pandemic related solid waste is in addition to the normal Municipal and medical solid waste and warrants scientific management. The pandemic has created supply chain problem of necessary resources (plastic based products) while also affecting the management of solid waste at downstream end. The sudden change in volume and composition of medical and municipal solid wastes has shocked the waste management facilities, a problem that needs urgent attention **[45]**.

Best practices in solid waste management are crucial for maintaining both public health and the quality of the eco-system. It is important to mention here that, during the initial phase of lockdowns, municipal solid waste collection halted or operated at reduced capacity **[76].** The lower volume of waste collected will subsequently be reported; hence any assessment made from short-term waste collection data can't reflect the waste generation curve of the entire pandemic. Additionally, the trend will also vary from locality to locality, depending upon waste management infrastructure, for example, household waste in Italy and China were observed to be lower while higher for Thailand during the lockdown. **[36]**. The thumb rule that 'in a linear economy as the consumption of material in society increases, the subsequent waste generation will also increase' shall be used for making a predictive assessment. "The demand for medical, commercial and household plastic products is expected to keep increasing as the number of COVID-19 cases increases and the waste generation curve will match global pandemic curve" **[45].**

***a**. Volume of solid waste*: Due to the pandemic, medical waste management facilities are dealing with an unprecedented increase in the volume of medical waste. The staggering amount of medical and municipal solid waste being generated daily, penetration of pandemic, infection, and hospitalization rate, and with no vaccine in sight, COVID-19 led waste management is a global issue that requires urgent attention. To stop the spread of disease, quarantine and isolation facilities are being set up at village, town and city level, where asymptomatic people, people with travel history, and incoming travelers are being accommodated for a certain period before allowing them to assimilate with society. Given the number of such centers, the high number of inhabitants, and the duration of stay, the total waste generated from these facilities is very high.

Given the severity of the pandemic and vast campaign for adopting better hygiene practices and preventive measures, it's very hard to have an estimate on the volume of PPE being disposed of daily. Here we try to propose a rough estimation on daily PPE generation from public settings and in medical facilities by modifying the equation proposed in the literature **[32].**

$$N = P * AR * DPU \qquad (1)$$

N= Total number of PPE waste generated daily in a town, city village.
P= Total Population of habitat (Rural/Urban, Developing/Developed)
AR = Factor to account for the type of habitat (Rural/Urban, Developing/Developed).

Setting	**Value of AR (between 0 to 1)**
Rural area in a developing country	0.6
Urban area in a developing country	0.8
Rural area in a developed country	0.8
Urban area in a developed country	1

DPU = Daily PPE Use (varies depending upon local regulation, and public motivation). It is recommended to change PPE (masks and gloves) daily, in such cases DPU = 2.

To calculate the total number of PPE waste generated daily in a country, total waste generated in a rural area and urban area of the country needs to be calculated separately using urban or rural population of the country and daily PPE used. The summation of rural and urban PPE waste generated will yield statistics for the whole country.

***b**. Change in Composition:* Solid waste management facilities deal with municipal waste and medical waste separately but due to pandemic, clinical waste from quarantine center is being mixed with municipal solid waste, which is not only dangerous but also putting strain on MSW management plants **[77]**. The pathogenic nature of clinical and medical solid waste requires a diligent approach in handling and disposal. The medical waste management facilities are capable of handling medical waste at the steady-state condition with a fixed volume of waste being generated periodically (Daily/weekly/monthly) and of pre-recognized composition. Any deviation from a normal condition such as a change in the volume, flow rate, or composition will lead to failure or ineffectiveness of the system. Most of the facilities are based on incineration or pyrolysis with sterilization where the medical waste is properly decontaminated and disposed of following very strict guidelines. The problem here is such guidelines are based on normal operating conditions but amid pandemic, the volume and composition of waste being generated daily are far more than the anticipated protocols, leading to the inability of facilities to deal with such wastes.

Similarly, with the onset of pandemic, the plastic composition in municipal solid waste changed rapidly due to the extensive use of PPE, sanitizers, and SUPs. Making lots of variations in the composition and volume of plastic waste generated daily. The waste handling and management (medical, clinical, and municipal solid) procedures are being modified to accommodate the effect of the pandemic. These modifications are directly changing the composition of wastes. The most common modifications that can be seen globally are treating recyclable products such as plastic bags and food containers in clinical solid waste as hazardous waste.

***c**. Infectious nature of the pandemic related solid waste:* The COVID-19 related solid waste poses a threat to the health of those who work on the frontline in the waste management sector such as trash collectors, transporters, rag pickers, street cleaners, waste segregator, etc. During the onset of the pandemic, emphasis was put on containing the spread of the virus and treating patients, solid waste implications were largely overlooked. Initially, the waste management practices were put on hold and in few cases augmented. While in countries like Australia, China, the USA, India, France, Italy, New Zeeland, etc., concerned authorities have put forth guidelines for the management of medical and clinical solid waste due to the toxic nature of such waste, the municipal solid waste remains overlooked. The fact that virus can survive on various materials up to 3 days puts the lives of those involved in the waste management sector at risk due to strong possibility of secondary transmission **[78].**

The medical waste generated due to COVID-19 is being handled carefully and poses no additional problem relating to secondary transmission due to diligent approach being practiced for managing medical wastes. The clinical solid waste originating from quarantine centers, testing facilities, affected households poses the risk of contamination and is recommended to be handled through the same waste management procedure used for managing medical waste **[79-80].** The real concern here lies with municipal solid wastes originating from households, public places, workplaces, and leisure and entertainment centers. As per WHO and other national agencies, the number of asymptomatic COVID-19 cases keeps increasing which makes it tough not only for health professionals for early detection but also have grave consequences for those involved in waste management.

Asymptomatic patients may generate contaminated municipal solid waste, as the virus can sustain on various surfaces, and in absence of modified municipal solid waste management protocols, it poses a threat to waste workers in both formal and informal sectors and may spread the disease further in community. The municipal solid waste management procedures vary across countries. The waste generated from individual habitat is collected and transported to waste management facilities. The person responsible for door to door collection or kerb-side collection and transporting the waste can come in contact with the virus persisting on the surface of wastes if the waste is not handled safely. In waste facilities, waste is segregated for resource recovery (separating the material that can be recycled or reused). Those involved in doing these tasks are also at risk of contracting the virus. At treatment and disposal end, whether thermal decomposition method or landfills, personnel responsible to carry out these tasks may come in direct contact with the contaminated waste.

The body of scientific knowledge on the survival of viruses after waste disposal is very limited with the majority of them focusing on enteric viruses only **[81-83]** , though influenza virus (H6N2) inactivation rate was up to 600 days in landfills, confirming that virus was active after disposal **[84]**. Secondary transmission of COVID-19 virus from solid waste to workers and reactivation rate of virus from the waste has not been studied yet, hence the secondary infection from occupational sectors can't be overlooked and requires scientific studies on priority basis.

The problem increases manifold where the systematic waste management is not in practice. In developing countries, municipal solid waste is disposed directly at dumping sites, or in the low-lying fields, where it remains in an open environment, depending upon whether the dumped waste is collected and transported for final disposal, mostly in open solid waste dumping sites away from the community, or left there for a prolonged duration, in either case, there exists greater risk for both waste workers and common public. More so, in the informal waste management sector, where waste or rag pickers, who work on the frontline to keep our habitats free from waste, indulge in waste scavenging, collect, recycle and reuse various items from municipal solid waste or garbage, as part of their livelihood, are at most risk, as they don't have access to personal protective equipment for their safety. The most scavenged items are plastic-based, which find way back to the community when scavengers sell the recycled plastic either to recyclers or plastic product manufacturing industries. The spread of disease in this instance is not limited to those who are directly involved. They may get infected by the virus and go on with their daily life, starting to interact with their family members and the general public, putting the health of anyone who comes in contact with them at risk.

d. *Logistical and operational challenges:* The systematic solid waste handling and management infrastructure designed to handle steady-state volume are under overwhelming stress, due to an increase in waste volume, changing composition, and associated risk of secondary infection. While waste professionals tasked with dealing with medical and clinical solid waste have already undergone safety training programs to deal with hazardous and toxic waste, those involved in the MSWM sector require immediate training campaigns, to follow better safety protocols and practices to use PPE.

Additionally, the medical waste management facilities need to develop ways to not only deal with the increased volume and changing composition of waste, but also consider their possible impact on waste to energy plants. The plastic waste management facilities are based on the concept of optimum resource recovery and energy production. The integrated solid waste management involves aspects of traditional solid waste management strategies coupled with energy recovery facility such as incineration. The ISWM also uses landfills to dispose of residue from incineration plants obtained after energy recovery. Plastic has very high calorific value, making it fit for waste to energy conversion using incineration plants. The changing volume and composition of solid waste creates logistical and operational challenges. The energy recovery based waste management facilities are not able to operate efficiently due to the change in the composition of waste. If recycling and segregation are being avoided, it creates a fiscal and operational imbalance. The operation on ISWM relies heavily on the recycling of waste, which is no longer possible due to COVID-19. The mass migration of people in developing countries may decrease the available manpower. In absence of segregation, the cost of waste management may increase significantly.

6. Unsustainable Lifestyle changes and its impact on the circular economy, 3R practices, and waste management

Any disaster event, natural or man-made, changes public perception on various facets of human life and often creates a socio-cultural and environmental shift. This changed behavior is observed through increased concern for the environment, advocacy for a sustainable lifestyle, and shift in consumption pattern of resource (material & product). The direct impact of the pandemic on public health like an increase in cases, increased hospitalization and morbidity is unprecedented, though remains manageable. However, the indirect impact of the pandemic on public health and especially on the ecosystem can be far worse and sustain for a longer duration. The public fear and control measures put sustainability and environmental concerns on hold. The most significant impact can be seen on the circular economy, especially on 3R practices, where each of the components is now affected at the individual, community, and governmental levels.

a. *Reduce*: The advocacy and public motivation to reduce the waste to curb carbon and plastic footprint remains dormant amid ongoing pandemic. The increased risk perception among the public due to media coverage and social media discourse, government, and concerned agencies urging to adopt social distancing and better hygiene practices, and scientific knowledge on the role of fomites in spreading infectious disease motivated the public to change their sustainable lifestyle. The key observable changes are increased consumption of SUPs, disposable items, and PPE, and increased solid waste generation. For better hygiene practices, the public opted for PPE, disinfecting wipes, tissue papers, and sanitizer bottles. Lockdown and social distancing put restrictions on mobility, leading to increased consumption of SUP products such as plastic plates, cups, spoons, forks, water bottles, straws, and food packaging containers. Much of this waste is of single-use type, which will end up being part of the waste as their usability ends. The public information campaigns and advocacy to reduce SUPs in recent years which changed public perception is facing major setbacks as the public is more concerned about looming COVID-19 threat than to reduce their carbon footprint. The major governmental emphasis has been put on saving lives that require the utilization of resources on a global scale. To meet the demand for resources, governments directed industries to stop their production and start creating more medical resources. This changes the reduction policy and interferes with resource balance. The increased consumables, mostly toxic, will end up as waste, increasing the carbon, energy, and waste footprint.

b. *Reuse:* The ongoing pandemic has not only increased the consumption of SUPs but also created a public perception to avoid reusing many products that have come in contact with the outer environment. New studies infer that the SARS-CoV-2 virus persists and remains active on the surface of various materials that the public uses from several hours to several days. The survival time of SARS-CoV-2 on plastic, paper, and cardboard surfaces has been a matter of discussion on traditional and social media. Lack of governmental and authoritative guidelines to address the reuse practices, widespread penetration of information relating to transmission of the virus through fomite created panic causing behavioral and lifestyle changes. Fears of contamination and convenience, hamper with the reuse practices. The increased consumption of products results in visible changes that include disposal of reusable products such as cardboard, paper products, and especially plastic bags after a single-use, regardless of the reusable nature of the products.

c. Recycle: In a resource scarce world with increased waste generation, the sustainable lifestyle requires recycling of the waste products at individual and community levels which forms the foundation of a circular economy. Amid COVID-19, the recycling practices have either been reduced or completely stopped. The fear of secondary contamination from the waste creates fear among those waste workers that are tasked to segregate and recycle waste in the formal sector. Logistical problems, especially the reduced workforce is forcing waste management officials to prioritize treatment and disposal of the overwhelming volume of waste, which can't be stored for longer duration due to special constraints. Informal waste and rag pickers also stopped their activity as governments are restricting mobility.

Additionally, as the recycling plants are shut down or working at reduced capacity, the production of recycled waste remains low, and increased logistical and operational challenges increase the cost of recycled waste. The sectors that utilize recycled waste is also affected by the pandemic. Their remains lower demand for recycled waste as the industries are operating at lower capacity and fear of secondary contamination from recycled waste.

The recycling business is also affected by the reduced cost of virgin plastics. To cope up with the low demand for fossil fuel, the oil industries are increasing the production of plastics at a cheaper cost. As the cost of plastic drops, the cost of plastic products will also drop and during the glooming financial uncertainties, the people will opt for cheaper products even if it contributes to an increase in carbon footprint. Sectors that consume recycled plastic waste as feed material has now access to cheaper original plastic, it will be very crucial to see if these industries priorities environment or economy.

Lastly, the plastic lobbying industries are capitalizing on the pandemic to further push their agenda where they are trying to change the public perception at the public level and asking the government to consider a more lenient approach towards plastic consumption **[85].** The effect of the pandemic on circular economy is going to be long-term and rectification measures will depend upon how fast the public perception and unsustainable consumption patterns will change to a more sustainable and eco-friendlier one. Pushing the reset button to move towards a sustainable circular economy will require participation from all stakeholders.

7. Solution and suggestion for handling and management of COVID-19 related pandemic solid waste

Environmental and man-made disasters have a devastating impact on both biotic and abiotic components of the environment. Presently global and national authorities are struggling to contain the spread of the virus and finding a vaccine. The emphasis has been put on saving lives and on public health, overlooking the other far-reaching consequences to the environment. The control and preventive measures adopted by authorities are generating an overwhelming amount of solid waste, which if not handled efficiently will keep deteriorating the ecosystem. The global solid waste and plastic waste generation was all-time high even before the pandemic. Finding a scientific and sustainable way to manage solid and plastic waste has been the focus of attention among authorities and the scientific community.

The unprecedented amount of COVID-19 waste is far too much for waste management facilities to handle. To effectively manage this waste, the waste-management and handling procedure should be such that it involves all stakeholders and simultaneously address the future concerns of sustainability. Keeping the current COVID-19 crisis in mind, here we propose a Pandemic Solid Waste Handling and Management (PSWM) strategy that can be modified and used to manage solid waste (medical waste, clinical solid waste, and municipal solid waste) during any pandemic event. The strategy involves all stake-holders and addresses all key components of traditional solid waste management strategy, with special attention on the infectious nature of the waste.

During the onset of a pandemic event, the authorities should immediately identify waste generation points. Temporary waste collection centers can be set-up depending upon proximity to treatment facilities and the volume of the waste generated. Concerned authorities should issue public guidelines on how to collect and store generated waste at the point of generation (hospitals, quarantine centers, households, etc.). The guidelines should ensure minimization of the risk of secondary contamination. On waste handling end, training and educational campaign for personnel safety is the key to avoid waste personal from contracting the disease.

A proper waste collection, transportation, storage, treatment, and disposal strategy needs to be communicated to the waste management agencies. Increased financial and personnel support, distribution of PPE, sanitation liquids, and allocation of dedicated equipment must be priorities.

The ongoing pandemic highlights the crucial role that plastic plays in our life. On one hand, increased demand for PPEs and SUPs generates substantial amounts of plastic waste, and on the other hand, due to infection fears, the public is moving away from sustainable practices mainly 3R (reduce, reuse, recycle), further adding on to the plastic problem. The plastic footprint created during the pandemic event can be seen up to thousands of years and hence it requires special attention. The increased plastic waste can be good for waste to energy recovery facilities, but polluting emissions will also increase.

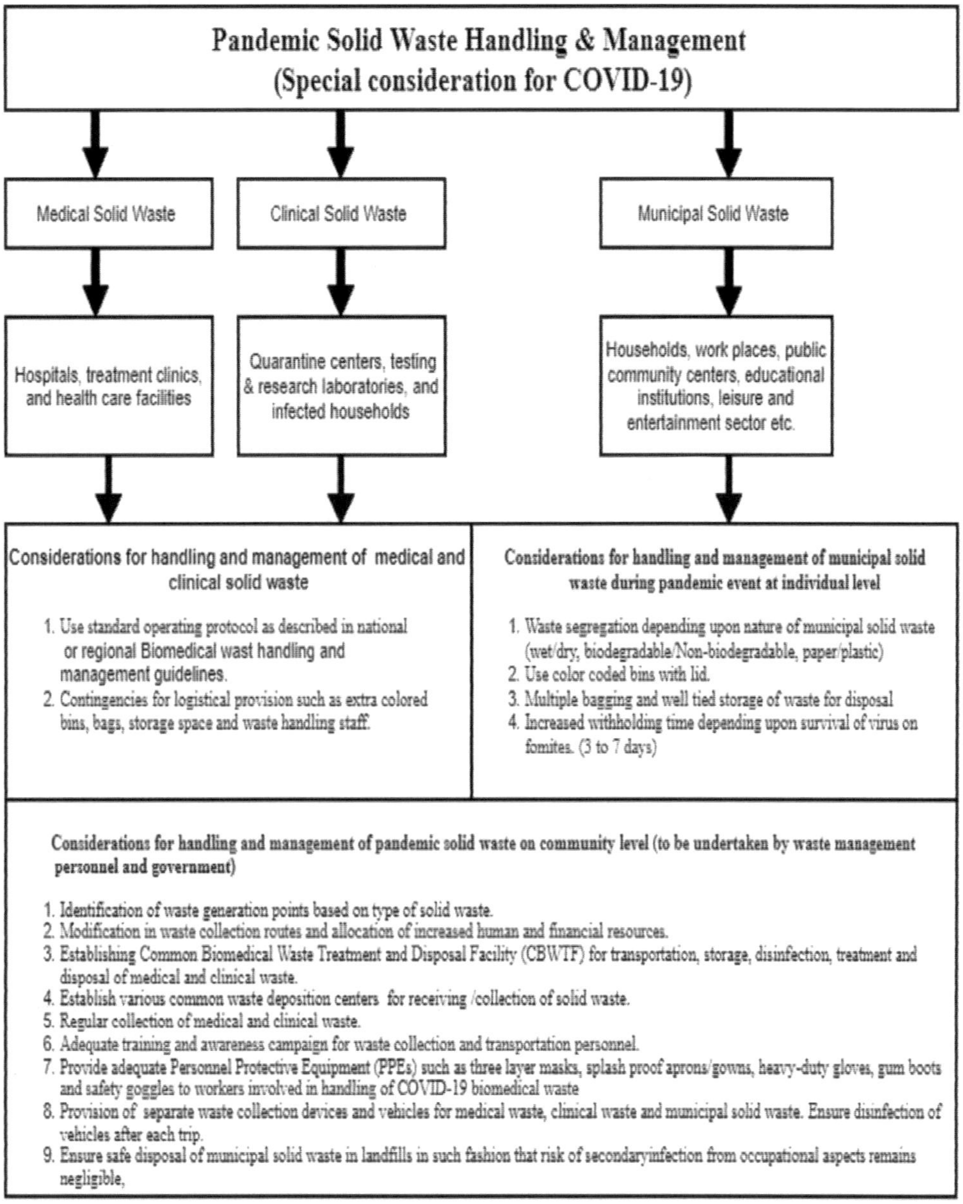

Fig.3. Proposed COVID-19 related pandemic solid waste handling and management. (Based on information obtained from [86-88]

The ISWM which is usually designed on certain assumptions (steady-state flow rate of waste generation and percentage of recycled and reuse) are no longer valid. The concept of **Plastic Waste Footprint (PWF)** can be used to track the role of a pandemic event on plastic waste management strategy. The concept of PWF which accounts for 3R practices makes it an effective tool in understanding the impact of a pandemic on sustainability practices. *"PWF is defined as the total mass of plastic waste used for a process, product or service minus amount of plastic avoided reused + recycled + reprocessed"* **[45]**. Calculating PWF before a pandemic, during a pandemic, and after pandemic will help us in assessing the waste management strategies severely hit by the pandemic and can be used to strengthen those facilities.

The waste management facilities are facing major operational challenge to accommodate the increased volume of solid waste that too of changed compositions. Clinical solid waste is to be treated as medical waste is hazardous and requires separate handling, treatment, and disposal methods. The already stressed medical waste treatment facilities require adequate modification and treatment alternatives other than existing facilities. In this regard, municipal solid waste incineration plant and cement plant can be used to reduce the load with appropriate modifications that can effectively reduce the dangers associated with toxicity. The incineration temperature to ensure the destruction of waste and pathogens varies with countries. Additionally, various countries (Germany, China, Norway, and Spain) have contingency plans already in place to cope with the surge in medical and clinical waste **[45]**.

Streamlining waste handling and management practices, making a circumstance (COVID-19) based substantial changes in existing solid waste management practices, advocacy for sustainable practices, public awareness, and safety campaigns can be way forward to maintain a circular economy and deal with the solid waste generated during the current pandemic. Appropriate awareness campaigns to counter plastic advocating lobby, deescalating the public fear, promoting safe 3R practices on the governmental level will ensure that behavioral changes won't be long-term, or else, it will be very challenging to return to sustainability, once this all over.

8. Status of solid and plastic waste handling and management in developed (high-income group) and developing (low-income group) countries

Global solid waste generation is all-time high with 2 billion tons of waste being generated annually. The annual waste generation rate is expected to rise 70 percent by 2050 if we keep business as usual. The figure calculated by **World Bank in 2018**, doesn't account for any unforeseen global disasters such as an ongoing pandemic. Additionally, the rate of solid waste generation along with solid waste handling and management policies that threaten our eco-system, varies from country to country, especially among developed and developing countries **[63].**

The severity of the global pandemic varies from country to country and accordingly the response to pandemic varies. As discussed earlier, the severity and level of response affect resource consumption and subsequent waste generation. It will be interesting to compare the status of pandemic solid waste generation from developed and developing countries and also the waste handling and management practices.

The virus first started from china and quickly spread across the developed nations such as Italy, Spain, Germany, and the USA, with Wuhan, New York, and Italy becoming the hotspots for COVID-19 by end of March 2020. The other nations, though comparatively less affected, undertook containment and precautionary measures. With the continued spread of the virus more countries like Russia, India, Brazil, Mexico, and Iran, etc. also became global hotspots. Reports of the second wave of disease in already affected china and South Korea are surfacing **[89]**. The solid waste generation trend will follow the spread of the virus with areas that are already affected generating more waste.

High-income countries generate (34 percent) nearly seven times as much waste generated by the lower-income group (5 percent) as shown in figure 4, confirming that advancement, development, and affluence generate more waste. The resource-rich developed countries have better infrastructure for solid waste management with optimum waste collection to efficient engineered waste treatment and disposal.

The integrated waste management including comprehensive collection of segregated waste is the foundation of waste management. Solid waste generating from the municipalities must be collected in entirety, to minimize the threat from mismanaged waste. Developed countries collect almost all the waste (100%) while developing countries are struggling with waste collection (30%). The management of solid waste also varies country wise. Traditionally, open dumping was practiced throughout the globe to deal with waste.

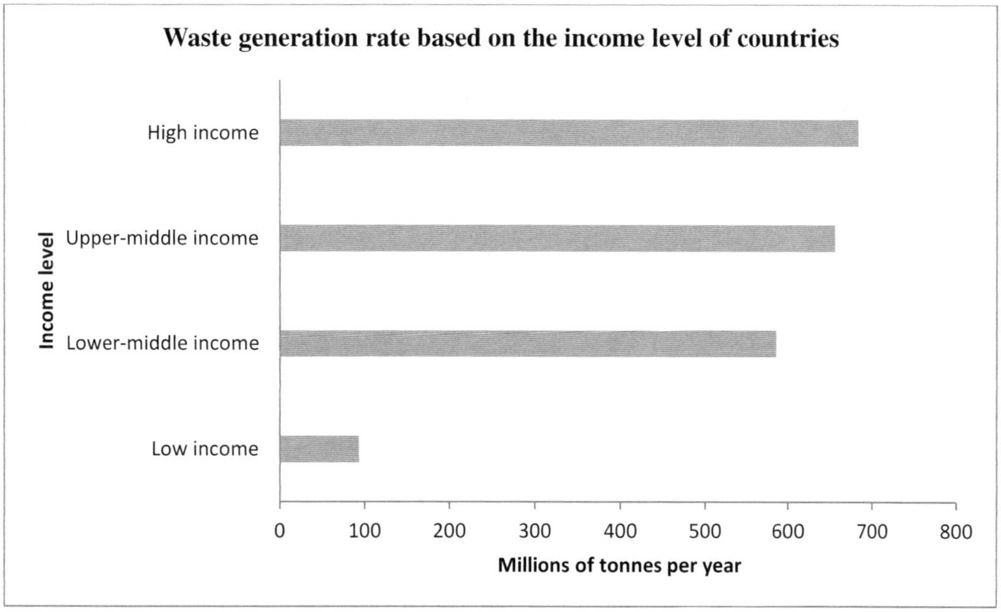

Fig. 4. Solid waste generation rate based on the income level of countries (based on data reported by **[63]**)

As the rate of waste generation increased and the scientific body of knowledge on waste management developed, landfills, incineration plants, and recycling methods started gaining popularity **[90]**. 40% of waste is disposed of in landfills while only 19% of the waste is recycled for material recovery. Energy recovery from waste to energy-based incineration plants though remains problematic and handles 11% of global solid waste **[63]**. "The Global Waste Index which ranks 36 countries within the Organization for Economic Co-operation and Development (OECD) on basis of most environmentally-friendly methods of waste management on per capita basis highlights that developed countries like Spain, Italy, Canada, USA, New Zealand and Israel not only generate an enormous amount of waste but also found to be less efficient with waste recycling and management, despite having better infrastructure" **[91].**

The COVID-19 related PSWM requires a joint effort from both developing and developed countries as the trans-boundary migration of contaminated waste is a possibility that cannot be overlooked. Developed countries with efficient collection and management systems with contingencies can handle the PSW by modifying the existing protocols to accommodate the diverse nature of pandemic waste and strict enforcement of waste disposal laws among the public.

Resource challenged COVID-19 affected developing countries are diverting their resources in healthcare infrastructure and sanitation drive. The manpower responsible for waste collection is mostly engaged in sanitizing the habitats creating logistical problems in waste management. Open dumping of solid waste is prevalent where landfills are not yet available or available in limited quantity and 93% of waste being dumped in open dumping sites, or local water bodies. This coupled with the possibility of secondary transmission due to pandemic affected waste raises an alarming concern. This may further worsen as the number of cases in developing countries keeps increasing. [**Table 2**] summarizes the top 10 countries with the highest number of reported COVID-19 cases and solid waste generation and recycling rate.

The amount of plastic waste generated throughout the years has threatened every aspect of our ecosystem. It has been estimated that 6300 million tonnes of plastics waste was generated between 1950 and 2015. 80% of plastic waste was mismanaged (figure 5) and directly discarded either in the environment or in landfills, where they persist for thousands of years **[92].**

Table 2. Top 10 countries with the highest number of reported COVID-19 cases and solid waste generation & recycling rate. **(Source: Data accessed on 15-June-2020 from** https://www.worldometers.info and [63])

S.No.	Country	Cases	Death	Population	MSW (tons/year)	Medical Waste (tons/year)	Plastic composition (%)	Recycling rate of municipal solid waste (%)
1	USA	2,142,224	117,527	330,912,290	258,000,000.00	2,600,000.00	12.90	34.60
2	Brazil	850,796	42,791	212,488,104	79,889,010.00	260,063.00	13.50	1.40
3	Russia	520,129	6,829	145,931,723	60,000,000.00	-	14.21	4.50
4	India	321,626	9,199	1,379,344,820	168,403,239.95	180,784.50	-	5.00
5	UK	294,375	41,662	67,869,522	31,567,000.00	4,218,205.00	20.20	27.30
6	Spain	290,685	27,136	46,753,986	20,151,000.00	2,990,346.00	9.00	16.80
7	Italy	236,651	34,301	60,465,633	29,524,000.00	16,071,024.00	11.60	25.90
8	Peru	225,132	6,498	32,948,535	8,356,711.13	7,176.00	10.52	4.00
9	Germany	187,423	8,867	83,771,837	51,046,000.00	8,821,344.00	13.00	47.80
10	Iran	184,955	8,730	83,939,011	17,885,000.00	1,609,650.00	8.50	5.00

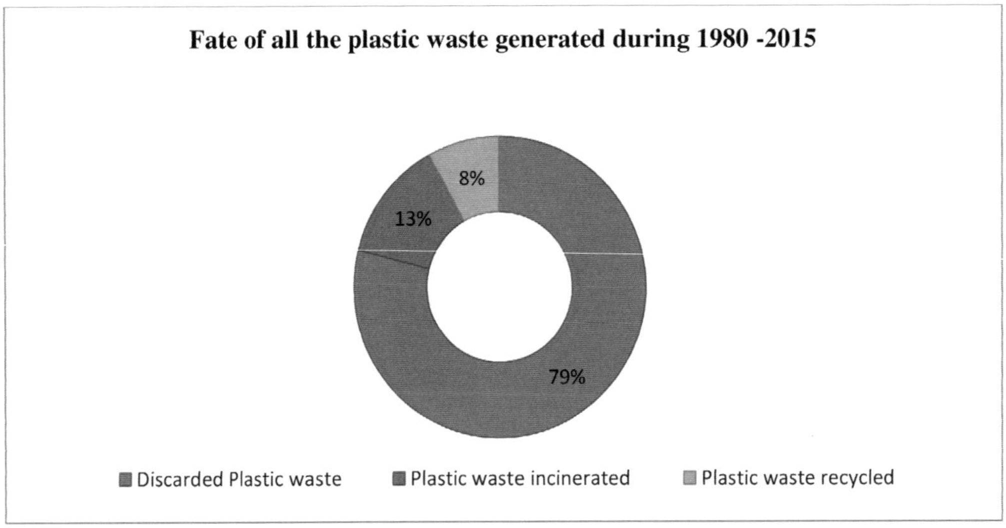

Fig. 5. Solid waste generation rate based on the income level of countries **(based on data reported by Geyer et. al., 2017)**

Besides the physical damage, they pose to wildlife and ecosystem, the indirect association with toxic chemicals and pathogens is threatening the biodiversity of environmental matrices. These mismanaged plastic waste also generate microplastics, small pieces of plastic with a size range below 5 mm. Besides plastic pollutant that is already present in an aquatic and terrestrial environment, an additional 5 to 13 million tons are being disposed of annually. This quantity will further increase due to ongoing pandemic which is generating a flux of plastic waste. The daily PPE consumption by the top 10 countries with the highest number of reported COVID-19 cases is calculated using equation 1 and shown in **table 3**. Based on our assumptions, 3.75 billion PPE waste is generated daily from 10 countries. The figure will likely to be lower, considering the acceptability rate varies with public perception.

Table 3. Top 10 countries with the highest number of reported COVID-19 cases and estimated daily PPE waste generation.

S. No.	Country	Population*	Urban Population (%)	Urban Population	Accept-ance Rate	Rural Population	Accep-tance Rate	Daily PPE Use	Daily PPE waste Generated
1	USA	330,912,290	83	273664463.8	1.00	57,247,826	0.80	2	6.39E+08
2	Brazil	212,488,104	87.1	185077138.6	0.80	27,410,965	0.60	2	3.29E+08
3	Russia	145,931,723	74.8	109156928.8	1.00	36,774,794	0.80	2	2.77E+08
4	India	1,379,344,820	34.9	481391342.2	0.80	897,953,478	0.60	2	1.85E+09
5	UK	67,869,522	83.9	56942528.96	1.00	10,926,993	0.80	2	1.31E+08
6	Spain	46,753,986	80.8	37777220.69	1.00	8,976,765	0.80	2	8.99E+07
7	Italy	60,465,633	71	42930599.43	1.00	17,535,034	0.80	2	1.14E+08
8	Peru	32,948,535	78.3	25798702.91	0.80	7,149,832	0.60	2	4.99E+07
9	Germany	83,771,837	77.5	64923173.68	1.00	18,848,663	0.80	2	1.60E+08
10	Iran	83,939,011	75.9	63709709.35	0.80	20,229,302	0.60	2	1.26E+08

The management of plastic wastes either through landfill or thermal methods, though generates energy, is not environmentally friendly due to emissions and scarcity of land. Life-cycle assessments of plastics management show that 3R based integrated plastic management practices are more economical end eco-friendly with significantly lower greenhouse gas footprint when compared with other methods. **[93-94].**

This is very crucial information for developing countries where informal plastic waste recycling is prevalent. Secondary transmission of the virus from dumped contaminated plastic waste especially gloves and masks can be threatening for those involved in such sectors. Governments of developing countries must take immediate action to address this problem. As the income level of the country increases, the resource utilization will further increase. The existing 'produce-consume-dispose' led linear economy is unsustainable in the future. The concept of a circular economy is being explored and there is an increased concern towards emissions from waste to energy plastic management plants, the advocacy for 3R based waste management keeps increasing. The sustainable circular economy to reduce carbon, energy, and plastic footprint coupled with more environment-friendly plastic waste treatment and disposal to reduce the emissions needs to be long term priority of government and industries with due consideration being given to the effect of the pandemic.

9. Future Research and Policy Priorities

COVID-19 which started from Wuhan has turned into a global pandemic on March 11, 2020. The widespread transmission around the globe has affected most of the nations in all continents. A pandemic has deleterious effects not only on public health but also on the environment. The priority concern right now is containment and eradication of the pandemic, ensuring the safety of public.

However, COVID-19 related control and mitigation strategies are generating a whooping amount of pandemic related solid waste, if not managed properly, not only poses the risk of the secondary outbreak but also has a long-term effect on the environment. The pandemic is also affecting the circular economy and all efforts made towards sustainability are in grave danger. A pandemic of such scale is a wakeup call for us and needs to be used to priorities the environment, ecosystem, and public health simultaneously. Keeping that in mind here we highlight a few important points that need to be a priority concern for researchers and policy makers.

a. *Pandemic solid waste handling and management guidelines*
The mismanagement and threat of solid waste generating during current pandemic calls for the creation of separate domain in the waste management field dedicated to biological disasters. Pandemic solid waste handling and management (PSWM) guidelines that focuses on dealing with medical, clinical, and municipal solid waste management should be prioritized. The guidelines must address the sudden shift in volume and composition of waste. Waste collection, transportation, storage, segregation, treatment, and disposal of pandemic solid waste should be such that it minimizes the risk of secondary contamination to the waste workers. Additional funds must be allocated to deal with logistical and fiscal challenges amid pandemic.

b. *Collaborative response*
The scale at which the current pandemic has spread calls for collaborative response encompassing all countries and regions on all fronts. The policy and guidelines to address pandemic solid waste should not be limited to local scale. The trans-boundary migration of pandemic related solid waste poses the risk of secondary transmission, migrates the disease to unaffected areas, and presents an eco-toxicological threat to the eco-system. The government and academia must work together to identify transport mechanisms through which COVID-19 related solid waste is migrating must be identified and preventive measures should be adopted to stop it.

c. *The utility of plastics during a pandemic and finding an alternative*
The current pandemic highlights the importance and dependence of plastics on making human life safe and comfortable. Additionally plastic forms the backbone of medical care. On the individual level, single-use plastic products have now become not only convenient but also safe to use. The skyrocketing demand for PPE is merely the tip of the iceberg. The world was already fighting with plastic pollution, this pandemic has both long-term and short-term impact on the way we perceive and consume plastic. The need of searching the alternative of plastic material is now more than ever. Collaborative studies on replacement of plastic used in the health-care industry with eco-friendly material must be undertaken by researchers. Additionally, a logistical and efficient framework needs to be prepared for the effective allocation of plastic resources keeping sustainability and public health priority in mind for any future events.

d. *Countering fake information and increased advocacy for a sustainable lifestyle and circular economy*
The sustainable circular economy is not some abstract concept and groundbreaking progress by re-imagining resource utilization with pollution mitigation is observable in many places. Advocacy and mass campaign for sustainable practices such as 3R yielded concrete public behavioral changes where people were more cautious of practicing recycling and reuse while avoiding consumption of single-use plastics. The pandemic halted this crucial progress and public perception changed towards single-use plastics. The spread of fake information and lack of prompt governmental response encourages behavioral changes in the consumption and recycling of plastics. Plastic industries are now lobbying against banning single-use plastics by citing results of studies on the survival of viruses on fomites **[85]**.

The argument made so far is unscientific and twisted to create a pro-plastic narrative to persuade people towards unsustainable consumption of plastics. The governments need to contain the spread of such fake news, to ensure that people don't fall back to the linear system economy in the post-pandemic world which is known to have the negative consequence to people, environment, and eco-system as a whole. The prompt governmental response for containment of the virus also suggests that the governments can address unsustainable practices. Detailed studies and campaigns on consumer (fresh plastics and recyclable plastic) end are required to transcend the benefits of circular economy and sustainable lifestyle. Lastly, a swift governmental response towards plastic recycling industries is also important to ensure that sustainable recycling practices are not compromised.

***e.** Contingencies and plan of action to deal with the indirect impact of a pandemic on waste management*

A pandemic or any disaster is known to bring societal and behavioral changes. The environmental and economic impact across different sectors and geographical scales is going to create many future challenges. Other indirect impacts include a change in consumption pattern and migration of the workforce. The consumption and behavioral changes amid pandemic may influence the environmental footprints such as carbon footprint, water footprint, and plastic waste footprint. These frameworks need to be reassessed and recalibrated to address the impact of the pandemic to gives us a better understanding of ways a pandemic affects the environment.

Additionally, pandemic related sudden mass migration puts pressure on the local eco-system. The scale and duration of such migration and impact on local environmental resources should be a research priority. Additionally, the mass migration will interfere with the social and economic dynamics at origin and destination end. For example, mass migration also reduces human resource in the waste management sector. Government and researchers need to handle this issue on a priority basis to manage the increased pandemic solid waste and social science and economic experts' needs to study the effect of migration in detail as the local economy is intertwined with the national circular economy and environment.

***f.** Creation, development, and application of scientific or engineering tools and frameworks to deal with pandemic related solid waste*

The severity of COVID-19 calls for the advancement of technologies to be used in fighting and controlling the pandemic. Contact tracing using a smart-phone application is an important tool that is being used by governments during this pandemic. Additionally, fab lab and 3D printing are being widely used at local levels for fabrication of face masks and medical items such as cotton masks, face shields, and ventilators for hospitals. The application of artificial intelligence and machine learning for optimum allocation of waste management resources during pandemic events will ease the burden and may eliminate possibility of mismanagement or error. From predicting and identifying the waste hot spots that require immediate collection, finding optimum waste collection frequency based on affected clusters, optimum pickup routes to efficient treatment and disposal methods can easily be accomplished. Sensor based monitoring, data collection, analytics and AI can further help in optimizing local and regional solid waste response strategies which will transcend the coordinated effort on national level. Using drones to identify pandemic waste, robotics for collection of contaminated wastes needs to be explored for feasibility. The data obtained from COVID-19 can be analyzed to create a technological framework that can be used during future pandemic or epidemic events.

10. Concluding remarks

The anthropogenic pollution of the environment is a priority concern for scientists, governments, and the public. Severe acute respiratory syndrome coronavirus 2 (SARS-CoV-2) pandemic has affected most nations throughout the globe. The severity and reach of the contagion resulted in prompt governmental response with various containment and mitigation measures regionally and globally. These measures brought short-term positive environmental changes. The long-term adverse environmental impact of this pandemic in terms of a sudden surge in increased solid waste generation requires immediate attention of all stakeholders.

The major composition of pandemic solid waste is comprised of plastic. The sudden surge in the volume of plastic waste due to increased consumption of PPE, medical consumables, and appliances and SUPs, along with unsustainable lifestyle changes, puts a burden on solid waste management facilities. Pandemic solid waste, such as gloves and masks, has already found its way to the environment. These pandemic related plastic waste, if not managed properly will further pollute environmental matrices as they are known to persist for a long time.

The plastic waste generation sources are highlighted and discussed in detail. Further, we examined the consumption pattern of various polymer types and identified the plastic types and sectors that will keep surging in near future. During the onset of the pandemic, the volume of medical solid waste and clinical solid waste increased drastically. This trend will continue as the number of cases and rates of hospitalization increase. The demand for PP, PE, PS, PET, nylon, and nitrile, commonly used to manufacture PPE, medical consumables, SUPs, packaging material, food containers, and bottles will keep increasing at a steady rate. The demand for plastic is expected to increase in medical, leisure, and online shopping sectors, this trend is also expected to maintain a steady pace in the post-pandemic world. Among various factors that affect the quantity of plastic waste, the duration of pandemic and unsustainable change in lifestyle seem to be dominant.

The solid waste management facilities were already operating at extreme conditions to cope with ever-increasing solid waste. The pandemic solid waste generates overwhelming stress on these facilities. The major challenges include an increase in volume and change in solid waste composition. The logistical and operational challenges require additional resources. There is an urgent need to formulate COVID-19 pandemic solid waste handling and management action plan, prioritizing the safety of waste workers.

Amid fear of contacting viruses from fomites, the sudden change in lifestyle, behavior, and consumption pattern puts sustainable circular economy on hold. On the governmental level, restriction on SUPs is either lifted or temporarily put on hold. The public is increasing the consumption of SUPs while avoiding 3R practices. Plastic industries are already lobbying against the existing ban or restrictions on plastics. All of these factors are collectively in play, which not only generates a whopping amount of plastic waste but also hampers with the efforts to shift from an unsustainable linear economy to a sustainable and environment-friendly circular economy. The delayed governmental response will not only decelerate the shift but all the progress made towards public awareness may be lost.

The solution lies in involving all the stakeholders and streamlining effective pandemic solid waste management response. Though developed countries are thought to be efficient in solid waste management, with engineered waste disposal systems, the progress made by developing countries can't be overlooked in this regard, given the dynamic nature of environmental pollutants and trans-boundary migration, the effective waste management system should start from regional level and streamlined to the global scale.

Prioritizing public health while ignoring environmental concerns may be a valid short-term response, but in the longer run, both aspects require equal attention due to their interdependence. The future governmental outlook and research priority can use the points discussed in the relevant section. In the post-pandemic world, creation of an environment friendly, energy-efficient, and wholesome earth should be the priority of government, public and scientific community. The post-pandemic world needs a global initiative that involves all stakeholders to rebuild mother earth.

***Author Contributions*:** Both authors have equal contribution in preparation of manuscript.

Funding: This research received no external funding.

Acknowledgments: The authors would like to acknowledge administrative and technical staff of National Institute of Technology, Raipur for their valuable support.

Conflicts of Interest: The authors declare no conflict of interest.

References:

1. World Health Organization. Basic protective measures against the new coronavirus. 2020 Apr;28(03).
2. Centers for Disease Control and Prevention. Coronavirus disease 2019 (COVID-19): symptoms of coronavirus. Atlanta: Centers for Disease Control and Prevention. 2020 Jun 4.
3. Chen S, Yang J, Yang W, Wang C, Bärnighausen T. COVID-19 control in China during mass population movements at New Year. The Lancet. 2020 Mar 7;395(10226):764-6.

4. Hellewell, J., Abbott, S., Gimma, A., Bosse, N. I., Jarvis, C. I., Russell, T. W., ... & Flasche, S. (2020). Feasibility of controlling COVID-19 outbreaks by isolation of cases and contacts. *The Lancet Global Health*. https://doi.org/10.1016/S2214-109X(20)30074-7
5. Bedford J, Enria D, Giesecke J, Heymann DL, Ihekweazu C, Kobinger G, Lane HC, Memish Z, Oh MD, Schuchat A, Ungchusak K. COVID-19: towards controlling of a pandemic. The lancet. 2020 Mar 28;395(10229):1015-8.
6. Lau H, Khosrawipour V, Kocbach P, Mikolajczyk A, Schubert J, Bania J, Khosrawipour T. The positive impact of lockdown in Wuhan on containing the COVID-19 outbreak in China. Journal of travel medicine. 2020 May 18.
7. Gupta S, Nguyen T, Raman S, Lee B, Lozano-Rojas F, Bento A, Simon K, Wing C. Tracking public and private responses to the COVID-19 epidemic: evidence from state and local government actions. American Journal of Health Economics. 2021 Sep 1;7(4):361-404.
8. Guerrieri V, Lorenzoni G, Straub L, Werning I. Macroeconomic implications of COVID-19: Can negative supply shocks cause demand shortages?. National Bureau of Economic Research; 2020 Apr 6.
9. Saadat S, Rawtani D, Hussain CM. Environmental perspective of COVID-19. Science of the Total environment. 2020 Aug 1;728:138870.
10. Sharma S, Zhang M, Gao J, Zhang H, Kota SH. Effect of restricted emissions during COVID-19 on air quality in India. Science of the Total Environment. 2020 Aug 1;728:138878.
11. Mahato S, Pal S, Ghosh KG. Effect of lockdown amid COVID-19 pandemic on air quality of the megacity Delhi, India. Science of the total environment. 2020 Aug 15;730:139086.
12. Dutheil F, Baker JS, Navel V. COVID-19 as a factor influencing air pollution?. Environmental pollution. 2020 Aug 1;263:114466.
13. Isaifan RJ. The dramatic impact of Coronavirus outbreak on air quality: has it saved as much as it has killed so far?. Global Journal of Environmental Science and Management. 2020 Jul 1;6(3):275-88.
14. Shrestha AM, Shrestha UB, Sharma R, Bhattarai S, Tran HN, Rupakheti M. Lockdown caused by COVID-19 pandemic reduces air pollution in cities worldwide.
15. Pandit R. Basic protective measures against the new coronavirus pandemic–COVID-19. J Manag Res Anal. 2020;7(1):1-2.
16. Centers for Disease Control and Prevention. Coronavirus disease 2019 (COVID-19): symptoms of coronavirus. Atlanta: Centers for Disease Control and Prevention. 2020 Jun 4.
17. Adams JG, Walls RM. Supporting the health care workforce during the COVID-19 global epidemic. Jama. 2020 Apr 21;323(15):1439-40.
18. World Health Organization. Water, sanitation, hygiene and waste management for COVID-19: technical brief, 03 March 2020. World Health Organization; 2020.
19. Daily Sabah, 2020, (www article), Expert warns of potential drought in Turkey due to increased water use amid COVID-19 outbreak. https://www.dailysabah.com/turkey/expert-warns-of-potential-drought-in-turkey-due-to-increased-water-use-amid-covid-19-outbreak/news (Accessed 13 June 2020).
20. Hill K, Zamyadi A, Deere D, Vanrolleghem PA, Crosbie ND. SARS-CoV-2 known and unknowns, implications for the water sector and wastewater-based epidemiology to support national responses worldwide: early review of global experiences with the COVID-19 pandemic. Water Quality Research Journal. 2021 May;56(2):57-67
21. WHO. 2020g. (www article) Shortage of personal protective equipment endangering health workers worldwide. https://www.who.int/news-room/detail/03-03-2020-shortage-of-personal-protective-equipment-endangering-health-workers-worldwide.(Accessed : 7 June 2020).
22. WHO. 2020f. *Rational use of personal protective equipment (PPE) for coronavirus disease (COVID-19): interim guidance, 19 March 2020* (No. WHO/2019-nCoV/IPC PPE_use/2020.2). World Health Organization
23. KDMARKETINSIGHT, 2020, (www article), Medical plastics market receives significant impetus during covid-19 outbreak; demand to sustain post pandemichttps://3wnews.org/news/873516/medical-plastics-market-receives-significant-impetus-during-covid-19-outbreak-demand-to-sustain-post-pandemic/. (Accessed : 7 June 2020).

24. Cooke SJ, Twardek WM, Lynch AJ, Cowx IG, Olden JD, Funge-Smith S, Lorenzen K, Arlinghaus R, Chen Y, Weyl OL, Nyboer EA. A global perspective on the influence of the COVID-19 pandemic on freshwater fish biodiversity. Biological Conservation. 2021 Jan 1;253:108932.
25. Allam Z. The second 50 days: a detailed chronological timeline and extensive review of literature documenting the COVID-19 pandemic from day 50 to day 100. Surveying the Covid-19 pandemic and its implications. 2020:9.
26. Spinazzè A, Cattaneo A, Cavallo DM. COVID-19 outbreak in Italy: protecting worker health and the response of the Italian Industrial Hygienists Association. Annals of work exposures and health. 2020 Jul 1;64(6):559-64.
27. KDMARKETINSIGHT, 2020, (www article), Medical plastics market receives significant impetus during covid-19 outbreak; demand to sustain post pandemichttps://3wnews.org/news/873516/medical-plastics-market-receives-significant-impetus-during-covid-19-outbreak-demand-to-sustain-post-pandemic/. (Accessed : 7 June 2020).
28. Cook TM. Personal protective equipment during the coronavirus disease (COVID) 2019 pandemic–a narrative review. Anaesthesia. 2020 Jul;75(7):920-7.
29. Bengali S. The COVID-19 pandemic is unleashing a tidal wave of plastic waste. The Los Angeles Times. 2020 Jun 13.
30. Ranney ML, Griffeth V, Jha AK. Critical supply shortages—the need for ventilators and personal protective equipment during the Covid-19 pandemic. New England Journal of Medicine. 2020 Apr 30;382(18):e41.
31. Nghiem LD, Morgan B, Donner E, Short MD. The COVID-19 pandemic: considerations for the waste and wastewater services sector. Case studies in chemical and environmental engineering. 2020 May 1;1:100006.
32. Nzediegwu C, Chang SX. Improper solid waste management increases potential for COVID-19 spread in developing countries. Resources, conservation, and recycling. 2020 Oct;161:104947.
33. Van Doremalen N, Bushmaker T, Morris DH, Holbrook MG, Gamble A, Williamson BN, Tamin A, Harcourt JL, Thornburg NJ, Gerber SI, Lloyd-Smith JO. Aerosol and surface stability of SARS-CoV-2 as compared with SARS-CoV-1. New England journal of medicine. 2020 Apr 16;382(16):1564-7.
34. Klemeš JJ, Van Fan Y, Tan RR, Jiang P. Minimising the present and future plastic waste, energy and environmental footprints related to COVID-19. Renewable and Sustainable Energy Reviews. 2020 Jul 1;127:109883.
35. Verma R, Vinoda KS, Papireddy M, Gowda AN. Toxic pollutants from plastic waste-a review. Procedia Environmental Sciences. 2016 Jan 1;35:701-8.
36. Prata JC, Silva AL, Walker TR, Duarte AC, Rocha-Santos T. COVID-19 pandemic repercussions on the use and management of plastics. Environmental science & technology. 2020 Jun 12;54(13):7760-5.
37. Eriksen M, Lebreton LC, Carson HS, Thiel M, Moore CJ, Borerro JC, Galgani F, Ryan PG, Reisser J. Plastic pollution in the world's oceans: more than 5 trillion plastic pieces weighing over 250,000 tons afloat at sea. PloS one. 2014 Dec 10;9(12):e111913.
38. Xanthos D, Walker TR. International policies to reduce plastic marine pollution from single-use plastics (plastic bags and microbeads): A review. Marine pollution bulletin. 2017 May 15;118(1-2):17-26.
39. Ilyas M, Ahmad W, Khan H, Yousaf S, Khan K, Nazir S. Plastic waste as a significant threat to environment–a systematic literature review. Reviews on environmental health. 2018 Dec 1;33(4):383-406.
40. Ziraba AK, Haregu TN, Mberu B. A review and framework for understanding the potential impact of poor solid waste management on health in developing countries. Archives of Public Health. 2016 Dec;74(1):1-1.
41. Sasao T. Cost and efficiency of disaster waste disposal: A case study of the Great East Japan Earthquake. Waste management. 2016 Dec 1;58:3-13.
42. Dugar N, Karanjit S, Khatiwada NR, Shakya SM, Ghimire A. Post-disaster waste management: lessons learnt from 2015 Nepal earthquake. InSustainable Waste Management: Policies and Case Studies 2020 (pp. 465-483). Springer, Singapore.

43. Kalina M, Tilley E. "This is our next problem": Cleaning up from the COVID-19 response. Waste Management. 2020 May 1;108:202-5.
44. Australian standard, (2018) *Management of clinical and related wastes.*
45. KDMARKETINSIGHT, 2020, (www article), Medical plastics market receives significant impetus during covid-19 outbreak; demand to sustain post pandemichttps://3wnews.org/news/873516/medical-plastics-market-receives-significant-impetus-during-covid-19-outbreak-demand-to-sustain-post-pandemic/. (Accessed : 7 June 2020).
46. Klemeš JJ, Van Fan Y, Tan RR, Jiang P. Minimising the present and future plastic waste, energy and environmental footprints related to COVID-19. Renewable and Sustainable Energy Reviews. 2020 Jul 1;127:109883.
47. Yuan L, Zhi N, Yu C, Ming G, Yingle L, Kumar GN, Li S, Yusen D, Jing C, Dane W, Xinjin L. Aerodynamic characteristics and RNA concentration of SARS-CoV-2 aerosol in Wuhan
48. Nelson KE, Williams CM. Early History of Infectious Disease: Epidemiology and Control of. Infectious disease epidemiology: theory and practice. 2014:3.
49. Gershon, R. R., Karkashian, C. D., Grosch, J. W., Murphy, L. R., Escamilla-Cejudo, A., Flanagan, P. A., ... & Martin, L. (2000). Hospital safety climate and its relationship with safe work practices and workplace exposure incidents. *American journal of infection control*, *28*(3), 211-221.
50. Gamage B, Moore D, Copes R, Yassi A, Bryce E, BC Interdisciplinary Respiratory Protection Study Group. Protecting health care workers from SARS and other respiratory pathogens: a review of the infection control literature. American journal of infection control. 2005 Mar 1;33(2):114-21.
51. Babanyara, Y.Y., Ibrahim, D.B., Garba, T., Bogoro, A.G. and Abubakar, M.Y., 2013. Poor Medical Waste Management (MWM) practices and its risks to human health and the environment: a literature review. *Int J Environ Ealth Sci Eng*, *11*(7), pp.1-8.
52. GlobalData Healthcare**, 2020, (www article)** No sign of relief for healthcare workers still struggling with PPE shortage. https://www.medicaldevice-network.com/comment/ppe-shortage-covid-19/
53. Dargaville T, Spann K, Celina M. Opinion to address the personal protective equipment shortage in the global community during the COVID-19 outbreak. Polymer degradation and stability. 2020 Jun 1;176:109162.
54. Press C. Coronavirus: The NHS workers wearing bin bags as protection. BBC News. 2020 Apr 5;5
55. Habib ZH, Alam AN. Biosafety Measures in the Laboratory during Handling of Specimens of Suspected COVID-19 Patients. Bangladesh Journal of Infectious Diseases. 2020 Apr 24:S66-8.
56. Hong, K. H., Lee, S. W., Kim, T. S., Huh, H. J., Lee, J., Kim, S. Y., ... & Kim, J. S. (2020). Guidelines for laboratory diagnosis of coronavirus disease 2019 (COVID-19) in Korea. *Ann Lab Med*, *40*(5), 351-360.
57. Richard VS, Kanchi M. A practical guide to a safe sequence of donning and doffing of personal protective equipment in relation to COVID-19. Journal of Cardiac Critical Care TSS. 2020 Aug;4(01):61-3.
58. Ma, Y., Lin, X., Wu, A., Huang, Q., Li, X., & Yan, J. (2020). Suggested guidelines for emergency treatment of medical waste during COVID-19: Chinese experience. *Waste Disposal & Sustainable Energy*, 1
59. Christopher. D. J., & Tharayan. P., 2020, (www article), Masks are mandatory for all now. https://www.thehindu.com/news/cities/chennai/masks-are-mandatory-for-all-now/article31400601.ece (Accessed: 1 June 2020).
60. Howard J, Huang A, Li Z, Tufekci Z, Zdimal V, van der Westhuizen HM, von Delft A, Price A, Fridman L, Tang LH, Tang V. An evidence review of face masks against COVID-19. Proceedings of the National Academy of Sciences. 2021 Jan 26;118(4).
61. M. & M., 2020a, (www article), Medical Plastics Market by Type (Engineering Plastics, HPP, Standard Plastics, Silicone, and Others), Application (Medical Disposables, Prosthetics, Medical Instruments & Tools, Drug Delivery, and Others), Region - Global Forecast to 2024. https://www.marketsandmarkets.com/Market-Reports/medical-plastics-market-83738633.html (Accessed : 1 June 2020).

62. M. & Mb., 2020b, (www article), COVID-19 Impact on Packaging Market by Material Type (Plastics/Polymers, Paper & Paperboard, Glass And Metal), Application (Healthcare, Food & Beverages, Household Hygiene, Beauty & Personal Care And Electrical & Electronics) And Region - Global Forecast to 2021.https://www.marketsandmarkets.com/Market-Reports/covid-19-impact-on-packaging-market-27035953.html (Accessed : 1 June 2020).
63. Kaza S, Yao L, Bhada-Tata P, Van Woerden F. What a waste 2.0: a global snapshot of solid waste management to 2050. World Bank Publications; 2018 Dec 6.
64. Rajendran K, Lin R, Wall DM, Murphy JD. Influential aspects in waste management practices. InSustainable Resource Recovery and Zero Waste Approaches 2019 Jan 1 (pp. 65-78). Elsevier.
65. Azevedo BD, Scavarda LF, Caiado RG. Urban solid waste management in developing countries from the sustainable supply chain management perspective: A case study of Brazil's largest slum. Journal of cleaner production. 2019 Oct 1;233:1377-86.
66. Ewing BT, Thompson MA. Dynamic cyclical comovements of oil prices with industrial production, consumer prices, unemployment, and stock prices. Energy Policy. 2007 Nov 1;35(11):5535-40.
67. Nicola M, Alsafi Z, Sohrabi C, Kerwan A, Al-Jabir A, Iosifidis C, Agha M, Agha R. The socio-economic implications of the coronavirus pandemic (COVID-19): A review. International journal of surgery. 2020 Jun 1;78:185-93
68. Kilian L, Park C. The impact of oil price shocks on the US stock market. International Economic Review. 2009 Nov;50(4):1267-87
69. Prasad. S., (2016), (www article), What type of relationship exists between prices of Crude Oil and Polyethylene? https://www.beroeinc.com/blog/crude-oil-polyethylene-prices/ (Accessed : 1 June 2020).
70. Greenwood. Al., (2020), (www article), Falling oil prices squeeze polymer margins. https://www.icis.com/explore/resources/news/2020/03/25/10486703/icis-webinar-falling-oil-prices-squeeze-polymer-margins (Accessed : 1 June 2020).
71. Plastic Expert, (2020), (www article), How oil prices affect recycling. https://www.plasticexpert.co.uk/oil-prices-plastic-recycling/ (Accessed : 1 June 2020).
72. Morath C, Weigand MA, Zeier M, Speer C, Tiwari-Heckler S, Merle U. Plasma exchange in critically ill COVID-19 patients. Critical Care. 2020 Dec;24(1):1-4.
73. Bomey N. Plastic bag bans reversed: states, cities, stores are suddenly banning reusable bags during coronavirus. USA TODAY. https://www. usatoday. com/story/money/2020/04/08/plastic-bag-bans-reversed-coronavirus-reusable-bags-covid-19/2967950001/.(Accessed 8 June 2020). 2020.
74. SCHLANGER. Z., 2020, (www article)_Will Coronavirus Be the Death or Salvation of Big Plastic? https://time.com/5831005/coronavirus-plastic-industry/ (Accessed: 1 June 2020).
75. Kissler SM, Tedijanto C, Goldstein E, Grad YH, Lipsitch M. Projecting the transmission dynamics of SARS-CoV-2 through the postpandemic period. Science. 2020 May 22;368(6493):860-8.
76. TOI 2020, (www artcle)After lockdown, waste collection must be on track. https://timesofindia.indiatimes.com/city/goa/after-lockdown-waste-collection-must-be-on-track/articleshow/75524544.cms (Accessed: 1 June 2020).
77. Mascarenhas. A., 2020, (www article), Covid waste overburdens common incineration facility. https://indianexpress.com/article/cities/pune/covid-waste-overburdens-common-incineration-facility-6421501/
78. Ajwani. R., 2020, (www article), COVID Crisis Could Worsen Until Waste-Workers Get PPE & Insurance. https://www.thequint.com/voices/opinion/solid-waste-management-biomedical-waste-covid-infection-crisis-health-disaster-waste-pickers-protection (Accessed: 1 June 2020).
79. WHO, 2020a. (www article), Declaration of Public Health Emergency of International Concern. https://www.who.int/emergencies/diseases/novel-coronavirus-2019/events-as-they-happen. (Accessed: 1 June 2020).
80. Babanyara, Y.Y., Ibrahim, D.B., Garba, T., Bogoro, A.G. and Abubakar, M.Y., 2013. Poor Medical Waste Management (MWM) practices and its risks to human health and the environment: a literature review. *Int J Environ Ealth Sci Eng*, *11*(7), pp.1-8.

81. Sobsey MD, Wallis C, Melnick JL. Studies on the survival and fate of enteroviruses in an experimental model of a municipal solid waste landfill and leachate. Applied microbiology. 1975 Oct;30(4):565-74.
82. Collins CH. Microbiological hazards of occupational needlestick and'sharps' injuries. J Appl Bacteriol. 1987;62:385-402.
83. Carducci A, Federigi I, Verani M. Virus occupational exposure in solid waste processing facilities. Annals of occupational hygiene. 2013 Nov 1;57(9):1115-27.
84. Graiver DA, Topliff CL, Kelling CL, Bartelt-Hunt SL. Survival of the avian influenza virus (H6N2) after land disposal. Environmental science & technology. 2009 Jun 1;43(11):4063-7.
85. Kaufan. L., 2020 (www article), Plastic Is the Hero of Coronavirus, Says the Plastics Industry. https://www.bloomberg.com/news/features/2020-06-08/is-plastic-the-coronavirus-hero-the-plastics-industry-thinks-so (Accessed: 1 June 2020)
86. WHO. 2020i, (www article),Water, sanitation, hygiene, and waste management for the COVID-19 virus: interim guidance.https://www.who.int/publications/i/item/water-sanitation-hygiene-and-waste-management-for-the-covid-19-virus-interim-guidance. (Accessed: 1 June 2020).
87. ACR, 2020, (www article), Municipal waste management and COVID-19. https://www.acrplus.org/en/municipal-waste-management-covid-19 (Accessed: 1 June 2020).
88. NCDC, 2020, Guidelines for Quarantine facilities COVID-19. https://ncdc.gov.in/WriteReadData/l892s/90542653311584546120.pdf (Accessed: 1 June 2020).
89. Gutiérrez J, García-Palomares JC. Transport and accessibility.
90. Hoornweg D, Bhada-Tata P. What a waste: a global review of solid waste management.
91. Sensoneo, 2019, (www article), Global Waste Index 2019. https://sensoneo.com/sensoneo-global-waste-index-2019/ (Accessed : 10 June 2020).
92. Geyer, R., J. Jambeck and K. Law (2017), "Production, use, and fate of all plastics ever made", Science Advances, Vol. 3/7, p. e1700782, http://dx.doi.org/10.1126/sciadv.1700782.
93. HPRC (2015), *Environmental Impacts of Recycling Compared to Other Waste Disposal Methods*, https://docs.wixstatic.com/ugd/49d7a0_6bb3ebb481ec49ceafef92f0b0ba010d.pdf (accessed on 28 March 2018).
94. Bernardo CA, Simões CL, Pinto LM. Environmental and economic life cycle analysis of plastic waste management options. A review. InAIP conference proceedings 2016 Oct 31 (Vol. 1779, No. 1, p. 140001). AIP Publishing LLC.

Noise Mapping during New Year Eve in selected residential zone of Raipur City

Vishal Kumar[1,a], A. V. Ahirwar[2,a], A.D. Prasad[2,a]
[1]Research Scholar, [2]Assistant Professor
[a]Civil Engineering Department, National Institute of technology Raipur

Abstract
Noise can be said to be invisible danger which present in our environment. Monitoring and mitigating of noise is very much important to build a health environment. This study mapped the noise levels increased during the New Year Eve 2022 of selected residential area of Raipur City. Increase in the levels of noise was seen during the Eve due to burning of crackers and use of loudspeakers during the celebrations. Five locations were selected for the study and noise map was prepared for pre and post New Year Eve 2022. From the map all the five locations showed higher noise levels (above 70dBA pre Eve and above 67dBA post Eve) than the prescribed limit (above 55 dBA day time and 45 dBA night time) of Central Pollution Control Board, New Delhi.
Keywords: Noise Map, Noise Pollution, New Year Eve.

1. Introduction

Sound levels which are unacceptable by the ear are defined as noise. Hampering mental health and disturbing the physical peace of human are among the common ill effects of noise[1][2]. Noise pollution is occurred in the environment mainly due to the vehicles and machines along with the human activities such as burning of crackers during different eve and functions[3]. Adverse health effect of on humans due to noise pollution was reported in 2011 by World Health Organization in seven categories. Sleep disturbance, hearing problem, mental health, cardiovascular, task performance, annoyance and social behaviour in negative were categorised in it[4]. In developing countries hearing impairments causing factor is environmental noise along with the occupational noise (WHO 1999). Exposed population of the residential zones are mostly affected by the cardiovascular disease due to higher levels of noise[5]. Blood pressure of higher noise levels exposed population also goes higher than the population exposed to lower levels of noise [6].

Duration of noise in the environment is lowest than the other pollutants of air and water. Major contribution of noise in the environment is done by the vehicles and there honking while running on road[7]. Number of vehicles are generally increasing due to the increase in population of the country [8].Noise map shows the distribution of noise levels in the environment due to the various sources at different locations in graphical form[9][10]. It is easy to understand the levels of noise in the single platform of various locations simultaneously. The distribution of noise can be prepared by the GIS software using interpolation[4][11]. The noise map helps the policy maker for better understanding the levels of noise in different zones of the selected area or the cities[12]. The main objective of this study is to measure the noise levels on New Year eve at selected residential zones with high population density and prepare the noise map in order to understand increase in noise levels from crackers and music systems. Also the noise levels were compared from the prescribed limit of Central Pollution control Board, New Delhi.

2. Study Area

Raipur city the capital of Chhattisgarh state is one of the cities with highest population density in the state. It is located at 22°33'to 21°14'N latitude and 82°06' to 81 ° 38'E longitudes with 40,63,872 population. The study was carried out in the selected residential areas of the city which are having high population density. The five locations: Tagore Nagar, KatoraTalab, Shankar Nagar, AwantiVihar and VIP Chowkwere selected for the study. Fifth location is having the highest number of hotels where the New Year Eve celebration was going on. The locations of the sampling stations are shown in figure 1.

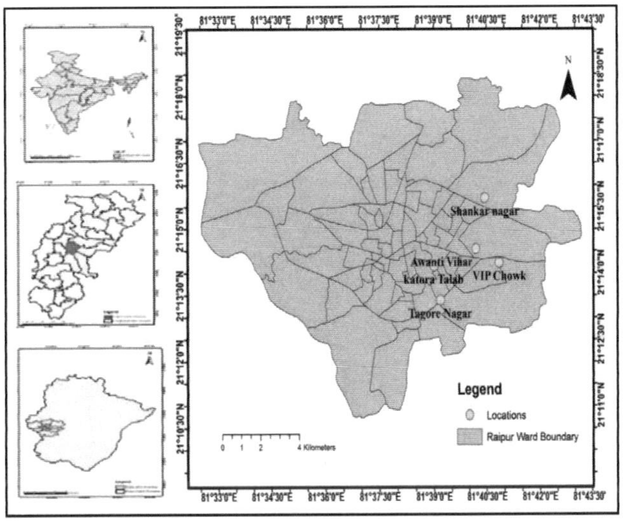

Figure1: Study Area Map of Noise Mapping during New Year Eve 2022

3. Methodology

Sampling was done with the help of Sound Level Meter (SLM-100, Class II Envirotech make) as shown in figure 2. The readings were recorded at four different locations during New Year Eve (31/12/2021 - 02/01/2022). The duration of sampling was taken from 7:00PM to 12:30PM (Tagore Nagar-(7:00 to 7:52) PM, KatoraTalab (8:00 to 8:52PM), Shankar Nagar (9:08 to 9:56 PM), AwantiVihar (10:08 to 11:00 PM) and VIP Chowk (11:20 to 12:28AM) measure the noise levels. At all the locations the coordinated of the locations were recorded. Instrument was set 1.5m above the ground and Leq values were recorded along with Lmax and Lmin. After taking the observations the data along with the coordinates was used in Arc GIS to make noise map. The methodology has been shown by the flow diagram in figure 3. Interpolation was performed in the software and noise maps were generated as shown in figure 5 and figure 6 respectively.

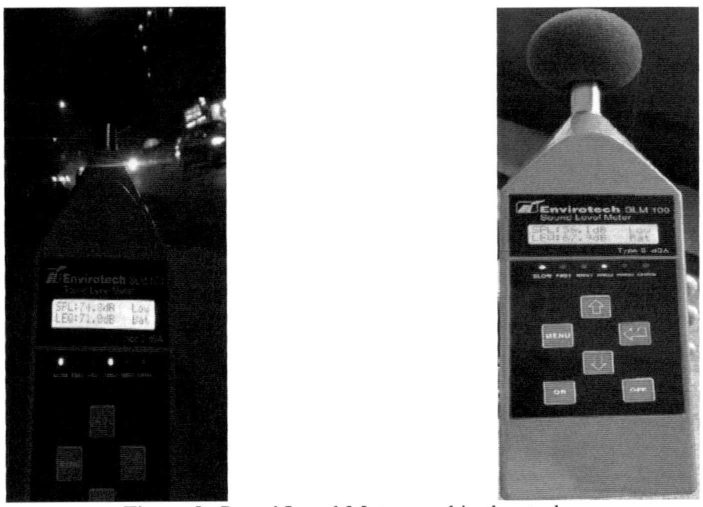

Figure 2: Sound Level Meter used in the study

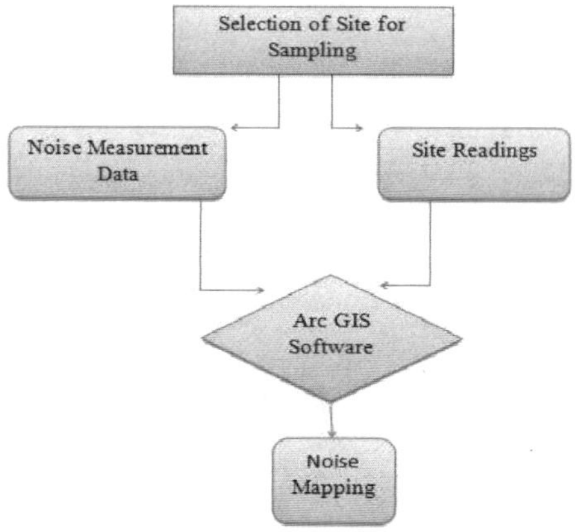

Figure 3: Flow Chart of Methodology

4. Result and Discussions

Average sound levels recorded is shown in the table 1 along with the CPCB prescribed limit. The Leq value of all the five locations is shown along with the Lmax and Lmin value. In all the locations the sound levels are higher on both the days however on 31st December it is higher due to the New Year Eve celebrations. Tagore Nagar is situated nearer to the National Highway so the noise levels on both the days were higher than prescribed limit of CPCB. At VIP Chowk difference of 6 dBA is observed due to the loud speakers and burning of crackers at mid night. A sudden increase in noise levels were observed from 11:59 PM to 12:20 AM on 31st December. Noise map shows the difference in noise level post New Year Eve. On comparing the levels of noise with prescribed limits(55dBA day time & 45 dBA night time) all the five locations are at higher end(above 72 dBA pre Eve and above 67dBA post Eve). The comparison has been shown in figure 4.

Table 1: Observed Noise Levels during the New Year Eve 2022

Zone	Location& Time	Average Sound Levels L_{eq}(dBA)			
		Latitude & Longitude of Study Station	31/12/2021	02/01/2022	CPCB Limit
Residential	Tagore Nagar (19:00-19:52)	(21°13' 23.2032") (81°39' 10.4076")	75.78 (max-90.50) (min- 47.20)	73.13 (max-88.70) (min- 61.40)	Day Time(55dBA) Night Time (45dBA)
	KatoraTalab (20:00-20:52)	(21° 14' 09.2832") (81° 39' 00.0072")	75.01 (max-95.50) (min- 54.60)	72.12 (max-92.40) (min- 48.60)	
	Shankar Nagar (21:08-21:56)	(21° 15' 41.9220") (81° 40' 26.3892")	73.23 (max-98.70) (min- 53.30)	70.30 (max-94.30) (min- 51.20)	
	AwantiVihar (22:08-23:00)	(21° 14' 32.8416") (81° 40' 10.9704")	72.22 (max-96.50) (min- 40.20)	67.77 (max-89.20) (min- 50.40)	
	VIP Chowk (23:20-00:28)	(21° 13' 23.2032") (81° 39' 10.4076")	73.47 (max-96.30) (min- 60.20)	67.92 (max-93.10) (min- 46.20)	

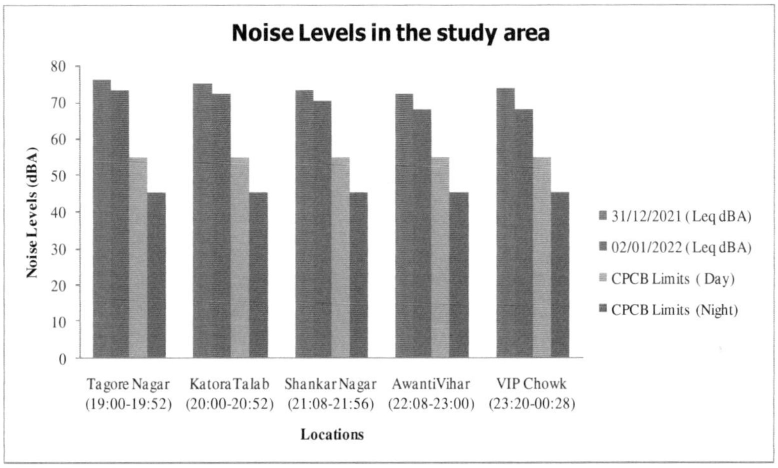

Figure 4: Comparative Study of Noise Levels During Eve

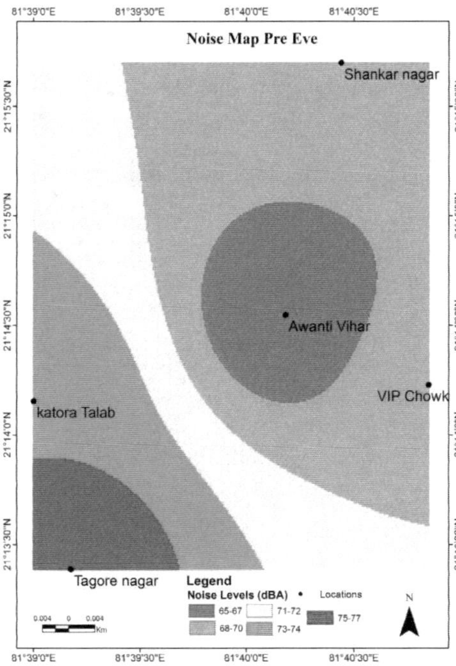

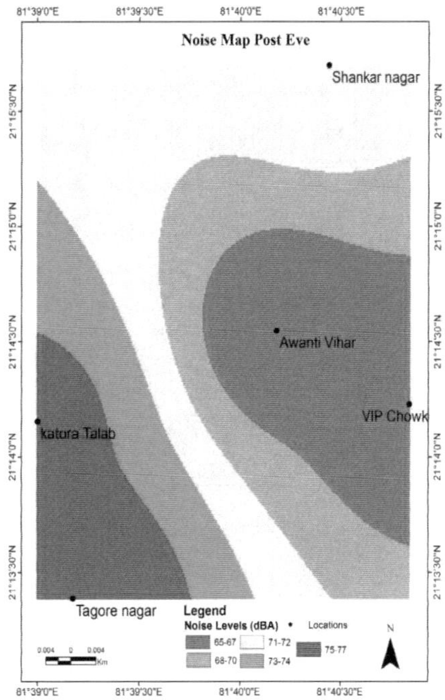

Figure 5: Noise Map 31st December 2021 Figure 6: Noise Map 2nd January 2022

5. Conclusion

From the observations we can say that all the selected locations shows higher noise levels than the prescribed limit of CPCB. However noise levels are on elevated side during the New Year Eve 2022. From the noise map it reveals that noise in the environment elevated during the Eve due to the burning of crackers and uses of loudspeakers at all the five residential areas of Raipur City. Government should make policy on uses of crackers and loudspeakers in order to mitigate the noise pollution in the residential areas of densely populated. Noise pollution shall be controlled to make our environment good for living a healthy life.

6. References

[1] K. Kalawapudi, T. Singh, J. Dey, R. Vijay, and R. Kumar, "Noise pollution in Mumbai Metropolitan Region (MMR): An emerging environmental threat," *Environ. Monit. Assess.*, vol. 192, no. 2, p. 152, Feb. 2020, doi: 10.1007/s10661-020-8121-9.

[2] J. Wu, C. Zou, S. He, X. Sun, X. Wang, and Q. Yan, "Traffic noise exposure of high-rise residential buildings in urban area," *Environ. Sci. Pollut. Res.*, vol. 26, no. 9, pp. 8502–8515, Mar. 2019, doi: 10.1007/s11356-019-04640-1.

[3] P. Bhagwat and P. M. Meshram, "Study of Noise Pollution During Ganesh Utsav in Yavatmal City," vol. 2, p. 4, 2013.

[4] E. A. Akintunde, J. Y. Bayei, and J. A. Akintunde, "Noise level mapping in University of Jos, Nigeria," *GeoJournal*, Jan. 2020, doi: 10.1007/s10708-019-10135-w.

[5] S. Stansfeld and C. Clark, "Health Effects of Noise Exposure in Children," *Curr. Environ. Health Rep.*, vol. 2, no. 2, pp. 171–178, Jun. 2015, doi: 10.1007/s40572-015-0044-1.

[6] S. Tabraiz, S. Ahmad, I. Shehzadi, and M. B. Asif, "Study of physio-psychological effects on traffic wardens due to traffic noise pollution; exposure-effect relation," *J. Environ. Health Sci. Eng.*, vol. 13, no. 1, p. 30, Dec. 2015, doi: 10.1186/s40201-015-0187-x.

[7] S. T. Leong and P. Laortanakul, "Monitoring and Assessment of Daily Exposure of Roadside Workers to Traffic Noise Levels in an Asian City: A Case Study of Bangkok Streets," p. 17.

[8] F. A. de Noronha Castro Pinto and M. D. Moreno Mardones, "Noise mapping of densely populated neighborhoods—example of Copacabana, Rio de Janeiro—Brazil," *Environ. Monit. Assess.*, vol. 155, no. 1–4, pp. 309–318, Aug. 2009, doi: 10.1007/s10661-008-0437-9.

[9] P. Das, S. Talukdar, S. Ziaul, S. Das, and S. Pal, "Noise mapping and assessing vulnerability in meso level urban environment of Eastern India," *Sustain. Cities Soc.*, vol. 46, p. 101416, Apr. 2019, doi: 10.1016/j.scs.2019.01.001.

[10] W. Yang, J. He, C. He, and M. Cai, "Evaluation of urban traffic noise pollution based on noise maps," *Transp. Res. Part Transp. Environ.*, vol. 87, p. 102516, Oct. 2020, doi: 10.1016/j.trd.2020.102516.

[11] N. Akhtar, K. Ahmad, and P. Alam, "Noise Monitoring and Mapping for Some Pre-selected Locations of New Delhi, India," *Fluct. Noise Lett.*, vol. 15, no. 02, p. 1650019, Jun. 2016, doi: 10.1142/S021947751650019X.

[12] R. Kumar, A. Mukherjee, and V. P. Singh, "Traffic noise mapping of Indian roads through smartphone user community participation," *Environ. Monit. Assess.*, vol. 189, no. 1, p. 17, Jan. 2017, doi: 10.1007/s10661-016-5741-1.

Estimation of crop water requirements within Wainganga sub-basin for Kharif and Rabi season using spatial analysis

Digambar S Londhe[1*], Mamta D Kapgate[1], Y B Katpatal[1] and M S Mukesh[1]

[1]Department of Civil Engineering, Visvesvaraya National Institute of Technology, Nagpur, Maharashtra, India.

*Corresponding author's e-mail: digambarlondhe@students.vnit.ac.in

Abstract. In India, largest user of the water is agriculture sector is the, so it is important to do the proper management of available water. Aim of this study to estimate crop water requirements (CWR) for Kharif and Rabi seasons within Wainganga sub-basin by using remote sensing and GIS technique. For this, reference evapotranspiration (ET_0) is estimated by using the Food and Agricultural Organization (FAO) Penman-Monteith method. Weather data from 8 weather stations has been collected to estimate ET_0. The crop coefficient (K_C) is estimated by using linear relationship with Normalized Difference Vegetation Index (NDVI). The MODIS NDVI dataset is used for calculation of crop coefficient. The effective precipitation (P_e) has been calculated to estimate CWR by using FAO recommended empirical method. The spatial variation maps for ET_0, K_C, actual evapotranspiration (ET_a) and CWR are generated using Inverse Distance Weightage (IDW) interpolation technique in ArcGIS software. The results show that ET_0 and ET_a are higher in Kharif season than the Rabi season. The K_C is also found higher in kharif than in rabi. Since the P_e is negligible in Rabi season, CWR in Rabi season is found to be higher than kharif season. The CWR in Rabi varies between 320 mm to 378 mm and in kharif season it varies between 94 mm to 263 mm. It is concluded that the seasonal estimation of CWR helps in understanding the peak water demand in that season in better way. It is required to provide sufficient irrigation to the crops in Rabi season especially as the agricultural production completely depends on the irrigation facilities in the study area because there is no rainfall in this season.
Keywords: Crop water requirement, kharif and rabi season, spatial analysis, reference and actual evapotranspiration, effective precipitation.

1. Introduction

Water is the most important component for the agriculture production. There is increase in demand of freshwater due to rising population, urbanization and industrialization. Due to global warming, temperature increases rapidly and hence, demand of freshwater increases as well. Due to the rise in temperature and variation in climatic factors, crop production and crop water requirements are significantly affected [1]. As the effective precipitation is the key factor for the crop water requirement estimation, it is important to analyse the effects of climate change on precipitation. There is need to get thorough knowledge about the climate change effect on reference evapotranspiration, actual evapotranspiration and effective precipitation to monitor and control the future agricultural policy, research and development [2].

In recent studies, researchers have estimated reference crop evapotranspiration by using many methods such as crop models based on climatic parameters or surface observations, remote sensing techniques and FAO Penman-Monteith method. Montazar et al 2020 [3] used residual of energy balance approach for estimation of ET_a and K_C in California with the aim to better manage limited water resources by gaining more accurate knowledge regarding evapotranspiration of crops.

Many researchers give the different models and methods for the crop water requirements. Li et al. 2020 [4] estimated irrigation water requirements based on GIS techniques and Penman-Monteith formula for three main crops (wheat, cotton and corn) during their growing periods in Xinjiang province, China. It was concluded that planting area of crops were generally more sensitive to irrigation water requirements than rising temperature. Wang et al. 2020 [1] used CROPWAT 8.0 to estimate crop water requirements in Heilongjiang Province, China. Mann-Kendall trend test was used to analyse the changing trend of ET_0, ET_a and P_e. Usman et al. 2021 [5] used surface energy balance system model to estimate Ea and FAO-56 Penman-Monteith method to estimate ET_0 and comparison between them.

The main objective of the study is to calculate the CWR within Wainganga river sub-basin, Nagpur, Maharashtra. For this weather data from 8 weather stations has been obtained and analysed. The ET_0 was calculated by using FAO Penman-Monteith method and K_C was estimated by using MODIS NDVI data. P_e was calculated by using FAO based empirical method. IDW method is used for interpolation of weather data in GIS environment. The spatial variation of K_C, ET_0 and CWR for kharif and Rabi season was also estimated.

The objectives of the present study are as follow:

1) To calculate the Crop water requirement of the study area for both Kharif and Rabi seasons for the years 2001, 2005, 2010 and 2013.

2) To calculate and analyse spatial variation of ET_0, K_C and ET_a in the study area.

3) To calculate effective precipitation within study area.

4) To analyse the spatial variation of crop water requirement in study area due to climate change for Kharif and Rabi seasons.

2. Study Area

Wainganga River sub-basin is located in Nagpur district, Maharashtra, India. It lies between 21°10'0"N to 21°40'0"N latitude and 78°30'0" E to 79°20'0" E longitude (Figure 1). The Wainganga River sub-basin has elevation range from 304 m to 467 m above mean sea level. The total area is 1677.2911 Sq. Km. Wainganga River basin within Nagpur District has 40 sub-watersheds [6], out of them 8 sub-watersheds are selected for study. The geographical location map of the study area is shown in Figure 1.

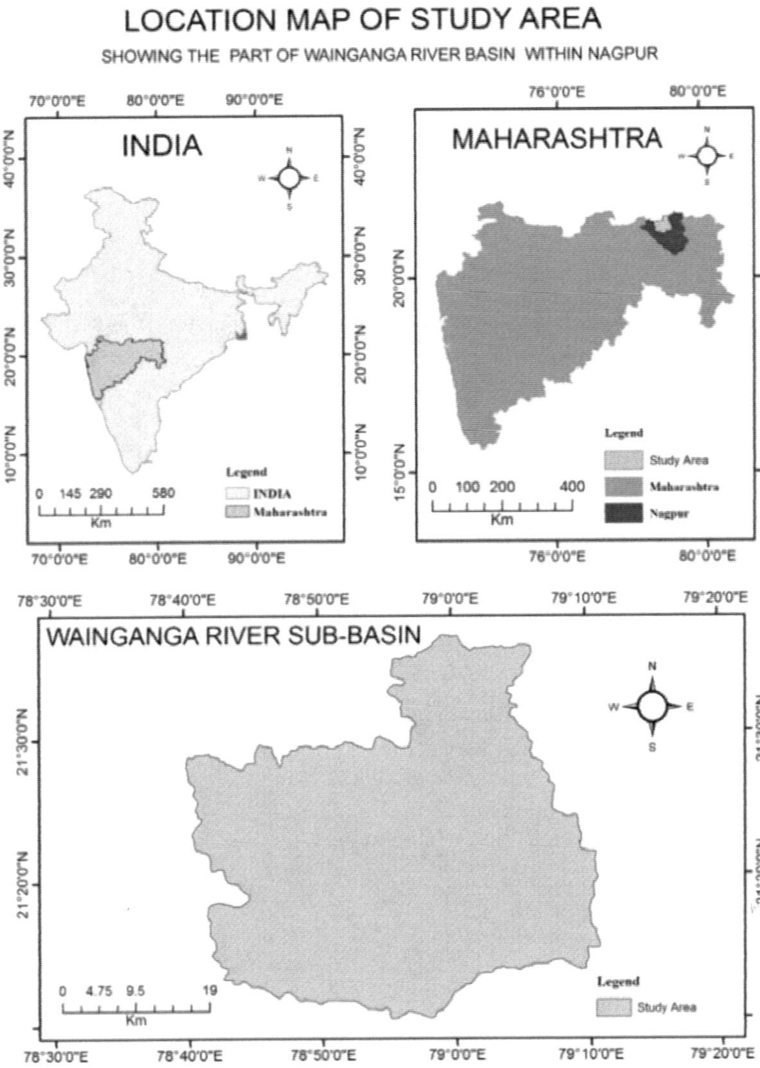

Figure 1. Location map of the Wainganga river sub-basin

3. Data Used

3.1 Meteorological Data

Meteorological data is obtained from CFSR (Climate Forecast System Reanalysis) which is coupled atmosphere, ocean and land system developed at NOAA-NCEP. The spatial resolution of CFSR data is 0.35^0. CFSR provides the maximum and minimum temperatures (°C), precipitation (mm), wind velocity (m/s), humidity (%) and solar radiation (MJ/m^2) values from 1979 [7]. Meteorological data obtained from CFSR is useful to estimate ET_0 by FAO Penman Monteith method.

3.2 MODIS NDVI Data

MODIS NDVI data was downloaded and analysed. There are two platforms from where datasets are collected viz. Aqua and Terra. MOD13Q1 data has spatial resolution of 250 m and temporal resolution of 16 days. The NDVI is a vegetation index, which gives the information about the actual growth condition of the vegetation. NDVI value varies between -1 to +1. During a year, minimum NDVI occurs in May and maximum NDVI occurs in September.

4. Methodology

The main objective of study is to estimate the CWR for the part of Wainganga sub-basin which is situated within Nagpur for kharif and rabi seasons. For this, precipitation, wind speed, relative humidity, solar radiations and temperature were collected. ET_0 was estimated by using Penman-Monteith method [8, 9]. The Penman-Monteith method is widely used all over the world for estimating ET_0. For the estimation of K_C, MODIS NDVI data is used. The ET_a was calculated by using K_c and ET_0. The P_e was calculated by using empirical formula which is suggested by FAO using total monthly precipitation. Finally, the CWR is estimated by using Et_a and P_e.

The overall methodology adopted for the calculation of CWR is given in flowchart (Figure 2) as follows:

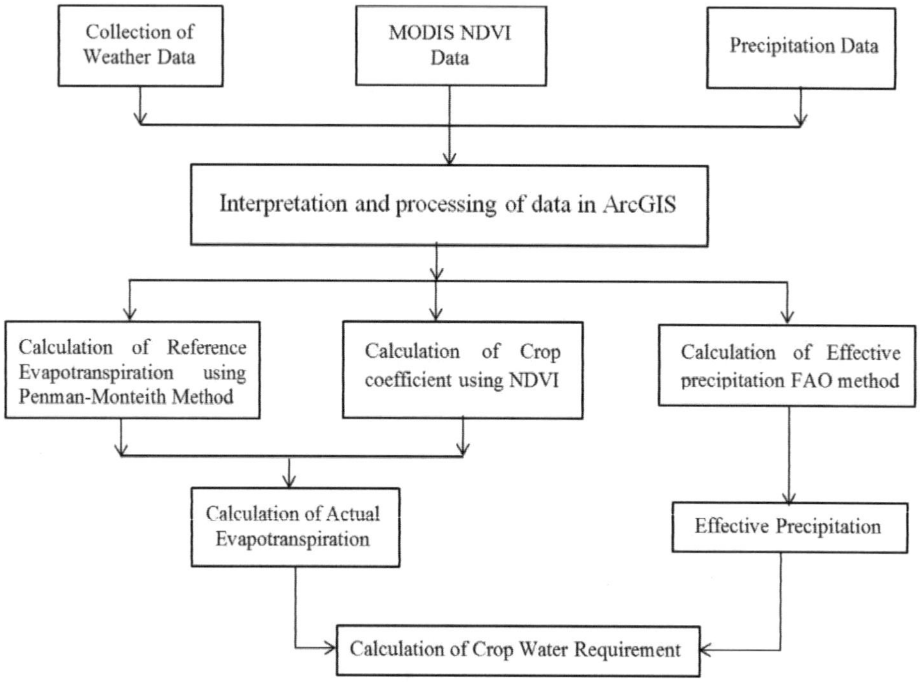

Figure 2. Overall methodology adopted for calculation of crop water requirements

4.1 FAO Penman-Monteith method for Calculation of ET_0

The FAO Penman-Monteith method is now mostly recommended method for estimation of ET_0 which is applicable in almost all regions and climatic conditions. The Penman-Monteith method uses precipitation, daily mean temperature, relative humidity, solar radiations and wind speed. Zotarelli, et.al (2010) [10] provides steps for calculation of ET_0. The equation for the estimation of ET_0 is given as Eq. 1.

$$ET_o = \frac{0.408\,\Delta\,(R_n - G) + \gamma\left[\frac{900}{T+273}\right] * u_2\,(e_s - e_a)}{\Delta + \gamma\,(1 + 0.34u_2)} \qquad \text{Eq. 1}$$

Where, ET_0 is reference evapotranspiration [$mmday^{-1}$];
R_n is net radiation at the crop surface [$MJ/m^2/day$];
G is soil heat flux density [$MJm^{-2}day^{-1}$];
T is mean daily air temperature at 2 m height [°C];
u_2 is wind speed at 2 m height [ms^{-1}];
e_s is saturation vapour pressure [kPa];

e_a is actual vapour pressure [kPa];
e_s-e_a is saturation vapour pressure deficit [kPa];
Δ is slope vapour pressure curve [kPa/°C] and
γ is psychrometric constant [kPa/°C].

4.2 Estimation of crop coefficient (K_C) by using satellite-based vegetation index

The crop coefficient is determined by using NDVI. The different models were developed to estimate crop coefficient directly with the help of NDVI. Kamble et al. (2013) [11] developed simple linear relation model for crop coefficient estimation. The linear relation between K_C and NDVI is given as Eq. 2.

$$K_C = 1.457\ (NDVI) - 0.1725 \qquad \text{Eq. 2}$$

Where, K_C - Crop Coefficient;
NDVI - Normalised Difference Vegetation Index.

4.3 Calculation of Crop water Requirements (CWR)

The crop water requirement is the actual water required to that crop during its growth period. The crop water requirement estimation formula [4] is given as follows,

$$CWR = ET_a - P_e \qquad \text{Eq. 3}$$

Where, ET_a- Actual evapotranspiration in mm/day;
P_e- Effective precipitation in mm.

The ET_a in mm for 3 months was calculated and also effective precipitation for 3 months for each season was calculated. The obtained CWR values were interpolated by IDW tool in ArcGIS software. The spatial variation maps of CWR for kharif and Rabi season were generated.

5. Results and Discussion

The estimated results of ET_0 obtained from the FAO Penman-Monteith method are analysed for kharif and rabi seasons for years 2001, 2005, 2010 and 2013. The K_C estimated by using MODIS NDVI data, ET_a by using ET_0 and CWR using ET_a and P_e within study area for kharif and rabi seasons are compared and analysed.

Statistical summary of the ET_0 (mm/day), K_C, ET_a ((mm/day)) and CWR (mm) over study area for kharif and rabi seasons are summarized in Table 1.

Table 1. Statistical summary of the ET_0 (mm/day), K_C, ET_a ((mm/day)) and CWR (mm) over study area for kharif and rabi seasons

	ET_0		K_C		ET_a		CWR	
	Kharif	Rabi	Kharif	Rabi	Kharif	Rabi	Kharif	Rabi
2001	4.686	3.931	0.739	0.371	3.460	1.457	201.555	359.723
2005	3.853	4.046	0.757	0.415	2.914	1.681	49.138	369.789
2010	3.841	3.606	0.775	0.506	2.976	1.824	38.985	327.015
2013	3.503	3.485	0.714	0.472	2.499	1.645	20.455	316.903

The statistical values of CWR in Table 1 indicate that the CWR for rabi season is much more as compared to the kharif season. This is because of the effective precipitation is zero in rabi season as there is no precipitation and more effective precipitation during kharif season due to monsoon rainfall. This affects the requirement of the crop water in the different seasons in the study area.

5.1 Spatial variation of ET_0 for Kharif and Rabi season

The accurate estimation of ET_0 is important for ET_a estimation. These weather data is collected from 8 weather stations for year 2001, 2005, 2010 and 2013 which is further interpolated in ArcGIS. The spatial variation maps are prepared for years 2001, 2005, 2010 and 2013. The spatial variation maps for kharif and rabi season for respective years are as shown in Figure 3.

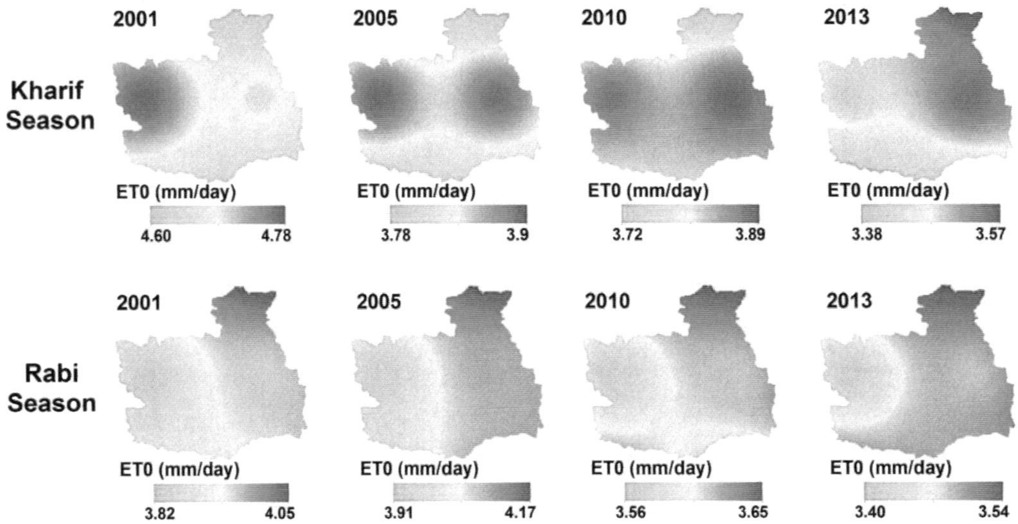

Figure 3. Spatial variation of ET_0 maps for kharif and rabi season for 2001, 2005, 2010 and 2013

The results show that the ET_0 varies in Kharif season ranges from 3.38 mm/day to 4.78 mm/day. The ET_0 value slightly decreases from 2001 to 2013 for kharif season. The rate of ET_0 is maximum at western side of Wainganga basin within Nagpur for year 2001 and then moved to the eastern part of the study area for year 2005. ET_0 rate is less in northern part of the study area.

Variation of ET_0 in Rabi season ranges from 3.4 mm/day to 4.17 mm/day. The rate of ET_0 is slightly increases from year 2001 to 2005 and then decreases for year 2010 and then again decreases for year 2013. The rate of ET_0 is slightly lower at western part of the study area. As the available moisture content in atmosphere is less in rabi season than kharif season, ET_0 rate is lower in Rabi season.

5.2 Spatial variation of crop coefficient (K_C) for Kharif and Rabi season

The K_C depends on the growth stages of the crop, canopy of crop and area under crop. The K_C is higher at peak growth stage and then decreases when the crop attends maturity. The MODIS data was processed in ArcGIS and K_C was calculated by using Raster Calculator tool in ArcGIS. Spatial variation maps of K_C are shown in Figure 4.

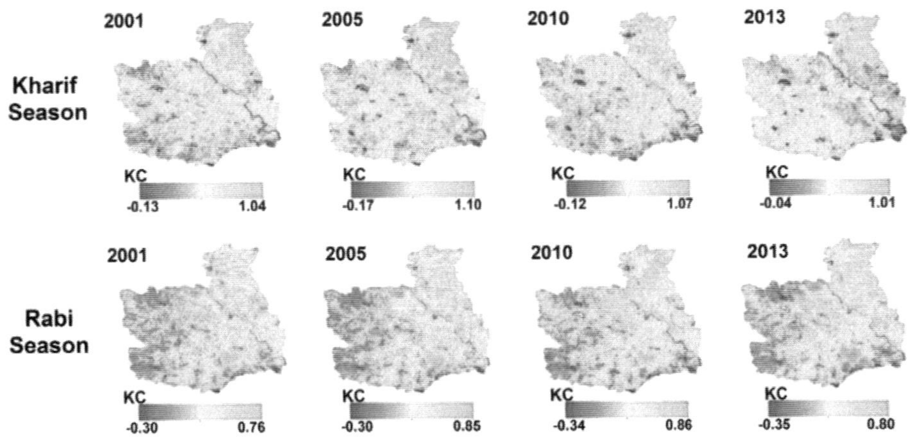

Figure 4. Spatial variation of K_C maps for kharif and rabi season for 2001, 2005, 2010 and 2013

Results of spatial variation of K_C show that the K_C value varies from -0.047 to +1.10 in kharif season. The negative values are indicating the presence of water body. The spatial variation maps of K_C indicate that in Rabi season K_C value ranges from -0.30 to +0.86. The result shows that the K_C is higher in Kharif season than Rabi season. The reason behind this may be the initial growth stage of the crop. The ET_a is affected by the K_C value. K_C is the simply ratio of ET_a to ET_0.

5.3 Spatial variation of ET_a for Kharif and Rabi season

The ET_a is the loss of water due to evaporation and transpiration from actual crop field. Spatial variation maps of ET_a are shown in Figure 5.

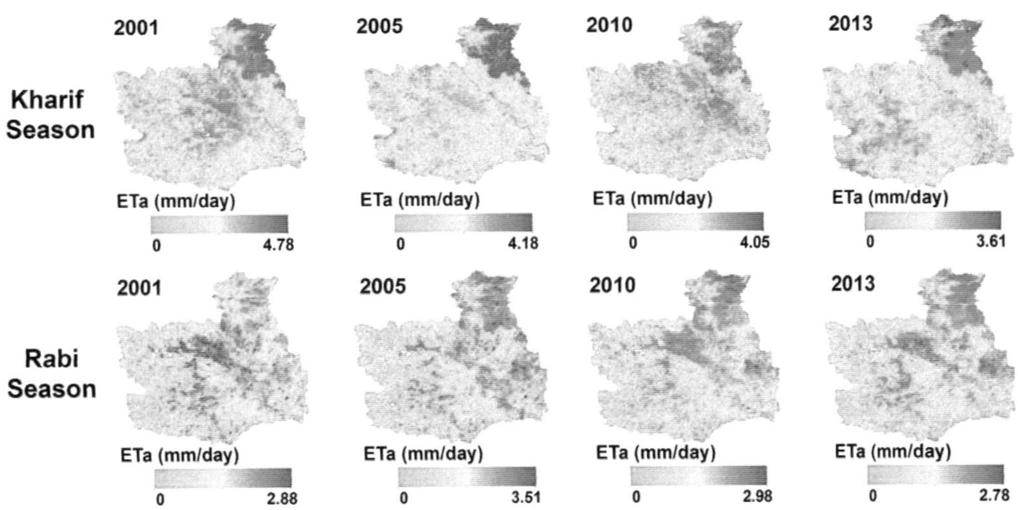

Figure 5. Spatial variation of ET_a maps for kharif and rabi season for 2001, 2005, 2010 and 2013

The results show that, ET_a value varies up to 4.78 mm/day in Kharif season and 3.51 mm/day in Rabi season. It is clearly revealed that ET_a is higher in kharif season than Rabi season. The maximum ET_a in kharif season was found to be 4.78 mm/day in year 2001, 4.18 mm/day in 2005, 4.05 mm/day in 2010 and 3.62 mm/day in 2013. Whereas in Rabi season, maximum ET_a was found to be 2.88 mm/day in year 2001, 3.51 mm/day in year 2005, 2.98 mm/day in year 2010 and 2.78 mm/day in year 2013.

5.4 Spatial variation of CWR for Kharif and Rabi season

The crop water requirement varies with the crop as well as for the different seasons. Here the spatial variation maps for CWR are prepared for kharif as well as rabi season for years 2001, 2005, 2010 and 2013. CWR for given study area was estimated with the help of actual evapotranspiration rate for 3 months in mm and effective precipitation for equivalent 3 months in mm. The spatial variation maps for CWR for kharif and Rabi season are given in Figure 6.

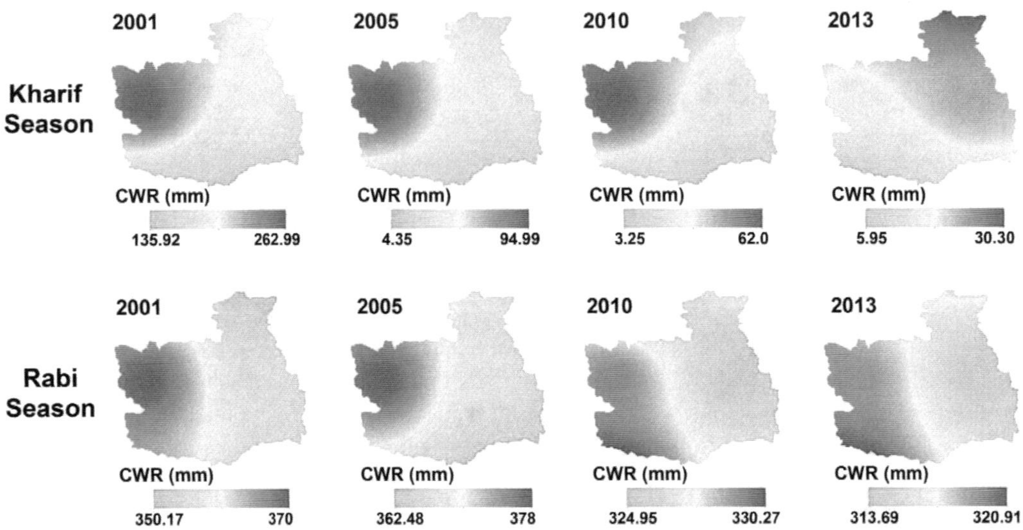

Figure 6. Spatial variation of CWR maps for kharif and rabi season for 2001, 2005, 2010 and 2013

The results show that the maximum CWR varies from 320 mm to 378 mm for Rabi season and it varies from 94 mm to 263 mm in Kharif season. The results show that water requirement for crops is maximum in rabi season as compared to Kharif season. The precipitation is the important affecting factor. During the Rabi season the precipitation is very less or negligible; so effective precipitation is found to be zero and hence, CWR is maximum in rabi season than kharif season. Also in Kharif season, the precipitation is quite sufficient to fulfil the water requirements of the crops.The results also show that the CWR in Rabi season decreases gradually from year 2001 to year 2013. This is because of the changing climate and other anthropogenic activities related to the agriculture.

6. Conclusion

The main objective is to estimation of crop water requirements within Wainganga sub-basin for kharif and rabi season. And the spatial variation maps for ET_0, ET_a and CWR are also prepared and analysed. The results show that ET_0 is lower in Rabi season than because of the variation in soil moisture content in Kharif and Rabi season.

The spatial variation maps of K_C show that value of Kc is higher in Rabi season than Kharif season. The K_C directly affects the actual evapotranspiration rate. The K_C was estimated by using MODIS NDVI data. As NDVI directly related to the crop coefficient, it represents the growth stages of crop and crop type. The ET_a map shows that ET_a is higher in kharif season than Rabi season as ET_0 is maximum in kharif season because of higher temperature.

The spatial variation maps of CWR shows that the CWR in Rabi season is higher than Kharif season. The maximum CWR found in rabi season which range from 320 mm to 378 mm and in kharif season, CWR is range from 94 mm to 263 mm. The rainfall reduces the amount of irrigation water requirement as water requirements is fulfilled by rainfall mostly. It is concluded that the seasonal estimation of CWR helps in better understanding the peak crop water demand. It is required to select sufficient irrigation discharge to the crop field in rabi season specially as the agricultural production is completely depends on the irrigation water because of no rainfall in this season. Hence, the remote sensing and GIS techniques proved to be most reliable technique as all the data required for CWR estimation is easily available and it can be used to generate spatial variation maps of all the parameters.

References

[1] Wang T, Du C, Nie T, Sun Z, Zhu S, Feng C, Dai C, Chu L, Liu Y, Liang Q 2020 Spatiotemporal Analysis of Maize Water Requirement in the Heilongjiang Province of China during 1960–2015 *Water* **12(9)** 2472

[2] Zhao J, Li H, Cai X, Chen F, Wang L, Yu D 2020 Long-term (2002–2017) impacts of Danjiangkou dam on thermal regimes of downstream Han River (China) using Landsat thermal infrared imagery *Journal of Hydrology* **589:**125135

[3] Montazar A, Krueger R, Corwin D, Pourreza A, Little C, Rios S, Snyder RL 2020 Determination of actual evapotranspiration and crop coefficients of California date palms using the residual of energy balance approach *Water* **12(8)** 2253

[4] Li Y, Wang H, Chen Y, Deng M, Li Q, Wufu A, Wang D, Ma L 2020 Estimation of regional irrigation water requirements and water balance in Xinjiang, China during 1995–2017 *PeerJ*. 2020 Jan 3 8:e8243

[5] Usman M, Ndehedehe CE, Manzanas R, Ahmad B, Adeyeri OE 2021 Impacts of climate change on the hydrometeorological characteristics of the soan river basin, Pakistan *Atmosphere* **12(6)** 792

[6] Chandan KS, Yashwant BK 2017 Optimization of groundwater level monitoring network using GIS-based geostatistical method and multi-parameter analysis: a case study in Wainganga Sub-basin, India *Chinese geographical science* **27(2)** 201-15

[7] Saha S, Moorthi S, Wu X, Wang J, Nadiga S, Tripp P, Behringer D, Hou YT, Chuang HY, Iredell M, Ek M. The NCEP climate forecast system version 2014 *Journal of climate* **27(6)** 2185-208

[8] Penman HL 1948 Natural evaporation from open water, bare soil and grass Proceedings of the Royal Society of London. Series A. Mathematical and Physical Sciences. 1948 Apr 22; **193(1032):**120-45

[9] Monteith JL 1965 Evaporation and environment In Symposia of the society for experimental biology 1965 (Vol. 19, pp. 205-234). *Cambridge University Press (CUP) Cambridge*

[10] Zotarelli L, Dukes MD, Romero CC, Migliaccio KW, Morgan KT 2010 Step by step calculation of the Penman-Monteith Evapotranspiration (FAO-56 Method) Institute of Food and Agricultural Sciences. University of Florida

[11] Kamble B, Kilic A, Hubbard K 2013 Estimating crop coefficients using remote sensing-based vegetation index *Remote sensing* **5(4)** 1588-602

Critical analysis of hydrological mass variations of northwest India

Mira Shivani S[1*], S Srivastava[2], A Singh[1]

[1]Centre for Wireless Network & Applications (WNA), Amrita Vishwa Vidyapeetham, Amritapuri, India
[2]Amrita School for Sustainable Development, Amrita Vishwa Vidyapeetham, Amritapuri, India

Corresponding authors e-mail: smirashivani@gmail.com, amiddids21011@am.students.amrita.edu, alkasingh@am.amrita.edu

Abstract. With current climate change, water availability is a huge concern. Ground-water (GW) is depleting at a steep rate globally and more specifically in Northwest India. Estimation and analysis of GW availability would be of great use for formulating a proper water management plan for the future. The study is carried out in the north-western part of India with the aim of generating a time series for total water storage (TWS) using different remote sensing and model-based data like GRACE, Landsat, MODIS. From the time series, it is clear-cut that, the TWS is showing a declining trend and this might be due to depletion of groundwater as, other variables like precipitation, evapotranspiration, soil moisture (obtained from MERRA-2 and CFSR data) are not showing any negative trend during the study period (2002 - 2021). Also, from the change detection analysis of land use land cover maps and crop yield statistics of water-intensive crops it can be concluded that croplands in the study area are increasing denoting the usage of water for irrigation at a large scale. Thus, better management of the groundwater is required for avoiding severe water scarcity in the future.

Keywords: GRACE, TWS, ground water, change detection, crop yield

1. Introduction

Groundwater is an essential resource for agricultural, industrial, and domestic water use and it's a primary source of freshwater [1]–[4]. GRACE (Gravity Recovery and Climate Experiment) data has played a key role in understanding the TWS, which integrates different compartments of water storage like surface water, soil moisture, and groundwater. Many studies have been carried out on GRACE to understand the status of groundwater [5]–[9], drought [10], urban groundwater variability [11] and solid earth [12]. Ahmed et al., (2021) has demonstrated the impact of anthropogenic activities on groundwater in Morocco and the need for taking sustainable measures. The major economy of India is from the agriculture industry and the production of water-intensive crops like paddy, wheat, cotton, sugarcane, has been suggested as one of the major reasons for groundwater depletion in the country, specifically north-western states [14]. Water intensive crops can cause a great impact on the rivers by reducing their run-off from the catchment and drawing large amount of groundwater resources (WWF, 2007)[15], [16]. Different climatic variables also influence TWS like, temperature, evapotranspiration rate, and precipitation. Another factor is population density, as it will exert stress on the available water resources in an area [17]. The study aims to analyse the behaviour and the Spatio-temporal trend of TWS from 2002 to 2021.

2. Study Area

The study area (figure 1) lies between 22°N to 30°N and 72°E to 84°E, covering the north western part of India (Rajasthan, Haryana, Punjab, Uttar Pradesh and Delhi). The region receives an average rainfall of 700mm annually and has rivers like Yamuna, Chambal, Son, etc., The groundwater in the region is withdrawn continuously for the purpose of irrigation exceeding the recharge rate [1], [2] which leads to depletion of groundwater in the region.

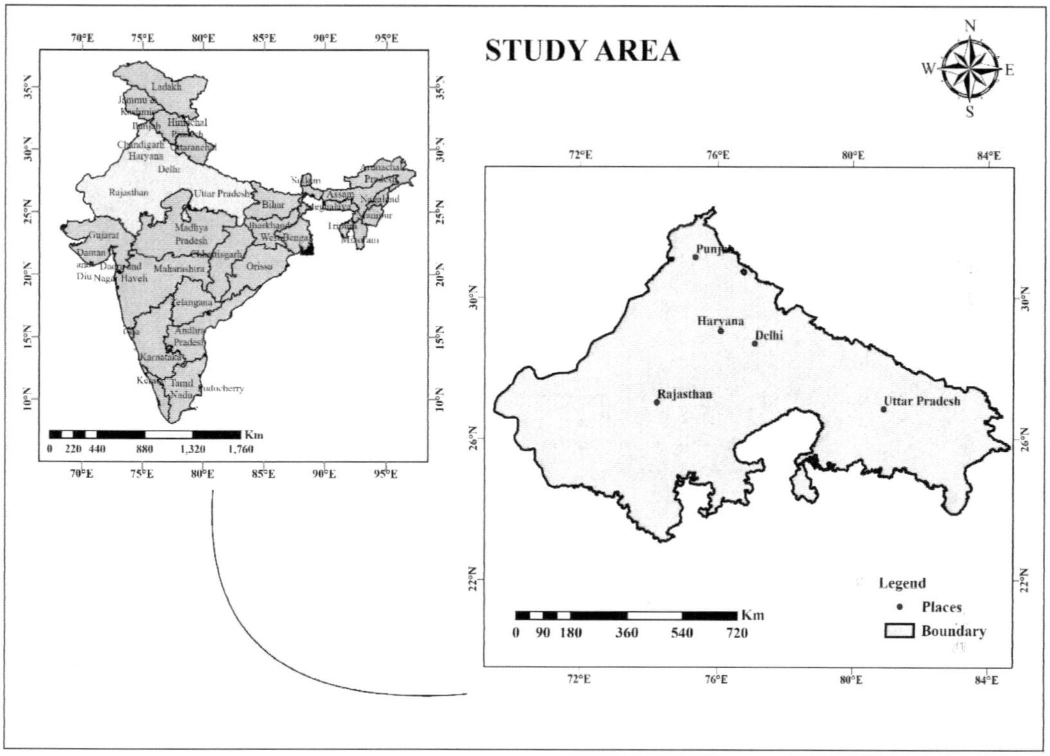

Figure 1. Study Area Map

3. Data and Methodology

3.1. Data

GRACE dataset (JPL GRACE Mascon solution) acquired for the study period (April 2002 to September 2021) has gridded monthly global water storage/height anomalies at 0.5-degree resolution. The crop yield data was obtained from National informatics centre's (NIC) area productive statistics report, and the land use land cover (LULC) acquired from decadal and MODIS' (Moderate resolution Imaging Spectroradiometer) MCD12Q1 products respectively. Other variables like temperature, evapotranspiration, precipitation, and soil moisture are analysed for the region with the help of MERRA-2 (Modern Era-Retrospective Analysis for Research and Analysis) Reanalysis data (temperature – evapotranspiration) and CFS (Climate Forecast System) Reanalysis data (precipitation – soil moisture) in climate engine. The resolution of MERRA-2 and CFS datasets are 0.5x0.625 degrees and 0.2 degrees respectively.

3.2. Methodology

The methodology (figure - 2) adopted for this study is as follows, GRACE data had gaps which were filled with the estimated seasonality and trend after which, the time series and spatial trend were analysed. LULC change detection analysis was carried out for the years 1985 – 2005; 2002 – 2021 with the decadal LULC map and MODIS MCD12Q1 products respectively. More importance was given to the features which got converted to croplands and built up. Then, the crop yield trend was derived from statistical analysis of the water-intensive crops (cotton, rice, sugarcane, and wheat) of the study area. Other variables such as temperature, evapotranspiration, soil moisture and precipitation were analysed in the climate engine research app.

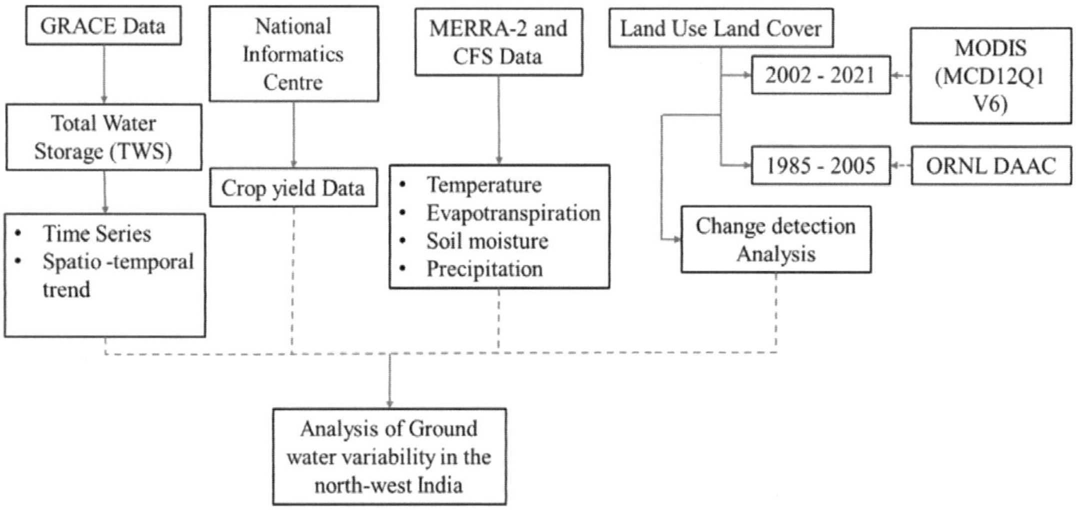

Figure 2. Methodology Flowchart

4. Results

The time series plot of GRACE data (figure 3a) clearly shows a declining trend in the TWS of the region, it is around +1cm in 2002 and significantly drops to -2 cm in 2021. The Spatio-temporal trend (figure 3b) of the TWS data depicts the strong drop in the parts of Rajasthan and Punjab. From the change detection analysis (figure 4) carried out for the year 2002 and 2021, the increase of croplands especially in the parts of Rajasthan can be witnessed which could be a reason for the negative trend of TWS as groundwater is extracted at a large scale for crop production. During 1985 – 2005, 66% barren lands, 21% scrublands, 9% waterbodies and 2% forest have been converted into croplands. Whereas, during the study period (2002 - 2021), 56% scrublands and 42% grasslands have been converted to croplands. Also, the crop yield data explicitly shows a rise in the water – intensive crops in the study area which is another crucial reason for groundwater depletion.

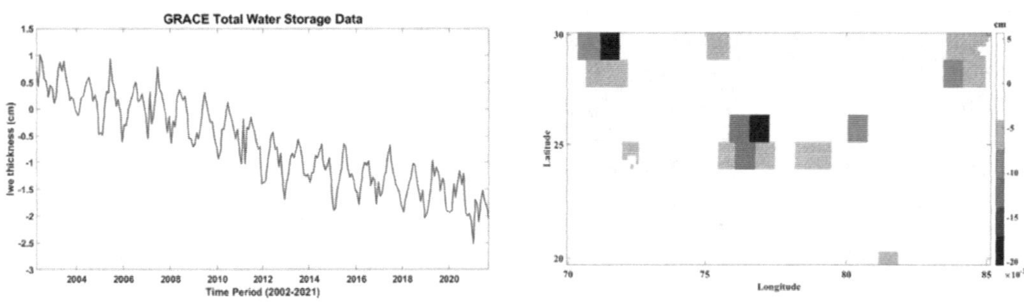

Figure 3. (a) GRACE Time series, (b) Spatio-temporal trend of TWS in the study area

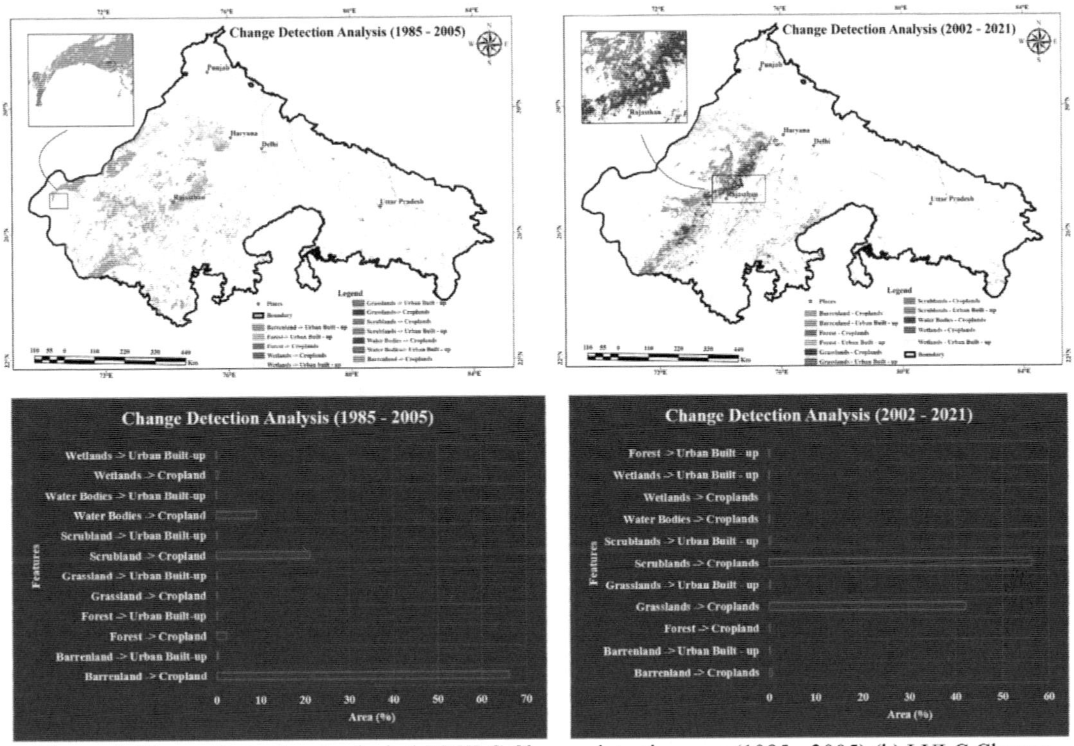

Figure 4. Change Detection Analysis (a) LULC Change detection map (1985 - 2005) (b) LULC Change detection map (2002 - 2021) (c) Graph showing the change of LULC (1985 - 2005) (d) Graph showing the change of LULC (2002 - 2021)

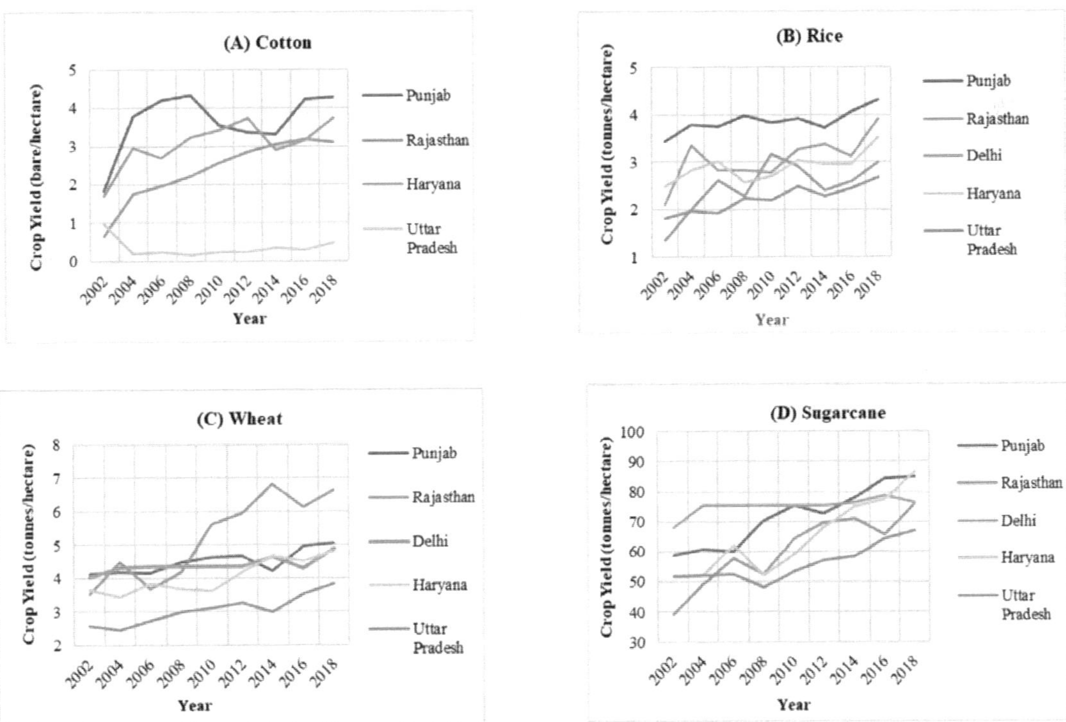

Figure 5. Annual Crop yield Trend in North-western states of India (a) Cotton (b) Rice (c) Wheat (d) Sugarcane

5. Discussion

GW depletion in north-western India has been exceeded over 150 percent, especially in the major crop-producing states like Haryana, Punjab, Rajasthan, and Uttar Pradesh as the groundwater abstraction is far more than its recharging pattern [18]. The annual crop yield trend for the period 2002 - 2018 in the north-western part of India (Haryana, Punjab, Uttar Pradesh and Rajasthan and Delhi) has been demonstrated and the statistics show a positive trend for four water intensive crops, cotton (figure 5a), rice (figure 5b), wheat (figure 5c) and sugarcane (figure 5d). The cotton yield in three states is increasing and Punjab stands the highest with 4.273 bare/ha of crop for the year 2018 while Rajasthan is showing a continuous growth and Uttar Pradesh has a decreasing trend and yield of 0.477 bare/hectare in 2018. The rice production is increasing in all the parts and the yield is highest in Punjab with 4.313 t/ha and lowest in Uttar Pradesh with 2.687 t/ha. In the case of wheat and sugarcane production, Rajasthan and Haryana stand at top with an annual yield of 6.639 t/ha and 86.62 t/ha respectively. Also, the other influencing factors of TWS such as, temperature, precipitation, evapotranspiration are not showing a negative trend (figure 6), denoting the decrease of GW as the reason for declining trend in TWS from GRACE. This study provides an outline of the declining trend in the GW availability of the region and future studies can be done using in situ GW data which will explicitly show the depletion of GW in the area.

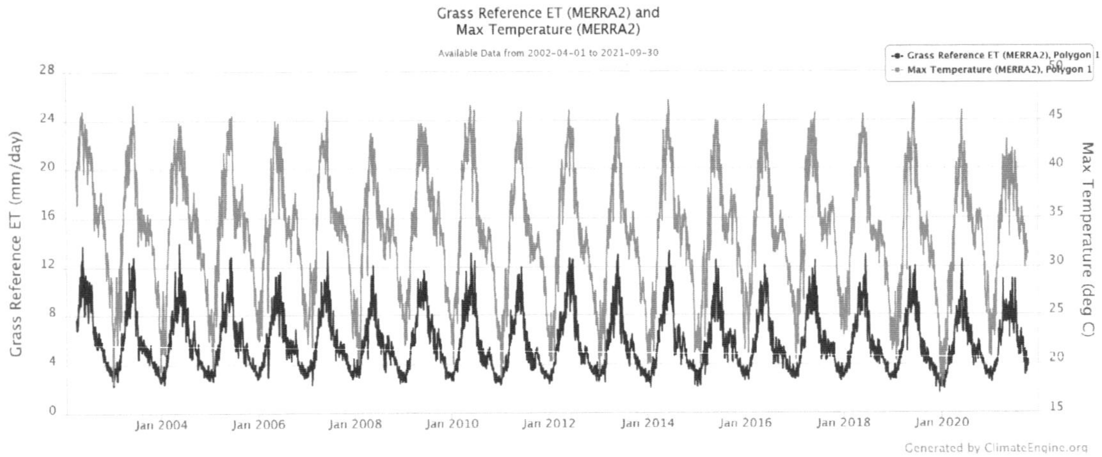

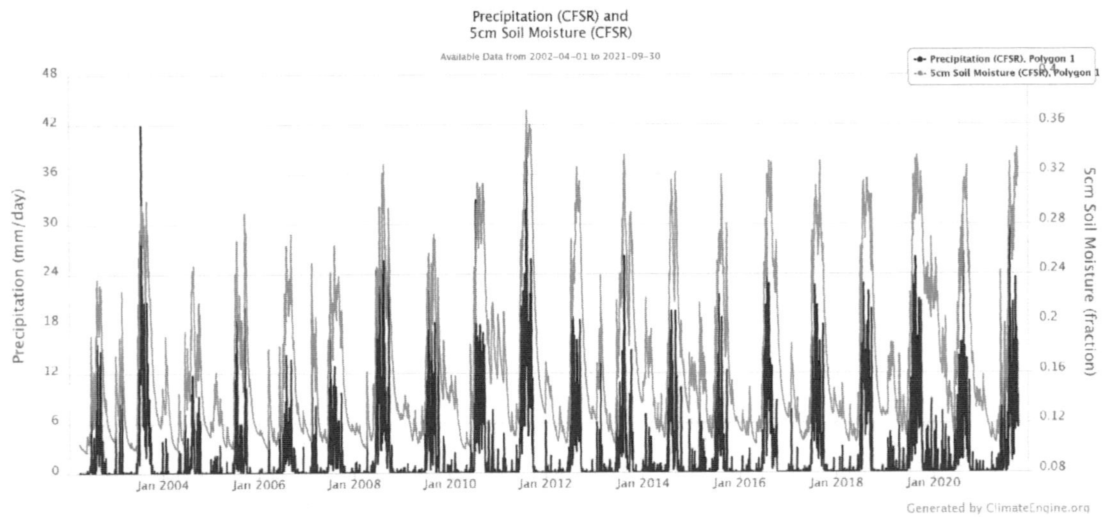

Figure 6. (a) Temperature – Evapotranspiration (b) Precipitation - Soil moisture

6. Conclusion

In all the crop yield plots, it is obvious that the crop production in north-western part of India is gradually increasing resulting in a growing trend of croplands (water-intensive crops) which is evident in the LULC maps also. Thus, it can be concluded that the declining trend of TWS, is due to over exploitation of groundwater in the region for the purpose of irrigation. On the other hand, population is shooting up and demands are escalating for food, land and water resources which urges for a better management of natural resources in order to lead a sustainable living.

References

[1] D. Long *et al.*, "Have GRACE satellites overestimated groundwater depletion in the Northwest India Aquifer?," *Sci Rep*, vol. 6, no. 1, p. 24398, Jul. 2016, doi: 10.1038/srep24398.
[2] M. Rodell, I. Velicogna, and J. S. Famiglietti, "Satellite-based estimates of groundwater depletion in India," *Nature*, vol. 460, no. 7258, pp. 999–1002, Aug. 2009, doi: 10.1038/nature08238.
[3] M. Kaviarasan, P. Geetha, and K. P. Soman, "GIS-Based Ground Water Quality Monitoring in Thiruvannamalai District, Tamil Nadu, India," in *Proceedings of the International Conference on Soft Computing Systems*, vol. 397, L. P. Suresh and B. K. Panigrahi, Eds. New Delhi: Springer India, 2016, pp. 685–700. doi: 10.1007/978-81-322-2671-0_65.
[4] M. Kaviarasan, P. Geetha, and K. P. Soman, "Multivariate Statistical Technique for the Assessment of Ground Water Quality in Coonoor Taluk, Nilgiri District, Tamilnadu, India," *Indian Journal of Science and Technology*, vol. 8, no. 36, Dec. 2015, doi: 10.17485/ijst/2015/v8i36/87535.
[5] W. M. Alley and L. F. Konikow, "Bringing GRACE Down to Earth: W.M. Alley and L.F. Konikow Ground Water xx, no. x: xx-xx," *Groundwater*, vol. 53, no. 6, pp. 826–829, Nov. 2015, doi: 10.1111/gwat.12379.
[6] A. Rateb *et al.*, "Comparison of Groundwater Storage Changes From GRACE Satellites With Monitoring and Modeling of Major U.S. Aquifers," *Water Resour. Res.*, vol. 56, no. 12, Dec. 2020, doi: 10.1029/2020WR027556.
[7] S. Kumar Joshi *et al.*, "Strongly heterogeneous patterns of groundwater depletion in Northwestern India," *Journal of Hydrology*, vol. 598, p. 126492, Jul. 2021, doi: 10.1016/j.jhydrol.2021.126492.
[8] A. Singh and F. Seitz, "Water storage variations in the Aral Sea from multi-sensor satellite data in comparison with results from GRACE gravimetry," in *2012 IEEE International Geoscience and Remote Sensing Symposium*, Munich, Germany, Jul. 2012, pp. 3042–3045. doi: 10.1109/IGARSS.2012.6350784.
[9] A. Singh, F. Seitz, and C. Schwatke, "Inter-annual water storage changes in the Aral Sea from multi-mission satellite altimetry, optical remote sensing, and GRACE satellite gravimetry," *Remote Sensing of Environment*, vol. 123, pp. 187–195, Aug. 2012, doi: 10.1016/j.rse.2012.01.001.
[10] A. Singh, J. T. Reager, and A. Behrangi, "Estimation of hydrological drought recovery based on precipitation and Gravity Recovery and Climate Experiment (GRACE) water storage deficit," *Hydrol. Earth Syst. Sci.*, vol. 25, no. 2, pp. 511–526, Feb. 2021, doi: 10.5194/hess-25-511-2021.
[11] A. Mohanavelu *et al.*, "Trends and Non-Stationarity in Groundwater Level Changes in Rapidly Developing Indian Cities," *Water*, vol. 12, no. 11, p. 3209, Nov. 2020, doi: 10.3390/w12113209.
[12] M. Mandea, V. Dehant, and A. Cazenave, "GRACE—Gravity Data for Understanding the Deep Earth's Interior," *Remote Sensing*, vol. 12, no. 24, p. 4186, Dec. 2020, doi: 10.3390/rs12244186.
[13] M. Ahmed, M. Aqnouy, and J. Stitou El Messari, "Sustainability of Morocco's groundwater resources in response to natural and anthropogenic forces," *Journal of Hydrology*, vol. 603, p. 126866, Dec. 2021, doi: 10.1016/j.jhydrol.2021.126866.

[14] P. Bhalla, "Impact of Declining Groundwater Levels on Acreage Allocation in Haryana," *Economic and Political Weekly*, Jan. 2007, Accessed: Feb. 19, 2022. [Online]. Available: https://www.academia.edu/1367095/Impact_of_Declining_Groundwater_Levels_on_Acreage_Allocation_in_Haryana
[15] C. S. Bahinipati and P. K. Viswanathan, "Can Micro-Irrigation Technologies Resolve India's Groundwater Crisis? Reflections from Dark-Regions in Gujarat," *International Journal of the Commons*, vol. 13, no. 2, pp. 848–858, Oct. 2019, doi: 10.5334/ijc.888.
[16] C. S. Bahinipati, V. Kumar, and P. K. Viswanathan, "An evidence-based systematic review on farmers' adaptation strategies in India," *Food Sec.*, vol. 13, no. 2, pp. 399–418, Apr. 2021, doi: 10.1007/s12571-020-01139-3.
[17] E. Hasan, A. Tarhule, Y. Hong, and B. Moore, "Assessment of Physical Water Scarcity in Africa Using GRACE and TRMM Satellite Data," *Remote Sensing*, vol. 11, no. 8, p. 904, Apr. 2019, doi: 10.3390/rs11080904.
[18] O. Singh, A. Kasana, and P. Bhardwaj, "Understanding energy and groundwater irrigation nexus for sustainability over a highly irrigated ecosystem of north western India," *Appl Water Sci*, vol. 12, no. 3, p. 44, Mar. 2022, doi: 10.1007/s13201-021-01543-w.

Physically based modelling techniques for landslide susceptibility analysis: A comparison

V Meena[1], S Kumari[2] and V Shankar[3]

[1]PG Student, Department of Civil Engineering, National Institute of Technology, Hamirpur, Himachal Pradesh, 177005, India

[2]Ph.D Scholar, Department of Civil Engineering, National Institute of Technology, Hamirpur, Himachal Pradesh, 177005, India

[3]Associate Professor, Department of Civil Engineering, National Institute of Technology, Hamirpur, Himachal Pradesh, 177005, India

Email: 20mce317@nith.ac.in

Abstract. In mountainous areas, landslides are the most common natural catastrophic event, due to these events thousands of people are killed each year, and millions of dollars are lost in property damage. Landslides are mainly induced by earthquakes, rainfall, or manmade activities. Several GIS techniques, such as synthetic aperture radar, ranging data from spaceborne, airborne, and ground-based platforms, optical, and light detection and various physically based models such as SINMAP, TRIGRS, SHALSTAB etc. have been widely used to study slope failures in recent years. Each of these techniques has advantages and limitations for susceptibility analysis of landslides. The current research focuses on landslide susceptibility models that are physically based, their parameterization and working principle. The study infers that TRIGRS is the most commonly used model for slope stability analysis, whereas GEO top model is the only 3D slope stability model which considers the spatial variation of soil parameters hence it can be considered as the most advanced physically based slope stability model.

1. Introduction

Landslides are by far the most serious environmental threats, with widespread social, economic, and environmental consequences [1]. Landslides are caused by the interaction of a number of complicated components including geology, geomorphology, topography, and seismic variables [2]. Geological factors include weak and soft ground material, jointed and fractured materials, rainfall, snowfall, earthquake, and many more while the geomorphological factors include slope, erosion, vegetation change, tectonic and volcanic uplift etc. Although natural hazards such as landslides are unavoidable, a firm grasp of the tendency and a scientific approach to anticipate the behavior of such events can be effective tools in reducing natural risk and susceptibility [1]. In recent years, GIS-based landslide susceptibility mapping has become increasingly common [3]. The development of GIS based technologies such as satellite imagery and DEM improved the landslide susceptibility as it would be very difficult to analyze the slope stability over a large area with field observations and it would be very difficult to determine parameters such as slope elevation and aspect, drainage network, catchment area, slope convexity with field observations but these parameters can be very easily determined using DEM and satellite data [1]. Landslide hazard mapping has been done in a number of locations worldwide, particularly in the last few decades. Some of the strategies for landslide mapping that are available include mapping based on past data, evaluation of landslide occurrence probability, deterministic and statistical analysis, numerical modelling etc. [4].

Statistical approaches are data-driven ways for eliminating subjectivity in weightage assignment and creating more predictable outcomes [5]. Statistical methods include techniques like Fuzzy logic, ANN analysis, LR analysis etc. To reduce subjectivity in the weight assignment technique, probabilistic approaches have also been applied in landslide hazard mapping research. Within a probabilistic framework, the spatial distribution of landslides in response to diverse cause factors is compared using this method [5]. Conditional probability model, WoE technique under Bayesian probability model, SHETRAN physically based model etc. are some of the techniques based on this approach [5]. Mass, energy, and momentum conservation are the core of deterministic models. Distributed hydrological and slope instability methods be situated in deterministic landslide hazard zonation to determine the global variance of level of groundwater, pore pressures, and safety factor [6]. Physically based models like SHALSTAB, Rainfall threshold models etc. follows deterministic approach. Numerical modelling is also very commonly practiced technology to determine factor of safety in slope instability analysis. Generally numerical modelling technologies are based on Limit equilibrium method, Finite difference method, Finite element method etc. GeoStudio-Slope/W, Plaxis-2D/3D, CHASM, SLIDE etc. are some software packages that follow numerical modelling approach.

Physical models for slope failure are based on simple mechanical rules and describe the physical processes that contribute to the landslide event [4], some of these models follow deterministic approach while others follow probabilistic approach. These models considers the spatial variation of ground water-rainfall response throughout the slope, and they don't require landslide data from long past, so they are commonly used in places with incomplete landslide inventory [4].The present study aims to compare the various physical models such as SLIP, SHETRAN, GEOtop-FS, SUSHI, SINMAP, SHALSTAB, TRIGRS that have been used in studies worldwide over the last decade (2010-2020).

2. Methodology

The study shows the comparison of physically based models for landslide susceptibility analysis. To conduct this study, we collected articles from reputed journals on physically based models for slope instability analysis from the period 2010-2020. The detailed methodology used in the study is shown in the flow chart below.

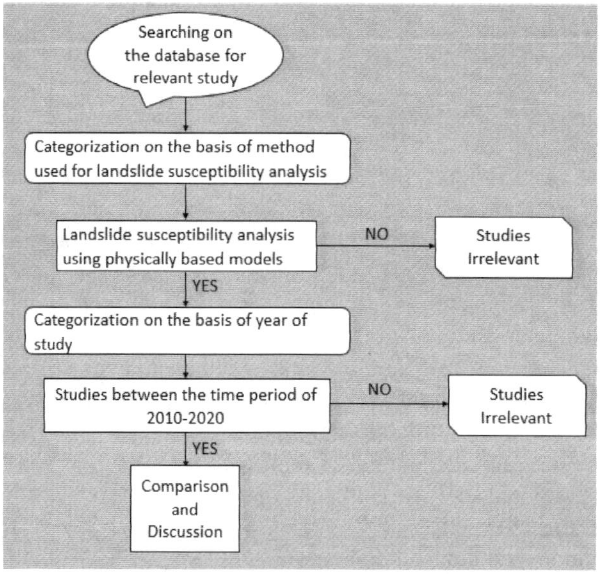

Figure 1. Flow chart of the methodology

2.1. Working principles of physically based models considered for the study

2.1.1. Stability index mapping (SINMAP). The stability model for infinite slope, used in the SINMAP technique, balances gravity's destabilizing components with friction and cohesion's restorative components sideways a probable plane of failure parallel to the earth's surface. To calculate the ground stability index (SI), SINMAP examines information such as geographic slope, watershed region, strength, climate etc. [7]. The possibility of any place to remain stable over the uncertainty ranges with considering uniform distribution of the parameters is defined as its SI. SI equals to 0 is considered as least stable whereas SI equals to 1 is the most stable [8–11].
The FoS of the model is simplified considering the wet and dry densities as equal (Eqn. 1).

$$SI = c' + \frac{cos^2\theta[\rho_s g(z - D_W) + (\rho_s g - \rho_w g)D_W] \tan\varphi}{z\rho_s g \sin\theta \cos\theta} \tag{1}$$

$c' = c_r + c_s$

TOPMODEL is the hydrological model used by the SINMAP technique [12]. Analysis done by SINMAP is based on some assumptions as follows:

- Topographical gradients were followed by subsurface flow at shallow depth. This suggests that the surface topography defines the area (*a*) making a contribution something to the flow at any point.
- Each point's lateral discharge q has a steady-state recharge R. (m/h)
- Lateral flux's measurements at each point is given by T.sinh, where T = K*z

$$w = \min\left(\frac{Ra}{T \sin\theta}, 1\right) \tag{2}$$

Eq. (2) is included into the dimensionless FoS to define the SI, which becomes:

$$SI = c + \frac{cos\theta\left[1 - \min\left(\frac{Ra}{Tsin\theta}, 1\right)\gamma\right] tan\varphi}{sin\theta} \tag{3}$$

Where, $c = c_r + c_s/(h\rho_s g)$, $\gamma = \rho_w/\rho_s$.

The DEM is used to determine the variables a and h, while the user provides the values for C, tan, γ, and R/T. SINMAP distinguishes between six different types of SI. SI >1.5 and SI >1.0 define the classes with normal stability, moderate stability, and quasi-stable classes; with the most cautious parameters in the indicated range, they illustrate locations that should not fail. FS<1.0 is found in the lower- and upper-threshold classes, with failure probabilities of lower than and more than 50%, respectively. These two classes, denoted by 1.0<SI<1.5 and 0<SI<1.0 respectively, indicates the minimum and maximum ground failure limits. The unconditionally unstable class has the highest chance of failure (>90 percent) within the stated range of parameters.

2.1.2. *Shallow Land sliding Stability Model (SHALSTAB).* SHALSTAB is developed by [13] and is a very common physically based model. This model conducts infinite-slope failure analysis when in steady-state circumstances. For slope analysis it incorporates a hill slope instability model with a hydrological model and Darcy's law to forecast 3-D variance of pore pressures [14–17]. SHALSTAB characterizes steady-state soil saturation depending on upslope impact region assessments using rainfall and topographical data, local slope and soil transmissivity, as well as hydrological model TOPOG established by O'Loughlin is used by SHALSTAB [18], [19]. Based on a saturated water flow parallel to the slope plane in a constant state, this model determines relative wetness (w):

$$w = \frac{Q}{T}\frac{a}{bsin\theta} \tag{4}$$

Where,
The drainage area and the width of the outflow barrier are represented by a and b, respectively, Q rainfall necessary to trigger landslides.

The SHALSTAB model forecasts the crucial rainfall needed to cause slope failure across a study area [13]. The model's design is a balance of hydrological in addition geo-mechanical elements. The SHALSTAB model solves the following equation to compute the essential precipitation for slope failure at each grid cell:

$$\log\left(\frac{Q}{T}\right) = \frac{sin\theta}{\frac{a}{b}}\left[\frac{c'}{\rho_W gzcos^2\theta tan\varphi} + \frac{\rho_s}{\rho_w}\left(1 - \frac{tan\theta}{tan\varphi}\right)\right] \tag{5}$$

This model's output is measured in mm/day of crucial rainfall, lower the values of output lower will be the intensity of rainfall required to trigger the slope failure and higher values indicate requirement of high intensity of rainfall for slope failure.

2.1.3. *Transient Rainfall Infiltration and Grid-based Slope-Stability (TRIGRS).* A well-known model is TRIGRS with a physical foundation for shallow landslide investigation. Various applications, such as landslide hazard mapping, have been tested using TRIGRS [20]. TRIGRS uses the input parameters: soil depth, slope, sub surface water table depth, saturated hydraulic conductivity (K_s) and steady (initial) surface flow to forecast in saturated conditions, the pore water pressure regime [21]. The sliding mass's depth is substantially lower than its length in TRIGRS 1-D infinite-slope instability model. The soil depth and physical attributes are uniform for every grid cell, and between grid cells, lateral stresses are neglected [22]. When calculating a safety factor with an infinite slope, the Taylor equation is utilised.

$$Fs(Z,t) = \frac{tan\varphi'}{tan\theta} + c' - \frac{\psi(Z,t)\gamma_W tan\varphi'}{\gamma_S Zsin\theta cos\theta} \tag{6}$$

Where, ψ (Z, t) is the ground water pressure head (ψ= u/γw), depending on Z is vertical direction coordinate and t is the time;
To get the safety factor above the saturated zone, the matric suction, ψ (Z, t) γ_w, is multiplied by χ, which can be approximated as:

$$\chi = \frac{\theta_v - \theta_r}{\theta_s - \theta_r} \tag{7}$$

2.1.4. *Systeme Hydrologique Europeen Transport (SHETRAN).* SHETRAN is a system having capability of modelling basin hydrology, sediment transport, and pollutant transport that is physically based and geographically dispersed [23]. Water flow, sediment transport, and pollutant transport are the three fundamental components of SHETRAN [24]. Using the 1-D FoS equation for infinite slope, to determine the frequency of landslides with a shallow depth, to mimic time-space changing soil saturation conditions, the hydrological model was utilised. This takes into account the influence of root cohesion and vegetation.

SHETRAN can model the hydrological response by accounting for spatial variations in land cover and soil parameters. SHETRAN is a hydrological model that now includes a landslide component. Root cohesion is taken into account, but it is usually time invariant. Despite the fact that the model can explain topographic convergence zones in theory but In practical, it is more likely to be used on unconfined debris flows and planar hillslope slides [24]. The SHETRAN model is being used to figure out where the crucial drenched soil depth is expected for a terrain failure to happen. If the time-varying drenched depth surpasses the limitation value during the simulation, the model predicts a landslide. The critical saturated soil depth spatial distribution efficiently offers a map of the failure chances distribution; areas with a higher critical depth have higher failure chances, according to the map, and a high slope angle is thought to be associated with a higher chance of failure [23].

2.1.5. *GEOtop-FS.* The GEOtop model is a spatially distributed and physically based three-dimensional (3-D) model that calculates the water, energy budgets [24-25]. It simulates unsaturated and saturated subsurface flows, channel flow, surface runoff and stormy fluctuations at the soil-atmosphere interface (e.g., temperature changes, earth temperature, etc.). The inputs of the model comprise maps (DEM, aspect, slope, landcover, soil form), rain, air temperature, and other climatological data are stored in meteorological files (such as relative humidness, wind velocity, shortwave radiation) if available, last but not least, other variables including parameters from Van Genuchten soil water retention curve, residual and saturated water content, numerical parameters for convergence criteria. Key output of the models includes soil suction and ground moisture map, sub-surface water depth maps at different levels, the infinite slope hypothesis is used to compute the FoS component.

The model is designed to work in a distributed environment and it calculates the Safety Factor for each pixel in the watershed, as well as each layer where the soil has been discretized [26].

$$F = \frac{tan\varphi'}{tan\beta} + \frac{2c'}{\gamma.z.\sin(2\theta)} - \sigma_s.\frac{tan\theta + cot\theta}{\gamma.z}.tan\varphi' \qquad (8)$$

Where, σ_S [kPa] is suction stress characteristic curve of the soil defined in eq. (9)

$$\sigma_s = \begin{cases} -(u_a - u_w) & if \ \ (u_a - u_w) \leq 0 \\ -\dfrac{\theta_v - \theta_r}{\theta_s - \theta_r}(u_a - u_w) & if \ \ (u_a - u_w) > 0 \end{cases} \qquad (9)$$

Soil wetness and pore water pressure multilayer calculation is provided using GEOtop and generates multilayer FoS maps.

2.1.6. *Probabilistic Infinite Slope Analysis model (PISA-m).* A DEM, maps of forest cover units, geotechnical soil units and data on geotechnical qualities and distributions for each map unit are all inputs into this model. [27]. Using β-PERT, triangular, uniform, and normal distributions [28]. Maps depicting the likelihood that the computed feature of static landslide safety is lower than the threshold value [Prob (FS<1)].

The infinite slope approximation is at the core of PISA-m, and is hence best suited to modelling the incidence of thin landslides in terms of length and width. For a forested infinite slope, the equation for the safety factor (FS) is used to govern the static function of the probabilistic model against sliding [27]

$$FS = \frac{c_r + c_s + [q_t + \gamma_m z + (\gamma_{sat} - \gamma_w - \gamma_m)H_w z]cos2\theta tan\varphi}{[q_t + \gamma_m z + (\gamma_{sat} - \gamma_m)H_w z]sin\theta cos\theta} \qquad (10)$$

Where, q_t is tree surcharge, γ_m, γ_{sat} are the soil moist and saturated unit weight respectively, H_w is the pore pressure coefficient.

2.1.7. *Saturated Unsaturated Simulation for Hillslope Instability (SUSHI).* The SUSHI model features two modules: a geotechnical component (Geo_SUSHI) for assessing slope instability and a hydrological component (Hydro_SUSHI) for investigating sub-soil water movement, with a focus on percolation processes and spatiotemporal changes in porewater pressure [29]–[31]. The Hydro_SUSHI module follows the Richards equation [29], stated as a function of the suction (ψ). The equation we get;

$$\frac{\partial}{\partial x}\left[K(\psi)\frac{\partial \psi}{\partial x}\right] + \frac{\partial}{\partial x}\left[K(\psi)\left(\frac{\partial \psi}{\partial z} - 1\right)\right] = C_{SU}(\psi)\frac{\partial \psi}{\partial t} \quad (11)$$

Where, $C_{SU}(\psi)$ is a simulation of water flow in both unsaturated and saturated zones, for unsaturated terrain, $K(\psi)$ is the hydraulic conductivity based on suction.

$$C_{SU}(\psi) = \frac{\theta}{\eta}S_s + C(\psi) \quad (12)$$

Where The slope of a soil–water characteristic curve is denoted by $C(\psi)$.

For unsaturated soils, the Geo SUSHI module employs the Fredlund and Rahardjo slope letdown equation, which is based on well-known General Limit Equilibrium methods.

$$\tau = c' + (\sigma - u_a)tan\varphi' + (u_a - u_w)tan\varphi_b \quad (13)$$

$$tan\varphi_b = \left\{tan\varphi'\left[\frac{\theta(\psi) - \theta_r}{\theta_s - \theta_r}\right]\right\} \quad (14)$$

where τ is shear strength, ϕ_b denotes the angle expressing the rate of increase in strength due to matric suction, σ is total normal stress and θ(ψ) is the water content shown by the suction level (ψ) retention curve.

The FoS is calculated using the GLE approach, which uses the summation of forces in two directions and moments around a generic point [30]. The FoS is the shear strength parameter reduction required to bring the soil mass into confined equilibrium along the anticipated slip surface.

$$S_m = \frac{\alpha\tau}{FS} \quad (15)$$

Where S_m is the shear force mobilized at the base of the slice, α is the sloping distance across the base of a slice.

2.1.8. *Shallow Landslides Instability Prediction (SLIP).* Developed by Montrasio and Valentino [32]. The factor of safety is calculated using a basic physically based model that takes into account the key factors that cause soil slip, such as the quantity of rain, slope geometry, and the condition, soil's mechanical and hydraulic properties. The model's hypotheses were created based on field observations; thus, it can forecast soil slipping in real-world situations. The hypothesis are as follows:

- The top layer of soil is partially saturated, but as you go deeper, the water content increases.
- The sliding surface is displaced in the soil's superficial layer and does not correspond to the soil-bedrock contact.

- The medium grade of slope is substantially bigger than the angle of shear strength of soil and the soil is generally consolidated, according to geometrical medium features.

The FoS derived using the limit equilibrium approach, is employed to assess the slope instability [33]. The Mohr-Coulomb strength criteria is used to analyze, stabilizing forces, it addresses the impact of surface saturation on soil cohesiveness, which is a key aspect in the triggering process [33], [34]. The factor of safety can be calculated as:

$$F_S = \frac{cot\theta . tan\varphi' . [\gamma + m.(\eta_w - 1)] + C'.\Omega}{\gamma + m.\eta_w} \tag{16}$$

Where;

$$\gamma = G_s.(1-\eta) + \eta . S_r \tag{17}$$

$$\eta_w = \eta .(1 - S_r) \tag{18}$$

$$\Omega = \frac{2}{sin2\theta . H.\gamma_w} \tag{19}$$

$$C' = [c' + c_\psi]L = [c' + A.S_r.(1-S_r)^\lambda.(1-m)^\alpha]L \tag{20}$$

$$m = \frac{\xi}{\eta H(1-S_r)} \sum_{i=1}^{\omega} h_i \exp[-K_T(t-t_i)] \tag{21}$$

Symbols in Eqs. (16-21) have following meanings:
The length of the soil slice is L, the saturated proportion of the soil layer in terms of its depth is m., H is the thickness of the potentially unstable layer, S_r is the degree of saturation of the terrain, G_s is the specific weight of the soil, c_ψ is the apparent cohesion given by the unfinished saturation of the surface; A, α and λ are limits for numerical calibration, K_T slope's discharge capacity, ξ is a runoff coefficient, t_i is the time in days to which the rainfall depth h_i resembles; and t is the time.

3. Comparison of models

Paulin et al. [12] used SINMAP model with 30-m, 10-m, 5-m and 1-m DEM and found that 1-m and 5-m maps gave better results than 10-m map and the 30-m map was considered as very less suitable for the prediction of landslide susceptibility analysis. From all the above studies [7-12] it was found that the ability of SINMAP for correctly predicting the unstable slopes depends on the quality of DEM used as input because in case of low resolutions more than 30-m the model fails to mark the unstable locations which shows very small failure and are very shallow.

TRIGRS models is used for numerous slope stability studies throughout the past decade and different researchers used the model under different input parameters. Sarma et al. [21] used input elevation maps with resolution of 1 and 1/3 arc-sec, Tran et al. [20] used a DEM of 1-m resolution and Ciurleo et al. [22] used the DEM with 5-m resolution. So, in the first study all the landslides were mapped correctly and efficiency reaching near to 90% in the second study also all the landslides were mapped good except few small one and it attained an efficiency of almost 85% but in the third study the efficiency dropped below 80% because in a 5-m DEM many shallow landslides were hard to map. It was found that the ability of TRIGRS to correctly model the landslide be subject to the quality of the input map and the preciseness of the calculated input parameters [19-21].

SHALSTAB is among the oldest physically based slope instability models, Vieira et al. [19] compared the prediction ability of SHALSTAB and TRIGRS under same conditions and it was found that with a 4-m DEM as input both models showed good results but TRIGRS performed comparatively better, this could be due to proper consideration of soil water characteristics and also the effect of water table in slope stability was considered by TRIGRS. Pradhan and Kim [8], Michel et al. [18] compared the landslide susceptibility mapping software SHALSTAB and SINMAP, the result showed that both the models matched the landslide scars but SHALSTAB showed better results than SINMAP. [13-18] used SHALSTAB for slope instability analysis and found that the property of the model to consider uniform spreading of the soil constraints throughout the area of study limits its ability to correctly map the unstable locations in case of locations with non-uniform variation of geotechnical and hydrological parameters.

Formetta et al. [26] used GEOtop model for slope stability analysis and claimed that this model considers the spatial variation of soil parameters, they studied that this model is based on 3-D Richard equations for stratified soils. They in their study found that the uncertainty of the soil parameters and quality of input maps mainly effects the efficiency of the model.

Two modules make up the SUSHI model for slope stability analysis. Hydro SUSHI is a model for investigating subterranean water circulation based on Richard's equation and Geo SUSHI is based on GLE methods and analyses slope stability using the Fredlund and Rahardjo equation [29]. This model can calculate actual evapotranspiration and its implications on summertime suction levels. [30-31].

Naidu et al. [27] performed slope stability analysis at 3 different locations using the PISA model. In the study they used 3 input maps namely the DEM, NDVI map and the soil map, with NDVI maps as input the effect of root cohesion and tree surcharge can be determined more efficiently with PISA model. Along with these maps this model also required the geotechnical and hydrological parameters of the area like any other slope stability model.

Montrasio et al. [27] used SLIP and TRIGRS model for slope stability analysis and the study area was located in Italy. They studied that SLIP model uses Mohr-Coulomb strength criterion for the calculation of FoS whereas TRIGRS was based on Iverson model and extended Richard's equation. TRIGRS model was found to be more efficient because TRIGRS considers the pressure due to groundwater head as a function of both vertical and temporal dimensions. Zizioli et al. [34] compared the model performance of SHALSTAB, SINMAP, TRIGRS and SLIP model. They concluded that SHALSTAB outputs effective rainfall whereas others output stability index and the FoS. The efficiencies of the models were nearly same SINMAP (79%), SHALSTAB (78%), TRIGRS (78%), SLIP (79%).

Zhao et al. [24] studied landslide susceptibility of Darjeeling Himalayas with the help of SHETRAN model. They concluded that SHETRAN model follows a probabilistic approach, the restricted possibility of a landslide occurring was calculated via Bayes' Theorem. The SHETRAN model was shown to be capable of accounting for the effect of foliage on slope instability in terms of evapotranspiration and canopy cover in the study. Abraham et al. [35] used SHETRAN model for slope stability analysis and discussed its limitations, which are, First, all upslope, downslope, and lateral barriers are ignored in the 1-D infinite-slope analysis. Second, a lack of data could hinder the model's capacity to capture local spatial inconsistency in soil characteristics and other factors that cause soil instability; additionally, the DEM's quality and resolution may limit the model's ability to interpretation for topographic controls on land sliding.

4. Discussion

4.1. Models used in landslide susceptibility analysis

Throughout the time span considered for the study (2010-2020) there were various technologies used to determine the susceptibility of a landslide. But as the study is only focused on physically based models there were total 8 models used to fulfil the objective of study, among which TRIGRS was most used model with 54% of studies using TRIGRS, followed by SINMAP (25%), SHALSTAB (14%) and SUSHI, SHETRAN, PISA-m, GEOtop, SLIP combinedly contributing to almost 7% of the studies.

IOP Conf. Series: Earth and Environmental Science 1032 (2022) 012033 doi:10.1088/1755-1315/1032/1/012033

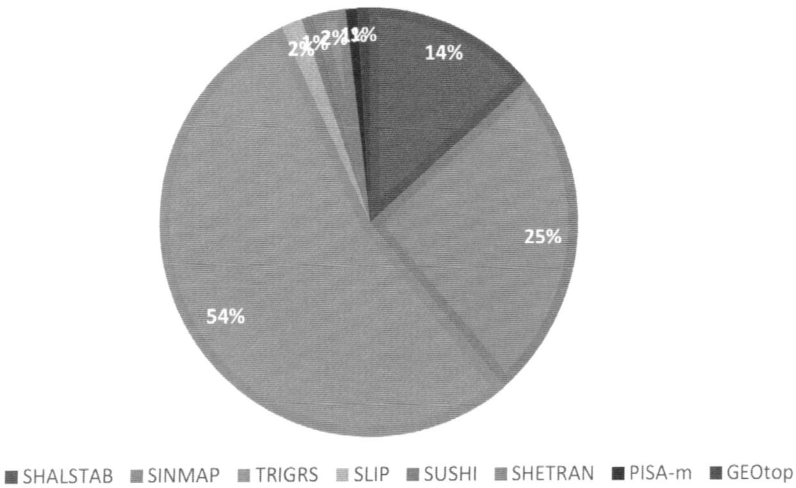

Figure 2. Studies with respect to model used.

4.2. *Input data required*

The variety of methodologies and models utilised to assess landslide susceptibility need a large number of data inputs. The cause and kind of landslide, as well as the characteristics of the investigated area, magnitude of the research, availability of data, and the technique utilised, all influence the factor selection [1]. Landslide susceptibility evaluation parameters are classified into four categories: geological, topographical, geotechnical, and ecological aspects [1].

Table 1. Input parameters required by different models

Model/Input Parameters	SHALSTAB	SINMAP	TRIGRS	SUSHI	GEOtop	PISA	SHETRAN	SLIP
Soil Depth	☑	☑	☑		☑		☑	☑
Hydraulic Conductivity	☑	☑	☑	☑	☑		☑	
Transmissivity		☑						
Recharge/ Rainfall		☑						
Effective Cohesion	☑	☑	☑	☑	☑	☑		☑
Internal friction angle	☑	☑	☑	☑	☑	☑		☑
Soil Density	☑	☑					☑	
Unit weight of soil	☑		☑			☑		
Soil Diffusivity			☑					
Saturated volumetric water content			☑	☑	☑			

Residual volumetric water content	☑	☑	☑		☑	
Soil Water Characteristic curve	☑	☑	☑		☑	
vanGenuchten-n			☑		☑	
vanGenuchten-alpha			☑		☑	
Depth of water table	☑					
Soil moisture content				☑		
Specific Gravity				☑		☑
Dry Unit weight of soil		☑				
Pore pressure				☑		
Void ratio				☑		
Saturated unit weight of soil		☑		☑		
Porosity						☑
Discharge capacity of the slope						☑
NDVI map				☑		
Evapotranspiration					☑	

Recent studies on landslide susceptibility analysis have considered a total of 25 factors (excluding the DEM) out of which cohesion and internal friction were the most considered limitations with 7 out of 8 models considering, soil depth and hydraulic conductivity were considered by 6 of the 8 models and NDVI index, Evapotranspiration and water table depth were the least considered parameters with only 1 of the 8 models considering each of them (Table 1).

5. Conclusion

The study presents a comparative analysis of various physically based slope stability models used for landslide susceptibility analysis. The study infers that TRIGRS, SHALSTAB and SINMAP contribute to almost 93 % of the studies in the past decade, this can be due to their efficiency along with opensource nature and simplicity in implementation as compared to the more advanced GEOtop model. The efficiencies of all these models be contingent on the correctness of input data used for the study (most precisely the quality of DEM), PISA-m and SNETRAN models are complex to use and have less efficiencies in varying soil conditions as compared to SINMAP, SHALSTAB and TRIGRS. Among all these models GEOtop was the only 3-D model so this model considers the spatial variation of the soil parameters and PISA model is the only model having NDVI map and soil map as input so this model can better explain the effect of tree surcharge and vegetation on slope instability.

Also, among the considered physically based models, internal friction angle and cohesiveness were the maximum considered limits for landslide susceptibility analysis and NDVI, water table depth and evapotranspiration were the least considered parameters. This could be because cohesion and inner friction are the first parameters affecting the soil strength most and NDVI, water table depth and evaporation being the least used parameters could be because of the complexity in determining these factors for a large area.

Symbols used and their meanings

c_r- Root cohesion
c_s- Soil cohesion
c'- Effective cohesion
θ- Slope angle
ρ_s- Wet soil density
ρ_w- Density of water
D_W - Vertical height of water table within soil layer
ϕ- Internal friction angle of the soil
T- Soil Transmissivity
w- Relative wetness
z- Soil thickness
g- Acceleration due to gravity
γ_w- the unit weight of groundwater
γ_s- the soil unit weight
χ- Bishop's effective stress parameter
θ_v- Volumetric water content
θ_r- Residual water content
θ_s- Water content at saturation
u_a- Pore air pressure
u_w- Pore water pressure
η- Porosity of soil
FS- Factor of safety

References

[1] H. R. Pourghasemi, Z. Teimoori Yansari, P. Panagos, and B. Pradhan, "Analysis and evaluation of landslide susceptibility: a review on articles published during 2005–2016 (periods of 2005–2012 and 2013–2016)," *Arabian Journal of Geosciences*, vol. **11**, no. 9. Arabian Journal of Geosciences, 2018, doi: 10.1007/s12517-018-3531-5.
[2] C. Zhao and Z. Lu, "Remote sensing of landslides-A review," *Remote Sens.*, vol. **10**, no. 2, pp. 8–13, 2018, doi: 10.3390/rs10020279.
[3] S. Lee, "Current and Future Status of GIS-based Landslide Susceptibility Mapping: A Literature Review," *Korean J. Remote Sens.*, vol. **35**, no. 1, pp. 179–193, 2019.
[4] S. D. Pardeshi, S. E. Autade, and S. S. Pardeshi, "Landslide hazard assessment: Recent trends and techniques," *Springerplus*, vol. **2**, no. 1, pp. 1–11, 2013, doi: 10.1186/2193-1801-2-523.
[5] D. P. Kanungo, M. K. Arora, S. Sarkar, and R. P. Gupta, "Landslide Susceptibility zonation (LSZ) mapping - a review," *J. South Asia Disaster Stud.*, vol. **2**, no. 1, pp. 81–106, 2009.
[6] M. T. J. Terlien, C. J. Van Westen, and T. W. J. van Asch, "Deterministic Modelling in Gis-Based Landslide Hazard Assessment," pp. 57–77, 1995, doi: 10.1007/978-94-015-8404-3_4.
[7] A. M. S. Pradhan and Y. T. Kim, "Application and comparison of shallow landslide susceptibility models in weathered granite soil under extreme rainfall events," *Environ. Earth Sci.*, vol. **73**, no. 9, pp. 5761–5771, 2015, doi: 10.1007/s12665-014-3829-x.

[8] J. Davis and L. Blesius, "A hybrid physical and maximum-entropy landslide susceptibility model," *Entropy*, vol. **17**, no. 6, pp. 4271–4292, 2015, doi: 10.3390/e17064271.
[9] A. Akgun and O. Erkan, "Landslide susceptibility mapping by geographical information system-based multivariate statistical and deterministic models: in an artificial reservoir area at Northern Turkey," *Arab. J. Geosci.*, vol. **9**, no. 2, pp. 1–15, 2016, doi: 10.1007/s12517-015-2142-7.
[10] M. L. Rabonza, R. P. Felix, A. M. F. A. Lagmay, R. N. C. Eco, I. J. G. Ortiz, and D. T. Aquino, "Shallow landslide susceptibility mapping using high-resolution topography for areas devastated by super typhoon Haiyan," *Landslides*, vol. **13**, no. 1, pp. 201–210, 2016, doi: 10.1007/s10346-015-0626-x.
[11] M. Lazzari and D. Gioia, "Regional-scale landslide inventory, central-western sector of the Basilicata region (Southern Apennines, Italy)," *J. Maps*, vol. **12**, no. 5, pp. 852–859, 2016, doi: 10.1080/17445647.2015.1091749.
[12] G. L. Paulin, M. Bursik, J. Lugo-hubp, and J. J. Z. Orozco, "Effect of pixel size on cartographic representation of shallow and deep-seated landslide , and its collateral effects on the forecasting of landslides by SINMAP and Multiple Logistic Regression landslide models," *Phys. Chem. Earth*, vol. **35**, no. 3–5, pp. 137–148, 2010, doi: 10.1016/j.pce.2010.04.008.
[13] D. R. Montgomery and W. E. Dietrich, "A physically based model for the topographic control on shallow landsliding," *Water Resour. Res.*, vol. **30**, no. 4, pp. 1153–1171, 1994, doi: 10.1029/93WR02979.
[14] A. M. S. Pradhan and Y. T. Kim, "Evaluation of a combined spatial multi-criteria evaluation model and deterministic model for landslide susceptibility mapping," *Catena*, vol. **140**, pp. 125–139, 2016, doi: 10.1016/j.catena.2016.01.022.
[15] F. Cervi, M. Berti, L. Borgatti, F. Ronchetti, F. Manenti, and A. Corsini, "Comparing predictive capability of statistical and deterministic methods for landslide susceptibility mapping: A case study in the northern Apennines (Reggio Emilia Province, Italy)," *Landslides*, vol. **7**, no. 4, pp. 433–444, 2010, doi: 10.1007/s10346-010-0207-y.
[16] M. Tiago Damas, V. Bianca Carvalho, F. Nelson Ferreira, O. Chisato, and M. David R., "Application of the SHALSTAB model for the identification of areas susceptible to landslides: Brazilian case studies.," *Rev. Geomorfol.*, vol. **19**, no. 1, pp. 136–144, 2017, doi: 10.21094/rg.2017.015.
[17] S. Sarkar, A. K. Roy, and P. Raha, "Deterministic approach for susceptibility assessment of shallow debris slide in the Darjeeling Himalayas, India," *Catena*, vol. **142**, pp. 36–46, 2016, doi: 10.1016/j.catena.2016.02.009.
[18] G. P. Michel, M. Kobiyama, and R. F. Goerl, "Comparative analysis of SHALSTAB and SINMAP for landslide susceptibility mapping in the Cunha River basin, southern Brazil," *J. Soils Sediments*, vol. **14**, no. 7, pp. 1266–1277, 2014, doi: 10.1007/s11368-014-0886-4.
[19] B. C. Vieira, N. F. Fernandes, O. Augusto Filho, T. D. Martins, and D. R. Montgomery, "Assessing shallow landslide hazards using the TRIGRS and SHALSTAB models, Serra do Mar, Brazil," *Environ. Earth Sci.*, vol. **77**, no. 6, pp. 1–15, 2018, doi: 10.1007/s12665-018-7436-0.
[20] G. L. Paulín *et al.*, "Landslide inventory mapping and landslide susceptibility modeling assessment on the SW flank of Pico de Orizaba volcano, Puebla-Veracruz, Mexico," *Zeitschrift fur Geomorphol.*, vol. **57**, no. 3, pp. 371–385, 2013, doi: 10.1127/0372-8854/2012/0097.
[21] C. P. Sarma, A. M. Krishna, and A. Dey, "Landslide hazard assessment of Guwahati region using physically based models," *6th Annu. Conf. Int. Soc. Integr. Disaster Risk Manag.*, no. December, pp. 1–13, 2015.
[22] M. E. Fauzan, A. Damayanti, and R. Saraswati, "Vulnerability assessment of landslide areas in Ci Manuk Upstream Watershed, Garut District, West Java Province," *Int. J. Adv. Sci. Eng. Inf. Technol.*, no. 1, pp. 219–226, 2020, doi: 10.18517/ijaseit.10.1.6755.
[23] J. C. Bathurst, C. I. Bovolo, and F. Cisneros, "Modelling the effect of forest cover on shallow

landslides at the river basin scale," *Ecol. Eng.*, vol. **36**, no. 3, pp. 317–327, 2010, doi: 10.1016/j.ecoleng.2009.05.001.

[24] B. Zhao, Q. Dai, D. Han, H. Dai, J. Mao, and L. Zhuo, "Probabilistic thresholds for landslides warning by integrating soil moisture conditions with rainfall thresholds," *J. Hydrol.*, vol. **574**, pp. 276–287, 2019, doi: 10.1016/j.jhydrol.2019.04.062.

[25] R. Rigon, G. Bertoldi, and T. M. Over, "GEOtop: A distributed hydrological model with coupled water and energy budgets," *J. Hydrometeorol.*, vol. **7**, no. 3, pp. 371–388, 2006, doi: 10.1175/JHM497.1.

[26] G. Formetta, V. Rago, G. Capparelli, R. Rigon, F. Muto, and P. Versace, "Integrated Physically based System for Modeling Landslide Susceptibility," *Procedia Earth Planet. Sci.*, vol. **9**, pp. 74–82, 2014, doi: 10.1016/j.proeps.2014.06.006.

[27] S. Naidu, K. S. Sajinkumar, T. Oommen, V. J. Anuja, R. A. Samuel, and C. Muraleedharan, "Early warning system for shallow landslides using rainfall threshold and slope stability analysis," *Geosci. Front.*, vol. **9**, no. 6, pp. 1871–1882, 2018, doi: 10.1016/j.gsf.2017.10.008.

[28] W. C. Haneberg, W. F. Cole, and G. Kasali, "High-resolution lidar-based landslide hazard mapping and modeling, UCSF Parnassus Campus, San Francisco, USA," *Bull. Eng. Geol. Environ.*, vol. **68**, no. 2, pp. 263–276, 2009, doi: 10.1007/s10064-009-0204-3.

[29] G. Capparelli and P. Versace, "FLaIR and SUSHI: Two mathematical models for early warning of landslides induced by rainfall," *Landslides*, vol. **8**, no. 1, pp. 67–79, 2011, doi: 10.1007/s10346-010-0228-6.

[30] G. Capparelli and P. P. Versace, "Landslide susceptibility from mathematical model in Sarno area," *Hydrol. Earth Syst. Sci. Discuss.*, vol. **10**, no. 10, pp. 12643–12662, 2013, doi: 10.5194/hessd-10-12643-2013.

[31] G. Capparelli and P. Versace, "Analysis of landslide triggering conditions in the Sarno area using a physically based model," *Hydrol. Earth Syst. Sci.*, vol. **18**, no. 8, pp. 3225–3237, 2014, doi: 10.5194/hess-18-3225-2014.

[32] L. Montrasio and R. Valentino, "A model for triggering mechanisms of shallow landslides," *Nat. Hazards Earth Syst. Sci.*, vol. **8**, no. 5, pp. 1149–1159, 2008, doi: 10.5194/nhess-8-1149-2008.

[33] L. Montrasio, R. Valentino, and G. L. Losi, "Towards a real-time susceptibility assessment of rainfall-induced shallow landslides on a regional scale," *Nat. Hazards Earth Syst. Sci.*, vol. **11**, no. 7, pp. 1927–1947, 2011, doi: 10.5194/nhess-11-1927-2011.

[34] D. Zizioli, C. Meisina, R. Valentino, and L. Montrasio, "Comparison between different approaches to modeling shallow landslide susceptibility: A case history in Oltrepo Pavese, Northern Italy," *Nat. Hazards Earth Syst. Sci.*, vol. **13**, no. 3, pp. 559–573, 2013, doi: 10.5194/nhess-13-559-2013.

[35] M. T. Abraham, N. Satyam, B. Pradhan, and A. M. Alamri, "Forecasting of landslides using rainfall severity and soil wetness: A probabilistic approach for Darjeeling Himalayas," *Water (Switzerland)*, vol. **12**, no. 3, pp. 1–19, 2020, doi: 10.3390/w12030804.

Spatial variation of Physico-chemical parameters and water quality assessment of urban ponds at Raipur, Chhattisgarh, India

Anjali Tamrakar[1], Kshitij Upadhyay[2] & Samir Bajpai[2]

1. Department of Environmental Science, Pt. Ravishankar Shukla University, Raipur, Chhattisgarh
2. Department of Civil Engineering, National Institute of Technology, Raipur, Raipur, Chhattisgarh, India.

Corresponding author: Kshitij Upadhyay

Email: kshitij.upadhyay111@gmail.com

Abstract-

Water bodies, natural or man-made, are under threat due to rapid urbanization. This research paper assesses the water quality of 20 ponds located in the vicinity of urban habitats of Raipur district, Chhattisgarh (India) using the Water Quality Index (WQI). The samples were analyzed for 16 physicochemical parameters in the post-monsoon winter season. The parameters were used to calculate WQI and the ponds were categorized accordingly for end use purposes, that is, drinking, irrigation, industrial, or unfit for use without treatment. The WQI of post-monsoon samples resulted in that 75% of the total pond water samples are in excellent condition, & 25% of the samples are unfit for various activities like drinking, irrigation, industrial & domestic. A high value of nutrients was observed in all the ponds, suggesting that the nutrient entering through surface runoff, solid waste disposal, and wastewater is entering the ponds. Our results conclude that all ponds require interventional steps to restore pond water quality and stop it from further degradation, determining which trophic state of the ponds may change in the future. The spatial monitoring of physical and chemical properties of pond water helps to predict, identify, and assess the natural condition of the urban ponds and their relation with the surrounding (environment) and helps in adopting rejuvenation measures.

Keywords- Pond Water, Physicochemical Properties, Water Quality Index (WQI), Raipur, Chhattisgarh.

1. Introduction-

Ponds are described as a small area filled with water, natural or man-made. They are smaller than a lake and are the habitat of various aquatic plants and animals [1]. Ponds are important hotspots for biodiversity. Healthy ponds are an indicator of a healthy environment, they provide safe havens to native wildlife, and they also help in nutrient cycling and remove environmental pollution [2].

Compared to other water bodies (oceans, rivers, lakes) that flow in nature, many ponds are more prone to pollution due to their stagnant nature [3]. The stagnant nature of a water body with high temperatures or high concentrations of phosphate and nitrates fuels the bloom of algae. Algae are always present in the water. The harmful algae bloom (HAB) has a direct correlation with the trophic state of the pond [4]. Excess fertilizer runoff caused by rainfall to rural, urban, and natural environments can remove nutrients such as phosphorus, nitrogen, and potassium from the water [5]. Agricultural waste (containing animal waste, bird droppings), industrial waste (chemical discharge and waste), and urban life (raw untreated wastewater and waste) also carry these nutrients. These nutrients along with warm temperatures and sunlight cause algal growth. Cyanobacteria is considered one of the HABs that releases toxins such as microcystin that enter the public drinking water system and kill the aquatic life [6]. Chlorophyll-a is also determined to know how much algae is present in the water source. Chlorophyll (a) is a pigment that is responsible for making plants and algae green in color. Chlorophyll (a) concentration can tell a lot about water quality, a water body with low productivity is deep and generally have good water quality and vice versa [7].

Hence, the productivity of the lake is used to make a quantitative and qualitative assessment of the water body based on the amount of biological activity happening in it. Trophic state index (TSI), a popular dimensionless parameter, is used to classify lakes or ponds individually based on the level of biomass and nutrients in the water body. Carlson proposed a widely used biomass-based TSI in 1977. It accounts for three important variables, chlorophyll-a, Secchi transparency depth, and total phosphorus concentration. The TSI value is calculated individually for each variable on a logarithmic scale and then averaged to give a final TSI value which indicates the level of biological activity in the particular water body. The TSI value is then used to label pond as oligotrophic (productivity level - less), mesotrophic (productivity level - moderate), and eutrophic (productivity level – very high) [8]. Many modifications have been proposed in the Carlson TSI by including or excluding other variables based on local geography [9, 10]. A major nutrient, Total Nitrogen (TN), is another key component that plays a vital role in the productivity of aquatic sources as it acts as a nutrient. The nutrient ratio (total nitrogen / total phosphorus) is used by the Florida TSI method to assess the limiting nutrient levels in the water body and TSI has been calculated accordingly [11, 12]. Long-term monitoring of biomass concentration, nutrient concentration is essential to establish levels of biological activity in the water bodies to ascertain their suitability to sustain aquatic life (plant, fish, wildlife) [13].

A safe water supply is essential for the survival of human species, still considered to be of least concern. Surface water bodies across the globe has shown a significant reduction in water quality since the last century [14]. Despite the ever-growing body of scientific findings that have established wastewater discharge (domestic, industrial), wastewater treatment plant effluent discharge, agricultural and surface runoff, and disposal of solid waste as the major drivers of surface water pollution, the condition of natural bodies continues to worsen. These anthropogenic activities lead to a change in the hydrological cycle and significantly affect the biogeochemical cycle in freshwater bodies [15, 16, 17].

Scientific findings suggest that loss of biodiversity due to changes in the landscape especially conversion of rural habitats, to urban habitats is directly responsible for ‘reduction of important ecosystem functions, such as productivity and material cycling’ [18].

The quality of the aquatic sources is interrelated with the variation in the land cover and associated land use in the respective watersheds. A good water quality lies in the vicinity of untouched forest watersheds. The water quality starts falling as the land use covers change. This has been observed as the rural landscapes’ changes from rural to urban through intense urbanization and industrialization. While the human endeavor and quest to advancement and pursuit of ease of life can’t be rationally stopped, the major task is balancing available and exploitable water resources and sustaining their quality [19].

The continuous degradation of lakes and ponds presents major challenges for restoration and will have significant ecological and material costs. Due to global warming and world population rising, the polluted water bodies will require increased conservation efforts to improve their quality. [20]. In third world countries which are transitioning from developing to developed economies like India, there remains a large inequality in water supply and sanitation. Lack of systematic and planned potable forced rural populations to look for alternative sources, while lack of proper education and health infrastructure, and employment aspiration have forced people to migrate to urban centers. The rapid industrialization and commercialization of rural habitats recently are transforming them into semi-urban and urban habitats [21]. This shift in land use and land cover has put the aquatic sources in the urban vicinity at increased risk of contamination. The urban ponds are susceptible to wastewater discharge, surface runoffs, street washing, and solid waste disposal. These ponds are also used for bathing and laundry purpose. Cattle owners use urban ponds to wash and cool off the cattle which introduces nutrients to the pond. Few urban pond waterfronts is a popular recreational and tourist hotspot, making them more susceptible to pollution.

Recent studies have shown that the water quality of Chhattisgarh is deteriorating [22]. The Raipur city of Chhattisgarh state is known for its high quantity of natural and manmade ponds [1]. The city is recently witnessing a habitat transformation. The spatial distribution of the physical and chemical properties of the pond water is limited, but few studies have shown it to be high, considering that most ponds lie in urban vicinity with poor sanitation practices. The main objective of this study was to assess the surface water quality of various urban ponds in Raipur City, as well as categorize their suitability using WQI.

WQI (Water Quality Index) is a mathematical model which provides information about water quality. It provides a single value for variable values of water quality parameters. It makes understanding water quality easy. WQI is of two types 1. Physicochemical indices, [2]. Biological indices. In this paper, we have used a physicochemical index to categorize ponds based on observed physical and chemical parameters. [23,24,25,26]. The study also examines the influence of land use patterns on the water quality of the ponds by collecting water samples from the ponds located in different land-use areas across Raipur city and making inter-comparison across the classified groups.

2. Materials and methods-

2.1 Location and geology of study area –

The 20 ponds studied here are located in heart of the industrial and residential area of Raipur city, which makes them prone to anthropogenic pollution. Raipur, the capital city of Chhattisgarh state, India falls under the Hirri sub-basin in the Chandi formation of Chhattisgarh basin. "The Chandi formation is further classified into two geological groups, the Raipur Limestone and the Deondongarh Shale. Raipur Limestone is known for grey, fine-grained, horizontally bedded, stromatolitic, massive limestone, high secondary porosity due to joints and karstification but with negligible primary porosity" [24, 25]. The tropical climate 0of Raipur city can be classified as sub-humid with average annual precipitation in range of 1300 to 1400 mm. Majority of the precipitation occurs in monsoon season (June to early October).

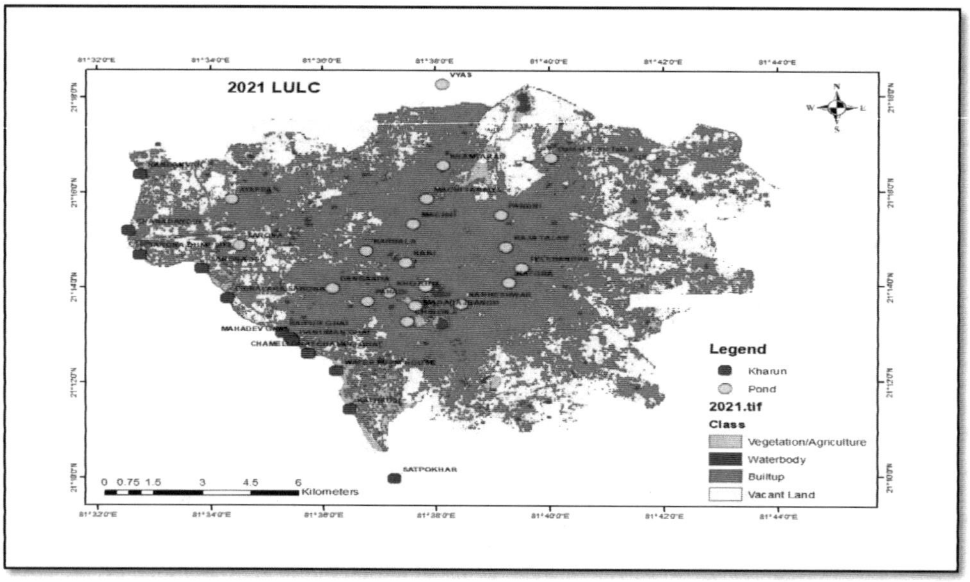

Figure 1. Land Use and land cover pattern of the study area along with the sampling location in Raipur city

2.2 Land use and land cover information-

Apparent changes in the land use pattern from forest, rural to urban and semi-industrial land transformation of Raipur can be observed through land-use and land–cover (LULC) maps. The city still lacks a proper sewerage system. The Land use and land cover image of 2021 is shown in fig. 1 while sampling locations are depicted in fig.2. A visible change in the land-use and land–cover (LULC) patterns can be deduced by comparing LULC 2013 to LULC 2021.

Table 1. Categorization of 20 ponds, sample code, accessibility to the public, and pond cleaning details.

	Classification of ponds in various groups based on habitat, land use and gentrification	Sample location code	Ponds studied	Accessibility to the public and frequency of pond cleaning
[A]	**Commercial / Industrial**	P1	Vyas pond	Accessible, no cleaning
		P2	Sarona pond	Accessible, no cleaning
		P3	Kari pond	Accessible, monthly cleaning
		P4	Khamtarai pond	Accessible, no cleaning
		P5	Pandri pond	Accessible, no cleaning
[B]	**Recreational**	P6	Budha Pond	Accessible, daily cleaning
		P7	Katora pond	Not accessible, no cleaning
		P8	Narheshwar pond	Accessible, no cleaning
		P9	Telebandha pond	Accessible, periodic cleaning
		P10	Danganiya pond	Accessible, no cleaning
[C]	**High-income group**	P11	Raja Pond	Accessible, no cleaning
		P12	Pahadi pond	Accessible, no cleaning
		P13	Karbala pond	Accessible, no cleaning
		P14	Kho kho pond	Accessible, no cleaning
		P15	Daldal seoni pond	Accessible, no cleaning
[D]	**Low-income group**	P16	Ayappa pond	Accessible, no cleaning
		P17	Chironji pond	Accessible, no cleaning
		P18	Macchhi pond	Accessible, no cleaning
		P19	Macchhi taraiya pond	Accessible, no cleaning
		P20	Mahrajbandh pond	Accessible, no cleaning

The 20 ponds considered in this study is divided into 4 category – High-Income group, Low-income group, Recreational/tourist, and the Commercial/Industrial zones. The detail of each pond with accessibility to general public, pond cleaning, and categorization is enlisted in [Table 1]. The classification and the selection of the ponds was done after preliminary survey of the ponds. Based on the reconnaissance survey a total of 5 ponds were selected from each category for this study.

A multitude of sources including unrestrained industrial growth, lack of wastewater collection infrastructure, gentrification, rapid urban migration, ineffective implementation of environmental and sanitation laws & practices, and lack of motivated participation from all the stakeholders involved in water quality management lead to have adverse impact on aquatic sources of Raipur city. Categorization of the pond is done based on the habitat and gentry surrounding the pond location, variations in the quality of water depend on the population living around it and the end use of the pond be the people living around it. Higher income groups are having different uses as compared to low-income group. As water bodies near low income group are being used more vigorously for multiple uses (bathing, irrigation, washing etc) as compared to high income group. This is the reason why we get different quality of water body when we studied those for this research.

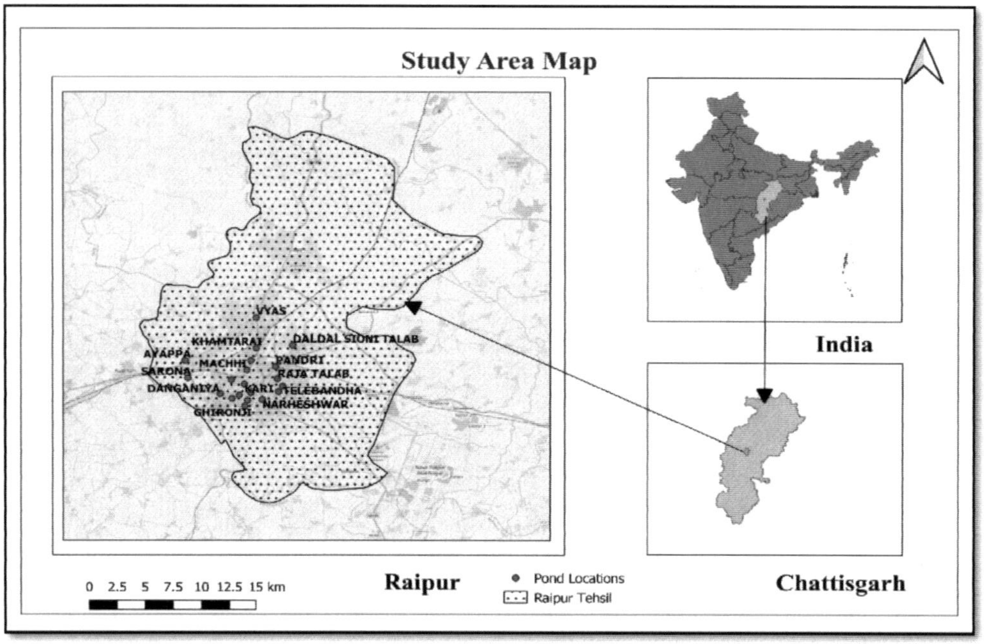

Figure 2. A map of the Raipur city depicting the location of the 20 ponds sampled in this study.

2.3 Sample collection–

Sample collection was done from 20 November to 10 December 2021 in the winter season. The average temperature during the sampling was 28° C. A total of 20 samples were collected and transported back to the laboratory for analysis. The images of 10 sampling location are shown in fig.3. The sample collection, preservation, and transportation were done strictly as described in the APHA manual. The collection of samples was done using plastic bottles. All the parameters are determined using standard procedures described in APHA [27]. All the results were collated with the WHO norms. Every sample was collected by taking all the necessary precautions. Every sample has been analyzed for 16 different physicochemical parameters for this research work and interpreted accordingly. All the samples were analyzed using standard methods as given in APHA manual. Samples were bought from the locations and were stored at below 4 degrees in deep refrigerator.

Figure 3. Photographs of the sampling sites.

2.4 Water Quality Index (WQI)-

Here, we have used a mathematical model called *water quality index* (WQI) based on the discrete parameter value of the water quality component. The quality of the water depends upon many variables which is commonly referred as water quality parameters whose values are compared with the standard permissible values prescribed by the governmental agency. A qualitative assessment and inter comparison of water sources become challenging due to the large numbers of involved water quality variables, and lack of uniform measurement units and relative importance of few variables like MPN, BOD, COD and nutrients etc. over others. The WQI address these challenges and provides a single value that represents the overall quality of the aquatic body at a certain location and time by incorporating all the variables that affects the water quality. With the central purpose of simplifying the complex datasets into an easily comprehensible and comparable value, the WQI is a very useful tool for all the stakeholders. In this paper we have studied the physicochemical parameters (Ph, Conductivity (μs/cm), Turbidity (NTU), Total Dissolved Solids (mg/L), Total Hardness as $CaCO_3$ (mg/L), Calcium Hardness as $CaCO_3$ (mg/L), Magnesium Hardness as $CaCO_3$ (mg/L), TKN (mg/L), Nitrate Nitrogen (mg/L), Phosphorus as PO_4^{3-} (mg/L), Chlorophyll-a (mg/m3), COD (mg/L), BOD (mg/L), Total Alkalinity (mg/L), Chlorides (mg/L), and DO (mg/L) [23].

Here we used Horton's WQI method which is discussed below.

$$\text{WQI} = \frac{\sum qn\, Wn}{\sum W}$$

Where, qn = quality rating assigned to the nth water quality parameter

Wn= unit weight of nth water quality parameter

<u>Quality Rating-</u>

Quantity rating can be expressed by using the formula given below-

$$q_n = \frac{(Vn - Vid)}{(Sn - Vid)} \times 100$$

Where,

Vn = Estimated value of the nth water quality parameter at a given sample location

Vid = Ideal value for nth parameter in pure water .(Vid for pH = 7 and 0 for all other parameters)

Sn = Standard permissible value of nth water quality parameter

Unit weight –

Unit weight can be calculated using the formula given below –

$$W = \frac{k}{Sn}$$

Where,

k = constant of proportionality and is calculated by using the formula –

$$k = \frac{1}{\frac{1}{Sn}}$$

Sn = Standard permissible value of nth water quality parameter

The individual pond is rated using the WQI values created using equations discussed above and the vale is compared with the standard WQI values, which gives us the overall status of the pond water and also suggest the possible usage of the pond water **[Table. 2]**.

Table 2. Categorization of WQI value, the corresponding status of water body, and suggestions for the possible usage of water. [58]

S.No.	WQI	Status	Possible Usage
1	0-25	Excellent	Drinking. Irrigation. Industrial
2	25-50	Good	Domestic, Irrigation, Industrial
3	51-75	Fair	Irrigation And Industrial
4	76-100	Poor	Irrigation
5	101-150	Very Poor	Restricted Use for Irrigation
6	Above 150	Unfit For Drinking	Proper Treatment Required Before Use.

3. Results-.

3.1 Physico-chemical parameters of the pond water -

The concentration of 16 physical and chemical parameters of 20 ponds studied here 9 is shown in fig 4, 5 & 6. All the ponds can be easily compared. Location variations in water revealed all the maximum and minimum values of all the physicochemical parameters which have been studied and calculated for this research. The results obtained from the physiochemical analysis are discussed below.

Ph values (Fig.4(a)) of these ponds were alkaline which is relatable with biological activity in each pond. Mahrajbandh pond showed the highest pH value, as the pond receives a constant influx of domestic wastewater, there is a good amount of phytoplankton population. Algal growth means higher photosynthesis, CO_2 depletion due to increased photosynthetic activities, and interaction with continuous vehicular emissions so these can be a very reason for higher ph. It has been observed that the average pH value in the commercial and industrial zone is more alkaline as compared to the other subgroups which were expected as these ponds are prone to deposition of vehicular emissions and atmospheric fallouts containing emissions from the industries.

The clarity of the water is very important, as it is the very first sign of good water. According to International Standard Organization (ISO) turbidity is caused due to the presence of suspended and colloidal matter, inorganic or organic in nature which reduces the transparency of liquid. Turbidity was measured by using the instrument called Nephelometer for this study. The range varies from 1.4 NTU (Kari Pond) to 20.8 NTU (Narheshwar pond) (Fig.4(b)). Ponds under the recreational group are more turbid, having a value of 15.38 NTU which is quite higher and the reason which has been observed is the aesthetic implications of the water supplies in those locations. Turbidity >5 NTU is usually acceptable for consumers. Raipur pond locations result in a higher average turbidity value of 9 NTU which is very high and it needs to be monitored as it indicates pollution triggered by environmental events (eg. Earthquakes, flood, fire, etc.) or anthropogenic activities or ingress of contamination due to infrastructures (Fig.4(b)) [31,32].

Dissolved oxygen concentration (mg/l) followed throughout the investigation period as it is a direct indicator of an aquatic resources' ability to support aquatic life. Low DO level in water leads to the death of aquatic organisms. The lower the concentration the higher the stress. If DO value of the water falls below 4mg/l, the survival chances of organisms will reduce. It was analyzed by using Winkler's method. The DO values of the samples varied from minimum (Mahrajbandh pond) – 1.7 mg/l to maximum (KhoKho pond)- 10.7 mg/l (Fig.4(c)). Higher values of phosphorus and nitrogen in Mahrajbandh pond supported the conclusion about its lower DO value. As high nutrients can lead to excessive plant growth and result in a decline in DO. Mainly phosphorus and nitrogen are responsible for the lower DO value. Uncontrolled plant growth leads to algal bloom, is often due to fertilizer run-offs causing a phenomenon called eutrophication. Eutrophication is mainly due to heavy nitrogen-enriched fertilizer runoffs leading to imbalance causing the death of aquatic life and creation of zones where survival of aquatic life is not possible. So, it can be reduced by taking preventive methods to reduce fertilizers or prevention of any type of plantation near highly eutrophic water bodies such as Maharjbandh pond (Fig.4(c)).

Electrical conductivity reflects the ability of water to conduct electricity. This electrical conductivity is due to the substances dissolved in the water which breaks down into positively and negatively charged ions. Electrical conductivity varies from 361µs/cm (Telebandha) to 922 µs/cm (Vyas) (Fig.4(d)). Vyas pond being on the commercial/ industrial location is having interaction with industrial runoffs that means dissolved salts and inorganic chemicals, conductivity increases. The electrical conductivity of low-income group ponds - (721 µs/cm) is much higher than other subgroups. Ponds in this group have access to many agricultural run-offs, i.e dissolved chemicals summarizing the very reason for higher electrical conductivity. Raipur ponds are having electrical conductivity of 634 µs/cm which is double the standard electrical conductivity value. Higher electrical conductivity is not a necessarily cause of concern due to its lack of direct health impact but dissolved insoluble solids may lead to higher water hardness and alkalinity (Fig.4(d)).

Total Dissolved Solids of water is the amount of dissolved inorganic salts and organic matter present in water. The major constituents dissolved in water are calcium, magnesium, sodium, potassium cations and carbonate, hydrogen carbonate, chloride, sulfate, and nitrate anions. Total Dissolved Solids range from 232 mg/l (Telebandha pond) – to 922 mg/l (Vyas Pond). Total dissolved solids in ponds in commercial/ industrial groups have higher TDS values as compared to other subgroups (Fig.5 (a)).

Alkalinity is the capacity to neutralize the acid present in water. Alkalinity in water sources is basically due to carbonates, bicarbonates, and hydroxide so it is considered as a sign of the presence of these constituents in the water body. (American Public Health Association, 1995). Water sources studied for this research range from Chironji pond (40 mg/l) to Raja Pond (304 mg/l) (Fig. 5 (b)).

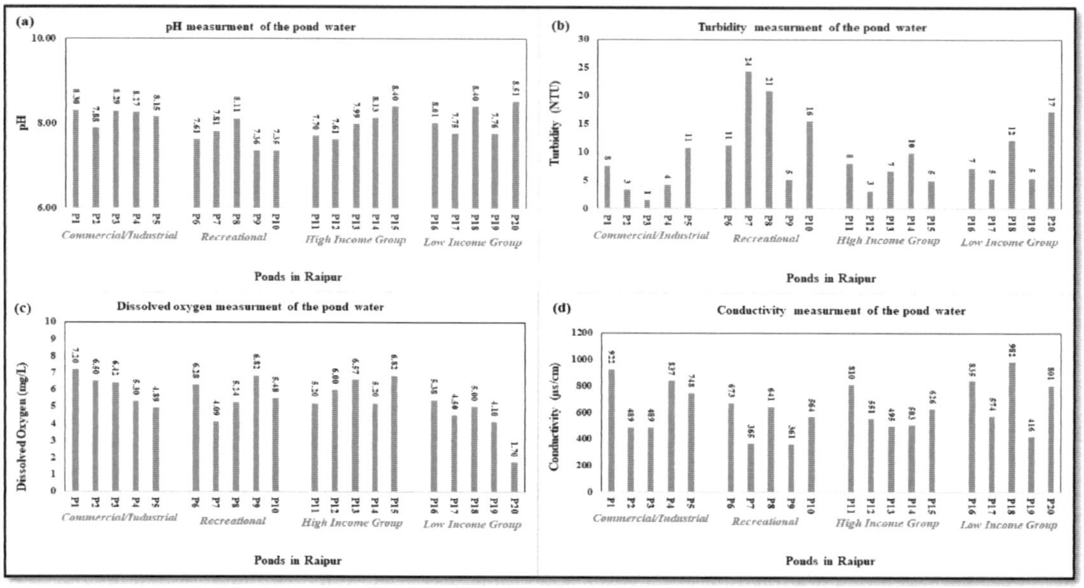

Figure 4. Spatial distribution of **(a)** pH, **(b)** Turbidity, **(c)** Dissolved oxygen, and **(d)** Conductivity in the surface water of urban ponds.

Phosphate measurements showed the recorded minimum value of 1.70 mg/l in (Telebandha pond) and a maximum value of 22.62 mg/l (Mahrajbandh pond). Phosphorus is an important nutrient for the growth of organisms. It is a growth-limiting nutrient so if the amount of phosphate speeds up due to continuous discharge from agricultural runoffs, industrial wastes, or raw or treated wastewater it will lead to eutrophication. Ponds in the high-income group have the highest phosphorus value as compared to other subgroups. Proving the very reason for higher phosphorus concentration i.e runoff from urban areas & lawns, leaking septic systems, etc. (Fig.5(c)).

High chloride concentration in water or wastewater causes harm to metallic pipes and structures, and plants. It is increasing due to an increase in the number of industries. It varies from 46 mg/l (BudhaPond) to 378 mg/l (MacchiTaraiyya) (Fig.5(d)).

Nitrogen-Nitrate concentration is normally in trace amount in surface and groundwater but it can reach to higher concentration due to agricultural runoffs industrial waste, human & animal waste discharge. It varies from 8.20 mg/l (Macchhi pond) to 51.20mg/l (Mahrajbandh pond).(Table2). The maximum nitrogen nitrate value was of ponds under the low-income group. TKN was measured from 1.20 mg/l (Kho Kho pond) to 28 mg/l (Mahrajbandh pond) (Fig.6(a)) [37].

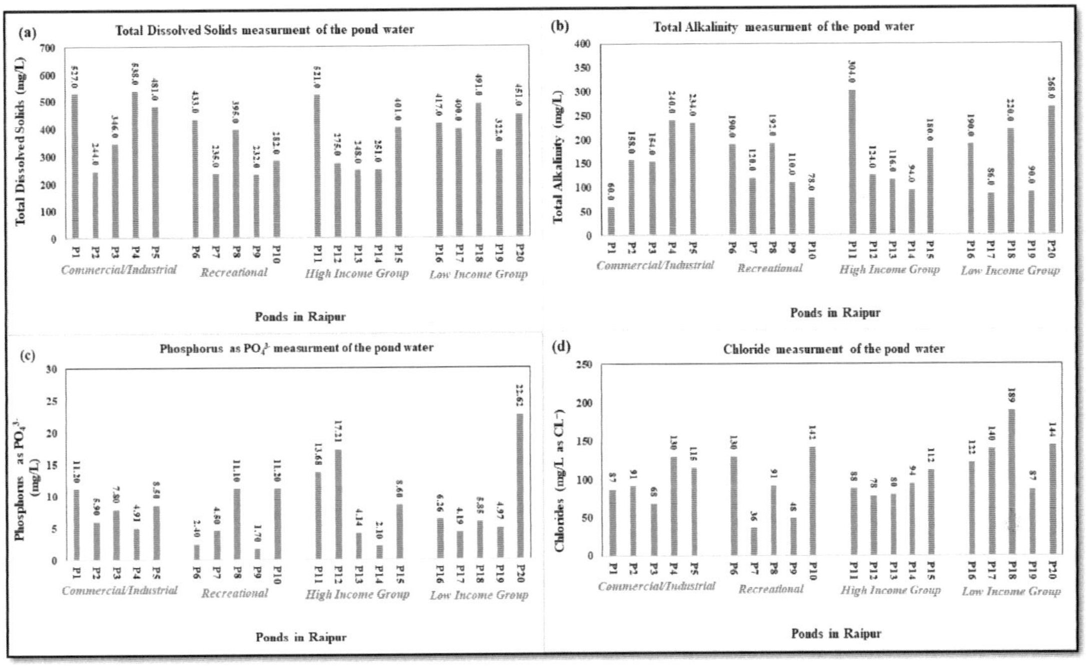

Figure 5. Spatial distribution of **(a)** TDS, **(b)** TA, **(c)** Phosphate, and **(d)** Chloride in the surface water of urban ponds.

Biological Oxygen Demand (BOD_5) and Chemical Oxygen Demand (COD) are the major physicochemical parameters measured for the evaluation of organic contamination pollution in water. The minimum COD value estimated is 36 mg/l (Kho Kho pond) and the maximum is 240 mg/l (Pandri pond) Higher values of COD mean a greater amount of oxidizable organic material in the water body which will reduce DO level. Pandri being in the commercial/ industrial location has higher industrial run-offs leading to an increase in COD values. The maximum COD and BOD values are of ponds that are under the low-income group. BOD_5 value estimated is a minimum of 10 mg/l (Telebandha pond) and the maximum is 142 mg/l. The COD/BOD ratio is very crucial parameter which tells us about the nature and draggability of the organic matter present in the water body. A high COD value reflect that major portion of the organic matter is better suited for chemically oxidation rather than biological oxidation. This organic matter often involves organic chemicals which are very toxic to the aquatic organisms. Thus, COD test is very essential to get comprehensive picture on the nature of organic matter which otherwise left unaccounted for in BOD test. (Fig.6 (b)) [35,36].

Hardness is observed to be higher in ponds of low-income groups. according to industries, a good water hardness level should not increase above 100 mg/l but these water sources have higher values of total hardness which is making the water unfit for consumption (fig.6(D)) [23].

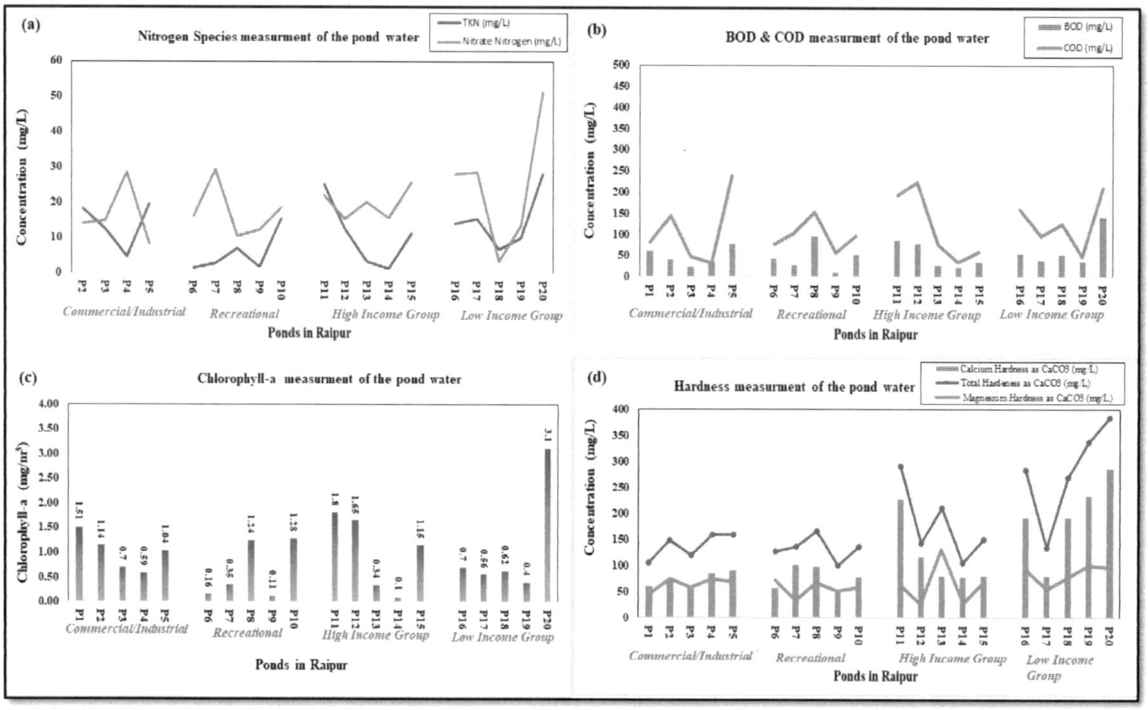

Figure 6. Spatial distribution of **(a)** Nitrogen species, **(b)** BOD & COD, **(c)** Chlorophyll-a, and **(d)** Hardness in the surface water of urban ponds.

3.2 Water quality assessment of 20 ponds using Horton's WQI -

16 Physicochemical parameters that were analyzed for 20 different ponds during the post-monsoon season were used to calculate WQI using Horton's method. Our investigations during post monsoon winter season shows that 75% of the pond water samples are in excellent condition, & 25% of the samples are unfit for various activities like drinking, irrigation, industrial & domestic. The index result constitutes the water quality status of the water samples [23,24].

Table 3. WQI values of the pond water of 20 ponds, the corresponding status of water body, and suggestion for the possible usage of water.

Sample Location Code	Wn*Qn	Water Quality Status	Possible Usage
P1	120.087976	Unfit for consumption	Proper treatment required before use
P2	60.57507927	Poor	Irrigation
P3	78.63236564	Very Poor	Restricted use for irrigation
P4	51.29076094	Poor	Irrigation

P5	85.13283493	Very Poor	Restricted use for irrigation
P6	29.10530265	Good	Domestic, Irrigation, Industrial
P7	46.25664794	Good	Domestic, Irrigation, Industrial
P8	114.7554323	Unfit for consumption	Proper treatment required before use
P9	19.23633221	Excellent	Domestic, Irrigation, Industrial
P10	112.2820819	Unfit for consumption	Proper treatment required before use
P11	132.7713479	Unfit for consumption	Proper treatment required before use
P12	166.4802608	Unfit for consumption	Proper treatment required before use
P13	43.4523277	Good	Domestic, Irrigation, Industrial
P14	24.07641117	Excellent	Drinking, Irrigation, Industrial
P15	87.52276089	Very Poor	Restricted use for irrigation
P16	158.8067259	Unfit for consumption	Proper treatment required before use
P17	44.5769208	Good	Drinking, Irrigation, Industrial
P18	60.85167751	Poor	Irrigation
P19	51.09124894	Poor	Irrigation
P20	217.8025077	Unfit for consumption	Proper treatment required before use

According to WQI calculated for various ponds, Mahrajbandh pond, Ayappa pond, Narheshwar pond, Tanzania, Raja Pond, Vyas Pond, Kari Pond, Sarona Pond, and Khamtarai Pond have water quality index above 100,they come under category – unfit for drinking according to the WQI ranges [table(3)]. The very reason behind the value being this high is that they are in industrial location & near high way having continuous exposure to chemicals, dust & waste disposal. This leads to an increase in nutrient concentration, toxicant concentration making them unfit for various activities.

Table 4. WQI values of the pond water of 20 ponds, the corresponding status of the 20 ponds

S.no.	WQI	Status	Ponds
1	0-25	Excellent	P14, P9
2	26-50	Good	P17, P13, P6, P7
3	51-75	Fair	NIL
4	76-100	Poor	P19, P18, P2, P4
5	101-150	Very poor	P15, P3, P5
6	157	Unfit for drinking	P20, P16, P1, P8, P10, P11, P12

4. Discussion-

The stagnant inland surface water sources are frequently subjected to an abrupt change in various environmental stressors due to changes in anthropogenic activities like habitat change, gentrification, industrialization, commercialization, or recreational activities. Urban ponds in Raipur city are facing immense pressure from increasing population density and gentrification which directly affects the pond water quality. The urban population of Raipur city increased from 1,139,172 in 2011 to 1,702,000. The population is increasing from an average rate of 3.5 % annually since 2010 [41].

A detailed study on the spatial and temporal variation of pond water quality is pertinent for conservation and rejuvenation efforts. Moreover, qualitative assessment of the pond water quality using WQI and TSI presents a summarised overview of the current status of the ponds which helps the authorities and masses for planning water quality management efforts. The present study is a part of a project where the long-term assessment of the inland water quality of Raipur city is being carried out. Here we present the physiochemical assessment of pond water in the post-monsoon winter season. Our findings confirm that the ponds lying in low-income group gentry, where the wastewater effluent is being discharged in the ponds are exhibiting high levels of nutrients, BOD, and COD. This is in line with the findings of earlier studies that have attributed the indiscriminate release of wastewater in the inland surface water body to their failing water quality [42, 43, 44].

The aesthetic and intrinsic value of the overall pond ecosystem decreases when the wastewater is discharged into the pond. Recreational ponds like Telebandha Pond which used to receive wastewater went through a restoration process. When the wastewater drains were stopped entering the pond, the water quality improved, thereby improving the aesthetic value. Now the pond serves as a waterfront and major recreational hotspot for the people of the entire city.

Variations in turbidity values followed variation in land use pattern with the maximum average value obtained in the vicinity of the recreational zone and minimum in the industrial and commercial zone. This variation may probably occur due to the direct disposal of solid waste and food waste by tourists. A slight high value of TH, MH, and CH was observed which can be explained by the dissociation of calcium carbonate, the release of calcium and magnesium from aquatic biota, and the use of calcium in aquaculture. The Calcium ion was observed to be higher than magnesium which can be attributed to the difference between the chemistry of calcium and magnesium mainly water solubility and precipitation rate. Further, the variation in the divalent cation species can be attributed the fact that aquatic flora have different utilisation rate for them.

The positive correlation between chlorophyll-a and dissolved oxygen suggests one possibility that increased chlorophyll-a values and related photosynthesis activities play a crucial role in the measured values of dissolved oxygen, particularly during winters. Significant increases in DO of surface water bodies are the result of phytoplankton growth activities. The low DO values in the LIG ponds can be attributed to wastewater discharge where older ponds still receive wastewater from the closed wastewater drains through seepage. The outlet drains are connected to the ponds at the bottom level which is manually not possible to observe [43].

A significant spatial variation in the nutrient levels exists within the subgroup and across the subgroup which is very alarming. The average nitrogen concentration in the ponds of LIG gentry is higher than others, however, ponds in the recreational zones have somewhat lower values. Maharajbandh pond is showing high nutrient values as the pond is constantly receiving sewage discharge from the Raipur city (Table 6). The Raja Pond in the HIG gentry is also exhibiting higher nutrient values still when compared with Maharajbandh Pond, the latter is greater which can mainly be attributed to factors such as difference in sewage discharge volumes, and quality of the wastewater.

Table 5. Nutrient and Chlorophyll-a value (average) across 4 pond categories.

Pond Classification	TKN (mg/L)	Nitrate Nitrogen (mg/L)	Total Nitrogen (mg/L)	Phosphorus as PO43- (mg/L)	Chlorophyll-a (mg/m^3)
Commercial/Industrial	15.2	17	32.2	7.662	0.996
Recreational	5.708	17.4	23.108	6.18	0.628
High Income Group	10.674	19.86	30.534	9.146	1.008
Low Income Group	14.78	24.94	39.72	8.778	1.076

The increase in the nutrient levels especially nitrate could be due to occasional additions of the organic manure and other fish foods that are added to the ponds by fishermen or by the fish waste product discharged by them into the water. The decrease or low nutrient values reflect the possibility that these ponds are not receiving any form of sewage or receiving waste that have less concentration of nutrients [46,47].

The high value of the phosphate that is observed in all 3 pond groups except those in recreational habitat could be due to fertilizers, manure, street washing, and runoffs. Groundwater table polluted with phosphate could also be responsible for elevated phosphate values **(Table. 5)**. Ponds like Pahadi Pond are recharged by pumping groundwater which may contain phosphorus. The excessive use of fertilizers in urban habitats in public and private parks and gardens is introducing phosphorus to the groundwater. Another possibility for the elevated phosphorus might be due to a process called sediment reflux which is very high during the winter season. The sampling was carried out at the onset of the winter season so were through the process known as 'winter stratification' where the water in the 'hypolimnion layer' with high phosphorus content may get mixed with the 'epilimnion layer'.

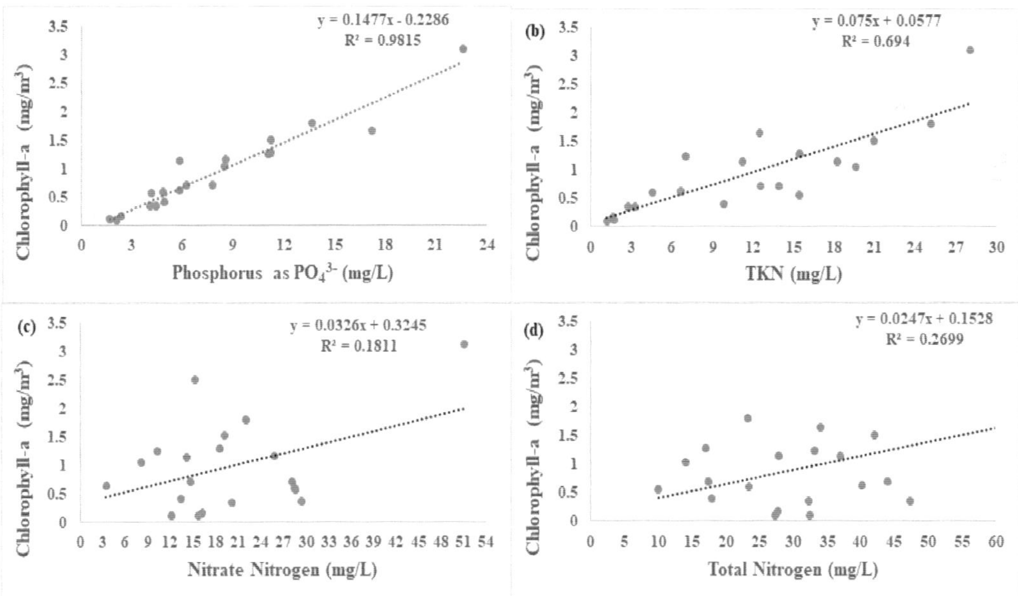

Figure 7. Correlation between Chlorophyll-a and **(a)** Phosphorus, **(b)** TKN, **(c)** Nitrate-Nitrogen, and **(d)** Total Nitrogen in the surface water of urban ponds.

On the contrary, the depletion of phosphate concentration could be due to its consumption by the phytoplankton family or due to the introduction of different particles having a high affinity for phosphorus, promoting phosphorus accumulation in the bottom sediments Phosphate in freshwater habitats is correlated with dissolved oxygen and phytoplankton. A significant negative correlation between phosphate and chlorophyll-a was observed in this study which confirms that phosphorus concentration decreased with an increase in the phytoplankton family [48-51].

As far as the correlation between Chlorophyll-a value with the nutrient is concerned, a high correlation was observed. A significant correlation between TKN and Chlorophyll-a is also observed which indicates that the ponds are receiving fresh sewage and high decay of the organic matter is happening in the ponds **(fig. 7)**. The ponds lying in the recreational zones are showing low nutrient value when compared with other groups. However, when compared with WHO standards for surface water, they are significantly higher. These ponds are tourist hotspots where people spend a considerable amount of time.

During sampling, a high volume of MSW was observed floating in surface water. Further, these ponds are prone to receiving kitchen waste from the food stalls catering the tourists which plausibly may introduce nutrients in the pond. One key point the authors would like to mention is that the ponds with a high recreational value that remains in the center of attraction receive much attention from municipal authorities where ponds like Budha and Telebandha enjoy periodic cleaning of the floating solid waste.

All the ponds except those in the recreational zone receive runoffs from street washing. These contain a high level of nutrients and organic matter. The street cattle and those from dairy farms discharge excreta and urine on the road surface which along with the organic matter in form of food waste discarded on the street gets mixed with street washing and enter the ponds.

Many criteria have been established that aid in classifying ponds and lakes using water transparency (turbidity, Secchi depth) algal biomass (chlorophyll-a), and nutrients (nitrogen and phosphorus species).
The nutrient ratio (Total-N/Total-P) is widely used to elucidate the concentration of phytoplankton and nutrient availability. Total Phosphorus act as a limiting nutrient in ponds and lakes where TN/TP value is greater than 7, whereas below 7 indicates total nitrogen as a limiting factor for algal growth. For general use, TN/TP value less than 10 indicates a nitrogen shortage, and higher than 20 a phosphorus shortage. Lower TN/TP ratios occur in eutrophic lakes and are higher in mesotrophic and oligotrophic lakes.
The urban ponds of Raipur city are showing a large variation in nutrient value. The TN/TP ratio varies between 1.62 and 10.50, while the average ratio is 5.02 which indicates TN to be the limiting nutrient.
The algal blooms which signify eutrophication was observed in many ponds during sampling which was further confirmed using Chlorophyll-a value [52-57].

The quality of surface water is affected by many physical and chemical variables and rating the ponds mathematically which reflects overall water quality by combing all the relevant variables of concern helps scientists to communicate the complex findings to the stakeholders. During the present study, sixteen parameters were studied out of which 15 parameters were considered for calculation of WQI of the pond water using Horton's method. The results indicate that the water quality of the pond ranged between Excellent (19.23%) at Telebandha pond and unfit for consumption (217%) at Mahrajbandh pond waters with being average of 86%.

The average WQI values of all the ponds were collectively classified as Poor and only fit for agriculture and irrigation purposes. The WQI allowed us to report and communicate our findings in a single number which enable stakeholders, policymakers, and even the public to evaluate the water quality and adopt preventive measures.

5. Conclusion-

In this study, variations in the water quality of different ponds in Raipur, Chhattisgarh, were examined in the post-monsoon winter season to observe and analyse the extent of water pollution. A successful attempt was made to identify the ponds that are most affected by the anthropogenic activities and then correlated with the habitats where they are located. It was observed that most of the water pollution is caused by human-induced sources. This conclusion was made as it has been analysed that ponds near service roads, industries, agricultural fields, (Maharajbandh pond,Vyas pond, Kari pond, Khamtarai pond, Sarona pond) are more polluted and this was supported by the Water Quality Index (mathematical model) used for this study. Long-term monitoring of the water quality and trophic level of the urban pond is of utmost importance to adopt a comprehensive rejuvenation and conservation strategy. It was confirmed that ponds are having high ph and algal bloom, many ponds with acidic ph didn't show any alarming situation as they are in the range of ph 6.5 which is ok with pond water. For majority of ponds turbidity was found about 5 NTU and above which shows presence of phytoplankton and other suspended solids in water indicating the increased nutrient inflow. DO is another parameter showing the health of pond, majority of ponds showed DO above 3 which is acceptable and is due to presence of phytoplankton due to higher photosynthetic activity as most of the samples were taken during the day time. Conductivity is also high showing substantial number of dissolved solids. TDS is in the range 300-500 which is acceptable above that will be an issue showing higher amount of suspended dissolved solid which can be a reason for salinity in pond water. Some of the pond water showed low level of alkalinity consuming the acidic effluent coming from industries. Phosphorus and nitrogen concentration is also high in majority of ponds showing continuous flow of detergents, or fertilizers used in gardens etc shows higher growth of phytoplankton. This conclusion showed the continuous anthropogenic activities are the very reason for this state of ponds. . Long-term monitoring of the water quality and trophic activity of urban ponds is of utmost importance to adopt a comprehensive rejuvenation and conservation strategy.

References-

1. Swarnakar, A. K., &Choubey, S. (2016). Testing and analysis of pond water in Raipur City, Chhattisgarh, India. International Journal of Science and Research, 5(4), 1962-1965.
2. Nag, A. (2014). Physicochemical analysis of some water ponds in and around Santiniketan, West Bengal, India. International journal of environmental sciences, 4(5), 676-682.
3. Gupta, N., Yadav, K. K., Kumar, V., & Singh, D. (2013). Assessment of physicochemical properties of Yamuna River in Agra City. International Journal of ChemTech Research, 5(1), 528-531.
4. Hecky, R. E., & Fee, E. J. (1981). Primary production and rates of algal growth in Lake Tanganyika 1. Limnology and Oceanography, 26(3), 532-547.
5. Lu, H., Yin, C., Wang, W., & Shan, B. (2007). A comparative study of nutrient transfer via surface runoff from two small agricultural catchments in north China. Environmental Geology, 52(8), 1549-1558.
6. Fried, S., Mackie, B., & Nothwehr, E. (2003). Nitrate and phosphate levels positively affect the growth of algae species found in Perry Pond. Tillers, 4, 21-24.
7. Björn, L. O., Papageorgiou, G. C., & Blankenship, R. E. (2009). A viewpoint: why chlorophyll a?. Photosynthesis research, 99(2), 85-98.
8. Fairchild, G. W., Anderson, J. N., & Velinsky, D. J. (2005). The trophic state 'chain of relationships' in ponds: does size matter?. Hydrobiologia, 539(1), 35-46.
9. Crossetti, L. O., & de M Bicudo, C. E. (2008). Phytoplankton as a monitoring tool in a tropical urban shallow reservoir (Garças Pond): the assemblage index application. Hydrobiologia, 610(1), 161-173.
10. Sharma, M. P., Kumar, A., & Rajvanshi, S. (2010). Assessment of trophic state of lakes: a case of Mansi Ganga Lake in India. Hydro Nepal: Journal of Water, Energy and Environment, 6, 65-72.
11. Kratzer, C. R., & Brezonik, P. L. (1981). A carlson-type trophic state index for nitrogen in Florida lakes 1. JAWRA Journal of the American Water Resources Association, 17(4), 713-715.
12. Sigua, G. C., Williams, M. J., Coleman, S. W., & Starks, R. (2006). Nitrogen and phosphorus status of soils and trophic state of lakes associated with forage-based beef cattle operations in Florida. Journal of environmental quality, 35(1), 240-252.
13. Bhateria, R., & Jain, D. (2016). Water quality assessment of lake water: a review. Sustainable Water Resources Management, 2(2), 161-173.
14. Gupta, A. P., Ranga, M. M., & Banerjee, A. Physico-chemical Properties of MoulviBandh, Ambikapur, Sargujawith reference to Water Quality.

15. Keller, C., Guntzer, F., Barboni, D., Labreuche, J., & Meunier, J. D. (2012). Impact of agriculture on the Si biogeochemical cycle: input from phytolith studies. Comptes Rendus Geoscience, 344(11-12), 739-746.
16. Chen, B., Zhang, X., Tao, J., Wu, J., Wang, J., Shi, P., ... & Yu, C. (2014). The impact of climate change and anthropogenic activities on alpine grassland over the Qinghai-Tibet Plateau. Agricultural and Forest Meteorology, 189, 11-18.
17. Doi, H., Katano, I., Negishi, J. N., Sanada, S., & Kayaba, Y. (2013). Effects of biodiversity, habitat structure, and water quality on recreational use of rivers. Ecosphere, 4(8), 1-11.
18. Hill, M. J., Biggs, J., Thornhill, I., Briers, R. A., Gledhill, D. G., White, J. C., ... & Hassall, C. (2017). Urban ponds as an aquatic biodiversity resource in modified landscapes. Global change biology, 23(3), 986-999.
19. Balvanera, P., Pfisterer, A. B., Buchmann, N., He, J. S., Nakashizuka, T., Raffaelli, D., & Schmid, B. (2006). Quantifying the evidence for biodiversity effects on ecosystem functioning and services. Ecology letters, 9(10), 1146-1156.
20. Jenny, J. P., Anneville, O., Arnaud, F., Baulaz, Y., Bouffard, D., Domaizon, I., ... & Weyhenmeyer, G. A. (2020). Scientists' warning to humanity: rapid degradation of the world's large lakes. Journal of Great Lakes Research, 46(4), 686-702.
21. Rahman, A., Jahanara, I., & Jolly, Y. N. (2021). Assessment of physicochemical properties of water and their seasonal variation in an urban river in Bangladesh. Water Science and Engineering.
22. Dixit, A. K. (2015). Study of physico-chemical parameters of different pond water of Bilaspur District, Chhattishgarh, India. Environmental Skeptics and Critics, 4(3), 89.
23. Akhtar, N., Ishak, M. I. S., Ahmad, M. I., Umar, K., Md Yusuff, M. S., Anees, M. T., ... & Ali Almanasir, Y. K. (2021). Modification of the water quality index (WQI) process for simple calculation using the multi-criteria decision-making (MCDM) method: a review. Water, 13(7), 905.
24. Adelagun, R. O. A., Etim, E. E., & Godwin, O. E. (2021). Application of Water Quality Index for the Assessment of Water from Different Sources in Nigeria. Promising Techniques for Wastewater Treatment and Water Quality Assessment, 267.
25. Haingotseheno, R., Chrysostome, R. J., Robert, R., & Tojonirina, A. R. WATER QUALITY INDEX (WQI) CALCULATION FOR THE EVALUATION OF PHYSICO-CHEMICAL QUALITY OF RAINWATER COLLECTED IN RESERVOIRS FULL OF SAND (RFS).
26. Suneetha, M., Sundar, S. B., & Ravindhranath, K. (2015). Calculation of water quality index (WQI) to assess the suitability of groundwater quality for drinking purposes in Vinukonda Mandal, Guntur District, Andhra Pradesh, India. Journal of Chemical and Pharmaceutical Research, 7(9), 538-545.
27. Bodhankar, N., & Chatterjee, B. (1994). Pollution of limestone aquifer due to urban waste disposal around Raipur, Madhya Pradesh, India. Environmental Geology, 23(3), 209-213.
28. Dar, F. A., Arora, T., Warsi, T., Devi, A. R., Wajihuddin, M., Grutzamer, G., ... & Ahmed, S. (2017). 3-D hydrogeological model of limestone aquifer for managed aquifer recharge in Raipur of central India. Carbonates and Evaporites, 32(4), 459-471.
29. Mukherjee R, Sahoo M, Naik KC (2011) Raipur City, Chhattisgarh. In: Groundwater scenario in major cities of India. Central Ground Water Board, pp 188–195
30. W.E.F. American Public Health Association (APHA) (2017). Standard Methods for the Examination of Water and Wastewater (23rd edition), American Water Works Association (AWWA), Washington DC, USA.
31. Lambrou, T. P., Anastasiou, C. C., & Panayiotou, C. G. (2009, September). A nephelometric turbidity system for monitoring residential drinking water quality. In International Conference on Sensor Applications, Experimentation and Logistics (pp. 43-55). Springer, Berlin, Heidelberg.
32. World Health Organization. (2017). Water quality and health-review of turbidity: information for regulators and water suppliers.
33. Garg, S. K., &Bhatnagar, A. (1996). Effect of varying closes of organic and inorganic fertilizers on plankton production and fish biomass in brackish water fish ponds. Aquaculture research, 27(3), 157-166.
34. Boyer, J. N., Kelble, C. R., Ortner, P. B., & Rudnick, D. T. (2009). Phytoplankton bloom status: Chlorophyll a biomass as an indicator of water quality condition in the southern estuaries of Florida, USA. Ecological indicators, 9(6), S56-S67.
35. Prambudy, H., Supriyatin, T., & Setiawan, F. (2019, October). The testing of chemical oxygen demand (COD) and biological oxygen demand (BOD) of river water in Cipager Cirebon. In Journal of Physics: Conference Series (Vol. 1360, No. 1, p. 012010). IOP Publishing.

36. Lee, J., Lee, S., Yu, S., & Rhew, D. (2016). Relationships between water quality parameters in rivers and lakes: BOD5, COD, NBOPs, and TOC. Environmental monitoring and assessment, 188(4), 1-8.
37. Nordin, R. N., Pommen, L. W., & Meays, C. L. (2009). Water quality guidelines for nitrogen (nitrate, nitrite, and ammonia). Water Stewardship Division, Ministry of Environment, Province of British Columbia, Canada, 1-29.
38. Khan, R., &Jhariya, D. C. (2017). Groundwater quality assessment for drinking purpose in Raipur city, Chhattisgarh using water quality index and geographic information system. Journal of the geological society of India, 90(1), 69-76.
39. Jena, V., &Sinha, D. (2017). Physicochemical analysis of ground water of selected areas of Raipur city. Indian J. Sci. Res, 13(1), 61-65.
40. Dwivedi, S. L., &Pathak, V. (2007). A preliminary assignment of water quality index to Mandakini River, Chitrakoot. Indian Journal of Environmental Protection, 27(11), 1036.
41. Web article https://www.census2011.co.in/census/city/280-raipur.html
42. Shiddamallayya, N., & Pratima, M. (2008). Impact of domestic sewage on fresh water body. Journal of environmental biology, 29(3), 303.
43. Kotut, K., Krienitz, L., & Muthuri, F. M. (1998). Temporal changes in phytoplankton structure and composition at the Turkwel Gorge Reservoir, Kenya. Hydrobiologia, 368(1), 41-59.
44. Otokunefor, T. V., &Obiukwu, C. (2005). Impact of refinery effluent on the physicochemical properties of a water body in the Niger delta. Applied ecology and environmental research, 3(1), 61-72.
45. Kant, S., Mandal, N. K., &Dubey, M. K. (2019). Impact of anthropogenic activities on physicochemical properties of water and sediment soil of a perennial pond of Godda District (SantalPargana), Jharkhand. Indian J Environ Sci, 23(2), 69-74.
46. Morée, A. L., Beusen, A. H. W., Bouwman, A. F., & Willems, W. J. (2013). Exploring global nitrogen and phosphorus flows in urban wastes during the twentieth century. Global biogeochemical cycles, 27(3), 836-846.
47. Okuku, E. O., Ohowa, B., Mwangi, S. N., Munga, D., Kiteresi, L. I., Wanjeri, V. O., ... & Kilonzo, J. (2011). Sewage pollution in the Coastal waters of Mombasa City, Kenya: A norm Rather than an Exception. International Journal of Environmental Research, 5(4), 865-874.
48. Soltan, M. E., Moalla, S. M. N., Rashed, M. N., & Fawzy, E. M. (2005). Physicochemical characteristics and distribution of some metals in the ecosystem of Lake Nasser, Egypt. Toxicological & Environmental Chemistry, 87(2), 167-197.
49. Romo, S., Miracle, M. R., Villena, M. J., Rueda, J., Ferriol, C., & Vicente, E. (2004). Mesocosm experiments on nutrient and fish effects on shallow lake food webs in a Mediterranean climate. Freshwater Biology, 49(12), 1593-1607.
50. Wu, X. Y., & Yang, Y. F. (2010). Accumulation of heavy metals and total phosphorus in intensive aquatic farm sediments: comparison of tilapia Oreochromis niloticus× Oreochromis aureu, Asian seabass Lates calcarifer and white shrimp Litopenaeus vannamei farms. Aquaculture Research, 41(9), 1377-1386.
51. Elçi, Ş. (2008). Effects of thermal stratification and mixing on reservoir water quality. Limnology, 9(2), 135-142.
52. Attayde, J. L., &Menezes, R. F. (2008). Effects of fish biomass and planktivore type on plankton communities. Journal of Plankton Research, 30(8), 885-892.
53. El-Serehy, H. A., Abdallah, H. S., Al-Misned, F. A., Al-Farraj, S. A., & Al-Rasheid, K. A. (2018). Assessing water quality and classifying trophic status for scientifically based managing the water resources of the Lake Timsah, the lake with salinity stratification along the Suez Canal. Saudi Journal of Biological Sciences, 25(7), 1247-1256.
54. Garg, S. K., &Bhatnagar, A. (1996). Effect of varying closes of organic and inorganic fertilizers on plankton production and fish biomass in brackish water fish ponds. Aquaculture research, 27(3), 157-166.
55. Egbueri, J. C. (2020). Groundwater quality assessment using pollution index of groundwater (PIG), ecological risk index (ERI) and hierarchical cluster analysis (HCA): a case study. Groundwater for Sustainable Development, 10, 100292.
56. El-Serehy, H. A., Abdallah, H. S., Al-Misned, F. A., Al-Farraj, S. A., & Al-Rasheid, K. A. (2018). Assessing water quality and classifying trophic status for scientifically based managing the water resources of the Lake Timsah, the lake with salinity stratification along the Suez Canal. Saudi Journal of Biological Sciences, 25(7), 1247-1256.

57. Daniel, T. C., Sharpley, A. N., Edwards, D. R., Wedepohl, R., & Lemunyon, J. L. (1994). Minimizing surface water eutrophication from agriculture by phosphorus management.
58. Abbasi T, Abbasi SA. Water quality indices. Elsevier; 2012 Jun 13.

Exploring temporal variation of $PM_{2.5}$ and PM_{10} and their association with meteorological data in Raipur, Chhattisgarh

P P Khobragade[1,a] and A V Ahirwar[2,b,*]

[1]Research Scholar, Department of Civil , National Institute of Technology Raipur, G.E.Road, Raipur - 492 010, Chhattisgarh, India
[2]Assistant Professor, Department of Civil Engineering, National Institute of Technology Raipur, G.E.Road, Raipur - 492 010, Chhattisgarh, India

Email: khobragadepallavi19@gmail.com; [b]avahirwar@nitrr.ac.in

Abstract: While significant efforts have been made to mitigate the negative health impacts of particulate matter, there are limited statistics on particle exposure in Raipur. A study was conducted to look at the short-term relationships between particulate matter ($PM_{2.5}$ and PM_{10}) and meteorological variables in Raipur. The current research was based on an experimental study conducted at Chhattisgarh's NIT Raipur. During the pre-monsoon, monsoon, post-monsoon, and winter seasons of 2021, a total of 125 air specimens were gathered from the campus. A respirable dust sampler and a fine particulate sampler were used to detect PM_{10} and $PM_{2.5}$ on the building's terrace. To assess the associations between $PM_{2.5}$ and PM_{10} particulates and both sampling season and site, Pearson's correlation study was used. The monthly mean mass concentrations of $PM_{2.5}$ and PM_{10} in Raipur ranged from 28.0-334.0 μg/m3 to 56.0-448.0 μg/m^3, respectively. Post-monsoon PM concentrations were highest, followed by pre-monsoon, winter, and post-monsoon. On more than 92 percent of days, $PM_{2.5}$ concentrations exceeded NAAQS standards, while PM_{10} concentrations exceeded restrictions on 90 percent of days. There were seasonal changes identified in the relationships between PM concentrations and meteorological variables.

Keywords: $PM_{2.5}$, PM_{10}, back-trajectory analysis

1. Introduction

Atmospheric aerosol has an impact on local and worldwide climate directly by absorbing and scattering arriving and departing solar radiation, ultimately by affecting cloud microphysical characteristics and semi-directly by reducing cloud covering [1]. Aerosols are the mixture of solid and/or liquid atoms in the environment, which are generated by a variety of natural and anthropogenic activities, including industrial emissions, power stations, burning of biomass, automobile emissions, forest fires, refuse burning, volcanic eruptions, and marine salt sprays, among others. In 2018, 97 percent of cities in low and middle-income nations failed to fulfil WHO air quality guiding principles (www.who.int/airpollution/data/cities/en/), indicating a serious problem [2]. In light of the aforementioned consequence, numerous researchers have investigated the particulate concentration and seasonal deviation of atmospheric aerosols in the India [3 ,4], specifically in central India, and concluded that $PM_{2.5-10}$ is emitted primarily due to soil particle re-suspension and $PM_{2.5}$ emits majorly from anthropogenic activities [5].

Because PM can enter the body through the respiratory tract, its size has significant toxicological and regulatory implications [6]. Epidemiological studies have shown that disclosure to fine as well as coarse particulates causes a growth in lung cancer, mortality to cardiopulmonary diseases and increased respirable morbidity [6, 7]. Outdoor air pollution is responsible for 3.3 million premature deaths global each year, with the figure expected to twice by 2050 if the problem is not addressed [8]. Air pollution has been shown to have a significant influence on lung surfactant structure as well as lung injury. As a result, air pollution increases the risk of respiratory toxicities, asthma, and additional chronic ailments in humans [9].

[10] identified meteorological elements as one of the key factors inducing urban air quality. Temperature, relative humidity, and wind speed and direction are considered major parameters since they can affect the dispersion process, elimination mechanisms of particles and emergence of atmospheric particulates [11,12], thus playing a significant role in limiting the concentrations of air pollutants [12,13]. Furthermore, rainfall may have a varying influence. Several studies have found that climatic variables can have an impact on ambient pollutant levels [8, 12]. The scattering, alteration, and exclusion of air contaminants from the atmosphere are the most essential functions of meteorology in this context [14].

Understanding the relationship between climatic conditions and particle variation is critical for eradicating particulate pollution by taking necessary actions. The objective of this study is to identify the influence of meteorological variables on fine as well coarse aerosols. Back-trajectory study was executed using the Hysplit Single Particle Lagrangian Integrated Trajectory model to better analyze the transboundary migration of particulates from diverse places.

2. Methodology

i. Site description

Raipur, the state capital of Chhattisgarh, is one of central India's most important industrial and commercial centres. Raipur district is industrially developed, with room for innovative companies and up-grading of prevailing ones [1]. Raipur has a tropical wet and dry climate with mild temperatures excluding pre-monsoon, when the peak temperature might reach 39 degrees Celsius. The winter season lasts from December-January, with lowest mean temperature. The samples were collected on the terrace of NIT Raipur institution at a height of 303.36 metres above sea level, far from concentrated pollutant emissions. The study site is cautiously chosen on a rooftop with no substantial vehicle or additional pollution within a 200-meter radius.

ii. Particulate sampling

PM_{10} specimens were acquired using a respirable dust sampler and $PM_{2.5}$ samples were collected using a fine particulate sampler (APM 550) (Envirotech instruments Pvt. Ltd., New Delhi, India) on 8" x 10" Whatman glass fibre tissues and a 47mm diameter PTFE tissue at the National Institute of Technology in Raipur, Chhattisgarh respectively. $PM_{2.5}$ and PM_{10} samples were acquired from 1st January 2021 to 15th December 2021. The sampling location was chosen to be distant from every direct concentrated pollutants within a 200-meter radius of the instruments. Prior to and following sampling, the filter tissues were kept in desiccators for 24 hours to remove any available moisture. A balance (Precisa-92SM-202A) with 0.01mg resolution was used to weigh the dehydrated filter papers. $PM_{2.5}$ and PM_{10} were calculated gravimetrically by dividing the variation between the filter paper's initial and final weights prior to and after experimenting by the sampled air volume. When moving the paper from the lab to the field, precautions were made by keeping it in covers to elude infection, and coding was done on top of it. The filter papers were collected and kept in the freezer until they were analyzed chemically in the lab. During the sampling period, meteorological elements such as temperature, wind speed, precipitation, and relative humidity were gathered using worldweatheronline.com's most recent weather forecasts.

3. Results and Discussions

i. Seasonal variation of particulates

The entire study period was divided into categories based on the many types of human activities and associated possible influence on aerosol mass concentrations and chemical compositions, as well as meteorological changes. $PM_{2.5}$ and PM_{10} mass concentrations ranged from 28.0-334.0 µg/m3 to 56.0-448.0 µg/m3 on a daily basis (Figure. 1). The mean yearly range of $PM_{2.5}$ and PM_{10} was around 135.04±79.4 µg/m3 to 225.37±104.6 µg/m3, which is much greater in comparison to the National Ambient Air Quality Guidelines set by the Central Pollution Control Board being three and four times higher than the USEPA's standards. The elevated particulate concentration in the Raipur area can be attributed to a cluster of industrial sectors that create significant particulates and have high pollution levels. $PM_{2.5}$ concentrations surpassed NAAQS limits on more than 92 percent of days and PM_{10} surpassed limits on 90 percent of days, posing a threat to human health, animal life, the climate, and historical sites [15]. Both the $PM_{2.5}$ and PM_{10} concentrations were observed to be minimum in the post monsoon season. $PM_{2.5}$ concentrations were reported to be 196.2 µg/m^3 during the post-monsoon season, higher than 117.5 µg/m^3 reported during the monsoon. $PM_{2.5}$ concentrations averaged 131.5 µg/m^3 during the pre-monsoon season. During the post-monsoon, pre-monsoon, and winter seasons, $PM_{2.5}$ concentrations are 1.7 times, 1.5 times, and 1.6 times greater than that reported in the monsoon season. During post-monsoon PM_{10} concentrations was observed to be 325.7 µg/m^3 whereas in monsoon the average particulate level was 182.2 µg/m^3. The average concentrations of PM_{10} for the pre-monsoon season were 239.7 µg/m^3. PM_{10} concentration is 1.4 times during post-monsoon and pre-monsoon and 1.6 times during winter time concentrations than the concentration observed during monsoon.

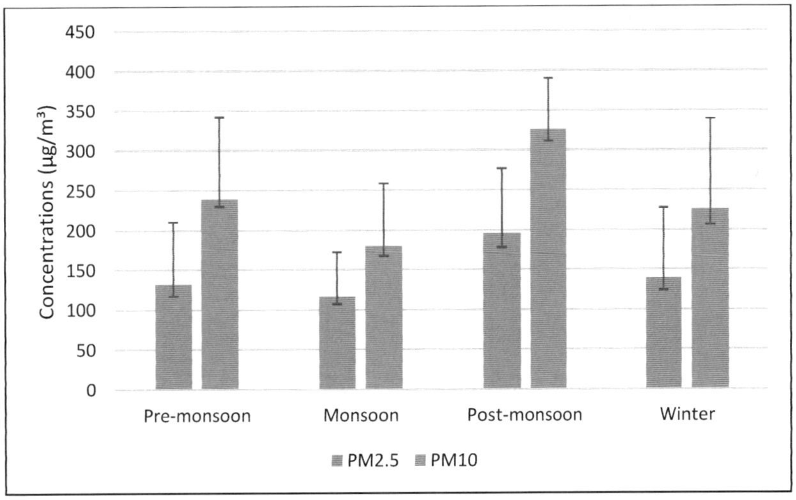

Figure 1: Seasonal variation of $PM_{2.5}$ and PM_{10} concentrations during January 2021 to December 2021

ii. Impact of meteorological criteria on particulate variation

Figure. 2 shows the variation in the daily concentration of PM_{10} at the sampling site with changes in the meteorological environment including temperature, wind speed, relative humidity and precipitation. Considerable seasonal changes have been observed in earlier studies also [16, 17, 18]. Seasonal $PM_{2.5}$ and PM_{10} correlation shows significant relationship among them in all the seasons. The gravimetric concentration level of $PM_{2.5}$ and PM_{10} changed in a similar way throughout time. During the study period, four and three different peaks in daily $PM_{2.5}$ and PM_{10} levels were identified. The average wind speed during monsoon and pre-monsoon is lower (15 km/h and 12 km/h) resulting in lower average $PM_{2.5}$ and PM_{10} concentrations (117.5 and 131.5 μg/m^3). The average wind speed observed during monsoon is 15 km/h, which is the highest of the year, with precipitation significantly lowering particulate levels in the city via a wet removal mechanism. After the post-monsoon period in October, when biomass/wood burning/combustion activities begin in the city as the temperature lowers, PM_{10} concentrations begin to rise, peaking in November and December.

Seasonal differences in climatic circumstances, as well as varying potential particulate sources, are the main causes of this discrepancy [19]. Maximum PM_{10} concentrations during the post-monsoon period may come from combustion combined with favourable meteorological circumstances that result in a lower boundary layer height, slower diffusion due to which pollutants remain in the atmosphere for extended periods of time. During monsoon, days during higher rainfall exhibited lowest particulate concentrations.

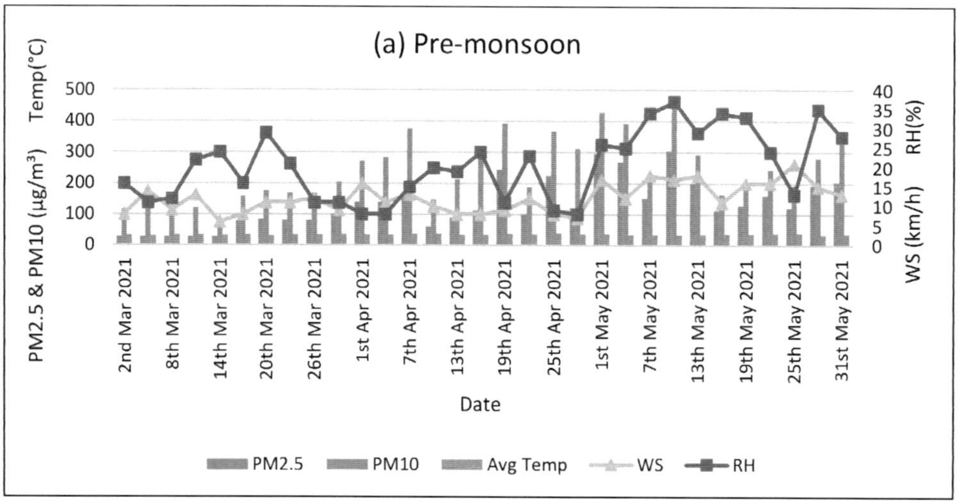

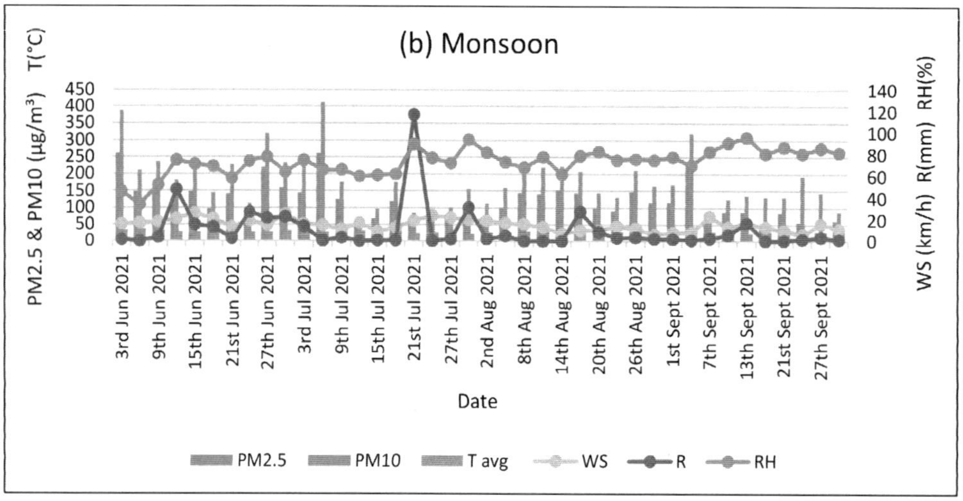

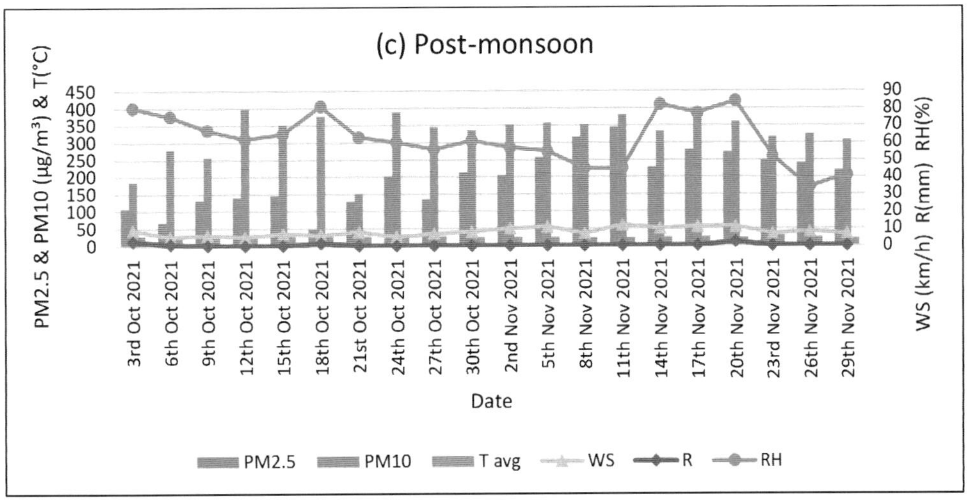

All the peak values of $PM_{2.5}$ and PM_{10} were observed in January month. Through winter and post-monsoon, the wind speed (8 km/h) was lowest when the particulate concentrations were observed to be highest. Diwali is one of India's most significant religious festivals, and it is observed for several days in October and November each year across the country, with illuminations and the lighting of firecrackers. High levels of respirable PM are caused by firework show emissions and Diwali festivities are associated through the greatest levels [20]. Also, during the months of October through January, when temperatures and wind speeds are lower, higher $PM_{2.5}$ and PM_{10} concentrations are seen [21]. During the monsoon season, heavy rain washes away pollutants, lowering particle concentrations in the area [22]. Temperature inversions occur during the summer, resulting in pollutant accumulation. Temperature, wind speed, precipitation, and relative humidity all play a role in PM_{10} fluctuations.

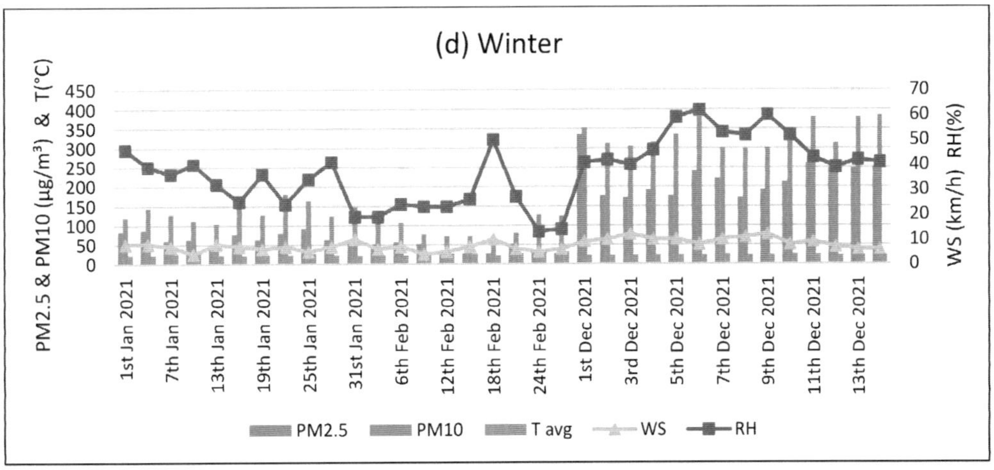

Figure: 2 $PM_{2.5}$ and PM_{10} variation with meteorological factors during (a) Pre-monsoon (b) Monsoon (c) Post-monsoon (d) Winter

Table 1 shows the findings of a Pearson correlation analysis of $PM_{2.5}$ and PM_{10} with other environmental variables such as meteorological parameters. $PM_{2.5}$ and PM_{10} shows a significant correlation. $PM_{2.5}$ concentrations were shown to have no correlation (0.04 and 0.01) with ambient temperature and wind speed for the yearly values. However, lower wind speeds showed a negative relationship with coarse particle concentrations (-0.32), indicating extended standstill periods at lower wind speeds [19]. Relative humidity exhibited negative correlation with both fine and coarse aerosols as humidity suppress the particulates minimizing their concentrations during monsoon. The temperature ranged from 6 to 41 degrees Celsius, with a relative humidity of 8 to 96 percent. The wind rose diagrams plotted throughout different seasons of the study period are shown in Figure 3. Throughout the three seasons, the overall wind speed fluctuated between 4 and 26 km/h. During post-monsoon and winter seasons, the wind blew from the northeast, in pre-monsoon the wind blows from southwest whereas in the monsoon, it blew from the southeast.

Table 1: Pearson correlation analysis of $PM_{2.5}$ and PM_{10} with meteorological factors during 2021

	PM2.5	*PM10*	*Temp*	*WS*	*RH*
PM2.5	1				
PM10	0.88	1			
Temp	0.04	0.18	1		
WS	0.01	-0.03	0.14	1	
RH	-0.16	-0.32	-0.69	0.13	1

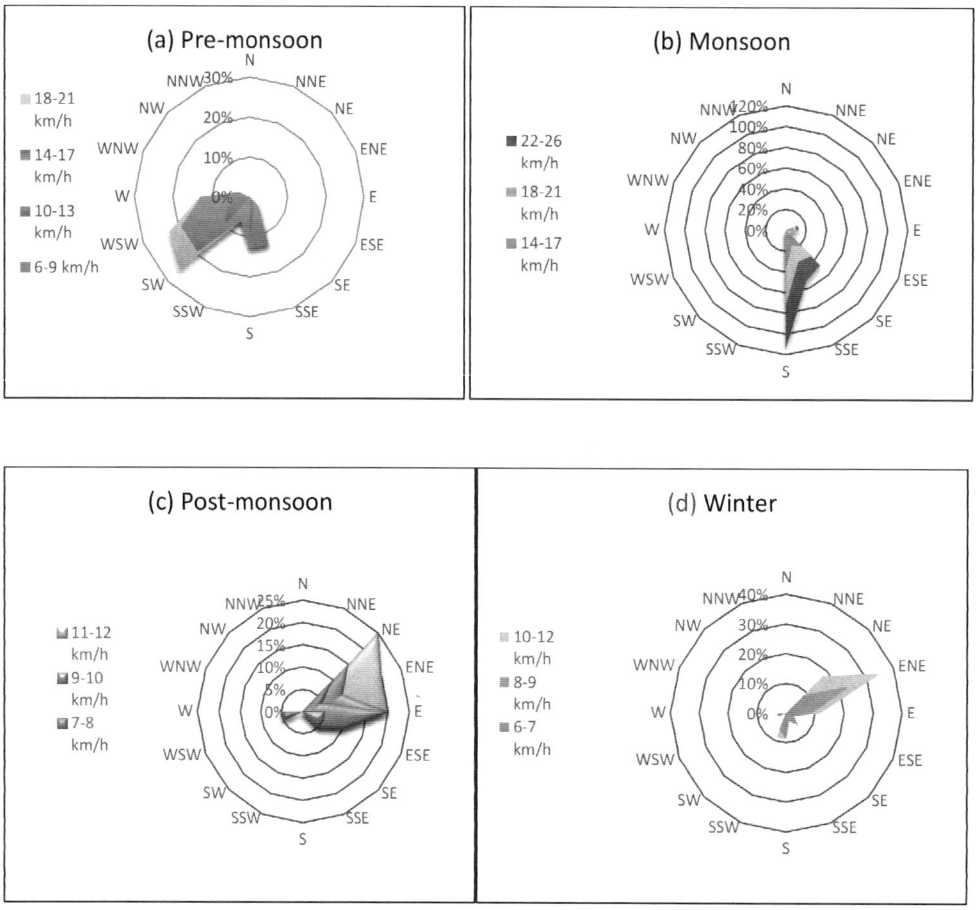

Figure 3: Wind rose diagrams plotted during (a) pre-monsoon (b) monsoon (c) post-monsoon and (d) winter seasons

iii. $PM_{2.5}$-PM_{10} relation

The relationship between PM_{10} and $PM_{2.5}$ concentrations was studied using coincident daily averages. For the studied year, $PM_{2.5}$ concentrations follow the same pattern as PM_{10}, however their means are clearly lower. Figure 4 displays a concentration scatterplot of PM_{10} vs. $PM_{2.5}$ concentrations measured in Raipur in 2021. The concentrations of PM_{10} and $PM_{2.5}$ are highly linked (r = 0.76). The PM ratio (PM= $PM_{2.5}$ / PM_{10}) was calculated and found to be 0.65, indicating that fine aerosols are present in significant quantities at our site.

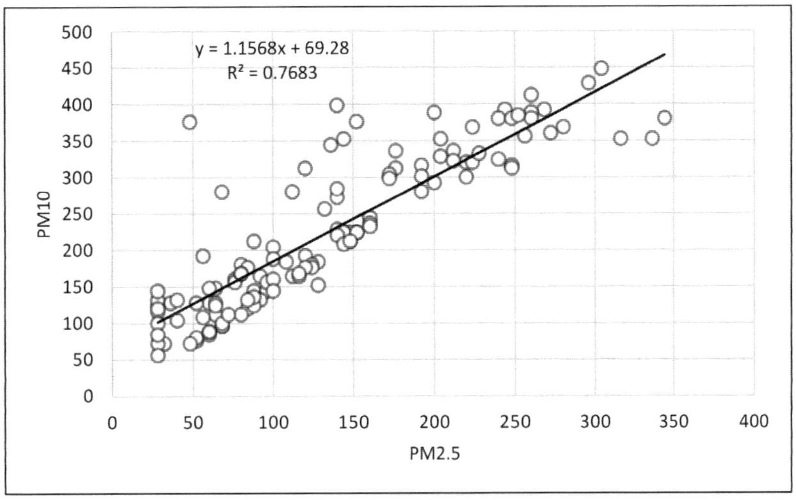

Figure: 4 Scatter plot of PM_{10} vs. $PM_{2.5}$ concentrations measured at Raipur

iv. Back-trajectory analysis

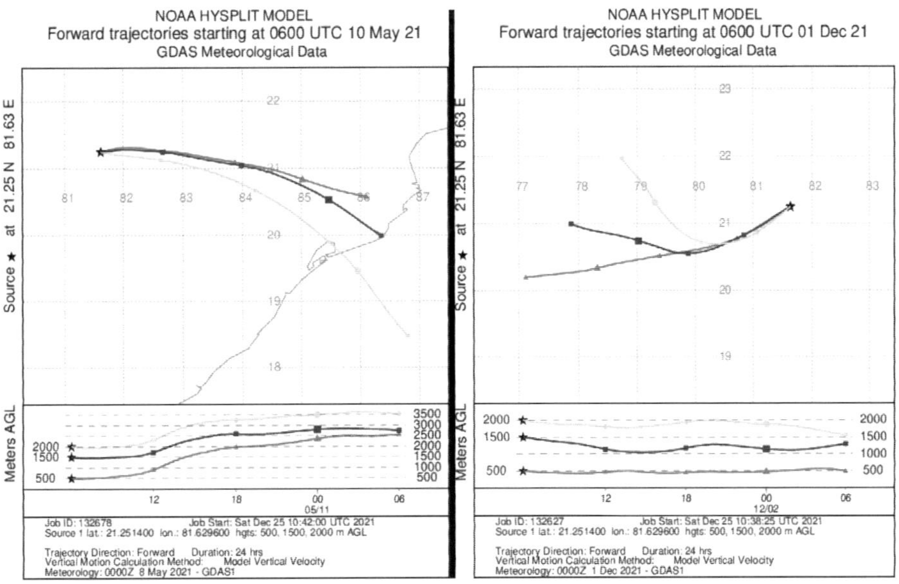

Figure 5: Back-trajectory plots using HYSPLIT NOAA

Stable air masses at the lower boundary level denoted local sources, whereas air masses at the upper boundary level with strong wind speeds denoted air masses from beyond. As illustrated in Figure 5, back-trajectories were computed at (21°15'00" N and 81°37'48"E) in the study region Raipur during periods of greatest concentration. Backward trajectories were computed for 120 hours at 500m, 1000m, and 1500m above ground level. During phases of maximum concentration detected in the winter season, the local sources were associated with NE winds with air masses at lower levels. According to the trajectory study, higher amounts of pollutants develop at lower levels, which could be attributed to anthropogenic activities during festivals. The highest $PM_{2.5}$ and PM_{10} concentration was reported on May 10th (448 $\mu g/m^3$ and 304 $\mu g/m^3$) during pre-monsoon period when the wind route was observed from west. In winter, highest $PM_{2.5}$ concentrations were noticed on December 1st and highest PM_{10} concentrations were observed on December 15th 2021 when the wind direction was observed from east. This suggests that, despite the greater wind speeds, the pollutant concentration was substantially higher, indicating the presence of powerful polluting sources, which could be local wood/coal burning sources used for heating and cooking during the winter.

4, Conclusion

Based on daily PM levels reported at the sampling station and concurrent estimation of surface meteorological factors, this paper examined the temporal variation of $PM_{2.5}$ and PM_{10} values in Raipur from January 2021 to December 2021, along with their interactions with different meteorological factors. From January to December 2021, the monthly mean mass concentrations of $PM_{2.5}$ and PM_{10} in Raipur ranged from 28.0-334.0 $\mu g/m^3$ to 56.0-448.0 $\mu g/m^3$, respectively. Post-monsoon PM concentrations were highest, followed by pre-monsoon, winter, and post-monsoon. Due to the inoculation of coarse particulates into the environment from earth crust in the area by heavy winds, PM_{10} levels were at their extreme in the zone in pre-monsoon. $PM_{2.5}$ concentrations were higher in the pre-monsoon season although the wind speed was higher as the region is surrounded by industrial areas. $PM_{2.5}$ showed no significant correlation with temperature (0.5) and wind speed (0.01) whereas PM_{10} showed no relation with temperature (0.04) and was negatively associated with wind speed (-0.03). Relative humidity was negatively associated with both $PM_{2.5}$ (-0.16) and PM_{10} (-0.32) indicating that moisture helps in the wet removal of both coarse and fine aerosols. The back-trajectory analysis depicts the local polluting sources throughout higher concentrations both during the pre-monsoon and post-monsoon season indicating the presence of regional polluting sources in the area. $PM_{2.5}$ / PM_{10} ratio concluded the presence of significant fine aerosols at the location. Overall, the $PM_{2.5}$ concentrations surpassed NAAQS limits on more than 92 percent of days and PM_{10} surpassed limits on 90 percent of days, posing a threat to human health. Seasonal differences in the correlations among PM concentrations and meteorological factors were discovered. The impacts of radiation, planetary boundary layer height, air temperature, relative humidity and wind speed must all be considered while studying PM pollution.

5. References

[1] A. V. Ahirwar and S. Bajpai, "Seasonal variability of TSPM, PM10 and PM2.5 in ambient air at an urban industrial area in eastern central part of India," p. 11.

[2] P. P. Khobragade and A. V. Ahirwar, "Source apportionment of SPM by positive matrix factorization and PM $_{2.5}$ analysis in an urban industrial area," *WJE*, vol. ahead-of-print, no. ahead-of-print, May 2021, doi: 10.1108/WJE-11-2020-0550.

[3] S. Tiwari *et al.*, "Black carbon and chemical characteristics of PM10 and PM2.5 at an urban site of North India," *J Atmos Chem*, vol. 62, no. 3, pp. 193–209, Apr. 2009, doi: 10.1007/s10874-010-9148-z.

[4] P. Pandey *et al.*, "A study of urban heat island and its association with particulate matter during winter months over Delhi," *Science of The Total Environment*, vol. 414, pp. 494–507, Jan. 2012, doi: 10.1016/j.scitotenv.2011.10.043.

[5] D. K. Deshmukh, M. K. Deb, D. Verma, and J. Nirmalkar, "Seasonal Air Quality Profile of Size-

Segregated Aerosols in the Ambient Air of a Central Indian Region," *Bull Environ Contam Toxicol*, vol. 91, no. 6, pp. 704–710, Dec. 2013, doi: 10.1007/s00128-013-1121-5.

[6] D. K. Deshmukh, M. K. Deb, and S. L. Mkoma, "Size distribution and seasonal variation of size-segregated particulate matter in the ambient air of Raipur city, India," *Air Qual Atmos Health*, vol. 6, no. 1, pp. 259–276, Mar. 2013, doi: 10.1007/s11869-011-0169-9.

[7] Y. Zhu *et al.*, "Spatiotemporal variation in residential PM2.5 and PM10 concentrations in China: National on-site survey," *Environmental Research*, vol. 202, p. 111731, Nov. 2021, doi: 10.1016/j.envres.2021.111731.

[8] I. Kayes, S. A. Shahriar, K. Hasan, M. Akhter, M. M. Kabir, and M. A. Salam, "The relationships between meteorological parameters and air pollutants in an urban environment," *GJESM*, vol. 5, no. 3, Jul. 2019, doi: 10.22034/GJESM.2019.03.01.

[9] M. Das, A. Das, R. Sarkar, P. Mandal, S. Saha, and S. Ghosh, "Exploring short term spatio-temporal pattern of PM2.5 and PM10 and their relationship with meteorological parameters during COVID-19 in Delhi," *Urban Climate*, vol. 39, p. 100944, Sep. 2021, doi: 10.1016/j.uclim.2021.100944.

[10] A. Manju *et al.*, "Spatio-seasonal variation in ambient air pollutants and influence of meteorological factors in Coimbatore, Southern India," *Air Qual Atmos Health*, vol. 11, no. 10, pp. 1179–1189, Dec. 2018, doi: 10.1007/s11869-018-0617-x.

[11] T. Mukta, M. Md. Hoque, Md. Sarker, M. Hossain, and G. Biswas, "Seasonal variations of gaseous air pollutants (SO2, NO2, O3, CO) and particulates (PM2.5, PM10) in Gazipur: an industrial city in Bangladesh," *AET*, vol. 6, no. 4, Oct. 2020, doi: 10.22104/aet.2021.4890.1320.

[12] H. Zhang, Y. Wang, J. Hu, Q. Ying, and X.-M. Hu, "Relationships between meteorological parameters and criteria air pollutants in three megacities in China," *Environmental Research*, vol. 140, pp. 242–254, Jul. 2015, doi: 10.1016/j.envres.2015.04.004.

[13] S. K. Goyal and C. V. Chalapati Rao, "Assessment of atmospheric assimilation potential for industrial development in an urban environment: Kochi (India)," *Science of The Total Environment*, vol. 376, no. 1–3, pp. 27–39, Apr. 2007, doi: 10.1016/j.scitotenv.2007.01.067.

[14] X. Li, Y. Ma, Y. Wang, N. Liu, and Y. Hong, "Temporal and spatial analyses of particulate matter (PM 10 and PM 2.5) and its relationship with meteorological parameters over an urban city in northeast China," *Atmospheric Research*, vol. 198, pp. 185–193, Dec. 2017, doi: 10.1016/j.atmosres.2017.08.023.

[15] W. Kliengchuay *et al.*, "Particulate matter (PM10) prediction based on multiple linear regression: a case study in Chiang Rai Province, Thailand," *BMC Public Health*, vol. 21, no. 1, p. 2149, Dec. 2021, doi: 10.1186/s12889-021-12217-2.

[16] V. Murari, N. Singh, R. Ranjan, R. S. Singh, and T. Banerjee, "Source apportionment and health risk assessment of airborne particulates over central Indo-Gangetic Plain," *Chemosphere*, vol. 257, p. 127145, Oct. 2020, doi: 10.1016/j.chemosphere.2020.127145.

[17] O. Ramírez, A. M. Sánchez de la Campa, F. Amato, R. A. Catacolí, N. Y. Rojas, and J. de la Rosa, "Chemical composition and source apportionment of PM10 at an urban background site in a high–altitude Latin American megacity (Bogota, Colombia)," *Environmental Pollution*, vol. 233, pp. 142–155, Feb. 2018, doi: 10.1016/j.envpol.2017.10.045.

[18] G. S. Saggu and S. K. Mittal, "Source apportionment of PM10 by positive matrix factorization model at a source region of biomass burning," *Journal of Environmental Management*, vol. 266, p. 110545, Jul. 2020, doi: 10.1016/j.jenvman.2020.110545.

[19] A. Srivastava, S. Gupta, and V. K. Jain, "Source Apportionment of Total Suspended Particulate Matter in Coarse and Fine Size Ranges Over Delhi," *Aerosol Air Qual. Res.*, vol. 8, no. 2, pp. 188–200, 2008, doi: 10.4209/aaqr.2007.09.0040.

[20] J. Nirmalkar, D. K. Deshmukh, M. K. Deb, Y. I. Tsai, and K. Sopajaree, "Mass loading and episodic variation of molecular markers in PM2.5 aerosols over a rural area in eastern central India," *Atmospheric Environment*, vol. 117, pp. 41–50, Sep. 2015, doi: 10.1016/j.atmosenv.2015.07.003.

[21] M. A. Zoran, R. S. Savastru, D. M. Savastru, and M. N. Tautan, "Assessing the relationship between surface levels of PM2.5 and PM10 particulate matter impact on COVID-19 in Milan, Italy," *Science of The Total Environment*, vol. 738, p. 139825, Oct. 2020, doi: 10.1016/j.scitotenv.2020.139825.

[22] R. Banoo, S. K. Sharma, R. Gadi, S. Gupta, and T. K. Mandal, "Seasonal Variation of Carbonaceous Species of PM10 Over Urban Sites of National Capital Region of India," *Aerosol Sci Eng*, vol. 4, no. 2, pp. 111–123, Jun. 2020, doi: 10.1007/s41810-020-00058-2.

Kodagu disaster (Floods - Emphasis on Catchment Fragmentation Index and unscientific land usage) analysis using GIS.

Kushal M[1,a*] and Sangita Deb Barman[2,b]

[1]UG Student, Department of Civil Engineering, Indian Institute of Engineering Science and Technology, Shibpur, West Bengal - 711103, India.

[2]Faculty, Department of Civil Engineering,Indian Institute of Engineering Science and Technology, Shibpur, West Bengal - 711103, India.

Email: [a] kushim250299@gmail.com[b]dbsangita@gmail.com

Abstract. India is a densely populated country, natural hazards like floods and landslides cause a great impact on peoples' life and property as well. The floods and landslides in Kodagu, Karnataka was not only because of unexpected heavy rain but has various man-made causes like unscientific land-usage modifications. This unscientific land usage and linear constructions like roads railways, etc. have caused catchment fragmentation, hindered or blocked the natural streamflow, and resulted in localized water stagnation, resulting in floods and landslides. This paper defines an index-catchment fragmentation index to generalize the causes of both unscientific land usage and linear constructions (mainly roads).Catchment fragmentation causes variations in soil moisture content, flow path, flow velocity, flow connectivity, as well as on concentration time. This paperuses the Multi-Criteria Decision Making-Analytical hierarchy process (MCDM-AHP) to create a flood hazard zonation map and also aims at predicting the hazard and vulnerability zones and hence triesto guide on correcting the unscientific land usages.

Keywords:MCDM-AHP, Catchment fragmentation index, Unscientific land usage.

1. Introduction

A natural hazard is a damaging phenomenon within a specified period and region due to a set of existing or predicted conditions that can cause loss of life as well as property [1].Floods are the overflow of water beyond its natural limits, mainly due to heavy rainfall or snowmelt and sometimes because of collapsing of the dam. Apart from these reasons it is also caused by deforestation, climate change, etc.

The study deals with interpreting and analyzing the reasons behind devastating floods and landslides of Kodagu district in Karnataka of the year 2018 and 2019.Though the floods were mainly due to heavy rainbut were also greatly influenced by anthropogenic activities which tend to disturb the equilibrium of the catchment. So, a zonation map is required considering the combined effect of all the influencing factors of the potential hazard. According to National Research Council, 2005; Rao, predicting the intensity and duration of the short burst intense rainfall is difficult, which often also triggers floods as well as landslides in slopy terrain. Kodagu faced around 130633.88 hectaresof cropland, INR 50,672 lakh of worth crops was destroyed, the main economic source of Coorg people. An estimated INR 415 crore loss due to damage to the roads, bridges, and infrastructure [2,3].

A zonation map is required before any disaster management or mitigation studies. As the regions of high probability zones can be concentrated upon, thus increasing the focus and reducing the time as well as investment in disaster management. The zonation maps can be generated by various techniques like -inventory-based probability studies or deterministic or heuristic or statistical techniques [4]. The Multi-criteria decision analysis method- analytic hierarchy process (AHP) is used in this study considering the heterogeneity of the attributes of all the criteria, belongs to heuristic analysis. One of the main advantages is that this method avoids the generalization of data and considers the uniqueness of the particular event, its region [6], and causative factors. This study mainly deals with the effect of anthropogenic activities which have increased the probability of such events and a flood hazard zonation map is prepared using ArcGIS and M algorithm as proposed by Satty [5].

2.Study area

Kodagu is considered one of the ecologically sensitive areasof Western Ghats of Karnataka (Figure 2), with rare and rich fauna and flora. Kodagu extends from 11.56°N-12.52°N and 75.22°E-76.12°E. The toposheet numbers of the region used are 48P/10, 48P/7, 48P/11, 48P/15, and 48P/12. During the monsoons of 2018, Kodagu has experienced very heavy rainfall. Madikeri and Somwarpet taluk had received cumulative rainfalls of 4692 mm and 2701 mm respectively, breaking the 87-year-old record of the year[7]. Though it was concluded that heavy rain was the cause but it is also debatable that anthropogenic pressure was the main reason for the floods. There is a lot of lessons that this disaster teaches us and help in future planning of our urban infrastructures which should not disturb the natural ecosystem and its processes.

The study area experienced heavy rainfall (Monsoon) which resulted in floods and landslides in August. The daily rainfall trends of the month August 2018 and August 2019 (collected from Karnataka State Natural Disaster Monitoring centre) is plotted to show the variation of such an event against a normal scenario. There was around a 400% increase in rainfall against the normal rainfall, for the days from 13th to 18th August 2018 [12], (Figure 1).

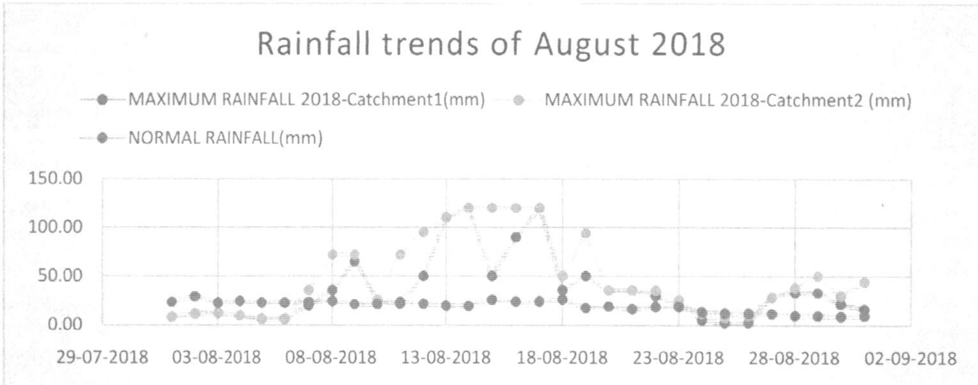

Figure 1. Rainfall in the month of August 2018 of the 2 catchments against the normal rainfall

The Kodagu district (Figure 3) is delineated and the 2 catchments (Figure 4) are considered for the study. Catchment 1 was the most affected by floods in the year 2018 and 2019, hence this catchment is chosen to study the floods in this paper. The landslide of catchment 2 was analysed in the previous part of this work [8].

Land-use-Land-cover (LULC) determines how the rainfall is portioned into the runoff, or whether it is infiltrated or evaporated [10].The catchment area showed that urbanized areas and the area under the plantations increased to about 20%, while the Forest area decreased to about 40% [9, 11], (Figure 4), which is especially alarming for an ecologically sensitive area like Kodagu. If the catchment is not having the capacity to retain rainwater, there would be no percolation into the ground, most water would flow into oceans, causing floods in the monsoon followed by droughts because of unreplenished groundwater. Along with flash floods, the district also witnessed landslides.

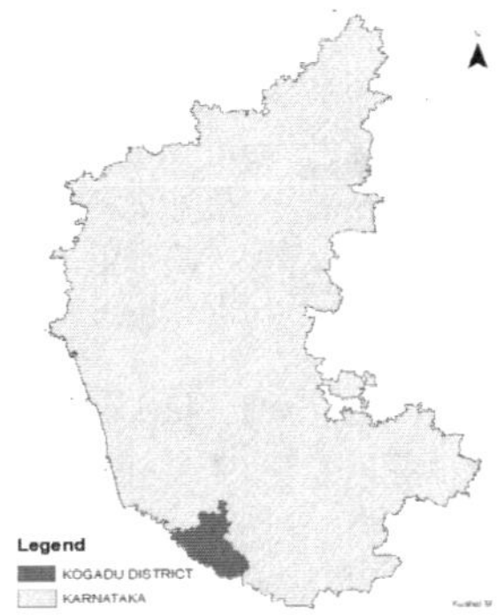

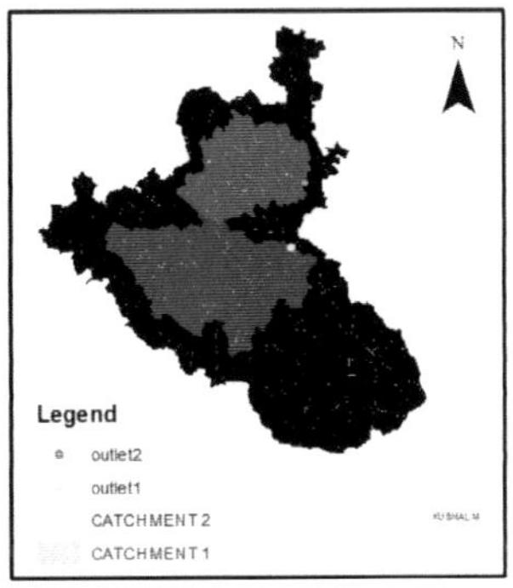

Figure 2. Karnataka state map **Figure 3.** Kodagu district map**Figure 4**. Delineated catchments

LULC of the year 2016 is considered, just before the floods and landslides, to prevent erroneous results due to changes in the landscape after the disaster.

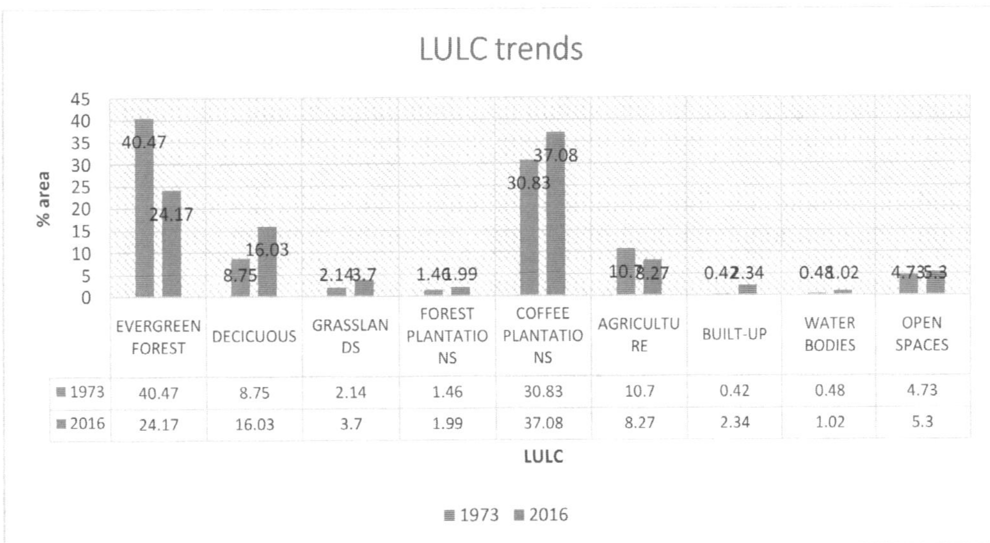

Figure 5. LULC trends of Kodagu for the years, 1973 and 2016

2.1. Data used

1). DEM data from Google earth pro.

2). LANDSAT 8 images from earth explorer-USGS.

3). Shapefiles of the district and taluk boundary from Karnataka GIS portal.

4). Digital Soil Map of the World prepared by the Food and Agricultural Organisation (FAO) has been used for the identification of the Hydrological Soil Groups (HSG) present in the study area.

5). Annual rainfall data from the IMD website and daily rainfall from Karnataka State Natural Disaster Monitoring Centre (KSNDM) [9].

6). Geomorphology, geology, road network, lineament shapefiles from Bhukosh website.

ArcGIS software is used in delineating the watershed, finding the catchment area as well as analyzing the AHP results using the tools of the software.

3. Analysis of disaster trends for the year 2018-19

By comparing the Land-use land cover trends of Kodagu district with previous disasters f 2018-19 (flood and landslide), a general conclusion can be drawn about the landslide and flood-affected areas.

- The urban settlements of the entire district are clustered about the state highway of Madikeri taluk. The map (released by the Government of Karnataka) shown above shows the landslide in the year 2018 and 2019 were clustered around the state highway of Madikeri taluk (Figure 5, Figure 7) and near plantations of Harangi basin of Somwarpet taluk [13,14]. The slopes created due to road construction if left unsupported can lead to slope instability causing landslides which are enhanced during rain. The stream network is hampered, thus leading to waterstagnation which causes more infiltration, decreasing the angle of friction leading to landslides
- The floods of the year 2019 and 2018 (Figure 6) also show a general trend as they are concentrated around urban settlements and plantations of Virajpete and Gonikoppa (Figure 7), even though they are away from the banks of the river north of them. This can infer that localized blocking of drainage and localized water stagnation due to slope modifications, unscientific land usages like plantations, construction of roads [11], clustering of settlements disturbed the catchment and no alternative methods were used to restore the blocked drainage. The terrace farming at the plantations is believed that have caused localized water stagnation.

Catchment 1 is considered for analysis, as it was most prone to floods.

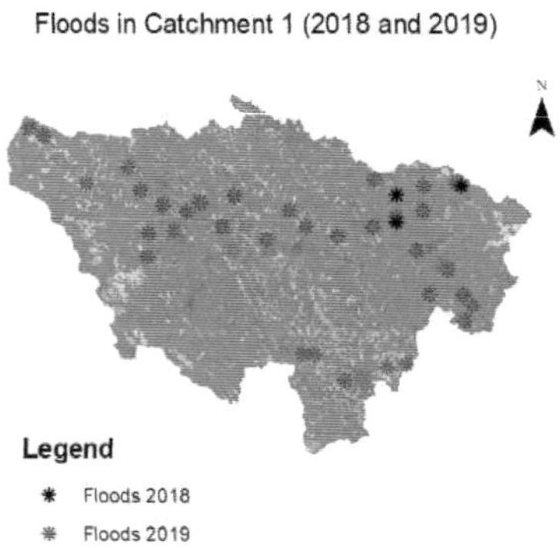

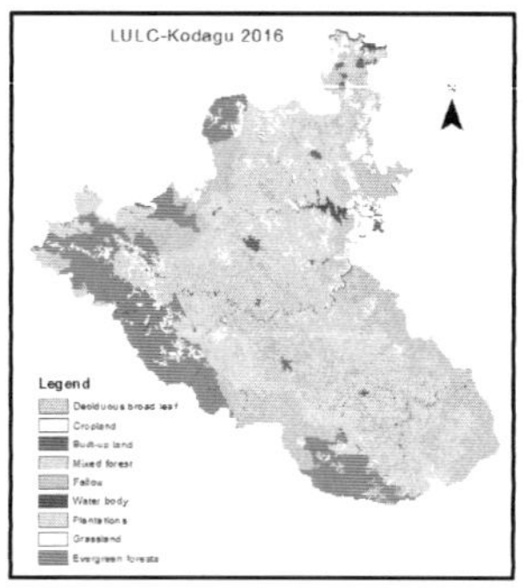

Figure 6. Floods of 2018 and 2019**Figure 7.** LULC map of 2016

4. Methodology

The Multi-Criteria Decision Making (MCDM)-Analytical Hierarchy Process (AHP) is used for generating the probable flood hazard zones [15] of Catchment 1. Considering the wide range as well as the heterogeneity of the attributes and multiple criteria, the MCDM-AHP method is chosen to predict the most probable zones of the hazard with the theoretical background as well as inventory studies of the different criteria considered for the Kodagu region.

- This method compares two criteria at a time through a pairwise comparison matrix as per users' decision based on theoretical background and the location-specific inventory studies; in which the user compares the two criteria based on the relative contribution to floods ranging from one to nine, where one means both criteria in pairwise comparison are equally contributing, whereas nine means one of the criteria is extremely dominating the other (1. equally contributing; 3. slightly more contribution; 5. Quite more contribution; 7. more contribution; 9. Extremely dominant in contribution; 2, 4, 6 and 8 are intermediate values) [5, 15, 16, 17]. After solving, we get relative priority weightage and ranking of the different criteria considered. The AHP method was used to conclude the contribution of each criterion to a particular hazard.

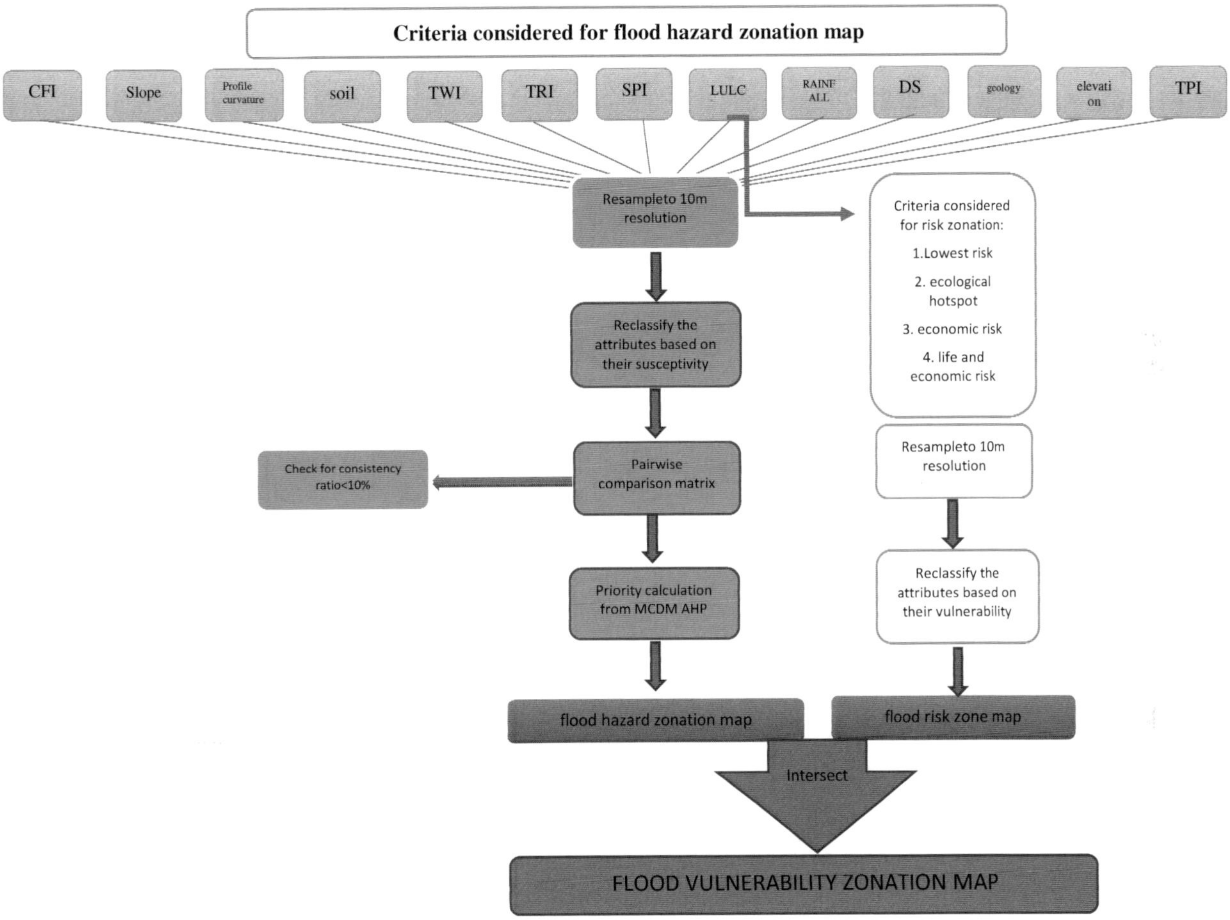

Figure 8. flow chart representing the methodology

- The thematic maps are in raster format and they are *Resample* to UTM zone 42 coordinate system and pixel resolution of 10m. The relative rank of different attributes of a particular attribute has been assigned using *Reclassify tool* of ArcGIS, based on its probability of occurrence of floods, using scale 1-9, (1- low probability of floods, 9-highest probability of floods).
- The advantages of the AHP method are, it allows identifying inconsistency of the ranks given to the pairwise comparisons using CR (consistency ratio) as an indicator. If CR <10%, it indicates a good level of consistency [5,18,19], but these ranks are assigned based on users' decisions.

(CR-consistency ratio, CI-consistency Index, RI-random index)

$$CR = \frac{CI}{RI} \quad (1)$$

Table1. Values of Random Index (RI) concerning

n	1	2	3	4	5	6	7	8	9	10
RI	0	0	0.58	0.89	1.12	1.24	1.32	1.41	1.45	1.49

$$CI = \frac{\lambda max - n}{n - 1} \quad (2)$$

Where λmax is the principal eigenvalue and n is the number of criteria. Random consistency index (RI) (Table 1) average of randomly generated reciprocal matrices (Saaty 1980).

- Based on priority obtained from the matrix calculations, the attribute Reclassified thematic maps are multiplied with their respective priority percentage (Equation 4) to obtain flood hazard zonation maps (Figure8).

4.1. The criteria considered for the study of the zonation map of the flood are:

4.1.1. Catchment Fragmentation Index (CFI)

This index depicts the macro-level effects of unscientific land usage and practices. These changes in LULC have led to catchment fragmentation which causes variations in soil moisture content, flow path, flow velocity, flow connectivity, water stagnation as well as concentration time.The deforestation and conversion of forests to plantations, agricultural lands, or grasslands has led to intensified and sudden flood peaks. The flow connectivity is getting disturbed due to roads and urban infrastructures. The stream networks have been hampered by unscientific land usages like plantations, and linear constructions like the transport network or the power lines which led to fragmentation of catchment thus blocking the stream networks and natural drainage systems.

This can infer that localized blocking of drainage and water stagnation due to slope modifications, unscientific land usages like plantations, construction of roads [11], clustering of settlements disturbed the catchment and no alternative methods were used to restore the blocked drainage.

Catchment Fragmentation i(CFI)(Equation 3, Figure 16)- has been defined to depict the extent of catchment fragmentation. The index has considered the blockage and hindrance of the stream network due to terrace farming, urban area, and linear constructions mainly roads (Figure 10). The main river network intersection is not considered since it's not hindered due to the presence of bridges and culverts. Thus the 15000<flow<2000(Figure 11, Figure 12, Figure 13) is considered to study the hindrance (Figure 14, Figure 15).

$$\boldsymbol{CFI} = \boldsymbol{BLOCKAGE\ DUE\ TO\ LINEAR\ CONSTRUCTIONS} + \boldsymbol{BLOCKAGE\ DUE\ TO\ UNSCIENTIFIC\ LULC(TERRACE\ FRAMING\ AND\ URBAN\ AREA)} \quad (3)$$

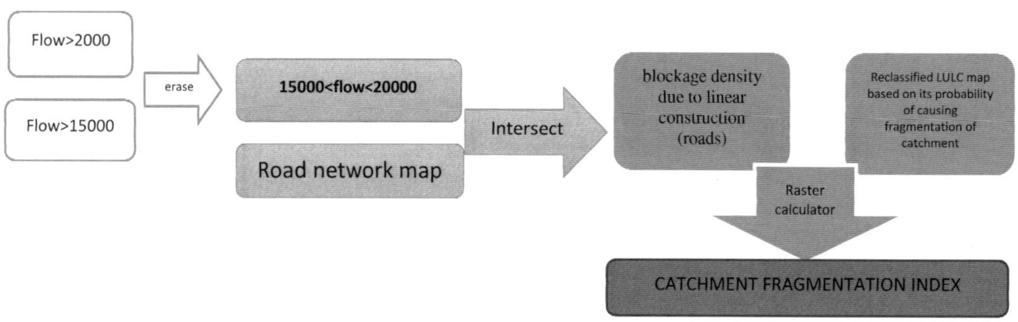

Figure 9. Flow chart showing Catchment fragmentation index methodology

If the road network increment and LULC changes can be predicted, the susceptibility of the catchment to flood and landslides in the future thus can also be predicted.

4.1.2. Slope

The slope is a topographic feature (Figure 17) that represents the steepness i.e rate of change of elevation, generally, the steepness increases the probability of the flood also decreasing at that particular region. The slope was classified into 5 classes-Very steep(>45°), steep(35°-45°), moderately steep(25°-35°), rolling(10°-25°), undulating(2°-10°) and flat(0-2°). At steep slopes, there are low chances of water stagnation, lower infiltration, and higher velocity of water, and regions downstream can face higher flood peaks due to steep slopes [20,26].

4.1.3. Slope curvature

Curvature is the second derivative of the surface. The PLAN CURVATURE is perpendicular to the direction of maximum slope. If it is positive surface is laterally convex, if negative the surface is laterally concave and a zero indicates the surface is linear. PROFILE CURVATURE (Figure 18) is parallel to the slope. If it is positive the surface is upwardly concave (flow is accelerated), if the negative surface is upwardly convex (flow is deaccelerated) and a zero indicates the surface is linear (ESRI). Profile curvature is considered for the study of floods [20,26].

4.1.4. Soil

Soil data is collected from the Hydrologic soil group (Figure 19) [23]. The soils present are 3,4,13,13.

3- moderately high runoff potential (<50% sand and 20-40% clay)

4- high runoff potential (<50% sand and >40%clay)

13- high runoff potential, unless drained (<50% sand and 20%-40%clay)

14- high runoff potential unless drained (<50% sand and >40%clay)

4.1.5. Topographic wetness index (TWI)

TWI (Figure 20) represents the tendency of water accumulation in an area with elevation difference and hence the soil moisture. It is a function of slope and upstream contributing area per unit width orthogonal to flow direction.

$$TWI = \ln\left(\frac{\text{SCA}}{\tan\varphi}\right) \qquad 4$$

SCA is a specific catchment area, assuming properties of soil and rock mass are the same SCA represents a tendency to receive water while slope represents the tendency to drain water.

Higher the TWI, the higher the probability of floods and landslides [20,26].

4.1.6. Topographic roughness index (TRI)

The topographic roughness index (Figure 21) indicates the amount of elevation difference between the pixels of DEM(Digital Elevation Model). Thus if TRI increases, the probability of occurrence of flood decreases [20,26].

4.1.7. Stream power index (SPI)

The SPI (Figure 22) is related to the geomorphology along a stream, which influences the loss of potential energy of the river along with its flow. It also depicts the errodibility capacity of the river. The higher the value of SPI, the more probable to floods [24].

4.1.8. LULC

The runoff in a catchment mainly depends on the amount of precipitation and land use land cover (LULC) (Figure 23) trends of that catchment [9]. Changes in LCLU are majorly influenced by anthropogenic activities more than natural changes. This greatly influences how rainfall partitions into Infiltration to groundwater recharge or runoff or evapotranspiration. The thick evergreen vegetation has deep root systems increasing slope stability. But most of the forests are transformed into plantations and urban settlements. The plantations in the study area followed the contour farming method which can be because of localized water stagnation [21]. LULC of the year 2030 (from CA Markov chain principle) is used for zonation map prediction.

4.1.9. Rainfall

The rainfall trends are shown in Figure 1 which shows the rare event of rain in the Kodagu region for August 2018 and this rare event is considered for the study (Figure 24).

4.1.10. Distance from stream

The distance from a stream (Figure 25) is calculated using the Euclidean distance tool of ArcGIS, the places closest to the river stream are most susceptible to floods [20,26].

4.1.11. Geology

The region consists of Charnockite, Sargur, Peninsular Gneissic complex, and Khondalite as its geological features (Figure 26). From the reports of previous floods, Khondalite, Peninsular Gneissic complex are most susceptible to floods followed by Charnockite and Sargur [22,25].

4.1.12. Elevation

Elevation of the region is shown (Figure 27). Relatively low-lying regions are most susceptible to floods [26].

4.1.13. Topographic position index (TPI)

TPI (Figure 28) compares the elevation of each pixel of DEM with the mean elevation of neighboring pixels. The extreme negative values indicate valleys, moderate negative values indicate lower slopes, zero indicates flat ground and a moderately positive value indicates upper slope, and extreme positive values indicate cliff edges or ridges [26].

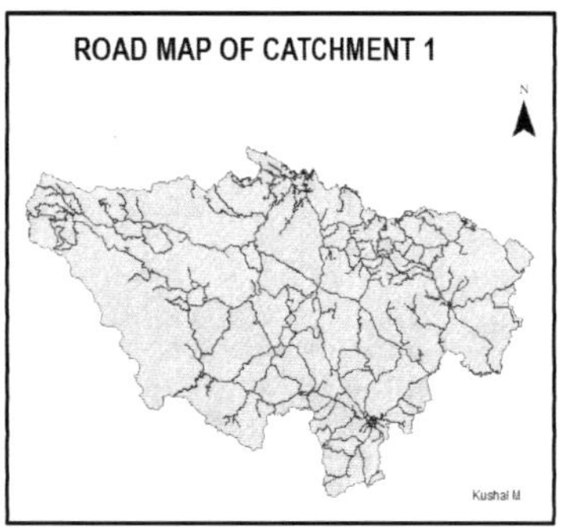

Figure 10. Road network map

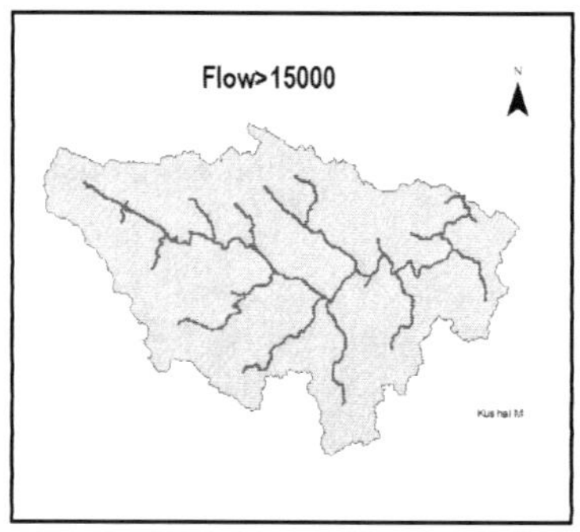

Figure 11. stream network with flow>15000

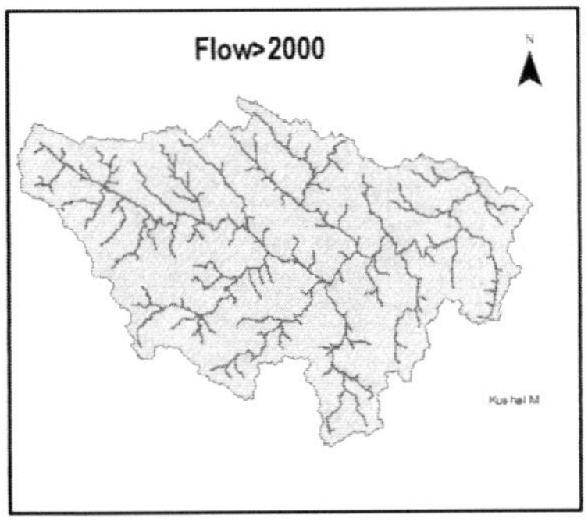

Figure12. flow>2000

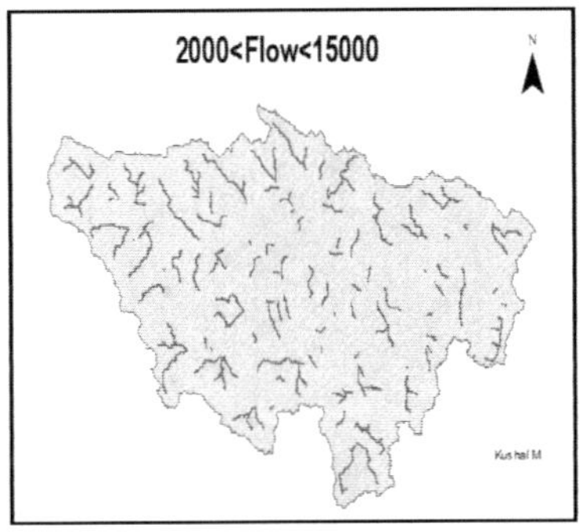

Figure 13. stream network with 2000<flow<15000

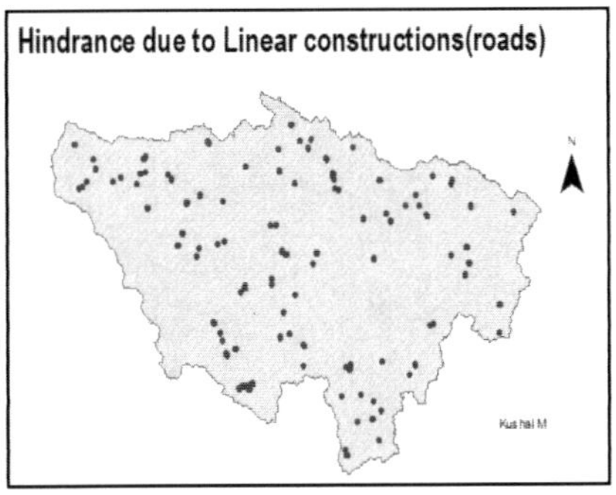

Figure 14. blockage points of stream

As no railways are present, main roads are considered as the linear construction causing an obstruction. 2000<flow<15000 is generated using overlay tools.

Hindrance points of a stream (2000<flow<15000) and roads network is generated using *intersect tool.*

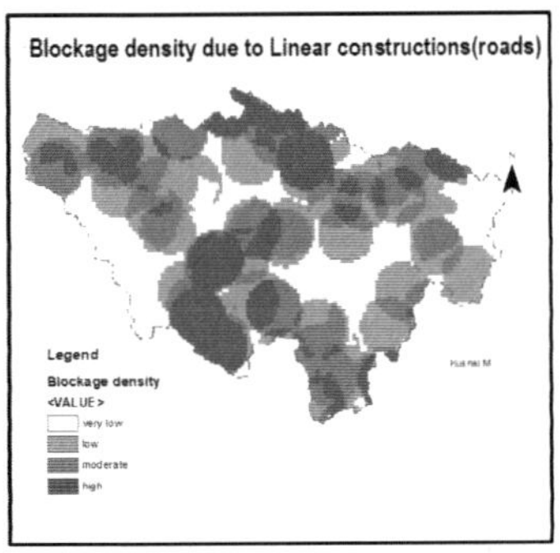

Figure 15. blockage density due to roads

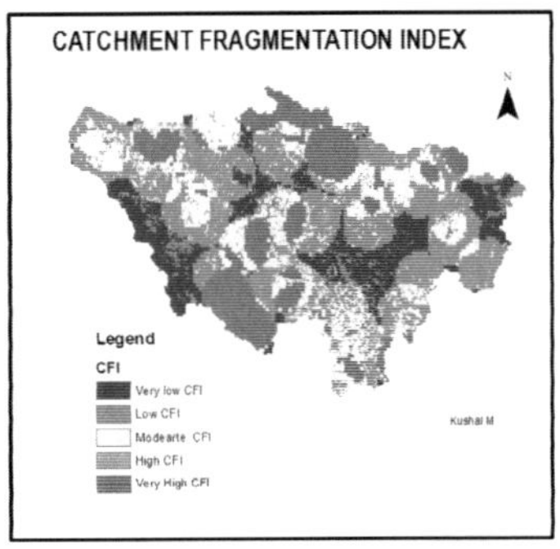

Figure16. Catchment fragmentationindex

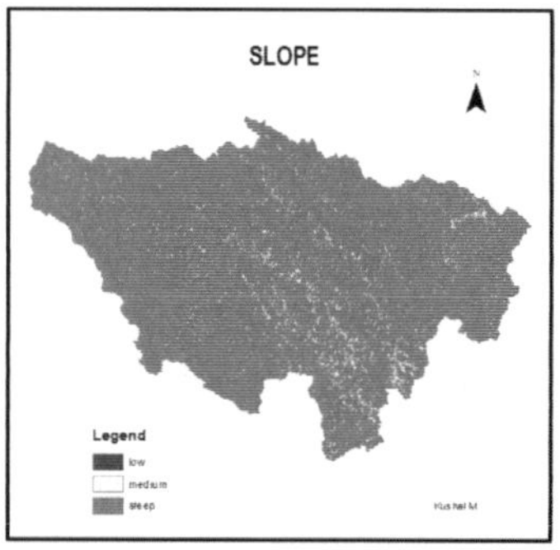

Figure 17. Slope variation map

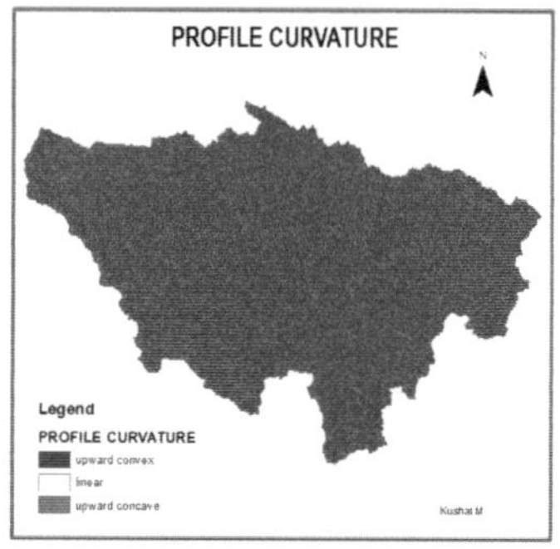

Figure 18. profile curvature map

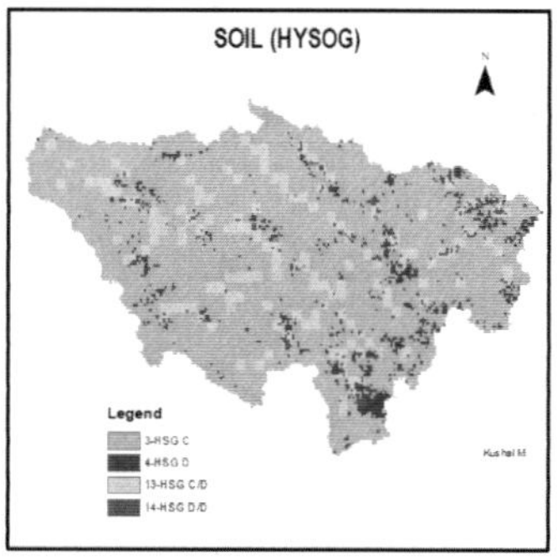

Figure 19. soil map

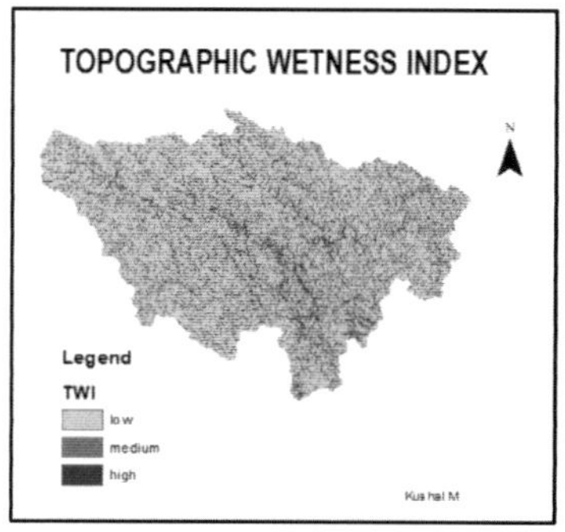

Figure 20. TWI variation map

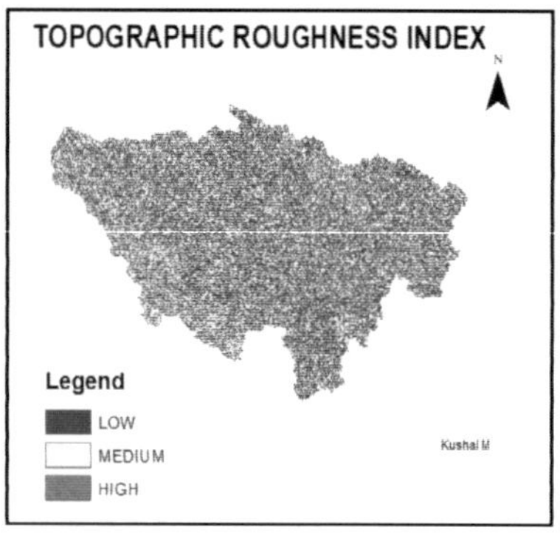

Figure 21. TRI variation map

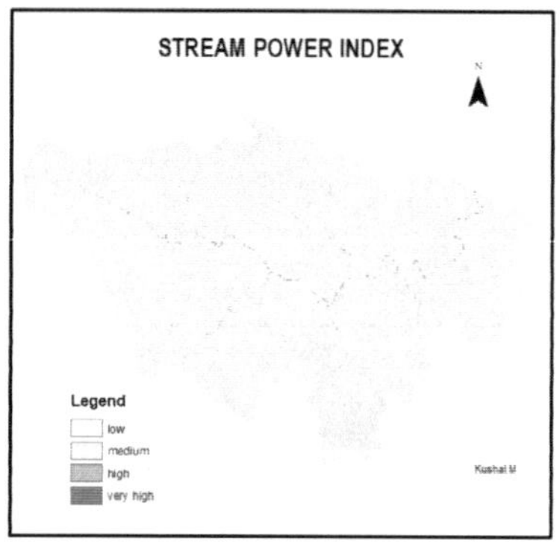

Figure 22. SPI variation map

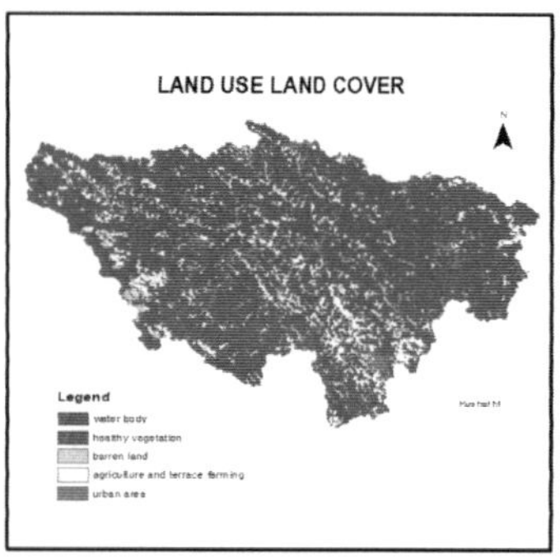

Figure 23. LULC 2030.

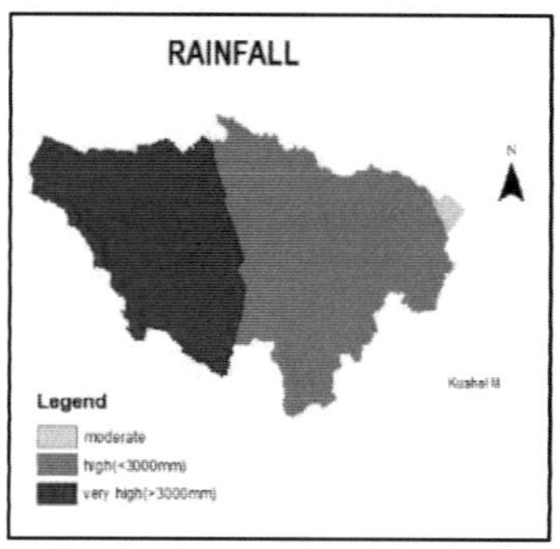

Figure 24. rainfall variation map

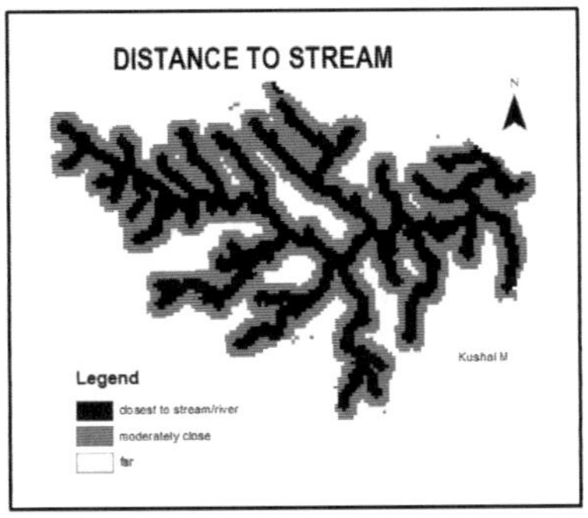

Figure 25. distance to streams

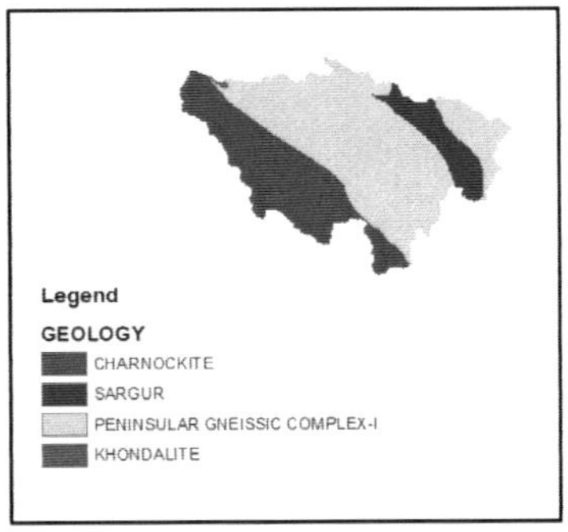

Figure 26 geology map

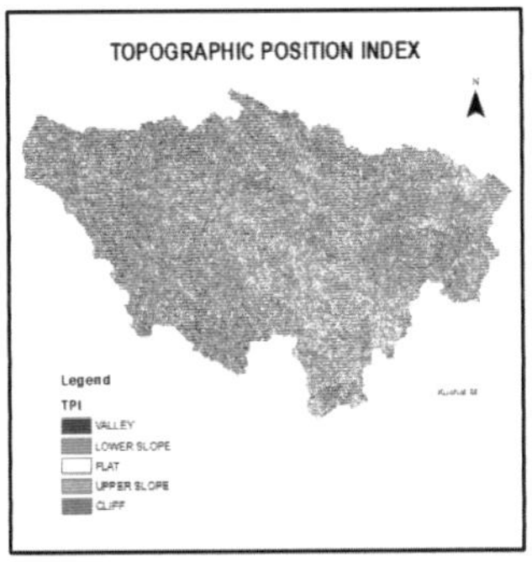

Figure 27. elevation map

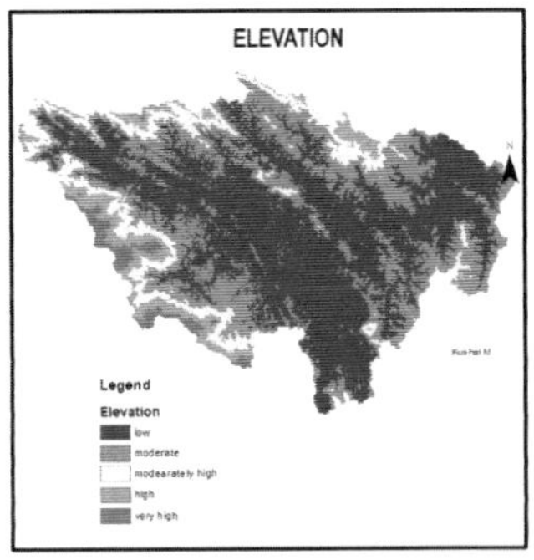

Figure 28. TPI map of catchment 1

5. Results and discussions

The pairwise comparison matrix of the AHP process is obtained (Figure 29).

There are 78 pairwise comparisons, the Principal Eigen Value is 14.543, and the Consistency ratio is 0.083(<0.1). Hence the AHP comparison results are consistent and reliable.

	1	2	3	4	5	6	7	8	9	10	11	12	13
1	1	7.00	7.00	6.00	2.00	5.00	1.00	2.00	2.00	0.50	6.00	6.00	4.00
2	0.14	1	2.00	2.00	0.33	0.33	0.17	0.25	0.33	0.12	4.00	4.00	0.33
3	0.14	0.50	1	1.00	0.25	0.25	0.11	0.14	0.14	0.12	2.00	3.00	0.25
4	0.17	0.50	1.00	1	0.14	1.00	0.14	0.14	0.50	0.14	3.00	2.00	0.20
5	0.50	3.00	4.00	7.00	1	5.00	1.00	3.00	2.00	0.17	3.00	3.00	2.00
6	0.20	3.00	4.00	1.00	0.20	1	0.14	0.17	0.50	0.14	4.00	4.00	1.00
7	1.00	6.00	9.00	7.00	1.00	7.00	1	1.00	3.00	1.00	6.00	6.00	4.00
8	0.50	4.00	7.00	7.00	0.33	6.00	1.00	1	1.00	0.14	6.00	6.00	1.00
9	0.50	3.00	7.00	2.00	0.50	2.00	0.33	1.00	1	0.12	6.00	6.00	1.00
10	2.00	8.00	8.00	7.00	6.00	7.00	1.00	7.00	8.00	1	9.00	7.00	5.00
11	0.17	0.25	0.50	0.33	0.33	0.25	0.17	0.17	0.17	0.11	1	0.33	0.25
12	0.17	0.25	0.33	0.50	0.33	0.25	0.17	0.17	0.17	0.14	3.00	1	0.25
13	0.25	3.00	4.00	5.00	0.50	1.00	0.25	1.00	1.00	0.20	4.00	4.00	1

Figure 29. Pairwise comparison matrix of the 13 criteria used for flood prediction

And the rank of each criterion (Table 2) and its priority percentage is as follows:

Table 2. rank and priority of different criteria

ATTRIBUTE	RANK	PRIORITY
CFI	3	13.8%
Slope	9	2.9%
Profile curvature	11	1.9%
Soil	10	2.2%
TWI	4	10.2%
TRI	8	3.9%
SPI	2	14.6%
LULC	5	9.1%
Rainfall	6	6.5%
DS	1	26.1%
Geology	13	1.4%
Elevation	12	1.7%
TPI	7	5.7%

Using the Raster calculator tool of ArcGIS, the flood hazard zonation map is prepared using the expression given below: the priority percentage is multiplied with the Reclassified thematic layers (all thematic layers used are with 10m resolution)

Flood hazard zonation map*= 0.138× (Reclassified CFI thematic map) + 0.029× (Reclassified slope thematic map) + 0.019× (Reclassified profile curvature thematic map) + 0.022× (Reclassified soil thematic map) + 0.102× (Reclassified TWI thematic map) + 0.039× (Reclassified TRI thematic map) + 0.146× (Reclassified SPI thematic map) + 0.091× (Reclassified LULC thematic map) + 0.065× (Reclassified rainfall thematic map) + 0.261× (Reclassified DS thematic map) + 0.014× (Reclassified geology thematic map) + 0.017× (Reclassified elevation thematic map) + 0.057× (Reclassified TPI thematic map)* 5

The flood hazard zonation (Figure 30) so generated is also validated with the previous events of 2018 and 2019of catchment 1.

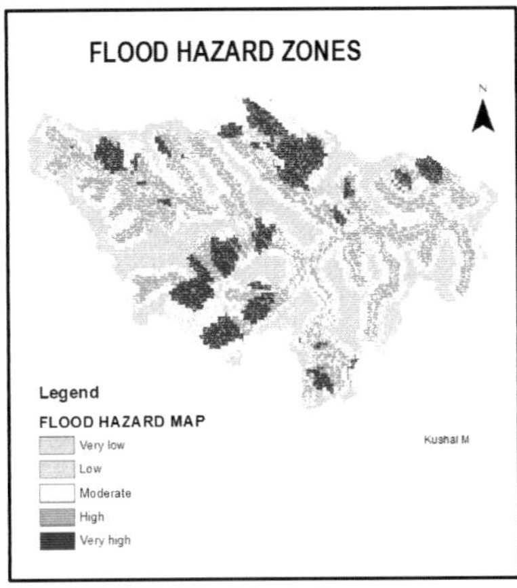

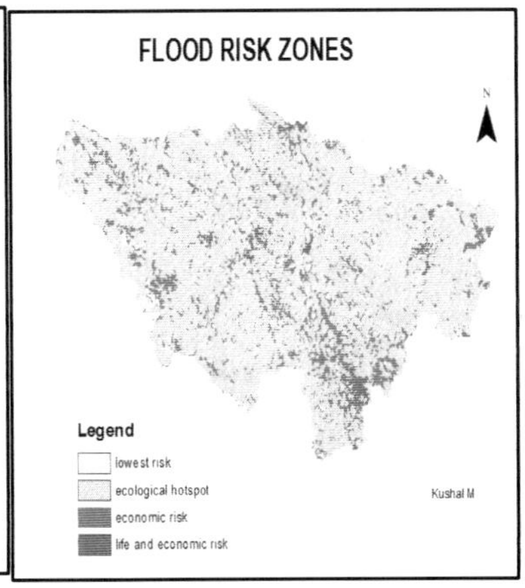

Figure 30. flood hazard zonation mapfor 2030.**Figure 31.** flood risk zonation map of

Considering the consequences after floods, the region is divided into risk zones- socio-economic risk zones are urban areas and most vulnerable to floods as they cause loss of life and property (Figure 31). The next most vulnerable region is the agriculture and plantation regions as they cause great economic losses and Kodagu being the ecologically most sensitive region, the forest region is considered vulnerable as well as there could be a threat to rich and rare fauna and flora. The least risk zones are barren land.

Using *Overlay-tools*, the risk and hazard zones are combined to get FLOOD VULNERABILITY ZONES (Figure 32).

The flood vulnerability zonation map divides the area into five different zones of different susceptibility and vulnerability classes- *very low, low, moderate, high, and very high-risk*. From the priority obtained from MCDM-AHP, we can also conclude that the man-made causes like LULC-urban settlements and plantations, road network density also contributed to triggering the floods and landslides, and these areas (the urban settlements and agriculture lands) are more vulnerable as it can cause loss of life and property.

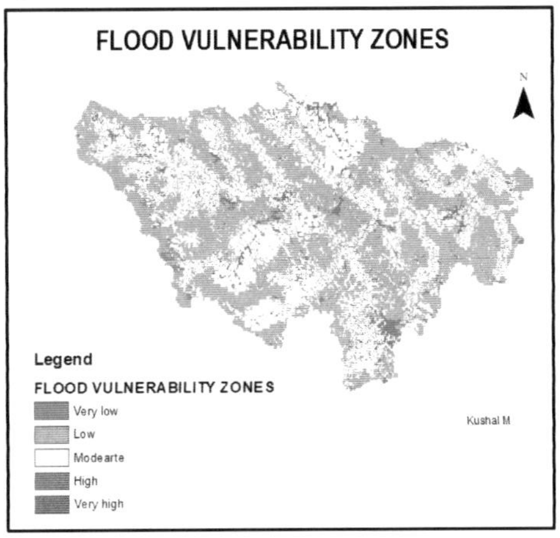

Figure 32. flood vulnerability zonation map of catchment 1 for 2030.

6. Conclusions

Catchment 1 of the Kodagu region was most susceptible to frequent flooding and the most common triggering factors are heavy rainfall, terrain, and mainly the anthropogenic pressure which disturbed the equilibrium of the catchment fragmentation as depicted by the catchment fragmentation index. In the present study, flood hazard zonation mapping for the year 2030 has been carried out using multi-criteria analysis (AHP-weighted) and CA-Markov chain principle with the help of remote sensing and GIS and ArcGIS software. The result was further validated by comparison with past events of 2018 and 2019. Future expansion of road networks, railways, plantations, or urban settlements should be done scientifically and no future increase in catchment fragmentation should be allowed, in unavoidable cases necessary steps to be taken to assure the equilibrium in the catchment and its hydrological processes.

7. Acknowledgements

This study is a part of the B.Tech project thesis and was done under the guidance of Dr. Sangita Deb Barman.

References

[1] Deoja, B., Dhatal, M., Thapa, B. and Wagner, A. 1991Mountain Risk Engineering Handbook Part II—*Application. International Centre for Integrated Mountain Development, Kathmandu***II.**

[2] "Kodagu floods: Karnataka's coffee-growing region stares at heavy losses after rains, landslides damaged plantations". *Firstpost, Indo-Asian news service,* 26th August 2018.

[3] Dr. Dayanand K.C 2019Analysis of August 2018 Disaster in Kodagu: An Overview*IOSR journal of humanities and social sciences***24 (8),** Ser. 4 (August. 2019) 43-48

[4] GUZETTI, F., CARRARA, A. AND CARDINALI, M. and REICHENBACH, P. 1999Landslide hazard evaluation: a review of current techniques and their application in a multi-scale study, central Italy. *Geomorphology*, **31**181–216

[5] SAATY, T.L. 1980The analytic hierarchy process: planning, priority setting, resource allocation". *McGraw-Hill International Book Company, New YorkMcGraw-Hill* 287

[6] Van Westen CJ, Ghosh S, Jaiswal P, Martha TR, Kuriakose SL 2011From landslide inventories to landslide risk assessment; an attempt to support methodological development in India In:*Proceedings of the second world landslide forum*, Rome*Terrigenous mass movements* 239-283

[7] Prof. RashmaJain2019Disaster Analysis of Kodagu District using Geomatics Technology *IJERT***8(6).**

[8] Kushal M 2022 Disaster (Floods and landslides due to unexpected heavy rainfall) analysis using GIS*Roorkee Water Conclave 2022.*

[9] D.K.Barik 2021Impact of Temporal Variation of Land Use/Land Cover and Precipitation on Runoff Estimation in an Ungauged Hilly Watershed in Kodagu *Watershed Management 2020, ASCE*pp 265-277.

[10] Deepak Khare, Lakwinder Singh& Mahesh Kumar Jat 2017 Impactofland usechange on urban runoff: A case studyof Jaipurcity, India*Proceedings of the 37th IAHR World Congress Kuala Lumpur, Malaysia*

[11] T.V. Ramachandra et al 2019Visualisation of impacts due to the proposed developmental projects in the ecologically fragile regions- Kodagu district, Karnataka Progress *in Disaster Science***3**

[12] Revenue DepartmentSEEKING CENTRAL ASSISTANCE FOR RELIEF AND EMERGENT WORKS DUE TO FLOOD AND LANDSLIDES IN MALNAD AND COASTAL DISTRICTS OF KARNATAKA DURING SOUTH-WEST MONSOON 2018 Kodagu: *Government of Karnataka.*

[13] Geological Survey of IndiaLandslides in Kodagu and western Ghats2018, 2019*GSI, Karnataka.*

[14] District Disaster Management authority, KodaguDisaster management report2019 Government of Karnataka

[15] Varsha Hassan Vishwanath, Brian Tomaszewski 2018 Flood Hazard, Vulnerability and Risk Assessments for Uttarakhand State in IndiaWiPe Paper 2018 *Geospatial Technologies and Geographic Information Science for Crisis Management (GIS) Proceedings of the 15th ISCRAM Conference – Rochester, NY, USA May 2018 Kees Boersma and Brian Tomaszewski, eds.*

[16] Danumah J. H., Odai, S. N., Saley, B. M. et al.2016*Geoenvironment Disasters,* **3**10

[17] Kazakis, Nerantzis, Kougias, Ioannis, Patsialis, Thomas December 2015Assessment of Flood Hazard Areas at a Regional Scale Using an Index-Based Approach and Analytical Hierarchy Process*Application in Rhodope – Evros Region, Greece, Science of The Total Environment,* 538, 555 – 563.

[18] Chen Y, Khan S, Paydar Z 2010A fuzzy GIS-based spatial multi-criteria evaluation framework for irrigated agriculture*Irrig Drain***59(2)**174–188.

[19] Feizizadeh B, Roodposhti MS, Jankowski P, Blaschke T 2014A GIS-based extended fuzzy multi-criteria evaluation for landslide susceptibility mapping*Computer Geoscience***73**208–221

[20] JOUR, Talukdar, Swapan, et al 2021Flood susceptibility modeling using advanced ensemble machine learning models*Geoscience Frontiers***12**

[21] M. Rogger1, M. Agnoletti, A. Alaoui, J. C. Bathurst, G. Bodner, M. Borga, V. Chaplot, F. Gallart, G. Glatzel, J. Hall1, J. Holden, L. Holko, R. Horn, A. Kiss1, S. Kohnova, G. Leitinger, B. Lennartz, J. Parajka1, R. Perdigao, S. Peth, L. Plavcova, J. N. Quinton, M. Robinson, J. L. Salinas1, A. Santoro, J. Szolgay, S. Tron, J. J. H. van den Akker, A. Viglione, and G. BloschlLand use change impact on floods at catchment scale*AGU publications*

[22] Janardhana, Vinutha, D.N 2007Landslides in Coorg District of Karnataka State *IJIRSET***10.15680**

[23] Palchaudhuri, MoumitaBiswas, SujataApplication 2016Application of AHP with GIS in drought risk assessment for Puruliya district, India*Natural Hazards***84JO**

[24] Pierluigi De Rosa 2019Stream Power Determination in GIS: An Index to Evaluate the Most 'Sensitive 'Points of a River *MDPI*

[25] Ramasamy, SM Gunasekaran2020*Geomorphology* and Landslide Proneness of Kerala, India A Geospatial StudyJournal-*Landslides***18**

[26] Sonali Bhandari 2014A comparative analysis between the Extreme flood events that wreaked havoc in Jammu and Kashmir in September 2014 and the unusual confluence of monsoons that triggered mammoth-scale destruction in Uttarakhand in June 2013, *15th Esri India User Conference 2014*

Advancement of Environmental Concerns Through Cli-Fi: James Bradley's *Clade* as a Prototype

Dr Anil Manjhi[1]

[1]Department of Basic Sciences & Humanities, Govt. Engineering College Raipur, India

E-mail: manjhi.anil3@gmail.com

Abstract. Climate Change due to anthropogenic factors and its repercussions in the form of disastrous environmental calamities is the burning issue in the current scenario. People of different fields are trying every possible means to bring awakening in the populace. The field of literature is also making contributions to this noble cause. Cli-fi or Climate Change Fiction is the newfangled branch of Ecocriticism from the field of literature, which is entirely dedicated to dealing with and bringing up climate change problems, global warming and likewise. The impact of this genre of literature has brought up the positive impact on the wider range of the population in simple language as compared to that of the technical language. Thus, this research study aims to bring forth the advancement of environmental concern employing Cli-fi or Climate Change Fiction by analysing James Bradley's work *Clade* as a prototype and inclusively quoting other key literary works in this genre and their primary contributions made in this field of operation. The narrative technique with vivid descriptions, the mapping of the psychological awakening of emotions towards environmental concern and the portrayal of environmental calamities will pave the way and establish the advanced approach of literature for environmental concern and climate change.

Keywords: Climate change, Cli-fi, ecocriticism, environmental concerns, Clade

1. Introduction

On September 23, 2019, Greta Thunberg; a young girl from Sweden roared in the UN Climate Action Summit organized in the city of New York against the Government and non-government entities for

> 'Stealing her hopes and childhood with mere false words leading to suffering of the people and death of the people. The entire ecosystem is collapsing. We are on the point of annihilation, and all you can speak about is wealth and unending economic expansion', 'How dare you!' [1]

Baffled over the non-fulfilment of the promise of reducing the carbon emissions by the government and non-government entities, she emphasized to recall their commitment and highlighted the purview of the aftermath which is approaching soon. Ever since the attention of the global population surprisingly again focused on the existential conundrum of Climate Change, to which Mr. Guterres; the UN Secretary-General supported exclusively by reaffirming it to be of anthropogenic nature by articulating 'We are putting life itself at jeopardy if we do not immediately alter our way of life' [1].

This saga of 'to do or not to do' is not new. The concern for climate change dates back to 1988 when the Inter-governmental Panel on Climate Change (IPCC) was formalized after facing the ramifications of greenhouse gases [2]. The Kyoto Protocol was adopted on December 11, 1997, and it mandated the UNFCCC to restrict and reduce GHG (Greenhouse Gas) emissions in accordance with agreed-upon targets. Main thrust was given in The Paris Agreement was signed at COP 21 on December 12, 2015, with the intention of achieving global warming below 2°C, optimally 1.5°C, compared to pre-industrial levels. Discussing Climate the first question arises what does one mean by this terminology? According to United Nations, 'Climate change refers to the long-term temperature variations and changes in weather patterns...due to the burning of fossil fuels such as coal, gas and oil' [3]. They are defined as 'changes in the atmosphere, as well as interactions between the atmosphere and several other chemical, biological, geological, and geographical variables within the Earth system, generated by periodic variations in planet Earth's climate' [4]. We, the humans are the primary culprit behind these major alterations resulting in Global warming, massive pollution, species extermination and biodiversity erosion, hazardous methane emissions, thawing icebergs, flooding, severe droughts, wildfires, and forest degradation [5]. So Climate Change has caught the eyes of researchers, political leaders and writers as well.

2. Ecocriticism & Cli-Fi

The field of literature as being the counterpart of the processes and events happening in the Society didn't remain untouched by the fire of environmental studies. It grew a new branch – Ecocriticism: 'an investigation into the connection between literature and the physical environment' [6]. Although William Rueckert came up with the term in his *Literature and Ecology: An Experiment in Ecocriticism*, but Cheryll Glotfelty is accredited for ecocriticism. This ecocritical branch gave birth to a new shoot specially dedicated to the Nobel cause of Climate change named 'Climate Change Fiction' in short popular as 'Cli-Fi'. This term was introduced by Dan Bloom; a journalist. According to him, 'Cli-fi' is a new name for novels, short stories, and films dealing with climate change and global warming issues: 'Cli' is from the word 'climate', while 'fi' is from the word 'fiction' forming 'Cli-fi' [7].

There are two ecocritical perspectives to Cli-fi literature: one is to conceive of ways to deal with climate change, whereas the latter is concerned with understanding climate change as a cultural phenomenon [8]. Translations of many works around the world like Iceland, Spain, Netherlands, Norway, Germany and Finland have contributed to the expansion of the Cli-Fi, which efficiently illustrates the ecological, societal and psychological effects. Some of the great works in this series are *Clade* by James Bradley (2015), *The Island Will Sink* (2016) by Briohny Doyle, Cat Spark's *Lotus Blue* (2017) Cli-fi is the fiction which depicts utter annihilation of the future and portrays dual perspectives: to begin it exhibits the societal, economical, and ecological life deteriorating, and the next exhibits the potential to withstand them and acclimatize to the new environment [9]. The goal of the Cli-fi is to assist the readers to comprehend the effects of the changing climate on individuals and communities by providing real information that will help them better understand what will happen in the future. As it reflects humanity's difficult relationship with the earth, Climate change fiction must address the human-caused component of global warming. It also discusses the role of humans in dealing with this unusual calamity [10]. Yet ultimately it teaches and entertains simultaneously.

3. James Bradley's *Clade*: As a Prototype

James Bradley, a novelist and critic of Australian origin has devoted his major concern to climate change. His *Clade* (2017) is analysed on the scale of Cli-fi as a prototype that depicts the story from the present to the future. It tells the narrative of Adam Leith, a scientist of climatology, Adam's friends and family members who are stuck in a world ravaged by the changing climate, including coastal floods, general flooding, wildfires, droughts, massive devastation, and worldwide pestilence. The title of the novel *Clade* is derived from *Klados* which signifies a faction of animals and plants that have a shared ancestor and all that descends from it. In this novel the common ancestor of his family and friend is

Adam. The novel resembles honeycomb structuralism as it exactly has ten chapters like ten compartments, which gives it an ecological linking. Further, the root of the climate change issue is defined by Adam's coworkers: 'We don't change because we don't believe in the problem, at least not at the deep, intuitive level we need to,' he would add. 'When it's right in front of us, we can see it and understand what it implies; we know we have to change. But as soon as we're away from it our old thinking reasserts itself, our desire to reproduce, to build power' [11].

3.1. Vivid Depictions

Bradley powerfully shows the current and future effects of climate change. When Adam goes to see his daughter, Summer, he tells her about the flood that has swept England:

> 'At first, all that is visible is the rising water, the refuse floating by. But then he hears a rushing noise, accompanied by creaking and grinding like the wind. And then, at the street's end, he sees it. It seems so improbable it's difficult to be afraid, for it is as if the water is pouring towards them in a sloping wall, a liquid hill that moves faster than any of them could run, a wave that does not end but comes and comes and comes' [12].

In a chat between Amir, a Dhaka-based physician and beekeeper, and Ellie, Adam's former better half, Bradley brings up the topic of crumbling of the honeybee colony and states that 'None of the other collapses has been comparable. It started in Europe and expanded in less than a year to America and Asia. We were safe for a while, but now the colonies here are also withering' [11]. Bradley also addresses the issue of population displacement. Amir's residence in Australia is unlawful; he and his wife and children were moved to camps after Bangladesh's government fell, but he fled following the demise of his spouse and daughter. Ellie is his narrator, and he tells her his narrative as she spent a long time in the camps and managed to evade one day. She had no idea where to go. She was 'heartbroken, depressed, and almost catatonic' [11]. Then Bradley describes a pandemic that starts in China and is triggered by the Acute Viral Respiratory Syndrome (AVRS) virus by stating that the media blackout in China remained in force, but the images that were leaking out were utterly horrifying consisting of overcrowded hospitals, corpses in the streets. This condition continues all over Asia, as well as America and Europe [11]. Furthermore, many species declined Birds/Bananas/Tigers/Frogs/Bees/Coffee/Polar bears/Coral and many got saved: Seeds/Elephants/Dolphins/Each other [11]. Climate change's terrible implications are accurately depicted.

3.2. Surfacing of Emotions

Climate Change triggers the emotional aspect of the characters in the novel *Clade*. This emotional surfacing encircles the key point that the Reader-Response Connection ties the audience to the central concern of the effect of climate change. The emotions of the characters are being felt by the readers and it creates a cognitive effect on the minds of the readers. This emergence of psychological feelings in the context of the characters in the novel lay a Catharatic effect leading to purgation and moral awakening towards Environment and nature. Several instances support the psychological emergence of emotions. Summer seems to be upset. She and Noah are found moving to England after leaving Australia. She appears to be uneasy and worried in a chat with Adam as a result of the impacts of the changing climate. When Adam inquires as to whether they have suffered any serious losses as a result of the flood. Her answer is, 'Stuff? Some. But people died, Dad. A lot of them. Or had you forgotten that?' [11]

Noah also has autism. He's always worried concerning the climatic conditions. According to Summer, matters of the flood, water and storm, stresses Noah out and makes him worse. He becomes quite agitated and 'cannot stop talking about it' [11]. After the flood, Adam, Summer, as well as Noah are confronted with the following terrifying images. They stumble upon the bodies in the wreckage on several occasions. A guy is stuck against a wall by an automobile, a lady is down in a pit, and a girl is entangled in a toppled tree. They initially tried that Noah shouldn't see all these gruesome sights, but later on, confronting the next corpse, they abandoned the idea of covering his eyes as it would be in vain [11]. Even Bradley acknowledges his desire to create emotional upheaval, stating that he sought to express a sense of loss associated with the extinction of landscapes, species, and possibilities of mankind.

3.3. The Portrayal of Environmental Calamities

The Environmental calamities have been depicted in their cruellest form with their resulting aftermath creating a deep reflective concern on the minds of their reader. As Bradley puts it,

> 'The rainfall that tends to arrive in July or August certainly didn't come, leaving the subcontinent in severely hot weather. The crop crisis resulted in food shortages and it led to severe hunger. Then, in November, horrific rains and floods killed over a million people and forced another 100 million to flee from their homes. Ultimately, the economy crumbled in the sequel, leading to severe joblessness, which has recently ignited rioting in Calcutta and Mumbai' [11].

He is able to sketch the climate-related generational transitions. Within four generations, the narrative universe shifts from a familiar location to one in which eco-collapse which has harmed both biological and adoptive generations descended from scientist, Adam. The story begins in Antarctica, where he is concerned about his artist wife, Ellie, and their in-vitro fertilisation procedure. Bradley wisely employs the circumstances with which the 21st-century readers are acquainted. The climate becomes a global level intimidating component in every topography: the absent monsoon, the melting of glaciers at Antarctica, the increase in asthmatic attacks, the abrupt deaths of birds and fish, the unsuccessful international climate talks and so on [11]. It generates contemporary apprehensions in the psyche of the readers, linking them to the real world resulting in didactic outcomes [13].

3.4. The Ramified Beauty in Destructed Environment

The scenarios of environmental degradations are portrayed in the sense of denial-acceptance. A literary and psychological style of depicting human nature of not taking seriously climate change and its after-effects is by shedding off themselves of the bitter reality of the havoc which was fruition of their deeds. The characters tend to find beauty in changes caused by the destruction. Maddie, Ellie's step mum, sees elegance in the wrecked seashore, despite the fact that the high water has dragged away constantly most of the portion of the sea beach. She believes that the beach's obliteration contains something exquisite [11]. Summer also mentions the beauty of the stars after the storm that rips England apart, stating, 'They are incredibly bright,' on which his father (Adam) affirms, 'With the power off, there are no lights to interfere with them' [11].

The cacophony seems amazing to Li Lijuan, Noah's nurse's daughter. She articulates that things were moving in the foliage, birds shouting and singing. Even the light had changed, becoming thicker and more oozing with smoke and colour. It was incredible [11]. This controversial beautification of the ruins of the wrath of climate change, puts one to ponder, really is it possible to find light in darkness or life in death? The contrast created by Bradley reveals the true human nature of being more selfish and less optimistic as instead of curating the anthropogenic measures, humans try to find out their profits. Akin to those steps of the developed countries in context to carbon emission reductions i.e. hearing everything but doing nothing.

4. Conclusion

Humans urged for development to sustain the continuity of their species but they never thought of the repercussions. Industrial development brought climatic developments but in a negative way. The periodicity of natural events changed, nature changed, the atmosphere changed, and changed the climate. What has not changed is the long sleep of ignorance of human beings. Thus, writers of literature took this noble aim of breaking this prolonged sleep by writing Cli-Fi to enlighten the audience on the intricacies of climate change and the disastrous outcomes and finally alarms the reader by giving an imaginary glimpse into the upcoming future. Bradley masters this aim as well as this genre and his *Clade* becomes a masterpiece of this forte. We have everything in it. The byzantine relationship of the characters with the environment, the physical and geological dilemmas, the emotional hullaballoo emerging out of the havoc caused by climate change and enjoyment of beauty in destruction and aftermath are such features that draw a cautionary note for the public: Beware! and establishes Cli-Fi as the literary tool that advances environmental concerns. Cli-Fi reiterates the words of Ms Greta Thunberg – 'We (Environment) will be watching you (human)' of the anthropogenic injustices caused on us by you and avenge by all means.

References

[1] Anon 2019 Greta Thunberg tells world leaders 'you are failing us', as nations announce fresh climate action *UN News*
[2] IPCC and WMO 1992 *Climate change: the 1990 and 1992 IPCC assessments, IPCC first assessment report overview and policymaker summaries and 1992 IPPC supplement* (Geneve: IPCC)
[3] Nations U What Is Climate Change? *U. N.*
[4] Anon climate change | Definition, Causes, Effects, & Facts | Britannica
[5] Jordan D J 2014 Climate Change Narratives in Australian Fiction.pdf
[6] Glotfelty C and Fromm H 1996 *The Ecocriticism Reader: Landmarks in Literary Ecology* (University of Georgia Press)
[7] Anon INTERVIEW: Dan Bloom on CliFi and Imagining the Cities of the Future | Smart Cities Dive
[8] Johns-Putra A 2016 Climate change in literature and literary studies: From cli-fi, climate change theater and ecopoetry to ecocriticism and climate change criticism *WIREs Clim. Change* 7 266–82
[9] Crownshaw R 2017 Climate Change Fiction and the Future of Memory: Speculating on Nathaniel Rich's Odds against Tomorrow *Resil. J. Environ. Humanit.* **4** 127
[10] Mehnert A 2016 Climate Change Fictions in Context: Socio-Politics, Environmental Discourse and Literature *Climate Change Fictions: Representations of Global Warming in American Literature Literatures, Cultures, and the Environment* ed A Mehnert (Cham: Springer International Publishing) pp 21–52
[11] Bradley J 2015 *Clade* (Australia: Penguin Books)
[12] Khalaf A M R 2020 Aesthetic Consolation: A Possible Remedy for Solastalgia in James Bradley's Clade (2017) *Engl. Lang. Lit. Stud.* **10** 54
[13] Lovell S, Thomas B and Wickham O 2019 Neither fish nor fowl: Travelling across genres and disciplines through 21st century Australian Cli-Fi *TEXT*

Coastal Vulnerability Assessment: A case study of the Ratnagiri coast, Maharashtra, India

A B Yadav[1], P C Mohanty[2] and A Singh[1]

[1]Center for Wireless Networks & Applications (WNA), Amrita Vishwa Vidyapeetham, Amritapuri, India

[2]Indian National Centre for Ocean Information Services (INCOIS), Ministry of Earth Sciences, Hyderabad 500 090, India

Corresponding author e-mail: yadavaparna896@gmail.com

Abstract. The coastal zone is a vulnerable habitat that needs extra caution to protect ecosystems. Coastal systems are increasingly threatened by possible climate change consequences, as evidenced by the Intergovernmental Panel on Climate Change's consecutive assessments . Increased tropical storm occurrences in recent years, in addition to the devastation caused by the tsunami in December 2004, have highlighted the necessity of analyzing the coast's susceptibility to flooding-induced hazards to get a better knowledge of the factors that generate various hazards and, as a result, reduce the after-effects of future occurrences. The Ratnagiri coast in Maharashtra is prone to erosional hazards, periodic land rehabilitation, and sudden rises in sea level. The main objective is to calculate the CVI (Coastal Vulnerability Index) for the Ratnagiri coast. To analyze the vulnerability of the coastal region, eight risk parameters were used, viz. shoreline change rate, coastal elevation, sea level change rate, coastal slope, tide range, significant wave height, coastal geomorphology, and tsunami arrival height. The coastal vulnerability map was created by categorizing the numerous coastal portions into three vulnerability groups: high, medium, and low, which will aid coastal residents in risk mitigation in the future.

Keywords: Coastal Vulnerability Index (CVI), DSAS, Ratnagiri, Coastal hazards.

1. Introduction

A vulnerability assessment is the procedure of finding, assessing, and monitoring risk. The coastal zone is a sensitive habitat that necessitates extra caution to protect ecosystems and human activities. As a result of various human activities, several coastal regions are experiencing severe environmental issues like coastal flooding, dune deterioration, and coastal erosion [1]. Inundation can worsen coastal erosion by moving submerged materials offshore, as well as prolong the impacts of coastal flooding by allowing storm waves to act. Increased tropical storm occurrences in recent years, as well as the terrible impact of the December 2004 tsunami. Cyclone Phyan (2009) caused massive damage to the property of coastal districts of Maharashtra. Also, since 1891, Cyclone Nisarga (2020) [2] has been the most severe and powerful tropical cyclone to hit the Maharashtra coast. These events have highlighted the necessity of analyzing the coast's susceptibility [3]. The Ratnagiri coastline, which stretches across India's west-central region, possibly vulnerable to rapid erosion. Most of the sites along the coast are already under threat from repeated storm floods and severe coastal erosion [4], [5].

CVI is the indicator of all the physical and geological risk factors in the coastal zone region (oceanic and landward hazard parameters). It synthesizes with individual parameters and creates a single Vulnerability Index, which leads to easy interpretation and identification of coastal risk zones along the coastline [6],[7]. It can help coastal managers, researchers, administrators, and industries for future infrastructure or other developments along the coastal area. Thieler and Hammar Klose's (2000) [8] approach is used to calculate CVI using geographical information systems (GIS). In India, a few attempts have been made to map the vulnerable areas along the Orissa coast [9], Ratnagiri coast [10], Andra Pradesh coast [11], and Mangalore coast [12]. The main aim of this work is to identify and prioritize the places on the Ratnagiri coast that are most susceptible to the adverse consequences of coastal hazards.

2. Study Area

The study was conducted along the coastline of Ratnagiri district, on the Arabian Sea coast of Maharashtra, India, which lies between 16.30°N - 18.0°N and 73.02°E - 73.53°E as shown in figure 1. The shoreline of the study area is about 276 km, which is distributed along five talukas, namely Ratnagiri, Dapoli, Guhagar, Rajapur, and Mandangad.

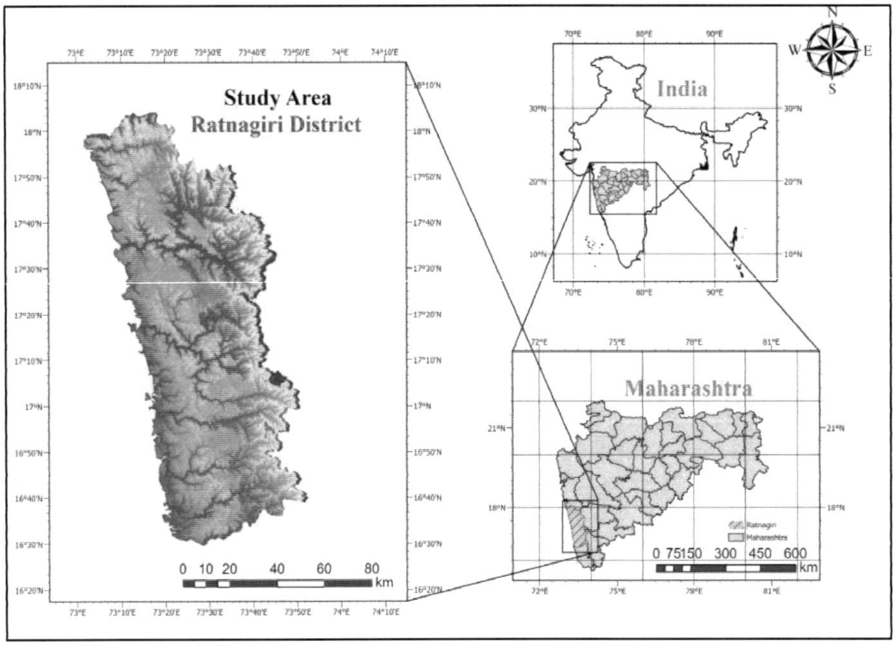

Figure 1. Study Area Map

3. Materials and Methodology

3.1. Data Collection

The majority of these parameters are dynamic along the coast and necessitate a large amount of data from various sources (Table 1). They are based upon data from geographic information systems, remote sensing, and numerical models.

Table 1. Data-sets and their respective sources.

Sr. No	Datasets	Sources
1.	Shoreline Change rate	USGS Earth Explorer, Landsat 7 and Landsat 8
2.	Sea level change rate	AVISO Altimetry
3.	Coastal slope	GEBCO (General Bathymetric Chart of Oceans)
4.	Coastal regional elevation	SRTM DEM
5.	Tidal range	WXTide Software
6.	Coastal Geomorphology	Geological Survey of India (GSI) Bhukosh web service
7.	Significant wave height	INCOIS
8.	Tsunami arrival height	INCOIS

A Landsat picture (path147-row 48) encompasses the whole Ratnagiri coastline, and data were gathered from ETM+ and OLI/TIRS sensors with 30 m resolution from 2001 to 2021. While downloading data, tidal levels were taken into consideration. To eliminate the influence of the tide, all satellite images taken for this study correspond to approximately equal tide levels (±0.5).

3.2. Methodology

The methodology adopted in this study is based on Gornitz's [13] Coastal Vulnerability Index and the work of Thieler and Hammer-Klose (2000) [8] and Doukakis (2005) [14]. Vulnerability is an internal risk factor that relates to a subject's or system's intrinsic predisposition to be impacted or prone to damage when exposed to a hazard [11]. As a result, eight geophysical parameters were used to compute the CVI: shoreline change rate, coastal slope, coastal regional elevation, sea-level change rate, tidal range, significant wave height, coastal geomorphology, and tsunami arrival height. The factors were categorized into three categories based on their risk level: low, moderate, or high. The flow chart (figure 2) below depicts the complete methodology used in the calculation of CVI.

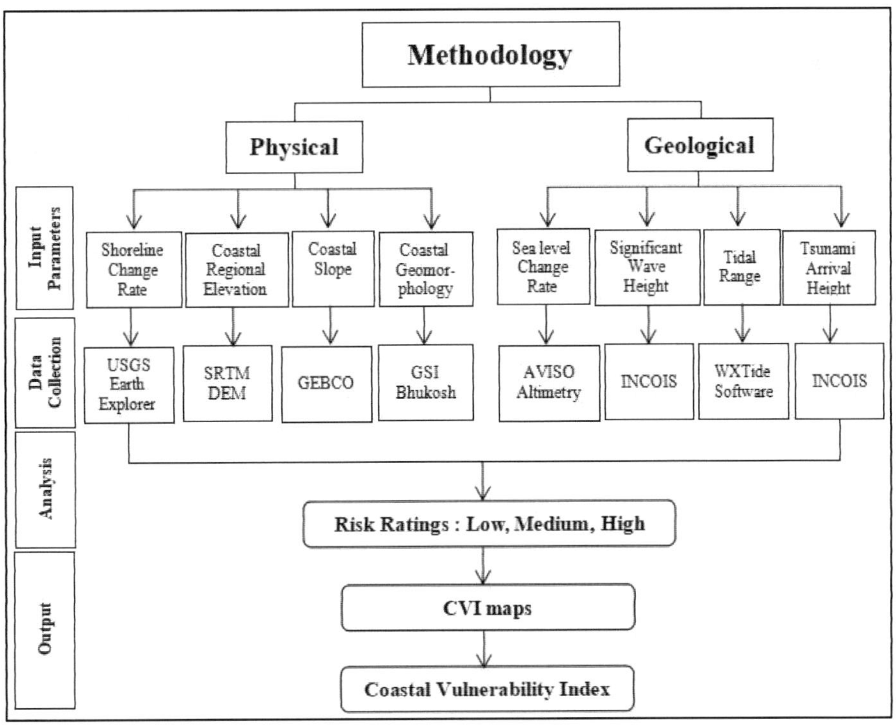

Figure 2. Methodology flowchart

3.2.1. Shoreline Change Rate. The shoreline change rate is obtained using the Digital Shoreline Analysis System (DSAS) toolbar, which is an extension of ArcGIS. The Landsat images obtained from the USGS Earth Explorer were projected to UTM WGS-1984 zone 43N. The extension includes three main aspects. The first step is to define the baseline, then to create transects at precise intervals along the coastline, and finally to determine the change rate using various statistical approaches. Shorelines were digitized manually using ArcGIS 10.5 for each year. The latest digitized shoreline was used to establish a baseline using a 100 m buffer. More than 2000 transects were created using a baseline at a 100 m interval. Each transect has a unique ID. Transects emerging from the baseline intersect all coastlines. Finally, the linear regression method was used to estimate the shoreline change rate throughout the entire coast of Ratnagiri. DSAS determines whether the shoreline change involves accretion or erosion, which depends upon the locations of coastlines with respect to the baseline. DSAS calculates the rate in numerical form. It's erosion if it has a negative value or accretion if it has a positive value.

3.2.2. Coastal Slope. The ratio of the change in height to the lateral distance measured between any two points along the shore is referred as coastal slope. The coastal slope is an important component in determining how vulnerable a coastal region is to inundation from severe storms and tsunamis [15]. A line's slope represents the steepness. The slope was calculated using GEBCO data with one-minute grid resolution. The study area's slope values are determined in degrees, and risk ratings are given. The slope values are computed in degrees, and risk ratings are assigned.

3.2.3. Coastal Regional Elevation. It is the average elevation of any region above the mean sea-level. It is crucial to study this parameter in detail to determine and quantify the amount of land affected by future rising sea levels. High-elevation coastal regions will be classified as less vulnerable areas, whereas low-elevation areas will be highly vulnerable. The Shuttle Radar Topography Mission (SRTM) data with a resolution of 30 m is used to compute elevation values.

3.2.4. Sea-Level Change Rate. Sea level changes are differences in mean sea level that are estimated by tide gauging stations. According to the IPCC, every millimeter increase in mean sea level results in a 1 m recession of the shoreline. According to Unnikrishnan et al. (2007) [16], there are two types one is global (mean sea level change), while the second is local sea level rise. As a result, sea-level rise is a significant phenomenon that, by combining global and localized processes, influences climate change-induced coastal hazards.The SLR data was acquired from AVISO altimetry data. Sea level anomaly (SLA) time-series data from 1993 to 2020 was obtained. The graph was plotted between SLA and the year and the slope was calculated in mm/yr.

3.2.5. Coastal Geomorphology. The scientific study of landforms is referred to as geomorphology. Geo implies "earth," morpho denotes "shape," and Logy means "discourse" or "science." The morphological structure of the coastal area is essential in evaluating the consequences of susceptibility. Coastal geomorphology studies give a fundamental knowledge of the coastal region. Coastal geomorphology was obtained using 1:250 K geomorphological maps from the Geological Survey of India (GSI) Bhukosh web services. The regions obtained in the study area were highly dissected plateaus, moderately dissected plateaus, a pediment pediplain complex, and coastal plains.

3.2.6. Tidal Range. The height difference between high and low tide is referred to as the tidal range. It is compelled by the moon's and sun's gravitational pull. One year of data for 2021 was acquired from the Windows Unix Programme for Tides (WXTide 32) software, which represents projections about the tidal range for tide stations around the study area. From the standpoint of vulnerability, there is an evident tendency to categorize coastal regions with a higher tidal range as extremely susceptible, while coastal regions with a lower tidal range are deemed less susceptible. The maximum and minimum tide amplitudes were calculated for the Rajapur river entry and Bassein.

3.2.7. Significant Wave Height (SWH). Significant wave height is used to assess the susceptibility of shorelines as an alternate option to wave energy. An increase in wave height results in land loss due to increased erosion and flooding along the coast. The higher the wave height, the more the region is vulnerable and vice-versa. The maximum value of significant wave height data was extracted from wave watch III at 0.5-degree spatial resolution during the 2018 to 2021 period.

3.2.8. Tsunami Arrival Height (TAH). The wave characteristics are affected by seismic source factors, bathymetry, seashore morphology, and coastal land elevation. These surges produce an inundation of seawater into land for up to 1 km or more, leading to damage to infrastructure and human life. The maximum value of tsunami wave height is simulated from all the historical tsunami-generated focal mechanism parameters for your academic project work. A risk rating will be provided for the entire study region based on the run-up values.

Table 2. Criteria for Risk Assessment

Datasets	Risk Ratings		
	Low (1)	**Moderate (2)**	**High (3)**
Shoreline Change Rate (m/y)	> 1	1 to -1	< -1
Coastal Slope (degree	>1	0.2 to 1	≤ 0.2
Coastal regional elevation (m)	> 6	3 to 6	≤ 3
Sea Level Change Rate (mm/yr)	-	0.64-0.68	-
Significant wave height (m)	-	0.47-0.74	-
Tidal range (m)	< 2	2 to 3.5	> 3.5
Tsunami arrival height (m)	< 2.3	2.3 to 3	> 3
Coastal Geomorphology	Highly dissected plateaus	Moderately dissected plateaus	Pediment pediplain complex and coastal plain

3.2.9. Coastal Vulnerability Index (CVI). It is computed by integrating different risk variables into a single index, as shown in equation (1). Each risk parameter is classified into one of three risk categories. A rating of 3 is given if the risk is severe, a rating of 2 is assigned for moderate risk, and a rating of 1 is given for low risk for a certain shoreline. Risk ratings are provided for each of the variables listed above. Table 2 shows the risk ratings that have been assigned to each variable. The CVI is determined by dividing the square root of the ranking variable by the total number of variables [17]. The CVI is calculated using the field calculator in ArcGIS based on the risk value provided as input parameters. The CVI value was determined for several stretches of the shoreline based upon its percentile value. If the value is less than the 25th percentile, it is considered low risk; if it is between the 25th and 50th percentile, it is considered moderate risk; and if it is more than the 50th percentile, it is considered high risk.

$$\mathbf{CVI} = \sqrt{[(\boldsymbol{P} * \boldsymbol{Q} * \boldsymbol{R} * \boldsymbol{S} * \boldsymbol{T} * \boldsymbol{U} * \boldsymbol{V} * \boldsymbol{W})/\mathbf{8}]} \quad (1)$$

where, P = Shoreline change rate, Q = Coastal slope, R = Coastal regional elevation, S = Coastal geomorphology, T = Sea level change rate, U = Significant wave height, V = Tidal range, W = Tsunami Arrival Height.

4. Result and Discussion

The current study's findings suggest a high risk of vulnerability along the Ratnagiri coast's low-lying sections and a low risk along rocky beaches for all CVI criteria. The assessment and ranking are discussed below.

4.1. Shoreline Change Rate

The Ratnagiri Coast's change rate was analyzed using a DSAS technique. The shoreline change analysis for the entire Ratnagiri coast for the study duration shows that around 48 km (18%) of coastline falls into a high-risk category, with an erosion rate of 5.32 m/yr and around 150 km (54%) remaining steady. Around 78 km (28%) of coastline recorded accretion falls under the low-risk category, with an accretion rate of 2.43 m/yr (figure 3a).

The research also identified high-risk vulnerable places along the beaches, creeks, and estuaries of Kelshi, Jaigad, Gavakhandi, Bhatye, Guhagar, Jaitapur, and Harnai, whereas low-risk zones such as cliffs and shoals are geologically exposed. As a result, the pace of change along the Ratnagiri coast indicates a moderate risk of coastal vulnerability.

4.2. Coastal Slope

According to the findings, approximately 222 km of coastline have low-risk ratings with a coastal slope greater than 1.0°, 52 km of coastline have moderate risk ratings, with values between 0.2° and 1.0°, and the remaining 3 km of coastline have low-risk ratings, with a slope less than 0.2° (figure 3b).

4.3. Coastal Regional Elevation

According to the current study, 178 km of coastline have high-risk ratings, with coastal elevations ranging from 0 to 3 m. A medium-risk rating of between 3 and 6 m has been observed along a 27-km stretch of shoreline. A total of 70 km of shoreline has low-risk ratings, with elevations exceeding 6 m (figure 3c).

4.4. Coastal Geomorphology

The present study shows 213 km of the study region is made up of coastal plain and pediment pediplain complexes with high-risk ratings. A total of 62 km of coastline has a medium risk rating, having moderately dissected plateaus. About 1 km of shoreline has a low-risk rating due to highly dissected plateaus (figure 3d).

4.5. Significant Wave Height

The mean significant wave height was found to be between 0.47 - 0.74 m in the current study. The entire coastline is categorized as moderately vulnerable (figure 4a).

4.6. Tidal Range

The current study shows 141 km of shoreline stretch have a moderate risk of tide ranges greater than 2 m. With a tide range of less than 2 m, about 135 km of coastline has low-risk ratings (figure 4b).

4.7. Tsunami Arrival Height

According to the current study, around 4 km of shoreline has a high-risk rating of greater than 3 m. A total of 9 km of shoreline stretch has a moderate risk of value ranging from 2.3 to 3.0 m. A total of 264 km of coastline has low-risk ratings ranging from 0 to 2.3 m (figure 4c).

4.8. Sea Level Change Rate

According to the current study, the 276 km of Ratnagiri coastline is at moderate risk, with sea-level change rates ranging from 0.64 to 0.67 mm/yr (figure 4d).

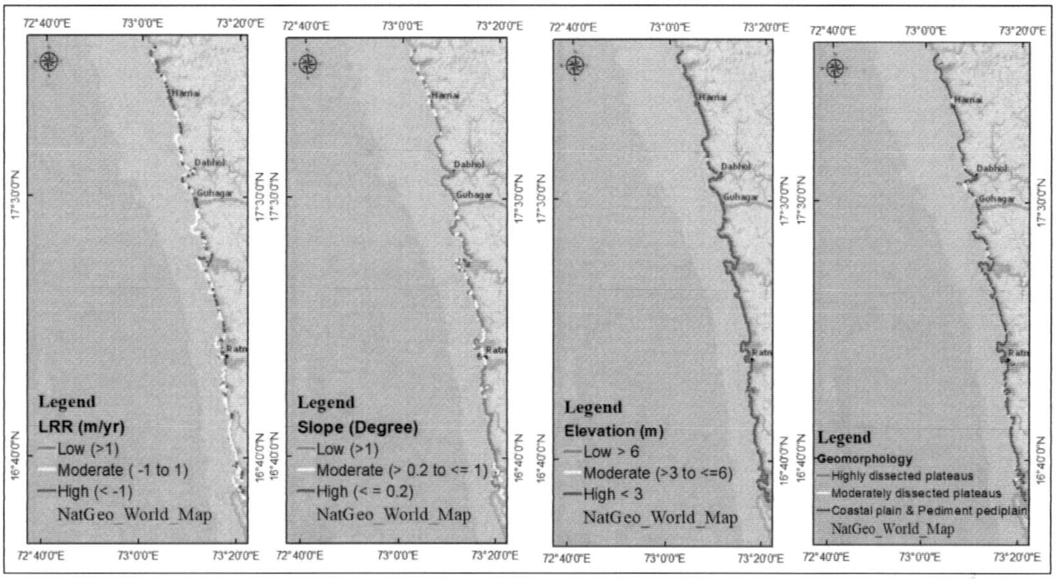

(a) (b) (c) (d)

Figure 3. (a) Risk categories for the Shoreline change rate (LRR), (b) Risk categories for the Coastal Slope, (c) Risk categories for the Coastal Regional Elevation & (d) Risk categories for the Geomorphology

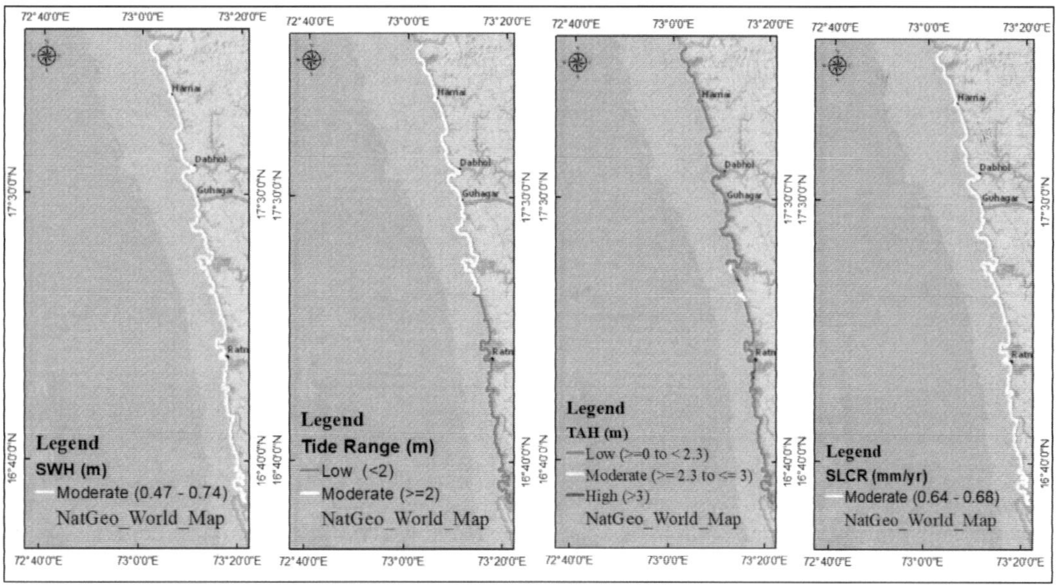

(a) (b) (c) (d)

Figure 4. (a) Risk categories for the significant wave height (SWH), (b) Risk categories for the Tide Range, (c) Risk categories for the Tsunami Arrival Height (TAH) & (d) Risk categories for the Sea level change rate (SLCR)

4.9. Coastal Vulnerability Index

The coastal lengths of Ratnagiri are categorized from low to high based on the vulnerability of risk parameters. The CVI is computed, and the vulnerable regions along the coastline are highlighted on the map (figure 5). The CVI ranged from 1 to 9 throughout the Ratnagiri coastline. The CVI values are 2.5 and 5, respectively, in the 25th and 50th percentiles. Regions of the shoreline with an index value of between 1 and 2.5 are considered less vulnerable, those between 2.5 and 5.0 are regarded as moderately vulnerable, and those with a value greater than 5.0 are considered extremely vulnerable. As a result, around 114 km of Ratnagiri's coastline stretch are less susceptible. About 143 km of Ratnagiri's shoreline are at medium risk, while 20 km are at high risk. The coastline stretch along Velas, Kelashi, Aade, Anjarle, Harnai, Guhagar, Palshet, Velneshwar, Jaigad shows a highly vulnerable area.

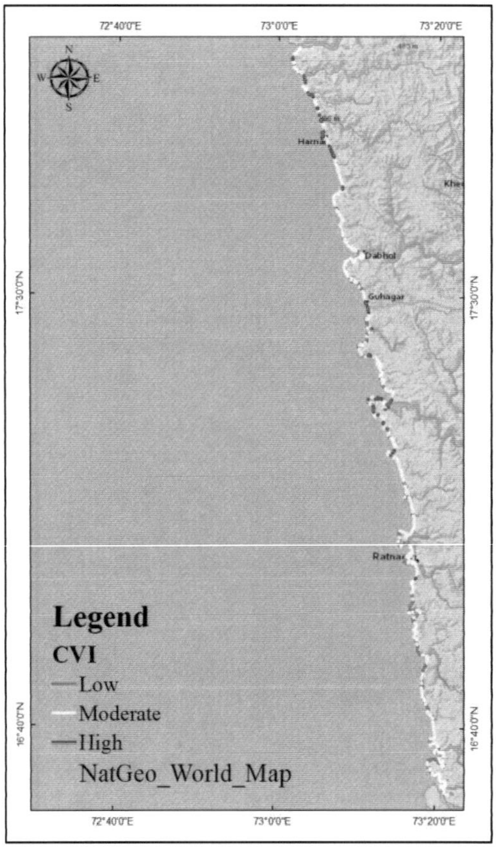

Figure 5. Risk classes for the Coastal Vulnerability Index

5. Conclusion

These coastal vulnerability maps serve as an indicator of risk to coastal geomorphic landforms along the Ratnagiri coast. The study area's segmentation aids in determining the degree of vulnerability on a regional scale. This study demonstrates comprehensive approaches to guide the adaptation process, as well as visualizations that will improve local government decision-making in response to climate change[18]. The study yielded new insights into the concept of vulnerability, which consists of three risk rating factors: low, medium, and high. Geophysical data were combined to create a single indication that can help prioritize and focus adaptation actions. The studies illustrate and propose larger implications for other groups and circumstances confronting comparable issues outside of the case study location.

IOP Conf. Series: Earth and Environmental Science 1032 (2022) 012038 doi:10.1088/1755-1315/1032/1/012038

References

[1] Satish A and Geetha P 2021 *4th International Conference on Electrical, Computer and Communication Technologies, ICECCT* **I** pp 1-7
[2] Mishra M, Kar D, Santos C A G, da Silva R M and Das P P 2022 *Marine pollution bulletin* **174** p 113262
[3] Mahendra R S, Mohanty P C, Kumar T S, Shenoi S S C and Nayak S R 2010 *Italian Journal of Remote Sensing* **42** pp 67-76
[4] Singh A, Seitz F and Schwatke C 2012 *Remote sensing of environment* **123** pp 187-195
[5] Prasad G, Rajesh R, and Arun K 2020 In *Proceedings of the International Conference on Industrial Engineering and Operations Management Dubai, UAE*
[6] Reddy D, Parthasarathy V and Rao S N 2017 In *Proceedings of the 18th ACM International Symposium on Mobile Ad Hoc Networking and Computing* pp 1-2
[7] Manasa K S S Geetha P 2022 *Lecture Notes in Networks and Systems* **314**
[8] Thieler E R and Hammar-Klose E S 2000 *National assessment of coastal vulnerability to sea-level rise; preliminary results for the US Pacific Coast* (No. 2000-178)
[9] Kumar T S, Mahendra R S, Nayak S, Radhakrishnan K and Sahu K C 2010 *Journal of Coastal Research* **26** pp 523-534
[10] Thakare L M and Shitole T A 2021 *Journal of Coastal Research* **37** pp 421-432
[11] Vivek G, Mahendra R S, Mohanty P C, Kumar T S and Nanda S 2016 *International Journal of Remote Sensing Geo-science (IJRSG)* **5** pp 1–7
[12] Hegde A V And Reju V R 2007 *Journal of Coastal Research* **23** pp 1106–1111
[13] Gornitz V M, Daniels R C, White T W and Birdwell K R 1994 *Journal of coastal research* pp 327-338
[14] Doukakis E 2005 *European Water* **11** pp 3-7
[15] Singh A, Reager T and Behrangi A *Hydrol. Earth Syst. Sci.* **25** pp 511–526
[16] Bindoff N L, Willebrand J, Artale V, Cazenave A, Gregory J M, Gulev S, Hanawa K, Le Quere C, Levitus S, Nojiri Y and Shum C K 2007 Observations: oceanic climate change and sea level
[17] Pendleton E A, Thieler E R and Williams S J 2005 *Coastal vulnerability assessment of golden gate national recreation area to sea-level rise* Woods Hole MA USA: US Geological Survey p 27
[18] Dhivvya J P, Rao S N and Simi S 2017 In *Proceedings of the 18th ACM International Symposium on Mobile Ad Hoc Networking and Computing* pp 1-2

3D Monitoring & its Significance in Tunneling: Special Emphasis on Tunnel T-74R of USBRL Project, India

A. Ahmed[1*], S. Mishra[2], A. K Singh[3], MD A. Azad[4]
[1]Deputy General Manager/Geology, ICT
[2]Associate Professor, G.D Goenka University
[3] Managing Director/ PMT Infrascience
[4] Department of Geology (CAS) University of Delhi-110007
*E-mail: aejazsz89@gmail.com

ABSTRACT

Nowadays, Tunnels are the major transportation network throughout the world. The importance of tunnel and its associated system also increases as the demand of Tunnel construction increases day by day. It has its significant importance in water conveyance networks as well as communication networks. As it is spreading fast over the world, the assessments of deformations are becoming an important part of tunneling system. The paper deals with the 3D monitoring system for the safety perspective and to make proper judgement about the condition of tunnels related to the information obtained in the measuring equipments. It also enables to react under various rock setting with the precise and reliable monitoring results. As we observed in the T-74R tunnel, where deformation of the tunnel's support system is a natural occurrence, the T-74R's main support system was also affected. The groundmass is prone to deformation due to its poor state and proximity to the principal thrust. The finding of the T-74R tunnel utilising 3D monitoring devices is important. They have provided the information about actual ground behaviour conditions.

*Keywords:*3D Monitoring, Tunneling, Measurements, Instrumentation

1.0 INTRODUCTION

The USBRL Project's Tunnel T-74R is situated in Banihal, Ramban district of Jammu & Kashmir. From 125+300 to 133+910, it runs through metamorphic rocks from the Ramsu and Machal formations in the Higher Himalayan tectonic belt, which is separated from the Lesser Himalayan tectonic belt by the Main Central Thrust, or MCT (see Fig. 1). However, because of the poor and unpredictable geological conditions, tunneling is a challenging activity in the Himalayan area [1-3]. The groundmass underwent multiple periods of deformation owing to the collision of the Indian and Eurasian plates, weakening the rocks of the sub- and lower Himalayas [4]. Construction on or through this distorted groundmass is thus difficult [5].

As a young fold mountain, neotectonic activity is significant, causing delays in building and damage during operation. So, the New Austrian Tunneling Method (NATM) is used a lot to build safe tunnels in these situations [6]. NATM is founded on the notion of tunneling by recognising the ground's reaction to an underground opening and mobilising it's self-supporting

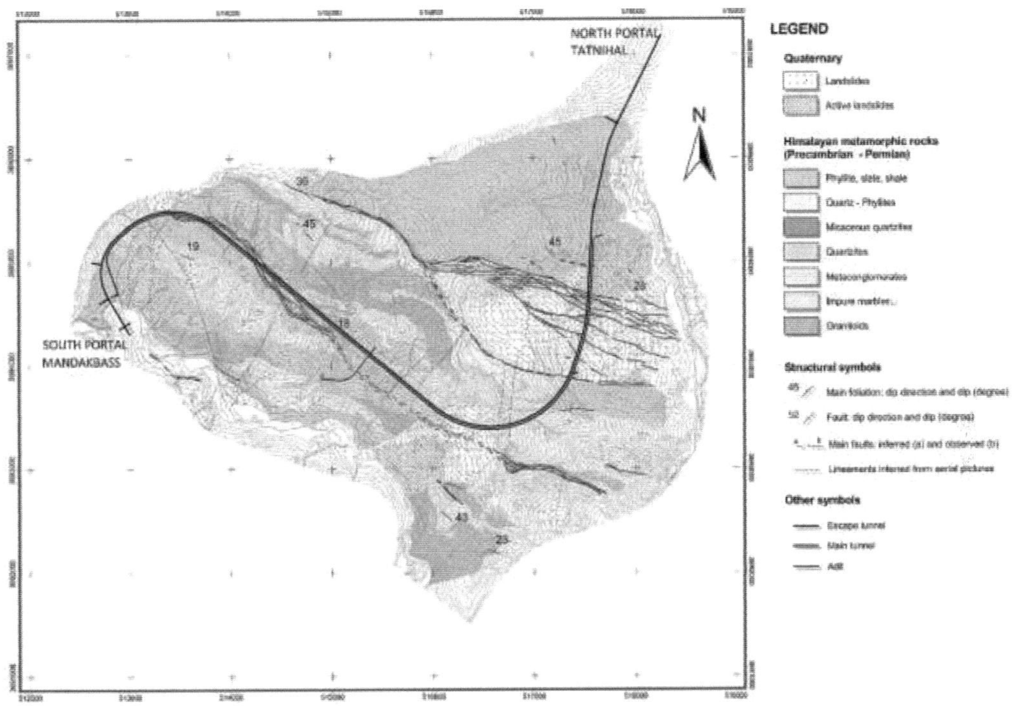

Fig. 1 Geological map of the region with the tunnel alignment superimposed

Capacity [6-8]. Creating a hole in the earth disturbs the balance of in-situ tensions. The openings surrounding ground reaches equilibrium by deforming and eventually becomes self-supporting. The final lining of tunnels cannot be done until the earth reaches equilibrium. So, to monitor the ground's activity, 3D monitoring points and geotechnical sensors are used [9]. The tunnels are designed for the predicted ground type and behaviour. However, while tunneling, particularly in the Himalayas, the encountered ground type and activity may change. The 3D monitoring system from USBRL Project Tunnel T-74R is shown here, along with its implications for understanding ground behaviour and a comparison between predicted and actual deformation.

Accidents in tunnel while under construction occur often. As a result of advances in database and communication technologies, a 3D tunnel construction monitoring system has been developed. The technology allows construction workers to examine 3D simulations in multiple settings. Construction progress simulation, automated alert warning and trend projection are all included in the 3D interactive operation. The method provides improved safety and visual construction site monitoring. It also improved the construction progress and offered solid technical assistance for the Tunnels' safety construction. 3D Monitoring is a novel approach that monitors tunnel deformation, surface settlements, ground water table depletion, face and unsupported ground instability, daylighting collapse, support failure, etc.

These deformations inside the tunnel causes damages and even collapse of tunnel in the future. So, in order to protect and to increase the life span of tunnels, such parameters are monitored by an advanced technique known as3-Dimensional Monitoring systems.

There are various reasons why 3D monitoring should take place in every tunneling system.

Some of the observations are listed below:

- To ensure the tunnel's safety and to validate design assumptions
- To ensure the tunnel's safety and to validate design assumptions
- Successful tunneling requires the ability to respond accurately and reliably to the diverse rock conditions.
- Any instrument must be dependable, simple, straightforward, and quick to install, use, and calibrate without interfering with building activity.
- The monitoring equipment must be long-lasting and resistant to damage before to and during installation.

2.0 Methodology

There are four parts to the methodology used in this study: (a) preparation of a database and a geological longitudinal profile of the tunnel that depicts the ground condition with overburden and support classes, (b) estimation of anticipated deformation at various zones along the study area's entire length, (c) analysis of actual 3D monitoring data from the tunnel, and (d) comparison of anticipated deformation with measured deformation. Primarily, ground behaviour is determined by the groundmass's geomechanically characteristics, which is why 3D monitoring points (bi-reflex targets) have been deployed (Fig. 2). We utilised the parameters from Ch. 125+712 to 127+662 in the Main Tunnel (South Portal) and Ch. 132+750 to 131+760 in the Main Tunnel (North Portal) for analysing ground and support system interaction (Fig. 3). The geological longitudinal profile and datasheet contain information on the litho-contacts, joints, faults, shear seams/zones, etc. that were acquired in the field during daily face logging. Additionally, this includes the projected and actual water conditions, overburden, etc. (Fig. 3 & Table 1).

During the excavation process for a tunnel, the ground's equilibrium is disrupted, resulting in deformation. The rock mass begins to deform half a tunnel diameter ahead of the advancing face and achieves its maximum deformation half a tunnel diameter behind the face [10]. Analysis of deformation incorporates notions of rock support interactions with rocks. Shotcrete, steel elements, and rock bolts were used to calculate the support pressure for each major support class evaluated in this study. The deformation zone (rp) forms around the aperture of the excavation in the plastic zone. The radius of the tunnel, overburden stress, and the geomechanical behaviour of the rockmass all have a role in determining the plastic zone's size. When the tunnel's principal support (Pi) is less than a critical support, rockmass failures occur (Pcr). The equations provided by Hoek and Brown [11] and Brady and Brown [12] were used to estimate Pi, Pcr, rp, and inward radial displacement (μ). (Tables 2 and 3) A Microsoft Excel spreadsheet has been produced, and the graphs have been displayed using both Microsoft Excel and Python software.

Deformation data has been acquired from the regular 3D monitoring reports, which show the real deformation. Each segment of Tunnel T-74R has five bireflex targets mounted. Instruments such as the TS-11 were used to achieve an accurate zero reading after they were implanted in shotcrete (total station). Continuous three-dimensional displacement (X, Y, and Z) measurements were made on a regular basis (on a daily basis). Using the same reference points, an interconnected observation system was constructed that allowed for more thorough monitoring of the area. Points having a maximum displacement rate of less than a millimetre per month were utilised as reference points. The data was analysed and the deformation was examined in relation to the zero-reading recorded soon after the installation of the bi-reflex targets. The AAA Chart (Awareness, Alarming, and Action Limits) approach has been utilised to verify the area's stability after the processing of data. Graphs showing the same data in multiple ways, such as time-displacement diagrams and distance-displacement diagrams (here the second one has been used).

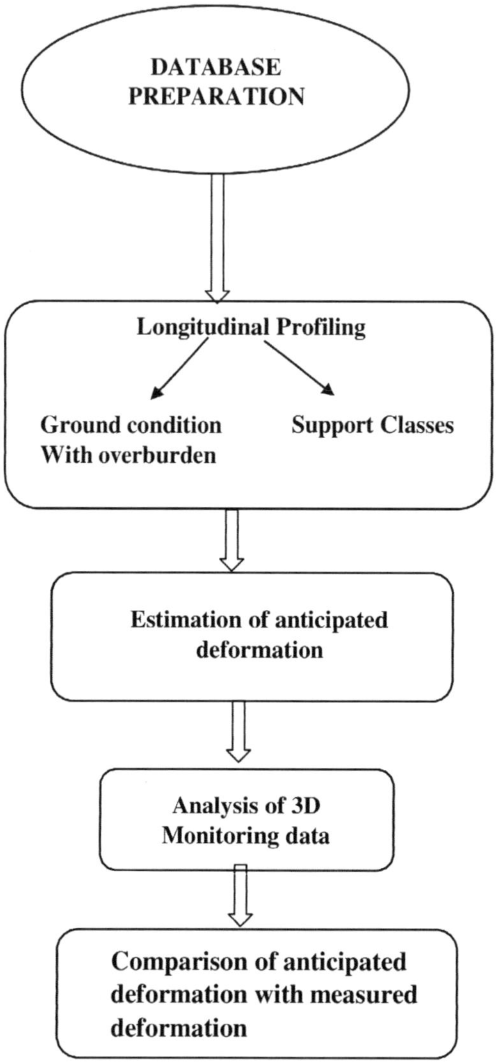

Flowchart showing methods adopted for tunneling

Finally, the predicted and actual deformation data were compared with the actual geological condition found during digging. Attempts have been made to determine the source of deformation in each zone and also to analyse differences between expected and actual deformation data for that specific geological zone, if there are any such differences. To account for both groundmass conditions and the overburden pressure, this has been taken into account. Because this is a tectonically active area, it's possible that the geological forces had a significant impact on the differences that were discovered.

Table 1: Geotechnical properties of rockmass

Zone No.	Tunnel meter		Length (m)	Chainages (Km)	UCS (Mpa) sig cm	Friction angle (θ) (deg)	Cohesion (c) (Mpa)	Modulus (E) (Mpa)	Poisson's ratio (v)	Rockmass UCS (σcm) (Mpa)	Rockmass constant (k)	Critical pressure (pcr) (Mpa)
	Start	End										
22	1450	1500	50	TM :126788	43	28.93	1.30	5844.32	0.3	4.41	1.94	6.65
22	1450	1500	50	TM :126800	43	28.93	1.30	5844.32	0.3	4.41	1.94	6.65
23	1500	1550	50	TM :126811	43	29.22	1.32	6190.62	0.3	4.50	1.95	6.18
23	1500	1550	50	TM :126817	43	29.22	1.32	6190.62	0.3	4.50	1.95	6.18
23	1500	1550	50	TM :126822	43	29.22	1.32	6190.62	0.3	4.50	1.95	6.18
23	1500	1550	50	TM :126829	43	29.22	1.32	6190.62	0.3	4.50	1.95	6.18
23	1500	1550	50	TM :126836	43	29.22	1.32	6190.62	0.3	4.50	1.95	6.18
23	1500	1550	50	TM :126846	43	29.22	1.32	6190.62	0.3	4.50	1.95	6.18
23	1500	1550	50	TM :126860	43	29.22	1.32	6190.62	0.3	4.50	1.95	6.18
24	1550	1600	50	TM :126871	40	28.93	1.21	5636.77	0.3	4.10	1.94	5.87
24	1550	1600	50	TM :126878	40	28.93	1.21	5636.77	0.3	4.10	1.94	5.87
24	1550	1600	50	TM :126886	40	28.93	1.21	5636.77	0.3	4.10	1.94	5.87
24	1550	1600	50	TM :126896	40	28.93	1.21	5636.77	0.3	4.10	1.94	5.87
24	1550	1600	50	TM :126907	40	28.93	1.21	5636.77	0.3	4.10	1.94	5.87
25	1600	1650	50	TM :126920	40	28.93	1.21	5636.77	0.3	4.10	1.94	5.46
25	1600	1650	50	TM :126933	40	28.93	1.21	5636.77	0.3	4.10	1.94	5.46
25	1600	1650	50	TM :126945	40	28.93	1.21	5636.77	0.3	4.10	1.94	5.46
25	1600	1650	50	TM :126957	40	28.93	1.21	5636.77	0.3	4.10	1.94	5.46
26	1650	1700	50	TM :126976	40	28.93	1.21	5636.77	0.3	4.10	1.94	4.88
26	1650	1700	50	TM :126990	40	28.93	1.21	5636.77	0.3	4.10	1.94	4.88
26	1650	1700	50	TM :127004	40	28.93	1.21	5636.77	0.3	4.10	1.94	4.88
31	1900	1950	50	TM :127261	45	29.51	1.40	6708.20	0.3	4.81	1.97	3.29
32	1950	2000	50	TM :127301	43	29.51	1.34	6557.44	0.3	4.60	1.97	3.28
34	2050	2100	50	TM :127387	44	29.51	1.37	6633.25	0.3	4.70	1.97	3.88
36	2150	2200	50	TM :127472	43	29.51	1.34	6557.44	0.3	4.60	1.97	3.64
36	2150	2200	50	TM :127487	43	29.51	1.34	6557.44	0.3	4.60	1.97	3.64
36	2150	2200	50	TM :127502	43	29.51	1.34	6557.44	0.3	4.60	1.97	3.64
37	2200	2250	50	TM :127515	43	29.51	1.34	6557.44	0.3	4.60	1.97	3.60

Table 2: Estimation of Anticipated deformation in South Portal

S. No.	Chainage	Max. Tunnel Displacement Anticipated (mm)	Tunnel	Overburden (m)
1	TM :126788	21.48	South Portal	754
2	TM :126800	21.48	South Portal	754
3	TM :126811	17.98	South Portal	716
4	TM :126817	17.98	South Portal	716
5	TM :126822	17.98	South Portal	716
6	TM :126829	17.98	South Portal	716
7	TM :126836	17.98	South Portal	716
8	TM :126846	17.98	South Portal	716
9	TM :126860	17.98	South Portal	716
10	TM :126871	18.60	South Portal	671
11	TM :126878	18.60	South Portal	671
12	TM :126886	18.60	South Portal	671
13	TM :126896	18.60	South Portal	671
14	TM :126907	18.60	South Portal	671
15	TM :126920	16.30	South Portal	631
16	TM :126933	16.30	South Portal	631
17	TM :126945	16.30	South Portal	631
18	TM :126957	16.30	South Portal	631
19	TM :126976	13.48	South Portal	575
20	TM :126990	13.48	South Portal	575
21	TM :127004	13.48	South Portal	575
22	TM :127261	6.70	South Portal	444
23	TM :127301	6.78	South Portal	438
24	TM :127387	8.41	South Portal	500
25	TM :127472	7.86	South Portal	473
26	TM :127487	7.86	South Portal	473
27	TM :127502	7.86	South Portal	473
28	TM :127515	7.80	South Portal	470

Table 3: Estimation of Anticipated deformation in North Portal

S. No.	Chainage	Max. Tunnel Displacement Anticipated (mm)	Tunnel	Overburden (m)
1	CH :131764.497	2.67	North Portal	N/A
2	CH :132340.182	1.02	North Portal	118
3	CH :132351.366	0.42	North Portal	162
4	CH :132368.000	0.42	North Portal	118
5	CH :132378.000	0.42	North Portal	118
6	CH :132395.085	0.42	North Portal	92
7	CH :132409.300	0.42	North Portal	92
8	CH :132472.720	0.64	North Portal	108

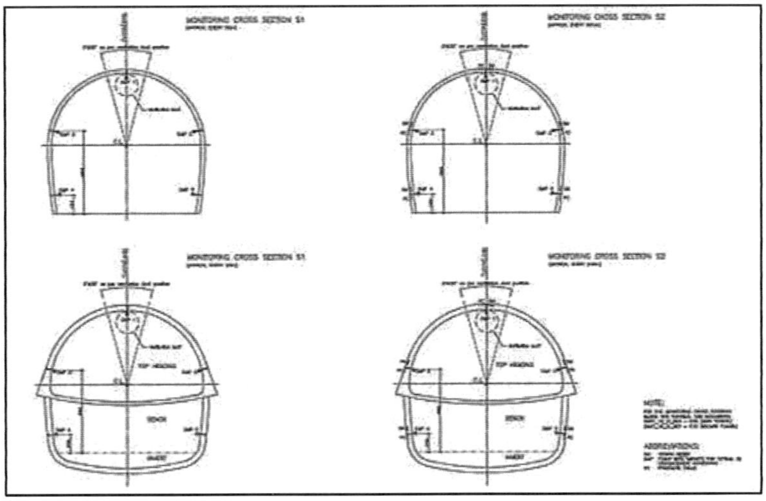

Fig. 2: Schematic diagram showing arrangement of 3D monitoring points in X-section

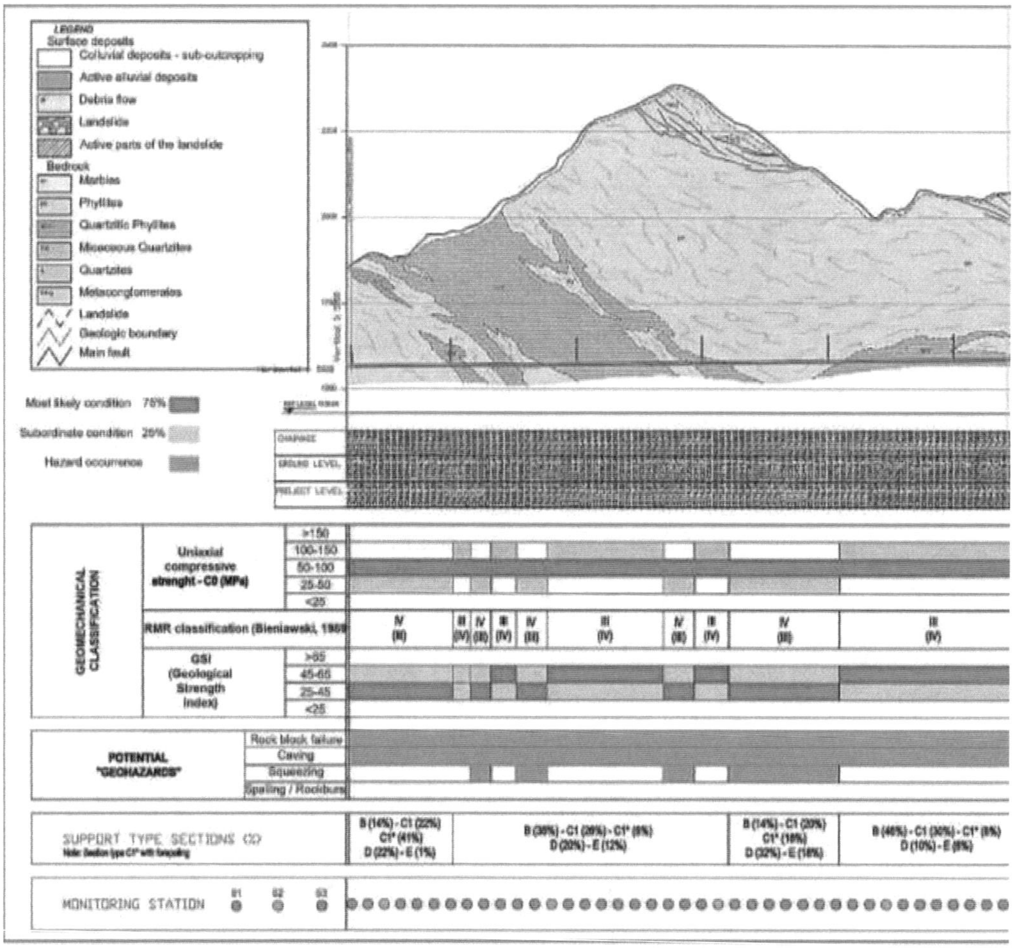

Fig. 3: Geological L-section of anticipated and actual geology with highlighted study area portion

3.0 Result and Discussion

In the longitudinal geological profile, there are several lenses of meta-conglomerate phyllite with some areas of medium-coarse grained quartzitic phyllite. In tunneling, there is one main fault with several shear seams (Fig. 3). The rockmass UCS varied from 25 to 100 Mpa, although local fault and shear zones exhibited UCS of 1–5 Mpa. The joint kinematic analysis revealed that the failure modes were mostly planar and wedge. Tunnel T-74R has four to five significant unavoidable collapse zones. The geological profile study revealed that all strongly deformed zones were linked to faults, shear zones, etc. However, variations in geomechanical behaviour between phyllite and quartzitic phyllite, as well as water intrusion, caused certain failures. It is computed that the support pressure, deformation radius, and radial displacement are all equal over the whole tunnel zone (Tables 2, 3, 4, and 5). The primary rock support class adopted at the analysed sites in Tunnel T74R is C1*/C1F (89.07%), with subordinate B (3.18%), C (7.29%), and D (7.29%). (0.44 percentage).Class B had a support pressure of 0.70 Mpa, Class C had a pressure of 1.30 Mpa, Class C1F had a pressure of 1.45 Mpa, and Class D had a pressure of 1.79 Mpa (Table 4 & 5).T74R's South Portal has 39 zones, each 50 m long, ranging from Ch.125+712 to Ch.127+662. The overburden at South Portal varies from 325 to 803 m. As a result, the in-situ stress is 6.5–16.06 Mpa (maximum at zone 20). The rockmass found in this location is fair, with a GSI ranging from 48 to 60. South Portal's rockmass has a GSI of about 50, which is on the lower side, in zones 19 and 20 where there is the highest overburden and in-situ stress. Its radius is 4–6.5 m (maximum at zone 19), and its estimated inward displacement is 2.3–24.00 mm (Table 2). The North Portal has 19 zones, each 50 m long, from Ch.132+750 to Ch.131+760. Overburden varies from 52 m (zone 5) to 254.2 m (zone 14). The in-situ stress varies depending on the overburden and rockmass. North Portal has lower overburden and in-situ stress than South Portal. The radius of the deformation zone is 4 to 4.13 m at the North Portal, indicating the absence of a deformation zone encircling the tunnel and very minor inward movement of the main tunnel support, which is 3.17 mm at zone 14. (Table 3). Tunnel T-74R's North and South Portal stresses were derived using the predicted GSI along the tunnel [13]. However, this assessment during the building stage allows us to evaluate the support systems built according to the rockmass encountered. Strain is the tunnel closure to tunnel diameter ratio. This stress is believed to be the product of tunnel depth and rockmass unit weight. During the tunneling process, no substantial stresses were recorded along the tunneling zone. A rapport of plastic zone and tunnel diameter with a ratio of rockmass strength and in-situ stress has been displayed (Fig. 4 & 5). These figures reveal a negative association, i.e., low rockmass increases the plastic zone under high in situ stress. In South Portal-MT, the ratio of plastic zone to tunnel radius is roughly 4, whereas in North Portal-MT, it is around 1.5. Tunnel closure has been predicted for the encountered rockmass support system based on the radius of the plastic zone surrounding the excavated region and the rockmass strength under specific in-situ stress. This study examined the link between tunnel closure and diameter ratios, as well as rockmass strength and in-situ stress (Fig. 6 & 7). No tunnel closure was emancipated in North Portal-MT, but in South Portal-MT, a small amount (max. ratio of 0.0025) was emancipated using the established support system in the encountered rockmass. The largest tunnel closure was expected in the 0.4 ratio region, where the rockmass was in poor condition and the in-situ stress was significant (Fig.9).

3D monitoring sites built at South Portal recorded data from chainages 126+788 to 126+822 with comparable overburden (716 m to 754 m). From chainage 126+822 to 126+886, there were several ongoing displacement fluctuations. From chainage 126+886 to 126+907, there is a sharp climb (50-150 mm) of 21 m in length. Maximum displacement was 225-250 mm between chainages 126+940 and 126+950, where the overburden was 631 m high (Fig. 8).The largest displacement was 26.6 mm at chainage 131+764.497, but at chainage 132+340.182, where the overburden was 118 m high, the displacement was practically the same (30 mm). 132+340 chainage, 182 mm displacement, 19.9 mm at 162 m overburden, 40.2 mm at 118 m overburden. Despite the same height of overburden, maximum displacements increased by 14.3 mm at only 10 m. A maximum displacement of 80 mm was found around chainage 132+390. When the overburden was only 92 m high, the maximum displacement of structures dropped suddenly, followed by an increase in displacement at the same overburden height. The findings obtained by calculating the projected deformation of the implemented support system theoretically and seeing the actual deformation on the installed deformation monitoring targets are considerably different (Table 6; Fig.10).

Table 4 : Support Pressure Estimation in South Portal

Zone No.	Tunnel meter		Length (m)	Chainages (Km)		Pull length	Actual Support	Max. overburden (m)	Support pressure (pi) (Mpa)
	Start	End		Start	End				
1	400	450	50	125+712	125+759.3	3.1	C1(F)	325	1.33
2	450	500	50	125+761.5	125+809.2	2.8	C1(F	329	1.34
3	500	550	50	125+810.7	125+859.2	1.6	C1(F)	340	1.44
4	550	600	50	125+860.7	125+909.2	2	C1(F)	353	1.40
5	600	650	50	125+910.6	125+959.5	2.5	C1(F)	375	1.36
6	650	700	50	125+962	126+008	3.7	C1(F)	398	1.31
7	700	750	50	126+010.4	126+058.5	3.7	C1(F)	410	1.31
8	750	800	50	126+061.3	126+108.8	4	C1(F)	458	1.31
9	800	850	50	126+110.8	126+160	4	C1	476	1.31
10	850	900	50	126+163	126+210	4.5	C1	505	1.30
11	900	950	50	126+212	126+260.5	3	B	545	0.73
12	950	1000	50	126+262.5	126+309.5	2	C	580	1.40
13	1000	1050	50	126+311	126+360.5	1.5	C1	644	1.45
14	1050	1100	50	126+362	126+410.5	2.3	C1	687	1.37
15	1100	1150	50	126+412	126+460	1.5	C1	702	1.45
16	1150	1200	50	126+461.5	126+509.5	1.5	C1	721	1.45
17	1200	1250	50	126+511	126+560.5	1.5	C1	743	1.45
18	1250	1300	50	126+562	126+610	1.5	C1	768	1.45
19	1300	1350	50	126+611.5	126+659.5	1.5	C1	798	1.45
20	1350	1400	50	126+661	126+710.5	1.5	C1	803	1.45
21	1400	1450	50	126+712	126+760	1.5	C1	791	1.45
22	1450	1500	50	126+761.5	126+810	1.5	C1	754	1.45
23	1500	1550	50	126+811	126+860	1.5	C1	716	1.45
24	1550	1600	50	126+861.5	126+910	1.8	C1	671	1.41
25	1600	1650	50	126+911	126+959.5	1.5	C1	631	1.45
26	1650	1700	50	126+960.75	127+009.5	1.3	C1	575	1.49
27	1700	1750	50	127+010.75	127+060.5	1.5	C1	558	1.45
28	1750	1800	50	127+062	127+109.75	1.5	C1(F)	545	1.45
29	1800	1850	50	127+111	127+157.75	1.8	C1(F)	520	1.41
30	1850	1900	50	127+160.75	127+210.25	3	C1	479	1.34
31	1900	1950	50	127+211.75	127+260.5	1.5	C1	444	1.45

Zone No.	Tunnel meter		Length (m)	Chainages (Km)		Pull length	Actual Support	Max. overburden (m)	Support pressure (pi) (Mpa)
	Start	End		Start	End				
32	1950	2000	50	127+262	127+310	1.5	C1	438	1.45
33	2000	2050	50	127+311.5	127+359.5	1.5	C1	463	1.45
34	2050	2100	50	127+361	127+410.5	1.5	C1	500	1.45
35	2100	2150	50	127+412	127+460	1.5	C1	500	1.45
36	2150	2200	50	127+461.5	127+510.057	1.8	C1	473	1.41
37	2200	2250	50	127+512	127+559.5	1.9	C1	470	1.40
38	2250	2300	50	127+561	127+610.5	1.5	C1	483	1.45
39	2300	2350	50	127+612	127+657	3	C1	492	1.34

Table 5: Support Pressure Estimation in North Portal

Zone No.	Tunnel meter		Length (m)	Chainages (Km) Katra		Pull length	Actual Support	Max. overburden (m)	Support pressure (pi) (Mpa)
	Start	End		Start	End				
1	0	50	50	132+738.8	132+690	4.5	C1	129	1.30
2	50	100	50	132+6888	132+639.5	4	B	129	0.71
3	100	150	50	132+637	132+589.5	3.5	C1	129	1.32
4	150	200	50	132+587.5	132+540	3.5	C1F	110	1.32
5	200	250	50	132+538.5	132+489	2.5	C1F	52	1.36
6	250	300	50	132+488	132+436.1	3	C1 F	108	1.34
7	300	350	50	132+435.1	132+386.5	1.1	C1 F	92	1.54
8	350	400	50	132+385.5	132+336.3	1.45	C1 F	118	1.46
9	400	450	50	132+335.3	132+286.4	1.5	C1 F	162	1.45
10	450	500	50	132+285.1	132+235.9	1.25	D	196	1.80
11	500	550	50	132+234.9	132+186.1	1.5	C1 F	218	1.45
12	550	600	50	132+184.6	132+136.6	1.5	C1 F	239.6	1.45
13	600	650	50	132+135.1	132+087.1	1.5	C1 F	253.1	1.45
14	650	700	50	132+085.6	132+036.1	1.5	C1 F	254.2	1.45
15	700	750	50	132+034.6	131+985.6	2	C1F	251.7	1.40
16	750	800	50	131+984.1	131+934.6	1.5	C1	250	1.45
17	800	850	50	131+933.1	131+884.1	1.5	C1	243	1.45
18	850	900	50	131+882.6	131+833.1	1.5	C1 F	232.5	1.45
19	900	950	50	131+831.6	131785.1	1.5	C1F	235	1.45

Table 6: Comparison chart of anticipated and actual deformation at important section

S. No.	Chainage	Max Displacement (mm)	Max. Tunnel Displacement Anticipated (mm)	Tunnel	Overburden
1	TM :131764.497	26.6	2.67	North Portal	N/A
2	TM :132340.182	28.5	1.02	North Portal	118
3	TM :132351.366	19.9	0.42	North Portal	162
4	TM :132368.000	40.2	0.42	North Portal	118
5	TM :132378.000	14.3	0.42	North Portal	118
6	TM :132395.085	80.2	0.42	North Portal	92
7	TM :132409.300	25.3	0.42	North Portal	92
8	TM :132472.720	16.04	0.64	North Portal	108
1	TM :126788	20	21.48	South Portal	754
2	TM :126800	21	21.48	South Portal	754
3	TM :126811	58	17.98	South Portal	716
4	TM :126817	68	17.98	South Portal	716
5	TM :126822	129	17.98	South Portal	716
6	TM :126829	123	17.98	South Portal	716
7	TM :126836	147	17.98	South Portal	716
8	TM :126846	119	17.98	South Portal	716
9	TM :126860	123	17.98	South Portal	716
10	TM :126871	110	18.60	South Portal	671
11	TM :126878	83	18.60	South Portal	671
12	TM :126886	106	18.60	South Portal	671
13	TM :126896	85	18.60	South Portal	671
14	TM :126907	52	18.60	South Portal	671
15	TM :126920	139	16.30	South Portal	631
16	TM :126933	218	16.30	South Portal	631
17	TM :126945	210	16.30	South Portal	631
18	TM :126957	257	16.30	South Portal	631
19	TM :126976	44	13.48	South Portal	575
20	TM :126990	24	13.48	South Portal	575
21	TM :127004	71	13.48	South Portal	575
22	TM :127261	38	6.70	South Portal	444

S. No.	Chainage	Max Displacement (mm)	Max. Tunnel Displacement Anticipated (mm)	Tunnel	Overburden
23	TM :127301	39	6.78	South Portal	438
24	TM :127387	26	8.41	South Portal	500
25	TM :127472	26	7.86	South Portal	473
26	TM :127487	63	7.86	South Portal	473
27	TM :127502	33	7.86	South Portal	473
28	TM :127515	48	7.80	South Portal	470

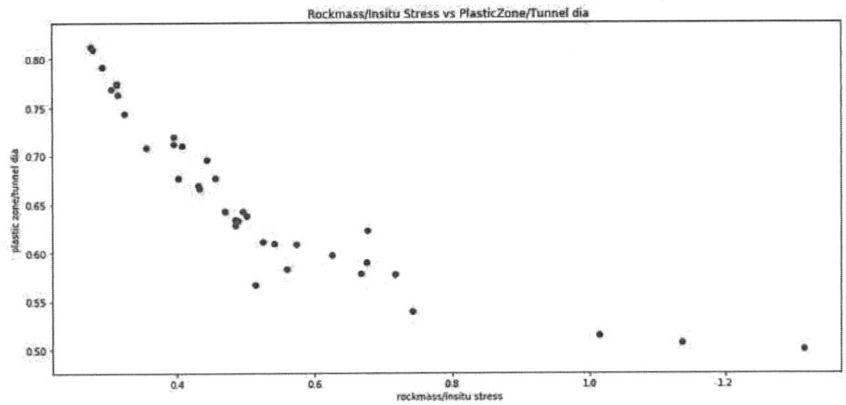

Fig. 4: Plastic zone and tunnel diameter vs rockmass strength and in-situ stress in South Portal

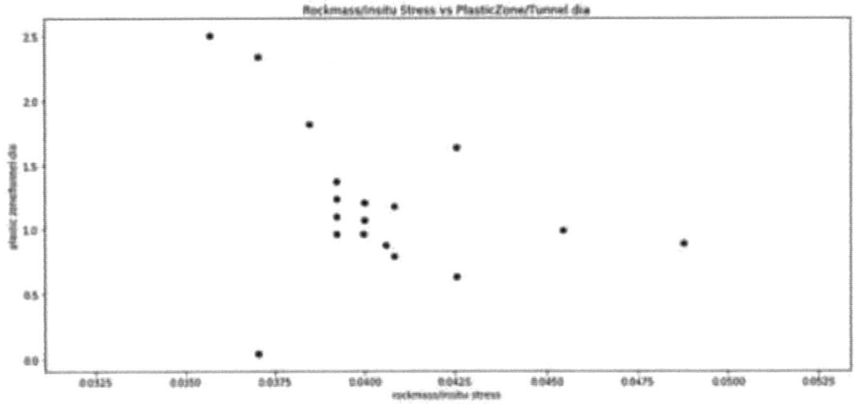

Fig. 5: Plastic zone and tunnel diameter vs rockmass strength and in-situ stress in North Portal

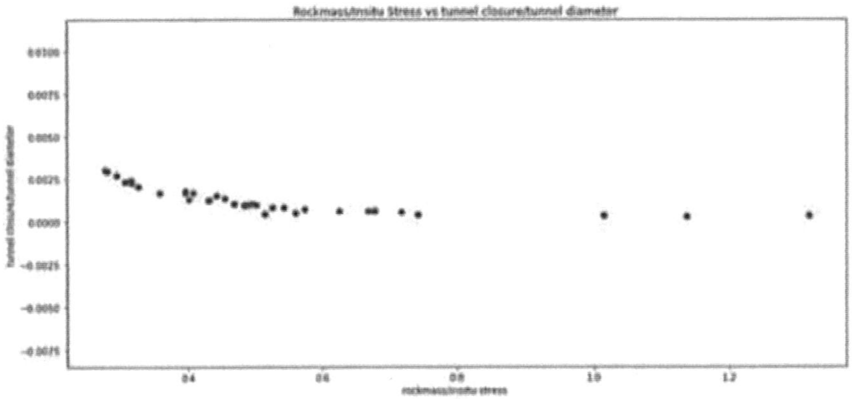

Fig. 6: Tunnel closure and tunnel diameter vs rockmass strength and in-situ stress in South Portal

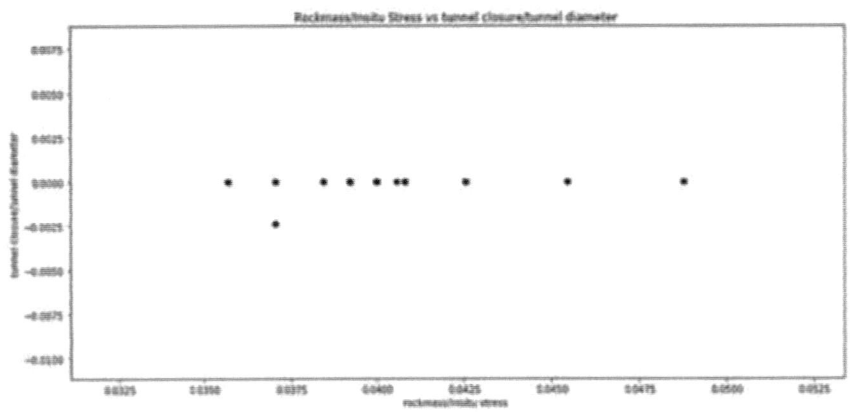

Fig. 7: Tunnel closure and tunnel diameter vs rockmass strength and in-situ stress in North Portal

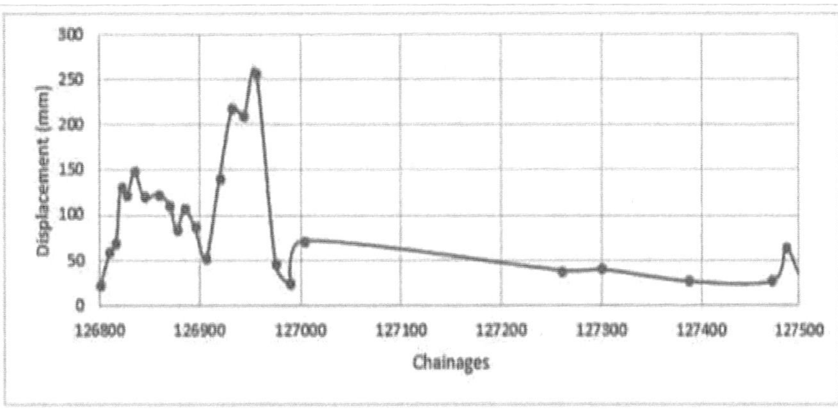

Fig. 8: Displacement of monitoring targets due to deformation at different chainages at South Portal of T74R

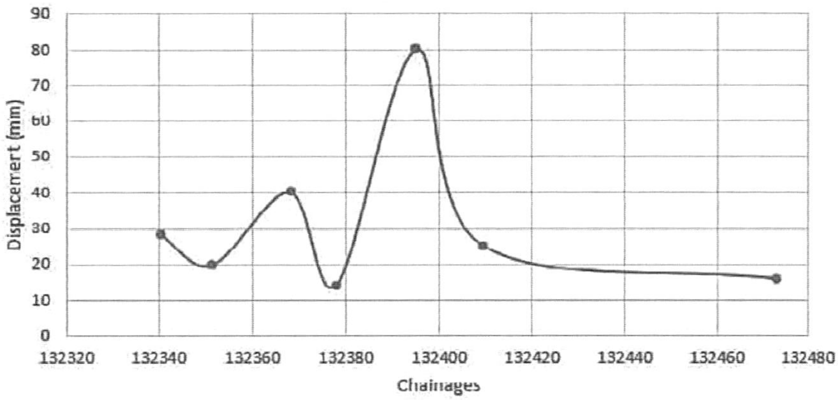

Fig. 9: Displacement of monitoring targets due to deformation at different chainages at North Portal of T74R

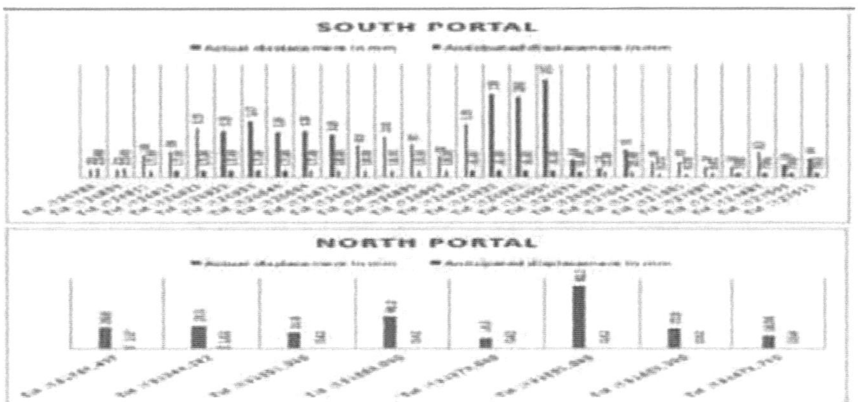

Fig. 10: Comparison of actual and anticipated displacement curvers of South & North Portal of T74 R

4.0 Conclusion

Tunnel T74R is located in one of the most seismically active areas of the Jammu & Kashmir Himalaya, therefore deformation of the tunnel support system was unavoidable. A poor groundmass near the principal thrust is prone to deformation. Our study compares the predicted and actual deformation of the support system based on ground conditions. The actual deformation monitoring data of T74R is found to be considerably different from the calculated deformation based on formulas, with the former revealing more deformation than expected for the installed support system in encountered groundmass. The discrepancy was largely due to not considering elements like horizontal/vertical stress ratio locally, shear strength characteristics of discontinuities in rockmass, and blasting effect. Besides technical concerns, various execution aspects also play a role in this variation. Also, the value of 3D monitoring resides in determining the real behaviour of the groundmass, as opposed to calculating it theoretically. Determination of shear strength parameters of joint in-fill material and joint data is required, as is the use of 3D monitoring data in final designs.

Acknowledgement:

The author wishes to express his gratitude to Ircon International Limited and Indian Railway (Northern Railway) for allowing him to work on the geological and geotechnical aspects of Tunnel T-74R's construction. It's all a component of my doctoral research.

References

[1] Mauriya, V.K., Yadav, P.K. and Angra, V.K. 2010 Challenges and strategies for tunneling in the Himalayan region, *Indian Geotechnical Conference-* Geo trendz, IGS Mumbai Chapter and IIT Bombay, p. 93-96

[2] Dhang, P.C. 2016 Tunneling in Lesser Himalaya, Jammu and Kashmir, India with special emphasis on tectonic mélange, *Journal of Geological Society of India*, **vol. 88**, p. 593-602,

[3] Dhang, P.C. 2017 Tunneling in overstressed rock: Example from lesser Himalaya of Jammu and Kashmir, India, *Proceeding: Conference on Tunneling in Himalayan Geology,* Tunneling Association of India, Jammu, 11th-13th October, 2017, **vol**. 340, p.378-389

[4] Jain, A.K., Dasgupta, S., Bhargava, O. N., Patel, R., 2016 Tectonics and Evolution of the Himalayas Deformation pattern in Kumaun, Lesser Himalaya, India *Proceedings of the Indian National Science Academy* **vol. 82** III.

[5] Daraei, A., Zare, S. 2019 Modified criterion for prediction of tunnel deformation in non-squeezing ground conditions. **J. Mod. Transport**. **vol. 27**, p-11–24

[6] Ahmad, A. & Dhang, P. C. 2019 Methodologies for the geological and geotechnical works during the construction of railway tunnels in Himalayan region, India, USBRL technical magazine, *Himprabhat*, V.XIII, p.32-40

[7] Shrestha, K.K & Panthi, K.K. 1970 The Use of Self Supporting Capacity of Rock Mass for Sustainable Hydropower An Analysis of the Middle Marsyangdi Headrace Tunnel, Nepal, Kathmandu University School of Education in Hydro Nepal *Journal of Water, Energy and Environment Journal of Water, Energy and Environment*, **Vol. 6**, p 18-26

[8] Nitschke, A., McRae, M., Ramirez, I. and Gall, V. 2014 NATM Excavation and Support Design and Construction of the Caldecott Fourth Bore.

[9] Karakus, M. and Fowell, R. 2004 An insight into the New Austrian Tunneling Method (NATM), Rockmec'2004-VIIth *Regional Rock Mechanics Symposium, Sivas*, Turkey

[10] Hoek, E. 1999 Support for very weak rock associated with faults and shear zones, Distinguished lecture for the opening of the International *Symposium on Rock Support and Reinforcement Practice in Mining, Kalgoorlie, Australia*, pp.1-20

[11] Hoek, E., & Brown, E. T. 1980 Empirical strength criterion for rock masses. *Journal of the geotechnical engineering division, 106* **vol 9**, p-1013-1035.

[12] Brady, B. H. G., & Brown, E. T. 1985 Energy changes accompanying underground mining. In *Rock Mechanics* Springer p. 240-259

[13] Hoek, E. and Marinos, P.G. 2010 Tunneling in overstressed rock, Rock Engineering in Difficult Ground Conditions- Soft Rocks and Karst- Vrkljan (Ed.), Taylor & Francis Group, London, p. 49-60.

An Experimental Study of concrete by Partial Replacement of Fine Aggregate with Bakelite Waste

R R Bhopi[1,a] **and M Sinha**[2,b,*]

[1]Research Scholar, Department of Civil Engineering, Pillai HOC College of Engineering and Technology, Rasayani-410207, Maharastra, India

[2]Assistant Professor, Department of Civil Engineering, Pillai HOC College of Engineering and Technology, Rasayani-410207, Maharastra, India

Email:[a]bhopi.ravina27@gmail.com; [b]madhusinha13@gmail.com

Abstract. Bakelite is a industrial waste produce in manufacturing process like in Auto Industries etc. Nowadays the waste increases rapidly because of these modern living and thus creates a waste management problem. It is necessary to overcome such waste related problem in a meaningful manner. The purpose of this study is to find whether the Bakelite gives us that much Compressive strength as compared to conventional M25 grade concrete by replacing fine aggregate with Bakelite waste. So, this research examines the utilization of waste Bakelite in concrete with varying percentage like 0%,17%,20%30%,40%,50%,60% and 70%. Different percentages of Bakelite are added and tests like the slump cone test, compaction factor test were conducted to investigate the fresh properties like workability and compression test were performed to find out the 7, 14 and 28 days compressive strength. It is found that the replacement of fine aggregate with Bakelite can be done up to 20% without compromising the compressive strength of the M25 grade concrete.

Keywords: Bakelite, Compressive Strength, Fine Aggregate.

1. Introduction

1.1 General

Bakelite is a kind of thermoset plastic that is the product of an elimination reaction where phenol reacts with formaldehyde[1]. Bakelite is mostly used in kitchenware appliances for heat resistance, automobile parts, and telephone casings. The reason behind providing the heat resistance is that Bakelite cannot be remelted into another product[11]. Just like every coin has two sides, Bakelite has it too. Advantages and disadvantages go hand in hand. There are a few disadvantages, that include the presence of methyl and ethyl alcohol[1,2]. These chemicals are harmful to the environment and also can lead to severe health issues. Disposal of Bakelite into the water bodies is also creating a hazardous environment for the aquatic ecosystem[6]. To avoid such health hazards there is an efficient alternative used which acts as a partial replacement for the sand in concrete i.e the Bakelite granules (Figure 1).

Figure 1. Waste Bakelite as a fine aggregate

Bakelite or polyoxybenzylmethylenglycolanhydride happens to be the very first plastic, produced from synthetic components[2]. It is also known as the thermosetting formaldehyde resin which is the product of the condensation reaction of phenol with formaldehyde[4]. A Belgian- American chemist named Leo Beakeland is the one who developed the product in the year 1907 [3]. Later in December 1909, Bakelite was patented. This invention acted as a boon for electrical insulators, the radio, telephone casings, kitchenware, jewellery, pipe stems, children's toys, and firearms[14]. Vital properties of the product were the electrical conductivity and heat resistance that eventually widened the usage value.

The objectives of the research are to determine % of fine aggregate which can be replaced by Bakelite waste without compromising the compressive strength of the conventional concrete of the grade M25, to study the effect of Bakelite waste on the slump value, compaction factor and compressive strength of the M25 grade concrete, utilization of waste Bakelite to reduce the dumping problem (Figure 2).

Figure 2. Procedure of making Bakelite granules

1.2 Literature Review

Sakthi Sasmitha, Dr. R. N. Uma (2019), carried out a study on concrete and paver blocks. Here, the coarse aggregate was replaced by Bakelite. The study stated the properties, strength, and use of Bakelite as a construction material in paver blocks of gradeM25. The optimum modifier content of waste Bakelite came out to be 8% for paver blocks and 35% for solid blocks. This study concluded that it will help to reduce the cost of construction to some extent and will solve the problem of waste plastic disposal.

Clement M, Rohini K, Krishna Kumar P, Boopathi M (2018) carried out a study on the structural behavior of cement concrete made by using partial replacement of coarse aggregate with Bakelite by casting 21 cubes of size 150 mm x 150 mm x 150 mm with different percentages of waste Bakelite. The cubes made with different percentages of waste Bakelite and conventional concrete were both compared to check their strength for 7, 14, and 28 days respectively. So was the fine aggregate, coarse aggregate, Bakelite, and cement tested based on the ISI standards. 3 cubes of M20 grade were cast by replacing 5, 10,15, 20, 25, and 30 percent of Bakelite. This was compared with M20 grade concrete which was made with natural coarse aggregates. The experiment came to a conclusion where the waste Bakelite up to 10% can be used as a substitute for aggregates in concrete. It will not hamper the compressive strength. It has also stated that it will always act as a better choice to avoid waste disposal and environmental hazard.

Arun Raja (2016) did a study on the mechanical and flexural strength of specimens by using E-plastic waste (Bakelite) in concrete as a partial replacement of fine aggregate. E- plastic waste was used instead of fine aggregate. The percentage of Bakelite used was 2%, 4%, 6%, 8%, and 10% by weight of total fine aggregate. The results stated that E-plastic waste up to 8% can act as a good substitute for fine aggregate in concrete. It will not compromise the compressive and flexural strength.

Pugal (2015) did a study on plastic waste as a coarse aggregate for structural concrete. The replacement Percentage used were 0%, 5%, 10%, and 15% by the total weight of coarse aggregate. The conclusion stated that the compressive strength and split tensile strength which contained plastic aggregate were retained in comparison. A decrease in strength was seen when the plastic content was greater than 20%.

2. Materials and Methodology

2.1 Materials

The materials (Figure 3) used for M25 grade concrete are as follows:

- Cement: The cement used for this study is JSW ordinary Portland cement (OPC) of 53-grade.[9]
- Fine aggregates: The grit powder sieve through BSS10 and retained on BSS20 was used as fine aggregate.
- Bakelite: The replacement for fine aggregate i.e. Bakelite. The Bakelite powder sieve through BSS10 and retained on BSS20 was used.
- Coarse aggregate: The 20mm size crushed stones were used.
- Water: Normal daily used water was used for mixing and curing.

Figure 3. Ingredients of concrete

2.2 Methodology

Following the literature research, a significant methodology was prepared to establish the objective of the project. This study is also focused on the effectiveness of Bakelite used in a construction project [7]. In Figure (4) step by step methodology of work is explained.

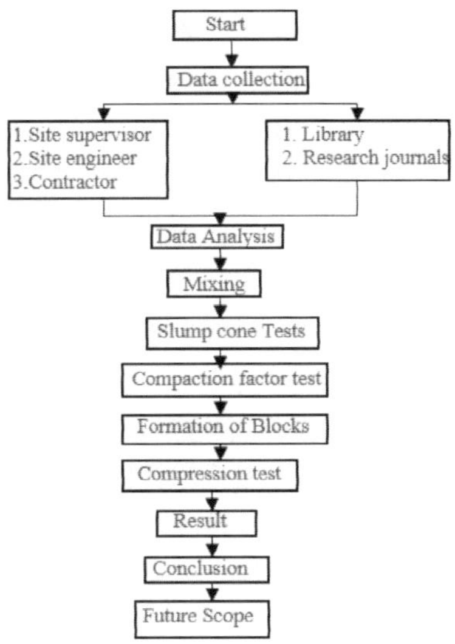

Figure 4. Methodology Flowchart

2.2.1 Data Collection

The data about the use of Bakelite in concrete is fetched from the site engineer, site supervisor, and contractor. For a detailed study of concrete technology and the use of Bakelite in concrete, books were referred from the college library. For some latest context, we referred to sample research papers like International Research Journals of Engineering and Technology (IRJET), Elsevier, MDPI, etc.

2.2.2 Data Analysis

The data collected was analysed thoroughly. Vital aspects like the quantity of material used, aim, results, conclusions were studied properly. Based on this context, the quantity of Bakelite to be used in concrete was decided. This overall collected data turned out to be our project topic.
Calculation of quantity of materials (Table 1) of M25 grade concrete: The quantities of cement, fine aggregate, Bakelite, coarse aggregate and water were calculated.

Dimensions of concrete block: Length=0.15m
Breadth=0.15m
Height=0.15m
Wet volume of blocks = 0.15*0.15*0.15*3
= 0.010125cu m
Dry volume of blocks = 1.54*0.010125
= 0.0155925cu m
Grade of concrete=M25
Ratio of cement: sand: aggregate =1:1:2
Sum of ratio =4
Quantity of cement= (1/sum of ratio) *density of cement*dry volume of blocks
Quantity of sand= (1/sum of ratio) *density of sand*dry volume of blocks
Quantity of aggregates = (2/sum of ratio) *density of aggregate*dry volume of blocks
Water cement ratio=0.45
Quantity of water required=0.45*quantity of cement

Table 1 . Quantities of materials

Sr.no.	Percentage of Bakelite to be used (%)	Quantity of Bakelite (kg)	Quantity of sand actually used(kg)	Quantity of cement (Kg)	Quantity of Aggregates (Kg)	Quantity of water required
1	0	0	5.6523	5.6133	12.0842	2.5259
2	17	0.9608	4.9615	5.6133	12.0842	2.5259
3	20	1.1304	4.5218	5.6133	12.0842	2.5259
4	30	1.6956	3.9567	5.6133	12.0842	2.5259
5	40	2.2609	3.3913	5.6133	12.0842	2.5259
6	50	2.8261	2.8261	5.6133	12.0842	2.5259
7	60	3.3913	2.2609	5.6133	12.0842	2.5259
8	70	3.9565	1.6956	5.6133	12.0842	2.5259

2.2.3 Mixing and Testing

After calculating all the materials quantity of all samples are mixed in tray (except water) and then water is added to it and again mixed thoroughly. The concrete mixture is then added in slump cone to perform slump cone test. After performing slump cone test, the same mixture is then used for compaction factor test. The same mixture was used to form blocks of size 150mm x150mm x 150mm [10,12]. After performing all the tests the result were tabulated.

2.2.4 Slump cone test

Slump cone test (Figure 5) is done to determine the consistency or workability of concrete. It is done at construction site or at laboratory to check the uniform quality of concrete during construction [13]. It is helpful to confirm whether the correct water quantity has been added to the mix or not. (Figure 7) shows the slump value (Table 2) of concrete. The mixing is done as per mix. Indian standards IS:1199-1959 . It is observed that workability of concrete decreases when the Bakelite added in the mix. For the conventional mix, 60 mm slump value is observed and from sample 1 to sample 7 the slump value decreases with increase in percentage of Bakelite.

2.2.5 Compaction factor test

The compaction factor test (Figure 6) is executed to estimate the degree of workability of fresh concrete. It is followed as per the Indian standard IS:1199-1959. It is more precise than slump cone test and particularly helpful for concrete mixes of very low workability [13]. (Figure 8) shows the compaction factor value (Table 3) of concrete as per mix. It is observed that, conventional concrete achieved 0.90 compaction factor, which is having medium workability. Whereas, sample-1 achieved maximum compaction factor of 0.958 which treated as good workability. Least compaction factor, 0.769 is for sample-7. On the basis of compaction factor results, it is observed that, as the fine aggregates contents in concrete are reduced by Bakelite, workability of concrete decreases.

2.2.6 Compressive strength test

Compressive strength (Figure 5) is the mechanical test which determines the maximum amount of compressive load a block or cube can carry before breaking. The samples were tested in compression testing machine after specified curing period for different percentage of Bakelite replacement. Compressive strength (Table 4) (Figure 9) of concrete is investigated for all the 8 specimens as per Indian standards IS:516-1959 at 7, 14 & 28 days curing period [17].

Figure 5. Slump cone test and compression test

Figure 6. Compaction factor test

3. Result and Discussion

The results obtained after performing the various tests on concrete are as follows:

3.1 For Slump cone test

The values of the slump are mentioned for all samples in the tables given below.

Table 2. Observations for slump cone test

Sample no.	Conventional Mix	1	2	3	4	5	6	7
% of replacement	0	17	20	30	40	50	60	70
Slump value(mm)	60	65	62	48	32	25	20	15

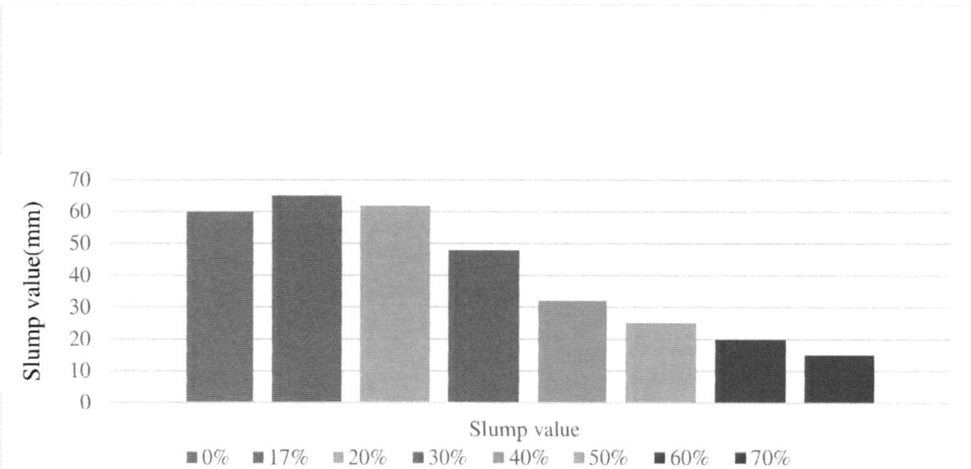

Figure 7. Comparison of slump value

3.2 *For Compaction factor test*

Compaction factor = (Weight of the partially compacted concrete with cylinder)/ (Weight of the fully compacted concrete with cylinder)

Table 3. Observations for compaction factor test

Sample no.	Conventional Mix	1	2	3	4	5	6	7
% of replacement	0	17	20	30	40	50	60	70
Compaction factor	0.90	0.958	0.922	0.877	0.869	0.841	0.822	0.769

The below figure shows the comparison of compaction factor values of all samples of concrete used in the experiment.

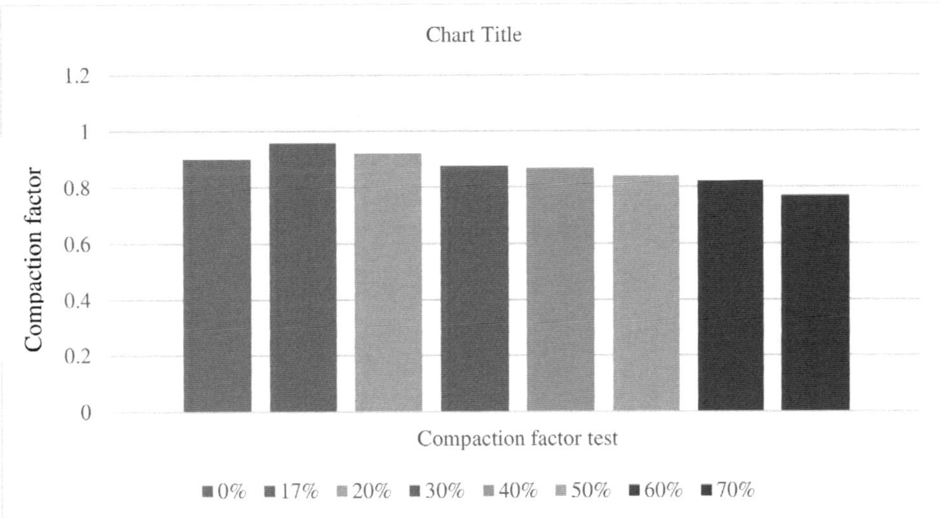

Figure 8. Comparison of compaction factor

3.3 *For Compression Test*

Compressive strength = load / cross-sectional Area

Compressive strength in N/mm2

Table 4. Calculation for compression test

Sample no.	Conventional Mix	1	2	3	4	5	6	7
% of replacement	0	17	20	30	40	50	60	70
7th day	17.77	20	13.33	18	12.22	13.33	11.11	7.11
14th day	19.55	20	20	20	16	18.22	19.11	12.88
28th day	26.44	26.44	26.44	24.44	22.88	24.66	25.05	17.33

The below figures shows the comparison of compressive strength of all samples of concrete used in the experiment.

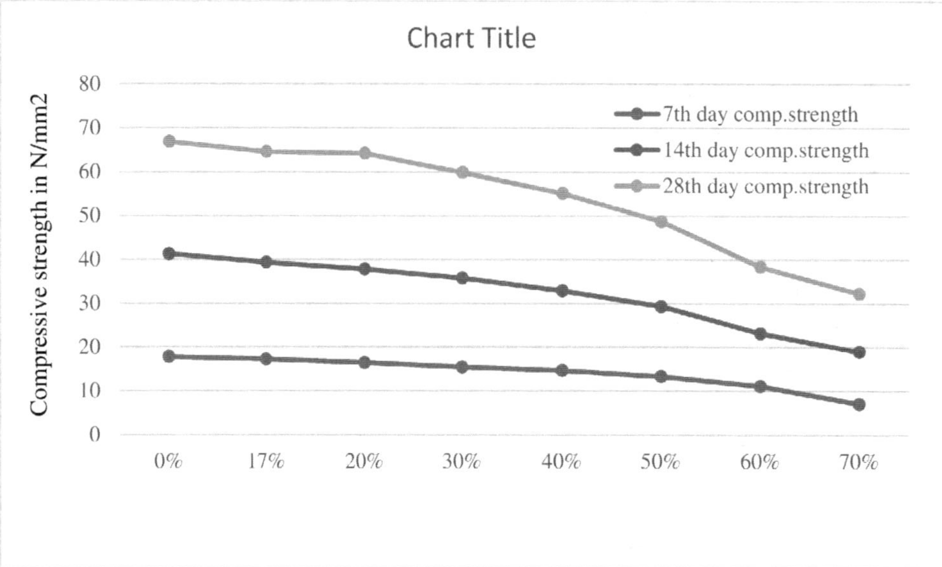

Figure 9. Compressive strength in N/mm2

4. Conclusion

The conclusions can be drawn from the experimental investigation are when the percentage of Bakelite is increased the value of slump decreases. There is a decrease in the compaction factor when we increase the Bakelite percentage. We can replace the fine aggregate upto 20% with Bakelite waste by weight in the concrete of M25 grade without compromising compressive strength. As the percentage of Bakelite is increased the concrete becomes lighter in weight.

REFERENCES:

[1] Nopagon U and Seree T, 2017 Case study The transformation of waste Bakelite to replace natural fine aggregate in cement mortar Case Studies in Construction Materials 120-133

[2] S Sakthi Sasmitha and Dr R N Uma, 2019 Waste Bakelite as Partial Replacement of Coarse Aggregate in Building Blocks *International Research Journal of Engineering and Technology (IRJET)* **6**(**5**) 4466-4472

[3] Clement M, Rohini K, Krishna Kumar P and Boopathi M, 2018 An experimental study on structural behavior of Cement Concrete manufactured by partial replacement of Coarse Aggregate with Bakelite waste *International Journal of Scientific Development and Research (IJSDR)* **3**(**10**) 29-36

[4] S Kongkarat, R Khanna, P Koshy, P O'Kane and V Sahajwalla, 2011 Use of waste bakelite as a raw material resource for recarburization in steelmaking processes Steel Res. Int. **82** (**10**) 1228–1239

[5] D Crespy, M Bozonnet and M Meier, 2008 100 years of bakelite the material of a 1000 uses chemiker angew Chem. Int. Ed. **47** 3322–3328

[6] S Dharanidharan, N Srivithya and N Meena, 2015 Experimental study on the flexural behavior of E-waste plastics in concrete *IJESRT* **4** (**11**) 660– 669

[7] Malek Batayneh, Iqbal Marie and Ibrahim Asi, 2006 Use of selected waste materials in concrete mixes.

[8] Mariaenrica Frigione , 2010 Recycling of PET bottles as fine aggregate in concrete

[9] Phaiboon Panyakapo and Mallika Panyakapo, 2007 Reuse of thermosetting plastic waste for lightweight concrete

[10] Arun Raja, P Kumar, Dr.M S Hameed and S K Raghunath, 2016 Study on flexural behavior of concrete by partially replacing fine aggregate with E-Plastic waste *International journal of Engineering Reasearch & Technology (IJERT)* **5**(**11**) 579-582

[11] S Sakthi Samitha, Dr.R N Uma, 2018 A critical review on the application of bakelite as a partial replacement of fine and coarse aggregate *IJSART* **4**(**11**) 174-178

[12] Murali K and Sambath K, 2020 Transformation of waste Bakelite into concrete and solid Blocks *IOSR Journal of Mechanical and civil engineering* **17**(**2**) 31-34

[13] Bhoopathi M, Clement M, Rohini K and Krishnakumar P, 2018 An experimental study on structural behavior of cement concrete manufactured by partial replacement of coarse aggregate with Bakelite waste *International journal of scientific development and research* **3**(**10**) 29-36

[14] P Dharani, R N Uma, 2018 Utilization of Bakelite and Waste plastic in flexible pavement construction-A Review *IJSART* **4**(**11**) 168-173

[15] A Kumar, S Chaudhary and A Ahamad, 2019 Effect of Bakelite plastic admixture as a coarse aggregate in a concrete *International Research Journal of Engineering and Technology (IRJET)* **6**(**3**) 3895-3899

[16] IS:1199-1959, Indian Standard Methods of sampling and analysis of concrete, Bureau of Indian Standards, New Delhi, 1959

[17] IS:516-1959, Indian Standard Method of Tests for Strength of Concrete, Bureau of Indian Standards, New Dehli, 1959

Retrieval of aerosols extinction coefficient from calipso satellite observations: a case study over Bhubaneswar

Sushree Sasmita[1], Dudam Bharath Kumar*[2],

School of Civil Engineering,

Kalinga Institute of Industrial Technology Deemed to be University (KIIT-DU), Bhubaneswar-750124, Odisha, India

*Corresponding Author

Email-ID: dudam.kumarfce@kiit.ac.in

Abstract. In order to examine the effects of aerosols on climate, knowledge on vertical distribution of aerosol in the lower atmosphere is essential. We analysed aerosol extinction coefficient obtained from the Cloud-Aerosol Lidar Infrared Pathfinder Satellite Observations (CALIPSO) on board Lidar at campus-3 of KIIT-DU Bhubaneswar during July 2018 to June 2019. Seasonal average (± standard deviation) of extinction coefficient for the winter (December-February: DJF), summer (March-May: MAM), monsoon (June-August: JJA) and post-monsoon (September-November: SON) are found to be 0.22±0.13, 13.08±0.07, 0.27±0.05, 0.15±0.14 km^{-1}, respectively. To differentiate the aerosol loading between the surface and elevated layer, we averaged extinction values for below and above 1km height. At surface level, the values were found to be maximum in post monsoon (0.54±0.13 km^{-1}) and minimum (0.14±0.03 km^{-1}) in winter. Whereas at elevated level, the values exhibit maximum in monsoon (0.45±0.45 km^{-1}) and minimum (0.11±0.02 km^{-1}) in post-monsoon. The vertical profile of aerosol can be used to anticipate aerosol extinction simulations over a variety of time periods and in regions where the CALIOP cannot reach.

Keywords: CALIPSO; extinction coefficient; vertical aerosol profile; surface level; elevated level

1. Introduction

By absorbing and dispersing incoming solar energy in the atmosphere, aerosols contribute to surface heating and cooling. Rapid industrialization and urbanization are responsible for the increase of aerosol content. To evaluate the effects of aerosols on the atmosphere, columnar distribution of particles is required. Aerosols, along with cloud characteristics and the Earth's radiation budget, play an important role in climate change [1]. The optical characteristics of elevated aerosol layers can be affected by long-range transport at any place [2][3]. Researchers investigated seasonal variability of vertical distribution and extinction profiles using the CALIOP level 3 product [4][5]. It provides global aerosol source attribution [6] and vertical distribution of aerosols relating to atmospheric circulation [7]. They reported about the monthly and seasonal values of aerosol extinction coefficient of backscatter at 532 nm similar to our study. The difference of aerosol concentrations differs at different altitudes due to difference in columnar levels of aerosol at various heights [8]. [9] have discussed about the presence of dust layers in altitude of 4 to 6 km in north western India analysing about the vertical profiles of aerosol extinction. This effect is necessary to analyse the impact on climate change and effect on environment [10]. Similar observations have been reported by [11] and [12]. [13] have analysed about higher concentration of aerosol in winter [14] in Indo-Gangetic plain as observed in our study. [15] have analysed about higher aerosol values in winter were due to crop burning. The goal of this study is to look into the vertical distribution of aerosols throughout the seasons and determine the impact of meteorology on aerosol and climatic impact at a given place. Bhubaneswar is one of the most rapidly rising cities in terms of infrastructure, technological innovation, greenery, and attempts to achieve smart city status.

2. Methodology

The Level 2.0 data of extinction coefficient was obtained from repository of onboard Lidar based satellite of Cloud-Aerosol Lidar Infrared Pathfinder Satellite Observations (CALIPSO) observations. The horizontal resolution of the satellite observations was 40 km resolution and 0.06 km vertical resolution for profile products at all altitudes.

2.1. *Study site*

Study site is situated at a latitude of 20.35°N and longitude of 85.81°E in campus-3, KIIT-DU Bhubaneswar at a height of 10m above ground level. Detailed description about the site is explained elsewhere [16].

2.2. *Back trajectory analysis*

The 7-day back-trajectory information was obtained from NOAA-Hybrid Single Particle Langrangian Integrated Trajectory (HYSPLIT) model by using the surface and elevated level back trajectory at heights of 10, 100, 500, 1000, 2500 and 5000 metre respectively to know the direction of the trajectory in all seasons.

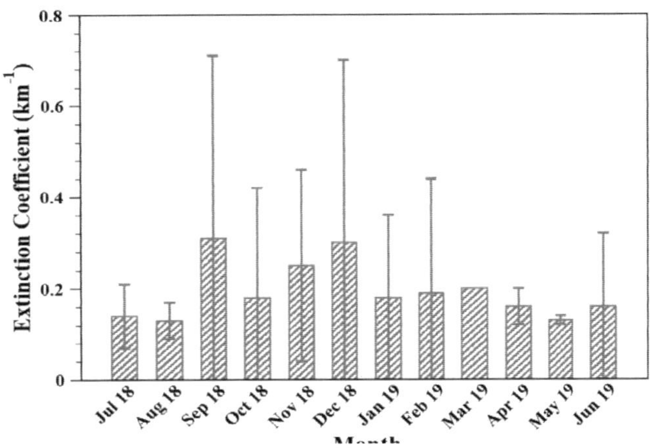

Figure 1. Monthly average extinction coefficient observed from CALIPSO satellite retrievals at study location during July 2018 to June 2019

3. Results and discussion

Month-wise aerosol extinction coefficient, that represents aerosol attenuation property, for four seasons in Bhubaneswar between July 2018 to June 2019 are as shown in Figure 1. September received highest attenuation (0.31±0.4 km^{-1}) suggesting that this month receives highest air pollution at surface level after the intense monsoon season ends. On the other hand, monsoon month August receives lowest attenuation (0.13±0.04 km^{-1}) due to the rainout or washout in the atmosphere. Seasonal average of extinction coefficients exhibits highest in post-monsoon (0.28±0.04 km^{-1}) followed by winter (0.22±0.07 km^{-1}), summer (16±0.04 km^{-1}) and lowest in monsoon (0.14±0.01) as referred as in Figure 2. The accumulation of aerosols at surface during post-monsoon season makes it highly polluted, particularly at surface level (0.54 km^{-1}) the extinction value shows 5 times higher than that at elevated level (0.11 km^{-1}).

Analysis of level wise aerosol attenuation and back-trajectory analysis is required to know about the possible transport and possible emanation of emissions from various origins, respectively. Surface and elevated level aerosol attenuation were categorized through averaging the extinction coefficient values for below and above 1 km, respectively (refer Figure 3). At surface level, peak value was observed in post-monsoon with 0.54±0.13. It indicates that the season after the monsoon such as post-monsoon receives relatively higher aerosol loading at surface level compared at

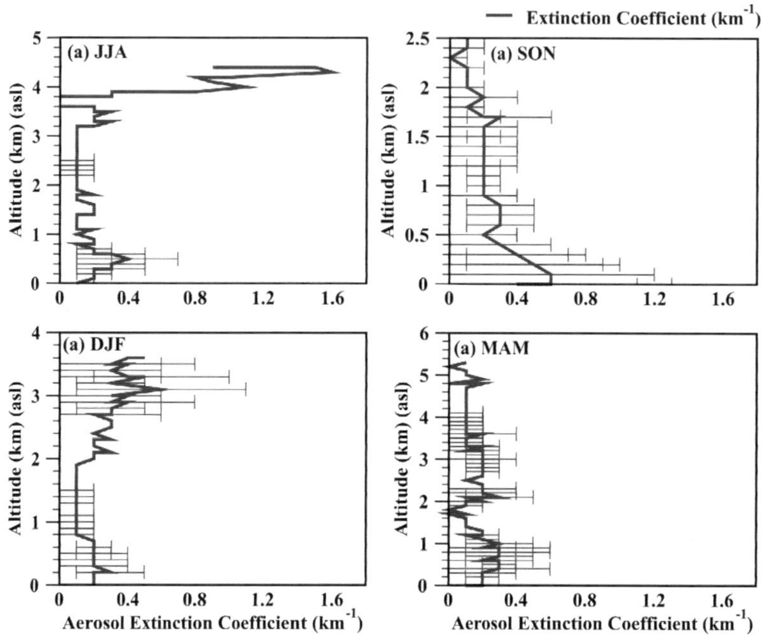

Figure 2. Vertical distribution of seasonal average (Standard deviation) of extinction coefficient observed from CALIPSO satellite retrievals at study location during July 2018 to June 2019

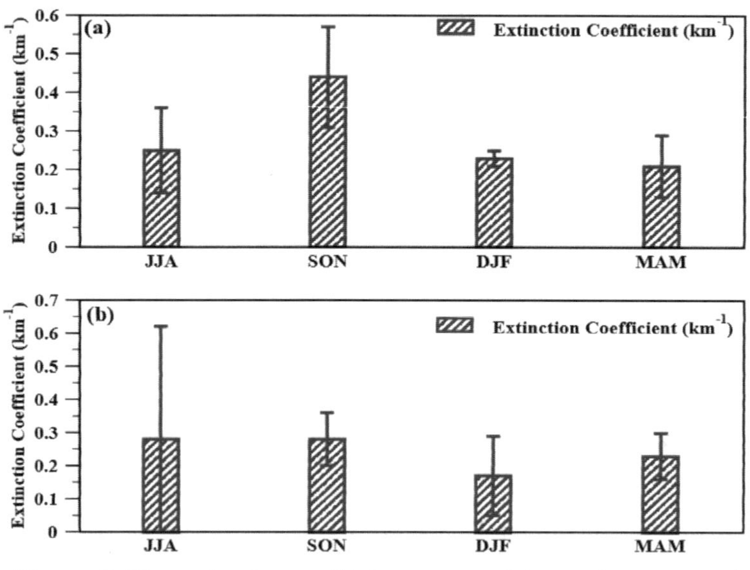

Figure 3. The extinction coefficient averaged for each season at (a) surface level: below 1 km (b) elevated level: above 1km that observed from CALIPSO satellite retrievals at study location during July 2018 to June 2019

elevated level. Whereas, a lowest value observed in winter with 0.14±0.03 km^{-1}, owing to the accumulation of aerosols at the surface followed by less convective transport in winter provides less aerosol loading, particularly at surface level. It is also supported by air-mass trajectories that exhibits scattered origins of pathways at elevated level compared of surface level where accumulation of pathways over possible source regions of eastern India and central India (refer Figure 4 and Figure 5). This indicates that elevated level transport is predominant in winter. This is reasoned by the trajectory analysis where airmass pathways traversed from elevated level (<1km) spent more time over aerosol hot spots of India including Indo-Gangetic region and central India followed by long-range sources such as west Asia (refer Figure 4 and Figure 5).

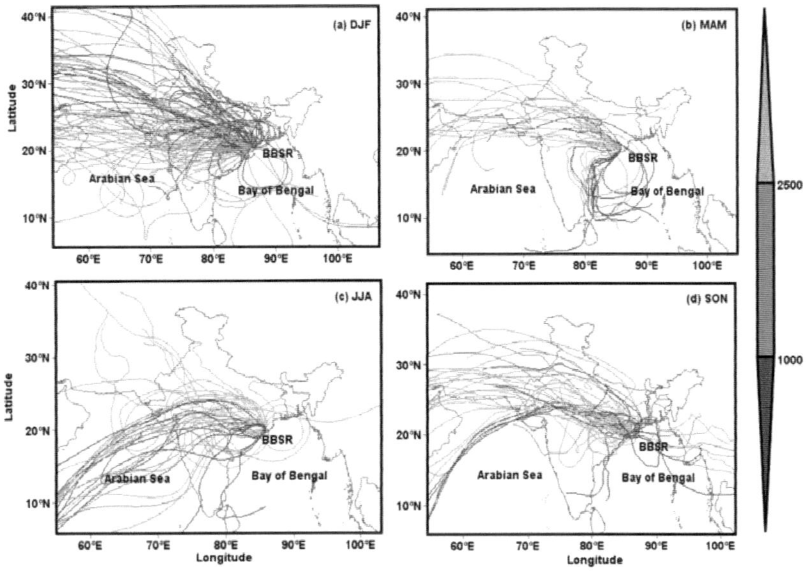

Figure 4. Air-mass pathways traversed from elevated levels (1000, 2500 and 5000m) of the receptor (study) location towards backward source origins observed at study location during July 2018 to June 2019

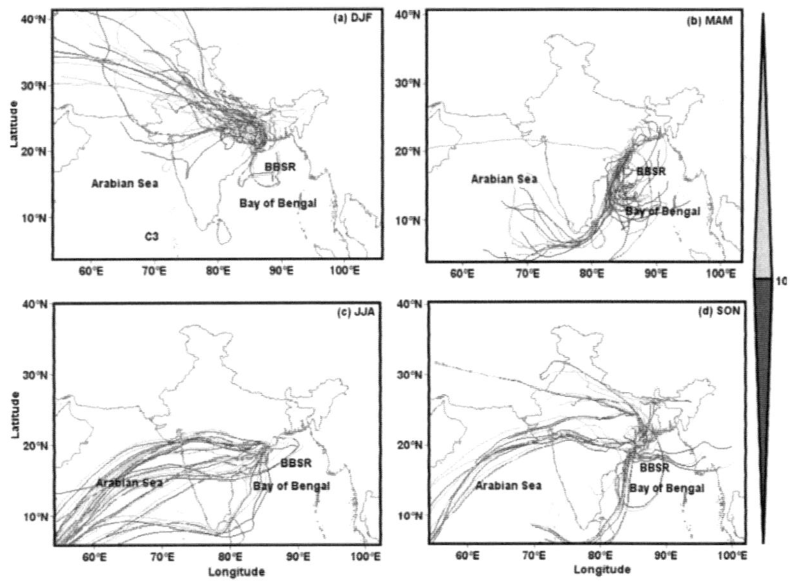

Figure 5. Same as Figure 4, but from surface level (10, 100 and 500m).

4. Conclusion

We conclude that the satellite observations of aerosol attenuation (e.g., extinction coefficient) showed highest in post-monsoon, which received aerosols at surface level relatively higher than that at elevated level. Back trajectory analysis was also corroborated with the satellite observations. Local weather effect was observed in monsoon season, where seasonal average extinction coefficient value was less that likely to receive highest rainfall in entire year.

Acknowledgments

We acknowledge Kalinga Institute of Industrial Technology-Deemed to be University (KIIT-DU) Bhubaneswar for its continuous support for providing computational laboratory facility.

5. References

[1] Boucher O, Randall D, Artaxo P, Bretherton C, Feingold G, Forster P, Kerminen VM, Kondo Y, Liao, H, Lohmann, U, Rasch, P, Satheesh, SK, Sherwood, S, Stevens, B, Zhang, XY 2013 *Clouds and aerosols, Climate Change 2013: The Physical Science Basis*. Contribution of Working Group I to the Fifth Assessment Report of the Intergovernmental Panel on Climate Change, 571-657

[2] Niranjan K, Madhavan B L, Sreekanth V 2007 micro pulse lidar observation of high-altitude aerosol layers at visakhapatnam located on the east coast of india Geo. phys. Res. Lett. 34 1-5.

[3] Satheesh S K, Vinoj V, Moorthy K K 2006. Vertical distribution of aerosols over an urban continental site in india inferred using a micro pulse lidar Geophys. Res. Lett. 33 2–6

[4] Huang L, Jiang J H, Jason L, Tackett H S, Rong, F 2013 Seasonal and diurnal variations of aerosol extinction profile and type distribution from calipso 5-year observations J. Geophys. Res.: atmospheres 118 4572–96

[5] Xu C, Ma Y, You C, Zhu Z K 2015 The regional distribution characteristics of aerosol optical depth over the tibetan plateau Atmos. Chem. Phys. 15 12065–78

[6] Prijith S S, Babu S S, Lakshmi N B, Satheesh S K, Moorthy K K 2013 meridional gradients in aerosol vertical distribution over indian mainland: observations and model simulations Atmos. Environ. 125 337-45

[7] Alizadeh-Choobari, O Sturman A, Zawar-Reza P 2014 A global satellite view of the seasonal distribution of mineral dust and its correlation with atmospheric circulation Dynam. Atmos. Oceans 68 20–34

[8] Muller D, Franke K, Wagner F, Althausen D, Ansmann A, Heintzenberg J, Verver J, 2001. Vertical profiling of optical and physical particle properties over the tropical indian ocean with six-wavelength lidar-seasonal cycle J. Geophys. Res. 106 567-75

[9] Gogoi M M, Moorthy KK, Kompalli SK, Chaubey JP, Babu SS, Manoj MR, Nair VS and Prabhu, TP 2014 Physical and optical properties of aerosols in a free tropospheric environment: results from long-term observations over western trans-himalayas Atmos. Environ. 84 262–74

[10] Mishra A K, Shibata T 2012 Climatological aspects of seasonal variation of aerosol vertical distribution over central indo-gangetic belt (igb) inferred by the spaceborne lidar caliop Atmos. Environ. 46, 365–75

[11] Liu D, Wang Z, Liu Z, Winker D, Trepte C 2008. A height resolved global view of dust aerosols from the first year CALIPSO lidar measurements J. Geophys. Res. 113 1-15

[12] Omar A, Liu Z, Vaughan M 2010 Extinction-to-backscatter ratios of saharan dust layers derived from in situ measurements and calipso overflights during namma Journal of Geophy. Res. 115 1-21
[13] Dey S, Tripathi, SN, Singh, RP 2004 Influence of dust storms on the aerosol optical properties over the indo-gangetic basin Journal of Geophysical Research 109 1-13
[14] Sasmita S, Kumar D B, Babu P 2022 Assessment of sources and health impacts of pm10 in an urban environment over eastern coastal plain of india, Environ. Challenges 7 100457
[15] Badarinath KVS, Kiran TR, Prasad VK 2006 Agriculture crop residue burning in the indo-gangetic plains- a study using irs-p6 awifs satellite data Current Science 91 1085-86
[16] Sasmita, S, Kumar D B 2021 Seasonal variability of satellite-derived aerosol optical depth in smart city bhubaneshwar (ICRDSI Conf. 11th-13th July 2019) Recent Developments in Sustainable Infrastructure (Bhubaneswar: Springer Science) **85**.

Appendix

Figure A1. Scaled, color-modulated, altitude-time images of CALIPSO for aerosol subtype for (a) monsoon (b) post-monsoon (c) winter (d) summer

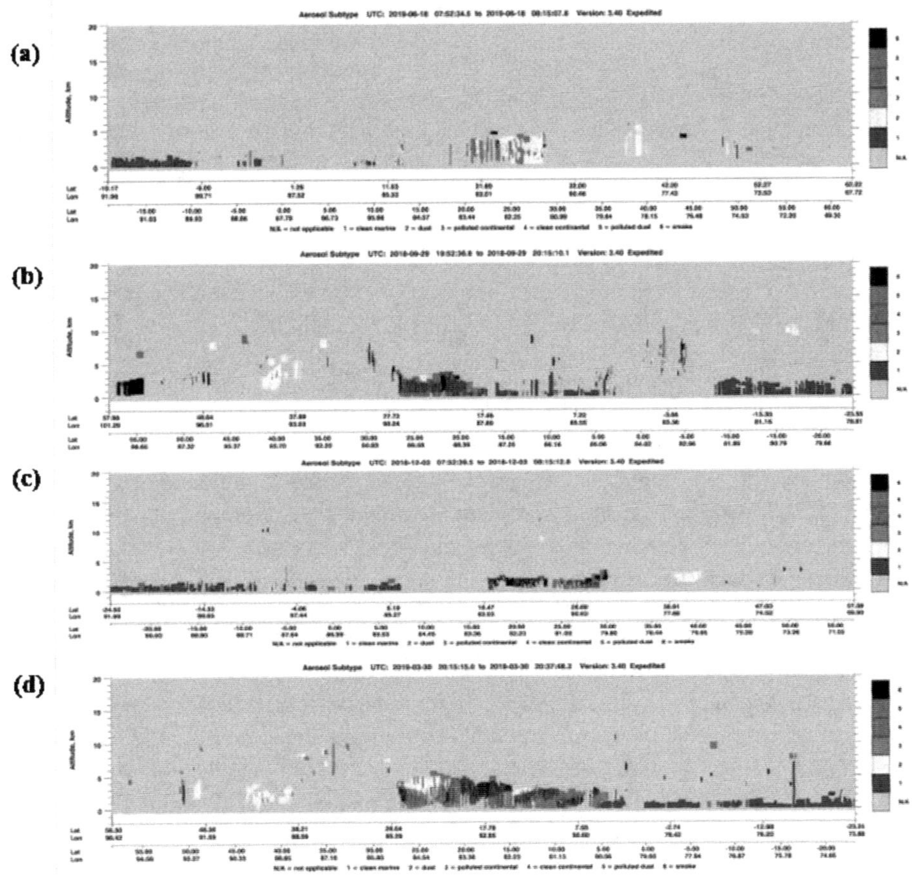

Impact of directional effect of strong ground motion on scenario-based earthquake hazards: preliminary results

A Kumar[1], **S Pasari**[1], **A Mehta**[2] **and H Verma**[1]
[1] Mathematics Department, Birla Institute of Technology and Science, Pilani, Jhunjhunu – 333031, Rajasthan, India
[2] Civil Engineering Department, Institute of Infrastructure Technology, Research and Management, Ahmedabad – 380026, Gujarat, India

Abstract. Scenario-based earthquake hazards are useful for social planning and disaster mitigation. In this study, general attenuation properties of earthquake events are analysed empirically with respect to the direction of seismic rupture from the epicentre. The study is primarily focused on presenting a relationship between fault source characteristics and the most credible direction of any earthquake that occurs at that source. Since such a direction is not only a function of source but also is dependent on site parameters, several ground motion prediction equations are utilised in congregation with methods to evaluate site parameters. The method involves a graphical relationship between scenario spectral ordinates and polar coordinates to estimate the most credible direction for that scenario. An analysis illustrating the method is presented here for the Himalayan megathrust fault, the Main Boundary Thrust.

1. Introduction

Earthquakes are one of the most prominent natural hazards that cause countless destruction to life and property both socially and economically. They often pose a high-level threat to all kinds of civil structures be it buildings, bridges, or dams. The impact of destruction is not confined at the vicinity of the source but goes as far as 300-400 km (Villaverde 2009). To reduce earthquake losses due to ground failure, several methods have been proposed to understand earthquake dynamics and their perspective location, date and time, intensity, and nature of devastation. In this direction, seismic hazard analysis, a term associated with the selection of ground motion characteristics based on historical information, geological data, statistical inferences, and empirical correlations, is often carried out. It involves both deterministic and probabilistic hazard analysis including scenario-based earthquakes, and provides information of conditional probability of earthquakes (exceedance probability curves), interevent time analysis, peak ground accelerations (PGA), and peak ground velocities (PGV) at the target areas (Villaverde 2009). The basic inputs are the types of geological faults and seismic sources, recurrence patterns, medium characteristics, and the attenuation relationships to estimate the ground motion for a given source-to-site distance and size of the seismic event. In this regard, most credible direction (MCD) is defined as the direction where PGA or spectral ordinates' amplitudes attenuate with the slowest rate as compared to other directions of attenuation. The concept of MCD not only is useful to identify high risk areas but also is important for structural engineers to effectively design seismic-resistant structures. In a recent study on the effect of ground motion direction on the response of engineering structures (Sun et al. 2016), the authors have used various statistical techniques and coordinate transformations to demonstrate that peak and phase of the ground motions are quite different in different angles. Similarly, during the 2013 Lushan earthquake in China, it was observed that the difference in structural internal force between the maximum ground motion direction input and the input of original seismic record is immense (Nie et al. 2018). Therefore, response spectrum analysis of various hypothetical (scenario-based) earthquakes in the vicinity of large-scale structures of national importance such as Taj Mahal, Qutub Minar, Fetehpur Sikri, and Red Fort could be of major help in improving the structural strength of monuments by identifying the most credible directions of perspective earthquakes originated in and around the active Himalayan seismic belt.

In view of the above, the present study provides a relationship between fault source characteristics and the MCD of scenario-based earthquakes. We compare various source dependent parameters such as fault type, fault orientation, direction of rupture, depth of rupture, magnitude of earthquake, and site dependent propagation medium characteristics to find out appropriate attenuation equations applicable to the model, usually representable with the help of spectral ordinates such as spectral acceleration, spectral velocity, PGA, and PGV. We thereafter formulate equations to evaluate associated parameters for site including different kinds of distances and fitting them with V_{S30} (the mean velocity of shear-wave to a depth of 30 m) profiles. Finally, we utilise attenuation equations and the corresponding regression coefficient tables to construct hypothetical earthquake scenario for the study site and use them to evaluate the MCD for the perspective earthquake.

2. Model formulation

The proposed model involves discussion about attenuation equation, site parameters, and model assumptions as presented below.

2.1. Attenuation equation

Ground motion intensity at a certain site is often correlated to earthquake magnitude, hypocentral distance, fault type, fault orientation, and local soil conditions (Villaverde 2009). It can be estimated with the help of attenuation relationships. Most attenuation relationships are developed using regression analysis, using data from previously measured strong ground motions. The functional form of equations is usually selected to reflect the underlying mechanism that gives rise to the ground motions (Villaverde 2009). The functional form is often expressed as:

$$\ln(Y) = C_1 + C_2 M + C_3 \ln(R + C_4) + C_5 R + C_6 f_1(\text{source}) + C_7 f_2(\text{soil}) \quad (1)$$

where, the symbols have the following meanings:

Y: Spectral ordinates / PGA / PGV

M: Magnitude of earthquake

R: Distance to source

f_1 (source): Dependent on characteristics of the fault responsible for the earthquake

f_2 (soil): Parameters (like V_{S30}) that are dependent on the site where the ground motion is being evaluated

C_i: Regression coefficients

Using previous earthquakes' response spectrum, regression models are developed to estimate regression coefficients for different time periods to essentially estimate spectral ordinates for any given values of the above-mentioned parameters. In this study, as the results are based on the active tectonic region of Himalaya, we consider the ground motion prediction equation (GMPE) provided by Akkar et al. (2010). The parameters for this model is described by the region type (active tectonic region), intensity measure types (PGA, PGV), site parameters (V_{S30}), rupture parameters (rake angle and earthquake magnitude), and distance measure type (Joyner-Boore distance, R_{JB}). A schematic diagram of an individual seismic source with its other properties is presented in Figure 1.

2.2. Site parameters

This step involves defining distance metrics and rupture related parameters for an assumed site. These distance measures are: (1) The straight distance between a site and the rupture plane (the rupture distance, R_{RUP}), (2) the shortest distance from a site to the surface projection of the rupture surface (the Joyner-Boore distance, R_{JB}), and (3) The shortest horizontal distance from a site to the strike line defined by extending the fault trace to infinity in both directions (the site coordinate, R_X); note that values on the hanging wall are positive and those on the foot wall are negative. The three rupture parameters for a fault are the dip angle (δ), width of down-dip rupture (W), and rupture from depth to top (Z_{TOR}) (Kaklamanos et al. 2011; Demircioglu et al. 2018). A realization of these three distances and three rupture parameters are illustrated in Figure 2(a) at a hanging wall site.

In order to evaluate the most credible directions, hypothetical earthquakes were generated over a distance of 100 km from the centre of the projection of top edge of fault line. Thus, the footwall (FW) belongs to the first two quadrants $(0 \le \theta \le \pi)$, whereas the hanging wall (HW) belongs to the next two quadrants $(\pi \le \theta \le 2\pi)$ as illustrated in Figure 2(b). Several mathematical equations to relate the parameters may be derived from the illustration of Figure 2 (Villaverde 2009), though we have purposefully skipped such descriptions here.

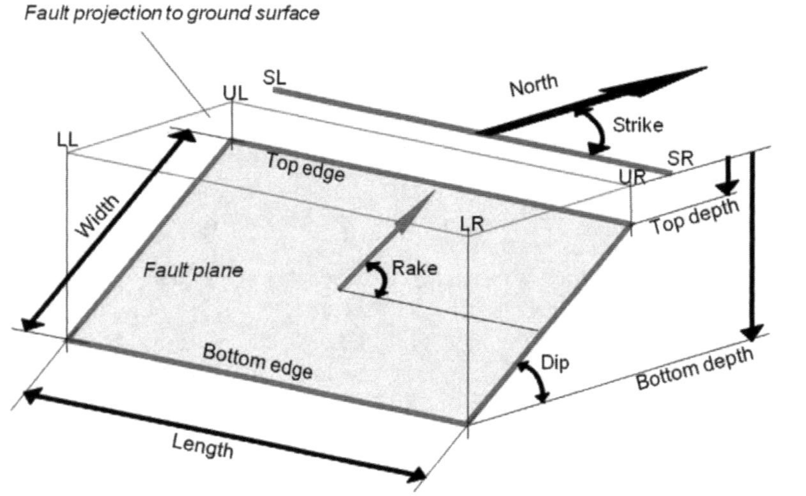

Figure 1. Schematic diagram of individual seismic source with its other properties; top depth and bottom depth are the top and bottom depths of the considered fault; LL: lower left; LR: lower right; similarly, for UL, UR, SL, and SR (Basili et al. 2008).

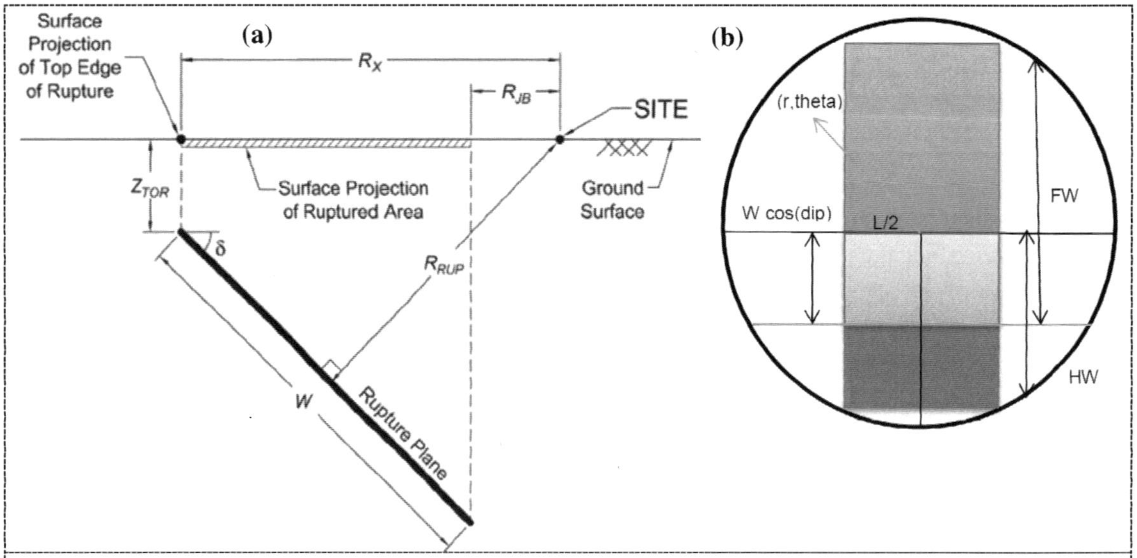

Figure 2. (a) The schematic representation of the three distance metrics and three parameters associated with the fault rupture for a hypothetical site; (b) top view of the fault plane represented in polar coordinates (r,θ); L is the length of the fault (Kaklamanos et al. 2011).

2.3. Assumptions for analysis

To develop the MCD for a scenario-based event, we consider an earthquake of moment magnitude *Mw* 6.85 and the fault parameters corresponding to the Himalayan Boundary Thrust (MBT). The three rupture parameters for the MBT are $\delta = 29^o$, $W = 10$ km, and $Z_{TOR} = 3$ km. The length of the fault is more than 2000 km, with rake angle 94° and strike angle 290° (Sharma et al. 2020). However, for computational simplicity, we consider the fault length as 40 km.

With the above considerations, a circular mesh of 100 radii was created, each with concentric circle of a fixed radius *i*, containing either $100(i+1)$ points in case of PGA or $50(i+1)$ points in case of PGV, distributed uniformly on the corresponding concentric circles. A MATLAB code was developed for constructing mesh-grids, whereas Python scripts were written for performing natural interpolation on CSV data. The results are provided in the next section.

3. Results and discussions

As mentioned earlier, we consider the GMPE given by Akkar et al. (2010) for a constant V_{S30} profile of 1200 m/s associated with a rocky soil base, and the associated spectral ordinate is taken to be peak ground acceleration. With this setting, the PGA at grid points for the MBT is pictorially represented in Figure 3(a), whereas the MCD for the PGA at grid points with constant V_{S30} profile for the MBT is shown in Figure 3(b).

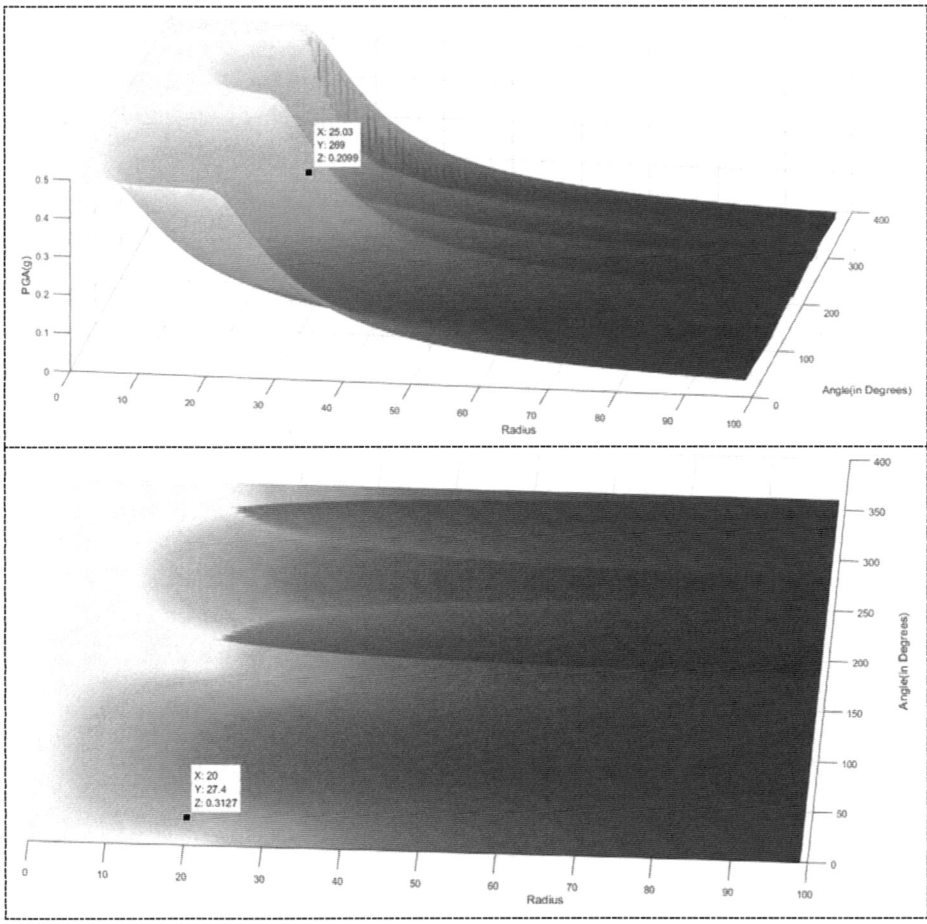

Figure 3. (a) The PGA at grid points with constant V_{S30} profile of 1200 m/s for the MBT and (b) the MCD for the PGA at grid points with constant V_{S30} profile for the MBT.

From Figure 3(b), it is observed that the model generates three ranges of directions measured at a distance of 20 km from the centre in which peak ground acceleration attenuates with the slowest rate. Moreover, we note that there are two directions of steepest descent (attenuation) in between these three ranges of MCDs. Also, we notice another interesting phenomenon at a distance of 22.14 km from centre in which the directions fork into two parts at angles of 334°and 200°. Moreover, the first fork at 200° furthers dissects itself into directions that curve the converge to 187° and 260°, whereas the other fork at 334° furthers dissects itself into directions that curve the converge to 281° and 354°. Therefore, we observe that PGA attenuation varies with different angles and radial distances, indicating the dependency between direction of propagation and earthquake origin characteristics.

4. Summary and conclusions

In this study, we have presented a relationship between fault source characteristics and the most credible direction of any earthquake that occurs at the source. For this, we have discussed attenuation equations, site parameters, and associated model assumptions corresponding to the Main Boundary Thrust. We derived peak ground acceleration and the most credible direction for the peak ground acceleration for a hypothetical earthquake. It was observed that the implemented model generates three ranges of maximum credible directions with a forking nature at two places. Therefore, the present study has successfully implemented the directional effect of strong ground motion.

Although we have derived surface curves assuming constant parameters for sites, a similar real-time MCD estimation can be performed by considering the actual soil profiles around the source. The analysis can also be carried out for different response time periods just by changing regression coefficients for the appropriate frequency. This will lead to obtain credible directions for the respective response time periods for a particular fault to enable risk analysis of selected monuments of national importance.

5. References

[1] Akkar S, Sandıkkaya MA and Bommer JJ 2014 Empirical ground-motion models for point- and extended-source crustal earthquake scenarios in Europe and the Middle East, *Bulletin of Earthquake Engineering* **12**, 359–387.

[2] Basili R, Valensise G, Vannoli P, Burrato P, Fracassi U, Mariano S, Tiberti MM and Boschi E 2008 The database of individual seismogenic sources (DISS), version 3: summarizing 20 years of research on Italy's earthquake geology, *Tectonophysics* **453**, 20–43.

[3] Demircioglu MB, Sesetyan K, Duman TY, Can T, Tekin S and Ergintav S 2018 A probabilistic seismic hazard assessment for the Turkish territory: part II—fault source and background seismicity model, *Bulletin of Earthquake Engineering* **16**, 3399–3438.

[4] Kaklamanos J, Baise LG and Boore DM 2011 Estimating unknown input parameters when implementing the NGA ground-motion prediction equations in engineering practice, *Earthquake Spectra* **27**, 1219–1235.

[5] Nie G, Zhang C, Dai J and Liu K 2018 Seismic damage investigation and seismic performance study of space double-layered lattice structure, *Journal of Performance of Constructed Facilities* **32**: 04018003.

[6] Sharma Y, Pasari S, Ching KE, Dikshit O, Kato T, Malik JN, Chang CP and Yen JY 2020 Spatial distribution of earthquake potential along the Himalayan arc, *Tectonophysics* **791**: 228556.

[7] Sun M, Fan F, Sun B and Zhi X 2016 Study on the effect of ground motion direction on the response of engineering structure, *Earthquake Engineering and Engineering Vibration* **15**, 649–656.

[8] Villaverde R 2009 *Fundamental Concepts of Earthquake Engineering*, CRC Press.

Acknowledgments

The authors thankfully acknowledge anonymous reviewers for this useful comments and suggestions. We are also thankful to Integrated Research on Disaster Risk, International Center of Excellence (IRDR ICoE Taipei), Taipei and International Science Council Regional Office for Asia and Pacific (ISC ROAP) for the financial support through seed grant for 2018 TC-EHRA. We also acknowledge the financial support from SERB-MATRICS grant.

IOP Conf. Series: Earth and Environmental Science 1032 (2022) 012043 doi:10.1088/1755-1315/1032/1/012043

Assessment of active tectonics in the Siwalik basin around the Subansiri river, NE India.

Archana Singh*[1], Bijoylakshmi Gogoi[1], Devojit Bezbaruah[1] and Neelratan Singh[2]

[1]Department of Applied Geology, Dibrugarh University, Dibrugarh, India

[2]CSIR- Central Institute of Mining and Fuel Research, Nagpur, India

*Corresponding author, e-mail: **itsarchana.singh456@gmail.com**

Abstract. In this paper we evaluate the morphotectonics of the area around the Subansiri river in north-eastern India. Research focuses on understanding the impacts on areas experiencing active tectonic deformation. It also explains the variety of kinematics observed in the area from the Miocene to the present. The tectonics can explain some structures related to the extensional tectonics predominant in the region and the development of the minor and major morph structures driven by faults with different kinematics. We developed a morphotectonic evolutionary model for the area based on the morphotectonic analysis and geological mapping. Morphotectonic indices used to evaluate the tectonic activeness in this study are Mountain Front sinuosity Index, Valley Floor Width to Valley Height Ratio, Asymmetry Factor, Transverse Topographic Symmetry Factor, Mountain Front Steepness Index, Basin Shape Index and Stream Length Gradient Index. There are numerous lineaments present in the study area which are trending in NW-SE and NE-SW direction.

1. Introduction

The Himalayas, the Mishmi Hills, the Naga Hills and the Dhubri fault surrounds the Assam block from all sides. The Himalayan range, the world's youngest mountain range, was created when the plates of the Indian and Eurasian continents collided around 50 million years ago [25]. There have been several major tectonic events in the area. Micro-tectonic data has been gathered in several Himalayan locations during the twentieth century and used to estimate the area's activity level. The Himalayan Frontal Thrust appears to be one of the most active zones in the Himalaya, as evidenced by the upliftment of terrace deposits, the creation of steep scarps, and the upliftment of Siwaliks over recent alluvium. Because the current morphology of the mountain ranges is the outcome of both tectonic and erosional action, we can regard recent and active tectonics as the principal mechanism leading to the rock upliftment [28, 2]. Geomorphic landforms including tilted fans, fluvial terraces, unpaired terraces, aberrant drainage patterns, mountain front sinuosity, and drainage basin asymmetry are analysed in tectonic geomorphology [27, 28, 5, 6, 7, 36, 37, 33, 25, 21, 11, 12].

We are interested in understanding the morphotectonic evolution and the current state of tectonic activity. The study focuses on drainage network, topography and geomorphic appearance. Tectonic activity influences drainage patterns and river behaviour [19, 35, 38] which may be quantified as well as qualitatively stated [18, 22]. Drainage patterns are susceptible to folding and faulting in tectonically active areas. This process accelerates river incision, asymmetries, complex drainage geometry, and river deflections [11]. Assessing earthquake risks necessitates a thorough examination of active tectonics, particularly in locations where there has been a lot of tectonic activity recently. [22]. It is difficult to identify a specific region for quantitative studies of active tectonics on a regional scale [13]. Furthermore, this study will investigate whether the tectonic activity classification developed when working in semi-arid or arid climates can be applied to the region with subtropical or tropical climates.

2. Study area

The current research area includes sections of Assam's Lakhimpur and Dhemaji district and Arunachal Pradesh's Lower Subansiri, Kamle and Lower Siang district (Figure1). The toposheet numbers 83I/2, 83I/3, 83I/6, 83E/15 and 83E/14 are used to carry out the study. We did the field investigation on the Assam-Arunachal border route sector. It covers the Lower Subansiri dam site region, the Gerukanala river section, the Dirpai nala river section, the Dakhin Rupahi hamlet in the Subansiri River downstream, and the Jiadhol River section.

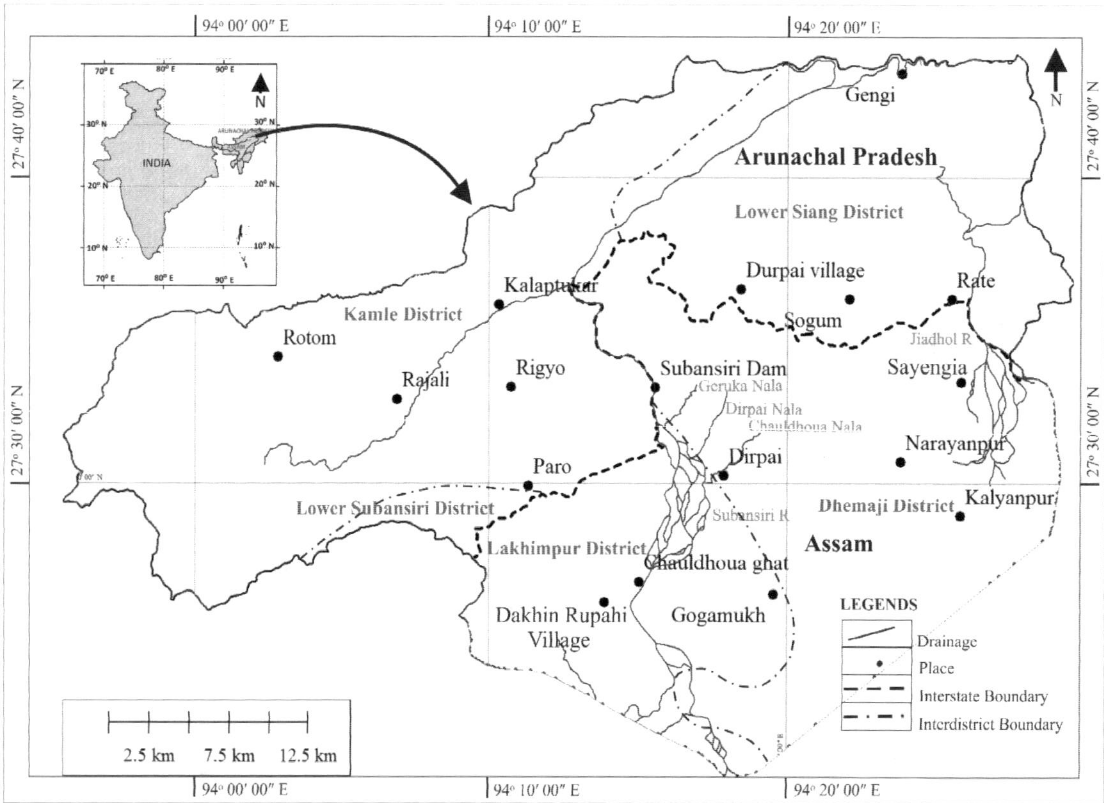

Figure 1. Location Map showing major rivers and places in the study area.

3. Geology

On account of its structural characteristics and stratigraphy, the Arunachal Himalaya is categorized into two tectonic groups, which is the Frontal Fold Belt (FFB) with sediments and rocks of the Siwalik Group, that constitutes the Sub-Himalaya, and the second one is the Main Himalayan Belt (MHB) with sediments and acid magmatic rocks of the Paleoproterozoic to Middle Eocene period which forms the

Lesser and the Higher Himalaya. These two belts are divided by the Main Boundary Fault (MBF) [23]. [19] Karunakaran and Ranga Rao (1979) of ONGCL, based on lithology, classified the Siwalik rocks in Arunachal Pradesh into three units (Table 1) (Figure 2).

Table 1: Stratigraphic succession of the Siwalik.

Age	Lithotype
Recent to Sub-Recent	Alluvium and older deposits
Plio- Pleistocene	Kimin Formation: soft sandstone, shales, siltstone and conglomerates, etc.
Middle to Upper Miocene	Subansiri Formation: coarse to medium grained, massive to poorly bedded, pebbly micaceous soft friable sandstone, shale with coal streaks.
Middle to Lower Miocene	Dafla Formation: fine to medium grained silty Sandstone, shale, claystones, mudstone and carbonaceous shale.

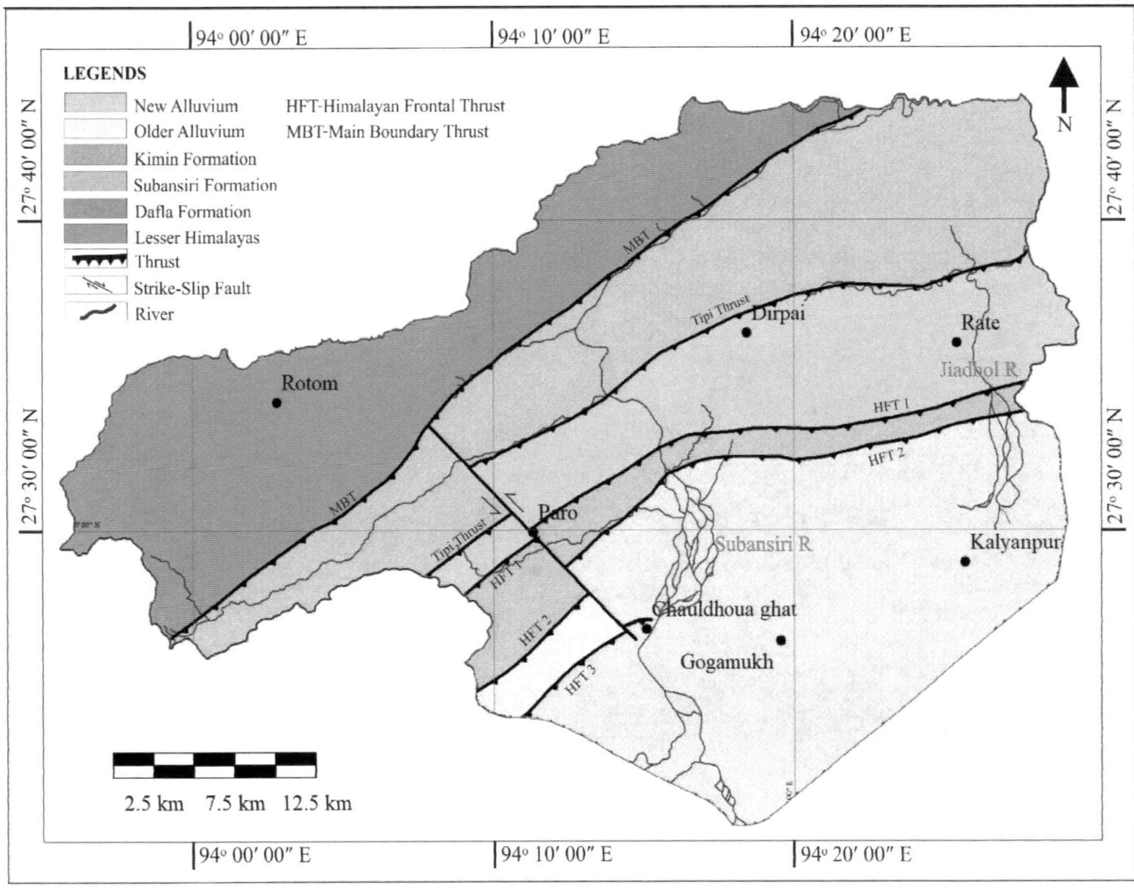

Figure 2. Geological map showing major thrust and lithologies present in the study area

We studied sections of Subansiri Dam, Gerukanala Section, Dirpai River Section, Jiadhol River Section, and the Paharpur area of landslides and the Dakhin Rupahi Village. The dam site at Subansiri has a section of road that runs along the Siwaliks. The lithology consists of alternate bands of sandstone and conglomerate. The high-level river terraces in this area lie unconformably above the rocks of the Siwalik group. These sections are characterized by several small-scale faults (Figure 3).

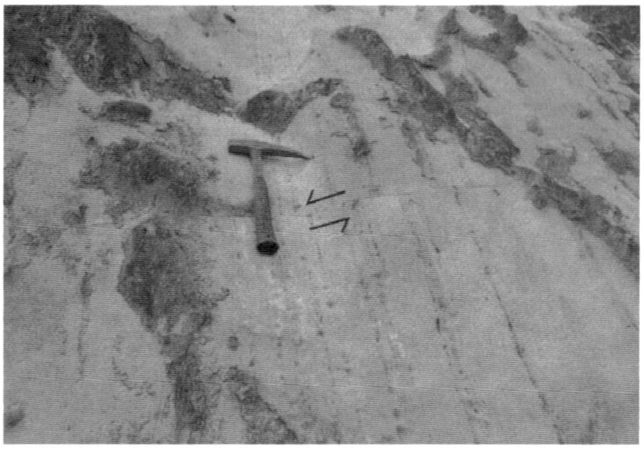

Figure 3. Small scale fault nearby Subansiri River.

There are alternate bands of sandstone and boulder beds exposed in the Gerukanala section. The Gerukanala is one of the tributaries of the Subansiri River. The sandstone present is greyish in colour and has friable characteristics and belongs to the Kimin Formation of the Siwalik Group. High-level river terrace deposits with a horizontal disposition overlie this deposit (Figure 4). It is evident that the conglomerate bands are brecciated, which is indicative of a thrust zone marked by the designation HFT-1 on the geological map (Figure 5). Several unconformities (Figure 6), minor faults and a fault surface (Figure 7) are exposed in the Gerukanala river bank. HFT-1 is the name given to the entire fault zone that separates the Subansiri and Kimin formations.

Figure 4. Kimin Formation overlain by terrace deposit.

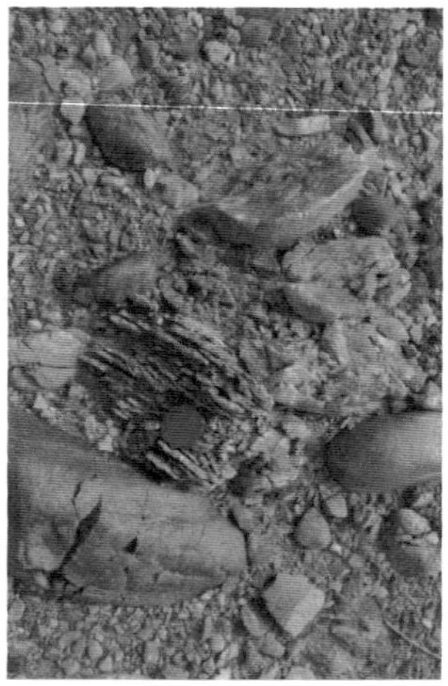

Figure 5. Crushed pebble on the Gerukanala river section

Figure 6. Unconformity in Gerukanala river section

Figure 7. Fault surface in the Gerukanala river section

There are terrace deposits 5.3 meters thick in the Dirpai nala river section. It is one of the tributaries of the Subansiri River. In the base of the terrace deposit, boulders, cobbles, and pebbles have been deposited, which are overlain by the cross-bedded sand unit. There are low hills towards the north and a flat, featureless plain towards the south, which are part of the Subansiri floodplain. The abrupt topographical change in the region is evidence of the presence of the thrust (Figure 8), marked as HFT-2 on the geological map.

Figure 8. Sharp evidence of HFT on the Jiadhol river section

Figure 9. Crushed zone along the Jiadhol river section

A terrace deposit can be found in the Jiadhol River section at its mouth. It is also the tributary of the Subansiri River. It is known as Kumotiya River in Assam from Gogamukh. A zone of brecciate and shear sandstone exists north of the terrace deposit, indicating the presence of HFT-1 (Figure 9). The rocks of the Subansiri Formation are found north of the thrust zone. The Kimin sandstones are exposed in a landslide zone at Sayengia and Paharpur, west of the Jiadhol River section (Figure 10). Sandstone, siltstone, and shale make up the lithology, interspersed with a conglomerate substrate. The clasts in the conglomerate strata are mostly quartzite, with sandstone boulders as subordinates.

The Kimin Formation is unconformably overlain by river terraces where we find channel fill deposits (Figure 11). The presence of sandstone boulders within the Kimin Formation indicates that the Dafla and Subansiri Formations were uplifted and became the source during the deposition of the Kimin

Formation (Figure 12). The topography abruptly changes to featureless plains to the south of this exposure, indicating the presence of HFT-2 nearby.

Figure 10. The Kimin Formation near Sayengia and Paharpur

Figure 11. Channel fill deposit above the Kimin Formation

Older alluvium and recent deposits can be found near Dakhin Rupahi village, along the Subansiri Riverbank. The bank stratigraphy comprises lithified conglomeratic bed at the base followed by a yellowish-brown colour older alluvium, which is overlain by recent deposits of Subansiri River. The recent deposit consists of fine-grain sand, overlain by a silt layer of around 20 cm thickness. Downstream of this location, the recent deposit near the riverbank has been eroded and exposed older alluvium. Here, over the older alluvium, we find lots of upright tree trunks, which indicates the presence of a forest buried under recent deposits (Figure 13). The forest submerged under the flood during the 1950 great Assam earthquake, as per the information provided by the area's residents. The great Assam earthquake of 1950 caused great destruction, a dam formed by landslides eventually burst and caused the great flood in the Subansiri river [17, 34]. The historic flood of 1950 contributed so much sediment that it deposited up to 1 m of coarser sediment comprising pebbles and coarse sand over the earlier flood plain alluvium, as can be seen in Chauldhoa ghat, Assam [17].

Figure 12. Pink colored sandstone present in the Kimin Formation.

Figure 13. Tree trunks on the Subansiri river buried by flood after 1950 earthquake.

4. Materials and Methodology

Through the use of various morphotectonic indices, the paper examines the area's active tectonic activity. Morphotectonic indices taken for the study are Mountain Front Sinuosity Index (Smf), Valley Floor Width to Valley Height Ratio (Vf), Asymmetry Factor (Af), Transverse Topographic Symmetry Factor (T), Mountain Front Steepness Index (S), Basin Shape Index (Bs), Stream Length Gradient Index (SL) and Relative Tectonic Activity Index (Iat). We calculated the morphotectonic indices for four mountain fronts and 32 watersheds within the study area (Figure 15). The study uses topographic maps by carrying out extensive field surveys and laboratory work. The general methodology applied in the study is given in the flow chart (Figure 14). Exposed rock outcrops are studied during field investigation using geological field tools such as Brunton Compass, hammer and GPS. Laboratory investigation involves geological maps preparation, map digitization and morphotectonic index calculation. ASTER-GDEM of 30m resolution is used to obtain the elevation data required during the preparation of geological profiles, cross-sections and morphotectonic indices calculation.

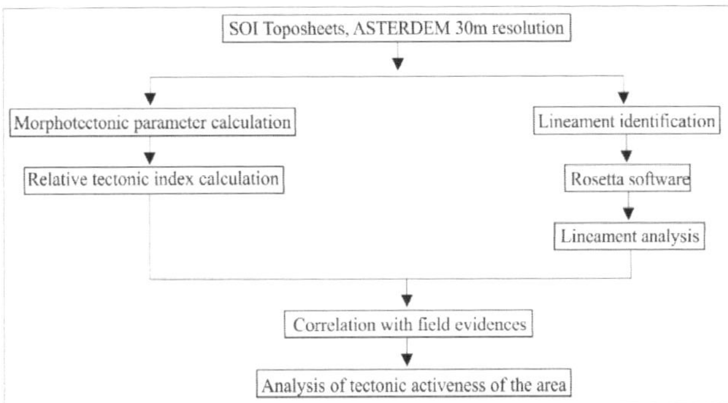

Figure 14: Methodology flow chart of the study.

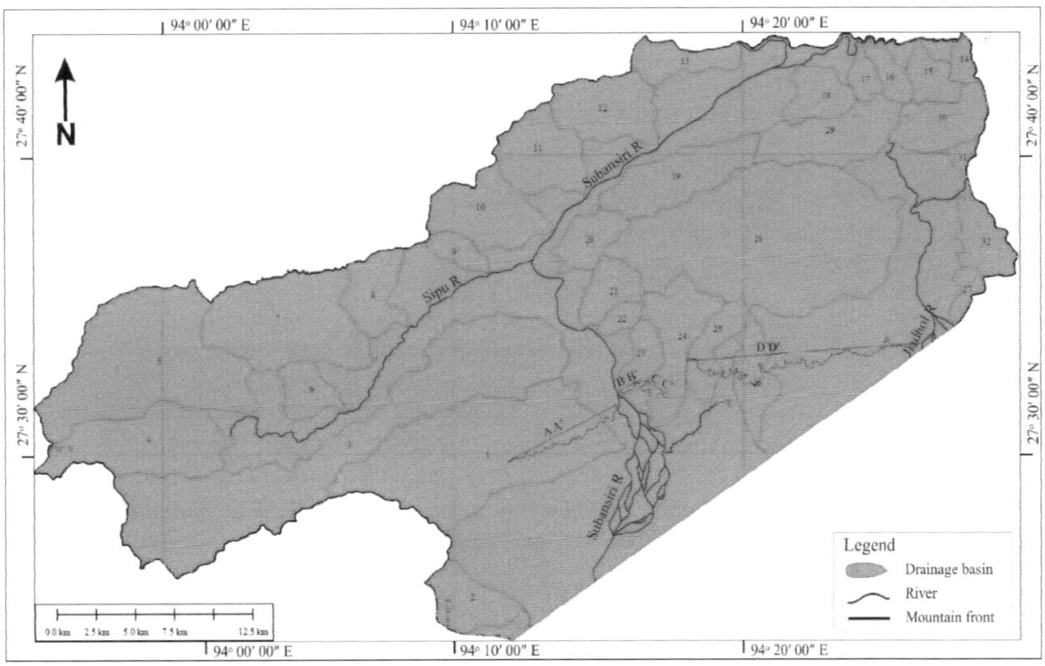

Figure 15. Drainage basin/watershed and Segments of the mountain front.

5. Morphotectonic indices and results

5.1. Mountain front sinuosity index

Many mountains are characterised by straight or curved fronts due to a constant struggle between active tectonic pressure and erosion. Tectonic activity produces a majority of straight mountain fronts; these faults can be evaluated to derive basic information about the tectonic activity and long-term deformation [39]. Mountain front Sinuosity Index (Smf) measures how much erosional modification has taken place on tectonic structural features or topographic escarpments bounded by fault [4, 6, 7, 22, 31, 34, 39].

It is described as the ratio between Lmf and Ls where Lmf is the length between two points of the mountain front along its base where there is a change in slope from hills to plain, and Ls is the straight length between the two ends of mountain front [6, 7] and is expressed in the formula:

Smf= Lmf/Ls

The Smf index in this study is measured at four fronts A A', B B', C C'and D D'. These fronts are separated by Dulang River, Subansiri River, Gerukanala, Dirpai river and Jiadhol river. The obtained Smf values are analysed and shown in Table 3 as per the ranges given by various authors. The ranges given by different authors are presented in Table 2.

Table 2: Tectonic activity classes and their ranges given by different authors.

Authors	Tectonic classes	Smf ranges	Vf ranges
Bull & Mcfadden (1977)	1	1.0 to 1.6	0.05-0.9
	2	1.4 to 3.0	0.5- 2.0
	3	1.5 to >5	>2.0
Rockwell et al. (1984)	active fronts	1.01- 1.34	0.43-1.91
	less active fronts	1.57- 2.72	
Silva et al. (2003)	1	1.17- 1.53	<0.5
	2	1.8- 2.30	0.3-0.8
	3	2.8- 3.5	>0.7(mostly 0.8-1.2)

Table 3 The Mountain front sinuosity values and their tectonic activity class.

Front	Smf (avg)	Tectonic activity class		
		Bull and McFadden (1977)	Silva (2003)	Rockwell (1984)
A A'	1.381	1	1	1
B B'	1.382	1	1	1
C C'	2.066	2	2	2
D D'	1.975	2	2	2

The mountain front analysis shows front A A', B B' as active and C C', D D' as less active. None of the segments came out as inactive. The ranges given by all the mentioned authors suitably indicates the area's tectonic activeness.

5.2. Valley Floor Width to Valley Height Ratio (Vf)

The Vf Index is calculated at the valleys at some distance upstream from the mountain front. It can be counted as an essential stability indicator. It signifies the dissimilarity between valleys with V and U shapes. The V-shape valleys indicate high tectonic activity and show active uplift, whereas U-shape valleys are broad-floored due to prominent erosion in the lateral sides up to the adjoining hillslopes, and it displays low tectonic activity. Vf with high value is related to U shape valley and low uplift rates, while low Vf values represent deep V-shaped valleys with actively incising streams linked with uplift [4, 6, 7, 8, 21, 30, 32, 37]. The Vf index is expressed by the formula [7],

$V_f = 2V_{fw}/ \{(E_{ld} - E_{sc}) + (E_{rd} - E_{sc})\}$ Where V_{fw} is the valley floor width measurement, elevation of the left and right valley divide is E_{ld} and E_{rd} and E_{sc} is the valley floor elevation.

The Vf index is calculated at twelve locations in the respective mountain fronts segments. Two transect distances are taken to determine the valley profiles, one at 250 meters [34] and another at 1 km upstream of the mountain fronts [7]. The calculation results at two different transect distances can be analyzed and compared in terms of upliftment, incision, and erosion.

For the valley transect profile taken at 250m upstream in the river from the mountain front. The Vf class at front A A' is moderately active, highly active at front B B', C C' and D D'. Similarly, we calculated the Vf index for the Subansiri River, and the result infers that it is a class 3 U type valley indicating moderate tectonic activity. The Vf index measured for Jiadhol river indicates V shaped valley and high tectonic activity class. For the valley transect profile taken at 1km upstream in the river from the mountain front. The Vf class at front A A', B B', C C' and D D' is highly active. It is a class 1, V type valley for the Subansiri River and Jiadhol River, indicating high tectonic activity.

5.3. Asymmetry Factor

The watershed development in a tectonically active terrain is affected by the ongoing active deformation. Streams in the watershed follow distinct patterns and geometries that can be qualitatively and quantitatively described [18, 22]. To identify tectonic tilting transverse to flow at watersheds and larger scales, the asymmetry factor was originated [18, 22]. As with most morphotectonic indices, the Af is most effective where the same lithology underlies all the watersheds. This method assumes that asymmetry doesn't occur due to lithologic formations such as dipping sedimentary layers or the climate in a region [22]. The formula obtains the Asymmetry Factor (AF) of a watershed, AF = 100 (Ar /At), Where, At denotes the total watershed area and area of the watershed to the right (facing downstream) of the trunk stream is Ar.

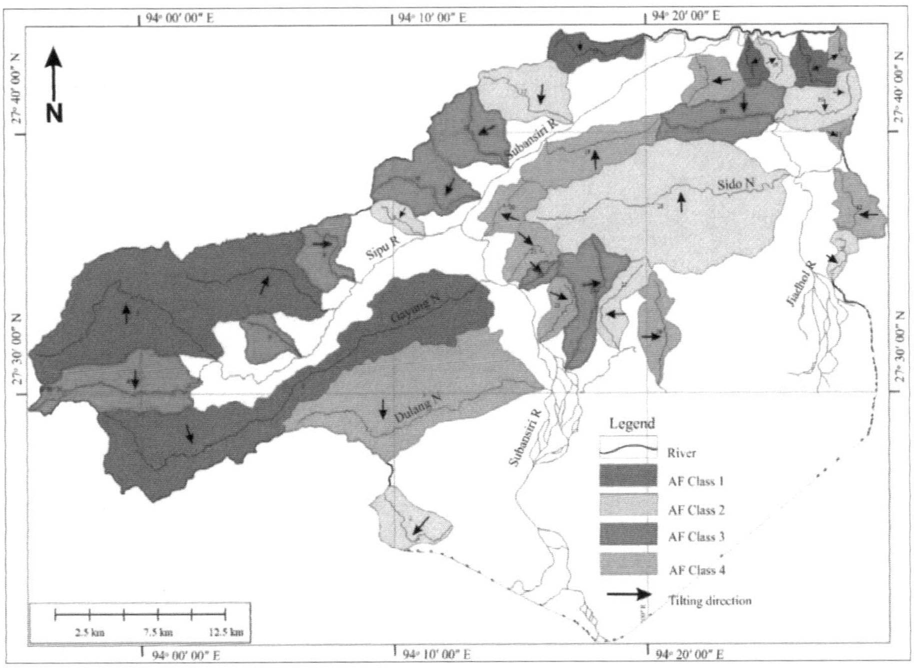

Figure 16. Asymmetry map of the study area.

The Af values for the watersheds range from 1.11 to 30. We have used the absolute Af values to determine the degree of asymmetry for each watershed. The absolute Af are categorized into four classes: symmetric watershed where Af is less than 5, gently asymmetric watershed where Af ranges between 5 to 10, moderately asymmetric watersheds where Af ranges between 0 to 15 and strongly asymmetric watersheds where Af is greater than 15 [27]. Many watersheds have Af values greater or less than 50, which indicates tilt in those watersheds due to tectonic activity. There is prominent tilting in the E, NE and SW directions. Out of 32 watersheds selected for the study, 10 watersheds are strongly asymmetric, 8 is moderately asymmetric, 8 is gently asymmetric, and the rest 6 are symmetric watersheds (Figure 16). It can be concluded that Af shows the area to be tectonically active.

5.4. *Transverse Topographic Symmetry Factor (T)*

T evaluates asymmetry and determines the tilting direction based on the deviation from the watershed midline of the trunk stream. It is computed with the formula T = Da / Dd, Where Dd is the distance taken from the watershed midline to the watershed divide and Da is the distance taken from the watershed midline to the midline of the stream. The T values show the shifting at a right angle to the watershed axis. In other words, it is a vector whose direction and magnitude ranges from zero to one, corresponding to the perfect asymmetric watershed [8, 11, 12, 22].

The T value ranges from 0.05 to 0.99, showing highly active tectonic activity. With increasing T values, the watershed becomes increasingly asymmetric. Changes in slope caused by tilting of the watershed in a particular direction cause the stream to shift in that direction and the length of the tributaries to increase. As we observe the trend of the streams in all the watersheds, the main channel in each watershed follows an E, NE and SW direction.

5.5. *Basin Shape Index*

An index of basin shape (Bs) may indicate the horizontal projection of a basin or watershed [9, 30]. It is calculated with the formula Bs = Bl / Bw, Where Bl is the length between the highest point to the mouth of the watershed, and the width of a watershed at its widest point is Bw. In tectonically active areas, watersheds are elongated in shape and perpendicular to the topographic slope [7, 30]. So, it is possible to find the rate of active tectonics through Bs.

It is found in the study that most of the watersheds in the region are circular. Four watersheds are semicircular, one is elongated, and all the rest are circular. In this region, the Bs index is not indicating tectonic activity. So, it can be concluded that the Bs index is not applicable to find the tectonic activeness in the Eastern Himalaya because of its weak and friable lithology and high precipitation level.

5.6. *Mountain Front Steepness Index*

By examining the slope in the frontal portion of the mountains, the Mountain Front Steepness Index (Ms) detects the transition in topography in frontal regions of mountainous ranges from hill to plain. The cross-section of the mountain front produced from a DEM is used to calculate this index. A tectonically active mountain front will have a steeper mountain front than a less active one if the region's lithology, climate, and vegetation are all consistent. Ms is defined as Ms = Y/X, Where the vertical distance between the hill's highest elevation points and the point where the slope breaks are Y and the horizontal distance between the hill and the plain where remarkable slope changes occur is denoted by X.

The mountain front steepness index is calculated in this study by obtaining perpendicular cross-sections of the mountain fronts A A', B B', and D D'. The Ms value for the front A A', B B', and D D' is 0.31, 0.42 and 0.23. The Ms values indicate that segment D D' is more active than the other two segments.

5.7. *The Stream Length-Gradient Index (SL)*

The Stream Length-Gradient index can be used to evaluate any divergence in river longitudinal profile stability caused by tectonic, climatic, or lithological factors. [17]. Hack (1973) defined the stream length gradient index as SL = (ΔH / ΔL). L Where ΔH is the elevation change, ΔL is the length of the reach, the ratio of ΔH/ΔL is the channel slope, and the channel length between the midpoints of the reach to the divide is L. Because the SL index is sensitive to variations in channel slope, it may be used to assess the link between the tectonic activity, rock resistance, topography, and stream length. [3, 17, 22, 40]. The required parameters for calculating the SL index are obtained from topographic maps. [3, 17, 22].

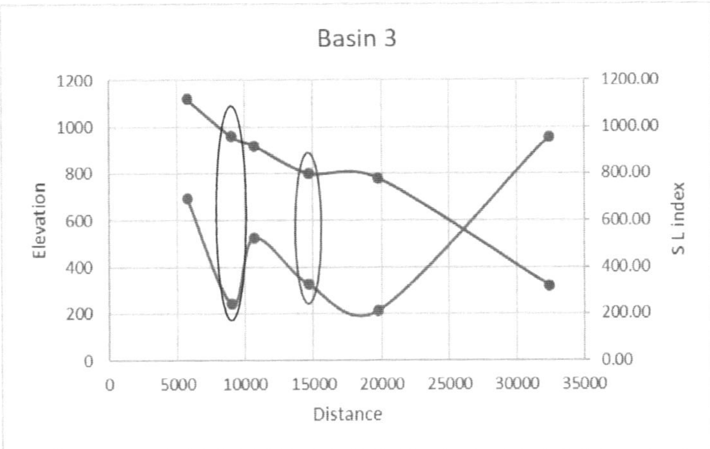

Figure 17 Following Graph showing relation between Slope and SL index of a watershed. (Slope in the graph is represented in brown line and SL index in blue line)

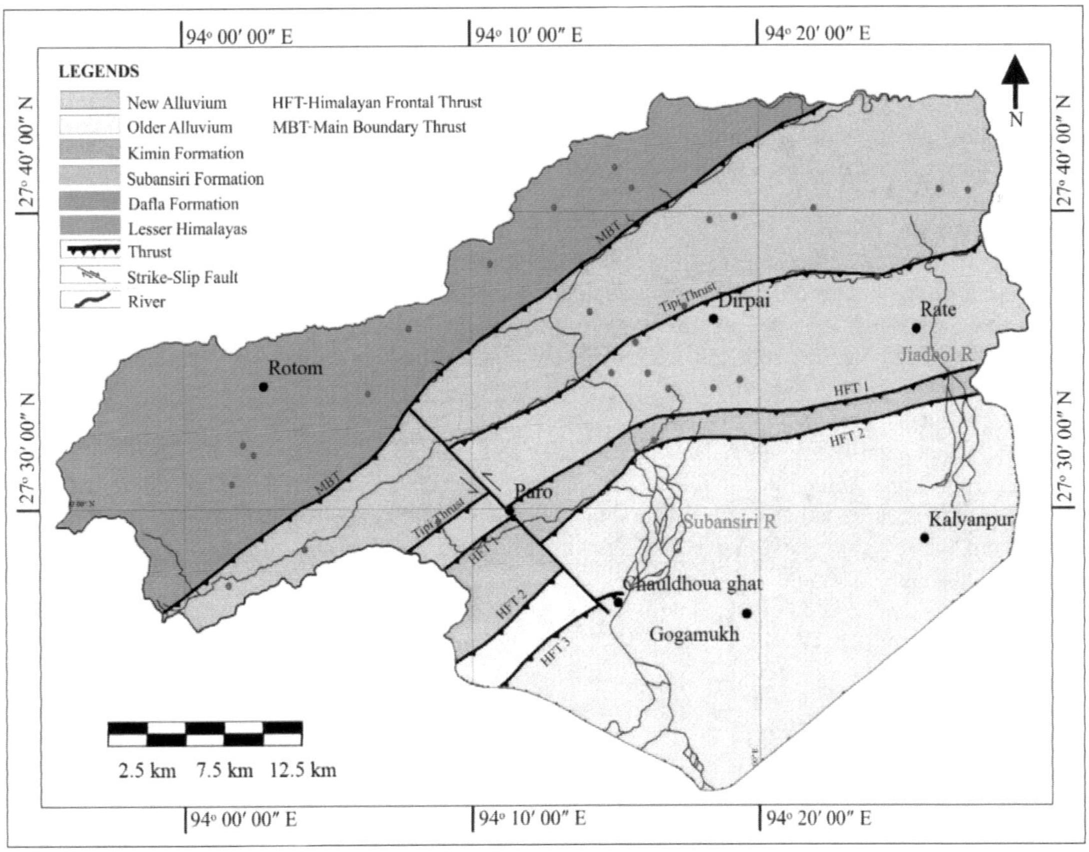

Figure 18: Map showing S L points in the area.

For watersheds of third or higher-order, the SL index is measured. The SL graphs are made by plotting the SL values against L, and the longitudinal profile is prepared by plotting elevation against L. distance from the source to the point of measurement in the river profile (Figure 17). From the overlay of the SL index with the longitudinal profile, it is observed that the anomalous value of the SL index lies close to the break in slope in the longitudinal profile. When these anomalous points are marked in the geological map (Figure 18), many Sl points are noted to fall right over the fault or lined parallel to the major faults in the vicinity. It can therefore be said that all of the primary faults in the study region are active.

5.8. Relative tectonic activity index (Iat):

We use the approach described by [13] to calculate an index that represents relative tectonic activity over a specific region in this study. The different indices are grouped into three categories, with class one indicating high activity and class four indicating low activity (Table 4). Iat is calculated by taking the average of tectonic activity classes of different morphotectonic indices (S/n). It is divided into four classes to determine the degree of tectonic activity where class 1 denotes very high tectonic activity (between 1 to 1.5), class 2 denotes high tectonic activity (S/n> 1.5 but < 2), class 3 denotes moderate tectonic activity (S/n>2 but < 2.5), and class 4 denotes low active tectonics (S/n> 2.5).

Table 4: Morphotectonic indexes classification for Iat analysis [13].

Class	Smf	Vf	S.L.	Af	Bs
1	< 1.1	<0.5	>/500	Af>/65 or <35	>4
2	1.1- 1.5	0.5-1	300\<	35\<Af<43 or 57\<Af<65	3-4
3	>1.5	>300	<300	43\<Af<57	<3

Iat values in the study area are mainly class 2 and class 3. The data indicates that the area is experiencing high and moderate tectonic activity.

6. Lineament Analysis:

Fractures, jointing, and other linear geological processes that occur everywhere from the terrain surface to potentially large depths are shown as lineament on the terrain surface. [13]. Lineament is a straight or slightly curving feature topographically manifested as depression, aligned depression or ridge. It is more intense in tectonically active zones than in tectonically quiescent zones. An important factor in the lineament study is the trend of the lineament. If the trend of the lineaments is correctly plotted on the Rose diagram, it can provide information about the principal stress and deformation direction of an area. The trend of the lineaments near the study area is determined and plotted on the Rose diagram to determine the stress directions. Analyses of lineaments in areas with poor exposures and dense vegetation can often reveal evidence of tectonic activity. The major lineaments of the study area are traced from the toposheet and AsterDEM 30m resolution. The map showing the locations of these lineaments is shown in Figure 19.

According to the Rose plot created with Rozeta Software, the predominant trend of lineaments in all three segments is in the NW-SE and NE-SW directions, similar to the region's major thrust and cross faults. (Figure 20).

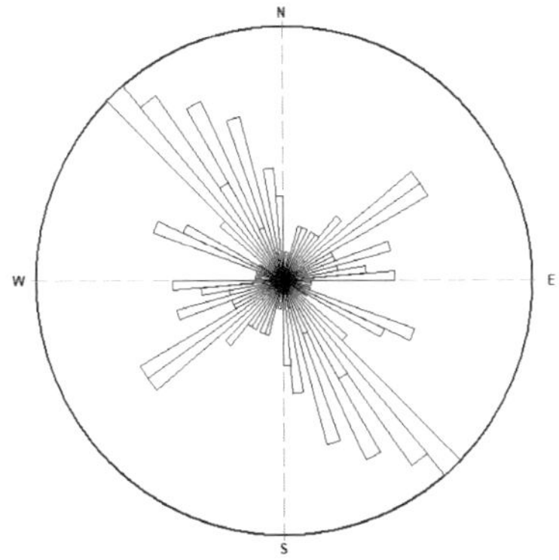

Figure 19. Bi-directional rose diagram showing direction of lineaments identified in the study area.

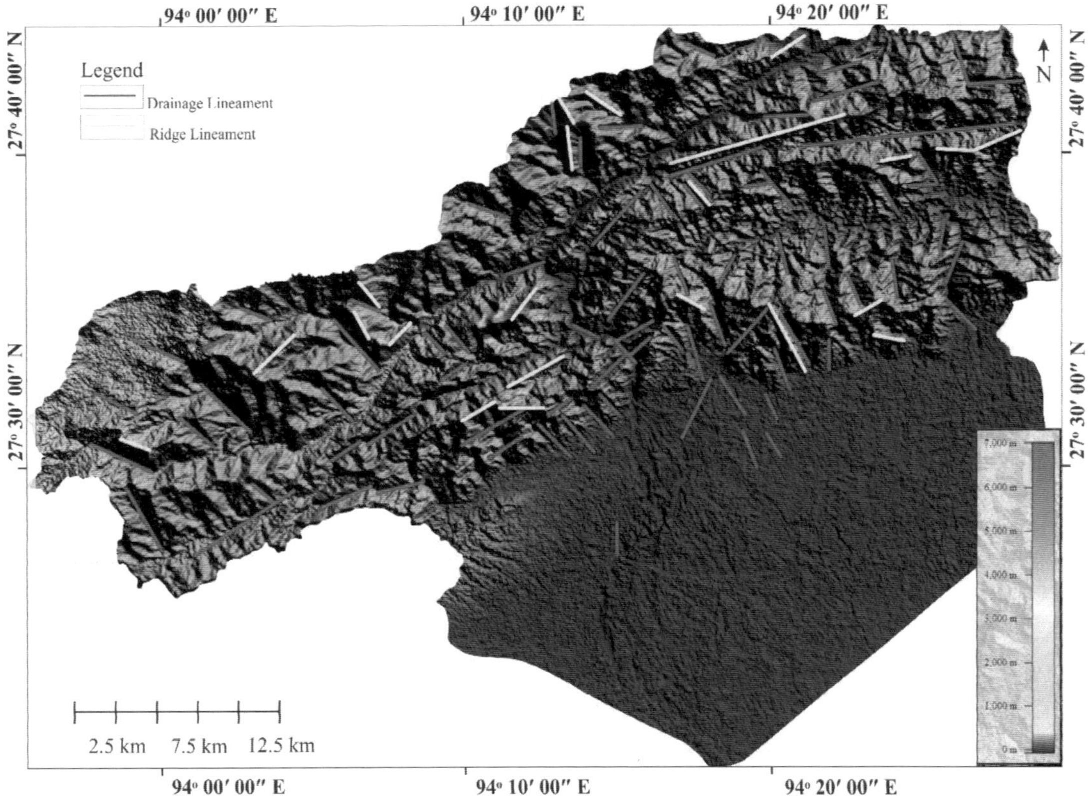

Figure 20. Lineation map of the area.

7. Morphotectonic evolution of the area

Along the whole length of the Himalayan strike, there are four critical structural breaks [24] that contribute to the Himalayan Mountain belt's tectonic setting. They are MFT, MBT, MCT and ITS (Indus Tsangpo suture).

The moderate to steep northerly dipping Main boundary thrust (MBT) separates the Lesser Himalaya's sedimentary strata from the Sub Himalaya. The Himalayan frontal thrust is the southern limit of the Himalayan orogeny. The Siwaliks are Late Tertiary sediments found over the whole length of the Himalayan foothills and deposited between the MBT and MFT. The most common lithology in the Sub Himalaya is sandstone.

In the study area, the overriding of thrust sheets due to intense compression probably occurred during the early Miocene, which resulted in the formation of a foredeep. In this foredeep basin, the Siwalik sediments are deposited (Figure 21). There is continuous North-South compression in the Himalayan frontal part, which has led to the development of several thrusts. Due to the constant compression during this orogeny, thrusts propagated southward into the basin and uplifted some portion of the foreland basin. The ongoing tectonic deformation has resulted in the southward propagation of numerous branching faults that belong to the fault system like MBT and HFT. The branching faults are the manifestation of an imbricate thrust fault in the region. The development of the thrust system is aided by basal detachment. MBT may be originated in the early Eocene age and separated the Gondwana group of rock from the Siwaliks. HFT is a prominent branching fault in the study area.

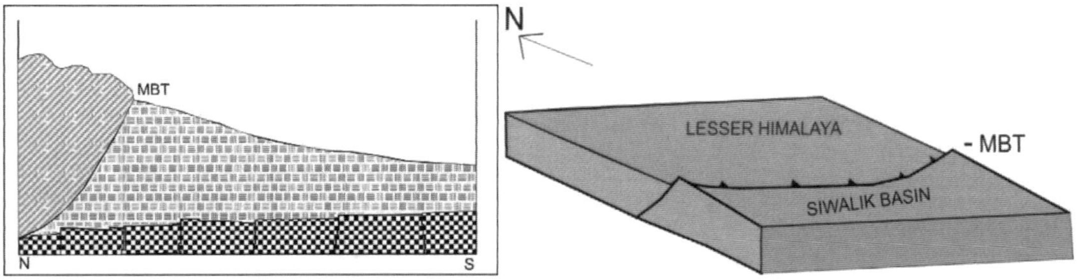

Figure 21. Development of the Siwalik basin.

The HFT-1 in the study area probably originated in the late Pliocene when the foreland basin compressed intensely. With the development of HFT-1, part of the Siwalik basin containing Dafla and Subansiri formation uplifted. It resulted in the southward migration of the foredeep. In this foredeep, Kimin Formation got deposited along with some alluvium (Figure 22). HFT-1 uplifted the middle Siwalik rocks and led to the development of the Outer Himalaya Ranges. Further movement along the frontal fault led to the deformation of the Kimin Formation, and they thrusted over recent alluvium (Figure 23). This episode involved the activation of another fault, which branched out of HFT-1 and was named HFT-2.

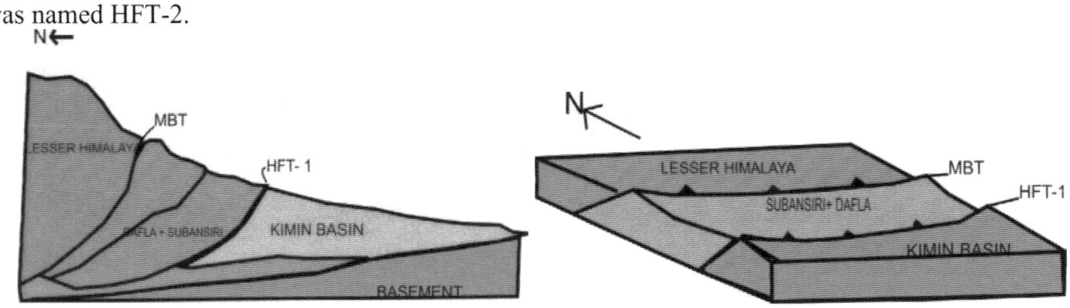

Figure 22. Upliftment of Dafla and Subansiri Formation and development of Kimin basin.

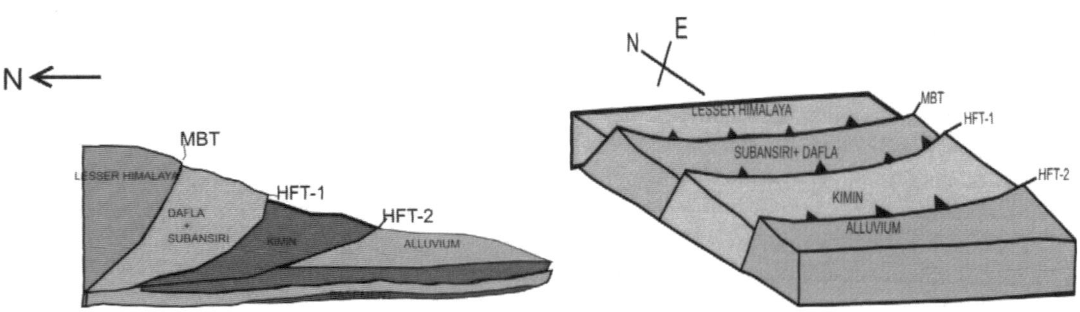

Figure 23. Upliftment of Kimin Formation and Development of foreland basin where alluvium is deposited.

In the study area, we observe significant differences in the lithology present in the right and left banks of the Subansiri river. The dip direction, dip amount, tilting of the bed, fold axis of the beds variy for both the river bank. The dipping of beds on the left bank is towards the SW and NE, whereas on the right bank, it is towards the W and S. Fold axis orientation is also different in both the bank. The right bank's folding is more intense than the left banks, indicating the presence of differential compression in the area. It is probably because of the difference in resistance for thrust movement in both the banks. These field evidence indicate the occurrence of a cross fault across the Dulung River in the study area. HFT-2 has been displaced by this cross fault, which indicates its age is late Pleistocene to post Pleistocene in age. The evolutionary model is prepared for the study area and shown in Figure 24.

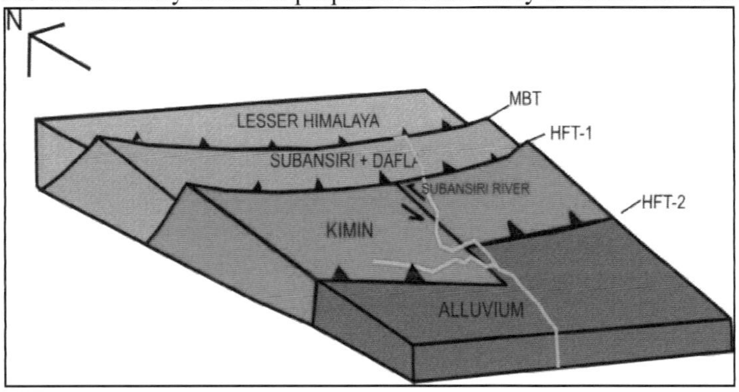

Figure 24. Block diagram shows the evolutionary model of the study area.

8. Conclusion

From studying seven morphotectonic indices, it can be said that the area is tectonically active and continuously undergoing tectonic deformation. The anomalous points identified by Stream Length Gradient Index falls along or near the MBT and HFT showing active tectonics. Therefore, we can say that these thrusts are still active. The relative tectonic index is calculated, and the parameters are classified into four classes. It indicates that the area is high to moderately active. In the following discussion, the ranges of every single morphotectonic parameter are defined on the basis of research conducted in the arid and semiarid regions. Considering the geographical location of the study area, which is subtropical-tropical climate zone, it receives very high precipitation (2000mm to 5000mm) annually, so rivers have a higher erosive power. The eastern Himalaya's frontal portion consists primarily of rock from the Siwalik Group, which is weak and friable. It is easy for these rocks to erode. This may be why the drainage basins attain a circular shape rapidly. Hence, the range for the Basin shape Index indicating tectonic classes for Eastern Himalaya needs to be revised. The lineament study shows that most features align parallel to MBT and HFT or the cross-fault in the study area. As a result, tectonics is crucial in the evolution of lineaments in the research domain. An evolutionary model of the study area is introduced to understand its tectonic evolution. It is observed that there is a development of splays from HFT, which is HFT 1, HFT 2 and HFT 3. So, it can be concluded that the Himalayan Mountain front could be in a state of southward expansion.

Reference:

[1] Adams J 1980 *Active tilting of the United State mid-continent; geodetic and geomorphic evidence* Geology **8** 442-446
[2] Andermann C, Gloaguen R 2009 *Estimation of erosion in tectonically active orogenies. Example from the Bhotekoshi catchment, Himalaya (Nepal)* International Journal of Remote Sensing

30 3075-3096

[3] Azor A, Keller E A, and Yeats R S 2002 *Geomorphic indicators of active fold growth: South Mountain-Oak Ridge anticline, Ventura basin, southern California* Geological Society of America Bulletin **114** 745–753

[4] Bull W B 1977a *The alluvial-fan environment* Progress in Physical Geography **1**(2) 222-270

[5] Bull W B 1977b *Tectonic geomorphology of the Mojave Desert* U. S. Geological Survey Contract Report 14-08-0001-G-394 Office of Earthquakes, Volcanoes, and Engineering, Menlo Park, California 188

[6] Bull W B 1978 *Geomorphic tectonic activity classes of the front of the San Gariel Mountains, California* U.S. Geological Survey Contract Report 14-08-001-G-394 California 59

[7] Bull W B, McFadden L D 1977 *Tectonic geomorphology north and south of the Garlock fault, California* In: D.O. Doehring (Ed.), Geomorphology in Arid regions. Proc. 8th Annual Geomorphology Symposium State University of New York Binghamton 115-138

[8] Burbank D W, Anderson R S 2001 *Tectonic Geomorphology* Malden, Massachusetts: Blackwell Science Inc

[9] Cannon P J 1976 *Generation of explicit parameters for a quantitative geomorphic study of Mill Creek drainage basin* Oklahoma Geology Notes **36** (1) 3–16

[10] Cox R T 1988b *Evidence of Quaternary ground tilting associated with the Reelfoot Rift zone, northeast Arkansas* Southeastern Geology **28** 211-224

[11] Cox R T 1994 *Analysis of drainage-basin symmetry as a rapid technique to identify areas of possible Quaternary tilt-block tectonics: an example from the Mississippi Embayment* Geological Society of America Bulletin **106** 571-581

[12] Cox R T, Van Arsdale R B and Harris J B 2001 *Identification of possible Quaternary deformation in the northeastern Mississippi Embayment using quantitative geomorphic analysis of drainage-basin asymmetry* Geological Society of America Bulletin **113** 615-624

[13] Cracknell A P, Hayes L W B 1993 *Introduction of Remote Sensing* Taylor and Francis, London.

[14] El Hamdouni R, Irigaray C, Fernandez T, Chacón J, and Keller E A 2008 *Assessment of relative active tectonics, southwest border of Sierra Nevada (Southern Spain)* Geomorphology **96** 150-173

[15] Gomez B, Marron D C 1991 *Neotectonic effects on sinuosity and channel migration, Belle Fourche River, Western South Dakota* Earth Surfaces Process and Landforms **16** 227-235

[16] Goswami U, Sarma J N and Patgiri A D 1999 *River channel changes of the Subansiri in Assam, India.*Geomorphology **30** 227-224.

[17] Hack J T 1973 *Stream-profiles analysis and stream-gradient index* Journal of Research of the U.S. Geological Survey **4** 421-429

[18] Hare P H, Gardner T W 1985 *Geomorphic indicators of vertical neotectonism along converging plate margins, Nicoya Peninsula, Costa Rica.* In: Morisawa M., Hack J.T. (Eds.) Tectonic Geomorphology Allen and Unwin Boston 75–104

[19] Holbrook John, Schumm S A 1999 *Geomorphic and sedimentary response of rivers to tectonic deformation: a brief and critique of a tool for recognizing subtle epeirogenic deformation in modern and ancient settings* Tectonophysics **305** 287-306

[20] Karunakaran C, Ranga rao A 1979 Misc. Publ. Geol. Surv. India **41** 1–66

[21] Keller E A 1986 *Investigation of Active Tectonics: Use of Surficial Earth Processes* In: R.E. Wallace (Ed.) Active Tectonics National Academy Press Washington DC 136-147

[22] Keller E A, Pinter N Eds. 2002 *Active Tectonics: Earthquakes, Uplift, and Landscape* 2nd ed. Prentice Hall Upper Saddle River N J 362

[23] Kumar G 1977 *Geology of Arunachal Pradesh* Geological society of India Bangalore

[24] Mahadevan T M 1994 *Deep continent structure of India: A review* Memoir 28. xv + Bangalore Geological Society of India 569

[25] Mayer L 1986 *Tectonic geomorphology of Escarpments and mountain fronts* In: R.E. Wallace (Ed.) Active Tectonics National Academy Press Washington D C 125-1135

[26] Molnar P, Tapponnier P 1975 *Cenozoic tectonics of Asia: Effects of a continental collision. Science* **189** 419-426

[27] Nakata, T 1972 *Geomorphic history and crustal movements of the foothills of Himalayas* Tohoku University Institute of Geography Sendai 111-118

[28] Nakata T 1989 *Active faults of Himalaya of India and Nepal* Geological Society of America Special Paper no. **232** 243-264

[29] Perez-Pena J V, Azanon J M, Azor A, Delgado J, Gonzalez F, Lodeiro 2009 *Spatial analysis of stream power using GIS: SLk anomaly maps* Earth Surface Processes and Landforms **34** 16-25

[30] Ramírez-Herrera M T, 1998 *Geomorphic assessment of active tectonics in the Acambay Graben, Mexican volcanic belt* Earth Surface Processes and Landforms **23** 317–332

[31] Rockwell T K, Keller E A, Johnson D L, 1984 *Tectonic geomorphology of alluvial fans and mountain front near Ventura, California* In: Morisawa, M., Hack, J.T. (Eds.) Tectonic Geomorphology Proceedings of the 15th Annual Binghampton Geomorph. Symp. Sep. 1984 Allen and Unwin Boston 183–208

[32] Sarma J N 1993 *Asamar Nad-Nadi* Asom Sahitya Sabha Jorhat 187–193.

[33] Schumm S A 1986 *Alluvial River response to active tectonics* In: R.E. Wallace (Ed.) Active Tectonics National Academy Press Washington D C 80-84

[34] Silva P G, Goy J L, Zazo C, Bardají T 2003 *Fault-generated Mountain fronts in southeast Spain: geomorphologic assessment of tectonic and seismic activity* Geomorphology **50** 203–225

[35] Sinha Roy S, 2001 *Neotectonic significance of longitudinal river profiles: An example from the Banas drainage basin, Rajasthan* Jour. Geol. Soc. India **58** 143-156

[36] Valdiya K S 1986 *Neotectonic activities in the Himalayan belt* Proc. Internal. Symp. Neotectonics in South Asia Survey of India Dehradun 241-251

[37] Valdiya K S 2003 *Reactivations of Himalayan Frontal Fault: Implications* Current Science **85** 1031-1040

[38] Valdiya K S, Narayana A C 2007 *River response to neotectonic activity: Example from Kerala India* Jour. Geol. Soc. India **70** 427-443

[39] Wells S G, Bullard T F, Menges T M, Drake P G, Karas P A, Kelson K I, Ritter J B, Wesling J R 1988 *Regional variations in tectonic geomorphology along segmented convergent plate boundary* Pacific Coast of Costa Rica. Geomorphology **1** 239–265

[40] Zovoili E, Konstantinidi E. and Koukouvelas I K 2004) *Tectonic Geomorphology of Escarpments: The Case of Kompotades and Nea Anchialos Faults* Bulletin of the Geological Society of Greece **36** 1716-1725

Assessment of Air Pollution before and during COVID-19 Pandemic Lockdown in Chhattisgarh State, India

H Rajput[1] and S Barde[2]
[1]Research Scholar, MSIT, MATS University, Raipur (C.G), India
[2]Professor, MSIT, MATS University, Raipur (C.G), India
rajputhansa1991@gmail.com
v.snehabarde@gmail.com

Abstract. Air pollution is known to be the most common problem worldwide. Most cities have air pollution, and new pollutants are being added to the atmosphere, making it more toxic. Air quality prediction is a complex task because of the dynamics, volatility, and high temporal and spatial variability of pollutants and particles. Intending to control the expansion of COVID–19, the Indian government has announced a strict and total lockdown beginning March 25, 2020, in all countries, besides essentials. The goal of this research work is to find out the effect of lockdown on air quality in the four cities of Chhattisgarh state: Raipur, Bilaspur, Durg-Bhilai, and Korba. For that, we are using data from the official site of the Chhattisgarh Environment Conservation Board (CECB). The data is divided into two phases before the three months of lockdown (December 2019 to February 2020) and during the three months of lockdown (March to May 2020). We compared the data and performed a detailed analysis using SPSS software of the impact of COVID-19 on air quality and, as a result, found a major change in the Air Quality Index (AQI), geographical circulation levels, and concentration of various parameters: PM_{10}, $PM_{2.5}$, NO_2, SO_2, and CO.
Keywords: Air Pollution, IoT, Covid-19, Air Quality Index

1. Introduction

The new Corona-virus disease COVID–19 coming from SARS-CoV-2 Virus (Severe Acute Respiratory Syndrome Corona-virus 2) was foremost recognized in December 2019 in Wuhan, China [2, 3]. Next to March 2020, the disease had spread quickly to other countries, and a pandemic was proclaimed on March 12, 2020 (WHO 2020). COVID–19 had afflicted 110 million people worldwide as of February 18, 2021, and killed over 2.45 million people [2, 3, 4, 7].

India has imposed a national lockdown for several months, starting March 25, 2020, to stop the spread of COVID–19 infections. Except for required services, all transportation (roads, planes, railroads), schools, universities, industrial plants, hotels, restaurants, retail malls, and markets were closed during the lockdown (CPCB, 2020). The National Air Quality Index (NAQI) proclaimed by the Central Pollution Control Board (CPCB) poses various economic challenges. Nevertheless, it suggested a positive result. Several studies have reported the short-term and long-term changes in air quality, as well as the impact of COVID–19 dictated restrictions in Indian cities. Unlike most of the published work and research on Indian cities, the current study reports the findings on COVID-19 induced changes in air quality throughout both the lockdown and unlocking periods. In addition, we also report both periods have many changes and outcomes [1]. The improvement and enhancement inspired us to further investigate COVID–19 effects on air quality in four districts of Chhattisgarh State.

It is widely agreed, the growth of human population especially the rapid modernization, economic development, and co-modification of Chhattisgarh state resulted in a deterioration of the quality of air [11]. Chhattisgarh has been declared one of the most polluted states in the world (WHO, 2016). Air pollution was found to be less than or under the acceptable limits of air quality criteria set by the Central Pollution Control Board (CPCB, 2009) even in the most polluted cities. To better understand the effects of lockdown, some research has been done on air pollution. To understand the details of the air quality changes, we have divided the whole period into two stages: Before Lockdown and During Lockdown phase [8].

This research will assist us in evaluating the lockdown's impact on the targeted five pollutants (PM_{10}, $PM_{2.5}$, NO_2, SO_2, and CO). All industrial, vehicle and commercial activities were prohibited during the lockdown period. As a result, this is a perfect situation in which to investigate their direct impact on pollution reduction and their role in raising air pollution [10].

2. Research study area

The research area encompasses Chhattisgarh state, which includes four cities, namely Raipur, Bilaspur, Durg-Bhilai, and Korba (Figure 1). These cities are among Chhattisgarh's most developed and densely populated. They are also experiencing rapid urbanisation and industrialization, as well as increased socio-economic activity. However, the problem of air pollution is increasing due to rapid development. As a result, cities are facing serious air pollution problems and challenges, thus Chhattisgarh state has been named one of the most polluted states in the country.

S.No.	State	Cities S.No.	Cities
1	Andhra Pradesh	1	Guntur
		2	Kurnool
		3	Nellore
		4	Vijaywada
		5	Vishakhapatnam
2	Assam	6	Guwahati
		7	Nagaon
		8	Nalbari
		9	Sibsagar
		10	Silchar
3	Chandigarh	11	Chandigarh
4	Chhattisgarh	12	Bhilai
		13	Korba
		14	Raipur
5	Delhi	15	Delhi
6	Gujarat	16	Surat
		17	Ahemdabad

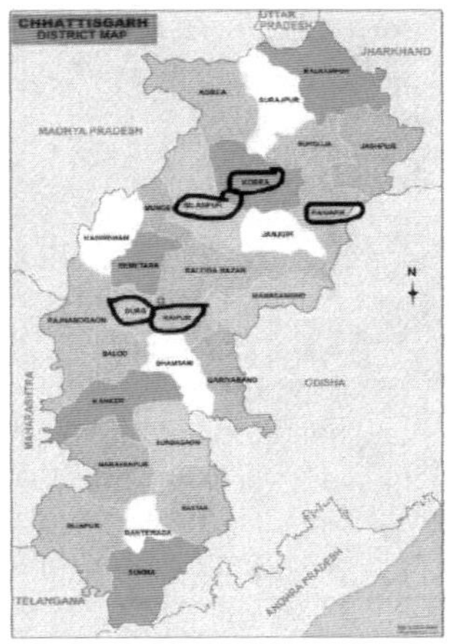

Figure 1.a) Polluted states of India (Source: WHO Report 2014 – 2018) b) in map black circles showing study area of Chhattisgarh state (Source: infoandopinion.com).

Raipur

Raipur city is jammed in the middle of 2 industrial buildings on its east and west edges (extended at distances of about 20–30 km) in the company of a few notable industries, namely Century Cement Factory, Monnet Sponge Iron and Power Plant, Bhilai Steel Plant, Larson & Toubro Cement Plant, Jamul Cement Factory, Woolworth Textile Plant and others. Within 30 kilometers of the circle, there are numerous agro forestry plants, electrical product manufacturing factories, fertilizer plants, and oil extraction plants. This massive coal-burning industry generates a lot of fly ash. The air distribution method has an impact on the local environment and the surrounding area [12, 13].

Figure 2.a) Waste product of chemical factory have been thrown in open field in Lalpur area of Raipur b) Carbon spread in all over the city in the form of smoke c) Heavy traffic in many areas of the city (Source: oneindia.com)

Bilaspur

In the case of pollution in the country, Justice Bilaspur has now come to the top. Bilaspur is also known as the center of India's power generation hub, among other industries. Comparable to Raipur, Bilaspur has witnessed numerous changes since the creation of the state. Alike another city in Chhattisgarh, Bilaspur city is also home to many coal and iron factories. Iron and Steel production is a result of speedy development, which has resulted in environmental issues due to the massive amount of waste produced and its management implications. The steel industry is the largest consumer of energy and natural resources, emits massive amounts of air pollutants, and consequently causes substantial local and worldwide environmental challenges.

Figure 3.a) Smoke billows out of a cement plant at Barmana in Bilaspur (Source: tribune.com) b) A toxic haze has blanketed the streets of Bilaspur since morning time c) Dust from construction work.

Durg-Bhilai

The Durg-Bhilai city area in Chhattisgarh state is home to the state's new administrative capital, which is connected by an expressway that serves as an industrial corridor and is situated between the edge of the Mahanadi River in the North and Deccan Plateau in the South Chota Nagpur. Raipur, Durg-Bhilai (RDB) is the largest steel manufacturing hub, with many foundries, metal-alloy factories, steel casting, and rolling work, iron and sponge factories, cement production, and formalin manufacturing hub in India. India's largest and most productive integrated steel plant is the Bhilai Steel Plant [12, 13].

Figure 4.a) Alarming levels of air emissions b) Open Waste Burning c) Smog and dust.

Korba

Korba, placed inside the northern 1/2 of Chhattisgarh, quarters to many coal-floor thermal power plants, inclusive of National Thermal Power Corporation (NTPCR) and Chhattisgarh State Electricity Board (CSEB). Korba has 14 thermal power plants with a turn-out of 8,625 MW accounting for 17.26 percent of the country's total coal production. Annually, the Korba plant produces over 1 lakh metric tons of fly ash. These power plants produce 6090 (MW) of electricity and supply electricity to Korba and its surrounding areas. This is one of the reasons to make the city a pollution hub. Coal transportation in open trucks is frequent in Korba, which contributes to poor air and water pollution. Korba is one of the most severely contaminated locations in the country, according to the Central Pollution Control Board's (CPCB) Comprehensive Environmental Pollution Index Report [12].

Figure 5.a) In Korba, Transportation of coal in open trucks is very common (Source: Mayank Aggarwal/Mongabay-India) b) Coal mines that are just a few minutes away from the main Korba city (Source: Mayank Aggarwal/Mongabay-India) c) Black Dust in local area and on the roadside is a daily occurrence.

3. Data collection

The data on air pollutants were collected from the official site of the Chhattisgarh Environment Conservation Board (CECB). CECB's headquarters are in Raipur, with few regional offices in significant cities like Raipur, Bilaspur, Durg-Bhilai, and Korba. The board recently promoted Raipur Laboratory to Central Laboratory and Research & Development Center. Each laboratory has high-precision laboratory equipment for examining and measuring a wide range of environmental samples. The data is divided into two phases: before lockdown (December 2019 to February 2020) and during lockdown (March to May 2020). Data were obtained for four cities from 15 monitoring stations: 6 from Raipur, 1 from Bilaspur, 4 from Durg-Bhilai, and 4 from Korba, respectively.

CECB maintains an air pollution monitoring system in Chhattisgarh, together with those used in this study, through an air quality monitoring network called the Ambient Air Quality Monitoring System (AAQMS). Intending to completely understand the influences of procedures made during the phase of lockdown on air quality, this study took into 5 most relevant pollutants were, and these air pollutants incorporate particulate matter (PM_{10} and $PM_{2.5}$), nitrogen dioxide (NO_2), sulfur dioxide (SO_2), and carbon mono-oxide (CO).

4. Result Analysis

The air quality in Chhattisgarh state is being investigated in four major districts. The real-time data set were divided into two phases, i.e. before three months of lockdown and during three months of lockdown. For identifying the variation among different parameters, SPSS software is used as a tool for the calculation of pollutants (minimum, maximum, mean, and standard deviation) shown in below tables, for plotting graphs the ROC curve is used. The phase-wise variations of the pollutants for four cities are shown below.

Table 1.a) Raipur: Descriptive Statistics before Lockdown (December 2019 to February 2020).

Parameter	N	Minimum	Maximum	Mean	Std. Deviation
PM_{10} ($\mu g/m^3$)	55	45.00	92.00	67.4436	25.04423
$PM_{2.5}$ ($\mu g/m^3$)	55	40.37	90.54	63.7476	14.80779
NO_2 ($\mu g/m^3$)	55	10.41	33.96	27.2200	4.69800
SO_2 ($\mu g/m^3$)	55	8.59	23.59	19.1773	4.16274
CO (mg/m^3)	55	10.59	23.59	11.4927	5.84078
Valid N (listwise)	55				

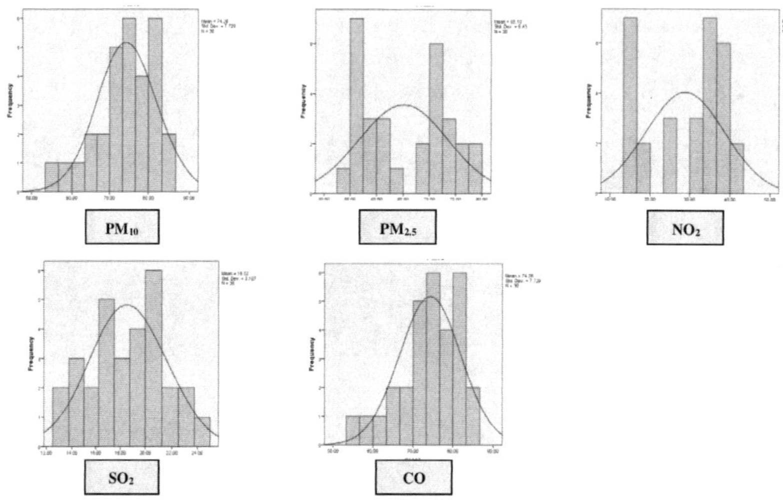

Figure6. a) ROC curve: Mean and standard deviation of PM_{10}, $PM_{2.5}$, NO_2, SO_2 and CO (pollutants are in units of μg/m^3 and CO is expressed in units of mg/m^3) levels in Raipur before lockdown.

Table 1.b) Raipur: Descriptive Statistics during the Lockdown (March to May 2020).

Parameters	N	Minimum	Maximum	Mean	Std. Deviation
PM_{10} (μg/m^3)	47	30.50	61.35	46.1540	5.37943
$PM_{2.5}$ (μg/m^3)	47	30.50	61.35	45.2679	5.55641
NO_2 (μg/m^3)	47	7.50	28.63	25.3636	3.86007
SO_2 (μg/m^3)	47	4.75	17.63	7.0919	1.96676
CO (mg/m^3)	47	4.30	17.45	7.2577	1.77024
Valid N (listwise)	47				

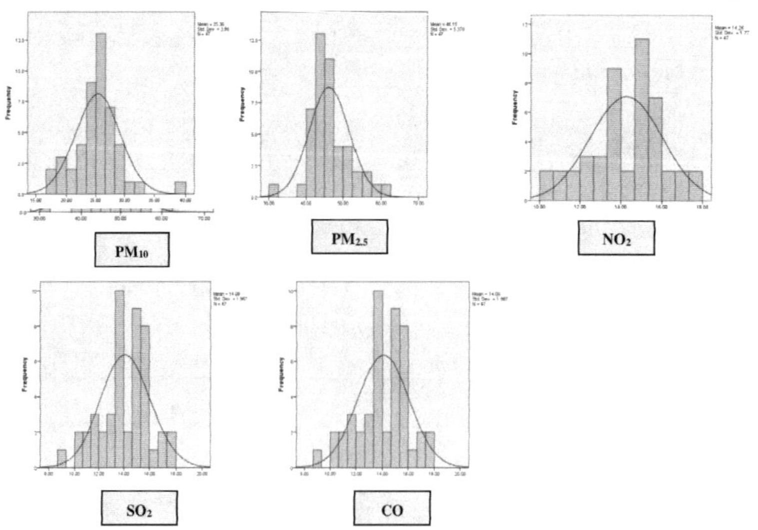

Figure6. b) ROC curve: Mean and standard deviation of PM_{10}, $PM_{2.5}$, NO_2, SO_2 and CO (pollutants are in units of μg/m^3 and CO is expressed in units of mg/m^3) levels in Raipur during lockdown.

Table 2.a)Bilaspur: Descriptive Statistics before Lockdown (December 2019 to February 2020).

Parameters	N	Minimum	Maximum	Mean	Std. Deviation
PM_{10} (μg/m^3)	25	48.50	63.50	56.9160	4.04925
$PM_{2.5}$ (μg/m^3)	25	48.50	63.50	57.3240	3.90099
NO_2 (μg/m^3)	25	10.50	13.30	12.1600	.68860
SO_2 (μg/m^3)	25	3.50	6.80	5.4200	.78049
CO (mg/m^3)	25	3.50	6.80	67	.64420
Valid N (listwise)	25				

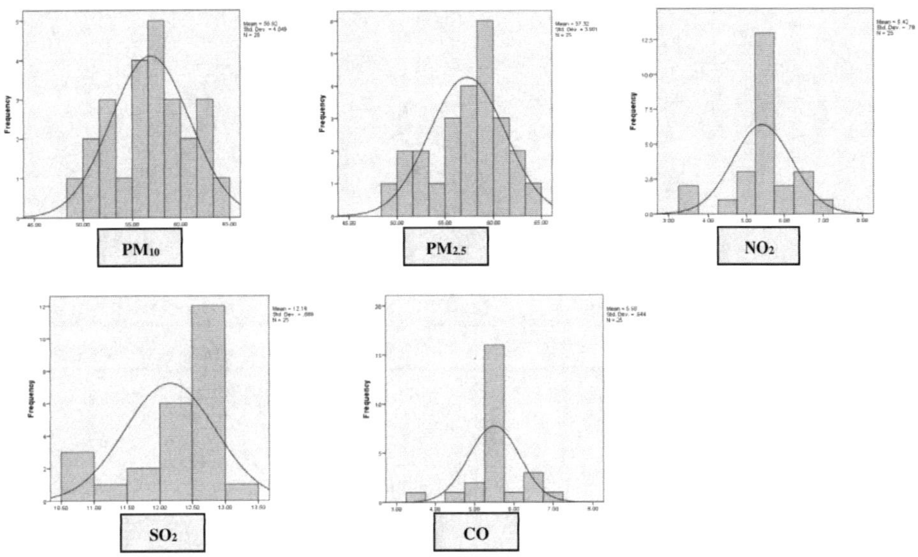

Figure7. a) ROC curve: Mean and standard deviation of PM_{10}, $PM_{2.5}$, NO_2, SO_2 and CO (pollutants are in units of μg/m^3 and CO is expressed in units of mg/m^3) levels in Bilaspur before lockdown.

Table 2.b)Bilaspur: Descriptive Statistics during the Lockdown (March to May 2020).

Parameters	N	Minimum	Maximum	Mean	Std. Deviation
PM_{10} (μg/m^3)	12	33.70	63.00	55.0333	7.27728
$PM_{2.5}$ (μg/m^3)	12	33.70	63.00	55.9167	7.62935
NO_2 (μg/m^3)	12	11.20	17.00	14.3750	2.00097
SO_2 (μg/m^3)	12	3.30	8.00	6.1333	1.63892
CO (mg/m^3)	12	3.30	17.00	11.9833	5.45324
Valid N (listwise)	12				

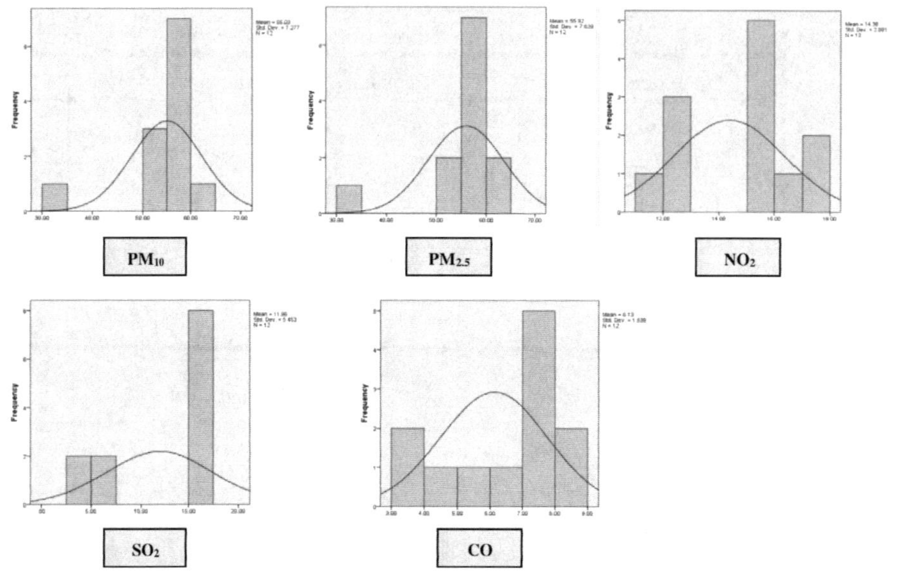

Figure 7.b) ROC curve: Mean and standard deviation of PM_{10}, $PM_{2.5}$, NO_2, SO_2 and CO (pollutants are in units of µg/m^3 and CO is expressed in units of mg/m^3) levels in Bilaspur during the lockdown.

Table 3.a) Durg-Bhilai: Descriptive Statistics before Lockdown (December to February 2020).

Parameters	N	Minimum	Maximum	Mean	Std. Deviation
PM_{10} (µg/m^3)	74	.00	92.00	68.4332	22.64548
$PM_{2.5}$ (µg/m^3)	74	40.37	90.54	62.6682	14.73104
NO_2 (µg/m^3)	74	10.41	23.96	17.1692	4.73712
SO_2 (µg/m^3)	74	4.59	23.59	9.1984	4.14219
CO (mg/m^3)	4.59	23.59	11.0930	5.70610	4.59
Valid N (listwise)	74				

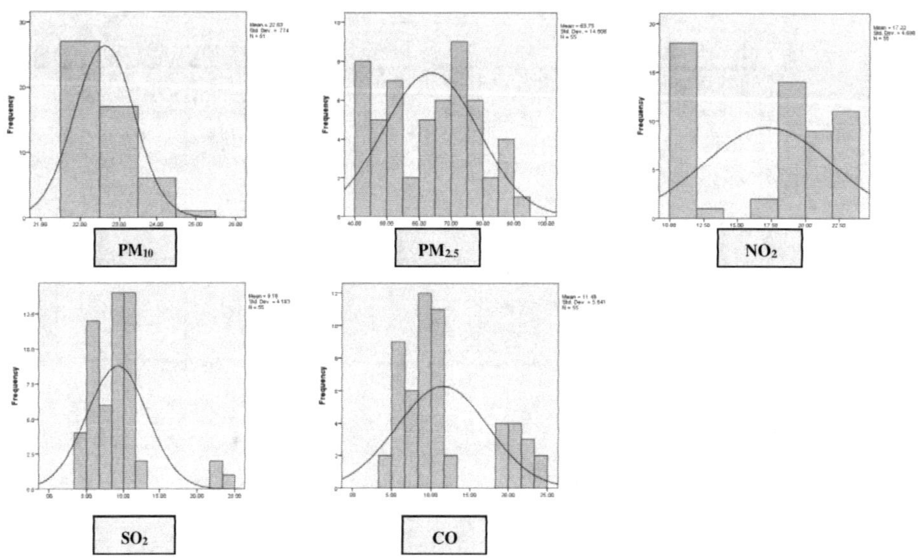

Figure 8.a) ROC curve: Mean and standard deviation of PM_{10}, $PM_{2.5}$, NO_2, SO_2 and CO (pollutants are in units of µg/m3 and CO is expressed in units of mg/m^3) levels in Durg-Bhilai before lockdown.

IOP Conf. Series: Earth and Environmental Science 1032 (2022) 012044 doi:10.1088/1755-1315/1032/1/012044

Table 3.b) Durg-Bhilai: Descriptive Statistics during the Lockdown (March to May 2020).

Parameters	N	Minimum	Maximum	Mean	Std. Deviation
PM_{10} (µg/m3)	52	8.75	83.46	50.5704	15.66342
$PM_{2.5}$ (µg/m3)	52	6.25	75.66	43.3710	15.98413
NO_2 (µg/m3)	52	6.66	47.94	11.5571	7.23068
SO_2 (µg/m3)	52	4.16	46.75	7.1708	5.93203
CO (mg/m3)	52	4.38	22.50	10.2719	5.21595
Valid N (listwise)	52				

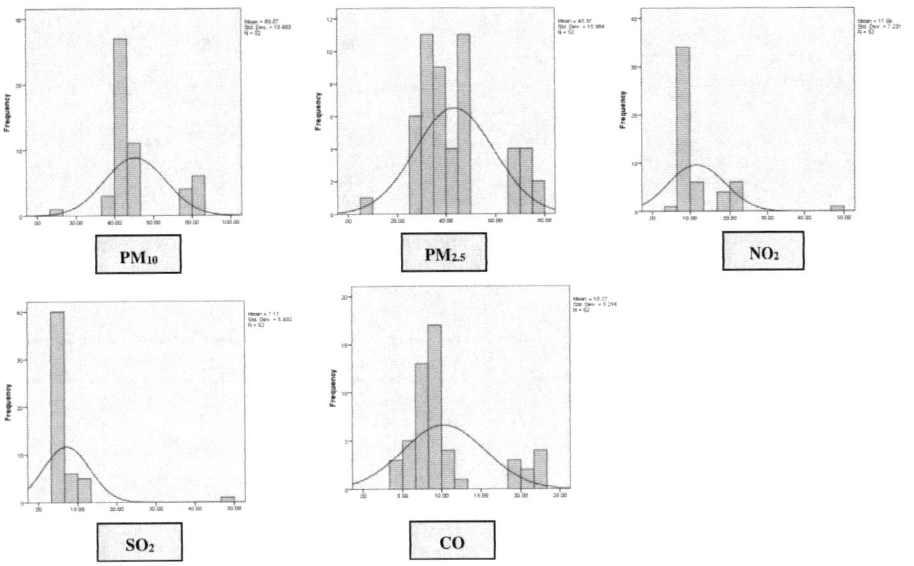

Figure8. b) ROC curve: Mean and standard deviation of PM_{10}, $PM_{2.5}$, NO_2, SO_2 and CO (pollutants are in units of µg/m^3 and CO is expressed in units of mg/m^3) levels in Durg-Bhilai during the lockdown.

Table 4.a) Korba: Descriptive Statistics before Lockdown (December to February 2020).

Parameters	N	Minimum	Maximum	Mean	Std. Deviation
PM_{10} (µg/m3)	51	41.00	66.00	56.3137	5.90081
$PM_{2.5}$ (µg/m3)	51	41.00	66.00	56.9608	5.57480
NO_2 (µg/m3)	51	22.00	25.00	22.8039	.82510
SO_2 (µg/m3)	51	10.00	14.00	10.6275	.74728
CO (mg/m3)	51	22.00	25.00	22.6275	.77358
Valid N (listwise)	51				

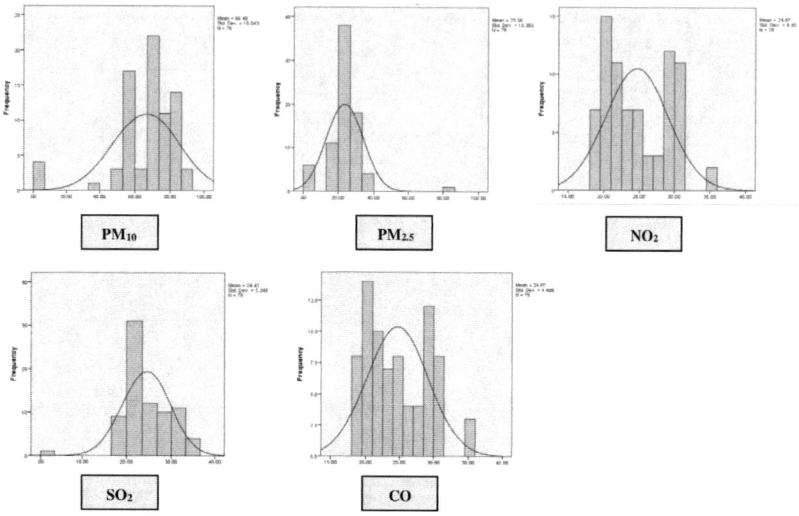

Figure 9.a) ROC curve: Mean and standard deviation of PM_{10}, $PM_{2.5}$, NO_2, SO_2 and CO (pollutants are in units of μg/m^3 and CO is expressed in units of mg/m^3) levels in Korba before lockdown.

Table 4.b) Korba: Descriptive Statistics during the Lockdown (March to May 2020).

Parameters	N	Minimum	Maximum	Mean	Std. Deviation
PM_{10} (μg/m3)	38	32.00	68.00	45.9737	10.07938
$PM_{2.5}$ (μg/m3)	38	17.00	56.00	39.1316	10.04092
NO_2 (μg/m3)	38	20.00	23.00	21.5526	1.03185
SO_2 (μg/m3)	38	6.00	12.00	9.2105	1.21161
CO (mg/m3)	38	20.00	24.00	21.8684	1.09473
Valid N (listwise)	38				

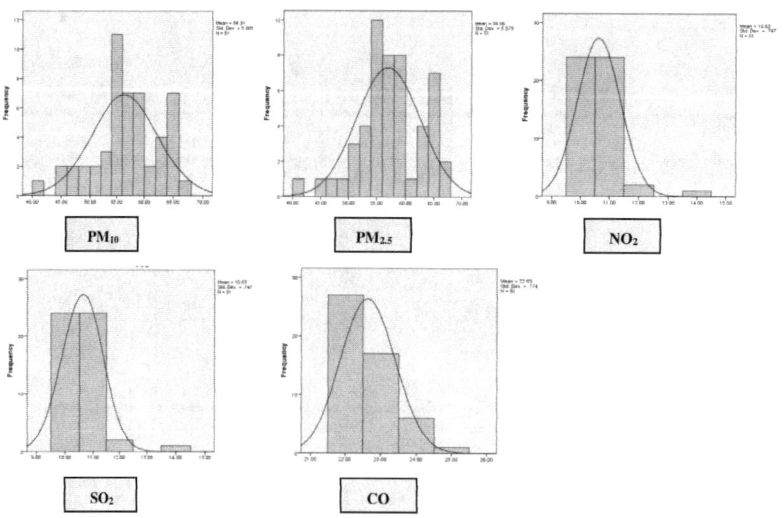

Figure 9.b) ROC curve: Mean and standard deviation of PM_{10}, $PM_{2.5}$, NO_2, SO_2 and CO (pollutants are in unitsof μg/m^3 and CO is expressed in units of mg/m^3) levels in Korba during the lockdown.

Before lockdown phase, the combined data of four cities are shown in (Table 5 and Figure 10) which depict the mean and standard deviation of pollutants (PM_{10}, $PM_{2.5}$, NO_2, SO_2 and CO). Before lockdown, the PM_{10} mean value was highest (74.26) in Raipur and minimum (56.91) in Bilaspur, PM $_{2.5}$ mean value was highest (65.1) in Raipur and minimum (23.58) in Korba, NO_2 mean value was highest (28.79) in Raipur and lowest in Bilaspur (12.16), SO_2 was highest (24.41) in Korba and lowest (5.42) in Bilaspur and CO was highest (24.47) in Korba and lowest (5.5) in Bilaspur.

Table 5. Before lockdown phase (December 2019 to February 2020): Mean and Standard Deviation of four cities (all pollutants are in units of $\mu g/m^3$ and CO is expressed in units of mg/m^3).

Parameter	City							
	Raipur		Bilaspur		Durg-Bhilai		Korba	
	Before Lockdown		Before Lockdown		Before Lockdown		Before Lockdown	
	Mean	Std.Dev	Mean	Std.Dev	Mean	Std.Dev	Mean	Std.Dev
PM_{10}(μg/m3)	74.26	7.72	56.916	4.04925	67.44	25.04	66.38	19.04
$PM_{2.5}$(μg/m3)	65.1	8.43	57.324	3.90099	63.75	14.8	23.58	7.90
NO_2(μg/m3)	28.79	9.86	12.16	0.6886	17.22	4.69	24.67	4.45
SO_2(μg/m3)	18.52	3.1	5.42	0.78049	9.18	4.16	24.41	5.34
CO(mg/m3)	23.89	8.52	5.5	0.6442	11.49	5.84	24.47	3.39

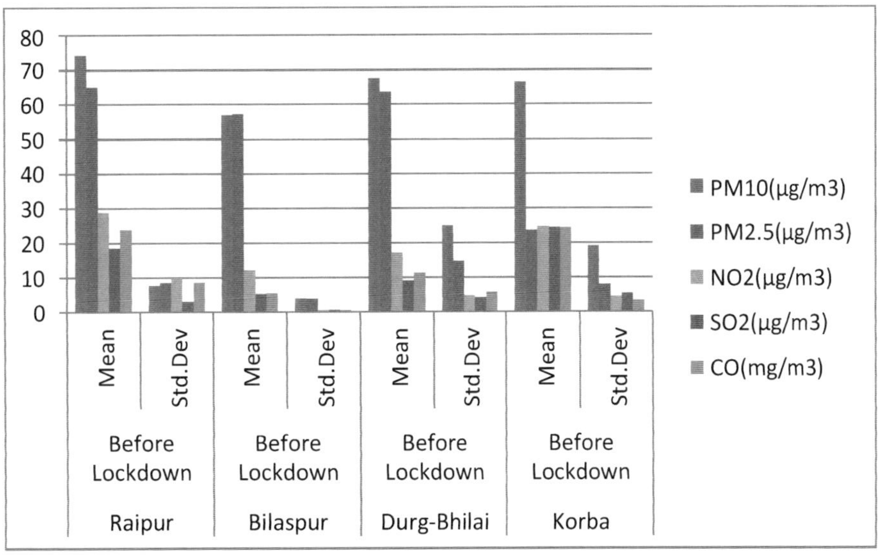

Figure10. Variation of air pollutants of four cities: Before Lockdown.

During the lockdown, the percentage fluctuation of the mean and standard deviation of pollutants in four cities is shown in (Table 6 and Figure 11). From the table, it can be observed that the PM_{10} mean value was highest (56.31) in Korba and lowest (46.15) in Raipur. $PM_{2.5}$ mean value was highest (56.96) in Korba and lowest (43.37) in Durg – Bhilai. NO_2 mean value was highest (25.36) in Raipur and lowest (11.56) in Durg – Bhilai. SO_2 mean value was highest (14.09) in Raipur and lowest (6.13) in Bilaspur. CO mean value was highest (22.63) in Korba and lowest (10.27) in Durg – Bhilai.

Table 6. During lockdown phase (March to May 2020): Mean and Standard Deviation of four cities (all pollutants are in units of $\mu g/m^3$ and CO is expressed in units of mg/m^3).

	City							
	Raipur		Bilaspur		Durg-Bhilai		Korba	
Parameters	During Lockdown		During Lockdown		During Lockdown		During Lockdown	
	Mean	std.Dev	Mean	std.Dev	Mean	std.Dev	Mean	std.Dev
PM_{10}(μg/m3)	46.15	5.37	55.03	7.27	50.47	15.66	56.31	5.9
$PM_{2.5}$(μg/m3)	45.26	5.55	55.92	7.26	43.37	15.98	56.96	5.57
NO_2(μg/m3)	25.36	3.86	14.38	2	11.56	7.23	22.8	0.82
SO_2(μg/m3)	14.09	1.96	6.13	1.63	7.17	5.93	10.63	0.74
CO(mg/m3)	14.25	1.77	11.98	5.45	10.27	5.21	22.63	0.77

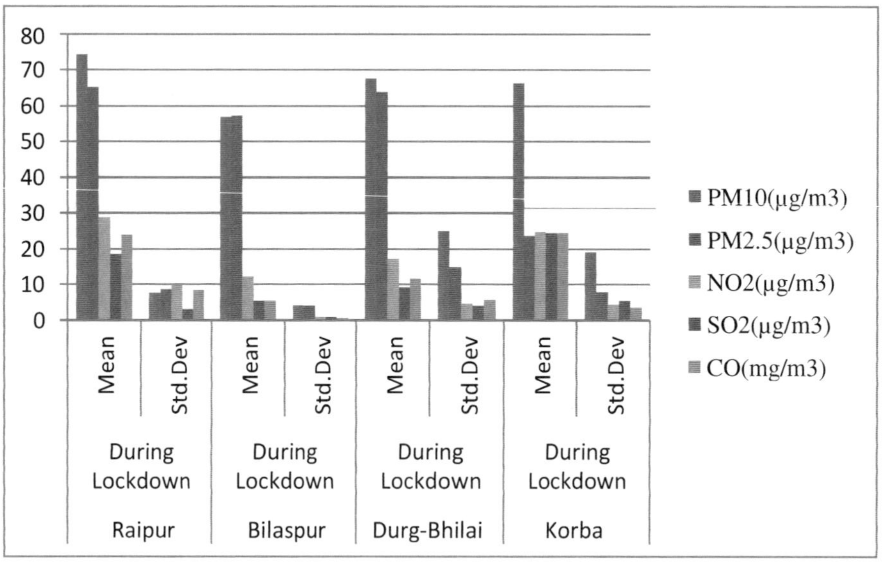

Figure11. Variation of air pollutants of four cities: During Lockdown.

5. Conclusion

The impact of pollution improvements due to decreased anthropogenic interference during the COVID – 19 epidemics in Chhattisgarh was investigated in this study. The study looked at the impact of COVID-19 on air quality in four main cities in Chhattisgarh. These cities have been rated as seriously polluted in Chhattisgarh state in recent decades. The current study, on the other hand, revealed a significant reduction in air pollution. We collected the data and separated it into two phases before three months of lockdown and during three months lockdown period. According to the acquired results, it was found that the concentration of all taken pollutants reduced dramatically due to several restrictive measures adopted by government authorities during the lockdown period. These reductions in pollutants are depicted in (Table 7 and Figure 11). Through these stages, a notable reduction in PM_{10} (41.81%) was observed, similarly $PM_{2.5}$ decreased by 37.08%, NO_2 decreased by 7.30%, SO_2 decreased by 3.43% and CO decreased by 5.53%.

As a result of several strict measures implemented during the lockdown, such as limiting public and private transportation, halting industrial and commercial activity, and so on, the concentration of several pollutants (PM_{10}, $PM_{2.5}$, NO_2, SO_2, CO) decreased dramatically, and this remarkable decline in the concentration of several pollutants can be seen. This demonstrates that there is ample scope to maintain ambient air quality with proper implementation of regulations, norms, and standards.

Reference

[1] Kumar D, Singh A et al 2021 COVID-19 driven changes in the air quality; a study of major cities in the Indian state of Uttar Pradesh *Environmental Pollution: Elsevier* **274** 1-10

[2] Liu F, Wang et al 2021 Effects of COVID-19 lockdown on global air quality and health, *Science of the Total Environment: Elsevier* **755** 1-13

[3] Nigam R, Pandya K et al 2021 Positive effect of COVID-19 lockdown on air quality of industrial cities (Ankleshwar and Vapi) of Western India *Scientific Reports* **4285** 1-12

[4] Karuppas M, Seshachalam S et al 2020 Air pollution improvement and mortality rate during COVID-19 pandemic in India: global intersectional study *Air Quality, Atmosphere & Health: Springer* **13** 1375–1384

[5] Menut L, Bessagnet B et al 2020 Impact of lockdown measures to combat Covid-19 on air quality over western Europe *Science of the Total Environment: Elsevier* **741** 1-9

[6] Bao R and Zhang A 2020 Does lockdown reduce air pollution? Evidence from 44 cities in northern China, *Science of the Total Environment: Elsevier* **731** 1-12

[7] Rodríguez D and Rodríguez L 2020 Air quality during the COVID-19: PM2.5 analysis in the 50 most polluted capital cities in the world, *Environmental Pollution: Elsevier* **266** 1-12

[8] Li L, L Qing L et al 2020 Air quality changes during the COVID-19 lockdown over the Yangtze River Delta Region, An insight into the impact of human activity pattern changes on air pollution variation *Science of the Total Environment: Elsevier* **732** 6- 11

[9] WHO (World Health Organization).World Health Organization Emergency Dashboard. https ://covid 19.who.int/regio n/searo /count ry/in. Accessed 8 July 2020.

[10] Garg A, Kumar A et al 2020 Comprehensive study on impact assessment of lockdown on overall ambient air quality amid COVID-19 in Delhi and its NCR, India *Journal of Hazardous Materials Letters* **2** 1-6

[11] Barde S and Rajput H 2018 An Idea to design a system to detect Air pollution in different area *International Journal of computer science and Engineering* **6** 1034-1036

[12] Tiwari M, Bajpai S et al 2016 Air and Leaching Pollution Scenario by Iron and Steel Plants, in central India *Elixir Pollution* **101** 44011 -44017

[13] Nishad K, Jawahar P et al 2019 Particulate Matter Source Contributions for Raipur-Durg-Bhilai Region of Chhattisgarh, India *Aerosol and Air Quality Research* **19** 528 -540

Igneous rock classification using Convolutional neural networks (CNN)

Surendra Patro[1,a], Dr. D.C. Jhariya[2,b,*], Dr. Mridu Sahu[3,c], Dr Pankaj Dewangan[4,d] Dr P. Y. Dhekne[5,d]

[1]B. Tech Mining Engineering Pre-Final Year, National Institute of Technology Raipur, G.E.Road, Raipur - 492010, Chhattisgarh, India
[2]Assistant Professor, Department of Applied Geology, National Institute of Technology Raipur, G.E.Road, Raipur - 492 010, Chhattisgarh, India
[3]Assistant Professor, Department of Information Technology, National Institute of Technology Raipur, G.E. Road, Raipur-492010, Chhattisgarh, India
[4]Associate Professor, Department of Mining Engineering, National Institute of Technology Raipur, G.E. Road, Raipur-492010, Chhattisgarh, India
[5]Associate Professor, Department of Mining Engineering, National Institute of Technology Raipur, G.E. Road, Raipur-492010, Chhattisgarh, India

Abstract –

This paper describes how convolutional neural networks are used to identify and classify igneous rocks (CNN). Igneous rocks are formed while still hot, hot magma crystallises and solidifies. Melt originates deep beneath the Earth's surface, amid active plate borders or hot zones, and then rises to the surface. There are also various kinds of igneous rocks, which are addressed throughout this work, distinguishing each one is a difficult feat in and of itself. Machine learning, which is fundamentally a three-layer neural network, is a subset of deep learning. These neural networks attempt to mimic conscious brain function of humans by letting it to "learn" from massive volumes of data, but they continue to fail. A Convolutional Neural Network (ConvNet/CNN) is a Deep Learning approach for assigning priority (learnable weights and biases) to various aspects/objects in a picture while also identifying them. Classification of images, audio and video segmentation, decision support systems, speech recognition, and image analysis are just a few of the applications for CNNs.

Keywords – Igneous Rocks, CNN (convolutional neural networks), layers, Python and Jupyter Notebook, training and testing.

1. Introduction –

Igneous rocks are formed by the cooling and crystallisation of molten rock, often known as magma. Magma is molten rock that exists deep within the ground. It is the source of all igneous rock. Because the planet was mainly melted when it originated, magma may be regarded as the beginning of the rock cycle. Igneous rocks' origins are documented in their composition. We can derive processes that occur within the earth and comprehend volcanic activities that occur on the planet's surface by carefully examining igneous rocks and interpreting the information they carry. We can learn about the igneous portion of geologic history by studying igneous rocks [1]. Igneous petrology is the study of the identification, classification, origin, development, and formation and crystallisation processes of igneous rocks [2].

Andesite, Diorite, Basalt and Obsidian

Igneous rock categorization and identification are critical because they can assist us understand the relative mineral richness of the rock. Based on their chemical/mineral composition, texture, and grain size, igneous rocks can be classified as felsic, intermediate, mafic, or ultramafic: Extrusive igneous rocks are fine-grained i.e., fine textured (infinitesimally small crystals) or amorphous, whereas intrusive igneous rocks are coarse-grained (tiny crystals) (no minerals; no crystalline structure). The porphyritic texture of volcanic rocks, notably felsic and intermediate, is characterised by visible crystals floating in a fine-grained groundmass [3].

Physically identifying each igneous rock by its texture, grain size, colours, faults, and patterns is a difficult process that necessitates the use of a skilled geologist with rock and mineral understanding. The act of identifying and classifying rocks may be arduous and time consuming for persons from other fields. This paper discusses a tool that can aid in the task's completion. We employed some specific deep learning methods to create this tool. Based on the image that the user uploads, our system offers an accurate forecast of rock [4].

The method of deep learning is a machine learning and artificial intelligence (AI) methodology that mimics human learning and biological neural networks, In data science statistical insights and predictive modelling, has embraced deep learning technology as an essential component. Data scientists in charge of acquiring, evaluating, and perceiving large amounts of data gain tremendously from deep learning because it improves and simplifies the task [7]. Deep Learning advancements in computer vision and image identification have been established and improved throughout time, mainly due to the use of a single algorithm: the Convolutional Neural Network.

A Convolutional Neural Network (ConvNet/CNN) is a Deep Learning system which can ingest a picture and assign relevance (learnable weights and biases) to various aspects/objects while also discerning between them. ConvNet/CNN requires substantially more pre-processing than other classification techniques. With enough training, ConvNets can learn these filters/characteristics [17], whereas simple approaches need filter hand-engineering. Other classification approaches need substantially higher pre-processing than a ConvNet. While foundational approaches need the hand-engineering of filters, CNN can acquire these filters/characteristics with enough training [17]. The extent of pre-processing necessary by a ConvNet is much less than that required by other classification algorithms. While basic approaches need hand-engineering of filters, CNN can learn these filters/characteristics provided with sufficient training [17]. The amount of pre-processing required by a ConvNet is significantly lower than that required by other classification techniques. While simple techniques need filter designing by hand, CNN can learn these filters/characteristics with enough training. The arrangement of the Visual Cortex influenced the construction of a CNN, which is similar to the connection pattern of Neurons in the Human Brain. Individual neurons can only respond to stimuli in the Receptive Field, a tiny portion of the visual field. If a set of comparable fields overlap, the entire visual region is covered. The CNN's role is to compress the images into a format that is easier to handle while keeping crucial properties for accurate prediction [17].

Literature Review –

Sharma N., Jain V., Mishra A., "An analysis of convolutional networks for image classification". This study looks at the performance of common convolutional neural networks (CNNs) for recognising objects in real-time image processing. [23].

Xin M. and Wang Y., "Research on image classification model based on deep convolution neural network". This paper explains architecture of image classification models based deep convolutional neural networks [24].

H. C. Shin; Holger R. Roth; Gao M.; Le Lu; Xu Z.; Nogues I.; Yao J., "Deep Convolutional Neural Networks for Computer-Aided Detection: CNN Architectures, Dataset Characteristics and Transfer Learning". Three crucial, but formerly unexplored aspects of using deep convolutional neural networks to solve computer-aided detection challenges are described in this research. To begin, the research and assessment of several CNN architectures are described. The models analysed ranging in size from 5,000 to 160 million variables, with varying amounts of layers. The effect of dataset scale and spatial image context on performance is then determined. [25].

JP Iddings, "The origin of igneous rocks". This paper discusses in depth analysis of igneous rocks – formation, features and types [26].

MacKenzie W.S., Donaldson C. H. and C. Guilford, "Atlas of igneous rocks and their textures". The textures of various volcanic and plutonic igneous rocks are discussed [27].

2. Methodology-

To predict and identify rocks the model first needs to be trained. For training, a good number of images with decent resolution and quality are required. The following flow chart explains the sequence of tasks we are ought to perform to create an accurate model.

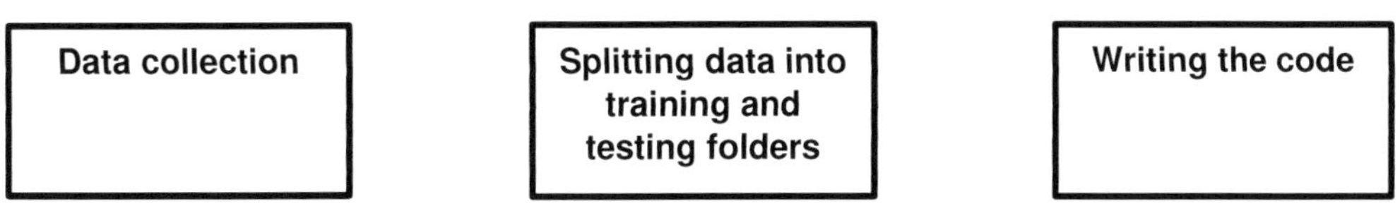

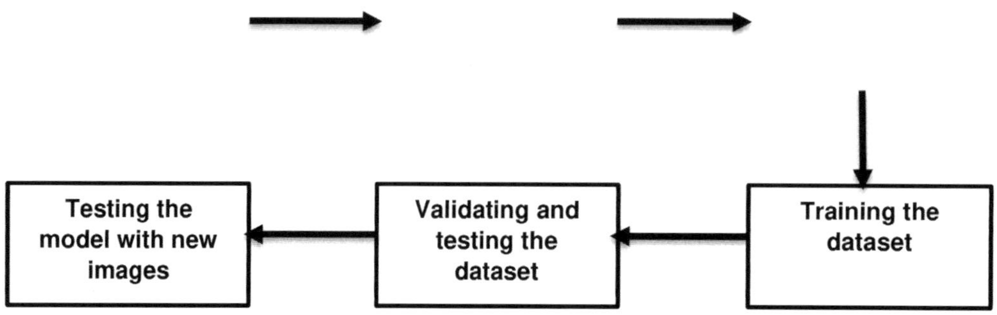

a. Data collection and data splitting –

Data collection is the process of obtaining, measuring, and interpreting precise information from a multitude of sources in order to explore solutions to examine situations, answer questions, and assess outcomes [12]. To train our model images of various igneous rocks are required. It is recommended to collect at least 150-200 distinct images of each class of igneous rocks. This paper aims to perform a multiclass classification of 4 types of igneous rocks; however, any number of igneous rocks can be considered if enough image data is available for each class. The four classes of Igneous rocks used in the scope of this paper are *andesite, basalt, diorite and obsidian.*

Training and testing folders

First create two folders in any directory and name them as "train" and "test". Create 4 folders of "andesite", "basalt", "diorite" and "obsidian" inside each of these training and testing folders. Split the total dataset of each class into a ratio of 80:20 in training and testing folders (*For example – If 200 images of Andesite are available, move 160 images to andesite folder present inside training folder and remaining 40 images to andesite folder of testing folder*). Once images are moved to their respective folders, the process of training and testing the model begins.

Inside the folders, images of the specific rock classes are stored. In this case its Basalt

b. Tools and techniques -

Python is the programming language utilised in this project. Jupyter Notebook, a cloud-based IDE, is used to develop the python code. Because it saves information on a cloud server and does all processing on the cloud, Jupyter notebooks allow you to interact with applications and APIs that may demand a high system specification (RAM, GPU).

Keras Sequential API is used to generate and train a model. This API simplifies the process of building and training with only a few lines of code. TensorFlow is accompanied by the Keras API. As a result, simply installing TensorFlow on your command prompt will enough [15].

TensorFlow is a machine learning framework that streamlines the entire process. It attributes a diverse ecosystem of tools, libraries, and community resources that enable academics to continue to progress the trailblazing techniques in machine learning and developers to instantaneously compose and implement machine learning optimization algorithms [7][13]. Google designed and published TensorFlow, a Python library for swift computationally intensive tasks [13].

Keras is an API that is based on Python-programming language. It is a deep learning API that operates atop TensorFlow machine learning framework. It was founded with a specific aim of enabling for speedy experimentation and research [8]. Keras uses many optimization methods to make high-level neural network API simpler and more performant. Keras is built on TensorFlow, an open-source machine learning framework. Keras is based on a straightforward framework for creating deep learning models with TensorFlow [7]. It's a simple framework that may be used on both the CPU and GPU. Established neural network deployment approaches either adds substantial latencies, cost a lot of money and time to incorporate into embedded code bases, or specifically supports a few numbers of model types.

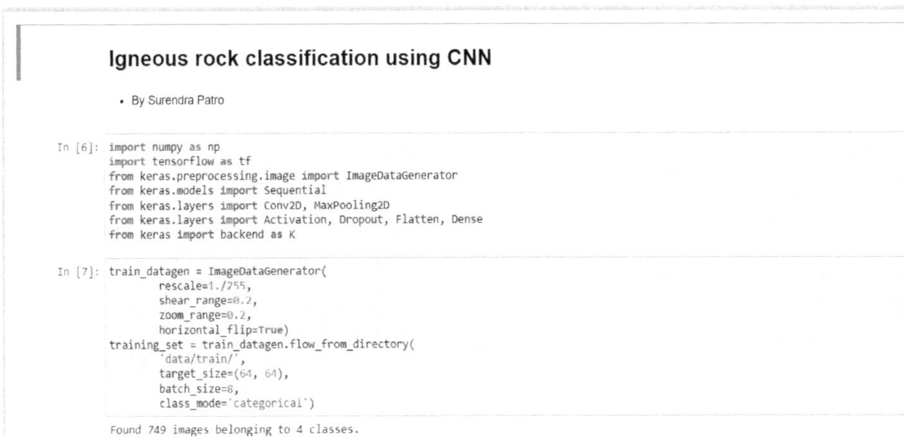
Igneous rock classification using CNN

- By Surendra Patro

```
In [6]: import numpy as np
        import tensorflow as tf
        from keras.preprocessing.image import ImageDataGenerator
        from keras.models import Sequential
        from keras.layers import Conv2D, MaxPooling2D
        from keras.layers import Activation, Dropout, Flatten, Dense
        from keras import backend as K

In [7]: train_datagen = ImageDataGenerator(
                rescale=1./255,
                shear_range=0.2,
                zoom_range=0.2,
                horizontal_flip=True)
        training_set = train_datagen.flow_from_directory(
                'data/train/',
                target_size=(64, 64),
                batch_size=8,
                class_mode='categorical')

        Found 749 images belonging to 4 classes.
```

Importing the libraries

Import the TensorFlow Platform and Keras API first. Data augmentation is a crucial aspect of the training process. For data augmentation, we use the "ImageDataGenerator" technique that comes with Keras. Then add the Sequential class to your project. The basic concept of Sequential API is to arrange the Keras layers in a sequential manner, hence the name. Most ANNs have layers that are arranged in a sequential sequence, with data flowing from one layer to the next in the specified order until it reaches the output layer. Then add layers for activation, dropout, flatten, and dense CNN. An activation function governs how the sum of weighted inputs is generated into an outcome from a network as well as nodes in a layer in a neural network. Dropout is a method of preventing a model beyond being overfit. Dropout converts the information the outgoing points of hidden nodes (neurons that comprises the hidden layers) to 0 at random throughout every cycle of the training procedure [8][15].

Training the model –

Once arranging the images in all folders is completed, we define two variables "train_datagen" and "training_set". The data gen variable calls out the ImageDataGenerator function which helps in the process of image augmentation. Second variable training set stores the path of the directory in which the training images are stored. These images can be stored locally as well in cloud storage platforms like Google drive and dropbox [11].

ImageDataGenerator( ) class has several arguments, *rescale* – This argument rescales the RGB pixel value into numbers between 0-1 (usually they're somewhere near 0 – 255), *shear range and zoom range*- The flow from directory method has following arguments – *The directory path, target size* – Pixel size of the image, *batch size* – Number of images to be taken in a single batch, *class mode*- Here categorical class mode as an argument is used since we have multiple classes.

Now the process of creating the model starts. To create a CNN model, we are going to add several layers to it using "keras.layers" and add() function. The CNN is made up of three types of standard layers they are-

Convolutional layers, Pooling layers, and Fully-connected (FC) layers. When these layers are combined, a standard CNN architecture is formed. Remaining two significant components are the activation layer and the dropout function [10]. A standard diagram of CNN architecture is shown below.

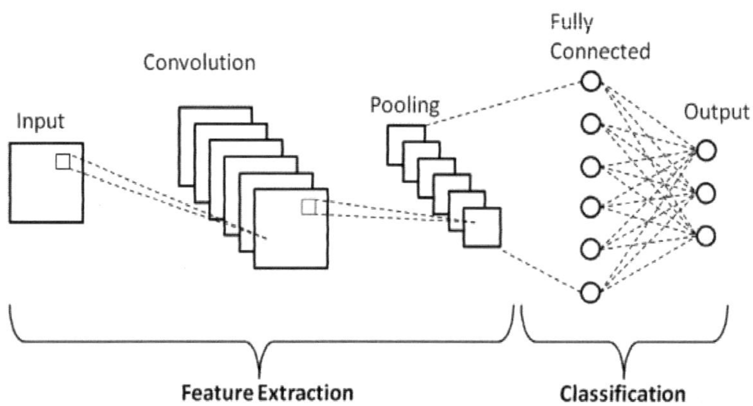

i. Convolutional Layer-

To detect picture characteristics, a convolution filter is applied to the image. The following is how it works:

A convolution is a function that multiplies a collection of weights by the neural network's inputs. Kernels or screens kernel (for 2D arrays of parameters) or a filter (for 3D structures) passes throughout an image several times during the multiplication process. The filtering is employed from right to left and top to bottom to cover the entire image. During the convolution, a mathematical operation known as the dot or scalar product is used. The weights are multiplied by each filter with various input values. The entire inputs are combined together to provide each filter position a unique value [18][19][21].

ii. Pooling Layer-

The pooling layers shrink the image over time, preserving just the most crucial details. For each set of pixels, the pixel with the peak score is kept (this is known as max pooling), or simply the average is kept (average pooling). By lowering the number of calculations and parameters in the network, pooling layers aid in the management of overfitting. [18].

iii. Fully Connected Layer-

A fully connected neural network is made up of layers that are all linked. A fully connected layer is expressed as a function ranging from R m to R n. Each input dimension determines the output dimension. "Neurons" is a term used to describe the nodes in fully connected networks. As a result, fully connected networks will be referred to as "neural networks" elsewhere in the literature [21].

iv. Dropout-

The Dropout layer, which helps minimise overfitting, changes input unit values to null at random with a frequency of specific rate at each step during training time. Inputs that aren't adjusted to 0 are escaladed up by 1/(1 - r), where r is that specific rate that was mentioned, so that the overall sum does not change its value [19].

```
In [7]: train_datagen = ImageDataGenerator(
                rescale=1./255,
                shear_range=0.2,
                zoom_range=0.2,
                horizontal_flip=True)
        training_set = train_datagen.flow_from_directory(
                'data/train/',
                target_size=(64, 64),
                batch_size=8,
                class_mode='categorical')

        Found 749 images belonging to 4 classes.

In [8]: test_datagen = ImageDataGenerator(rescale=1./255)
        test_set = test_datagen.flow_from_directory(
                'data/test/',
                target_size=(64, 64),
                batch_size=8,
                class_mode='categorical')

        Found 160 images belonging to 4 classes.

In [9]: cnn = tf.keras.models.Sequential()
        cnn.add(tf.keras.layers.Conv2D(filters=64 , kernel_size=3 , activation='relu' , input_shape=[64,64,3]))
        cnn.add(tf.keras.layers.MaxPool2D(pool_size=2,strides=2))
        cnn.add(tf.keras.layers.Conv2D(filters=64 , kernel_size=3 , activation='relu' ))
        cnn.add(tf.keras.layers.MaxPool2D(pool_size=2 , strides=2))
        cnn.add(tf.keras.layers.Dropout(0.5))
        cnn.add(tf.keras.layers.Flatten())
        cnn.add(tf.keras.layers.Dense(units=128, activation='relu'))
        cnn.add(tf.keras.layers.Dense(units=4 , activation='softmax'))
        cnn.compile(optimizer = 'rmsprop' , loss = 'categorical_crossentropy' , metrics = ['accuracy'])
```

Training the model and adding layers to it

v. Activation Functions-

Furthermore, the activation function can be considered as one of the most critical parts of the CNN model. These functions are employed to train themselves and estimate specific sort of network mode to node correlation, including continuous and complicated. It defines which model characteristics should be transmitted forward and which will be returned at the network's end, in simple words [12][20].

Note: The activation function employed is RELU, or rectified linear activation function, is a piecewise function that returns 0 if the input is negative and returns 1 if it is positive. Because it is easier to make it learn and often produces better quality results, it has transformed itself into the preferred activation function for several modelling techniques [13].

A nonlinear activation function is applied to the outcomes of a linear operations, such as convolution. Regardless of the fact that relatively smoother nonlinear functions like the sigmoid or hyperbolic tangent (tanh) function were usually prominent as mathematical representations of actual neuron behaviour, the rectified linear unit (ReLU) is now the most prominent nonlinear activation function [22].

f(x) = maximum (0, x)

Testing and validating the data-

Testing data can help us understand whether our model is performing correctly on newer data. In this line of code, we fit the trained CNN model with test data set. The *epoch* argument shows the number of cycles the data passes in and out of the neural network.

```
In [5]: cnn.fit(x = training_set , validation_data = test_set , epochs = 10)

Epoch 1/10
94/94 [==============================] - 32s 337ms/step - loss: 1.0939 - accuracy: 0.4967 - val_loss: 0.6575 - val_accuracy: 0.
7125
Epoch 2/10
94/94 [==============================] - 17s 184ms/step - loss: 0.6692 - accuracy: 0.7303 - val_loss: 0.6004 - val_accuracy: 0.
7750
Epoch 3/10
94/94 [==============================] - 18s 195ms/step - loss: 0.4574 - accuracy: 0.8224 - val_loss: 0.3686 - val_accuracy: 0.
8750
Epoch 4/10
94/94 [==============================] - 19s 198ms/step - loss: 0.3582 - accuracy: 0.8678 - val_loss: 0.2257 - val_accuracy: 0.
9500
Epoch 5/10
94/94 [==============================] - 17s 179ms/step - loss: 0.2345 - accuracy: 0.9132 - val_loss: 0.3150 - val_accuracy: 0.
9312
Epoch 6/10
94/94 [==============================] - 17s 182ms/step - loss: 0.2339 - accuracy: 0.9212 - val_loss: 0.2814 - val_accuracy: 0.
9375
Epoch 7/10
94/94 [==============================] - 17s 181ms/step - loss: 0.2136 - accuracy: 0.9346 - val_loss: 0.3758 - val_accuracy: 0.
9062
Epoch 8/10
94/94 [==============================] - 17s 183ms/step - loss: 0.2026 - accuracy: 0.9359 - val_loss: 0.2667 - val_accuracy: 0.
9563
Epoch 9/10
94/94 [==============================] - 17s 182ms/step - loss: 0.1355 - accuracy: 0.9546 - val_loss: 0.3778 - val_accuracy: 0.
9375
Epoch 10/10
94/94 [==============================] - 17s 185ms/step - loss: 0.1010 - accuracy: 0.9720 - val_loss: 0.9549 - val_accuracy: 0.
8438

Out[5]: <keras.callbacks.History at 0x22708173cd0>
```

Testing data

The testing of our model is completed. The performance parameter shows an accuracy of 95%, which is great. *(Note – This accuracy might vary depending upon the type and quality of data).*

Model summary –

Here we show our model summary. We need model summary to confirm that everything (layers, arguments and other parameters) is as expected. [14][15]

```
In [6]: cnn.summary()

Model: "sequential"
_________________________________________________________________
 Layer (type)                Output Shape              Param #
=================================================================
 conv2d (Conv2D)             (None, 62, 62, 64)        1792

 max_pooling2d (MaxPooling2D  (None, 31, 31, 64)       0
 )

 conv2d_1 (Conv2D)           (None, 29, 29, 64)        36928

 max_pooling2d_1 (MaxPooling  (None, 14, 14, 64)       0
 2D)

 dropout (Dropout)           (None, 14, 14, 64)        0

 flatten (Flatten)           (None, 12544)             0

 dense (Dense)               (None, 128)               1605760

 dense_1 (Dense)             (None, 4)                 516

=================================================================
Total params: 1,644,996
Trainable params: 1,644,996
Non-trainable params: 0
_________________________________________________________________
```

CNN model summary

Performance parameters –

For performance parameter, two Line graphs are plotted. One is of *accuracy vs epochs* and other is of *loss vs epochs*. The graph shows that the validation and training accuracies were almost identical, with a combined accuracy of over 90%. The CNN model's loss is a negative trailing graph, indicating that the model is functioning as predicted, with the loss decreasing with each epoch [16][19].

IOP Conf. Series: Earth and Environmental Science 1032 (2022) 012045 doi:10.1088/1755-1315/1032/1/012045

```
In [9]: import matplotlib.pyplot as plt
        plt.plot(history.history['accuracy'])
        plt.plot(history.history['val_accuracy'])
        plt.title('model accuracy')
        plt.ylabel('accuracy')
        plt.xlabel('epoch')
        plt.legend(['train', 'test'], loc='upper left')
        plt.show()
```

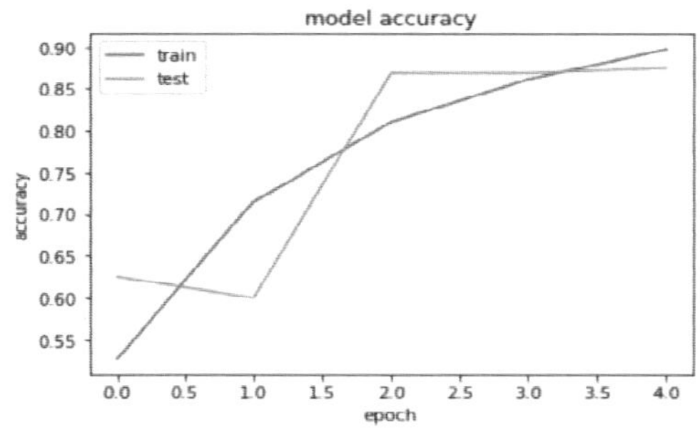


```
In [10]: # summarize history for loss
         plt.plot(history.history['loss'])
         plt.plot(history.history['val_loss'])
         plt.title('model loss')
         plt.ylabel('loss')
         plt.xlabel('epoch')
         plt.legend(['train', 'test'], loc='upper left')
         plt.show()
```

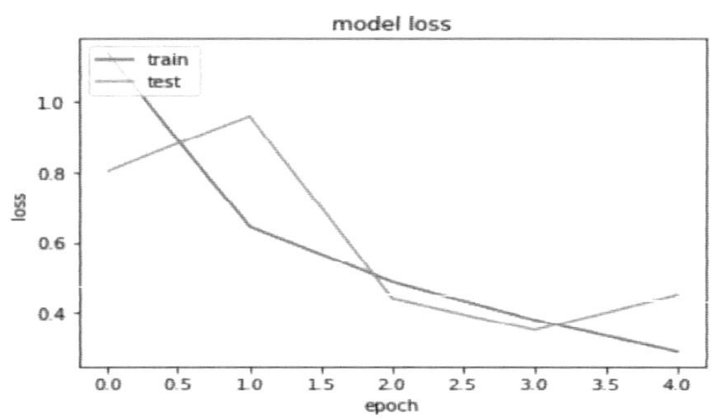

Results and Outcome –

After training, validating and testing, a CNN model is created which can identify igneous rocks. To test the model on new data, following lines of python code can be used. In this cell, an input box is provided where path location of any locally saved image is added. (*Note – make sure the new image data is locally saved in your device*)

User input

```
In [*]: from keras.preprocessing import image
        user_input = input('Enter your image link - ')
        test_image = image.load_img(user_input, target_size=(64,64))
        test_image = image.img_to_array(test_image)
        test_image = np.expand_dims(test_image,axis=0)
        result = cnn.predict(test_image)
        training_set.class_indices

        Enter your image link -
```

After entering the image path -

User input

```
In [38]: from keras.preprocessing import image
         user_input = input('Enter your image link - ')
         test_image = image.load_img(user_input, target_size=(64,64))
         test_image = image.img_to_array(test_image)
         test_image = np.expand_dims(test_image,axis=0)
         result = cnn.predict(test_image)
         training_set.class_indices

         Enter your image link - image3.jpg

Out[38]: {'andesite': 0, 'basalt': 1, 'diorite': 2, 'obsidian': 3}
```

Once the image path is entered into the input box shown above. The image data is passed through the neural network, which predicts the rock type.

The uploaded test image is showcased -

Uploaded image

```
In [39]: from IPython import display
         display.Image(user_input, width = 256, height = 256)

Out[39]:
```

And finally, the model predicts an outcome. The out is usually in a form of an array. A simple if-else python statement can solve this. Finally, the desired result is obtained.

Result

```
In [40]: if result[0][0]==1:
             print('The uploaded image is of andesite')
         elif result[0][1]==1:
             print('The uploaded image is of basalt')
         elif result[0][2]==1:
             print('The uploaded image is of diorite')
         elif result[0][3]==1:
             print('The uploaded image is of obsidian')

         The uploaded image is of diorite
```

In this case the Igneous rock is **diorite!**

Conclusion –

Classification of igneous rocks using CNN algorithm can help us automate many aspects of geological surveys. First organise the image dataset into two folders "training and testing". After this the model is provided with several layers like convolutional layer, pooling layer and activation function (In this case RELU). This CNN algorithm can classify images with considerably good accuracy. The four igneous rock classes we've taken for this paper are

Andesite, basalt, diorite and obsidian. Once the user inputs a rock image to the algorithm, the result is predicted. In our case we provided an image of Diorite rock, which our model predicted very accurately. There is no restriction in the number of classes. However, it is important to organise the rock images in proper folders and images should be of good quality.

This tool makes geologist's life easy. Geological surveying can be automated by using these machine learning tools. This rock classification algorithm can also be used for surveying rocks and minerals on a larger scale and also for understanding underlying geological structures. The CNN image classification algorithm can also be used for several other geological/mining problems like predicting the rock fragmentation, predicting the river basin sediments etc...,

References –

[1] MacKenzie, William Scott, Colin H. Donaldson, and Corin Guilford. Atlas of igneous rocks and their textures. Vol. 148. Harlow: Longman, 1982.

[2] Gill, Robin. Igneous rocks and processes: a practical guide. John Wiley & Sons, 2011.

[3] McBirney, Alexander R. *Igneous petrology*. Jones & Bartlett Learning, 1993.

[4] Marsh, Bruce D. "On some fundamentals of igneous petrology." *Contributions to Mineralogy and Petrology* 166.3 (2013): 665-690.

[5] Maaloe, Sven. *Principles of igneous petrology*. Springer Science & Business Media, 2012.

[6] Cox, Keith Gordon, ed. The interpretation of igneous rocks. Springer Science & Business Media, 2013.

[7] Abadi, Martín. "TensorFlow: learning functions at scale." Proceedings of the 21st ACM SIGPLAN International Conference on Functional Programming. 2016.

[8] Arnold, Taylor B. "kerasR: R Interface to the Keras Deep Learning Library." J. Open Source Softw. 2.14 (2017): 296.

[9] Learning, Deep. "Deep learning." High-Dimensional Fuzzy Clustering (2020).

[10] Lu, Dengsheng, and Qihao Weng. "A survey of image classification methods and techniques for improving classification performance." International journal of Remote sensing 28.5 (2007): 823-870.

[11] Kattenborn, Teja, et al. "Review on Convolutional Neural Networks (CNN) in vegetation remote sensing." ISPRS Journal of Photogrammetry and Remote Sensing 173 (2021): 24-49.

[12] Sapsford, Roger, and Victor Jupp, eds. Data collection and analysis. Sage, 1996.

[13] Abadi, Martín. "TensorFlow: learning functions at scale." Proceedings of the 21st ACM SIGPLAN International Conference on Functional Programming. 2016.

[14] Gulli, Antonio, and Sujit Pal. Deep learning with Keras. Packt Publishing Ltd, 2017.

[15] Moolayil, Jojo. "An introduction to deep learning and keras." Learn Keras for Deep Neural Networks. Apress, Berkeley, CA, 2019. 1-16.

[16] Albawi, Saad, Tareq Abed Mohammed, and Saad Al-Zawi. "Understanding of a convolutional neural network." 2017 international conference on engineering and technology (ICET). Ieee, 2017.

[17] Hussain, Mahbub, Jordan J. Bird, and Diego R. Faria. "A study on cnn transfer learning for image classification." UK Workshop on computational Intelligence. Springer, Cham, 2018.

[18] Bakhshi, Ali, Stephan Chalup, and Nasimul Noman. "Fast evolution of CNN architecture for image classification." *Deep Neural Evolution*. Springer, Singapore, 2020. 209-229.

[19] Jmour, Nadia, Sehla Zayen, and Afef Abdelkrim. "Convolutional neural networks for image classification." *2018 international conference on advanced systems and electric technologies (IC_ASET)*. IEEE, 2018.

[20] Sharma, Sagar, Simone Sharma, and Anidhya Athaiya. "Activation functions in neural networks." towards data science 6.12 (2017): 310-316.

[21] Ma, Wei, and Jun Lu. "An equivalence of fully connected layer and convolutional layer." arXiv preprint arXiv:1712.01252 (2017).

[22] Agarap, Abien Fred. "Deep learning using rectified linear units (relu)." arXiv preprint arXiv:1803.08375 (2018).

[23] Sharma, Neha, Vibhor Jain, and Anju Mishra. "An analysis of convolutional neural networks for image classification." Procedia computer science 132 (2018): 377-384.

[24] Xin, Mingyuan, and Yong Wang. "Research on image classification model based on deep convolution neural network." EURASIP Journal on Image and Video Processing 2019.1 (2019): 1-11.

[25] Shin, Hoo-Chang, et al. "Deep convolutional neural networks for computer-aided detection: CNN architectures, dataset characteristics and transfer learning." IEEE transactions on medical imaging 35.5 (2016): 1285-1298.

[26] Iddings, Joseph Paxson. The origin of igneous rocks. Vol. 12. Philosophical Society, 1892.

[27] MacKenzie, William Scott, Colin H. Donaldson, and Corin Guilford. Atlas of igneous rocks and their textures. Vol. 148. Harlow: Longman, 1982.

Classifying Minerals using Deep Learning Algorithms

Tajendra Singh[1, a], Dr. D.C. Jhariya[2, b, *], Dr. Mridu Sahu[3, c], Dr Pankaj Dewangan[4, d], Dr. P.Y. Dhekne[5, e.]

[1]B. Tech Mining Engineering Pre-Final Year, National Institute of Technology Raipur, G.E.Road, Raipur - 492010, Chhattisgarh, India
[2]Assistant Professor, Department of Applied Geology, National Institute of Technology Raipur, G.E.Road, Raipur - 492 010, Chhattisgarh, India
[3]Assistant Professor, Department of Information Technology, National Institute of Technology Raipur, G.E. Road, Raipur-492010, Chhattisgarh, India
[4]Associate Professor, Department of Mining Engineering, National Institute of Technology Raipur, G.E. Road, Raipur-492010, Chhattisgarh, India
[5]Associate Professor, Department of Mining Engineering, National Institute of Technology Raipur, G.E. Road, Raipur-492010, Chhattisgarh, India

Email: [a]singhtajendra01@gmail.com; [b]dcjhariya.geo@nitrr.ac.in; [c]mrisahu.it@nitrr.ac.in; [d]pankajnitrr@yahoo.co.in, [e]pdhekne@nitrr.ac.in

Abstract: A mineral is an inorganic substance that occurs in nature with specific chemical content and an ordered atomic positioning. Minerals are identified by their physical properties. Minerals' physical properties are related to their chemical composition and bonding. Quartz is extremely valuable economically. Valuable minerals and some examples of gemstones are citrine, amethyst, quartz with smoky texture and quartz of rose color can be said as rose quartz are some examples of gemstones. Sandstone, primarily composing quartz, is the most used building stone. Biotite has limited number of applications for commercial use. Deep learning is the subset of machine learning. It is based on self-learning and improvement through the examination of computer algorithms. TensorFlow library of machine learning combines a number of different algorithms and models which allows users to build deep neural networks for projects/model such as image recognition/classification and many more. Image Classification is the assignment of one label from a fixed set of categories to an input image. In this paper Convolutional neural networks (CNNs) are used primarily for image processing, classification, segmentation, and other auto-correlated data. This paper will explain the techniques and explanation for classifying minerals images using a deep learning algorithm called a convolutional neural network. Identifying minerals on a field is a tedious activity and requires a lot of information and conformation here with the help of deep learning algorithms we made a deep learning model which has all its feature already embedded in it that can be used to classify minerals with a reasonable accuracy furthermore in future it can be made more accurate and fit accordingly to the conditions.

Keywords: Minerals, Quartz, Biotite, Deep learning, TensorFlow, Keras, Convolutional neural network.

1. Introduction

Minerals either alone or in a combination called compounds forms the Earth. Mineral can be of a single element or can be compound.[1]. It is an inorganic substance that occurs naturally having specified chemical composition and ordered atomic structure.

Mineral Economics (ME) Division of the Indian Bureau of Mines was established in 1948. It was elevated to the rank of division of the department in 1956[2]. The Mineral Economics division provides information support and advisory services to the mineral industry and the government, mainly being focused on issues of marketing, mineral specifications, uses of minerals, mineral legislation, mineral resource inventory, mining leases and taxation, and many more.

In India, Metallic minerals are valued at Rs 54,000 crore, while for non-metallic minerals including minor minerals, are valued at Rs 61,000 crore, according to the report [3]. According to the reports of 2016-17 the value of metallic minerals such as ores of: iron, manganese, zinc, bauxite, copper, gold(native), chromite, and lead was around Rupees 40,000 crore.

Andhra Pradesh (AP), Chhattisgarh (CG), Goa (GA), Gujarat (GJ), Jharkhand, Karnataka, Madhya Pradesh (MP), Maharashtra (MH), Orissa (OR), Rajasthan (RJ), and Tamil Nadu(TN) are among the states with the most mineral reserves and contribute significantly to the country's mineral production [4].

India's major mineral resources constitute Iron Ore, Coal, Mica, Manganese ore, Bauxite Chromite, Diamonds, Limestone, and Thorium.[5].

Identification of Minerals on Field: Minerals can be distinguished by a variety of characteristics. Colour, streak, lustre, hardness, crystal form, cleavage, specific gravity, and habit are the most frequent characteristics used to identify minerals. Even when a geologist is out in the field, most of these can be easily analysed [6]. Mineral identification is required to identify rocks and can be used to understand the landscape also the geologic history of the area.

Deep learning: Is one of the types of machine learning in which computers tends to learn by doing, much like people do. It is also an important component of self-driven cars because it allows them to identify traffic signs and distinguish between pedestrians and obstacles on road [7]**.** In this, a computer model learns to perform classification tasks by taking input as pictures, text, or sound directly. These models may achieve the greatest accuracy of time also sometimes they can even outperform humans [8]. Multilayer neural network topologies and a large amount of labelled data are used to train models [9]. Depending on the type of the earth science exploration, a variety of algorithms may be used· For specific goals, number of algorithms may perform much more accurate than others.

CNN: Artificial neural networks (ANN) are good at solid classification and convolutional neural networks (CNN) are good at throwing light on photographs, but they are more computationally costly to train than support-vector machine (SVM) learning [10]. In recent decades, machine learning has grown in popularity as other technologies such as unmanned aerial vehicles (UAVs), ultra-high resolution remote sensing technology, and high-performance computer units have made vast, high-quality datasets and more powerful algorithms available [11].

The CNN is an artificial neural network that is being used for image identification and processing, because of its capacity to detect patterns in photos A CNN is a powerful tool, but training it takes millions of labelled data points [12]. If CNNs are to generate results quickly enough to be effective, they must be trained with high-power processors such as a GPU or an NPU.

CNN for Mineral Classification: Extracting characteristics from a picture (in this example of biotite and quartz) in order to discover trends in a dataset is known as image classification [13]. ANN for image classification would be highly computationally costly due to the large number of trainable parameters [14]. Let's say, if there are 50 X 50 image of a minerals and want to train our standard ANN to classify it as biotite or quartz, the trainable parameters are – (50*50). 2,50,302 image pixels multiplied by hidden layer + 100 bias + 2 = 2,50,302 image pixels multiplied by hidden layer + 100 bias + 2.

Problem Statement: Mineral exploration is costly and risky due to limited geologic knowledge and an incomplete understanding of the ore-forming processes that control mineral distribution over geologic space and over hundreds of millions of years of geologic time. Classifying them by the means of deep learning model will come out to be the great change in the mineral identification problem and have potential to overcome the difficulties during the process.

2. Literature Review

Folorunso, I. O., et al. "A rule-based expert system for mineral identification." Journal of Emerging Trends in Computing and Information. In this project, an expert system is used to identify minerals. The expert system's design and execution are based on the physical characteristics of the forty minerals studied as the knowledge domain. The inference engine is rule-based, as past research has shown to be more effective.

Mlynarczuk, Mariusz, and Marta Skiba in "The application of artificial intelligence for the identification of the maceral groups and mineral components of coal." Computers & Geosciences 103 (2017). It was demonstrated that artificial intelligence approaches may be used to successfully identify specific petrographic properties of coal.

Zhang, Ye, in. "Intelligent identification for rock-mineral microscopic images using ensemble machine learning algorithms." *Sensors* 19.18 (2019): Deep learning and convolutional neural networks (CNNs) provide a mechanism for rapidly and intelligently analysing mineral microscopic pictures. The transfer learning model of mineral microscopic pictures is developed in this study using the Inception-v3 architecture.

Maitre, Julien, Kévin Bouchard, and L. Paul Bédard. "Mineral grains recognition using computer vision and machine learning." *Computers & Geosciences* 130 (2019): Computer vision combined with machine learning can categorise minerals in sand grains, with mineral recognition results reaching 90% accuracy.

3. Methodology

For a good and accurate working model, it first needs to be trained on a dataset containing separate folder for each class as in this case "Biotite" & "Quartz", of good quality images that are free of background noises, so the first step is to collect the data that will be fed into the model.

3.1 Data Collection

Data collection is the systematic way/technique of collection and gathering of data in the systematic form with features of interest clubbed together so that it can be used for testing hypotheses and for evaluating the outcome of problems [15]. For gathering the image data of minerals, using "bingimagedownloader" api, for each class there are 199 images (Biotite and Quartz) so total is 199*2 which is 398. Make folder for each class and name them accordingly [16].

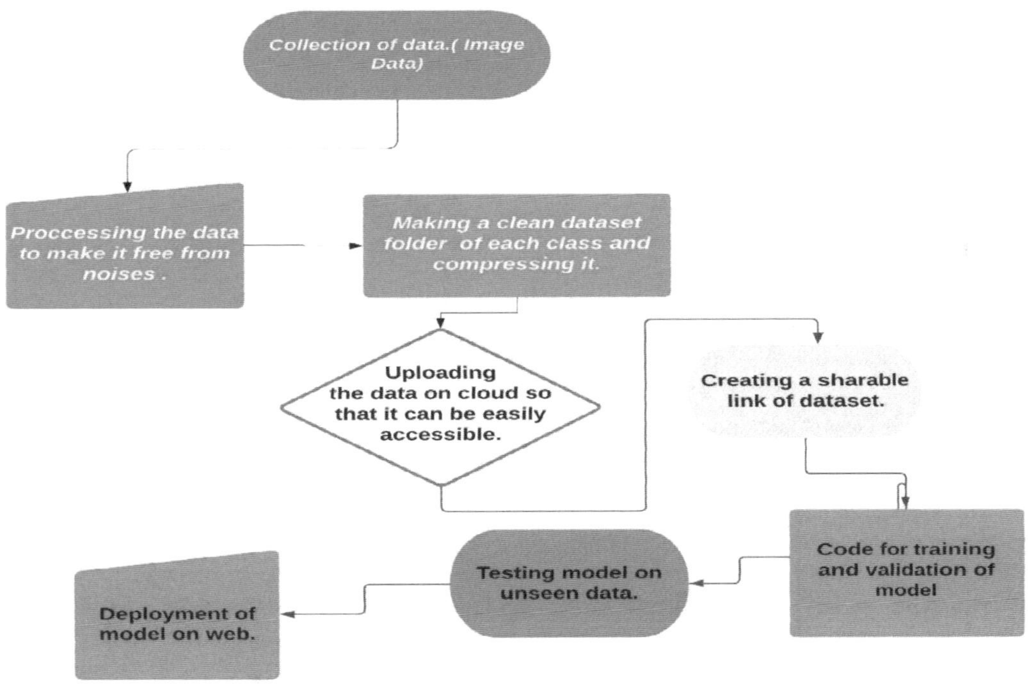

Figure 1: Process Flow Chart

Now, compress this folder and upload it to your cloud account so that it can be imported directly for use. For this "dropbox" is used as it creates direct shareable link.

Figure 2: Uploading the compressed folder

3.2 Tools used

For writing script in python google colab is used as development environment. It's a cloud-based integrated development environment [17]. Colab is thought to be more portable and simpler to use than Jupyter since it is easier to set up. Colab also facilitates team collaboration, which is not available with Jupyter [18].

For this project, Python syntax is used. As a framework, Keras is implemented, It is a Python-based high-level neural network API. But for lower-level tasks it cannot be utilized or function without a backend [19].

Keras Sequential API is used to generate and train a model. This API simplifies the process of building and training with only a few lines of code. TensorFlow is accompanied by the Keras API [20].

Step 1: Import the libraries you'll need.

For this project, Keras library is used for building and training of the deep learing model. Also used Matplotlib to visualise our dataset so that we can better grasp the photos we'll be dealing with. Opencv is another useful library for dealing with picture data.

```
[ ] import matplotlib.pyplot as plt
    import numpy as np
    import os
    import PIL
    import tensorflow as tf

    from tensorflow import keras
    from tensorflow.keras import layers
    from tensorflow.keras.models import Sequential
```

Figure 3: Importing Libraries

TensorFlow accepts tensor data forms, tensors are high-dimensional and multi-dimensional arrays. These multi-dimensional arrays are useful for taking in account with large volumes of data [21].

TensorFlow is based on graphs that depict data flows and have nodes and edges. Because the execution method is in the form of graphs, TensorFlow code is significantly easier to spread over a cluster of computers when using GPUs [22].

Keras is an API of high-level in deep learning has a use for creating neural networks developed by Google. Its script is written in Python and is made for simplification to the development of neural networks [23]. Backend calculation of several neural networks is also enabled in keras.

2nd Step of Methodology: Next, is to define the data's path, and will be downloading the data for the url that we had created before. Name the folder same as it was in compressed files("Minerals").

Code is shown below:

```
import pathlib
dataset_url = "https://dl.dropboxusercontent.com/s/w0jyd2f23mk6wu0/minerals2.tgz?dl=0"
data_dir = tf.keras.utils.get_file('minerals', origin=dataset_url, untar=True)
data_dir = pathlib.Path(data_dir)

Downloading data from https://dl.dropboxusercontent.com/s/w0jyd2f23mk6wu0/minerals2.tgz?dl=0
63012864/Unknown - 2s 0us/step
```

Figure 4: Importing the dataset

Next is checking if the data is imported correctly or not.

Code along with the output is shown below:

```
[ ] biotite = list(data_dir.glob('biotite/*'))
    print(biotite[0])
    PIL.Image.open(str(biotite[0]))

    /root/.keras/datasets/minerals/biotite/Biotite (175).jpeg
```

Figure 5: Dataset Output

The third step is to split the dataset. Setting up the training, validation and test sets has a huge impact on productivity. It is important to choose the validation and test sets from the same distribution and it must be taken randomly from all the data [24].

```
[ ] img_height,img_width=180,180
    batch_size=32
    train_ds = tf.keras.preprocessing.image_dataset_from_directory(
      data_dir,
      validation_split=0.2,
      subset="training",
      seed=123,
      image_size=(img_height, img_width),
      batch_size=batch_size)

    Found 371 files belonging to 2 classes.
    Using 297 files for training.

[ ] val_ds = tf.keras.preprocessing.image_dataset_from_directory(
      data_dir,
      validation_split=0.2,
      subset="validation",
      seed=123,
      image_size=(img_height, img_width),
      batch_size=batch_size)

    Found 371 files belonging to 2 classes.
    Using 74 files for validation.
```

Figure 6: Splitting the dataset into training and validation classes

Step 3: Create a visual representation of the data.

Let's take a look at the data and see what we're dealing with. Here, utilises "mat plot" library for plotting the total quantity of photos in both classes, and result can be seen below.

```
[ ] import matplotlib.pyplot as plt

    plt.figure(figsize=(10, 10))
    for images, labels in train_ds.take(1):
      for i in range(9):
        ax = plt.subplot(3, 3, i + 1)
        plt.imshow(images[i].numpy().astype("uint8"))
        plt.title(class_names[labels[i]])
        plt.axis("off")
```

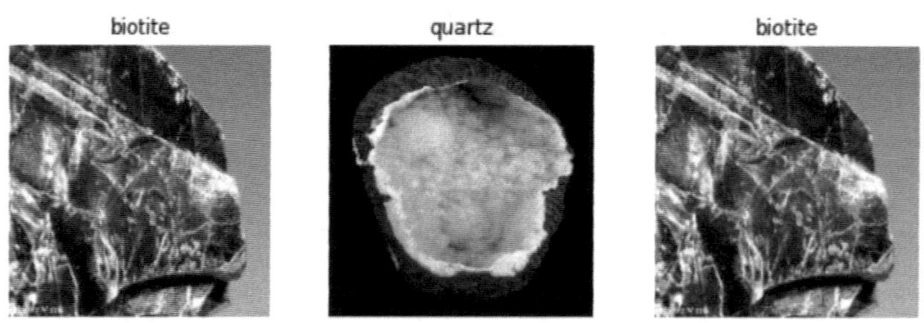

Figure 7: Checking the Split Dataset.

The next step is to Define the model:

Starting with the simple CNN model having three Convolutional layers and then three max-pooling layers [25]. Succeeding the third maxpool operation a dropout layer was inserted so that overfitting is avoided.

```
[ ] num_classes = 2

    model = Sequential([
      layers.experimental.preprocessing.Rescaling(1./255, input_shape=(img_height, img_width, 3)),
      layers.Conv2D(16, 3, padding='same', activation='relu'),
      layers.MaxPooling2D(),
      layers.Conv2D(32, 3, padding='same', activation='relu'),
      layers.MaxPooling2D(),
      layers.Conv2D(64, 3, padding='same', activation='relu'),
      layers.MaxPooling2D(),
      layers.Flatten(),
      layers.Dense(128, activation='relu'),
      layers.Dense(num_classes,activation='softmax')
    ])
```

Figure 8: Importing Functions

The input is half padded when padding == "SAME." Because the size of the output is same as of input (with stride=1), the padding type is termed SAME. Using the keyword 'SAME' ensures that the filter is applied to all of the input's items. Padding is usually set to "SAME" when training the model [26]. For subsequent computation, the output size is mathematically convenient [27].

Activation function is originally a basic function for turning the inputs into the outputs within a certain range of values, and is set to "relu." In contrast to other forms of activation functions, the sigmoid activation function accepts input and converts the output values between 0 and 1 in a unique fashion [28].

For compilation of the model optimization type is chosen as Adam optimizer, and for the loss function, the SparseCategoricalCrossEntropy is chosen. For obtaining the smoother curve learning rate was set to "$1x10^{-6}$" [29].

Here is the code:

```
model.compile(optimizer='adam',
              loss=tf.keras.losses.SparseCategoricalCrossentropy(from_logits=True),
              metrics=['accuracy'])#compiling the model
```

Figure 9: Compiling Model

This graph is showing the learning rate with the loss of pictures approximately=0.4% (which is a pretty less loss value and indicates an accurate result). Learning rate and loss graph was obtained to check if the training of the model was not up to the mark or not, it tells the rate at which model is being trained and also the loss of features due to the learning rate, a good learning rate is much needed so that loss of the features is less and also the training of the model is efficient.

```
LR Finder is complete, type {learner_name}.recorder.plot() to see the graph.
Min numerical gradient: 3.31E-06
Min loss divided by 10: 6.31E-08
```

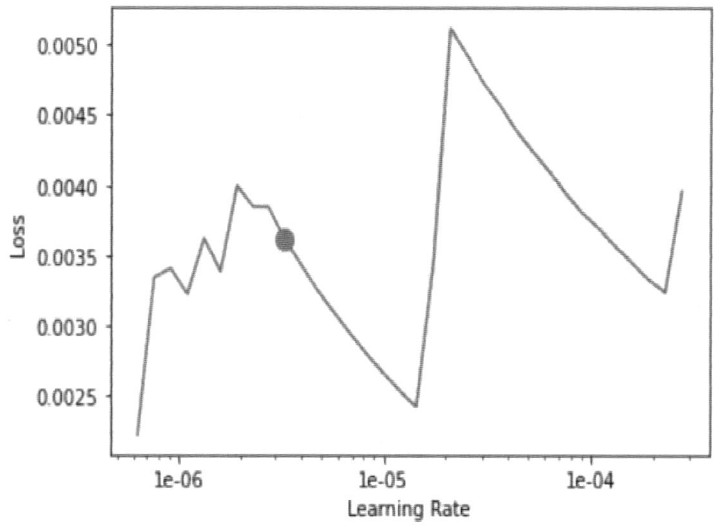

Figure 10: Learning Rate and Loss

The next step is to train our model, Now, because our learning rate is so low, let's train our model for 10 epochs.

```
[ ]  epochs=10
     history = model.fit(
       train_ds,
       validation_data=val_ds,
       epochs=epochs
     )
```

Figure 11: Setting Epochs for model fitting

On the next, the output is shown along with the time, loss, accuracy, valid loss and valid accuracy.

```
Epoch 1/10
/usr/local/lib/python3.7/dist-packages/tensorflow/python/util/dispatch.py:1082: UserWarning: "`sparse_categorical_crossentropy` received `from_logits=True`
  return dispatch_target(*args, **kwargs)
10/10 [==============================] - 11s 913ms/step - loss: 0.9854 - accuracy: 0.6263 - val_loss: 0.6184 - val_accuracy: 0.6892
Epoch 2/10
10/10 [==============================] - 10s 879ms/step - loss: 0.4211 - accuracy: 0.8148 - val_loss: 0.5981 - val_accuracy: 0.6757
Epoch 3/10
10/10 [==============================] - 10s 888ms/step - loss: 0.3654 - accuracy: 0.8283 - val_loss: 0.5964 - val_accuracy: 0.7703
Epoch 4/10
10/10 [==============================] - 10s 865ms/step - loss: 0.2595 - accuracy: 0.8855 - val_loss: 0.4804 - val_accuracy: 0.8378
Epoch 5/10
10/10 [==============================] - 10s 855ms/step - loss: 0.1875 - accuracy: 0.9428 - val_loss: 0.5079 - val_accuracy: 0.8108
Epoch 6/10
10/10 [==============================] - 10s 862ms/step - loss: 0.1028 - accuracy: 0.9832 - val_loss: 0.6542 - val_accuracy: 0.7973
Epoch 7/10
10/10 [==============================] - 10s 863ms/step - loss: 0.0757 - accuracy: 0.9697 - val_loss: 0.5101 - val_accuracy: 0.8378
Epoch 8/10
10/10 [==============================] - 10s 866ms/step - loss: 0.0479 - accuracy: 0.9899 - val_loss: 0.5676 - val_accuracy: 0.8784
Epoch 9/10
10/10 [==============================] - 10s 897ms/step - loss: 0.0277 - accuracy: 0.9966 - val_loss: 0.6210 - val_accuracy: 0.8649
Epoch 10/10
10/10 [==============================] - 10s 898ms/step - loss: 0.0088 - accuracy: 1.0000 - val_loss: 0.7544 - val_accuracy: 0.8649
```

Figure 12: Output of epochs

Further Step: Assessing the outcome:

Training and Validation accuracy and as well as training and validation loss was plotted.

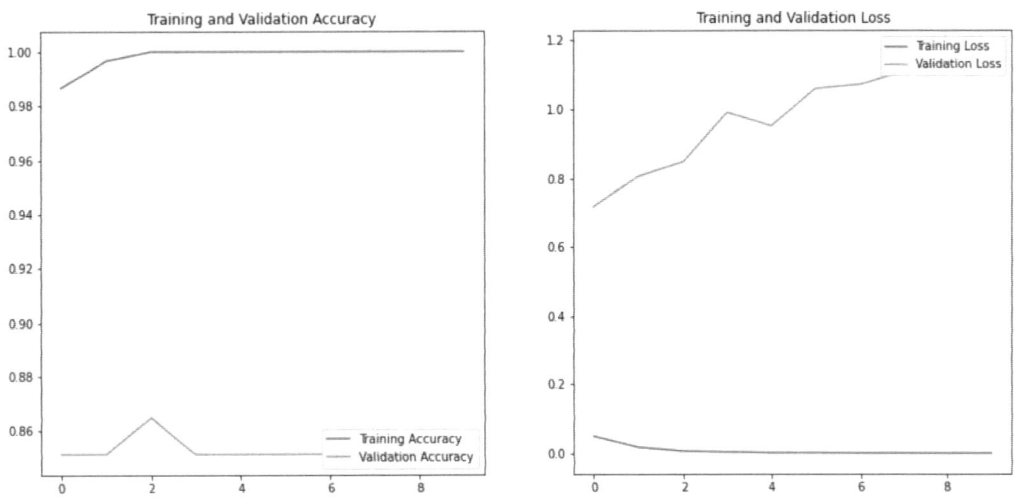

Figure 13: Training and Validation Illustration

Now for the model summary:

Keras makes it possible to summarise a model.

The textual summary contains information about:

• The layers of the model, as well as their arrangement.

• The ultimate output form of each layer.

• About parameters of each tier (weights)

Also, all the parameters in the model (weights). The summary is constructed by invoking the model's summary () method, which returns a string that may then be printed [30].

```
model.summary()

Model: "sequential"
_________________________________________________________________
 Layer (type)                Output Shape              Param #
=================================================================
 rescaling (Rescaling)       (None, 180, 180, 3)       0

 conv2d (Conv2D)             (None, 180, 180, 16)      448

 max_pooling2d (MaxPooling2D  (None, 90, 90, 16)       0
 )

 conv2d_1 (Conv2D)           (None, 90, 90, 32)        4640

 max_pooling2d_1 (MaxPooling  (None, 45, 45, 32)       0
 2D)

 conv2d_2 (Conv2D)           (None, 45, 45, 64)        18496

 max_pooling2d_2 (MaxPooling  (None, 22, 22, 64)       0
 2D)

 flatten (Flatten)           (None, 30976)             0

 dense (Dense)               (None, 128)               3965056

 dense_1 (Dense)             (None, 2)                 258

=================================================================
Total params: 3,988,898
Trainable params: 3,988,898
Non-trainable params: 0
```

Figure 14: Model Summary Output

Next step is to:

A function is defined which can accept images, and reshape those images to (180, 180):

```
def predict_image(img):
  img_4d=img.reshape(-1,180,180,3)
  prediction=model.predict(img_4d)[0]
  return {class_names[i]: float(prediction[i]) for i in range(2)} #function to reshape image so that it ca
n be fed into the input
```

Next step is to deploy the model on web:

For deployment of my deep learning model, gradio is used because, Gradio is a free, open-source Python toolkit that lets you quickly develop UI components for your machine learning model that are easy to use and customise [31].

IOP Conf. Series: Earth and Environmental Science 1032 (2022) 012046 doi:10.1088/1755-1315/1032/1/012046

Gradio makes it easier to use by integrating the GUI straight into your Python notebook.

Install gradio:

```
!pip install gradio #for deployment of model on web as demo.

import gradio as gr
```

Figure 15: Importing gradio

A simple pre-made web template for taking input from local host and to show the images along with prediction made by model.

Gradio can be used as a webpage or embedded in Python notebooks. A Gradio interface can build a public connection automatically. Furthermore, model result is displayed by taking an input picture, into the UI, it can be easily fed to the model by drag and drop function or by directly clicking the upload button.

4. Model Result

Figure 16: Simple UI for accepting images and giving output.

Images can be uploaded by drag and drop or can be uploaded directly:

The uploaded image can be seen the frame and can be edited in that frame also.

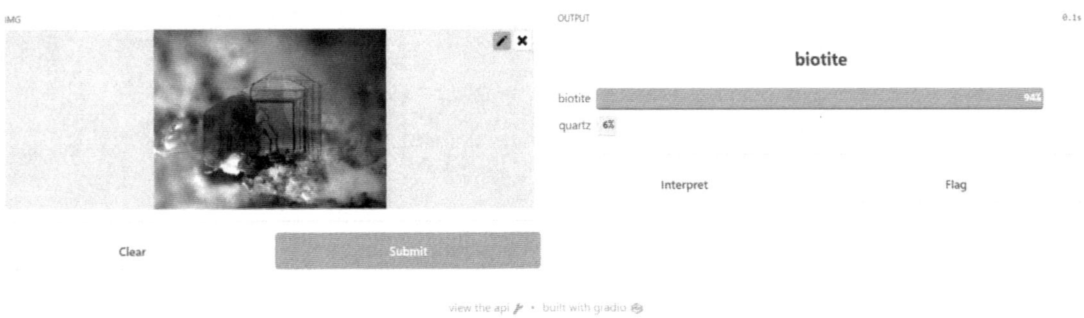

Figure 17: Prediction output for biotite image.

5. Conclusion

A deep learning model which can classify minerals using CNN can be deployed with reasonable accuracy. In this paper, employed Convolutional Neural Networks (CNN) for image categorization in this paper, and we used photos that got using the bingimagedownloader api for “Biotite” and “Quartz”. This data collection was used for both training and testing using CNN. It has a 98 percent accuracy rating. The processing time for these images is significantly long when compared to regular JPEG shots. Image classification results will be more accurate if clusters of GPUs are used to stack the model with extra layers and train the network with more picture data. The next update will concentrate on identifying large-scale coloured images, which will be highly valuable in the image segmentation process. Although it has a great accuracy but it can be better if a large number of noise free images being used for better precision and accuracy can be used classifying minerals.

References:

[1] Magnus, Albertus. *The Book of Minerals*. Dalcassian Publishing Company, 1967.

[2] Brown, J. Coggin. *India's Mineral wealth: A guide to the occurrences and economics of the useful minerals of the Indian Empire*. Oxford University Press, London, 1936.

[3] Kalra, G. D. "The Indian minerals industry—an overview." *Minerals and Energy* 7.2 (1990): 35-37.

[4] MINES, INDIAN BUREAU OF. "Indian Minerals Yearbook 2016." *Gov. India Minist. Mines Nagpur* 13 (2018): 1-17.

[5] Randive, Kirtikumar, and Sanjeevani Jawadand. "Strategic minerals in India: present status and future challenges." *Mineral Economics* 32.3 (2019): 337-352.

[6] Bowie, S. HU, and K. Taylor. *A system of ore mineral identification*. No. A/CONF. 15/P/42. Geological Survey, London, 1959.

[7] LeCun, Yann, Yoshua Bengio, and Geoffrey Hinton. "Deep learning." *nature* 521.7553 (2015): 436-444.

[8] Goodfellow, Ian, Yoshua Bengio, and Aaron Courville. *Deep learning*. MIT press, 2016.

[9] Bengio, Yoshua, Ian Goodfellow, and Aaron Courville. *Deep learning*. Vol. 1. Cambridge, MA, USA: MIT press, 2017.

[10] Shin, Hoo-Chang, et al. "Deep convolutional neural networks for computer-aided detection: CNN architectures, dataset characteristics and transfer learning." *IEEE transactions on medical imaging* 35.5 (2016): 1285-1298.

[11] Roska, Tamas, and Leon O. Chua. "The CNN universal machine: an analogic array computer." *IEEE Transactions on Circuits and Systems II: Analog and Digital Signal Processing* 40.3 (1993): 163-173.

[12] Wang, Jiang, et al. "Cnn-rnn: A unified framework for multi-label image classification." *Proceedings of the IEEE conference on computer vision and pattern recognition*. 2016.

[13] Sang, Xiancheng, et al. "One-Dimensional Deep Convolutional Neural Network for Mineral Classification from Raman Spectroscopy." *Neural Processing Letters* 54.1 (2022): 677-690.

[14] Lou, Wei, Dexian Zhang, and Richard C. Bayless. "Review of mineral recognition and its future." *Applied Geochemistry* 122 (2020): 104727.

[15] Weller, Susan C., and A. Kimball Romney. *Systematic data collection*. Vol. 10. Sage publications, 1988.

[16] Couper, Mick P. "The future of modes of data collection." *Public Opinion Quarterly* 75.5 (2011): 889-908.

[17] Carneiro, Tiago, et al. "Performance analysis of google colaboratory as a tool for accelerating deep learning applications." *IEEE Access* 6 (2018): 61677-61685.

[18] Kluyver, Thomas, et al. *Jupyter Notebooks-a publishing format for reproducible computational workflows*. Vol. 2016. 2016.

[19] Raschka, Sebastian. *Python machine learning*. Packt publishing ltd, 2015.

[20] Gulli, Antonio, and Sujit Pal. *Deep learning with Keras*. Packt Publishing Ltd, 2017.

[21] Dillon, J. V., Langmore, I., Tran, D., Brevdo, E., Vasudevan, S., Moore, D., ... & Saurous, R. A. (2017). Tensorflow distributions. *arXiv preprint arXiv:1711.10604*.
[22] Pang, Bo, Erik Nijkamp, and Ying Nian Wu. "Deep learning with tensorflow: A review." *Journal of Educational and Behavioral Statistics* 45.2 (2020): 227-248.

[23] Ketkar, Nikhil. "Introduction to keras." *Deep learning with Python*. Apress, Berkeley, CA, 2017. 97-111.

[24] Wei, Qiong, and Roland L. Dunbrack Jr. "The role of balanced training and testing data sets for binary classifiers in bioinformatics." *PloS one* 8.7 (2013): e67863.

[25] Jung, Wonkyung, et al. "Restructuring batch normalization to accelerate CNN training." *Proceedings of Machine Learning and Systems* 1 (2019): 14-26.

[26] Liu, Guilin, et al. "Partial convolution based padding." *arXiv preprint arXiv:1811.11718* (2018).

[27] Dumoulin, Vincent, and Francesco Visin. "A guide to convolution arithmetic for deep learning." *arXiv preprint arXiv:1603.07285* (2016).

[28] Sharma, Sagar, Simone Sharma, and Anidhya Athaiya. "Activation functions in neural networks." *towards data science* 6.12 (2017): 310-316.

[29] Jais, Imran Khan Mohd, Amelia Ritahani Ismail, and Syed Qamrun Nisa. "Adam optimization algorithm for wide and deep neural network." *Knowledge Engineering and Data Science* 2.1 (2019): 41-46.

[30] Shlezinger, Nir, et al. "Model-based deep learning." *arXiv preprint arXiv:2012.08405* (2020).

Delineation of Groundwater Potential Zones in Koyna River Watershed, Maharashtra using Remote Sensing and GIS.

R C Swain[1, a], D C Jhariya[2, b*], A Swain[3, c], Y K Mawale[4, d] and S Rana[5, e]

[1]M. Tech Student, Department of Applied Geology, National Institute of Technology Raipur, G.E. Road, Raipur - 492 010, Chhattisgarh, India

[2]Assistant Professor, Department of Applied Geology, National Institute of Technology Raipur, G.E. Road, Raipur - 492 010, Chhattisgarh, India

[3]M. Tech Student, Department of Applied Geology, National Institute of Technology Raipur, G.E. Road, Raipur - 492 010, Chhattisgarh, India

[4]Associate Professor, Department of Applied Geology, Sant Gadge Baba Amravati University, Amravati- 444602, India

[5]M. Sc. Student, Department of Applied Geology, National Institute of Technology Rourkela, Rourkela- 769008, India

Email: [a] romuswain@gmail.com; [b] dcjhariya.geo@nitrr.ac.in; [c]anshumanswain2598@gmail.com; [d]myadaorao@yahoo.co.in; [e]soumyarana33@gmail.com

Abstract.: Groundwater is considered as only source which provides water for meeting domestic, industrial and agricultural requirement. Continuous uses of water without any conservation cause lowering of Ground water level and thus living forms are facing many problems now-a-days. In current situation some existing wells getting dried up because of depletion of groundwater table as the natural groundwater recharge is not sufficient for requirements. As a result, the access to resources and their demand allows individuals to pick favorable locations for population increase. As the world's population grows, so does urbanization and groundwater exploration. As a rising metropolitan region, it is critical to analyze the present groundwater scenario of the Koyna River watershed. This study is simple to do with the use of remote sensing and geographical information systems (GIS) by cost-effectively incorporating factors impacting groundwater potential. We will combine numerous data sets and maps in this study, including satellite imaging, drainage, groundwater level, precipitation, LULC, slope, and its existing geological map. After including the groundwater potential, it was divided into five distinct zones: extremely poor, poor, moderate, good, and very excellent [1].

Keywords- Groundwater potential, GIS, RS, LULC, Koyna River, Thematic Layers.

1. Introduction

The Western Ghats are a hilly terrain that runs along the western border of the Indian Peninsula (hills) [4]. The Koyna River basin is located east of the primary ridge, which is locally known as the Sahyadri Hill Range (portion of the Krishna River basin). Mahabaleshwar,

Satara district, Western Maharashtra, India is the source of the Krishna River [7]. It emerges near Mahabaleshwar, a well-known Western Ghats hill station. Four tributaries feed the Koyna River [2]. All tributaries flow from the Western Ghats [mpcb.gov.in]. Unlike other Maharashtra rivers, the Koyna River flows north-south. This river basin is 2,036 km^2 in size and is located in the Deccan region of Maharashtra State, India. Geologist became concerned in this river basin after the 1967 Koyna earthquake. Despite these efforts, there have been few groundwater storage zone analyses in the river basin. The goal of this study is to define the river basin's hydro geologic framework, including aquifers, underground water occurrence, well efficiency, and aquatic habitat [2].

The physiography of the Koyna River basin is typical of the Deccan Plateau, with elevations ranging from 550 to 1,460 metres above sea level features a sub-dendritic drainage system with a lot of baseflow and steep slopes, therefore groundwater recharge is negligible. The river basin is well-drained, with a lot of surface runoff and a high gradient, as well as low absorption rate. [2].

The Koyna Dam, built on the Koyna River, is a significant supply of water for Maharashtra. It's a crumbled concrete dam with a height of 103.02 metres above the deep footings and 85.35 metres above the riverbed. It is 807.22 metres long in total.

A subtropical monsoon climate characterizes the Koyna River watershed. Rainfall averages more than 5000 mm in the River Course's watershed up to Helwak. The catchment area is 891.78 square kilometers from Koyna Dam to the mouth of the river. 120 TMC is the mean annual production with 75% reliability [2].

A number of techniques have been used for mapping groundwater potential zones. But recently, remote sensing along with Geographic Information System plays a vital role in order to provide information about potentiality, occurrence, mapping, problems regarding exploitation of groundwater. Exploration of groundwater with the help of remote sensing data leads to discover new information about inaccessible or remote areas. Remote sensing and GIS technique are not only accurate, cost-effective, time-saving but also, they provide reliability as well as detail interpretation of data. GIS refers to a set of technologies for gathering, storing, retrieving, processing, and presenting geographical data from the actual world [1].

The main application of using this advance technology to know; (i) amplifying the efficacy of planning and decision making that help in data distribution and controlling unwanted data base (ii) reducing replication ability to amalgamate information derived from many resources (iii) complex interpretation of geographical referenced data in order to create new information.

2. Study area

The Konya River Basin is located between 17.54 and 17.16 N and 73.42 and 74.06 E. The Konya River runs north-south for 65 kilometres, approximately parallel to the continental divide (Fig. 1). It runs from Mahabaleshwar to the hamlet of Helwak, passing via King Shivaji's fort of Pratapgadh. It swings steeply eastwards near Helwak, runs for approximately 56 kilometres, and meets the Krishna River at Karad. It's an odd confluence where two rivers meet head on. Preeti Sangam is the common name for this confluence.

The river runs through Patan and Karad talukas. The River Course's catchment area extends up to Helwak and has an average rainfall of more than 5000m. The river's catchment area up to Konya Dam is 891.78 square kilometres.

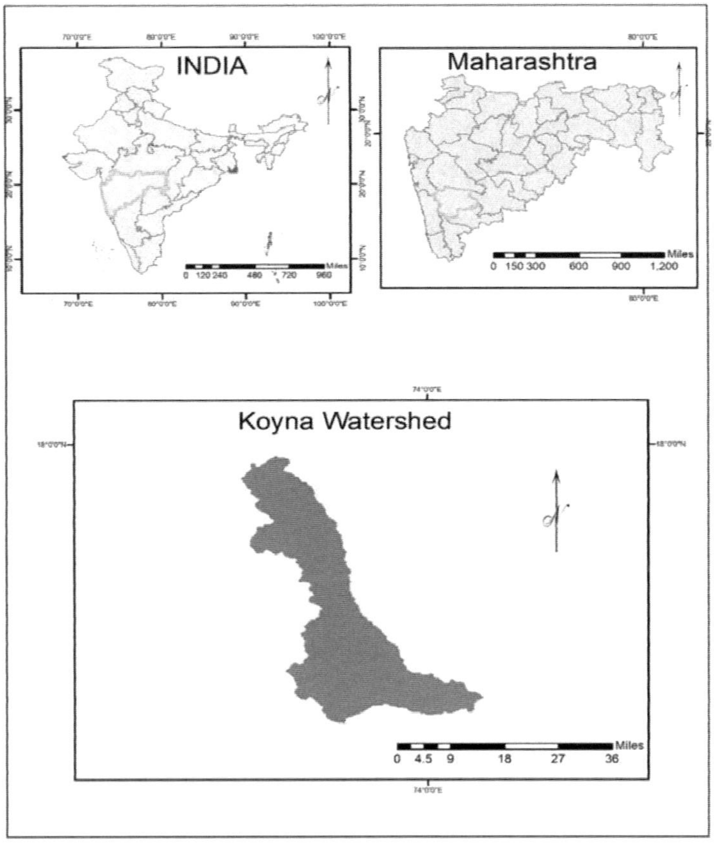

Figure. 1 Koyna Watershed

Table 1: date used in preparation of different maps

SL. No.	Layer Name	Source	Sites
1	Geology	Bhukosh	www.bhukosh.gsi.gov.in
2	Geomorphology- Landforms	Bhukosh	www.bhukosh.gsi.gov.in
3	Geomorphology- Units	Bhukosh	www.bhukosh.gsi.gov.in
4	Lithology	Bhukosh	www.bhukosh.gsi.gov.in
5	Drainage pattern	DEM	www.earthexplorer.usgs.gov
6	Lineament Density	Bhuvan	www.bhuvan.nrsc.gov.in
7	Precipitation	India-WRIS	www.indiawris.gov.in
8	Groundwater level	India-WRIS	www.indiawris.gov.in
9	Land Use and Land Cover	Sentinel-2	www.earthexplorer.usgs.gov
10	Slope	DEM	www.earthexplorer.usgs.gov

3. Methodology

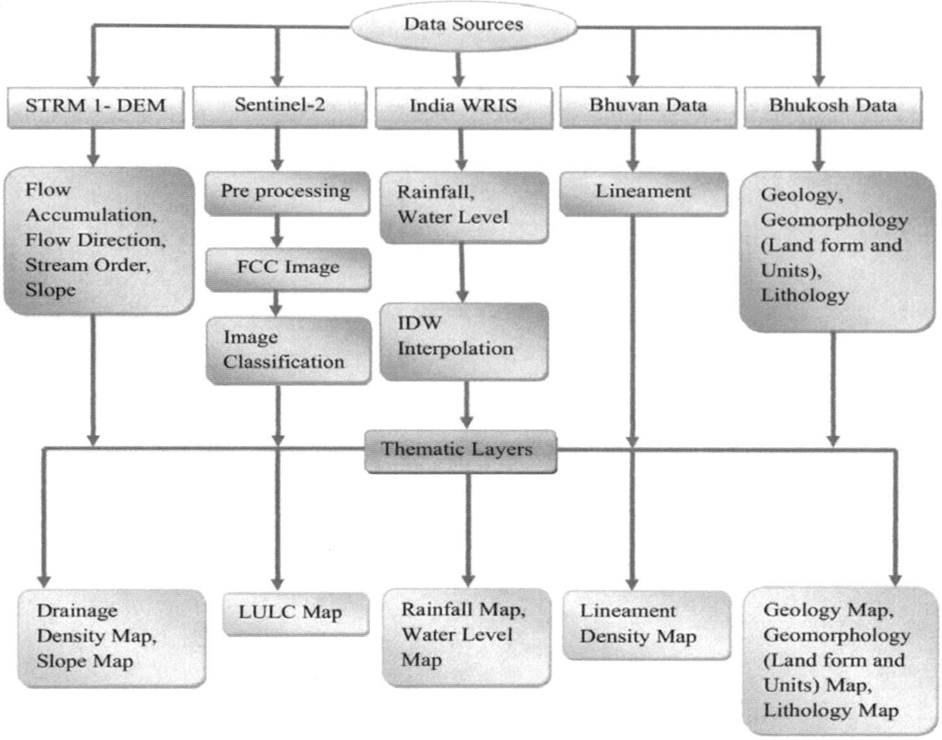

Figure. 2 Adopted Methodology Flow Chart

Systematic methodology adopted in the present study (Fig. 2). The identification and extensive research of ground water potential zones in the Koyna basin was accomplished through the use of GIS technology to analyse a variety of remote sensing data and numerous thematic maps. Some layers were created and used to identify recharging locations in our present investigation. SRTM data was used to create the drainage layer [1]. A proportion of the slope map was created in addition. Bhuvan.gov.in provided the geomorphology theme map. The Geological Survey of India provided this geology map. For ease of use in GIS settings, all of these data were geo-rectified and projected in WGS 1984 with UTM Zone 45 N [3]. Geology, geomorphology, LULC, gradient, drainage/water bodies, post-monsoon water level, and worn thickness & cracks occurring near the surface all determine the best site for groundwater recharge. By combining geographical data from the research region and utilizing an index overlay model in a GIS context, the best zone for groundwater recharge was discovered. In this paradigm, weightage for the theme was assigned on a scale of 1–10, while subclasses of the theme were assigned on a scale of 1–5. The map was then divided into five categories: very high, high, moderate, extremely poor, and poor.

4. Climate

The climate of the Koyna River basin is subtropical monsoon. The Indian rainy season, the most significant of India's three seasons, runs from June to September and accounts for around 88 percent of yearly rainfall. The winter months are November to February, while the summer months are March to May the Western Ghats' orographic influence has resulted in a wide variation of rainfall distribution in the region. Annual rainfall decreases steadily from the western to the eastern parts of the state. In the north, Mahabaleswar has the highest annual precipitation (6,024 mm), while in the southeast, Karad has the lowest (745 mm). The contribution of rainfall is just 8% in the winter and 4% in the summer [2].

5. Thematic Layers

Thematic layers that play significant role for GW occurrence and management of aquifer system in research area are as follows;

5.1 Lithology

Basaltic lava flows make up the lithology of the Koyna River basin. It was mostly discovered between the Cretaceous and the Lower Eocene epochs. Groundwater in this region is mainly unconfined in shallow aquifers, but semi-confined to confined in deeper aquifers. It is mostly related with Basalt. The Basaltic flow is divided into two trap units: a bottom massive unit and an upper vesicular unit. Containing 60 to 85 percent of the Basaltic flows, the enormous unit is the major trap unit. It's primarily fine-grained, thick, and compact, and the color ranges from greenish grey to dark grey [4]. It also behaves as an impermeable bed with negligible porosity and permeability (Fig. 3).

The vesicular trap unit is found at the top of each flow and accounts for 15–40% of the total basaltic flow volume. These are soft, fine-grained, and range in hue from greenish to brownish. The lava flows are porphyritic to varying degrees and are tholeiitic in nature (Geological Survey of India 1974). Laterites produce flat plateaus and tablelands at altitudes ranging from 975 to 1,400 meters above sea level as cap rocks over basalts. In the eastern and northern regions, the lateritic profiles are 12–30 m thick, whereas in the southern parts, they are 2–5 m thick. The color of laterites is often reddish-brown [2].

5.2 Geomorphology

In study of geomorphology of a particular area we mainly cover its landforms, topography and different geological features. But this is not always accessible to study these features of a particular area. So, the recent advancement in remote sensing and GIS helps to understand these features in a very effective way and by using this advanced technology, it is very easy to detect ground water potential zone of concerned area (Fig. 4). Various available features of an area can indicate good site for ground water availability [9]. We know that there are many geomorphic features are controlling factor for groundwater storage and its movements. We are focusing the detailed study of groundwater potential zone of (GWPZ) of koyna river.

We discovered inselberg, mesa, pediment, pediplain, plateau top, scarp, and butte morphological characteristics. Approximately 31% of the territory is made up of fragmented plateaus with steep slopes, whereas the remainder 28percent is somewhat low and residence to the general populace. Almost 6percent of the basin's water-spread area is covered by the Koyna reservoir. The western half of Mahabaleshwar is primarily wooded. The shattered plateau and hillsides, but from the other side, have a scant layer of soil and a limited vegetation, largely comprising of shrub, steppe, and a few distributed trees. The main valley is mostly cultivated and has deep soils and regolith. Lands that have been eroded are also common, especially where the fragmented plateau meets the main valley [2].

5.3 Slope

A Slope is the rise or fall of the land surface. It is proportional to the amount of water flowing through it. The Western half of the research region has the largest slope angles, according to the Koyna River basin's slope map [1]. This location (Fig. 5) has a slope angle of more than 15 °.

IOP Conf. Series: Earth and Environmental Science 1032 (2022) 012047 doi:10.1088/1755-1315/1032/1/012047

According to the investigation, high slopes may be found in upstream regions, while very low slope ratios can be found in downstream areas. In general, slopes range from the direction is northwest to southeast. As a result, all of the streams are determined to be flowing in these orientations.

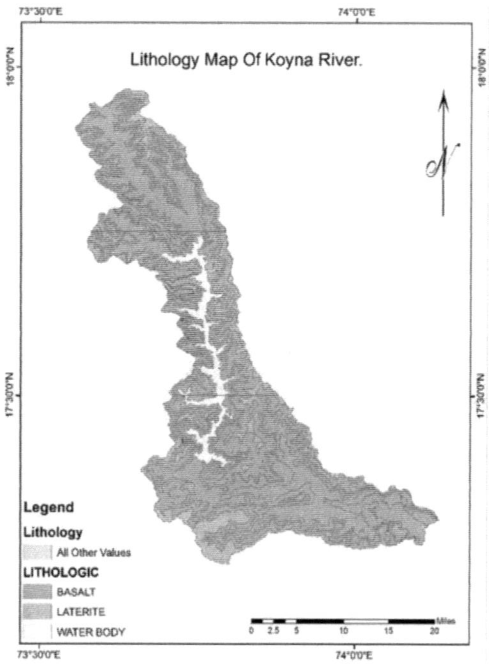

Figure. 3 Lithology Map

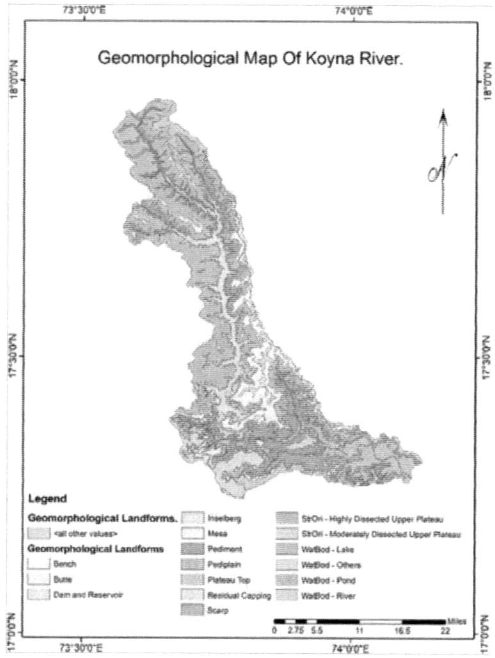

Figure. 4 Geomorphology Map

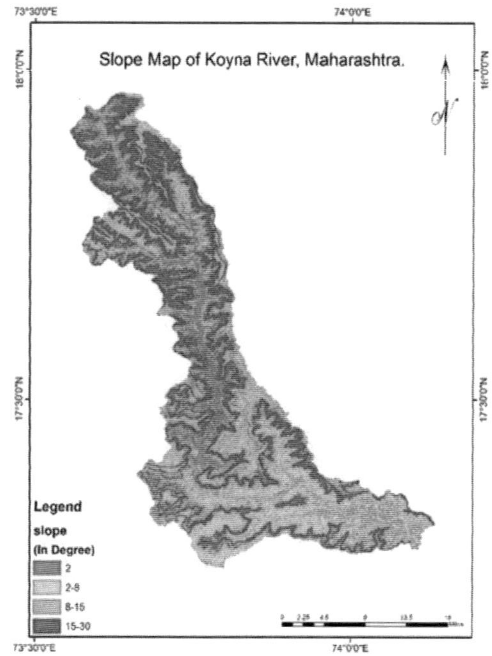

Figure. 5 Slope Map

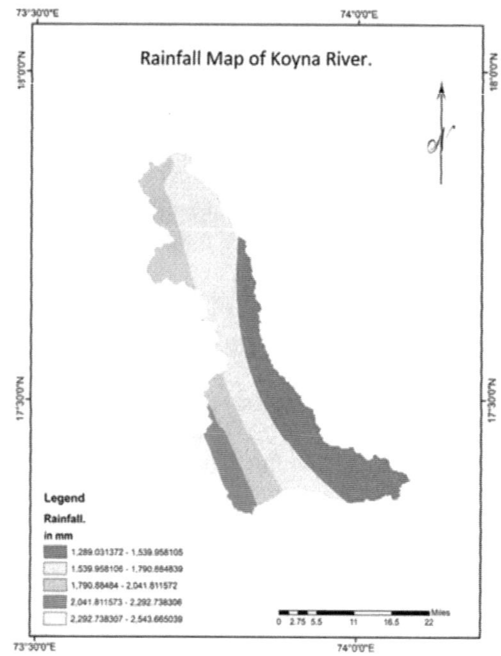

Figure. 6 Rainfall Ma

5.4 Rainfall

Our groundwater is recharged only by rainfall. The orographic effects of the Western Ghats causes a large variance in rainfall distribution in the Koyna River basin. Annual rainfall decreases steadily from the western to the eastern part of the state. Northern Mahabaleswar (6024 mm) receives the greatest annual average rainfall, while the southeast's Karad (745 mm) receives the least. In the winter, rainfall contributes just 8%, whereas in the summer it contributes 4% [2].

The rainfall map has been categorized into five classes in milli meters (Fig. 6), such as

Class 1- 989.56 to 1262.73, Class 2- 1262.73 to 1535.90, Class 3- 1535.90 to 1809.07, Class 4- 1809.07 to 2082.24, Class 5- 2082.24 to 2395.40.

5.5 Water Level

As per the map, groundwater may be found at both surface and subsurface depths in the Koyna River basin. Aquifers are divided in to two types, shallow aquifers and deeper aquifers. Dug wells seldom reach depths of more than 15 meters, and the majority are between two and eight meters. The diameter ranged from 2-10 meters, with 50% and diameter of 2-4 meters. Bore wells typically have a diameter of 150mm and are 30-80 meter deep, seldom going deeper than 90-100 meter; the majority are 50-70m deep (Fig. 7). Dug wells are used to access the lower aquifers whereas borewells are used to access the deeper water resources [2]. The major part of the area is located near Koyna River, so any further groundwater development to meet drinking water requirements can be suitable on a village-to-village basis.

5.6 Lineament

Usually, lineament-controlled terrains that are underlain by the zone of increased porosity and permeability. The lineament density of lineament map is closely related with groundwater occurrences and its distribution (Fig. 8). The surface interpretation of the underpinning geologic structure is lineament, which is a linear feature existing in the earth's surface. The presence of lineament implies favorable groundwater prospects since they operate as possible pathways for groundwater circulation [1].

5.7 Land Use Land Cover (LULC)

LULC is a significant indicator of human impact on groundwater resources. Groundwater storage and outflow, as well as infiltration, are influenced by land use patterns [5]. Forest land, nonagricultural land, and cultivable land are the three basic categories of land use. The Koyna River basin covers 280671 acres of territory from the two talukas of Patan and Karad in the Satara district. 38322 hectares are covered in forest, 21804 hectares are utilised for non -agricultural activities, and 220545 hectares are cultivable land (Fig. 9).

According to our observations, trees cover the majority of the land, whereas bareground constitutes a relatively small percentage of the total surface of the region.

5.8 Drainage Density

The DD of the Koyna River and its tributaries reveals that the drainage basin's general trend is from northwest to southeast. This corresponds to the altitude of the land surface, which maintains the same trend as inferred from the DEM data [12]. Because the waterways in the research region run from west to east, almost all of them meet on the right bank of the Krishna River (Fig. 10). Streams and rivers flow into the major river Krishna from the western mountainous region. The drainage density in the research region was divided into five categories. The maximum density was discovered at the edge of the catchment area, with a value of 40.83-190.82, and the lowest density was found in the heart of our catchment region, with a value of 640.78-790.77.

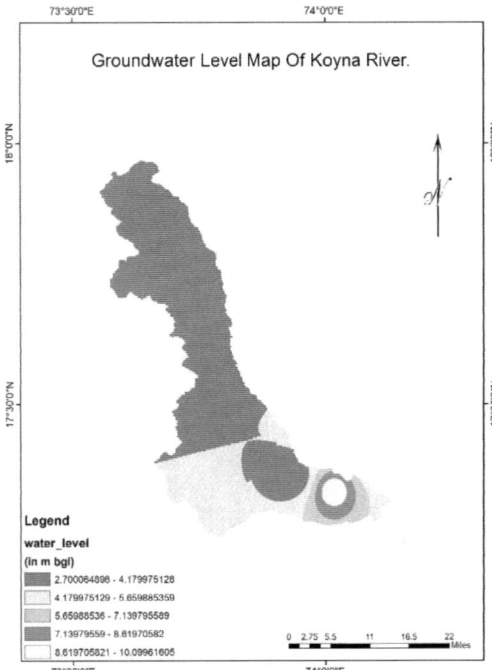

Figure. 7 Groundwater Level Map

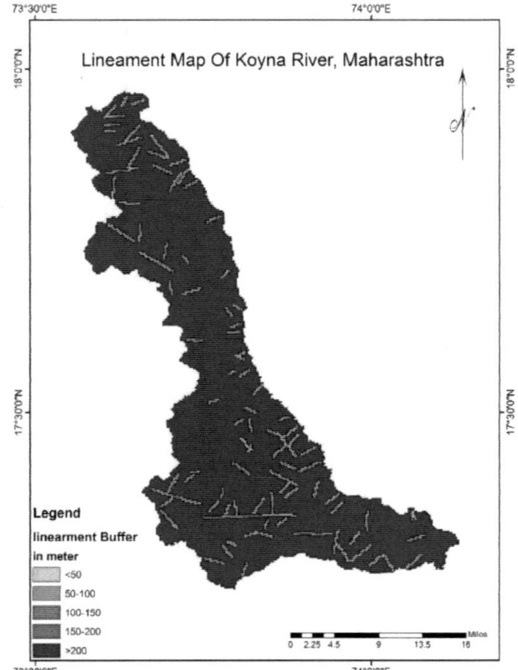

Figure. 8 Lineament Map

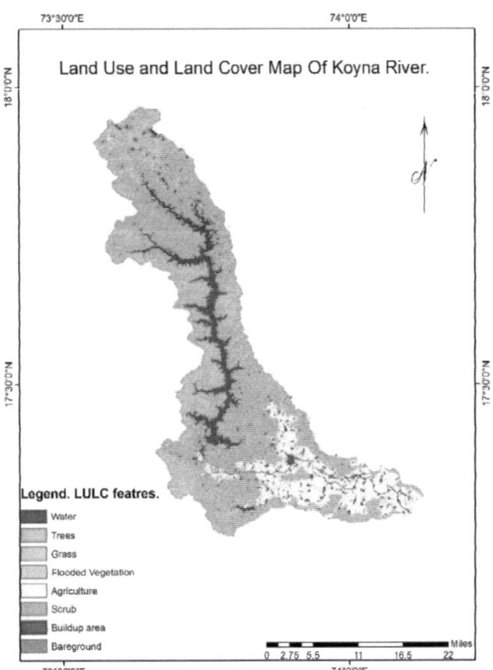

Figure. 9 LULC Map

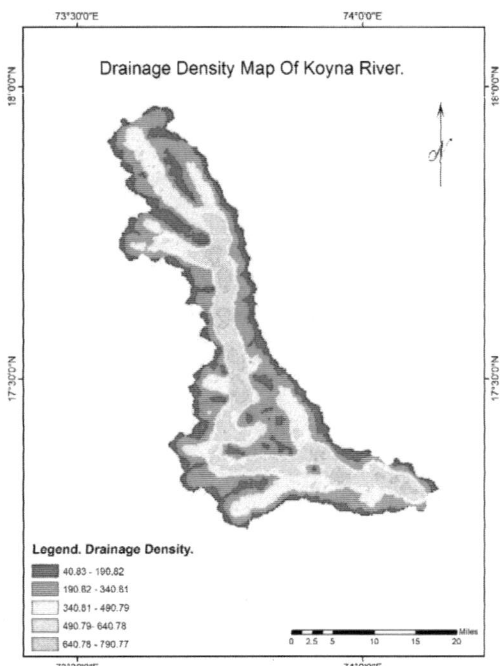

Figure. 10 Drainage Density Ma

Table 2. Weightage Table

Influencing Factors	Category (Classes)	Groundwater Potentiality	Rating	Weightage
Geology	Basalt	Poor	2	5
	Laterite	Good	4	
	Pediment	Good	4	
	Pedi Plain	Poor	2	
	Mesa	Very poor	1	
	Inselberg	Good	4	
	Bench	Poor	2	
	Butte	Very Poor	1	
	Dam and Reservoir	Poor	2	
	Lake	very Good	5	
Geomorphology	Pond	Moderate	3	8
	River	very Good	5	
	<50m	Very Poor	1	5
	50-100m	Poor	2	
	100-150m	Moderate	3	
Lineament Density	150-200m	Good	4	
	>200m	Very Good	5	
	>2*	Very Good	5	
	2-8*	Good	4	
Slope	8-15*	Moderate	3	
	15-30*	Poor	2	5
	2.700-4.179bgl	Very Good	5	
	4.179-5.659bgl	Good	4	
	5.659-7.139bgl	Moderate	3	

Groundwater Depth	7.139-8.619bgl	Poor	2	7
	8.619-10.099bgl	Very Poor	1	
Rainfall(mm)	1289.03-1539.95	Moderate	3	8
	1539.95-1790.88	Good	4	
	1790.88-2041.81	Moderate	3	

LULC	Scrub lands	Good	4	9
	Bare ground	Very Poor	1	
	Flooded Vegetation	Very Poor	1	
	Trees	Very Good	5	
	Grass	Poor	2	
	Build-ups	Poor	2	
Drainage Density (Km/ Km^2)	40.83-190.82	Very Good	5	6
	190.82-340.81	Good	4	
	340.81-490.79	Moderate	3	
	490.79-640.78	Poor	2	
	640.78-790.77	Very Poor	1	

6. Groundwater Potential Zone Map:

The relative importance of multiple themed layers and their relevant classes were used to develop the groundwater potential zone. The weighted sum technique was used to produce the groundwater prospecting zone study area. The potential zones were created using a weighted total with five distinct zones. Very poor, poor, moderate, good, and very good are the zones. According to the findings, the researched region has a significant groundwater potential zone (Fig. 11). According to a research of the groundwater potential map, the distribution of groundwater potential is influenced by rainfall, drainage density, slope, LULC, and other geomorphologic and geological factors [6].

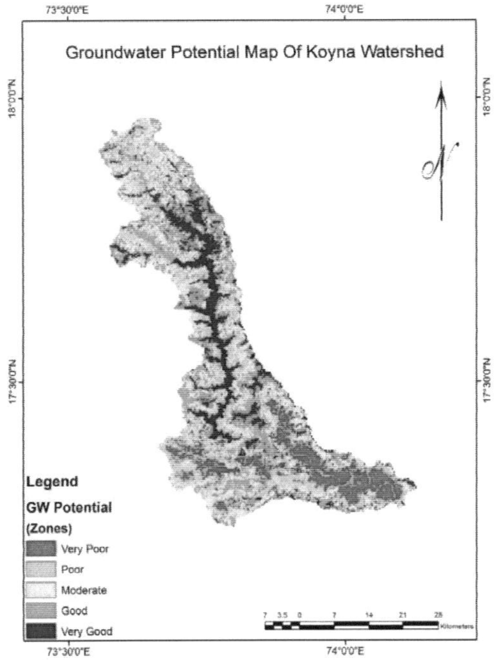

Figure.11 Groundwater Potential Map

7. Discussion

The Koyna watershed's final groundwater recharge zone shows extremely poor, poor, moderate, good, and very good. The presence of a very excellent recharging zone may be detected in a major section of the northern half of the research region, but it declines as one moves southern [11]. As a result, the majority of the northern and central parts of the Koyna River watershed have a good groundwater potential zone. Basalts cover the majority of our research region, with some Laterite thrown in for good measure. The intermediate groundwater potential zone is situated in the middle of the research region, whereas the poor zone is found in the southern section (Fig. 11).

8. Conclusion

The purpose of this study was to create a groundwater potential map for the Koyna River basin. The findings suggest that the most effective groundwater potential zone is located in the research area's northern half. Around here the dam, reservoir and crop land have high infiltration capacity and the area is gaining moderate to low rainfall and having lower lineament density. Because of its lithology and low lineament density, moderate rainfall, and inadequate drainage density, the northern half of the research region is most effective for groundwater potential. Geomorphology, rainfall, lithology, LU/LC, drainage density, lineament density, and slope (percent) all have a role in the presence and potential of groundwater in the research region, according to GIS analysis. To map the groundwater occurrence map, movement, and management, remote sensing and GIS techniques were utilized to merge distinct theme maps. From the above study, the utilization of RS and GIS techniques gives amazing tool to consider groundwater resources and represent a reasonable exploration plan for recharge of groundwater around there, and which can be utilized for improvement in the groundwater potential and holding for the study area. So, by doing this, results can be used to advancement of groundwater management for future.

References

[1] K Indhulekha, Jhariya Dalchand (2020). Delineation of groundwater potential zones in Samoda watershed, Chhattisgarh India, using Remote Sensing and GIS techniques. IOP Conference Series: Earth and Environmental Science. 597. 012007. 10.1088/1755-1315/597/1/012007.
[2] Naik, Pradeep & Awasthi, Arun & a v s s, Anand & Mohan, Prakash. (2001). Hydrogeologic framework of the Deccan terrain of the Koyna River basin, India. Hydrogeology Journal. 9. 243-264. 10.1007/s100400100123.
[3] Senthilkumar, M. & Gnanasundar, Devadasan & Arumugam, Rethinam. (2019). Identifying groundwater recharge zones using remote sensing & GIS techniques in Amaravathi aquifer system, Tamil Nadu, South India. Sustainable Environment Research. 29. 10.1186/s42834-019-0014-7.
[4] Naik, Pradeep & Awasthi, A. & a v s s, Anand & Behera, P.. (2009). Hydrogeochemistry of the Koyna River basin, India. Environmental earth sciences. 59. 613-629. 10.1007/s12665-009-0059-8.
[5] Roy, Annadasankar & Keesari, Tirumalesh & Sinha, U.K. & Sabarathinam, Chidambaram. (2019). Delineating groundwater prospect zones in a region with extreme climatic conditions using GIS and remote sensing techniques: A case study from central India. Journal of Earth System Science. 128. 10.1007/s12040-019-1205-7.
[6] Jhariya, Dalchand & Kumar, Tarun & Gobinath, M. & Diwan, Prabhat & Kishore, Nawal. (2016). Assessment of groundwater potential zone using remote sensing, GIS and multi criteria decision analysis techniques. Journal of the Geological Society of India. 88. 481-492. 10.1007/s12594-016-0511-9.
[7] Naik PK (1994) Hydrogeological investigations of the Deccan terrain of the Koyna sub- basin, India. PhD Thesis, University of Roorkee.
[8] Chow, V.T., D.R. Maidment, and L.W. Mays, Applied Hydrology, McGraw-Hill, Inc., New York, NY, 1988.

[9] Jhariya, Dalchand. (2019). Assessment of Groundwater Pollution Vulnerability Using GIS-Based DRASTIC Model and its Validation Using Nitrate Concentration in Tandula Watershed, Chhattisgarh. Journal of the Geological Society of India. 93. 567-573. 10.1007/s12594-019-1218-5.
[10] M., Rajasekhar & Kadam, Ajaykumar & Raju, Gadhiraju & Bhagat, Vijay. (2021). Identification of Artificial Groundwater Recharge Zones in Semi-arid region of Southern India using Geospatial and Integrated Decision-Making Approaches. Environmental Challenges. 5. 100278. 10.1016/j.envc.2021.100278.
[11] Naik, Pradeep & Dehury, Biranchi & Tiwari, Arun. (2007). Groundwater Pollution Around an Industrial Area in the Coastal Stretch of Maharashtra State, India. Environmental monitoring and assessment. 132. 207-33. 10.1007/s10661-006-9529-6.
[12] Bhunia, Gouri. (2020). An approach to demarcate groundwater recharge potential zone using geospatial technology. Applied Water Science. 10. 10.1007/s13201-020-01231-1.
[13] Darnell, W.L., W.F. Staylor, S.K. Gupta, N.A. Ritchey, and A.C. Wilber, "Seasonal Variation of Surface Radiation Budget Derived from ISCCP-C1 Data," J. Geophysical. Res., 97, 15741- 15760, 1992.

Research on shear strength of geopolymer concrete by using fly ash

Sachin A. Daur[1,a], Madhulika Sinha[2,b]

[1]Construction Management @P H C E T Engineering College, Rasayani, New Mumbai, India

[2]Professor @P H C E T Engineering College, Rasayani, New Mumbai, India.

Email: [a]sachindaur13@gmail.com; [b]madhulikas@mes.ac.in

Abstract: Cement is one of the most important construction materials. Cement is a backbone of civil engineering work. Development of whole world depends upon increasing infrastructure. For which cement is very necessary. Manufacturing of cement produce huge amount of carbon di-oxide and other greenhouse gases which is hazardous to the environment causes pollution. Hence it is very necessary to develop alternative binding material instead of cement for construction. Many researchers contribute their work on replacement of cement in concrete.

Keywords: Geopolymer, Fly ash, Alkaline solution, Compressive strength, Shear strength.

1. Introduction

Ash-based geopolymer concrete is a major change in the construction and concrete industry in the use of Portland cement and 100% cement material is an industrial product, i.e., fly ash. The adoption of fly-based concrete is an efficient use of the world's largest ash-filled flies as a result of coal-fired power plants.

In structural and mechanical engineering, the shear strength of a component is important in designing the size and materials to be used in the construction or construction of the component. In a reinforced concrete line, the main purpose of reinforcing bar stirrups is to increase the shear strength.

Cement is one of the most widely used materials in the world. It is an important ingredient in producing concrete. Over the past decades, more research has been done to replace cement in concrete even with debris and industrial products such as ash ash (FA) and ground granulated blast furnace slag (GGBS) to reduce the amount of greenhouse gases associated with cement production. In addition, the use of waste and industrial effluents will reduce the use of non-renewable natural resources in construction that makes concrete sustainable. The use of fly ash in the concrete industry not only has a positive impact on the environment but also enhances the resilience and economy of concrete production. However, many previous studies limit the use of additional cement materials to about 30% of cement content [1]. The whole development of a country depends on the amount of energy production and consequently its use as energy. Our country, India needs a lot of energy resources to meet the expectations of its people and its goal of becoming a developed nation by 2030. Mineral fuel plays an important role in meeting the demand for electricity generation. Coal is considered one of the major energy sources the world's richest fossil fuels. Globally, India ranks third in the world in terms of coal production and has the fourth largest coal-fired power plant (197 billion tons). It is estimated that 75% of India's total energy source is thermal when half of the coal is about 90%. About 600 million tons of coal is produced worldwide each year, with Fly ash generation estimated at 500 MT (60-78%) of total ash production [2].

1.1 The need to learn

1. Find another OPC method in concrete.
2. Produce a strong infrastructure that is able to design a life that is measured for hundreds of years.
3. Reduce CO2 emissions and produce eco-friendly concrete mix.

4. Protect water sources and areas above freshwater by removing fly ash dumps.
5. Producing a product that works well for the price.
6. Producing higher strength concrete than standard Portland cement concrete.

1.2 Purpose and performance of the work
1. Production of ash-based geopolymer concrete with varying percentage of GGBS.
2. To study the effect of GGBS 'expansion on its new and stronger geopolymer concrete features.
3. Testing the compressive strength and shaving power of geopolymer concrete with various percentages of GGBS.
4. Determining the total percentage of fly ash and GGBS in the result.

2. *Book responsibility:*
The purpose of this section is to review the previous work on geopolymer concrete by paying particular attention to material structures and their structural behavior.

2.1 *B. Laxman Raju et al. [1]* investigated the "Interface Shear of Fly ash and GGBS Based Geopolymer Concrete" This article presents an experimental study conducted to determine the shear strength of geopolymer concrete. In addition, it was proposed to evaluate the validity of existing statistics such as Birkeland and Birkeland, Mattock and the Design ACI 318 code, developed by standard Portland concrete to measure the shear volume of geopolymer concrete. Push off Samples were used to study the shear strength at the interface. Both reinforced concrete samples and reinforced concrete samples were used for the investigation. It was noted that the shear strength of geopolymer concrete is higher than that of OPC concrete. The presence of reinforced transversely shear plane produced an increase of about 29% against shear resistance against slipping.

2.2 *Satyajit kumar et al. [2]* Discuss "The investigation of fly ash polymer composite" Industrial waste, such as ashes that cause environmental problems, is widely used as a building material due to its low cost and easy availability. But the main disadvantage of these bricks is their low strength. Therefore, more research is being done to increase the strength of these bricks. Current research work is under way to develop a new systematic process of producing composite ashes that will have high compression strength. Here fly-ash is mixed with Cold setting resin in different proportions and water is treated at different temperatures to find a solution in the brick industry. Pressure strength, durability, water absorption, density and thermal performance of fly ash-resin powders obtained under the best test conditions are 11.24 Mpa, 47.37HV, 19.09% 1.68 g / cm3, and 0.055 W / mK respectively. Slide behavior is also under investigation. The structural correlations of these compounds are studied using X-ray diffraction, FTIR analysis and scan electron microscopy.

2.3 *Benny Josef et al. [3]* learned from "Interface shear strength of fly ash based geopolymer concrete." This current paper and experimental study conducted to evaluate the impact of composite content on the shear strength of a geopolymer concrete connector and compare it with that of standard Portland cement concrete. In addition, it is proposed to consider the suitability of the existing OPC concrete figures to assess the shear capacity of the geopolymer concrete. A push-off template was used to study the shear strength of the interface. Both reinforced and non-reinforced concrete were used in the study. It has been observed that the shear strength of geopolymer concrete is lower than that of OPC concrete and that the combined content of less than 65% in geopolymer concrete leads to a significant reduction in shear volume. A 50% reduction in value has been proposed to predict the shear strength of the geopolymer concrete.

2.4 U-Subhash V. Patankar et al. [4] Discuss "The mixed design of fly ash based geo-polymer" Geopolymer is a new development in the concrete world where cement is completely replaced by pozzolanic materials such as fly ash and is activated by highly alkaline solutions to act as a bond. in concrete mixing. In order to select the appropriate ingredients for geopolymer concrete to achieve the desired strength in the required performance, exploratory research has been conducted to plan the geopolymer concrete and the composite design process is proposed on the basis of quantity and fineness of fly ash, sum of water, grated finely ground aggregate, and fine to total aggregate proportion. Sodium silicate solution containing Na2O = 16.37%, SiO2 = 34.35% and H2O = 49.28% and sodium hydroxide solution with 13M concentration were maintained throughout the study. A water-to-geopolymer binder scale of 0.35, an alkaline-to-fly ash level of 0.35 and a sodium silicate-to- sodium hydroxide ratio of 1.0 were concentrated on the basis of performance and pressure cube pressure. The performance of the geopolymer concrete is measured by the use of a flow table and 150mm side cubes are cast and tested for compression strength after the specified heating time of the oven. The oven temperature is maintained at 60 ° C for 24 hours duration and tested 7 days after heating. It is recognized that performance results and pressures are closely related to the required level of performance and pressure. Therefore, the proposed method is used to design a standard and standard geopolymer concrete.

2.5 Rohit Zende et al. [5] read "research on fly ash and GGBS-based geo-polymer concrete under ambient concrete" Construction has been the most important human activity since mediaeval time. Concrete is widely used and reliable building materials. Some of the challenges in the construction industry are global warming and a shortage of building materials. One of the ways to change concrete materials is the use of geo-polymer which helps to apply a very small amount of cement to the concrete. This project deals study of the mechanical properties of geopolymer concrete with different mixes. In this study, Geopolymer concrete is produced by fly ash and sodium hydroxide and sodium silicate are used as binder. Fly ash is replaced by GGBS by 25%, 50% and 75% to improve various concrete structures. In this project, compound formation was performed at 11M and 13M sodium hydroxide concentration. A solution of 2.5 alkaline activator and an alkaline solvent fluid 0.40 was selected for this study. Template size 150x150x150mm cubes, 150x300mm cylinders and 500x100x100mm prisms are made and geo-polymer concrete templates are treated at ambient temperature for 7 days and 28 days. The treated specimens are then tested for compressive strength, dividing force and consecutive bending strength.

3. Experimental work and methodology:

3.1 Material used:

For the preparation of fly ash based geo-polymer concrete following ingredients are used as shown in figure 1.

Figure 1. Material required for Geo-polymer concrete

3.1.1 Fly ash:

In this project work fly ash as shown in figure 2 was obtained from Sofia Thermal Power Plant Nandgao Khandeshwar, Amravati district, Maharashtra state. The unprocessed fly ash having specific gravity of 1.84 calculated from pycnometer test.

Chemical composition of fly ash which is use as a binder in geo-polymer concrete as shown in table 1.

3.1.2 Ground granulated blast slag (GGBS):

Ground-granulated blast-furnace slag (GGBS) is obtained by quenching molten iron slag (a by-product of iron and steel-making) from blast furnace in water or steam, so produce a glassy, granulated product that is then dried and ground into a fine powder as shown in figure 3. Ground- granulated blast furnace slag is highly cementitious and high in CHS (calcium silicate hydrates) which is a strength enhancing compound which improves the strength, durability and appearance of the concrete.

GGBS is purchase from Magma iron Pvt. Ltd. Nagpur. It is stored in tight bags. The GGBS having specific gravity 2.88. Physical properties and chemical properties of GGBS are given in table 2 and table 3 respectively:

Table 1. Chemical composition of Fly ash

Sr. No.	Characteristics	Fly ash (% wt.)
1	Silica	55-65
2	Aluminium oxide	22-27
3	Iron oxide	5-7
4	Calcium oxide	5-7
5	Magnesium oxide	<1
6	Titanium oxide	<1
7	Phosphorous	<1
8	Sulphates	0.1
9	Alkali oxide	<1
10	Loss of ignition	1-1.5

Figure 2. Fly ash

Figure 3. GGBS

Table 2. Physical properties of GGBS

Sr. No.	Test conducted	Test result	Test method
1	Specific gravity	2.88	IS 1727-1967
2	Fineness- specific area in m^2/Kg by Blaine's permeability	320	IS 1727-1967
3	Residue on 45 microns	3	IS 1727-1967
4	Residue on 90 microns	7	IS 1727-1967

3.1.3 Fine Aggregate

Natural sand obtained from Wardha River was used as fine aggregate which was passing through 4.75 mm IS sieve is use in this project.

1. Specific gravity of fine aggregate is 2.62.
2. Sieve analysis was conducted to the fine aggregate.
3. Water absorption for the fine aggregate is 1%.
4. Fineness modulus for fine aggregate is 3.02.

5. Specific gravity of fine aggregate is 2.62.
6. Sieve analysis was conducted to the fine aggregate.
7. Water absorption for the fine aggregate is 1%.
8. Fineness modulus for fine aggregate is 3.02.

Physical properties of Fine aggregate are as shown in table 4.

Table 4. Physical properties of Fine aggregate

Sr. No.	Characteristics	Result
1	Aggregate type	Natural
2	Specific gravity	2.62
3	Fineness modulus	3.02
4	Water absorption	1%

Table 3. Chemical properties of GGBS

Sr. No.	Test conducted	Test result	Requirements as per IS 12089-1987
1	Insoluble residue (max)%	0.23	5
2	Manganese oxide (max)%	0.29	5.5
3	Magnesium oxide (max)%	6.06	17
4	Sulphide Sulphur (max)%	0.58	2
5	Glass content %	96	Min 85.0
6	$(CaO+MgO+1/3Al_2O_3)/(SiO_2+2/3Al_2O_3)$	1.14	≥1.0
7	$(CaO + MgO+ Al_2O_3)/SiO2$	1.1	≥1.0

Sieve analysis is done on fine aggregate to calculate fineness modulus as shown in table 5.

Table 5. Sieve analysis on fine aggregate

Sr. No.	Sieve size	Weight Retained	Cumulative weight retained (kg)	Weight passing (%)	Percentage weight retained (%)
1	10mm	0	0	1	100
2	4.75mm	0	0	1	100
3	2.36mm	0.1195	0.1195	0.8805	88.05
4	1.18mm	0.3255	0.445	0.555	55.5
5	600 micron	0.192	0.637	0.363	36.3
6	300micron	0.2055	0.8425	0.1575	15.75
7	150micron	0.137	0.9795	0.0205	2.05
8	Pan	0.0215	-	-	-
9	Total	1.00	3.0235		

Fineness modulus of fine aggregate= Cumulative weight retain / Sum of weight retain = 3.0235/1.00 = 3.0235.

3.1.4 Coarse aggregate:

The material which is retained on BIS test sieve number greater than 4.75mm IS sieve size of aggregate is termed as coarse aggregate. The broken stone is generally used as a stone aggregate. Locally available coarse aggregate having maximum size of 20mm are used and minimum size of 12 mm are used as a coarse aggregate. The aggregate used, having rounded shape and are smooth. Specific gravity of coarse aggregate calculated by pycnometer apparatus is obtained 2.71. Water absorption test carried out in lab on coarse aggregate and it is obtained 2%. Sieve analysis was carried out to find fineness modulus of coarse aggregate as shown in table 6 and obtained fineness modulus is 2.7045.

Table 6. Sieve analysis of coarse aggregate

Sr. No.	Sieve size	Weight Retained	Cumulative weight retained (kg)	Weight passing (%)	Percentage weight retained (%)
1	20mm	0.53	0.053	9.947	99.47
2	16mm	3.556	3.609	6.391	63.91
3	12.5mm	2.809	6.418	3.582	35.82
4	10mm	0.7515	7.1695	2.8305	28.3
5	4.75mm	2.626	9.7955	0.2045	20.45
6	Pan	0.288	-	-	-
7	Total	10	27.045	-	-

Fineness modulus of coarse aggregate = Cumulative weight retain / Sum of weight retain = 27.045/10 = 2.7045.

Physical properties of coarse aggregate are tabulated in Table 7 are as follows:

3.1.5 Sodium Silicates:

The sodium silicate was purchased from Rajas chemicals Khamgao at 30Rs per kg. A sodium silicate used in this project content Na_2O=9%, SiO_2=29%, Water=63%. Physical and chemical properties of liquid sodium silicate gel are as follows:

Chemical formula:	Na_2O* SiO_2*H2O
Appearance:	Jelly
Color:	Colorless
Molecular weight:	184-254 g/mole
Molar ratio:	1:28
Specific gravity:	1.5
Baume:	50
Na_2O:	5.45%
SiO2:	31%
Total solids:	49.50%
pH:	12.6

Table 7. Physical properties of coarse aggregate

Sr. No.	Characteristics	Result
1	Aggregate type	Crushed stone
2	Maximum size of aggregate	20
3	Specific gravity	2.71
4	Shape of aggregate	Rounded and smooth
5	Fineness modulus	2.7045
6	Water absorption	2%

3.2 Methodology

3.2.1 Mix design of geo-polymer concrete:

Geo-polymer is a new development in the world of concrete in which cement is totally replaced by pozzolonic materials like fly ash and activated by highly alkaline solutions to act as a binder in the concrete mix. For the selection of suitable ingredients of geo-polymer concrete to achieve desire strength at required workability, an experimental investigation has been carried out for the gradation of geo-polymer concrete and a mix design procedure is proposed on the basis of the quantity of fi neness of fly ash, quantity of fly ash, quantity of water, grading of fine aggregate, fine to total aggregate ratio. Sodium silicate with Na_2O=9%, SiO_2=29%, Water=63% and Sodium hydroxide solution having 11M concentration were maintained constant throughout the experiment.

The primary difference between geo-polymer concrete and Portland cement is the binder. To form geo-polymer-paste alkaline activator solution used to react with silicon and Aluminium oxides which are present in fly ash and GGBS. This alkaline activator solution helps to bind coarse aggregate and fine aggregate to form geopolymer mix. The fine and coarse aggregate occupy nearly 75% to 80% mass of geo-polymer concrete according to literature survey as geo-polymer concrete does not have any specific mix design. The fine aggregate was taken as 30% of total aggregate. The density of geo-polymer concrete is taken 2400Kg/m3. The workability and strength of concrete are influence by properties of materials that make geo-polymer concrete. Fly ash is replaced by GGBS in range of 20%, 40%, 60%, and 80%. The ratio of Sodium silicate to sodium hydroxide is 2.5 and is kept constant throughout this study. The ratio of alkaline activator to fly ash is 0.40 kept constant. In this project work we used 11M NaOH solution.

3.2.2 Preparation of alkali solution

Alkaline activation is a chemical process in which a powdery aluminosilicate such as fly ash is mixed with an alkaline activator to produce a paste capable of setting and hardening within a reasonably short period of time.

The role of alkali activator solution is to dissolve the reactive portion of source materials Si and Al present in fly ash and provide a high alkaline liquid medium foe condensation. The alkaline activator plays an important role in geo-polymer concrete. Alkaline activator plays a fundamental role in the hydration of binders. Due to their characteristics, mortars and alkaline activated concretes can show higher mechanical resistance result than those obtained by a concrete made with OPC. The preparation of solution is done by dissolving sodium hydroxide in water. The concentration of sodium hydroxide changes with molarity. The quantity of sodium hydroxide solution with a concentration of 11M is calculated. The mass of NaOH solids in solution varied depending on the concentration of the solution express in terms of molar, M. The NaOH solution with concentration of 11M consisted of 11*40=440gm of NaOH solids per liter of the solution, where 40 is the molecular weight of NaOH. The mass of NaOH solids was measured as 306gm per kg of NaOH solution of 11M concentration. The sodium hydroxide added to the water and stirred about fifteen minutes to gel cool down. Then sodium silicate is added to the solution. This solution is used after 24 hours of its preparation.

3.2.3 Materials

In the proposed mix proportioning method, low calcium processed fly ash of thermal power plant was used as source material. The laboratory grade sodium hydroxide in flake form (98 % purity) and sodium silicate (38 % solids) solutions are used as alkaline activators. Locally available river sand is used as fine aggregate and locally available less than 20 and greater than 12.5 mm sizes crushed basalt stones are used as coarse aggregates.

3.2.4 Parameters considered for Mix Proportioning of Geopolymer Concrete

For the development of fly ash based geopolymer concrete mix design method, detailed investigations have been carried out and following parameters were selected on the basis of workability and compressive strength.

3.3 Method proposed for mix proportioning

Based on the experimental investigation carried out in the present study the following mix proportioning method is proposed. Subhash V. Patankar discuss on "Mix design of fly ash based geopolymer concrete" following mix design steps are taken from his research [4].

Table 8. Water content per cubic meter of concrete.

Sr. No.	Degree of workability	Flow in percentage	Quantity of water required in Kg/m^3			
			Fineness of fly ash in m^2/kg			
			<300	300–400	400–500	>500
1	Low	0–25	80	85	100	110
2	Medium	25-50	90	95	110	120
3	High	50-100	100	110	120	135
4	Very high	100–150	120	130	140	160

3.3.1 Data Required for Mix Design

1. Characteristic compressive strength of Geopolymer Concrete (f_{ck})
2. Fineness of fly ash in terms of specific surface in m^2/kg
3. Workability in terms of flow
4. Oven curing (heating) 60 °C for 24 hr and tested after 7 days.
5. Fineness modulus of fine aggregate
6. Water absorption and water content in fine and coarse aggregate

3.4 Design steps

1. Selection of quantity of fly ash: From Rohit Zende [5] quantity of fly ash is taken to be 394.28 Kg/m^3 at solution to fly ash ratio is 0.40.

2. Calculation of the quantity of alkaline activators:
Calculate the quantity of alkaline activators considering:
Solution/Fly ash ratio by mass = 0.4,
i.e., Mass of (Na_2SiO_3 + NaOH) / Fly ash = 0.4 Mass of (Na_2SiO_3 + NaOH) / 394.28 = 0.4 Mass of (Na_2SiO_3 + NaOH) = 157.712 kg/m^3.
Take the sodium silicate-to-sodium hydroxide ratio by mass of 2.5 [5]
Mass of sodium hydroxide solution (NaOH)= 45.06 kg/m^3 Mass of sodium silicate solution (Na_2SiO_3) = 112.65 kg/m^3

3. Calculation of total solid content in alkaline solution:
Solid content in sodium silicate solution = (38/100) x 112.65 = 42.807 kg/m^3
Solid content in sodium hydroxide solution = (38.50/100) x 45.06 =17.34 kg/m^3 Total Solid content in both alkaline solutions = 60.140kg/m^3

4. Selection of water content: For medium degree of workability and fineness of fly ash of 394.28 m^2/kg, water content per cubic meter of geopolymer concrete is selected from Table 8 Water content = 95 kg/m^3.
5. Adjustment in water content:
For sand conforming to grading-I, correction in water content is taken from Table 9. Adjustment in

water content = -1.5%.

Total quantity of water required = 95 - (1.5/100) x 95 = 93.575 kg/m^3 Water content in alkaline solutions = 157.712 - 60.140 = 97.572 kg/m^3.

As required water present in alkaline solution therefore no need to add extra water in concrete.

6. Calculation of Fine and Coursed aggregate content

Total aggregate content = (Wet density of GPC) - (Quantity of fly ash + Quantity of both solutions + extra water; if any),

= 2400 - (394.28 + 157.712 + 0)
= 1848.008 Kg/m^3

Sand content = [Fine to total aggregate content in %] x [Total quantity of all in aggregate]

= [30 / 100] x 1848.008
= 554.4 Kg/m^3

Coarse aggregate content = [Total quantity of all in aggregate] - [Sand content]
= 1848.008 - 554.4
= 1293.608Kg/m^3

Quantity of materials required per cubic meter of geopolymer concrete is shown in Table 9.

Table 9. Mix design for geo-polymer concrete.

Sr. No.	Designation of mix	Replacement of fly ash with GGBS (%)	Fly ash (Kg/m^3)	GGBS (Kg/m^3)	Fine aggregate (Kg/m^3)	Coarse aggregate (Kg/m^3)	Alkaline solution NaOH (Kg/m^3)	Na2SiO3 (Kg/m^3)
1	M1	00	394.28	00	554.4	1293.6	45.06	112.65
2	M2	20	315.42	78.856	554.4	1293.6	45.06	112.65
3	M3	40	236.57	157.712	554.4	1293.6	45.06	112.65
4	M4	60	157.71	236.568	554.4	1293.6	45.06	112.65
5	M5	80	78.856	315.424	554.4	1293.6	45.06	112.65

3.5 Experimental method

In figure 4, showing a rough idea about experimental procedure of project carried out in lab.

***3.5.1** Preparation of alkaline activator solution:*

The alkaline activator is produced first before 24 hours of mixing. First Sodium Hydroxide of required molarity was prepared first manually Are as follows: 1. Take a required quantity of solid sodium hydroxide in a glass beaker which having 98% purity, 2. Add a water into the beaker that exactly make a 1 Lit solution, 3. Stir well that mixture about 4 to 5 minutes, 4. Keep this Jar for some time to become a cool, and 5. After preparing 11M Sodium Hydroxide solution add a Sodium silicate liquid into that solution as shown in figure 5. Stir this solution up to 4 to 5 minutes and keep it away up to 24 hours before mixing.

3.5.2 *Mixing:*

Different weight of percentage of fly ash (100%, 80%, 60%, 40%, and 20%) with (0%, 20%, 40%, 60%,

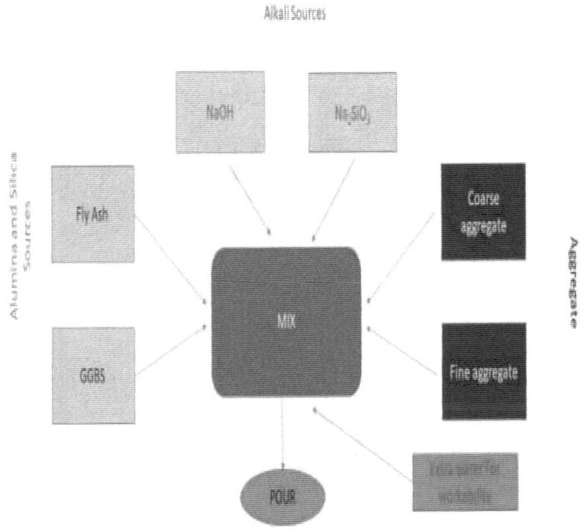

Figure 4. Flow chart of experimental procedure

Figure 5. Alkaline activator solution

and 80%) of Ground granulated blast slag was taken for geopolymer concrete mix M1, M2, M3, M4, and M5 respectively. Now weight the required quantity of Coarse and fine aggregate for mix. Before mixing all coarsed and fine aggregate were completely in dry state as shown in figure 6.

Figure 6. Weighted constituents of Geo-polymer Concrete: Fly Ash., Course Agg., Fine Agg., GGBS.

Different quantity of fly ash, GGBS along with alkali activator and aggregate were mix properly in a mixer are as follows:

1. Weight all quantity of mix properly on a digital weight machine first as shown in figure 7.
2. Separate all material properly such as we can handle it easily.
3. First add course and fine aggregate in to the mixer.
4. Mix it 2 to 3 minutes in a mixer properly. Keep in mind aggregate are completely in dry state.
5. Add Fly ash and GGBS in to the aggregate.

Figure 7. Weight of material

After fly ash and aggregate mix properly in to the mixer then add alkaline activator in a mix. Mix it properly up to next 5 minutes as shown in figure 8. Add water if required otherwise water is already present in alkaline activator.

Figure 8. Mixing of Geo-polymer concrete in mixer at lab

3.5.3 Compacting:

Properly mix Geo-polymer concrete is then poured first in slum cone test apparatus to check workability of geo-polymer concrete.

After completing slum cone test concrete in poured in to the moulds of casting into the lab. A (100 X 100 X 100) mm size moulds are used for compressive strength test. Concrete is placed in three layers and tamping is done for each layer by giving more than 25 blows, in order to get fully compacted geopolymer concrete specimens. Then the top surface is well finished. The sizes of the moulds used are cube (100 X 100 X 100) mm for compressive strength and 'L' shape moulds of [(150X150X150) -(60X90X150)] were used as shown in figure 9 for calculating shear strength of geo-polymer concrete.

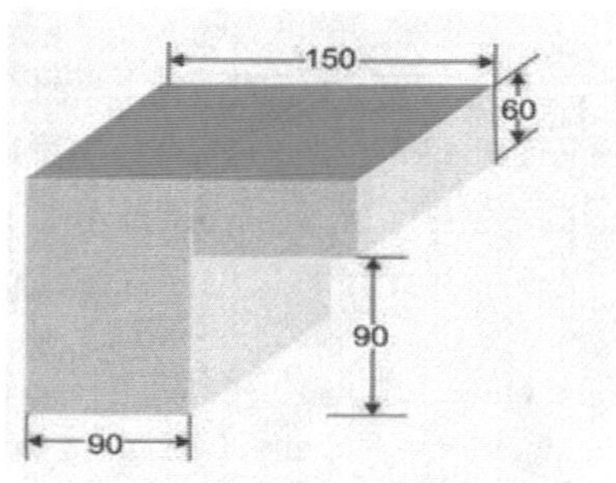

Figure 9. Shear specimen

For each mix six (100 X 100 X 100) mm size moulds were cast from which three for 7 days compressive test and three for 28 days compressive test and three 'L' size moulds of size [(150X150X150) mm - (60X90X150)] mm specially prepared moulds by adding (60X90X150) mm wooden block in standard (150X150X150) mm size moulds are cast for shear strength test as shown in figure 10. At the time of casting precaution is taken for casting those moulds such as exact "L" shape specimen obtained after demolding.

Figure 10. Casted geo-polymer concrete moulds in Lab

After 24 hours all the cubes are de-moulded and kept for curing at room temperature. As geo-polymer concrete content alkaline activator this activator produces a polymerization reaction with binder present in concrete.

To produce a polymerization reaction in concrete to get strength to geopolymer concrete does not required water for curing. Geo-polymer concrete required temperature curing for getting strength. But it is not possible to cure geo-polymer concrete at site for temperature curing hence, we do work on curing of geo-polymer concrete at ambient temperature. Hence, after demolded cubes are placed at room temperature for curing.

After 7 and 28 days later take compressive strength test of concrete and after complete gaining of concrete strength after 28 days perform shear strength of geopolymer concrete and result are noted.

For calculating shear strength, a specific 'L' shape cubes are prepared having dimensions [(150X150X150) - (60X90X150)]mm are cast in lab by providing (60X90X150) mm wooden block in (150X150X150) mm size cube on its one side as shown in figure 3.10.

Shear strength test assembly is kept on compressive test machine and apply load on shear block. Note the reading exactly on what amount of load the shear test specimen broken or crack on shear failure. Load obtained in test is not your final reading. The load obtained is divided into two equal parts due to shear force as shown in fig. 3.10.
Shear strength can be calculated by using formula:

Shear strength = Failure load/Area
Where, Failure load = Applied load/2.

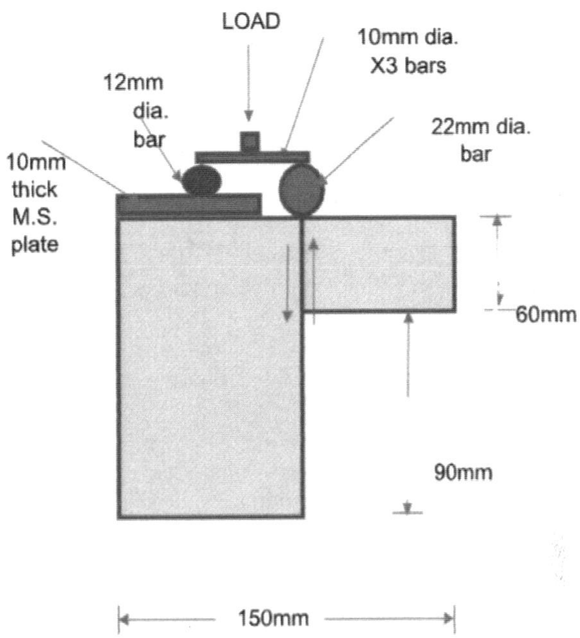

Figure 11. Shear strength test assembly

4. Result and discussion:

4.1 Determination of compressive strength
The compression test on cubes were conducted according to Indian standard specification (IS: 516-1959). After 28 days of casting cubes are taken for compressive strength test on compression testing machine in lab. Figures and tables show the compressive strength readings.

Compressive strength results of Geopolymer concrete at 7 and 28 days are as shown in table 10 and table 11 respectively. Graphical representation of compressive strength verses replacement of fly ash is as shown in figure 4.2. In this graph compressive strength values taken on Y- axis and replacement of fly ash with GGBS taken on X-axis.

Table 10. Compressive strength of geo-polymer concrete at 7 days at ambient temperature.

Sr. No.	Designation of mix	Compressive strength at 7 days of cube 1, 2 & 3 (Mpa)	Average compressive strength at 7 days (Mpa)
1	M1	7.30, 6.75, 7.16	7.07
2	M2	14.93, 16.82, 16.20	15.98
3	M3	20.48, 17.56, 21.31	19.78
4	M4	24.34, 20.11, 24.38	22.94
5	M5	20.36, 24.62, 27.36	24.11

Table 11. Compressive strength of geo-polymer concrete at 28 days at ambient temperature.

Sr. No.	Designation of mix	Compressive strength at 28 days of cube 1, 2 & 3 (Mpa)	Average compressive strength at 28 days (Mpa)
1	M1	10.21, 8.28, 11.16	9.88
2	M2	31.20, 33.69, 28.41	31.10
3	M3	33.73, 30.10, 35.63	33.15
4	M4	30.67, 36.42, 38.73	35.27
5	M5	43.16, 41.81, 38.26	41.07

Graph shows that as percentage of GGBS increase in concrete then compressive strength at 7 days increases polynomials over percentage replacement of fly ash by GGBS. Curve show that there is an up down or fluctuation in increment of compressive strength. Compressive strength of concrete at 28 days also shows polynomial curve means there is increment in compressive strength as GGBS increases in concrete but there are more fluctuations are between them. From above result and literature survey we can say that as percentage of GGBS increases in fly ash based geo-polymer concrete then its compressive strength increases [7].

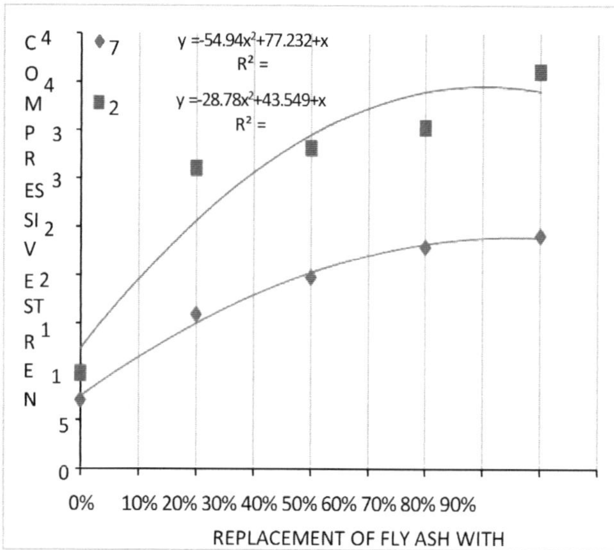

Figure 12. Graphical representation of Compressive strength of Geo-polymer concrete in ambient temperature with replacement of fly ash by GGBS.

4.2 Determination of shear strength

Due to covid-19 disease lockdown in India country all labs are closed therefore I unable to perform shear strength test. Therefore, from literature survey I found some results of shear strength of fly ash based geopolymer concrete. From literature survey it has been evaluate that shear strength from different standard codes and publisher gives a formula about shear strength of plain concrete in terms of compressive strength.

American concrete institute (ACI) committee 318 (2002) publish a building code requirement for structural concrete (ACI 318-02) mention shear strength formula for general plain concrete in chapter 22 are as shown in equation 1,

$$VC=0.166\sqrt{fck} \quad \text{..equation 1}$$

Where, V_C = Shear strength of concrete in Mpa, f_{ck} = Characteristic compressive strength in Mpa.
In designer's handbook of Eurocode-2 part 1.1, Design of concrete structures, Telford, London (1995) provides a formula for shear strength of plain concrete mention below in equation 2.

$$TRD=(0.25fctk,0.05)/Y \quad \text{..equation 2}$$

Where, T_{RD}= Shear strength of concrete in Mpa, $0.25f_{ctk}$, 0.05 = Formula for 5% fractile tensile strength = 0.7 f_{ctm}, f_{ctm} = Mean tensile strength = 0.3 $(f_{ck})^{0.67}$, Y_C = Factor of safety = 1.5.

In Canadian standard association (CSA) technical committee on reinforced concrete design (1994), Design of concrete structure A23.3-94 gives a formula for shear strength of plain concrete mention in equation 3.

$$V_C=0.2\sqrt{fck} \quad \text{..equation 3}$$

Where, V_C = Shear strength of concrete in Mp, f_{ck} = Characteristic compressive strength in Mpa.

Shear strength values of fly ash based geo-polymer concrete according to above formulae with reference to compressive strength of fly ash based geo-polymer concrete are as shown in table 12.

Table 12. Shear strength of concrete

Sr. No.	Designation of mix	Compressive strength at 28 days (Mpa)	Shear strength of concrete on the basis of compressive strength (Mpa)			Average shear strength (Mpa)
			ACI	CSA	EUROCODE -2	
1	M1	9.88	0.52	0.63	0.64	0.59
2	M2	31.10	0.92	1.11	1.40	1.14
3	M3	33.15	0.95	1.15	1.46	1.18
4	M4	35.27	0.98	1.19	1.52	1.23
5	M5	41.07	1.06	1.28	1.69	1.34

From table 12 we got a values of shear strength of geo-polymer concrete on the basis of different standard codes i.e., ACI, CSA, EUROCODE-2. From those values we take an average shear strength of all three values as a shear strength of geopolymer concrete. From above table we can say that as percentage of GGBS increases in geopolymer concrete shear strength of concrete also increases. When percentage of fly ash is 100% in geo-polymer concrete, we got 0.59 Mpa shear strength then as we increase GGBS percentage then shear strength increases for mix M2, M3, M4 & M5 is 1.14 Mpa, 1,18 Mpa, 1,23 Mpa and 1.34 Mpa respectively.

A Compressive strength verses shear strength of geo-polymer concrete variation showing graph is as shown in figure 13. A graph represents compressive strength of concrete in Mpa on X- axis and shear strength of concrete in Mpa on Y axis. In this graph blue line represents shear strength by ACI, dark red line represents shear strength by CSA, green line shows shear strength by using Eurocode-2 and from all this black line shows average shear strength geo-polymer concrete calculated from above code.

Above graph of compressive strength and shear strength shows there is a linear variation in shear strength with respect to compressive strength. From above graph we can say that as compression strength of concrete increases by day wise then there is also increment in shear strength of concrete.

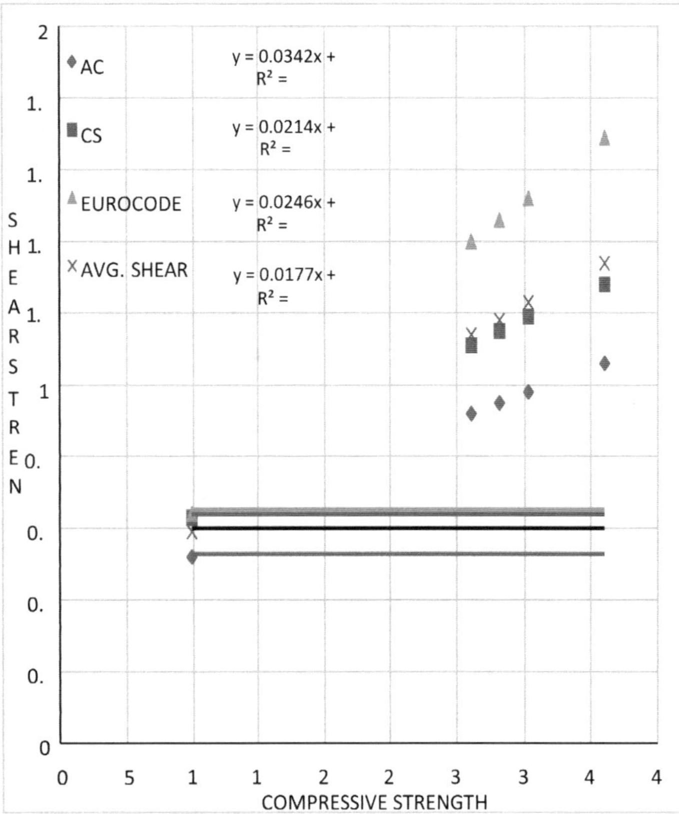

Figure 13. Graph showing variation between shear strength and compression strength.

5. Summary:

The use of geopolymer technology as a construction material in pre- cast industry applications like railway sleepers, sewer pipes, culverts and other prestressed concrete building components has been an increase in order to meet the demand of reducing the greenhouse gas emissions for sustainable development. This has led to an emphasis on information regarding the structural behavior and performance of fly ash-based Geopolymer concrete under aggressive Environments. However, a systematic compilation of the available information in literature clearly shows that the comprehensive investigation on the mix design procedure for High strength grade geopolymer concrete is almost nonexistent leading to uncertainty regarding its applicability. This made the author to study the development of unprocessed fly ash based geopolymer concrete and its behavior under ambient temperature.

As a preliminary investigation, studies were carried out to optimize the parameters affecting the properties of geopolymer concrete. Using the optimized parameters such as Alkaline liquid to fly ash ratio, ratio of Sodium silicate to sodium hydroxide, curing time, Curing temperature, Sodium hydroxide concentration, the Addition of extra water. The mix design was developed for geo-polymer concrete. Mechanical properties such as compressive strength, shear strength is determined in ambient temperature.

The outcome of the study shows the geo-polymer concrete is best replacement for the ordinary Portland cement and as the GGBS percentage increases in concrete increases strength properties and workability of concrete.

6. References:

[1]. B.Laxman Raju, L.Sudheer Reddy, Sumanth Kumar B "Interface Shear of Fly ash and GGBS Based Geopolymer Concrete" IJTIMES, ISSN: 2455-2585 Volume 5, Issue 04, April-2019.

[2]. Satyajit kumar "Investigation of fly ash polymer composite" Rourkela national technology center.

[3]. Benny JOSEPH, George Mathew "interface shear strength of fly ash based geopolymer concrete" International Journal of Engineering, 2013.

[4]. Subhash V. Ptankar "Mixed design of fly ash based geo-polymer" Conference paper2014 https://www.researchgate.net/publication/275340304.

[5]. Rohit Zende, Mamtha A. "Study On Fly Ash and GGBS Based Geopolymer Concrete under Ambient Curing" JETIR (ISSN-2349-5162), July 2015, Volume 2, Issue 7.

[6]. Dhanabal G., Drs. V. Sreevidya "Shear Power Features of Self Compacting Geopolymer Concrete Cast at Different Ages" International Journal of Recent Engineering and Management Research (IJLEMR) ISSN: 2455-4847, April 2019.

[7]. J. Guru Jawahar, D. Lavanya, C. Sashidhar "The effectiveness of fly ash and GGBS basedon geo-polymer concrete in the acid environment" International Journal of Scientific Research and Development (IJRSI), Volume III, Issued August 8, 2016.

[8]. B.H. Win "Properties of fly ash based geopolymer mortar", International Journal Of Engineering Research & Technology, vol.4, issue 07, July-2015.

[9]. Shaikh Usman "Energy lessons in geo-polymer concrete using fly ash and quarry dust", pace center of technology and science, vallur, 2015-2017.

[10]. Tanveer Singh Bains, Khushpreet Singh "Experimental Study on Geopolymer Concrete using Fly Ash, Bagasse Ash and Metakaolin with Pet Fiber".

[11]. International Journal of New Technology and Engineering Evaluation, Issued May 7, 2019.

Development of different hydrogeological influencing thematic layers and groundwater potential zones of Deo river watershed of Mayurbhanj district, Odisha, India

N K Mohanta[1], D C Jhariya[2], Y K Mawale[3] and M Sahu [4]

[1]M.Tech student, Department of Applied Geology, Geology, National Institute of Technology Raipur, G.E.Road, Raipur - 492 010, Chhattisgarh, India
[2]Assistant Professor, Department of Applied Geology, National Institute of Technology Raipur, G.E.Road, Raipur - 492 010, Chhattisgarh, India
[3]Associate Professor, Department of Applied Geology, Sant Gadge Baba Amravati University, Amravati- 444602, India
[4]Assistant Professor, Department of Information Tecnology, National Institute of Technology Raipur, G.E.Road, Raipur - 492 010, Chhattisgarh, India
Email: nandanmohanta1999@gmail.com, dcjharia.geo@nitrr.ac.in, mrisahu.it@nitrr.ac, myadaorao@yahoo.co.in

Abstract: Water is one of the most abundant renewable resources, and it is essential to all living things. Groundwater is one of the major sources of Water in the planet Earth. Continuous uses of water without any conservation cause lowering of Ground water level and thus living forms are facing many problems now-a-days. Due to increase in industrialisation in the state of Odisha water problem is rising its head. Using Remote Sensing and Geographic information system (GIS) methods, this research aids in the delineation of water potential zones in the Deo river watershed in Mayurbhanj district, odisha. For this, different thematic layers have been prepared and interpreted. This study may be helpful to monitor and conserve groundwater in near future.

Keywords: Groundwater, Remote sensing, GIS, Deo watershed

1. Introduction:

India is the world's largest consumer of groundwater. More than 60% of agricultural operations rely on groundwater, and more about 85% of people rely on it for potable water. Groundwater removal has caused water levels to rise and shallow groundwater harvesting structures to dry up eventually. The survey's goal is to delineate subsurface water for improved planning of the aquifer system in Mayurbhanj district, Odisha. Where groundwater deterioration has risen rapidly in recent decades. Groundwater levels have been lowered, with a long-term dropping trend and even wells drying up as a result of this persistent groundwater removal for diverse sectors (irrigation, household, and industrial)[14]. For calculating the groundwater resources, remote sensing techniques occupy a broad and inaccessible portion of the earth's surface in a short amount of time. For groundwater management and planning, detailed understanding of aquifers, replenishment, and potentiality is essential. Groundwater flow and distribution are governed by a number of elements, including rock types and geological qualities, elevation, , landform patterns, land use/land coverage patterns, drainage networks and climatic conditions. A GIS overlay analysis based on drainage density, lineament, slope, Land use / land cover (LULC), and geology was used to create the groundwater potential zone [1].

To generate thematic layers of the research region utilising satellite data from various thematic layer such as geology, landform, LULC, gradient, drainage network, precipitation, water level, lineament, and lithology [12]. To delineate the ground water potential zone of Deo watershed.

2. Research area:

The Deo river watershed (Figure 1) is located in Odisha's Mayurbhanj district. The catchment covers a total area of 456.07 km2 and includes one of Odisha's most hilly regions like similipal national park, Iron mines and agricultural lands. Accoding to Survey of India (SOI) Toposheets No. 73 J/18, Latitude 21°40'0" to 21°50'0" and Longitude 86°0'0" to 8°20'0". This region falls under tropical climatic areas, which supports the area with average rainfall of 1681.7 mm and mean temp 27.3°C.The southwest monsoon is the area's key source of precipitation [9].

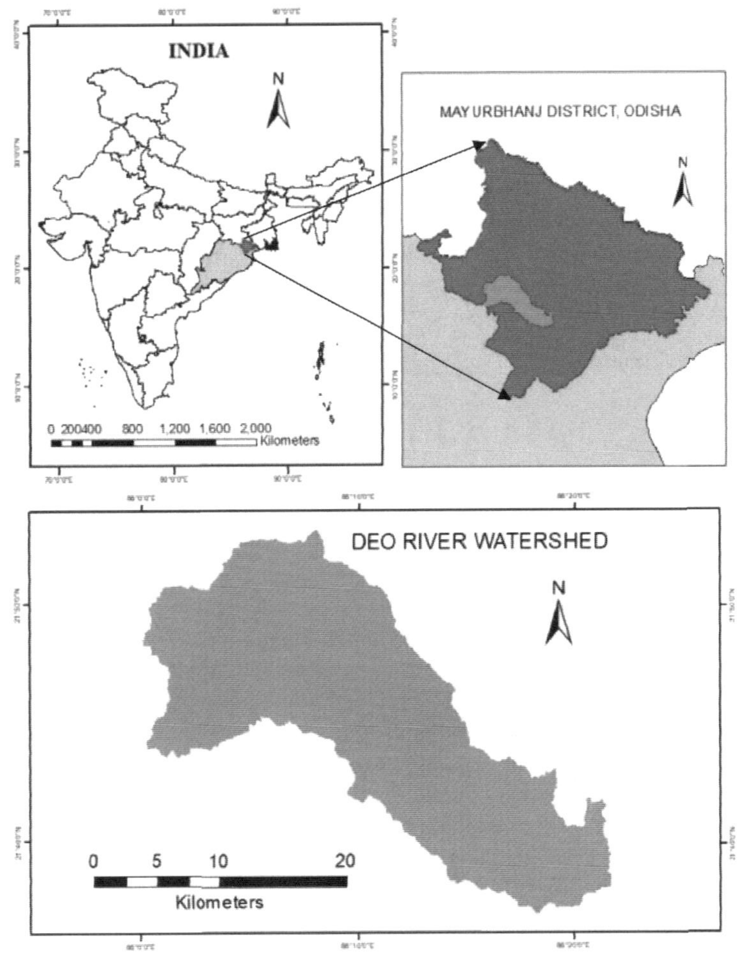

Fig.1. Research area location map

3. Data used:

The GIS software used to make all the thematic layers are ArcGIS and QGIS. Satellite data Used: Shuttle Radar Topography Mission (SRTM), Digital elevation model (DEM) imagery was used for the preparation of drainage and slope map. Sentinel-2 imagery was used to make LULC map of the study area. Bhukosh site data were used to prepare geology, geomorphology and lithology map [11]. India Water Resources Information System (India-WRIS) data were used for the precipitation and groundwater level maps. The source of data is described below **(Table-1).**

Table-1: date used in preparation of different maps

SL. No.	Layer Name	Source	Sites
1	Geology	Bhukosh	www.bhukosh.gsi.gov.in
2	Geomorphology- Landforms	Bhukosh	www.bhukosh.gsi.gov.in
3	Geomorphology- Units	Bhukosh	www.bhukosh.gsi.gov.in
4	Lithology	Bhukosh	www.bhukosh.gsi.gov.in
5	Drainage pattern	DEM	www.earthexplorer.usgs.gov
6	Lineament Density	Bhuvan	www.bhuvan.nrsc.gov.in
7	Precipitation	India-WRIS	www.indiawris.gov.in
8	Groundwater level	India-WRIS	www.indiawris.gov.in
9	Land Use and Land Cover	Sentinel-2	www.earthexplorer.usgs.gov
10	Slope	DEM	www.earthexplorer.usgs.gov

4. Methodology:

Groundwater potential zones in and around the Deo basin were identified using a GIS technology that included satellite-derived multiple themed images. The following layers were created and utilised to identify the groundwater potential zone in this investigation. SRTM data is process in ArcGIS and used to create the drainage density along with slope map was created [15]. Bhuvan.gov.in provided the geomorphology theme map which data is also process in ArcGIS. The GSI provide the geology map. For ease of use in GIS settings, all of these datasets were geo-rectified and projected in the WGS 1984 using Universal Transverse Mercator (UTM) Zone 45 N after that False colour composite (FCC) is generated by considering the band no. 8,3 and 2 from 12 band data then pixel processing then LULC map is prepared. From IWRIS site we found rainfall and post monsoon waterlevel data from different station in my research area then IDW interpolation process provide the rainfall and water level map. The flow chart depicts the approach used in this paper. Geology, geomorphology, LULC, slope, drainage system and water level are combining to give the best site for groundwater potential. Spatial analysis integration utilizing an index overlay approach in a GIS context was used to determine groundwater potential. Eight layers were combined in this design by giving relative weight to each theme map on a level of 1–10 and subcategories on a range of 1–5. The map was then classified as follows: very high, high, moderate, very poor, and poor [14]. As a result we found GW potential boundries.

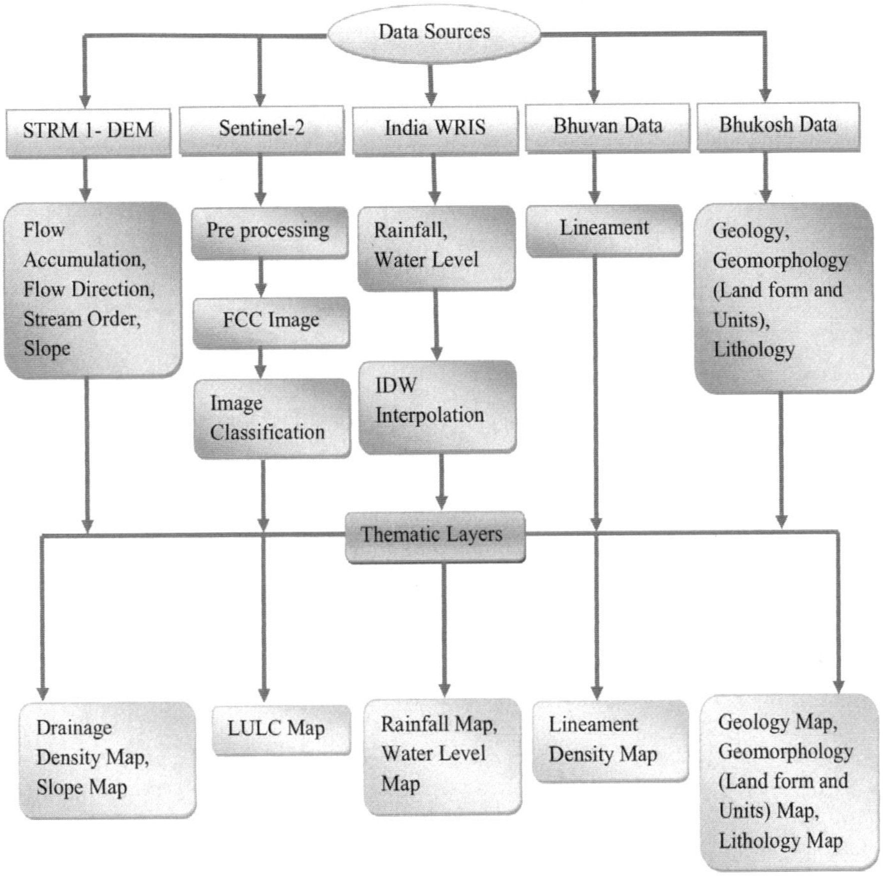

Figure-2.Methodology Flow chart for thematic layer preparation

5. Thematic layers

Thematic layers that play significant role for GW occurrence and management of aquifer system in research area are as follows;

5.1. Geology:

The study area comes under Singhbhum supergroup which is Proterozoic in age. Singhbhum supergroup divided into three groups. The study area comes under Singhbhum group and similipal group. Age of the study area is Meso-proterozoic to Neoproterozoic.

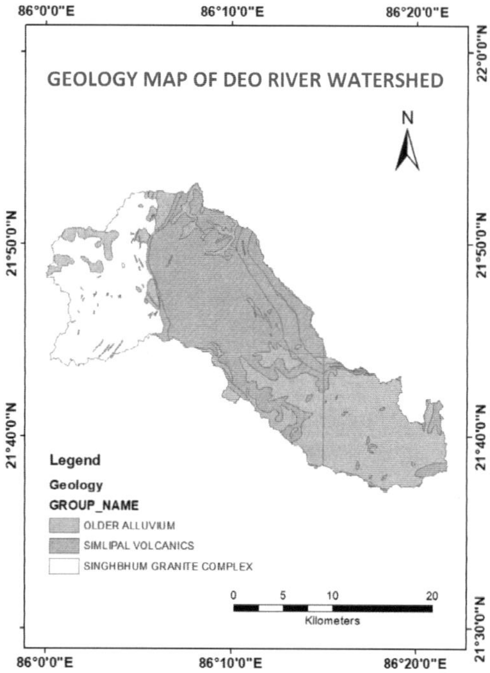

Fig.3. Geology map of Deo watershed

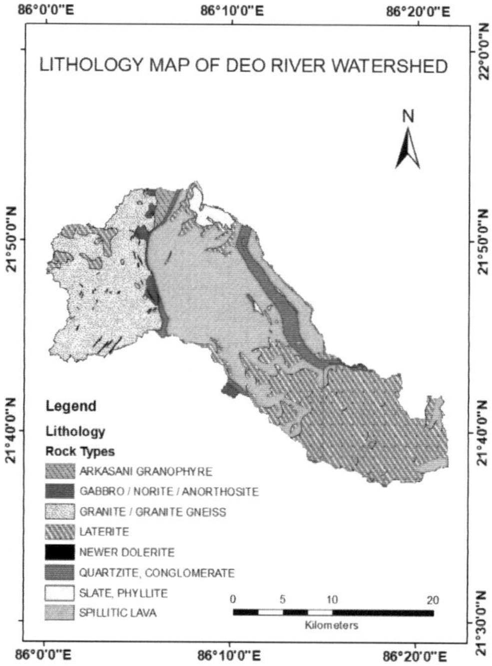

Fig.4. Lithology map of Deo watershed

5.2. Lithology

The study has a variable lithology which plays a crucial role to control the presence of groundwater [5]. As it represents a delta type environment here more sediment accumulates rather than rock mass. Granite, gabbro, quartzite, slate, phyllite, spilitic lava and lateritic soil cover most of the portion in the study area. These weathered zones can also hold a huge amount of groundwater. Towards the West part of my study area granitic rocks are dominant, but these have some linear features. So the regions may be categorized under medium recharge zones but in the central part covered with spilitic lava so this region is categorized under least recharge zone. The regions like dolerite dyke, Spilitic lava, quartzite, basic and ultra basic and ultramafic rocks have no porosity and permeability condition and Laterite acts as a good recharge zone. Hence these zones can be categorized under low recharge zones.

5.3. Geomorphology:

Recognition of various geomorphological feature are critical in assessing groundwater possibilities. Geomorphological variables such as gradient, degradation extent, and landform type all serve a part in defining a region's groundwater potency. The research region is split among thirteen primary landforms (according to the Bhukosh-Geological Survey of India categorization systems), with low-to-high dissected hills and valleys, as well as alluvial plain, pediment, and Pediplain being some of the most prominent characteristics. Dissected hills are generated by differential erosion and weathering processes and are made up of jointed and fractured granites and gneisses. Percolation of groundwater is possible due to the existence of fractures, joints, and topographic cuts, however as the slope rises, so does the probability of runoff, restricting the groundwater recharge process. As a result, low-dissected hills recharge groundwater greater efficiently than high-dissected hills. Pediplain in western part creates excellent recharging and buffer zones.

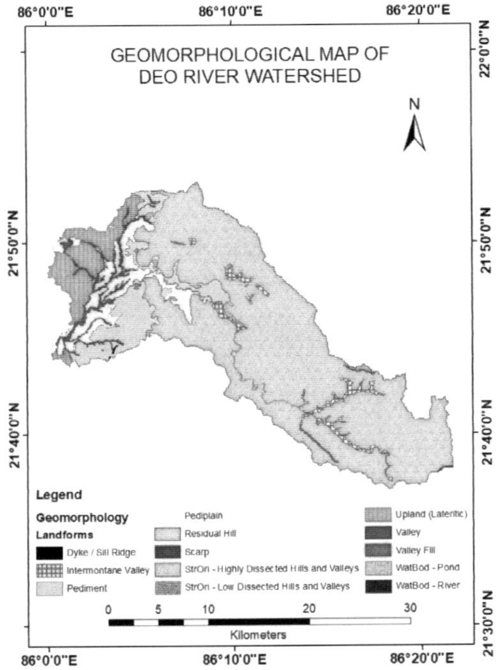

Fig.5. Geomorphology map of Deo watershed

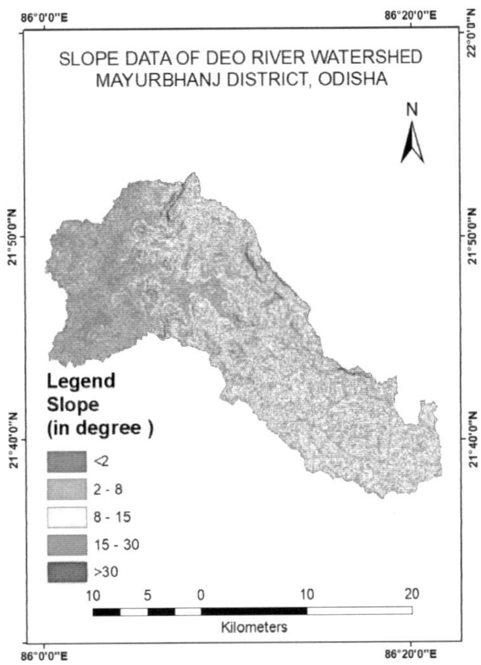

Fig.6. Slope map of Deo watershed

5.4. Slope:

Any terrain's slope is critical for permitting water to infiltrate the subsurface system. Runoff will be slower in areas with gentle slopes, giving rainfall more time to percolate, whereas rapid runoff occurs near steep slopes, allowing rainwater to enter for a shorter length of time. A medium to mild slope covers nearly half of the research area, indicating that moderate recharging is possible in these places. As per characteristics impacting groundwater recharge, holding, and recurrence, the grades were allocated to the relevant slope classes.

5.5. Drainage density:

Western part of research area is covered by high drainage density due to a deltaic environment. High drainage density indicates the underlying strata is more saturated hence more surface runoff occurs rather than percolation. In the low drainage density area (Southeast side) more infiltration occurs. The line density analysis technique was used to create the drainage density map. So the area with high drainage density more ground water potential and law recharge area [2].

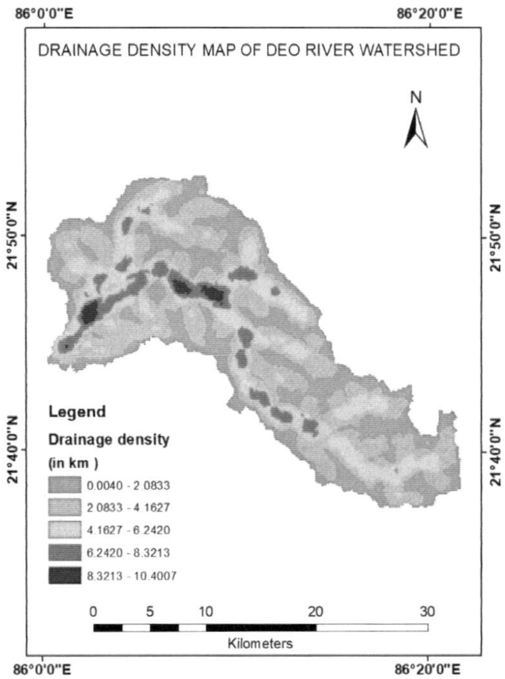

Fig.7. Drainage density map of Deo watershed

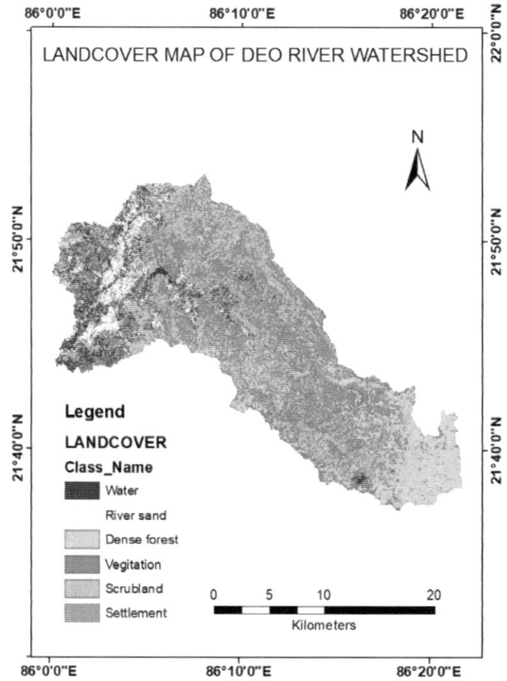

Fig.8. LU/LC map of Deo watershed

5.6. LULC:

LULC is a significant indicator of anthropogenic impact on groundwater resources. Groundwater absorption, preservation, and release are all influenced by land use patterns. Water bodies (rivers/canals/reservoirs), thick woods, vegetation, scrubland, towns, and river sand are the six forms of LULC in this area.

5.7. Lineament buffer:

The occurrence of lineaments implies favorable groundwater potential since they operate as possible channels for groundwater circulation. The zone of enhanced pore size distribution underpins lineament-controlled landscapes in general. As a result, lineament intensity mapping from lineament maps is closely linked to groundwater occurrence.

5.8. Water Level:

The water levels map was created using static water levels from pre-monsoon drilled wells (source: India WRIS) (figure-9). The water level is at its highest in the eastern half of the research area. Viz.12.44-15.31m and in northern part water level is minimum viz. 0.97-3.84m. Moderate water level occurs in central, western and southern of the study area viz. 3.84-6.70m, 6.70-9.57m and 9.57-12.44m.

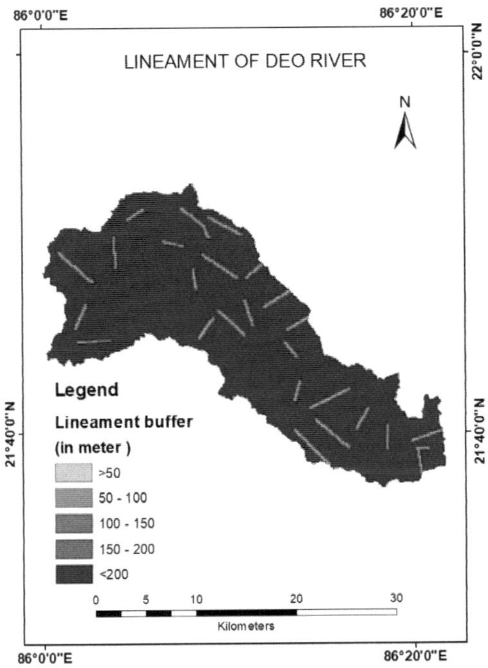

Fig.9. Lineament buffer map of Deo watershed

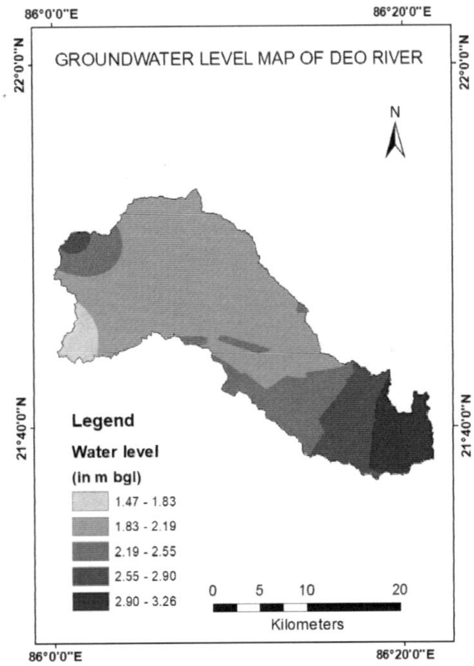

Fig.10. Water level map of Deo watershed

5.9. Rainfall:

The primary component of groundwater recharge is rainfall. On this study, a thematic map of rainfall was created using the Inverse Distance Weighted (IDW) approach in the Arc GIS software using multiple area points of data. The research region is classified into different rainfall zones. The rainfall total of 1533- 1568.58 mm falls into the moderate groundwater storage category. The area with rainfall between 1624.74 and 1656.32 mm are classified as Good. The groundwater potential of locations with rainfall ranging from 1656.32 to 1685.9 mm is rated as Very excellent. This is shown in the Fig.11.

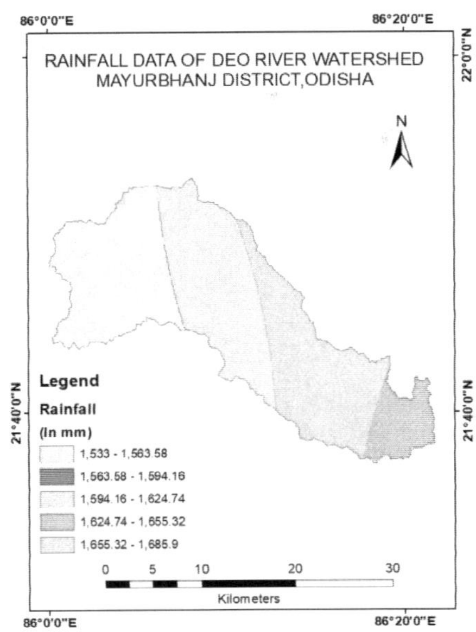

Fig.11. Rainfall map of Deo watershed

Table:2. Weightage table

Influencing Factors	Category (Classes)	Groundwater Potentiality	Rating	Weightage
Geology	Granite/ Granitic gneiss	Moderate	3	9
	Gabbro/ Anorthosite	Very Poor	1	
	Laterite	Good	4	
	Arkasani Granophyre	Poor	2	
	Slate, Phyllite	Poor	2	
	Newer Dolerite	Very Poor	1	
	Spilitic Lava	Very Poor	1	
	Quartzite, Conglomerate	Very Poor	1	
Geomorphology	Dyke/ sill ridge	Very Poor	1	8
	Highly Dissected Hills and Valleys	Very Poor	1	
	Low Dissected Hills and Valley	poor	2	
	Intermontane valley	Very Good	5	
	Pediment	Moderate	3	
	Pediplain	Good	4	
	Residual hill	Moderate	3	
	valley	Good	4	
	Scrap	Moderate	3	
	Lateritic Upland	Moderate	3	
	Valley Fill	Good	4	
	Water Body	Very Good	5	
Lineament Density	0-0.05 km/km^2	Very Poor	1	7
	0.05-0.1 km/km^2	Poor	2	
	0.1-0.15 km/km^2	Moderate	3	
	0.15-0.2 km/km^2	Good	4	
	>0.2 km/km^2	Very Good	5	
Slope	0- 2%	Very Good	5	5
	2- 8%	Good	4	
	8- 15%	Moderate	3	
	15- 30%	Poor	2	
	>30%	Very Poor	1	
Groundwater Depth	0.97-3.84m	Very Good	5	7
	3.84-6.70m	Good	4	
	6.70-9.57m	Moderate	3	
	9.57-12.44m	Poor	2	
	12.44-15.31m	Very Poor	1	
Rainfall	1533- 1563.58mm	Moderate	3	9
	1563.58-1594.16mm	Moderate	3	
	1594.16- 1624.74mm	Good	4	
	1624.14- 1655.32mm	Good	4	
	1655.32-1685.9mm	Very Good	5	

LULC	Water body	Very Good	5	4
	Dense Forest	Very Good	5	
	vegetation	Good	4	
	Scrubland	Poor	1	
	Sand	Good	4	
	settlement	Very Poor	1	
Drainage Density	0.004- 2.083 km/km^2	Very Good	5	6
	2.083-4.167 km/km^2	Good	4	
	4.167- 6.242 km/km^2	Moderate	3	
	6.242- 8.321 km/km^2	Poor	2	
	8.321- 10.4 km/km^2	Very Poor	1	

6. Groundwater potential zone:

Based on above 8 thematic maps groundwater potential map is made accoding to the GW potential it is divided into one of 5 groups: Very Good, Good, Moderate, Poor, and Very Poor [6]. The weights and ranks were determined using groundwater potentiality map - based features. All thematic maps are converted to raster format before being stacked using the weighted overlay method. Geomorphology and geology were given larger weights in the weighting process, whereas drainage density and lineaments were given lesser weights. Individual ranks for sub variables were assigned after weights were assigned to various factors. The characteristic of the high groundwater potentiality receives the maximum value, while the characteristic of the lowest groundwater potentiality receives the minimum. The greatest rank values are attributed to water bodies, while the lowest rank values are attributed to the built-up area in LULC. Sandstone and laterite have the greatest values in geology, whereas Granite, Gneiss, and volcanic rocks have the lowest. The extremely high lineament density groups are given higher rank values among the other lineament density groups because they have a higher possibility of groundwater penetration. Alluvial plain and valley has the maximum score of 5 in landforms, whereas hill has the lowest rank of 1. Table 2 summarises the overlay analysis. Using the Groundwater Potential Index (GWPI) algorithm in GIS, all of the theme maps were combined. The abovementioned approach is used to produce a complete groundwater potential map (Fig.12). In this paper, the groundwater potential zones were split into five categories: The five degrees of performance are Very Good, Good, Moderate, Poor, and Very Poor.

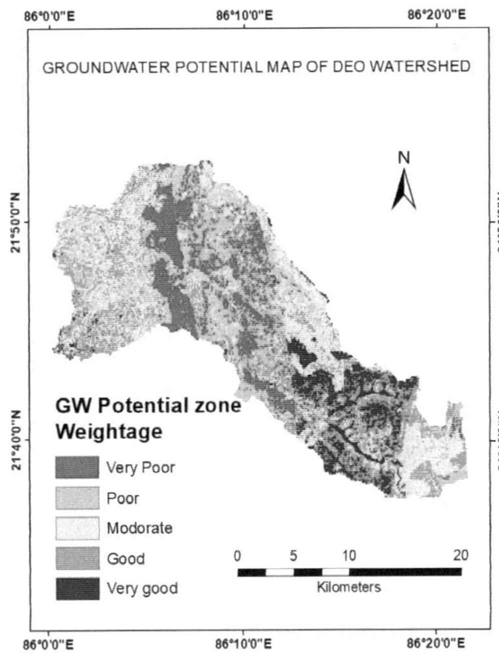

Fig.12.Groundwater potential map of Deo watershed

7. Conclusion:

This study revealed that GIS technology may be an incredible tool for analysing large volumes of cross-disciplinary data and making decisions in groundwater explorations. Several approaches have been proven useful and demonstrated for the monitoring of groundwater potential. For the delineation of potential zones, the weightage rating overlay methodology has shown to be quite beneficial. This groundwater potential data will be valuable in determining a good location for water extraction. This study found that remote sensing and GIS techniques are used efficiently to define groundwater resources map, which can be used to improve future groundwater studies in that region, as well as for other purposes such as identifying the locations of artificial recharge structures, tube well locations, and cost effective groundwater management for the benefit of society.

Reference:

[1] Ahirwar, Shobharam & Malik, Mohammad Subzar & Ahirwar, Rakesh & Shukla, J.. (2020). Application of Remote Sensing and GIS for Groundwater Recharge Potential Zone Mapping in Upper Betwa Watershed. Journal of the Geological Society of India. 95. 308-314. 10.1007/s12594-020-1430-3.

[2] Ajibade, Fidelis & Olajire, Olabanji & Ajibade, Temitope & George, Fadugba & Idowu, Temitope & Adelodun, Bashir & Opafola, Omobolaji & Lasisi, Kayode & Adewumi, J.. (2021). Groundwater potential assessment as a preliminary step to solving water scarcity challenges in Ekpoma, Edo State, Nigeria. Acta Geophysica. 69. 1-15. 10.1007/s11600-021-00611-8.

[3] Barik, Kamal Kumar & P.C., Dalai & S.P., Goudo & S.R., Panda & Nandi, Debabrata. (2017). Delineation of Groundwater Potential Zone in Baliguda Block of Kandhamal District, Odisha using Geospatial Technology Approach. International Journal of Advanced Remote Sensing and GIS. 6. 2068-2079. 10.23953/cloud.ijarsg.33.

[4] Benjmel, K., Amraoui, F., Boutaleb, S., Ouchchen, M., Tahiri, A. and Touab, A. (2020) Mapping of Groundwater Potential Zones in Crystalline Terrain Using Remote Sensing, GIS Techniques, and Multicriteria Data Analysis (Case of the Igherm Region, Western Anti-Atlas, Morocco). Water 2020,12,471;doi:10.3390/w12020471

[5] Bhattcharya, Raj & Das Chatterjee, Nilanjana & Das, Kousik. (2020). An integrated GIS approach to analyze the impact of land use change and land cover alteration on ground water potential level: A study in Kangsabati Basin, India. Groundwater for Sustainable Development. 11. 100399. 10.1016/j.gsd.2020.100399.

[6] Harini, P. & Sahadevan, DINESH KUMAR & Das, Iswar & Manikyamba, Prof. Dr. C. & Durgaprasad, M. & Nandan, Mogali. (2018). Regional Groundwater Assessment of Krishna River Basin Using Integrated GIS Approach. Journal of the Indian Society of Remote Sensing. 46. 10.1007/s12524-018-0780-4.

[7] Jhariya, D. C., Shrivastava, S., Nage, S. K. and Chatterjee, R. S. (2015) Integrated Remote Sensing and GIS approach to groundwater potential delineation in the Doon valley, Uttarakhand, India. SSARSC International journal of Geo-science and Geo-informatics, ISSN 2348-6198.

[8] Krishnamurthy, J., Srinivas, G.: Role of geological and geomorphological factors in groundwater exploration: a study using IRS LISS data. International Journal of Remote Sensing, 16, 2595-2618 (1995)
[9] Lekha K, Indhu & Jhariya, Dalchand. (2020). Delineation of groundwater potential zones in Samoda watershed, Chhattisgarh India, using Remote Sensing and GIS techniques. IOP Conference Series: Earth and Environmental Science. 597. 012007. 10.1088/1755-1315/597/1/012007.
[10] Mohanty, C. and Behera, S.C. (2010) Integrated remote sensing and GIS study for hydro geomorphological mapping and delineation of groundwater potential zones in Khallikote block, Ganjam district, Orissa; J. Indian Soc. Remote 38(2) 345-354, http://doi.org/10.1007/s12524-010-0023-9.
[11] Rahimi-Feyzabad, F., Yazdanpanah, M., Gholamrezai, S. *et al.* Institutional constraints to groundwater resource management in arid and semi-arid regions: a Straussian grounded theory study. *Hydrogeol J* **29,** 925–947 (2021). https://doi.org/10.1007/s10040-020-02283-y
[12] Roy, Annadasankar & Keesari, Tirumalesh & Sinha, U.K. & Sabarathinam, Chidambaram. (2019). Delineating groundwater prospect zones in a region with extreme climatic conditions using GIS and remote sensing techniques: A case study from central India. Journal of Earth System Science. 128. 10.1007/s12040-019-1205-7.
[13] Saraf, A.K., Choudary, P.R.: Integrated remote sensing and GIS for groundwater exploration and identification of artificial recharge sites. Int. J. Remote Sensing, 19, 1825-1841 (1998)
[14] Senthilkumar, M. & Gnanasundar, Devadasan & Arumugam, Rethinam. (2019). Identifying groundwater recharge zones using remote sensing & GIS techniques in Amaravathi aquifer system, Tamil Nadu, South India. Sustainable Environment Research. 29. 10.1186/s42834-019-0014-7.
[15] Ujjal Senapati and Tapan Kumar Das Journal: Arabian Journal of Geosciences, 2021, Volume 14, Number 11 DOI: 10.1007/s12517-021-07316-8This item cites DOI: 10.1007/978-3-030-62397-5_12

AUTHOR INDEX

Acharjee, Shukla 86
Agrawal, Ayush Kumar 261
Ahirwar, A V 385
Ahirwar, A.V. 332
Ahirwar, Ajay. V. 70
Ahmad, Ishtiyaq 267
Ahmed, A 424
Alok, Apurva 198
Ambadkar, Swati 145
Arora, Aman 287
Asudani, P 81
Azad, A 424
Azad, Md Alquamar 198
Baghel, Ankur 249
Bajpai, Samir 267, 306, 366
Bama, B Sathya 104
Barde, S 480
Barman, Sangita Deb 395
Bezbaruah, Devojit 463
Bhopi, R R 442
Das, Rajib Tarani 86
Daur, Sachin A. 534
Dave, P 53
Dewangan, Pankaj 493, 506
Dhekne, P.Y. 493, 506
Dilini, M.M.G.S. 86
Dixit, R. K. 280
Ganvir, P S 151
Ghose, D K 163
Gogoi, Bijoylakshmi 463
Goswami, Uttam 86
Guhey, Rajeeva 151
Gupta, Nidhi Rani 35
Haidar, Tanveer 1
Hari, K. R. 1
Hooda, Nishtha 35
Ingole, D K 217
Iyengar, B.K. 86
Janghel, Rekh Ram 173
Jena, S 227
Jhariya, D C 521, 550
Jhariya, D.C. 493, 506
Jibhakate, Tulshidas M. 138
Kannan, K Akash 104
Kapgate, Mamta D 337
Katpatal, Y B 337
Katpatal, Yashwant B. 138
Khobragade, P P 385
Killedar, D J 58
Killedar, D.J. 70
Kumar, A 458
Kumar, Dudam Bharath 452
Kumar, Nikhil 227
Kumar, Vishal 332
Kumari, S 353
Kushal, M 395
Londhe, Digambar S 337
Manjhi, Anil 410
Mawale, Y K 521, 550
Mawale, Y.K. 86
Meena, V 353
Mehta, A 81, 458
Mishra, Lipsa 227
Mishra, S 424
Mishra, V. K. 280
Mishra, Varun Kumar 112
Misra, Sagar 1, 112
Mohanta, N K 550
Mohanty, P C 415
Mukesh, M S 337
Mukherjee, A. 280
Naik, H K 227
Naredla, Pooja 23
Negi, R S 86
Nimbarte, G R 217
Pa, Yazhini Malar 104
Pandey, Manish 287
Parde, Divyesh 58
Pasari, S. 53, 81, 458
Pasari, Sumanta 261
Patel, M. D. 235
Patro, Surendra 493
Peddiwar, S T 217
Prabhakar, G. 280
Prasad, A.D. 209, 332
Praveenkumar, J 104
Qurashi, Jameel 287
Rajak, Tarun Kumar 12
Rajput, H 480
Ramchandani, Monica 173
Rana, S 521
Rathore, Purva 58, 70
Rathore, Yogesh Kumar 173
Ravinashree, A 187
Roomi, S Mohammed Mansoor 104
Sahare, Akansha 58, 70

Sahoo, A 163
Sahoo, M.M. 128
Sahu, M 550
Sahu, Mridu 493, 506
Sahu, R T. 209
Sangeetha, S 23
Sasmita, Sushree 452
Shankar, V 353
Sharma, Akriti 35
Sharma, Renu 35
Shekhar, Shubhendu 86
Shende, Ashay Devidas 128
Shivani, S Mira 346
Shuja, Mirza 287
Singh, A K 424
Singh, A 346, 415
Singh, Archana 463
Singh, H 209
Singh, Harvir 287
Singh, Neelratan 86, 198, 463
Singh, Priyanshi 173
Singh, S K 198
Singh, Tajendra 506
Sinha, M 442
Sinha, Madhulika 534
Sivapragasam, C 187
Soni, Sangeeta 287
Srivastava, S 346
Swain, A 521
Swain, R C 521
Swarnakar, Arvind Kumar 267
Tamrakar, Anjali 366
Tejaswini, Pallapothala 173
Tekade, Shrikant 128
Thakre, Ankush 12
Tiwari, R. N. 280
Tiwari, Rabindra Nath 112
Umare, Mahendra 145
Upadhyay, Kshitij 306, 366
Vasudevan, M 187
Verma, C. B. 280
Verma, H 458
Verma, Himanshu 261
Verma, M K 209
Verma, S K 209
Vishwakarma, Neeraj 1, 112
Yadav, A B 415